SUN OIL CO. LIBRARY

FEB 6 1969

MARCUS HOOK, PA.

DISCARD

JELKS BARKSDALE, Ph.D., Columbia University, is Associate Professor of Chemistry, Auburn University, and a consultant on titanium dioxide pigments. He has previously been a research chemist with the National Lead Company, Titanium Division. Dr. Barksdale has written widely on chemistry in general, including college textbooks, and titanium in particular, and holds a number of patents pertaining to titanium dioxide pigments.

# TITANIUM

Its Occurrence, Chemistry,
and Technology

**JELKS BARKSDALE**
Auburn University
Consultant on Titanium

SECOND EDITION

THE RONALD PRESS COMPANY · NEW YORK

Copyright © 1966 by
THE RONALD PRESS COMPANY

Copyright 1949 by
THE RONALD PRESS COMPANY

*All Rights Reserved*

No part of this book may be reproduced in any form without permission in writing from the publisher.

1
VR-VR

Library of Congress Catalog Card Number: 66-20080

PRINTED IN THE UNITED STATES OF AMERICA

To
Robbie Barksdale
and
William Jelks Barksdale

# Preface

The purpose of this book is to provide a comprehensive working reference on titanium for all those interested in the sources, chemistry, and technology of the element.

Since this book first appeared, many significant developments have been made in the field of titanium, and the literature has increased at an accelerated rate. Largely as a result of national defense demands coupled with government subsidies, titanium metal has been developed to the stage where many commercial plants have been built for its production. Civilian uses for titanium metal are increasing steadily.

Improved titanium dioxide pigments having the rutile crystal form and produced by modifications of the original sulfate process, which at the time of the First Edition had just been introduced commercially, have now largely replaced the older anatase-type pigments. During this period the commercial production of titanium dioxide pigments of the rutile type by the vapor phase oxidation of anhydrous titanium tetrachloride at a high temperature was accomplished. A number of plants now employ this process on a large scale.

This Second Edition emphasizes the more important recent pigment and metal developments, together with new occurrences and studies of ore deposits and advances in the chemistry of titanium.

JELKS BARKSDALE

Auburn, Alabama
   September, 1966

# Contents

| | | |
|---|---|---|
| 1 | The Discovery of Titanium | 3 |
| 2 | The Occurrence of Titanium | 5 |
| 3 | Geology and Mineralogy of Titanium | 10 |
| 4 | Mineral Deposits That Yield Titanium | 16 |
| 5 | Production and Imports of Titanium-Bearing Ores | 47 |
| 6 | The Chemistry of Elemental Titanium | 53 |
| 7 | Oxides and Hydroxides of Titanium | 67 |
| 8 | The Chemistry of Titanium Salts | 87 |
| 9 | Organic Compounds of Titanium | 128 |
| 10 | Chemical Analysis of Titanium and Its Compounds | 136 |
| 11 | Commercial Production of Titanium Metal | 145 |
| 12 | Leaching, Roasting, and Smelting of Ores | 187 |
| 13 | Titanium Sulfate Solutions for Pigment Manufacture | 213 |
| 14 | Hydrolysis of Titanium Sulfate Solutions to Produce Pigment | 256 |
| 15 | Filtering and Washing Hydrous Titanium Dioxide | 317 |
| 16 | Calcining, Milling, and Processing Titanium Dioxide to Produce Pigments | 324 |
| 17 | Sulfuric Acid Recovery | 361 |
| 18 | Hydrolysis of Titanium Tetrachloride to Produce Pigment | 372 |
| 19 | Production of Titanium Tetrachloride | 400 |
| 20 | Purification of Titanium Tetrachloride | 463 |

## CONTENTS

| | | |
|---|---|---:|
| 21 | Vapor Phase Oxidation of Titanium Tetrachloride to Produce Pigment | **480** |
| 22 | Special Processes of Pigment Manufacture | **509** |
| 23 | Chalking and Discoloration of Titanium Dioxide Pigments | **533** |
| 24 | The Titanium Pigment Industry | **568** |
| | Notes | **579** |
| | Index | **675** |

# TITANIUM

# 1

# The Discovery of Titanium

The element titanium was discovered by the Reverend William Gregor [1] in 1790. Gregor was born in the county of Cornwall, England, in 1762. He studied at Oxford, earning the A.B. degree in 1783 and the A.M. degree in 1786, both from St. John's College. His classical and mathematical attainments, which were of the highest order, procured for him the most distinguished honors at the university. In 1789 he was collated to the rectory of Deptford, and in 1793 he was presented to the vicarage of Bratton Clovelly in Devon. Soon afterward he transferred to the rectory of Creed, where he finally fixed his residence. In addition to his ecclesiastical duties he found time to continue his mineralogical and geological interests, and he was one of the founders and an active member of the Geological Society of Cornwall. The later years of his life were embittered by bodily suffering. He died of tuberculosis at Creed in 1817.

Berzelius,[2] in his letters, mentioned Gregor several times in connection with his discovery, analysis, and properties of minerals, and referred to him as a "celebrated mineralogist."

In 1791 Gregor communicated to the *Journal de Physique* the description and chemical analysis of a black magnetic sand found in the parish of Menaccan, six miles south of Falmouth, in Cornwall. The analysis showed almost 50 per cent of a white metallic oxide, up to that time unknown to chemists.[3] The sand was given the name *menaccanite* from the locality, and the new metallic oxide recovered from it was christened *menaccine* by Kirwan.[4]

Little interest was shown in the discovery until Klaproth,[5] in 1795, noticed the close agreement between Gregor's account of menaccine and the findings of his own investigations of the oxide extracted from "red schorl" (rutile) from Hungary. The identity of the two substances was soon established, and Klaproth, acknowledging Gregor's priority, applied the temporary name *titanium* to the new element with the comment: "Wherefore no name can be found for a new fossil (element) which indicates its peculiar and characteristic properties, in which position I find myself at present, I think it is best to choose such a denomination as means nothing of itself and thus can give no rise to any erroneous ideas. In consequence of this, as I did in the case of uranium, I shall borrow the name for this metallic substance from mythology, and in particular from the Titans, the first sons of the earth. I therefore call this metallic genus, titanium." This "temporary" name has been retained, however, and it now appears to have been an appropriate selection.

In those pioneer investigations the finely ground ore was leached with hydrochloric acid to remove the iron, the dried residue was fused with sodium carbonate, and the melt was taken up in dilute hydrochloric acid. On treatment with zinc or tin the solution assumed a bluish or reddish color, but on exposure to air it again became colorless. An attempt to concentrate the solution by heating it on a sand bath resulted in a turbid suspension resembling milk in appearance, which could not be cleared up by the addition of more acid. This property was the basis of the first analytical procedure developed; the solution was boiled to effect hydrolysis and the precipitate of hydrous titanic oxide was washed, dried, and weighed. Samples of the white oxide calcined alone became yellow, and in the presence of charcoal it became blue. It produced a yellow enamel color and was insoluble in acids, but it was soluble after fusion with an alkali. The new element showed a strong tendency to combine with oxygen.

Gregor's analysis of the black sand—oxide of iron 51.00 per cent, oxide of titanium 45.25 per cent, oxide of silicon 3.50 per cent, and oxide of manganese 0.25 per cent—corresponds to that of the mineral ilmenite.

It is interesting at this point to note that the fundamental chemical reactions on which the present-day titanium pigment industry is based were known by both these pioneer investigators before 1800, although it was not until 1918 that these pigments were available commercially on the American market.

# 2

# The Occurrence of Titanium

Titanium is ninth in abundance [1] of the elements making up the lithosphere, the crust of the earth, and accounts for 0.63 per cent of the total. It is exceeded in amount only by oxygen, silicon, aluminum, iron, magnesium, calcium, sodium, and potassium, and with the exception of these it comprises a larger proportion of the known terrestrial matter than all other metals combined, including zinc, copper, and lead, which are generally thought of as being common. Titanium is five to ten times as abundant as the well-known nonmetals, sulfur and phosphorus.

Titanium is a persistent constituent of practically all crystalline rocks and of sediments derived from them. Of 800 igneous rocks analyzed in the laboratories of the United States Geological Survey, 784 contained this element. It is present in most minerals, and it makes up the principal metallic constituent in an important group, the most common of which are ilmenite, rutile, arizonite, perovskite, leucoxene, and sphene or titanite.

Krugel and Retter [2] found 0.02 to 0.10 per cent titanium dioxide in phosphate rock from Florida, Curaçao Island, and North Africa. Apatite concentrate from the Chibinsk deposits of the Union of Soviet Socialist Republics contained 0.2 to 0.5 per cent. Forty-six samples of phosphate rock, representing practically all types available in the United States, showed 0.03 to 0.36 per cent.[3] The proportion was higher than that found in characteristic samples of bone ash. The element was detected in a radioactive bituminous mineral from the Boliden mine.[4] Spectrographic studies showed the presence of titanium in diamonds and in gold. Igneous rocks of China average 1.0 per cent of titanium dioxide.[5] Titanium dioxide has been reported in pyrite, quartz, diatomite, fluorite.

Dunnington [6] analyzed soils from many parts of the world and pointed out the wide distribution of titanium dioxide. From this work he estimated the average content of soils of Virginia as 1.57 per cent, of those of the remainder of the United States as 0.85 per cent; of Asia as 0.90 per cent; and of Europe as 0.54 per cent. Although data on the titanium dioxide content of soils from various parts of the world show wide variations, in a great majority of localities the values fall between the limits of 0.50 and 1.50 per cent. Frequently a considerable variation is also found in the strata or horizons of a given profile. Geilmann [7] detected this constituent in all soil samples he examined. The proportion was highest, up to 1.0 per cent, in clay soils, progressively less in loam and sand, and lowest in soils of the calcareous type. The titanium content of the principal soil zones of the European part of the Union of Soviet Socialist Republics was reported to vary from 0.24 to 0.79 per cent.[8] Arable or desert soil contains the lowest proportion. New Zealand soils [9] range from 0.25 to 1.5 per cent; the heavy loam and clay varieties contain the highest proportion. This constituent has also been reported in the soils of the Maritime Alps along the Mediterranean, in those of the cane-growing regions of India around Delhi, and in the soils of Brazil. The lateritic soils of Savaii Upolu, Western Samoa, contain an unusually high proportion,[10] from 7.8 to 12.0 per cent, and 2.33 per cent was reported in the soils of the historic island of St. Helena.[11] Although the average titanium dioxide content of soils is from 0.5 to 1.5 per cent, some Hawaiian soils average as much as 5 per cent. Local areas on mountainsides contain as much as 45 per cent titanium dioxide.

According to Charrin,[12] all clays contain titanium dioxide. Kaolin was found to average 0.28 per cent, and white clays 0.88 per cent.[13] Deposits in Wyoming [14] show an average of 0.23 per cent, and typical flint clays from Orange County, California,[15] have 1.0 to 1.2 per cent. Alabama clays [16] were found to contain titanium in the form of ilmenite, and this constituent could be removed by tabling. Twenty-seven samples from Finland [17] averaged 0.68 per cent titanium dioxide. A series of primary clays from the Union of Soviet Socialist Republics [18] contained from 0.22 to 1.35 per cent, and a similar series of secondary clays showed 0.94 to 2.72 per cent, probably as rutile. Kaolin from the Usol, Irkutsk District,[19] contains from 0.3 to 1.03 per cent. Similar proportions have been reported in the clays from San Paolo, near Rome,[20] and in the fire clays of Wales.[21] Much higher amounts, up to 15 per cent titanium dioxide, occur in the ceramic clays of Hawaii,[22] and from 11 to 27 per cent in the clays of the Sudetenland.[23] Isomatsu [24] reported 0.06 per cent of this oxide in bentonite from Japan. Titanium occurs as a minor constituent of most bauxites. For example, deposits in Bohemia [25] show 4.6 per cent of the dioxide, those in Brazil [26] 2 per cent, and those in the north Urals,[27] 2 per cent. Analysis of seven samples of Indian bauxite showed 2.0 to 3.2 per cent titanium dioxide.[28]

It occurs as sphene in English fuller's earth. Small proportions have been reported in the glauconites of Missouri. Aplite,[29] a ceramic raw material from Virginia, contains 23.8 per cent. Hart [30] found up to 0.34

per cent of the dioxide in Portland cement and 1.34 per cent in alumina cement. Titanium compounds are present in deep sea sediments and in material suspended in the waters of the seas and oceans, and in sea muds. The oxide was found in deep well borings at a depth of 4490 feet, and in the deep-sea dredgings of the *Challenger*.[31]

Titanium is a common constituent of coals. The ash of representative samples of anthracite coal from Pennsylvania [32] contained 1.0 to 1.8 per cent of the dioxide, and that of British coal [33] showed up to 3.0 per cent. This constituent has also been identified in the ash of German brown coal and of peat from North Carolina, and it has also been detected in the coals of the Charleroi basin in Belgium. The ash of five samples of crude oils used in Germany [34] averaged 0.1 per cent of this constituent. Titanium is present in sea water and in waters of the major streams of the world.[35] It has been reported in water below petroleum layers.

The titanium content of natural waters is of the order of $10^{-7}$ to $10^{-6}$ per cent.[36] It has been detected by spectrographic methods in sea water, in waters of the Rio de la Plata in Argentina, and in those of the Danube. Very small amounts have been reported in the waters of the hot springs of Japan, in many Spanish medicinal waters, and in the city water supply of Buenos Aires.

Noteworthy amounts of titanium dioxide were found in the volcanic dusts of Mont Pelée,[37] in the fumarole sublimation product from the crater of Vesuvius, and at Kilauea.[38] It has been detected in many meteoric stones.[39] The Perpeti meteoric shower [40] of May 14, 1935, showed 0.16 per cent; the Saratov meteorite,[41] which fell on September 6, 1918, contained 0.12 per cent in the magnetic fraction. A meteorite which fell on December 29, 1937, at Rangala, India,[42] contained 0.08 per cent, and another which fell in Ekely, Sweden,[43] on April 5, 1939, contained 0.22 per cent.

According to Lockyer,[44] the presence of titanium in the atmosphere of the sun was discovered by Thalen with the aid of the spectroscope as early as 1886. Later work confirmed the earlier findings and indicated that the element exists in the ionized form.[45] From the intensities of the lines of a number of elements, including titanium, the temperature of the solar atmosphere has been determined to be between 3500° and 5000° C.[46]

Lines corresponding to titanium have been observed in the spectra of a great many stars and satellites. Titanium is present in interstellar space.

The widespread distribution of titanium is by no means limited to rocks and minerals. Fifty samples of legumes, grains, vegetables, trees, and shrubs examined in the laboratories of the U.S. Department of Agriculture contained amounts up to 0.017 per cent.[47] Geilmann [48] examined twenty-three plants and plant parts, and of these cacao beans only showed no titanium. Potato tubers gave only a trace, and the ashes of the others contained up to 0.27 per cent of the dioxide. Green plant parts showed the highest proportions. Headden [49] reported 0.08 per cent titanium oxide in the ash of potato tubers and suggested that that in the ash of field-grown plants may have been derived from dust and soil blown upon them. The titanium content of tobacco is relatively high, while that of alfalfa is low.[50] Spectrographic examination revealed a higher percentage in the ashes of nodules than in the ashes of the roots of leguminous plants.[51] Titanium has been detected in samples of wire grass, in both cryptogamic and phanerogamic plants, and in the filter-press cake from a sugar factory on Manui Island, Hawaii. According to Bertrand and Voronca-Spirt,[52] most of the titanium in seeds is concentrated in the tegument. Buckwheat and leguminous plants were found to be relatively rich in this constituent and contained considerably higher proportions than ordinary cereals; peanuts and hazel nuts were much poorer, but more was detected in walnuts and coconuts. Foodstuffs, such as the edible part of fruits, the roots of carrots, and the tubers of potatoes were poor, while onions and similar bulbs contained distinctly higher proportions. In these products the titanium content varied from one part in 10,000,000 to one part in 200,000 parts of the fresh weight. The element has also been reported in cotton seed, in red clover, and in the ashes of sunflower seed, horse chestnut, walnut, pumpkin, peanut and wheat flour. Appreciable amounts were found in the ashes of various kinds of wood. Titanium is present in plants and trees, in the leaves of plants, in plant green parts, in clover, rye grass, in apples, and in yeast.

Baskerville [53] found that human flesh contains 0.0325 per cent of titanium dioxide, human bone a trace, beef flesh 0.013 per cent, and beef bone 0.0195 per cent. Amounts between 0.0015 and 0.011 mg per 100 g were reported in all organs of the human body,[54] the largest proportions being found in the lungs,[55] probably an accumulation from particles of titanium dioxide suspended in the air. It has also been detected in the bodies of the horse, calf, sheep, hog, and rabbit.[56] All animal tissues examined by Chuiko and Voinar [57] contained proportions ranging from 0.001 to 2.0 mg per 100 g. The lungs, hair, liver, and kidneys showed the higher proportions.

Titanium has been found in a variety of sea creatures. A number of specimens of fish [58] that were examined contained 0.3 to 0.9 mg per kg, and various varieties of shellfish contained even higher proportions.

The titanium content of land and marine plants and animals and of fresh-water organisms varies from 0.00001 to 0.01 per cent.[59] Plants contain a slightly higher proportion than animals. Small amounts have been detected in the teeth [60] of man and the lower animals. Berg [61] detected small proportions in many foods and in excreta. Spectrographic analyses of the ash of human and cow's milk from various countries have consistently shown the presence of titanium. Samples of cow's milk in the Union of Soviet Socialist Republics [62] contained 0.001 per cent. The percentages of ash and of titanium, respectively, in representative samples of human milk were 0.21 and 0.0136, in mare's milk 0.27 and 0.0178, and in cow's milk 0.64 and 0.0063.[63] Titanium has been detected in hen's eggs, in liver, in brain, in hair, nails and skin, in nerve cells, in kidney stones, in silkworm tissues, and in embryos.

The concentration of titanium in the suspended particles collected from the atmosphere of 20 American cities is below 1 gamma per cubic meter.[64]

# 3

# Geology and Mineralogy of Titanium

Titanium is a characteristic constituent of igneous and metamorphic rocks and of sediments derived from them. The chief mineralogical occurrences are as oxides, titanates, and silicotitanates, although the element is present in many silicates and less commonly in niobates and tantalates. In general, rock magmas rich in silica and poor in the base-forming elements, iron and magnesium, deposit their titanium component as oxide. Those containing relatively high proportions of calcium and silicon yield calcium titanate; those highest in iron and lowest in acid-forming oxides (silica) yield the iron titanates, ilmenite and arizonite. In silicates, titanium either replaces silicon in the acid radical or enters as a weak base-forming element. Titanium is often associated with magnetite and hematite, and it may make up a considerable proportion of such deposits.[1] Sundius[2] attributed the concentration of titanium in residual magmas to the presence of titanium tetrachloride and tetrafluoride, which hydrolyze to yield the dioxide, and to the occurrence of alkali silicates rich in titanium, which cause the formation of the alkali titanium silicates in the nepheline syenites. Lundegardh[3] studied the effect on the concentration of titanium of the time at which a mineral is formed during the crystallization differentiation.

**OXIDES**

As the stable dioxide, titanium occurs in nature in three crystal modifications corresponding to the minerals rutile, brookite, and anatase or octahedrite, although the latter two are relatively rare.[4] Rutile[5] crystallizes in the tetragonal system, commonly in prismatic crystals and often in slender acicular ones. It has a distinct cleavage, subconchoidal to uneven fracture, metallic-adamantine luster, hardness of 6.0 to 6.5, and specific gravity of 4.18 to 4.25. The color is usually reddish to brown, sometimes yellowish, bluish, or black, and rarely green. It is transparent to opaque and gives a pale brown streak. Iron and vanadium are usually present in proportions up to 10 per cent.

Brookite is of the orthorhombic crystal form. It has a specific gravity of 4.0, hardness of 5.5 to 6.0, and molecular volume of 20.0. The color is yellowish, reddish, brown, or iron black.

Anatase or octahedrite crystallizes in the tetragonal system. It has a specific gravity of 3.82 to 3.95, hardness of 5.5 to 6.0, and molecular volume of 20.5. The color varies from brown to indigo-blue to black. All three modifications in the pure state correspond to the chemical formula $TiO_2$.

Rutile is found in all types of rocks, igneous, metamorphic, and sedimentary. It is a constituent of eruptive rocks, pegmatite dikes, contact metamorphic deposits, regionally metamorphosed rocks, and veins both metalliferous and nonmetalliferous.[6] Although rutile occurs as a pyrogenetic mineral in eruptive rocks, it is more common in the metamorphics and is found in gneiss, mica schist, amphiboles, and phyllites,[7] and less commonly in granular limestones and dolomites.[8] In the crystallines it appears both as a pyrogenetic constituent and as an alteration product. In clastic deposits (sediments) the most important occurrences are in beach and stream sands. In minute crystals and small quantity the mineral is found in some of the dark micas, as large crystals in quartz veins closely related to pegmatites, and in schists to which siliceous solutions have carried the titanium. The known large deposits are in pegmatites or in very closely related basic rocks such as kragerolite, a rutile-plagioclase rock. Apatite and albite are common associates. Rarely rutile is found as a secondary mineral derived from ilmenite and titanite, and sometimes from certain of the ferromagnesian silicates, especially biotite. It is relatively very stable, but it sometimes alters to leucoxene along with ilmenite.

Brookite does not occur in fresh eruptive rocks, but is generally found in decomposed gneiss and quartz porphyry and in sedimentary deposits. It alters into rutile.

Octahedrite is never primary but is formed by the alteration of other titanium minerals. It has been observed under a great variety of conditions, as in granite, diabase, quartz porphyry, diorite, the crystalline schists, shale, sandstone, and limestone. Rutile is of pyrogenetic origin, while brookite and octahedrite occur only as secondary minerals.

All three modifications have been prepared synthetically. These experiments showed that rutile, the most stable form of titanium dioxide, was formed at the highest temperature; brookite at a temperature considerably lower; and anatase at a point still lower on the scale. These observations are in harmony with the known occurrences of the three species as rock-forming minerals in nature.

According to Walker,[9] the titanium in Transvaal lodestone is in the trivalent state. Because of this he suggested the name titaniferous magnetite, $(FeTi)_2O_3$. Rankama[10] reported that tantalum occurs in appreciable amounts in titanium minerals. Vanadium and chromium are also usually present.

## TITANATES

Of the many titanates, only a few are important industrially, particularly those of iron and calcium. Ilmenite[11] is a heavy mineral of metallic to submetallic luster, crystallizing in the rhombohedral class of the hexagonal system with specific gravity of 4.5 to 5.0, hardness of 5 to 6, and molecular volume of 30.4. The color is usually iron black, but it may be reddish to brown. The streak is black to reddish brown. It is opaque and slightly magnetic, cleavage is not developed, and fracture is conchoidal. The mineral is isomorphous with pyrophanite, manganese titanate, $MnTiO_3$. It is rarely seen in good megascopic crystals in rocks but usually occurs as embedded grains and masses or plates of irregular outline. The chemical composition of natural ilmenite is variable, with the result that its constitution has been much discussed although it is generally regarded as ferrous titanate corresponding to the chemical formula $FeTiO_3$. Proportions of iron higher than the theoretical may be due to admixed hematite or magnetite, while an excess of titanium may be due to the presence of rutile. Studies of ilmenite samples from different sources indicate that the ferric oxide is not in chemical combination, but rather is held mechanically as inclusions or intergrowths. Watson and Taber suggested an isomorphous series $(FeTi)_2O_3$ in which the metallic components may vary widely, even grading into hematite. Singewald[12] has held that the titanium in titaniferous iron ores may be present as an integral part of the magnetite molecule itself, although most authorities believe that these ores are composed primarily of magnetite or hematite, with intergrowths of ilmenite. The latter view is strengthened by the fact that high-grade concentrates can be prepared by mechanical means from many titaniferous iron ores.

From a study of X-ray absorption edges, Hamos and Shcherbina[13] obtained values for the titanium present corresponding to the quadrivalent state, and concluded that the chemical formula $FeTiO_3$ represents divalent iron and tetravalent titanium, although from a chemical standpoint these values might be three and three respectively. This is also substantiated by chemical analysis.

Ilmenite is widely distributed in nature. It is formed in igneous rocks, in pegmatites, in contact metamorphic deposits, under regional metamorphic conditions, and in deep veins.[14] The most important occurrences are in coarse-grained gabbros and anorthosite where the ilmenite is frequently segregated into large masses. The metamorphic rocks [15] which most commonly contain it are gneiss, mica schist, and amphibolite. The mineral appears in most igneous rocks and their lava forms, although its chief occurrences are in those varieties that are poor in silica. In addition to being an essential constituent of diabase, basalt, diorite, and gabbro, it is found in the acid crystalline rocks such as granite and syenite, in andesite and more basic lavas. Along with magnetite it is one of the first minerals to separate from magmas.[16] The large masses of ilmenite found in many parts of the world associated with more or less magnetite, olivine, pyroxene, and soda-lime feldspar are simply phases of the rock itself formed by concentration from the original magma. Near such ore bodies, consisting of a mixture of ilmenite and ferromagnesian silicates, the dark-colored basic components first increase and the feldspar finally disappears.[17] The mineral as flat, imperfect crystals occurs in small quantities in quartz veins, although its most common form in large deposits is as plates intergrown along the cleavage planes of magnetite. These plates are often microscopic in size. Ilmenite-magnetite and ilmenite-hematite in large bodies are found only with gabbros, anorthosite, and similar rocks.[18]

Pseudobrookite,[19] ferric orthotitanate, corresponding to the chemical formula $Fe_3(TiO_4)_3$, crystallizes in the orthorhombic system, has a specific gravity of 4.39, hardness of 6, and molecular volume of 127.5. The color varies from dark brown to black. It occurs as a rare accessory mineral in eruptive rocks of the andesite, trachyte, basalt, and nephelinite type.

Arizonite,[20] a ferric titanate having the chemical formula $Fe_2TiO_5$ is probably monoclinic. It has a dark steel-gray color, metallic luster, and brown streak. Thin sections, which are transparent, appear deep red in transmitted light. The mineral forms a large part of some sands classified as ilmenite, particularly those of Quilon, India, and Senegal, West Africa. Chromic oxide, usually present, may replace ferric oxide in the molecule.

Perovskite, calcium titanate, of the chemical formula $CaTiO_3$, crystallizes in the isometric system. The molecular volume is 34, specific gravity 4.0, hardness 5.5. The color is yellow, ranging through orange and brown to grayish black. It is found in both eruptive and metamorphic rocks. In the former it is associated with melilite, leucite, nepheline, and some peridotites, and is among the earliest secretions. Other occurrences are in chlorite schist, limestone, and quartz gneiss. It may be derived from titanite and may alter into titanic oxide.

Geikielite, magnesium titanate, of the chemical formula $MgTiO_3$, and pyrophanite, manganese titanate, having the chemical formula $MnTiO_3$, are rare as separate minerals but are of interest because they form intergrowths with ilmenite.

At the initial temperature of crystallization, 800° C, it is believed that much more than the normal 6 per cent of the ferric oxide can be accommodated in the ilmenite structure. However, as ordinary temperatures are reached, the ferric oxide in excess of 6 per cent cannot be accommodated in the structure and is ejected as hematite or magnetite. Much of the hematite and magnetite found intimately associated with the ilmenite has been formed in this manner and is referred to as exsolution hematite or magnetite. Minerals formed by exsolution can often by recognized by textural studies because the position of the exsolved mineral is usually determined by favorable crystallographic position within the host.

Although intergrowths of hematite or magnetite within ilmenite commonly result from exsolution, it is also possible to have any combination of these minerals form simultaneously and thus be intergrown from the original magma.

According to Buddington and Lindsley [21] the experimentally determined solubility of ilmenite in magnetite is too low to explain the formation of ilmenite-magnetite intergrowths of exsolution. It is shown experimentally that magnetite-ulvospinel solid solutions can be obtained from the ilmenite-magnetite intergrowth.

Information derived from chemical and X-ray studies indicates that the weathering of ilmenite proceeds according to the scheme; ilmenite, hydrated ilmenite, arizonite, leucoxene, rutile (or anatase or brookite).[22] This work confirms the existence of arizonite as a separate mineral having the formula $Fe_2O_3 \cdot nTiO_2 \cdot mH_2O$ where n is 3 to 5 and m is 1 to 2. Leucoxene which is widely distributed in nature is considered to be a solid hydrogel of titanium dioxide containing a variable but small amount of water. The formation of rutile is a complex process which cannot be attributed in its entirety to the rutilization of the ilmenite in the final stages of weathering but it must be associated with processes involving the colloidal nature of some of the weathering products.

Ilmenite ores may be listed in order of increasing degree of alteration as MacIntyre, North Carolina, Quilon, Brazil, and Florida.[23] Although more weathered than ore from Quilon, ore from Brazil is lower in titanium dioxide since less of its ferric oxide has been removed by leaching. Quilon, Brazil, and Florida concentrates all contain ore grains ranging in composition from ilmenite to highly altered products approaching pure titanium dioxide in composition. Arizonite is probably weathered ilmenite. Conclusive evidence is lacking for the existence of a compound $Fe_2O_3 \cdot 3TiO_2$. As the ilmenite is weathered, the iron content is oxidized to the ferric state and leached out. The crystalline ilmenite goes through an amorphous stage and finally the titanium dioxide recrystallizes in the rutile form. The final product of weathering is a gray colored, almost pure titanium dioxide mineral known as leucoxene.

The structural changes that take place in the weathering of ilmenite are ilmenite having the hexagonal crystal form, through an amorphous stage to rutile crystallizing in the tetragonal system. Chemical changes result in a decrease in the percentage of iron accompanied by an equivalent increase in the percentage of titanium dioxide. The final product of the weathering of ilmenite is a fine-grained, white to yellow mass called leucoxene which is essentially pure titanium dioxide. Leucoxene is for the most part masses of fine rutile needles. In leucoxene, rutile forms interesting pseudomorphs after sphene and ilmenite.[24] These pseudo ilmenites often have the aspect of well developed ilmenite, but are friable in their interior portions. X-ray diagrams show abundant rutile, only relict ilmenite and hematite as a new formation. The reactions causing the pseudomorphs are not a trivial diffusion of oxygen into the ilmenite, but a diffusion of ferrous ions outward along defect holes on the octahedron empty positions. Into these octahedral gaps titanium IV ions migrate and nucleate the crystallization of rutile. The resulting porous pseudomorph has a density of 3.3 as compared with 4.26 for rutile crystals, and a low rigidity. About one-fourth of the pseudomorph volume is pores.

Gruner[25] decomposed ilmenite into leucoxene and the accompanying iron oxides and hydroxides by exposure to hydrogen sulfide at elevated temperatures.

## SILICATES

Titanite or sphene is titanium calcium silicate that corresponds to the chemical formula $CaTiSiO_5$. It crystallizes in the monoclinic system, has a specific gravity of 3.54, hardness of 5.0 to 5.5, and molecular volume of 55.5. The color varies from yellow to green, red, brown, gray, or black. Pyrogenetic titanite is found along with the oldest secretions in the more siliceous rocks such as granite, diorite, syenite, and trachyte, and is abundant in phenolites and elaeolite syenites. It is also formed as an alteration product of rutile or ilmenite and is frequently associated with chlorite. Alteration of titanite into rutile has been reported.

The name silicoilmenite[26] has been suggested for a red-brown mineral forming polysynthetic intergrowths with ilmenite. It is found in the Ilmen Mountains, Union of Soviet Socialist Republics, and is reported to be a solid solution of a silicate or silica in ilmenite.

Complete pseudomorphs of anatase after sphene up to 4 cm in longest dimension occur in deeply weathered pegmatites 9 miles south of Roanoke, Virginia.[27] Thin, tabular crystals of anatase found on quartz and orthoclase in Austria are tetragonal with a = 3.79 and c = 9.51 Ångstrom units.[28] In a differential series of magma the titanium content decreases tenfold from gabbro diorites (0.66 per cent) to aplites (0.05 per cent).[29]

# 4

# Mineral Deposits That Yield Titanium

### Massive Ilmenite

In contrast to its widespread occurrence, only a few minerals bearing titanium, particularly ilmenite, rutile, arizonite, leucoxene, titanite, and perovskite, are known to occur in bodies large enough to be of economic importance.

**UNITED STATES**

Virginia. Pure ilmenite in place in rich minable ore bodies is found in western Virginia [1] in an unusual rock mass known as nelsonite. In it ilmenite and apatite, or rutile and apatite, are the essential materials. Phlogopite, quartz, pyrite, pyroxene, and hornblende are, however, frequently present as accessories. These nelsonite bodies, which have the form of dikes, are enclosed in a biotite schist of the Lovington granite gneiss. Apatite appears to have been the primary mineral and to have been replaced by ilmenite or rutile until in some places only small rounded pellets of the original mineral are left.

Another type of deposit is known in the same area in which the titanium minerals occur as disseminations in the anorthosite. Ross [2] concluded that both types were deposited by hydrothermal invading solutions believed to have been derived from a deep-seated, highly ferromagnesian rock, probably itself a differentiate from the same magma as the anorthosite. Ferrous iron was first abstracted from the carrying solutions to form ferromagnesian silicates and ilmenite. Magnesium and titanium traveled farther and deposited rutile and magnesium silicate. At greater distances the feldspar was completely altered without any marked change in its chemical composition. Davidson, Grout, and Schwartz [3] concluded that the ilmenite deposit was an irregular fingering dike intruded into anorthosite and later sheared and hydrothermally altered.

In adjacent Amherst and Nelson counties, in Virginia,[4] are nelsonite bodies large enough to be of economic value as producers of ilmenite. By far the largest of these yet discovered is located in Amherst County, across Piney River from the railroad station of the same name. This dike-like mass outcrops 70 to 80 feet high for a distance of 350 feet along the river in a bluff at the edge of an old terrace. The maximum surface width of 400 feet extends for 1300 feet to the southwest, then tapers within the next 600 feet to a width of less than 100 feet. A tail, 60 to 100 feet wide, extends for more than 1000 feet farther. Since the dike dips with the schists about 45 degrees to the southeast, the surface outcrop of 400 feet represents an actual thickness of the main body of about 283 feet. The tail is 28 to 71 feet thick. The ore, which has been proved to a depth of 400 feet, appears to continue deeper. Weathering has extended about 75 feet below the surface to such an extent that the nelsonite appears as a rather rusty rock, soft enough to be crushed in the hands. Deeper, however, it becomes harder. The ilmenite is a clear black, the apatite is comparatively clear, and there is enough chlorite to give the rock a greenish cast. Pyrite in small amounts is found along the cracks.

The original rock contains 18.5 per cent titanium dioxide, and after separating the apatite, which accounts for all the gangue material, it yields a concentrate containing 43.7 per cent titanic oxide.[5] The unweathered rock is said to be quite amenable to treatment, although the weathered material is more easily worked. This deposit, to a depth of 400 feet, contains 24,000,000 tons of nelsonite carrying more than 4,400,000 tons of titanium dioxide. For a number of years large quantities of apatite, obtained as a by-product in the ilmenite mining, have been employed as the raw material for the manufacture of primary calcium phosphate for use in the manufacture of baking powders.

About a mile and three quarters down Tye River, below Massie's Mill, there is a hard vertical dike of nelsonite 60 feet wide and 600 feet long. Another dike of about the same width, but shorter, occurs on Hat Creek. Two miles farther up the same creek is a dike that is reported to be 40 feet thick and of unknown but considerable length. These deposits are reported to contain 30 per cent titanium dioxide, and the available reserve has been estimated at 2,000,000 tons of such ore. In addition, many smaller dike-like bodies of nelsonite have been reported in both counties, as well as in Roanoke County.

**North Carolina.** A deposit of nearly pure ilmenite, slightly intermixed with talc and serpentine, occurs about one mile north of Finley, Caldwell County, North Carolina.[6] The ore body is lenticular in shape, 45 feet wide at the best exposure, and outcrops along the strike for half a mile. It is apparently a part of a peridotite intrusion along a fault plane dipping 35 degrees to the southeast. The country rock is sericite schist consisting of small particles of ore in a matrix made up chiefly of fibrous and scaly aggregates of chlorite, serpentine, and talc. A sample of the ore from this deposit analyzed 49 per cent titanium dioxide. Three fourths of a mile to the southeast a similar ore body outcrops for nearly three fourths of a mile. It is 25 feet thick but is not so rich as the first deposit.

## Ilmenite-Magnetite

New York. Very large deposits of ilmenite-magnetite occur in the Adirondack Mountains of New York State. Large bodies of ilmenite-magnetite carrying from 7 to 23 per cent titanium dioxide are inclosed in the country rock of anorthosites and dark gray gabbro or norite. The largest of these deposits referred to as Sanford Hill, MacIntyre, and Tahawas is in Essex County. As mined the ore contains 32 per cent ilmenite, 35 per cent magnetite, 10 per cent plagioclase feldspar, 8 per cent garnet, 8 per cent hornblende and pyroxene, 4 per cent biotite, 2 per cent spinel, 0.5 per cent pyrite, 0.3 per cent calcite and 0.2 per cent apatite.[7] By magnetic concentration, one ton of ore gives 0.2 ton of ilmenite containing 49 per cent titanium dioxide. The proved ore reserve of this body alone is 100,000,000 tons. Grinding to 20 mesh is required for effective separation of the mineral components.

Other but smaller deposits in the area are at Cheney Pond, Iron Mountain, and Calamity Mill Pond. Reserves of the Calamity Mill Pond deposit are given as 1,630,000 tons of titanium dioxide and the Iron Mountain deposit at 2,330,000 tons. Small deposits occur near Elizabethtown, Mineville, and Westport, also in Essex County.

After standing idle for many years, the Tahawas deposit was reopened in 1941 to supply the domestic demand for ilmenite formerly obtained from India but cut off by war conditions.[8] The ore, as mined, consists of a mixture of ilmenite, magnetite, feldspar, hornblende, pyroxene, garnet, and biotite, with a small amount of the vanadium mineral, coulsonite.[9] A magnetic concentrate, rich in iron and vanadium, is obtained from low intensity, wet-magnetic separation, and an ilmenite concentrate is obtained from the nonmagnetic tailings by hydraulic classifiers and tables. This ilmenite product, containing 38 per cent titanium dioxide, is improved in grade to 45 per cent on high intensity, dry-magnetic separators. The original ore contains 16 per cent titanium dioxide,[10] and in addition to the ilmenite concentrate, it yields magnetite containing 56 per cent iron with 0.12 to 0.17 per cent sulfur and only a trace of phosphorus.

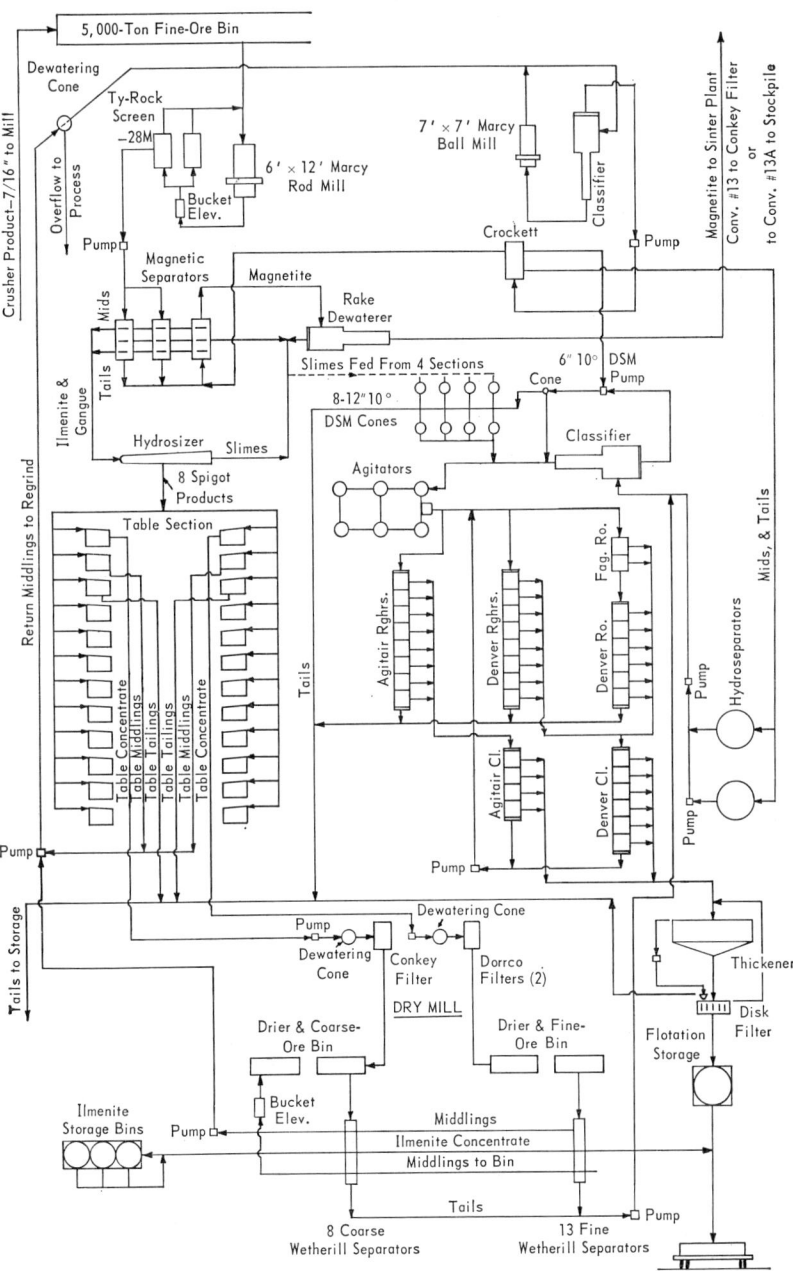

**Fig. 4–1.** Flowsheet of the ilmenite concentrating plant of National Lead Company at Tahawus, New York. (Reproduced from Bureau of Mines Information Circular 7791.)

**Minnesota.** Another extensive deposit of titaniferous iron ore is located in Lake and Cook counties, Minnesota.[11] The titanium content ranges from 3 to 20 per cent, although samples with a yield as high as 40 per cent have been obtained. Most of the ore-bearing rock contains very dark ferruginous minerals as gangue material, which closely resembles the ore. Few of the outcrops exceed 10 feet by 100 feet in area, although the number of small exposures is quite large.

**Rhode Island.** Another large deposit of such ore in the United States is in Cumberland, Rhode Island, where there is an outcrop 600 feet wide and 1500 feet long. A concentrate containing 22 per cent titanic oxide and 54 per cent iron can be produced by magnetic separation.

**Wyoming.** On Chugwater Creek, 8 miles west of Iron Mountain Station, Wyoming, on the Colorado and Southern Railroad, there is an extensive igneous dike of solid titaniferous magnetite one and one fourth miles long with an average width of 175 feet (100 to 200 feet), in anorthosite with some associated granite. Most of the analyses show 22 to 23 per cent titanium dioxide, although some much lower values have been reported.[12] Satisfactory concentration of the ore requires several steps. There are, in addition, several dike-like bodies in the vicinity.

Other smaller deposits in the Iron Mountain area include the Shanton deposit, containing 20 per cent titanium dioxide, and the Taylor deposit, containing from 5 to 31 per cent titanium dioxide. Estimates show the Iron Mountain deposit to contain 9,150,000 tons of ore equivalent to 1,278,000 tons of titanium dioxide. Resources of the entire Iron Mountain area are placed at 178,000,000 tons of ore containing 10 per cent titanium dioxide. This ore cannot be concentrated economically at the present time.

**Montana.** Composite drill cores from the Choteau titaniferous magnetite deposit in Teton County, Montana, analyzed 43.7 per cent iron, 7.2 per cent titanium dioxide, 0.05 per cent sulfur, 0.018 per cent phosphorus, 0.5 per cent manganese, 17.0 per cent silica, 1.6 per cent calcium oxide, 1.2 per cent magnesium oxide, and 5.2 per cent alumina.[13] Magnetic concentration of the 100-mesh material increased the percentage of iron to 60.6, but lowered the titanium dioxide content only to 6.6 per cent.

**California.** Large deposits of titaniferous magnetite are known in the San Gabriel Mountains, Los Angeles County, California. The principal occurrences are on the ridge south of Soledad Canyon, extending 5 miles from Lang to Russ Siding, and also in the area southwest of Mount Gleason and north of Tujunga Canyon. The ores of this region are associated with rocks of the gabbro family, and the largest bodies are contained in rocks composed chiefly of labradorite in coarsely crystalline, granitoid aggregates that have been called anorthosite by the Canadian geologists. (Anorthosite is a nearly pure plagioclase rock in which labradorite is the prominent feldspar.) These ores consist almost entirely of magnetite and ilmenite, with little visible gangue, and the titanium dioxide content is said to range from 11 to 25 per cent.

A study of the titaniferous magnetite deposits in the western San Gabriel Mountains shows that the minerals consist of an intergrowth of ilmenite and magnetite in association with hornblende and highly altered pyroxene.[14] Accessory minerals contain alumina, phosphorus, vanadium, and traces of beryllium. The large amounts of ilmenite heavy sand in the placer deposits of Sand and Pacoima canyons offer the greatest promise for commercial development.

**Colorado.** Titaniferous iron ores are found in Colorado at Caribou Hill, Boulder County, at Iron Mountain in Fremont County, and on Cebolla Creek, south of Powderhorn, in Gunnison County. Analyses of samples from these deposits show 8.5, 12.95, and 9 to 36 per cent titanium dioxide, respectively. The ore is not well adapted for magnetic concentration.

**Others.** Numerous smaller deposits of titaniferous magnetites, associated with anorthosite and believed to be magmatic segregations, occur in the Wichita Mountains of Oklahoma.[15] An average of five samples gave 7.31 per cent titanium dioxide, although values up to 16 per cent have been obtained. Large quantities of titanium ore, of satisfactory commercial grade, occur in Hidalgo County, New Mexico,[16] near the Arizona border. The ore body is a replacement deposit of contact metamorphic origin, and the areal extent averages 70 feet wide by 1000 feet long. Other occurrences of iron ores rich in titanium have been reported in North Carolina, South Carolina, Tennessee, and New Jersey.

**Quebec.** Both ilmenite-magnetite and ilmenite-hematite are found in many places in Canada, although the best known occurrences are at St. Urbain, Charlevoix County, Quebec. According to Gillson,[17] these ore bodies are formed by replacement of the anorthosite with which the ilmenite is associated and were deposited by gaseous or liquid solutions which soaked through the rock. The original solutions were derived from the magmatic reservoir. Similar deposits, but without rutile, occur at Ivry, Terrebone County, also in Quebec. Ores from both localities contain 20 to 30 per cent titanium dioxide, although several thousand tons, presumably concentrates, carrying 35 to 40 per cent, have been shipped to Niagara Falls, New York, for the manufacture of ferrocarbontitanium.

Bodies of titaniferous magnetites, seldom containing more than 15 per cent titanic oxide, are known to occur near Lake St. John on the Saguenay River, on Bay of Seven Islands, along Seine Bay, and in the Rainy Lake region, Quebec. Other apparently less important deposits have been reported in Renfrew, Hastings, and Hamilton counties, and in the Nipissig district, Ontario.[18] An ore body more important as a source of iron and containing less than 4 per cent titanium occurs in the Rocky Mountains near Burmis, Alberta. All these deposits are magmatic segregations genetically connected with basic eruptive rocks of the gabbro family and generally irregular and uncertain in their mode of occurrence, a characteristic of titanic iron ore deposits the world over. Large bodies of titaniferous magnetite are known in the western part of Newfoundland. The ore contains 4 to 16 per cent titanium and as much as 65 per cent iron.

In 1948 large deposits of ilmenite and ilmenite-bearing iron ores were discovered, in eastern Quebec. These deposits, reported to carry millions of tons of ore, are near Lake Allard, 400 miles northeast of Quebec City. An ore body of 125,000,000 tons was outlined by diamond drilling at Tio Lake.[19] Ore from this area developed jointly by the Kennecott Copper Corporation and the New Jersey Zinc Company is shipped by railroad 27 miles to Havre St. Pierre on the Saint Lawrence River. Then it is transported by ship 600 miles to Sorel for smelting in an electric furnace. The ore body consists of large masses of anorthosite rock, some of which are rich in ilmenite. Ore from this area contains 35 per cent titanium dioxide and 40 per cent iron. The ore in the 100,000,000 ton main body and in the 12,000,000 ton cliff ore body averages 32 per cent titanium dioxide and 36 per cent iron. Mining is confined to the main ore body, where 25 foot benches expose massive solid ilmenite faces with very little waste interspersed. Although the ore consists predominantly of ilmenite, it includes also about 15 per cent of hematite as submicroscopic lamellae within the ilmenite. This ore contains 35 per cent titanium dioxide. Estimated reserves vary from 112,000,000 to 200,000,000 tons of ore carrying 35,840,000 to 70,000,000 tons of titanium dioxide. This deposit has been developed extensively. In 1964, 545,000 tons of slag containing 70 per cent titanium dioxide were produced.

The Ivry deposit in Bresford Township, Terrebone County, is largely ilmenite with some included hematite and averages 32 per cent titanium dioxide. Available reserves are 320,000 tons of titanium dioxide.

**Alberta.** Low grade titaniferous magnetite deposits of sedimentary origin are present in the folded foothills area of southwestern Alberta.[20]

**Union of Soviet Socialist Republics.** With the exception of the Scandinavian Peninsula, Europe is not rich in titanium deposits, although there are apparently extensive deposits of low-grade ores in the Union of Soviet Socialist Republics. The main resources are the titaniferous magnetites in the Ilmen Mountains, a branch of the Urals, and it is from these mountains that the mineral *ilmenite* received its name. These ores, which contain 14 per cent titanium dioxide, 54 per cent iron, and 0.6 per cent vanadium pentoxide, are amenable to magnetic separation,[21] yielding a concentrate containing 42 per cent titanium dioxide and 37 per cent iron, and another product containing 64 per cent iron and 7 per cent titanium dioxide carrying 87 per cent of the original iron. Ore from the Kussinsk [22] area, containing iron 51.9, titanium dioxide 14.2, and vanadium pentoxide 0.66 per cent, ground to pass a 65-mesh screen and subjected to wet-magnetic separation, gave an ilmenite concentrate representing 19.5 per cent of the original weight, which contained 45.3 per cent titanium dioxide, 35.8 per cent iron, and 0.34 per cent vanadium pentoxide. The Ural deposits are reported to carry 400,000,000 tons of available ore. Other important occurrences [23] have been reported near Khibine, on the Kola Peninsula, and near Gatskavo, in the Ukraine.

Recent estimates of resources are 145,600,000 tons of titaniferous material containing 14,500,000 tons of titanium dioxide.[23a] The largest deposits in the Ural Mountains consist of ilmenite-magnetite as dikes in gabbro. Reserves have been placed at 144,900,000 tons of ore containing 13,800,000 tons of titanium dioxide. These deposits are described in Tables 4–1 and 4–2.

All basaltic rocks in the Urals can be divided into three groups according to the content of titanium dioxide in their titanomagnetites.[24] The first group contains 18 to 25 per cent titanium dioxide, which corresponds to a crystallization temperature of 1000 to 1100° C. In the second group of gabbro-diabases the titanomagnetite is a solid solution consisting of laminated intergrowths of ilmenite. These contain 12 to 14 per cent titanium dioxide, indicating a temperature of crystallization of 750 to 950° C. The low content of titanium dioxide (1 to 6 per cent) in the third group indicates a crystallization temperature of 400 to 700° C. In these rocks the accessory titanomagnetite is in the form of xenomorphic grains and consists of laminar intergrowths of ilmenite and spinel in the magnetite.

Alpine cleft hematites may contain up to 5 per cent titanium dioxide.[25] These hematites show precipitations of rutile. Ilmenite separations do not occur. There is an expansion of the hematite lattice because of the titanium content.

**Norway.** Extremely large deposits of massive ilmenite-magnetite ore are found in the southwest part of Norway between Egersund and Sogndal, and are known as Kohldahl, Kyland, Storgangen, Blaafjeld, and Lakesdal.[26] One of the largest and purest of those so far described is the Blaafjeld deposit, Jossing Fjord, in Sogndal, south of Egersund. This

## TABLE 4–1

Titanium Resources of the Union of Soviet Socialist Republics, in Tons

| Deposit | Titaniferous Material | TiO$_2$ Per Cent | TiO$_2$ Content |
|---|---|---|---|
| Pudozhgora, Kola Peninsula | 406,000 | 10.0 | 41,000 |
| Ukraine deposits, Sea of Azov sands at | | | |
| Belosarajskaja | 331,000 | 35.0 | 116,000 |
| Ganzukovskaja | 8,000 | 14.2 | 1,000 |
| Nogajskaja | 11,000 | 33.2 | 4,000 |
| Portovaja | 2,000 | 37.6 | 1,000 |
| Gakovske and Rudno-Gakovske | | | 492,000 |
| Ural Mountains | 144,880,000 | | 13,835,000 |
| Total | 145,638,000 | | 14,490,000 |

## TABLE 4–2

Titanium Resources of the Ural Mountains Region of the Union of Soviet Socialist Republics, in Tons

| Deposit | Titaniferous Material | TiO$_2$ Per Cent | TiO$_2$ Content | |
|---|---|---|---|---|
| Bilimbaevskoye | 3,400,000 | 3.5 | 119,000 | |
| Chernorechenskoye | 770,000 | 13.0 | 100,000 | |
| Kachkanar | 34,200,000 | 3.5 | 1,197,000 | |
| Kopanskoye | 17,600,000 | 11.0 | 1,936,000 | |
| Kusa (Kusin, Kusinska or Magnitnoye) | 66,800,000 | 14.2 | 9,486,000 | Concentrate contains 42% TiO$_2$ |
| Pervouralskoye | 11,000,000 | 3.0 | 333,000 | |
| Spomae | 110,000 | 6.5 | 7,000 | |
| Yubyshka Mountains | 11,000,000 | 6.0 | 660,000 | |
| Total | 144,880,000 | | 13,835,000 | |

ore contains from 40 to 45 per cent titanium dioxide and averages 42 per cent. The available reserves, to a depth of 100 meters (327 feet), have been placed at 3,000,000 long tons. It was on this deposit that the titanium pigment industry was started.

The Lakesdal deposit near by is estimated to contain 225,000 long tons of ilmenite-magnetite carrying 35 per cent titanium dioxide. These bodies are long dikes of ilmenite-magnetite from 50 to 300 feet wide and of unknown thickness in gabbro formation. The best known deposit of titaniferous magnetite is located at Rodsand, on the southeastern shore of Tingvoldfjord, and has been worked for the iron values. Run-of-mine ore contains 6.0 to 7.5 per cent titanium dioxide.

As mined the ore averages 17 per cent titanium dioxide and is concentrated to an ilmenite product containing 44 per cent titanium dioxide. The total resources of this area are placed at 225,703,000 tons of titanium dioxide.

Other deposits are at Tellnes and Rodsand. Elsewhere in Norway titaniferous deposits are common.

**Romania.** Concentrates of alluvial sands in Romania containing titanium dioxide 31, ferric oxide 38 to 42, and silica 11 to 14 per cent are employed for the manufacture of titanium pigments by the sulfate process.[27]

**Sweden.** The huge igneous sheets at Kirunavano, Sweden, contain a low percentage of titanium. Occurrences of titaniferous iron ore are known at Taberg. A deposit of ferro-orthotitanate at Sodra-Ulvon contains a large proportion of the mineral $Fe_2TiO_4$, not previously found in nature.[28]

**Finland.** The titanium ores of Finland have been described by Geiler.[29] Occurrences have also been reported in Transylvania and in the Carpathian Mountains.

**Africa.** There are extensive deposits of magnetic iron ore, containing as much as 14 per cent titanium dioxide, in the Republic of South Africa. Occurrences containing 10 per cent titanium dioxide and up to 40 per cent iron are found over a large area in the Transvaal, although the largest deposit is at Magnet Heights, north of Pretoria. Titaniferous magnetites, together with massive ilmenite, occur as veins and segregations in the norite of the Peninsula of Sierra Leone. Thicknesses of the former of more than 15 feet, and of the latter of 5 feet, have been observed. Occurrences have been reported in Rhodesia and in Dahomey. Ilmenite and rutile are associated with the diamond deposits of Sierra Leone.[30] Two deposits of titaniferous iron minerals of different types are known in Mozambique.[31] The larger consists of somewhat magnetic material, mainly titaniferous magnetite, and ilmenite with a titanium dioxide content of 10 to 13 per cent. Some goethite has been formed from the alteration of ilmenite. The smaller deposit contains ilmenite and magnetite in the form of grains 1 to 2 millimeters in diameter.

A deposit located near Betroka and consisting of an intimate mixture of ilmenite and hematite enclosed in gneiss contains 27 per cent titanic oxide; the quantity of the ore has been estimated at 3,500,000 tons. An occurrence of an ilmenite-magnetite mixture derived from gabbro is located in the valley of the Vongoabe River, a tributary of the Mahajlo. Extensive deposits of high-grade titaniferous magnetite occur at Liganga, in the Njombe district of Tanzania.[32] The reserve has been placed at 1,200,000,000 tons. Smaller deposits have been reported in The Congo.

Titaniferous iron ores of two varieties are found on the island of Madagascar.

**Asia.** Deposits of titaniferous iron ores are known to occur in Manchuria, Korea, Vietnam, and India. Large quantities of titaniferous magnetite have been reported in the Mysore State of India. These ores, which contain 11.6 per cent titanium, have not been utilized. Ilmenite, together with titaniferous magnetite, occurs in segregated masses in pegmatite veins in the Mellore District, Madras,[33] in commercial quantities. Similar deposits have been reported in the Arabian Desert.[34]

Extensive beds of titaniferous magnetite occur north of Port Stevens in New South Wales, and ilmenite and rutile are found in some permian sediments of Western Australia, Australia.[35]

## Beach and Buried Sands

As a result of the extreme stability of titanium minerals, particularly rutile, ilmenite, and arizonite, they are left behind as the country rocks are weathered and eroded away, and accumulate in the black sands of streams and beaches. These placer deposits are extremely important, since they are easiest and cheapest to work and have supplied a large part of the world production of titanium ores. Stream gravels are worked locally to a limited extent, but much the greater output is derived from ocean beaches. Natural concentration on some beaches has progressed to remarkably high degrees, for example, in the State of Travancore, India. In New South Wales, Australia, rutile, ilmenite, and zircon make up 75 per cent or more of the black sands. The mineralogical composition of the concentrate naturally depends upon the ratio of the heavy resistant minerals of the country rocks and upon the drainage area. For instance, it is impossible to get high-grade natural concentrations of ilmenite in localities in which the weathered rocks also yield large proportions of such minerals as garnet and magnetite, since these minerals accompany the ilmenite. Such a condition exists at Nome, Alaska, where large proportions of garnet are fed into the sea along with the ilmenite.

**UNITED STATES**

Sands at many places along the Atlantic, Pacific, and Gulf coasts contain titanium minerals, although the highest concentration is found along the east coast of Florida from the mouth of the St. Johns River to the town of St. Augustine. The occurrence consists of a complex black sand

## MINERAL DEPOSITS THAT YIELD TITANIUM

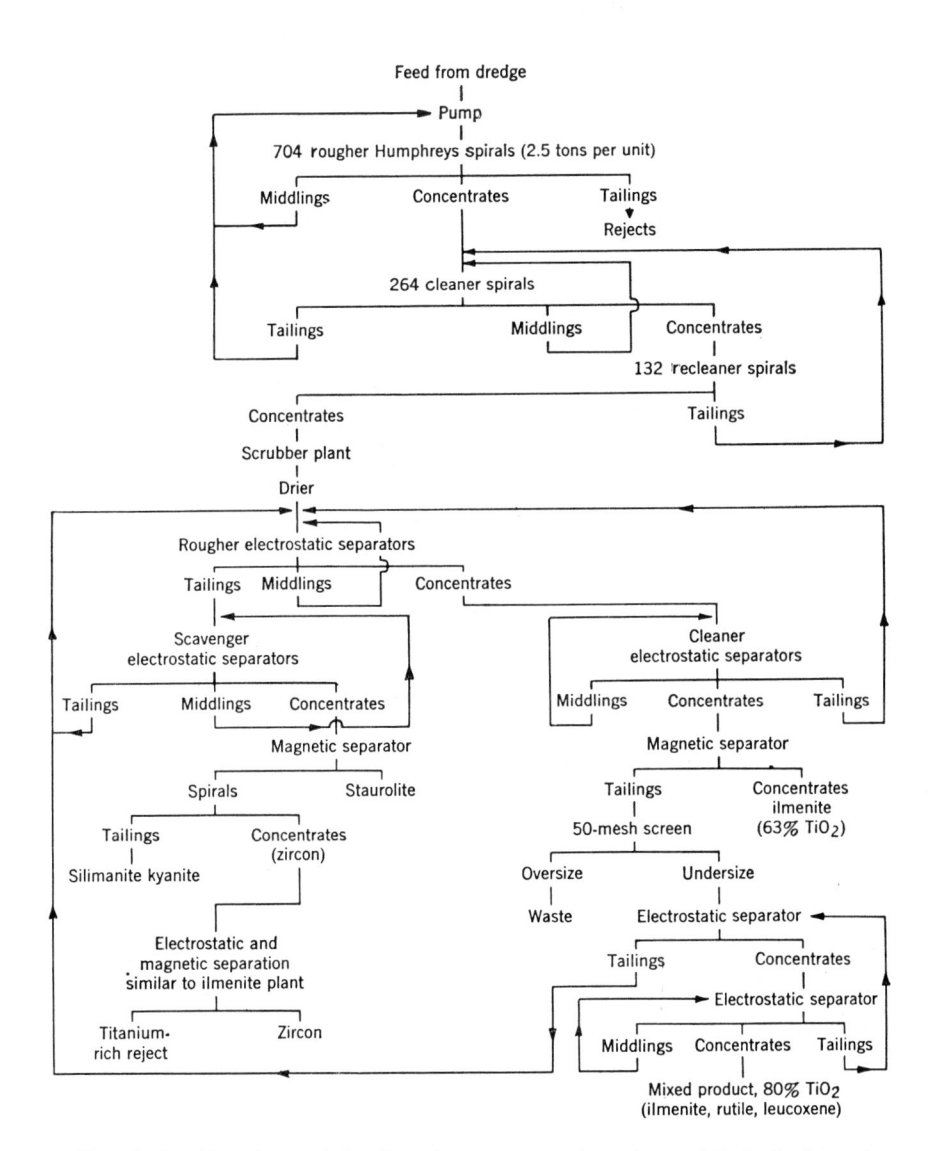

**Fig. 4–2.** Flowsheet of the ilmenite concentrating plant of E. I. du Pont de Nemours and Company at Trail Ridge, Florida. (Reproduced from Bureau of Mines Information Circular 7791.)

containing ilmenite, rutile, monazite, and zircon which have been concentrated naturally between low and high tides. A strip of the beach extending 3 miles north and 8 miles south of Mineral City, south of Jacksonville, was operated on a commercial scale for a number of years, beginning in 1918, but was abandoned in 1929. There were reports that the deposit was worked out. For a time ilmenite from this beach supplied the greater part of the requirements of the Titanium Pigment Company.[36] Around 30,000 tons of ilmenite and a few hundred tons of rutile were produced during this period. The principal beds occur on the back part of the beach at the foot of the dunes, and the heavy mineral content of the sands varies from 20 to 60 per cent.

**Florida.** Concentrations of heavy mineral sands of economic value, comprising 4.5 per cent of rutile and 45 per cent of ilmenite and leucoxene along with zircon, monazite, staurolite, garnet, and other resistant minerals, occur in buried sands, and in beach and dune deposits throughout the State of Florida. The richest developed areas are the Trail Ridge region east of Starke, the Highland district near Lawtey, an area 8 to 10 miles east of Jacksonville, and the beach and dune region near Vero Beach and Wobasso. The minerals are sedimentary in origin, their source being the Piedmont area of Georgia and the Carolinas.

Although buried titaniferous sands in Florida usually average less than 5 per cent heavy minerals, they are mined economically by dredging.[37] The sands mined by Humphrey Gold Corporation for the National Lead Company 10 miles west of Jacksonville contain 3 to 4 per cent heavy minerals. East of Starke on state-owned land Humphrey Gold Corporation operates a mine and concentration plant for E. I. du Pont de Nemours and Company. Here buried sands averaging 2 to 4 per cent heavy minerals are mined by a floating dredge. The products are an ilmenite concentrate containing 63 per cent titanium dioxide, a mixed ilmenite-leucoxene-rutile product of 80 per cent titanium dioxide, and concentrates of zircon and staurolite. Another mine of the same type is operated at Lawtey.

Extensive test drilling revealed that the buried sands of central Florida average 1 to 2 per cent heavy minerals and in some places exceed 3 per cent. Beach sands in the Pensacola Bay area contain 1 per cent ilmenite and 0.1 per cent rutile. Potentially commercial deposits were reported found on the Gulf Coast beaches near Panama City.

Analysis of drill hole samples in ten stabilized dune areas in central and northeast Florida showed an average of 1 to 2 per cent heavy mineral content although a few holes exceeded 3 per cent in grade.[38] The heavy minerals, including ilmenite, rutile, brookite, anatase, arizonite, leucoxene, and zircon, were deposited by wave action.

Heavy mineral concentrations occur in the Appalachicola shoals, Tarpon Springs, and Venice areas of the Florida west coast.[39] Deposits also occur in Baker, Bradford, Clay, and Duval counties.

The total reserves of all of the beach deposits of Florida have been estimated at 14 to 16 million tons of titanium dioxide.

Tanner, Mullins, and Bates [40] developed the hypothesis that heavy minerals carried by the Appalachicola River are being deposited on the rim of the river delta and that the heavy mineral content of the shoals increases significantly with depth.

**New Jersey.** A wooded area of alluvial quartz sand near Lakehurst in Ocean County, New Jersey, carries an economic concentration of heavy minerals.[41] This otherwise level tract is traversed by a ridge which has a maximum relief of 70 feet. The mineral bearing sands are at or near the surface and have an average thickness of 50 feet, although mining depths average about 25 feet. Dispersion of the heavy minerals in the quartz sand is erratic. For example, in two adjacent sample intervals of 2.5 feet, the heavy mineral content may range from 2 to 60 per cent. The entire sand body averages 0.5 per cent titanium dioxide, but taking the level to be mined, the concentration is 5 per cent. Reserves of heavy minerals amount to 2,000,000 tons. The heavy minerals are composed of ilmenite 82, leucoxene 3.25 to 4, rutile 1.5, and zircon 4 per cent with the remainder kyanite, sillimanite, and related minerals.

**Mississippi.** Quantitative investigations of the heavy mineral sand from recent beaches of the Gulf Coast of Mississippi and associated islands showed that ilmenite, kyanite, staurolite, and tourmaline are abundant.[42] This assemblage of minerals suggests metamorphic rocks as the original source material.

**Others.** The Bitterroot, Upper Clark Fork, and Jefferson River drainage areas of Montana contain ilmenite and other titanium bearing black sand minerals in amounts greater than 70 pounds per cubic yard.[43] Titaniferous black sandstones of Wyoming contain ilmenite, magnetite, and anatase as the principal ore minerals.[44] Monazite, rutile, and zircon have potential economic importance. Magnetic fractions show 5.6 to 71.3 per cent titanium dioxide. Large deposits of low-grade placer deposits containing ilmenite and other heavy black minerals were located in Idaho by subsurface exploration.[45] Spot samples from beaches along 220 miles of shoreline of the Alaska Peninsula indicate deposits containing up to 10 per cent recoverable titaniferous magnetite.[46]

**Arkansas.** Low-grade deposits of ilmenite-bearing sands occur in the southern part of Howard County, Arkansas.[47]

The beach sands of the entire Pacific Coast contain ilmenite in various proportions, together with magnetite, gold, and other heavy resistant minerals, although the richest strips appear to be in California. At Aptos, Santa Cruz County, an occurrence in the shape of an irregular crescent 50 feet wide extending for 200 feet along the foot of a detrital cliff carries layers of black sand, from the thickness of cardboard to 6 inches, which are separated by strata of lean sand throughout a profile of several hundred feet. The sand is said to average 16 per cent titanium dioxide and to yield a concentrate carrying 40 per cent of this constituent. Another extensive concentration is found south of Redondo, in Los Angeles County. This deposit consists of a lens-shaped body of the black sands from 14 inches to 8 feet in thickness, overlaid by gray and white quartz sand. The sands, as worked, contain 20 per cent titaniferous iron ore and magnetite mixed with silica, olivine, epidote, garnet, and zirconium silicate. Ilmenite has been produced sporadically from both these deposits.

Concentrations of ilmenite sands derived from nelsonite are known along stream courses in Virginia. Occurrences of minor importance have been described in Georgia, Texas, Maryland, Oklahoma, Missouri, Montana, Washington, and Alabama.

**Union of Soviet Socialist Republics.** Ancient marine ilmenite placer deposits occur in the territory of the western Siberian lowland.[48] The Tobal titanium deposit in the Turgai syncline in the Soviet Union was formed by wind rather than by water action. This deposit is broken into smaller units.[49] Ilmenite concentrate from the area contains 52.2 per cent titanium dioxide. Sands in the Zitomar district near Gakavske and Rudno-Gakavske contain 984,000 tons of ilmenite equivalent to 492,000 tons of titanium dioxide. Black sands of the Sea of Azov contain 14 to 38 per cent titanium dioxide. Resources are 352,000 tons of heavy sand containing 122,000 tons of titanium dioxide. Titanium bearing sands also occur in the Minus Delta, near Chutor Najdenovka and near Alabashly in the Caucasus.

**India.** Extensive beaches of highly concentrated titaniferous sands occur along the coast of the State of Travancore, which makes up the western part of the southern tip of India. Two strips along the shore line, one in the southern part of the state at Maravalakurichi and the other 80 miles to the north and 4 miles north of the town of Quilon are particularly rich. The productive section of the Maravalakurichi beach, arcuate in form, extends for 6000 feet from a rocky headland at Muttum westward toward the headland at Colachet. Sand has been mined from the locally rich areas on the beach front, where it has been naturally concentrated by the waves, and from buried, highly concentrated seams similarly formed at earlier stages in the development of the shore line. As mined the sand ranges from 50 to 70 per cent ilmenite and from 2 to 3 per cent rutile.

The second beach begins just south of Neendahara Inlet, 4 miles north of Quilon, and extends to Kayankulam Bar, 15 miles to the north. Deposits of economic value also occur in the dunes back from the shore north of Kovilthatam. Heavy minerals constitute 40 to 80 per cent of the beach deposits. The ratio of rutile to ilmenite is 1 to 25.

Other beach deposits having concentrations of titaniferous materials occur on the Chowghat-Ponnani coast, in the Tinnevelly district near the mouth of the Vaippar and Kallar rivers, near Tranquebar on the Madras coast, and along the Malabar coast. Reserves of the State of Travancore have been placed at 50,000,000 tons of ilmenite and 2,600,000 tons of rutile.

Rutile, ilmenite, and titaniferous magnetites are common in many primary deposits of the interior of the country. The Maravalakurichi concentrate as shipped, contains 54 per cent titanium dioxide, while that from Quilon is higher, around 60 per cent.

In Malaysia, a considerable proportion of ilmenite, along with zircon, accompanies the tin ore, cassiterite, in the placers and is obtained as a waste product from the mining operations. A large part of the ore comes from Perak, since the tin mining is concentrated there. At plants where magnetic separation is employed, almost pure ilmenite is produced incidental to the concentration of cassiterite. According to Hess,[50] the ilmenite is loaded on ships at a nominal cost and shipped to England for pigment manufacture. There are numerous deposits on the coast of Vietnam, and the magnetic sands of Ceylon are potential sources of ilmenite for commercial purposes.[51]

**Australia.** Workable deposits of ilmenite occur in Australia at Naracoopa, at the mouth of Fraser River on King Island, 60 miles north of Tasmania, in Bass Strait, and along the central eastern coast from Shoalhaven River, in New South Wales, toward Brisbane, in Queensland. The black sands of the Naracoopa beach, carrying ilmenite along with cassiterite, monazite, and gold, are 8 feet thick and yield a magnetic concentrate carrying 45 per cent titanium dioxide. Beach sands along the coast have been concentrated naturally by the action of waves, and these beaches have been worked intermittently for their tin, gold, and platinum content since the turn of the century. The largest deposits are north of Port Macquarie. Similar sands along the coast near Fitzroy, New Plymouth, New Zealand, contain 2.62 to 9.6 per cent titanium dioxide.[52] Water-sorted titaniferous iron sands on the beaches at Patea, South Taranaki, have been estimated at 12,250,000 tons.[53]

Remnants of old beaches extend a few feet above the present beach. According to Fisher,[54] reserves of ilmenite and rutile along the east coast of Australia are adequate for several years.

Japan. Japan has extremely large deposits of titaniferous magnetite sand derived from her many and widespread lavas and metamorphic rocks.[55] Such sands occur on most beaches of the main islands of Honshu, Kyushu, and Hokkaido. The richest deposits, which are on the east coast of Aomori and Iwate prefectures, Honshu, consist of thick layers of magnetic sand cemented with ferric oxide mud. The available reserves were placed by a government engineer at 10,000,000 tons, carrying, as mined, 20 to 30 per cent iron, 8 to 12 per cent titanic oxide, and 0.6 per cent vanadium pentoxide. These sands, which are amenable to concentration, give a product containing 60 to 65 per cent titanium dioxide. A strip 70 miles long and 20 miles wide, centering around Kuji, 100 miles from the north tip of the island of Honshu, has been raised above sea level and is blanketed with a layer of titanium-bearing sands averaging 18 feet in thickness. Minimum available reserves of this region have been placed at 200,000,000 tons of ore. From X-ray studies, Ando and Nitto[56] found that the Japanese ilmenite-rutile ores have the rutile structure and consist of a solid solution of rutile and mossite.

Natural concentrations of magnetic-iron sands containing ilmenite are found along the coast of the Island of Java.

Placer deposits of ilmenite, ilmenite-hematite, and ilmenite-magnetite of economic importance occur on the island of Hokkaido, Japan.[57] Fujiwara and Futa[58] investigated the titaniferous iron sands in Hokkaido and reported many chemical analyses. Some of these deposits are being worked for ilmenite.

Philippines. Seven areas of beach sand along the eastern coast of Surigao Peninsula, Philippines, have a total reserve of 143,000 metric tons of magnetite-ilmenite containing 50 to 56 per cent iron and 4.8 to 11.7 per cent titanium dioxide.[59]

## AFRICA

Senegal. Deposits of heavy black sands occur along the beach of Senegal, West Africa, from Rufisque to the north of the mouth of the Saloun River and near the mouth of the Casamance River.[60] The sands, which are of unknown but probably marine origin, are transported by high tides, particularly during the winter season. Zircon and quartz are mixed with the titanium-bearing component which, although generally called ilmenite, contains 55 to 60 per cent titanic oxide. Because of this chemical composition it is not a true ilmenite. Grains of quartz are coarser than those of zircon so that screening offers a partial separation, and screening combined with magnetic treatment effects complete separation.

Later investigations indicated that the titanium is present as the mineral arizonite. These ores contain a relatively large proportion of chromium which has replaced ferric iron. The deposits constitute a large potential source of ore containing around 52 per cent titanium dioxide. The chromium content by the conventional methods of manufacture, however, tends to give a discolored pigment.

**Republic of South Africa.** In many places along the southeastern coast of the Republic of South Africa the beaches contain deposits of heavy mineral sand. The best deposit, near Umgababa, contains 170,000 tons of rutile and 2,200,000 tons of ilmenite. The largest titaniferous deposits in the Republic of South Africa are in the Transvaal as ilmenite-magnetite layers in the Bushveld lopolith, a large saucer-shaped body of ferromagnesian igneous rocks. Although the titanium-bearing layers contain 8 to 24 per cent titanium dioxide, the material cannot be concentrated economically at the present time. The ilmenite is fine grained and intergrown with the magnetite. This ore deposit contains 400,000,000 tons of available titanium dioxide.

In South West Africa, 18 miles northeast of Karibab, is a small deposit of rutile.

**Mozambique.** Radioactive sands in Pebane, Mozambique, contain ilmenite 55 to 81, zircon 8 to 20, monazite 2 to 12, and rutile 1 to 3 per cent.[61] Ilmenite occurs in beach sands over a wide area of Kenya in concentrations up to 6.7 per cent titanium dioxide, but the average grade is less than 1 per cent.[62]

**Ivory Coast.** Ilmenite sands containing 37 per cent titanium dioxide occur along the Gulf of Guinea, Ivory Coast. Shaking screens yielded a concentrate containing 44.3 per cent titanium dioxide, 14.2 per cent silica, and 12.4 per cent zirconium oxide, and further magnetic treatment raised the titanium dioxide to 52 per cent and eliminated most of the other constituents. Other deposits have been reported in South Africa.[63] The black sands of Egypt are amenable to selective flotation.[64] Ilmenite and magnetite were removed by sodium sulforicinate with oleic acid in an alkaline solution. Concentrations of heavy black sands have also been reported in Gambia, on the Gold Coast, in Nigeria and South Rhodesia, and on the island of Madagascar.

## LATIN AMERICA

Brazil. Large beach deposits of ilmenite, mixed with monazite, zircon, garnet, and quartz sand occur along the southeast coast of Brazil.[65] The principal beaches are located at Guarapary and Boa Vista, in the State of Espirito Santo, in the vicinity of Prando, in the State of Bahia, and just south of Barra do Itabapoana, in the State of Rio de Janeiro. By a system of concentration and magnetic separation, an ilmenite product may be obtained containing around 50 per cent titanium dioxide. The deposits at Espirito Santo have been developed and worked sporadically. There is a very large quantity of available ilmenite on the Brazilian coast. According to Abreu,[66] these sands were derived from archean granites and gneisses. In places the titanium dioxide content of the sands may be as high as 43.6 per cent. A concentrate containing 71.6 per cent ilmenite, 6.0 per cent monazite, and 12.9 per cent zircon may be obtained by water tables and magnetic separation.

Argentina. Considerable quantities of beach sands are found, said to contain 24.7 per cent ilmenite, 29.2 per cent magnetic iron sand, and 27.9 per cent ferruginous black sand, not magnetic. The principal deposits extend southward from Meconchea through the Straits of Magellan. Similar occurrences have been reported in Guatemala and Mexico.

## EUROPE

Portugal. In Europe there is a workable deposit of ilmenite sands in the district of Castelo Branco, Portugal, and production from this area has amounted to a few hundred tons a year. Other occurrences have been reported in Italy, Sweden, Switzerland, Yugoslavia, and Czechoslovakia.

# Rutile

Deposits of rutile are known to occur in the United States in Arkansas, Florida, and Virginia, in Algeria, Australia, Brazil, Cameroons, Canada, Ceylon, Congo, Egypt, Equatorial Africa, Greenland, India, Mexico, Norway, Senegal, Sierra Leone, Republic of South Africa, and the Union of Soviet Socialist Republics. Of these deposits, those in Florida, Virginia, Australia, Brazil, Egypt, India, Norway, and Sierra Leone are the more important. Except for the deposits in Australia and Sierra Leone, the rutile is associated in small proportions with ilmenite, the primary mineral.

The world resources of rutile are given in Table 4–3.

**TABLE 4–3**

World Resources of Rutile

| Country | Titaniferous Material, Tons | Grade Per Cent TiO$_2$ | Rutile, Tons | |
|---|---|---|---|---|
| United States | | | | |
| Arkansas | 2,250,000–12,500,000 | 3.1 | 73,500–407,000 | Beneficiation not economical |
| Florida | 114,600,000 | 0.5–1.5 | 570,000 | Important producer of rutile concentrate |
| Virginia | 5,330,000 | 3.0 | 158,000 | |
| Australia | 10,503,000 | | 2,689,000 | Major source of rutile |
| India | 80,000,000 | | 2,000,000 | In Travancore State. Mined primarily for ilmenite. Rutile a secondary product. |
| Mexico | 27,500,000 | 17.5 | 5,052,000 | |
| Sierra Leone | | | 3,000,000 | Potential reserves 30,000,000 tons. |
| Republic of South Africa | 25,428,000 | | 175,000 | Beach deposit at Umgababa. Lode deposits in South West Africa. |

**Virginia.** According to Hess and Gillson,[67] the rutile deposit at Roseland, Virginia, is a broad pegmatite replacement of a peculiar aplite which was intruded into granite gneiss. In areas highly disturbed with wide and crossed cracks, the magmatic solutions penetrated more freely, and here the rutile formed in masses, some of which weighed several pounds. As in most replacement deposits of this type, the proportion of minerals varies from place to place. Both rutile and ilmenite in comparatively small grains follow the cracks through the pegmatites. This rock averages 4 to 5 per cent each of rutile and ilmenite and has been mined since the turn of the century. It is first crushed and separated on a shaking table to yield a concentrate of about equal proportions of the two minerals. The ilmenite is recovered from the ore by a magnetic separator yielding a rutile fraction containing 92.5 to 98 per cent titanium dioxide.

Areas of badly weathered pegmatites between Peers and Johnson's Spring, in Goochland County, and near Gouldin, in Hanover County, carrying dull-red rutile intergrown with ilmenite and white and green apatite, have been prospected, but the deposits proved too small for profitable development.

**Arkansas.** A large rutile deposit occurs within the Magnet Cove, Arkansas, intrusive complex in highly altered aegerine phenolite that is cut by rutile-bearing carbonate-feldspar veins.[68] Brookite occurs at the Magnet Cove area in two deposits in recrystallized novaculite associated with quartz crystals, taemiolite and rutile. These titanium deposits are believed to be of hydrothermal origin. The heavy minerals are concentrated by gravity methods and this concentrate is floated and jigged. After magnetic separation, a product containing 95 per cent titanium dioxide is obtained.[69]

**Florida.** The mineral concentrates from the black sands of east Florida, Mineral City workings, contain 6 per cent rutile. Other occurrences have been reported on Shooting Creek, North Carolina, in east Alabama, Georgia, New Jersey, Pennsylvania, South Carolina, South Dakota, and Washington.

**Quebec.** One of the ilmenite occurrences located 2 miles west of the village of St. Urbain,[70] Quebec, Canada, contains enough admixed rutile to make it of possible importance as a source of this ore alone. Rutile, accompanied with sapphirine and ilmenite inclosed in granular masses of ilmenite, has been intruded into anorthosite. The ore-bearing rock, of a brownish-black color, consists of granular ilmenite sprinkled with grains of orange-red rutile, feldspar like that in the enclosing anorthosite, biotite, sapphirine, and spinel. Microscopically, the ilmenite forms a background in which the other minerals lie. The size of the rutile particles ranges from mere specks to grains 3.5 millimeters in diameter and averages 0.6 millimeter. Rutile content of the rock varies from 8 to 20 per cent.

**Brazil.** Rutile deposits in Brazil occur in the states of Ceara, Goias, and Minas Gerais. Rutile- and ilmenite-bearing deposits of beach sand are known along the coast of Rio de Janeiro, Espirito Santo, and Baia. In Ceara alluvial and residual deposits contain about 1 per cent rutile and can be concentrated to a product containing 88 to 96 per cent titanium dioxide. Water for concentration is scarce in the area. In Goias, stream gravels and beach deposits of the Araguaia, Paranaiba, and Tocantins rivers contain rutile that can be recovered as a concentrate of 92 to 97 per cent titanium dioxide. The rutile is recovered as a by-product from the mining of the deposit for gold and diamonds. Occurrences of rutile in Minas Gerais are at Andrelandia, Araxa, Ayuruoca, Bon Jardin, Lima Durante, and Uberlandia. Most of the rutile occurs in residual or alluvial deposits overlying pegmatites and schist.

The best deposits in the State of Rio de Janeiro occur along the beach that extends southward 12 miles from the mouth of the Barra de Itabapoona River. In the State of Espirito Santo the best deposits are from south to north at Boa Vista de Siri; at Maratayses, between Barra do Itapemirim and Guarapury; at Cerapebus, Capubal, Jacareipe, and Boa Vista de Nova Almeida; near Ara Cruz; and near the mouth of the Rio Doce. In the State of Baia the best deposits occur along the southern coast at Guaratiba and Comoxatiba, just north of Caravellas.

Production from Minas Gerais ordinarily contains ilmenite, and the ore accordingly analyzes only 70 to 85 per cent titanium dioxide.[71] That from Goyaz is of a better quality, however, and runs up to 95 per cent or even more. These deposits have been worked on a limited scale for a number of years. At Bon Jardin [72] the rutile occurs contaminated with oxides of iron, most of which are in the forms of ilmenite, but after crushing the crude material to a particle size of 0.15 to 0.20 millimeter in diameter, magnetic separation gives a high-grade rutile concentrate. Samples from British Guiana showed 95 per cent titanium dioxide. Occurrences of minor importance have been reported in Mexico and Guatemala.

**Union of Soviet Socialist Republics.** Rutile occurs in the central Kyzyl-Kum.[73] The mineral is associated with ilmenite in the many and extensive deposits.

**Norway.** The important deposit at Kragero, in southern Norway, carries 10 to 15 per cent rutile in association with a moderately acid plagioclase known as kragerite. Productive rock occurs as streaks or stripes in foliated granite, with an enormous dike of granite-pegmatite on one side and of olivine-hyperite with amphibolite on the other, both of which are traversed by aplite-rutile veins. The rock mass yields a black rutile concentrated containing 97 per cent titanium dioxide and abnormally high proportions of chromium and vanadium. The most prominent megascopic constituents of the ore are light-gray and pinkish feldspars, and some rutile with a little quartz. The rutile grains rarely exceed 1 millimeter in diameter, and many are less than ½ millimeter.

**Greenland.** Rutile and titanite occur in the sands of East Greenland.[74]

**Africa.** Rutile is known to occur at a number of localities in Africa. The sands of Yaounde River, Cameroons,[75] contain commercial quantities, and a small production of rutile has been reported in recent years. It is also found in the former German colonies,[76] and the sands of Souf, south of Constantine, Algeria,[77] contain a considerable proportion of this mineral. Occurrences have also been reported in The Congo, West Africa, and South Africa.

**Sierra Leone.** An extensive rutile ore body lies on a strip of coastal plain 25 miles long and 16 miles wide in the southwest corner of Sierra Leone in west Africa. Potential reserves are 30,000,000 tons.[78] A richer deposit of 3,000,000 tons of rutile has been proved by drill holes. The ore will be mined with dredges. Indications are that rutile from this deposit will be sold in the open market.

**Senegal.** Rutile accompanies the ilmenite in the sand deposits that occur along the coast for 150 miles southeast from Dakar. In this area the best deposits are on the beach near Rufisque, in sand dunes between Joal and Palmarin, and on the beach at Casamance. The sands as mined generally contain 40 to 60 per cent ilmenite. Other deposits include titaniferous heavy mineral sands in the beaches between Dakar and St. Louis, and placer deposits in the Ketiou-Ko River.

**Madagascar.** Rutile is found in some abundance in Madagascar.[79] The most important occurrence is in the mica schists to the west of Ambstrofinandrahana and north of the Matsiatra River. It occurs as large crystals in the enclosing rocks, and small quantities of the loose mineral have accumulated in placers as a result of weathering. Ore from this deposit was used during the first World War for the manufacture of titanium tetrachloride to be used as a screening agent.

**Australia.** A major source of rutile in the world, Australia has reported resources of 2,689,000 tons of rutile and 4,289,000 tons of ilmenite in the many beach sand deposits, and 3,944,000 tons of titaniferous materials containing 420,000 tons of titanium dioxide in the primary deposits. The largest beach deposits are in Queensland, while the principal primary deposits are in New South Wales and Western Australia. In the beach deposits the ratio of rutile to ilmenite is about 2 to 3, while in deposits of other countries the ratio is about 1 to 20.

In New South Wales extensive beach deposits occur at Byron Bay and extend about 45 miles south from the Queensland border. There are also rich deposits at Swansea. In the interior several small bodies of titaniferous magnetite occur in sandstone near the junction of Williams and Karuh rivers.

Rutile and ilmenite sands occur at various places along a 50 mile strip of coastline in Queensland extending northward from the border of New South Wales. Also important deposits occur as much as 180 miles north of the border. There is a large deposit in the dunes on the main part of Standbroke Island.

In Western Australia, except for a ridge of titaniferous magnetite and hematite at Gabanintha, all the deposits are beach and dune placers. The principal of these are located at Bunker's Bay, Cheyne Bay, Doubtful Island Bay, Minninup, Wonnerup, and Gabanintha.

Residual deposits of rutile occur in the vicinity of Blumberg, Williamstown, and Mount Crawford in South Australia. Lenses of rutile occur in schists at Yankalilla.

The titanium resources of Australia are given in Tables 4–4, 4–5, and 4–6.

### TABLE 4–4
Titanium Resources of Queensland, in Tons

| Deposit | Grade Per Cent Heavy Minerals | Rutile, Tons | Ilmenite, Tons |
|---|---|---|---|
| Indian Head, Frazer Island | 65 | 16,500 | 54,500 |
| North Stanbroke Island | | | |
|   Beaches near north end | 12.4 | 174,700 | 309,100 |
|   Dunes in main part of island | 3.0 | 1,814,400 | 2,889,600 |
| Surfers Paradise to North Burleigh | 16.7 | 24,900 | 16,900 |
| North Burleigh to Burleigh | 12.6 | 42,200 | 28,800 |
| Burleigh | 15.7 | 3,500 | 5,500 |
| Palm Beach | 19.2 | 18,800 | 11,800 |
| Flat Rock Creek | 13.7 | 2,900 | 1,800 |
| Tegun–Billings | 17.9 | 9,500 | 7,500 |
|   Total | | 2,107,400 | 3,325,500 |

### TABLE 4–5
Titanium Resources of New South Wales, in Tons

| Deposit | Grade Per Cent Heavy Minerals | Rutile, Tons | Ilmenite, Tons |
|---|---|---|---|
| Tweed Heads to Fingal Point | 13.7 | 14,800 | 11,400 |
| Fingal Point to Cudgen Headland | 18.8 | 25,000 | 15,800 |
| Cudgen Headland to Norries Head | 17.2 | 250,200 | 152,000 |
| Norries Head to Hastings Point | 14.2 | 63,800 | 44,800 |
| Hastings Point to Potts Point | 29.3 | 27,800 | 17,300 |
| Potts Point to Brunswick Heads | 18.8 | 76,400 | 44,000 |
| Brunswick Heads to Cape Byron | 28.8 | 40,200 | 28,100 |
| Cape Byron to Broken Head | 10.7 | 62,200 | 41,700 |
| Broken Head to Lemon Head | 14.5 | 7,200 | 4,500 |
|   Total | | 567,000 | 359,600 |

### TABLE 4–6
Titanium Resources of Western Australia, in Tons

| Deposit | Rutile, Tons | Ilmenite, Tons |
|---|---|---|
| Bunker's Bay | 3,800 | 163,700 |
| Cheyne Bay | 1,300 | 31,600 |
| Doubtful Island Bay | 1,300 | 31,600 |
| Minninup | 200 | 22,400 |
| Wonnerup | 7,200 | 304,000 |
|   Total | 13,800 | 553,900 |

**Egypt.** In Egypt rutile is associated with the ilmenite occurrences in the sands of the Nile delta. Sands near Rosetta and Damietta contain the richer deposits. Reserves have been estimated at 15,000,000 tons of black sand. Other deposits occur in gabbro at Abu Ghalqua, near Wadi-Ranga, and inland from the Red Sea coast. Preliminary estimates indicate several million tons of titaniferous material, largely ilmenite.

**Flotation.** There has been developed a type of collector particularly applicable to the flotation of ilmenite and other minerals whose suspensoids in water are negative.[80] Characteristic examples of this type of reagent are the heavily loaded quaternary ammonium salts and cetyl pyridium bromide which give a positively charged surface-acting ion in aqueous solutions. The hydrocarbon mineral-oiling end of the molecule is the positive ion (micelle). Such substances are now generally referred to as cationic agents, in contrast to xanthates and the fatty acids which have the hydrocarbon chain in the anion and are known as anionic oilers. A concentration of iron and titanium compounds may be obtained by subjecting a pulp of the sand to froth flotation in the presence of wood creosote, the sodium salt of a sulfonated mineral oil, and oleic acid or a fatty acid soap.[81] Pickens[82] obtained an ilmenite concentrate containing 52.81 per cent titanium dioxide which represented a recovery of 56.35 per cent. The ore was ground with 1.5 pounds sodium hydroxide per ton, deslimed by decantation, and given a rough silicate float at 35 per cent solids by conditioning for 5 seconds and floating for 2 minutes with 0.06 pound per ton pine oil and 0.2 pound of the reaction product of thiourea and beta chloroethyl oleate, a reagent of cationic type. This concentrate was then floated.

At the Jacksonville, Florida, plant rutile, ilmenite, zircon, and monazite are recovered from a beach deposit containing about 2.5 per cent of these minerals.[83] The sand contains 4 per cent heavy minerals, of which 40 per cent is ilmenite, 4 per cent is leucoxene, 7 per cent is rutile, 11 per cent is zircon, and less than 0.5 per cent is monazite. The remaining group of heavy minerals is made up of various silicates including sillimanite, kyanite, staurolite, tourmaline, and garnet. In the ilmenite-rutile mill, ilmenite analyzing 60 per cent titanium dioxide is obtained as a magnetic fraction from the cleaned concentrate. The magnetite fraction contains rutile and some silicates. By passing this product through three cleaner units and one recleaner high-tension unit, a rutile fraction containing 92 per cent titanium dioxide is recovered. The final tailings from the ilmenite-rutile mill is pumped to the rougher bin of the zircon-spinel plant. For removal of the remaining titanium minerals the spiral concentrate is treated on two groups of four high-tension separators. The final concentrate is given two successive passes over an induced roll magnet.

The nonmagnetic fraction from this operation is a finished product containing 98 per cent zircon, less than 0.25 per cent titanium dioxide, and less than 1 per cent of either quartz or alumina. Fifteen per cent monazite, 40 per cent zircon, and the remainder various magnetic silicates comprise the magnetic fraction. This fraction is treated further to obtain a 95 per cent monazite product.

A flotation technique was developed to obtain bulk concentrates of ilmenite, rutile, sphene, and leucoxene from the anorthosite near Roseland, Virginia.[84] First the apatite was floated by using oleic acid as the collector in an alkaline pulp with dexterin as a titanium mineral depressant. From the apatite tailings, the titanium minerals could be floated by using fluosilicic acid to depress the remaining apatite, morin to depress ferromagnesian minerals, and oleic as the collector. A bulk titanium mineral concentrate containing 66.0 per cent titanium dioxide and 0.08 per cent phosphorus pentoxide was obtained from a sample of the unaltered rock, which contained 5.45 per cent titanium dioxide and 2.81 per cent phosphorus pentoxide. Overall recovery in the concentrate was 47.1 per cent of the total titanium dioxide in the ore. Magnetic separation of the ilmenite from the rutile in the concentrate was not satisfactory because of interlocking of the various titanium minerals. Consequently a sulfuric acid attack as used in pigment manufacture was used to dissolve the ilmenite and some of the leucoxene leaving the rutile as an enriched residue that could be further upgraded by hydraulic sizing or gravimetric separation.

Electromagnetic separation of composite ore samples from Nelson and Amherst counties, Virginia, recovers 83 per cent of the titanium dioxide as an ilmenite concentrate.[85] A flotation procedure employing oleic acid as an apatite depressant produces an overall recovery of 59.3 per cent on deslimed feed. By another process laurylamine hydrochloride removes silicates in an acid pulp followed by an alkali apatite flotation, with oleic acid as the collector and dextrin as the ilmenite depressant. The ilmenite rich tailing represents a titanium dioxide recovery of 66.2 per cent.

A saprolite deposit in Virginia 700 feet wide and 1.5 miles long contains 9 per cent titanium dioxide.[86] By flotation an ilmenite concentrate containing 45.5 per cent titanium dioxide is obtained. A fatty acid-fuel oil emulsion is an effective flotation collector.

Garnet, feldspar, hornblende, and augite are separated from ilmenite by froth flotation.[87] An aqueous slurry is formed of the material finer than 60 mesh, and adjusted with hydrofluoric acid to a pH of 3 to 6. Starch and a cationic amine flotation agent are added, after which the slurry is aerated to remove the silicate minerals in the froth. The slurry may contain 20 to 50 per cent solids and may be settled to remove slimes. Sulfuric acid may replace some of the hydrofluoric acid. From 0.5 to 20 pounds of starch is added per ton of ore to depress the ilmenite. As flotation agents, the higher aliphatic amines are particularly effective and 0.2 to 1 pound is used per ton of ore. Amine halide salts, quaternary ammonium salts of the higher aliphatic series, fatty amines, and fatty amine and rosin acetates may also be used.

For example, a 60 mesh ilmenite concentrate containing ilmenite 87, garnet 4, feldspar 2, hornblende 3, augite 3, and sulfides 0.5 per cent was first desulfurized by flotation with sodium carbonate and xanthate, and deslimed. Water was added to the residue to form a slurry containing 50 per cent solids. In a laboratory flotation machine 3.6 pounds of hydrofluoric acid per ton of ore was added, followed by aeration for 3 minutes. In the next step the mixture was agitated with 3 pounds of starch per ton of ore for 3 minutes followed by the addition of 0.28 pound of rosin amine D acetate per ton of ore and agitated for 5 minutes. At this stage the slurry was aerated and the silicate froth was removed. There were three other additions and froth flotations. The final tailings contained 49.6 per cent titanium dioxide and 3.4 per cent silicates, representing a removal of 79 per cent of the silicates and a 92.6 per cent recovery of titanium dioxide.

A frothy waste solution from the production of kraft pulp is used as the collector and the frother for recovering high titanium concentrates from titanium-containing iron sands by froth flotation in Japan.[88] Ilmenite is separated from apatite and other minerals by subjecting the pulp to froth flotation at a pH less than 6.8 in the presence of an anionic promoter such as sulfate and sulfonate-free higher fatty acids, rosin acids, naphthenic acid, tall oil, and sodium, potassium, and ammonium soaps of these materials.

Beach sands having an organic film are scrubbed with an alkali hydroxide prior to froth flotation to improve the recovery of titanium minerals.[89] One ton of sand as a slurry is treated with from 0.3 to 5 pounds of the hydroxide. After desliming and washing, flotation is accomplished in a medium containing an amine collector and a fluorine containing acid depressant at a pH of less than 5.

According to Lenhart,[90] spiral concentrators are being used to recover ilmenite, rutile, and zircon from beach sands. From this concentrate, an electrostatic machine separates the zircon and a magnetic separator removes the ilmenite.

A study was made to develop a procedure from the concentration of sands containing ilmenite, rutile, and zircon by flotation.[91] To achieve selectivity and complete extraction of valuable minerals in the main flotation, it is necessary to remove the slurry carefully from the initial sands with a preliminary cleaning of the surface of the minerals. According to one procedure, by taking advantage of the different flotation activities of the minerals, zircon and the titanium minerals were floated separately. In the other version all of the valuable minerals were floated together in a bulk concentrate. Flotation is conducted with alkali naphthenate in a neutral medium. Consumption is from 2 to 3 kg per ton. The titanium-containing minerals are suppressed with water glass. By further flotation the bulk concentrate may be separated into a zircon-epidote and an ilmenite-rutile fraction, followed by magnetic separation.

The entire complex of heavy minerals in titanium-zirconium sands has good floatability with collectors of the fatty acid type.[92] Aminosilicates and zircon have the highest floatability, and ilmenite and leucoxene the lowest. Sulfate and tallow oil are preferred as collectors.

In the Soviet Union the main concentration of titanium-zirconium sands from the sea is done principally on tables, jigs, and spirals.[93] For finishing the bulk concentrate, low capacity methods of electrostatic and electromagnetic separation are used. Bulk flotation is successful only after mud and clay have been completely removed from the product. A machine was designed which removes the mud in hydrocyclones in two stages without using a second pump. Oxidized petroleum previously saponified in a 10 per cent sodium hydroxide solution at 60 to 80° C for 1 hour gave good results as a replacement for oleic acid. Magnetic separation of the frothing product gave a high-grade concentrate containing 62 per cent zirconium oxide and 1.3 per cent titanium dioxide as well as a titanium dioxide concentrate.

A mixture of rutile and zircon after a reduction roast is separated and concentrated by magnetic separation in a strong field.[94] Applicability of this method is based on the presence of iron oxide inclusions in the rutile grains and their absence in the zircon grains. As a result of calcination the magnetic susceptibility of rutile increases while that of zircon remains constant. Reduction is carried out at 900° C.

Titanomagnetites are polymetallic ores containing iron, titanium, and vanadium. The magnetites containing iron and vanadium are separated by magnetic treatment. Then ilmenite containing iron and titanium are concentrated by selective flotation in which the impurities as cobalt pyrite are separated.[95] If the pulp solids are mechanically stirred and subjected to self-abrasion for 15 minutes in the presence of sodium fluosilicate and sulfuric acid as depressors, the ilmenite particles can be floated later with a higher yield of froth. Frothing agents are kerosene and tall oil. The tall oil contains 30 to 35 per cent fatty acids and the remainder, pitch tar derivatives. The concentrate obtained by hydrogravitation and electromagnetic methods from the detritic products of dacite contained titanium dioxide 38.0 to 40.7, ferrous oxide 34.2 to 7.4, ferric oxide 16.2 to 21.2, alumina 2.9 to 3.3, calcium oxide 2.4 to 2.3, manganese oxide 0.4 to 0.5, and silica 0.3 to 0.4 per cent.[96] This concentrate

## TABLE 4-7

Analyses of Ilmenite Ores

| Chemical Constituent | United States — Virginia — Piney River | Roseland | New York | Florida | California | Canada — Ivry | Bourget | Allard | Brazil |  |
|---|---|---|---|---|---|---|---|---|---|---|
| $TiO_2$ | 44.3 | 51.4 | 44.4 | 64.1 | 48.2 | 42.5 | 22.4 | 37.3 | 61.9 | 48.3 |
| FeO | 35.9 | 37.9 | 36.7 | 4.7 | 39.1 | 31.1 | 36.9 | 26.3 | 1.9 | 32.4 |
| $Fe_2O_3$ | 13.8 | 1.6 | 4.4 | 25.6 | 10.4 | 20.7 | 31.2 | 30.0 | 30.2 | 16.6 |
| $SiO_2$ | 2.0 | 4.6 | 3.2 | 0.3 | 1.4 | 0.88 | 1.0 |  | 1.6 | 1.3 |
| $Al_2O_3$ | 1.21 | 0.55 | 0.19 | 1.5 | 0.2 | 1.05 | 6.01 |  | 0.25 | 0.3 |
| $P_2O_5$ | 1.01 | 0.17 | 0.07 | 0.21 |  |  | 0.93 | 0.004 |  |  |
| $ZrO_2$ | 0.55 |  | 0.006 |  | 0.05 |  |  |  | 0.07 | 0.1 |
| MgO | 0.07 | 2.35 | 0.80 | 0.35 | 0.6 | 2.0 | 1.50 |  | 0.3 | 0.1 |
| MnO | 0.52 | 0.70 | 0.35 | 1.35 | 0.1 | 0.04 |  | 0.10 | 0.3 | 0.6 |
| CaO | 0.15 | 0.59 | 1.0 | 0.13 | 0.1 | 0.1 | 0.55 |  | 0.1 | 0.1 |
| $V_2O_5$ | 0.16 | 0.07 | 0.24 | 0.13 | 0.05 | 0.36 |  | 0.39 | 0.2 | 0.06 |
| $Cr_2O_3$ | 0.27 |  | 0.001 | 0.1 | 0.03 | 0.15 |  |  | 0.1 | 0.5 |
| $SnO_2$ | 0.001 | 0.02 | 0.001 |  | 0.001 | 0.001 |  |  | 0.001 | 0.003 |
| CuO | 0.0005 | 0.0005 | 0.004 |  | 0.005 |  |  | 0.46 |  |  |
| CoO | 0.005 |  |  |  |  |  |  | 0.018 |  |  |
| PbO | 0.005 |  |  |  |  |  |  |  |  |  |
| NiO | 0.05 |  |  |  |  |  |  | 0.04 |  |  |
| $WO_3$ |  |  |  |  | 0.005 |  |  |  |  |  |
| $ThO_2$ |  |  |  |  |  |  |  |  |  |  |
| $CeO_2$ | 0.02 | 0.01 | 0.002 |  |  |  |  |  |  |  |
| $U_2O_5$ |  |  |  |  |  |  |  |  |  |  |
| Au |  |  |  |  |  |  |  |  |  |  |
| Pt |  |  |  |  |  |  |  |  |  |  |
| Cb |  |  | 0.002 |  | 0.005 |  |  |  |  |  |

is used for the manufacture of titanium pigments. Cupferon combined with liquid technical soap and kerosene is used for the separation of sphene, titanomagnetite, and ilmenite from aegirite in the flotation of titanite-aegirite-nepheline tailings obtained in apatite production.[97] To increase the selectivity of the separation of ilmenite-magnetite concentrates by flotation, sodium humate is used as the depressor reagent for magnetite.[98]

Cationic collectors are more effective for ilmenite than anionic collectors.[99] Sodium alkyl sulfate and sulfurized oxidized oil recycle are effective collectors for ilmenite and magnetite in acid media. Separation of ilmenite from magnetite is possible at pH of 2.5 to 3. Preliminary separation of titanium-zirconium sands is done with screw separators.[100] The electric separation is carried out with drum separators of the corona type.

Australian sands consist primarily of quartz sand with up to 50 per cent of the mixture of heavy minerals.[101] The major components are zircon 25 to 60 per cent, rutile 25 to 50 per cent, and ilmenite 5 to 50 per cent. Primary concentration is by gravity using spirals or various

## TABLE 4-7 (Continued)

| Australia | Europe | | | | Asia | | | Africa | | |
|---|---|---|---|---|---|---|---|---|---|---|
| | Norway | Union of Soviet Socialist Republics | | Portugal | India Travancore M.K. | Quilon | Malaysia | Senegal | Sierra Leone | Rep. of S. Africa |
| 55.3 | 43.9 | 44.0 | 52.2 | 64.1 | 52.2 | 54.3 | 60.3 | 51.7 | 55.3 | 56.0 | 42.3 | 49.5 |
| 26.7 | 36.0 | 31.4 | 13.1 | 24.3 | 42.1 | 26.0 | 9.7 | 38.5 | 26.7 | 14.2 | 28.0 | 37.2 |
| 15.4 | 11.1 | 16.9 | 26.6 | | | 15.5 | 24.8 | 3.77 | 13.0 | 28.2 | 25.0 | 10.5 |
| 0.20 | 3.28 | 1.84 | 1.35 | 1.28 | 0.27 | 1.40 | 1.4 | 1.00 | 0.7 | 0.9 | | 0.76 |
| | 0.85 | | 1.50 | | 0.29 | 1.10 | 1.0 | 1.56 | 0.5 | 0.5 | | |
| 0.04 | 0.12 | 0.15 | | | 0.03 | 0.26 | 0.17 | 0.09 | 0.19 | 0.15 | | 0.02 |
| | 1.09 | | | | | 2.18 | 0.60 | | 0.10 | 2.37 | | 0.05 |
| 0.29 | 3.69 | 2.76 | 0.46 | 2.06 | 0.03 | 0.85 | 0.65 | 0.21 | 0.02 | 1.90 | | 0.59 |
| 1.64 | 0.33 | 0.72 | 1.88 | 2.71 | 5.00 | 0.40 | 0.40 | 3.15 | 0.70 | 0.20 | | 0.88 |
| 0.17 | 0.18 | | 0.51 | | 0.09 | 0.08 | 0.15 | | 0.50 | 0.10 | | |
| 0.19 | 0.20 | | | 0.16 | | 0.20 | 0.26 | 0.04 | 0.07 | 0.27 | 0.38 | |
| 0.03 | 0.03 | 0.05 | | 2.75 | | 0.07 | 0.14 | 0.02 | 0.03 | 0.23 | | 0.16 |
| | 0.05 | | | | | 0.06 | 0.01 | | 0.25 | | | |
| | 0.08 | | | | | 0.01 | | | | | | |
| | | | | | | 0.02 | 0.01 | | | | | |
| | | | | | | 0.02 | 0.01 | | | | | |
| | | | | | | 0.04 | 0.02 | | | | | |
| | | | | | | 0.005 | 0.004 | | | | | |
| | | | | | | 0.08 | | 0.03 | | | | |
| | | | | | | 0.08 | | | | | | |
| | | | | | | | | | | | | 0.33 |
| | | | | | | 0.006 | | | | | | |
| | | | | | | 0.02 | | | | | | |
| | | | | | | 0.1 | | | | | | |

forms of pinched sluice. The resulting concentrates are cleaned on wet tables. During these treatments leucoxene and impure rutile are eliminated. The separation of heavy minerals from each other is mainly based on differences in electrical and magnetic properties, although use is also made of surface properties, density, and grain size. In rutile concentrate the titanium-bearing grains are rutile, leucoxene, and ilmenite.

Quartz and garnet are first separated from titanium-bearing sands by tabling, after which the concentrate is separated magnetically into titanomagnetite, ilmenite, and monazite fractions with a residue of rutile, zircon, and nigrine.[102] The ilmenite concentrate is calcined in a reducing atmosphere to convert martite to titanomagnetite and again separated magnetically to yield a concentrate containing more than 50 per cent titanium dioxide. Zircon is recovered from the residue by froth flotation. A rutile concentrate containing 95 per cent titanium dioxide is obtained from the tailings by electric separation. Egyptian ilmenite is reduced with hydrogen in a three-stage process.[103] In the first stage hematite is reduced to ferrous oxide in the form of $2FeO \cdot TiO_2$ at 500 to 600° C. Then at

600 to 900° C metallic iron is liberated and titanium dioxide is reduced to a lower oxide, probably $FeTi_2O_4$. At very high temperatures $FeTi_2O_4$ is further reduced to $FeTi_2O_3$. The reaction products have greater magnetic properties than that corresponding to complete reduction of the iron oxide. Concentrates prepared by the magnetic separation of iron from the reduction product of ilmenite contain 50 to 80 per cent titanium III.[104] This lower valent titanium is oxidized.

In the Soviet Union electrostatic method is the principal process, and table and magnetic methods are auxiliary processes in separating rutile, ilmenite, and zircon.[105] In a review of the field, Finn [106] described the different types of titanium minerals, and the important primary and secondary titanium ore deposits throughout the world. Wet and dry gravity, magnetic, flotation, electrostatic, and metallurgical methods of beneficiation are described. Flow sheets and diagrams are given.

## ANALYSES OF ORES

**Ilmenite.** Based on the titanium dioxide content, ilmenite ores may be arranged roughly into four main groups, according to locality: Canada, with about 30 per cent titanium dioxide; Virginia, New York, Norway, and Union of Soviet Socialist Republics, with about 43 per cent of the dioxide; Florida, Virginia (Roseland), Maravalakurichi, Travancore, India, Malaysia, and Portugal, around 52 per cent titanium dioxide; Quilon, Travancore, India, Brazil, and Senegal, carrying approximately 60 per cent. Both the Quilon and Senegal ores usually classed as ilmenite contain a large proportion of the mineral arizonite.

The analyses of regular ilmenite concentrates from the more important deposits as collected from many but reliable sources are given in Table 4–7.

**Rutile.** Rutile concentrates, as marketed, usually average 94 to 98 per cent titanium dioxide. Representative analyses of samples from Australia showed 94 per cent titanium dioxide, from Brazil 85 to 95 per cent, from Cameroon 90 to 95 per cent, from Norway 90 to 93 per cent, and from Southwest Africa 94 to 98 per cent. A sample of rutile from Nelson County, Virginia, contained 99.28 per cent titanium dioxide, 0.40 per cent ferrous oxide, 0.28 per cent vanadium pentoxide, and 0.04 per cent chromic oxide. Another sample from Amherst County showed 99.7 per cent titanium dioxide, 0.20 per cent vanadium pentoxide, and 0.07 per cent chromium trioxide. Typical Australian rutile contains titanium dioxide 96.4, zirconium oxide 0.45, silica 0.35, ferric oxide 0.22, chromic oxide 0.19 and vanadium pentoxide 0.36 per cent.

# 5

# Production and Imports of Titanium-Bearing Ores

## PRODUCTION OF ILMENITE

The alluvial sands of Florida and the beach sands of the state of Travancore, India, and of Australia; the titaniferous magnetites of the Adirondack Mountains of New York State, of Canada, and of the Union of Soviet Socialist Republics; the lode deposits of Norway; and the nelsonite deposits of Virginia have supplied much the larger part of the world's production of ilmenite, although North Carolina, New Jersey, West Africa, and Malaysia have accounted for smaller but appreciable amounts.[1] Smaller tonnages have been produced in Brazil, Egypt, Portugal, Madagascar, and Finland.

**United States.** Significant production of ilmenite in the United States began along the east coast of Florida in 1922 to supply the needs of the Titanium Pigment Corporation. Until 1926 the output amounted to around 5000 tons annually, but it decreased to 3500 tons in 1927, and to 918 tons in 1928. Production remained well below 1000 tons a year until 1935, when there was a substantial increase in output from Virginia to supply the needs of the pigment plant at Piney River. In 1945 the expanded development, together with the pigment plant, was sold by the Virginia Chemical Company to the American Cyanamid Company. Annual production held around 15,000 tons until 1941, when deposits of ilmenite-magnetite in the Adirondack Mountains of New York were developed by the National Lead Company to supply the requirements of ilmenite formerly imported from India but cut off by war conditions. Output expanded rapidly. In 1944 a total of 200,000 tons of concentrate was obtained from 1,000,000 tons of ore mined, and in 1947 the production amounted to 300,000 tons of ilmenite concentrate. Production has increased steadily.

Late in December, 1947, E. I. du Pont de Nemours and Company announced a long-term lease of ilmenite-sand-bearing land in the north-central area of Florida and large-scale production began in 1948. Mines and concentrators are operated at Starke and Lawtey. Ilmenite from this area supplies the major requirements of the Company's pigment plants except the one in California. A fraction of weathered ilmenite known as leucoxene from the concentrators is used for the manufacture of titanium dioxide pigment by the chlorination process. Rutile is also recovered in the operations. Zircon is recovered as a valuable byproduct.

Ilmenite deposits at Finley, North Carolina, were developed in 1942 by the Glidden Company, and in 1944 the output amounted to 15,000 tons. In 1946 the production from this area reached 17,102 tons of ore, averaging 52.4 per cent titanium dioxide. No production was reported in 1962. The Glidden Company began construction of mining and processing facilities at its Lakehurst, New Jersey, ilmenite deposit in 1960, and production began in 1962.

In recent years the J. R. Simplot Company and Porter Brothers have produced ilmenite at Boise and Lowman, Idaho.

From a small beginning in 1922 production of ilmenite in the United States increased to 50,000 tons in 1925 but decreased to 1000 tons in 1930. From 1935 to 1940 production amounted to 15,000 tons a year. After a substantial increase in 1941, production reached 215,000 tons in 1944. In 1947 the total production of ilmenite in the United States was 336,533 tons. The output increased each year to 757,180 tons in 1957, decreased to 563,338 tons in 1958, but again increased to 807,725 tons in 1962. In 1964 production was 1,001,132 tons.

In 1962 ilmenite was produced by American Cyanamid Company at Piney River, Virginia; by E. I. du Pont de Nemours and Company at Starke and Lawtey, Florida; by M and T Chemicals, Incorporated, in Hanover County, Virginia; by National Lead Company at Tahawas, New York; by Titanium Alloy Manufacturing Division of National Lead Company at Skinner, Florida; by the Glidden Company at Lakehurst, New Jersey; by Florida Minerals Company at Vero Beach, Florida; by J. R. Simplot Company at Boise, Idaho; and by Porter Brothers Corporation at Lowman, Idaho.

India. The Travancore deposits of southern India were first worked for monazite in 1911 to supply the gas mantle industry, but not until 1922 was ilmenite saved when production was 406 tons. Except for a small recession in 1929, production steadily increased to 250,000 tons in 1940 to supply the expanding titanium pigment industry. As a result of war conditions and lack of shipping space, production decreased to 2000 tons in 1942, but again increased to 225,000 tons in 1946. Production reached a maximum of 376,321 tons of ilmenite in 1956 and decreased to 152,100 tons in 1962, and 11,849 tons in 1964.

**Australia.** As a by-product of the rutile demand for making titanium metal, ilmenite production in Australia increased rapidly from 600 tons in 1955 to 49,600 tons in 1957. Production reached 204,000 tons in 1962, and 343,500 tons in 1964.

**Norway.** The production of ilmenite from Norwegian deposits first reported in 1920 amounted to 67,194 tons in 1936 as a result of shipments to the National Lead plants in New Jersey and in Missouri. From 1938 to the close of World War II the entire production of slightly more than 60,000 tons annually went to Germany. In 1946 production was resumed by the National Lead Company. The total output increased steadily from 141,220 tons in 1953 to 343,723 tons in 1961, but decreased to 330,000 tons in 1962. In 1964 production reached 299,608 tons.

**Canada.** The production of ilmenite in Canada of a few thousand tons a year increased to 65,437 tons in 1943 as a result of war conditions but decreased to 7122 tons in 1947. In recent years the production reported by Canada has been largely electric furnace slag containing 70 per cent titanium dioxide produced from titanomagnetite. Production of slag increased from 151,176 tons in 1953 to 463,361 tons in 1961, with a break of only 161,312 tons in 1958. Output decreased to 301,448 tons in 1962. To produce this tonnage of slag 745,753 tons of ore were smelted. Production in 1964 was 544,721 tons.

**Malaysia.** Ilmenite production in Malaysia was 29,758 tons in 1953, 136,837 tons in 1956, 113,856 tons in 1962, and 144,754 tons in 1964.

**Finland.** Production of ilmenite in Finland increased from 3465 tons in 1953 to 117,384 tons in 1958, but decreased to 96,100 tons in 1962. In 1964 production increased to 127,937 tons.

**Egypt.** Egypt produced 3000 tons of ilmenite in 1958 and increased the output to 50,000 tons in 1961 and to 120,000 tons in 1962.

**Republic of South Africa.** The Republic of South Africa increased production of ilmenite from 3118 tons in 1957 to 99,009 tons in 1961 but dropped to 87,096 tons in 1962. No production was reported in 1964.

**Spain.** Annual production of ilmenite in Spain of less than 10,000 tons increased to 12,267 tons in 1960 and to 52,572 tons in 1962, but decreased to 44,357 tons in 1964.

**Other Countries.** Other producing countries of less importance in 1964 were Ceylon 50,880 tons, Japan (mostly slag) 2161 tons, and Senegal 2125 tons.

The world production of ilmenite not including the Union of Soviet Socialist Republics increased from 5000 tons in 1920 to 45,000 tons in 1930, 300,000 tons in 1938, 600,000 tons in 1947, 1,095,100 tons in 1953, 1,710,200 tons in 1958, 2,326,900 tons in 1961, and 2,295,100 tons in 1962. Production in 1964 was 2,576,400 tons.

## PRODUCTION OF RUTILE

**United States.** During 1962 rutile was produced by M and T Chemicals, Incorporated, Beaver Dam, Virginia, by Titanium Alloy Manufacturing Division of National Lead Company, Skinner, Florida, and by Florida Minerals Company, Vero Beach, Florida. The combined production of rutile concentrates amounted to 45 tons in 1925 and 500 tons in 1935. By 1947 the output had increased to 8562 tons. Production of 7406 tons in 1958 increased to 9981 tons in 1962, but decreased to 8062 tons in 1964.

**Australia.** Australia is by far the leading producer of rutile, with the United States a poor second. From 465 tons in 1938 output expanded to 4694 tons in 1941 and to 8843 tons in 1945. Most of the ore during the war years was shipped to the United States for welding rod coating. Production was accelerated by the demand for rutile for the manufacture of titanium metal and more recently by the commercial development of the chlorination process for the manufacture of titanium dioxide pigments. Production amounted to 7460 tons in 1947, 42,604 tons in 1953, 93,327 tons in 1958, 133,283 tons in 1962, and 201,522 tons in 1964.

**India.** From 117 tons in 1953 the output of rutile in India increased irregularly to 503 tons in 1958, to 1770 tons in 1962, and to 2062 tons in 1964.

**Other Countries.** Other producing countries of less importance in 1962 were Brazil with an output of 240 tons, Senegal 811 tons, and Egypt 1200 tons.

The world production of rutile other than the Union of Soviet Socialist Republics increased from 200 tons in 1930 to 19,588 tons in 1944, but decreased to 16,000 tons in 1947. Production in 1953 was 49,600 tons, in 1958 was 103,200 tons, in 1962 was 150,900 tons, and in 1964 was 211,600 tons.

**Composition of the Ores.** The rutile concentrates as shipped usually contain 93 to 94 per cent titanium dioxide, although values may range from 90 to 99 per cent.

## IMPORTS OF ILMENITE

Before World War II the domestic production of ilmenite supplied only a small part of the ore consumed in the titanium pigment and other industries. In 1929, the first year for which accurate figures are available, imports amounted to 15,685 tons. Imports increased steadily to 287,191 tons in 1939, then decreased to a low of 10,369 tons in 1942 because of lack of shipping space as a result of the war, but again increased to 242,826 tons in 1946. Total 1962 imports were 166,434 tons, and in 1964 totaled 119,819 tons—all from Australia, Canada, and India.

Prior to World War II, with the exception of 1935 and 1936, when 22,472 and 32,126 tons, respectively, were obtained from Norway, India supplied practically the entire amount. Imports from India amounted to 147,005 tons in 1953 and 212,479 tons in 1958 but decreased to 44,666 tons in 1961. No ilmenite was imported from India in 1962, but in 1964 imports amounted to 11,200 tons.

During the war years 1943 and 1944 imports of ilmenite from Canada were 65,437 and 32,580 tons, respectively, but dropped back to the more normal value of 6981 tons at the close of the war in 1945. In 1953 a total of 139,585 tons of high titania slag was imported from Canada. The value increased to 217,762 tons in 1957 but decreased to 108,493 tons in 1962, and 91,497 tons in 1964.

Significant amounts of ilmenite from Australia began in 1958, when 22,736 tons were brought in. The amount increased to 57,941 tons in 1962, but decreased to 17,122 tons in 1964.

## IMPORTS OF RUTILE

Imports of 364 tons of rutile in 1938 increased rapidly during the war years to 9635 tons in 1943, but dropped to 3304 tons in 1945. Practically all of this amount came from Australia and Brazil. From 16,098 tons in 1953, imports from Australia increased to 36,507 tons in 1958, decreased to 35,542 tons in 1962, but increased to 110,981 tons in 1964. Imports from the Republic of South Africa increased from 274 tons in 1959 to 1450 tons in 1951 but decreased to 424 tons in 1962. Total imports in 1964 amounted to 110,981 tons, all from Australia.

## PRICE

The price of ilmenite containing 60 per cent titanium dioxide in 1964 was 23 to 26 dollars a long ton f.o.b. Atlantic seaboard. Rutile containing 94 per cent titanium dioxide f.o.b. Atlantic seaboard was 104 dollars a ton.

## CONSUMPTION

In 1964 the United States consumed 980,426 tons of ilmenite, 128,203 tons of high titania slag, and 79,446 tons of rutile. These values compare with 731,424 tons of ilmenite, 117,581 tons of slag, and 21,677 tons of rutile consumed in 1958.

Of the total ilmenite consumption in 1964 the manufacture of titanium pigment took 977,178 tons, 576 tons went into welding rod coatings, 2625 tons were used in alloys and carbides, and 47 tons went into ceramics. In the same year 128,100 tons of high titania slag were consumed in the manufacture of pigments, and 103 tons went into welding rod coatings. Of the 79,446 tons of rutile consumed, 58,112 tons went into the manufacture of titanium metal, and 19,847 tons were used in welding rod coatings. Ceramics took 674 tons and fiber glass accounted for 813 tons.

# 6

# The Chemistry of Elemental Titanium

## PRODUCTION

The production of pure elemental titanium is fraught with many difficulties because of its high melting point and strong affinity for nitrogen and carbon as well as for oxygen. Although many attempts were made by the earlier investigators with varying degrees of success, probably the first to prepare the pure metal was Hunter [1] in 1910. Previous attempts to prepare the metal from its compounds usually resulted in nitrides, carbides, or cyanonitrides which, because of their metallic luster and appearance, were often mistaken for the element. Among the pioneer investigators were such well-known scientists as Klaproth,[2] Berzelius,[3] Wollaston,[4] Wohler,[5] and Deville.[6] In 1887 Nilson and Petersson [7] obtained a product of 94.7 per cent purity by reducing titanium tetrachloride with sodium in an airtight steel cylinder. The chief impurity was oxygen, and since this was considered to be combined as titanium monoxide, the sample contained only 78.9 per cent metallic titanium. Perhaps the nearest approach to the pure element, prior to the work of Hunter, was that of Moissan [8] who reduced titanium dioxide with carbon in a lime crucible at the temperature of a powerful electric arc. The primary product contained 5 per cent carbon, but by reheating with added titanium dioxide the amount was reduced to 2 per cent. Since the carbon was probably in chemical combination, the material was far from pure titanium metal. Two museum samples [9] examined some years later proved to be of high purity, but contained titanium carbide and iron as cell walls surrounding pure titanium crystals. Huppertz [10] is reported to have made a high-grade product, relatively free from oxygen and nitrogen, by an electrolytic process employing titanium dioxide in a calcium chloride electrolyte. The cell was a strontium metal furnace.

In his earlier work, Hunter [11] attempted to reduce sodium titanofluoride with potassium in an iron cylinder, but the best samples contained only 73.2 per cent titanium. Potassium and barium titanofluorides gave no better results. Reduction of the oxides with carbon, after the method of Moissan, yielded a product containing 4.6 per cent carbon. Finally, following the method of Nilson and Petersson, and exercising extraordinary care to exclude air from the apparatus, metallic titanium practically free from impurities was obtained. The titanium tetrachloride, prepared by chlorinating titanium carbide, was carefully purified by shaking with mercury and sodium amalgam followed by redistillation in an atmosphere of nitrogen to give a colorless constant boiling liquid. This purified titanium tetrachloride, together with the theoretical amount of sodium, was transferred to a steel bomb of 1000 ml capacity, capable of withstanding an internal pressure of 80,000 pounds. A typical charge consisted of 500 g titanium tetrachloride and 246 g metallic sodium. On heating the system to dull redness, the reaction took place with explosive violence and was almost instantaneous. The cooled and washed product consisted of small metallic beads, together with a minor proportion of the same material in the powder form. Two analyses of the beads gave 99.9 and 100.2 per cent of titanium, and no trace of iron or sodium was found. Ninety per cent of the titanium in the original charge was recovered as metal. These results were confirmed by Lely and Hamburger,[12] who reduced the tetrachloride with sodium in a steel bomb and fused the metal in an electric vacuum furnace. In a modification of this process, titanium tetrachloride was heated in a closed vessel with sodium, in the presence of a minor proportion of titanium dioxide or titanium nitride or cyanonitride.[13] The pure metal was produced in Germany by heating the tetrachloride with sodium at 800° C under a flux of sodium chloride and potassium chloride.[14]

An amorphous product of 85.7 per cent purity was prepared by reducing potassium titanofluoride with sodium.[15] After pressing the metal into sticks and fusing it in a vacuum furnace, the titanium content was found to be 97.5 per cent. According to Vaurnasos,[16] titanium tetrachloride is hard to reduce to the metal with potassium. A lower chloride formed, but on long heating the metal was obtained. Elemental titanium was also obtained by introducing the tetrachloride in vapor form into a bath of fused potassium chloride maintained at 700° to 900° C and covered with a molten layer of an alkali metal [17] in an inert atmosphere. Kroll [18] reduced titanium tetrachloride with pure magnesium in a molybdenum-lined, electrically heated crucible at 1000° C in an atmosphere of argon. The metal was separated from excess magnesium salts by leaching and acid treatment; no alloy of the two metals formed. The powdered titanium was compressed into bars and melted in a special electric vacuum apparatus. After this treatment, the product was easily hot-rolled and a strip less than one millimeter thick could be bent cold without fracture.

Attempts were made by Ruff and Brintzinger [19] to reduce oxides of titanium with sodium, calcium, and sodium-calcium alloy in a wrought-iron bomb. While sodium reduced the oxides at 900° to 950° C rather ineffectively, calcium was more effective, and a 30 per cent sodium-calcium alloy gave best results, yielding at this temperature 82 to 88 per cent recovery. Titanium dioxide was reduced to the metal by heating with calcium and a powdered, fused, 75 to 25 mixture of calcium chloride and barium chloride in an electric furnace in an atmosphere of argon at temperatures up to 750° C.[20] After a second reduction the product was ground and treated with water and strong hydrochloric acid. All efforts to remove the small proportion of lower oxides failed.

By carrying out the aluminothermic reaction in a hot electric furnace, Lohmann [21] obtained the pure metal. Merle [22] carried out the reaction between purified powders of titanium oxide and aluminum in a centrifuge, to effect better separation and recovery. Iwase and Nasu [23] found that only after mixing titanium dioxide with lampblack did reduction with atomic hydrogen and formation of carbide take place. Pure metal was produced only with difficulty. The dioxide was not reduced directly with atomic hydrogen.

Stahler and Bachman [24] obtained metallic titanium of 94 per cent purity by distilling the dichloride in an atmosphere of hydrogen at 1100° C. The tetrachloride mixed with hydrogen was reduced on passing through an iron tube at 900° C, but the inner walls of the tube combined with the liberated titanium to form an alloy.[25]

According to van Arkel and de Boer,[26] titanium prepared by reducing the tetrachloride with sodium always contains traces of oxides and nitrides, and electrolysis of double fluorides produces no better result. These workers found that metal could be produced in very pure form by heating in an evacuated vessel a compound which dissociates into the metal and a gas at a temperature below the melting point of the metal. Titanium tetraiodide fulfills these requirements. Its vapors were heated in a glass chamber, freed of air, at 650° C, and a layer of titanium was deposited on a filament (tungsten) located in the upper part of the vessel and heated electrically at 2000° C. Iodine was liberated and recombined with impure titanium powder placed on the bottom of the glass chamber so that the process proceeded until the supply of anode titanium was used up. By this method the metal was not fused and did not come in contact with impurities. The iodide gave better results than the tetrachloride, since it dissociates at a lower temperature. Titanium produced in this manner is of the highest degree of purity.

After a survey of the entire field, Fast [27] prepared titanium in a relatively impure form by the reduction at 1000° C of potassium titanofluoride with 1 per cent excess sodium, or of sodium titanofluoride with sodium in 10 per cent excess. Fluorides were extracted from the reaction product, the mass was boiled with strong sodium hydroxide solution, and the surface oxidation products were removed by treatment with dilute hydrochloric acid. Pure samples were obtained by the reduction of the tetrachloride with sodium at 700° to 800° C, or at even lower temperatures, in the presence of potassium chlorate to initiate the reaction.

Pure titanium was also obtained by the thermal decomposition of the iodide. The metal reacted with iodine vapor at room temperature and formed the dark-red vapor of the tetraiodide at 200° C, which was decomposed by a tungsten wire heated to at least 1100° C to effect deposition of the titanium. The nature of the deposit depended both upon the characteristic of the core and the temperature of the filament, but in general it was smoother at 1300° than at 1100° C. Deposition increased as the temperature of the vapor inside the closed vessel rose to 150° C, but at 200° C the reaction, $TiI_4 + Ti \rightleftarrows 2TiI_2$, became a factor. At 250° C the reaction, $2TiI_2 \rightleftarrows Ti + TiI_4$, became important. Deposition became slight at 300° C and ceased at 350° C. All iodine was believed to have been converted to $TiI_3$.

The metal may be produced by reducing the finely divided dioxide with natural gas.[28] It was first produced in large quantities by reducing the tetrachloride with magnesium in an atmosphere of argon or helium.[29] Purified titanium tetrachloride was dripped onto molten magnesium at 700° C, the resulting mixture was leached with dilute hydrochloric acid, and the powder residue was pressed into shape at 1000° C in vacuum. After considering virtually every process that had been proposed for the production of ductile titanium, Dean and others [30] concluded that this method of Kroll was the most practical for large-scale operation and developed a number of modifications. The modified method was employed by the United States Bureau of Mines to produce titanium metal on the pilot-plant scale by reduction of the tetrachloride with magnesium. After critically studying a number of methods, Hanna and Wormer [31] likewise concluded that a modification of the Kroll process was best suited for producing titanium needed in compounding getting mixtures. The tetrachloride was reduced with magnesium at 800° to 950° C to produce a product of not less than 98 per cent purity. Three methods were used by the Germans during World War II to produce the metal.[32] Potassium chloride, sodium chloride, and metallic sodium were heated in a covered iron pot to 800° C, and titanium tetrachloride was introduced, with stirring. The reaction product was cooled, removed from the pot, and treated with hydrochloric acid to separate the titanium as a residue. By another method, purified hydrogen was passed through a vaporizer containing titanic bromide into a quartz tube heated by tungsten wires at 1200° to 1400° C. As titanium built up on the wires, an increase in current was necessary to maintain decomposition temperature. According to a third process, titanium dioxide was heated with calcium hydride in a tube furnace made of molybdenum. The reaction mass was contained in shallow boats pushed into the tube heated at 700° C. Reaction was complete in ½ minute, and the boats were discharged at the opposite end of the tube. This product was treated with boiling formic acid, which dissolved the calcium oxide, leaving pure titanium as a gray powder. A similar process was employed by Alexander,[33] who recommended that the powder be packed in an inert gas such as argon or helium, to prevent oxidation.

The change in free energy, $\Delta F.° = 175{,}600 - 29.1\,T$, of the decomposition of titanium tetrachloride, $TiCl_4 \rightarrow Ti + 2Cl_2$, shows that the reaction is possible only at an unobtainable high temperature. Reduction of the tetrachloride with hydrogen, $TiCl_4 + 2H_2 \rightarrow Ti + 4HCl$, $\Delta F.° = 87{,}000 - 35.8\,T$, should be possible around 2000° C. The reaction for the reduction of titanium tetrachloride with sodium, $TiCl_4 + 4Na \rightarrow Ti + 4NaCl$, $\Delta F.° = -226{,}200 - 65.2\,T$, and with magnesium, $TiCl_4 + 2Mg \rightarrow Ti + 2MgCl_2$, $\Delta F.° = -129{,}200 - 45\,T$, are exothermic, and provide satisfactory methods for producing titanium. Both these methods have been employed commercially, but the magnesium process has gained wider application.

In the magnesium reduction process the unreacted magnesium and the magnesium chloride are either leached from the titanium by hydrochloric acid or removed by vacuum distillation. Refrigeration is employed to keep the temperature of the hydrochloric acid leach solution below 25° C since solution of the magnesium chloride is exothermic. Furthermore the liberated hydrogen dissolves in the spongy titanium. Employing sodium as the reducing agent, the chloride produced must be maintained in the liquid state, otherwise its formation on the surface of the sodium soon brings the reaction to a halt. The small temperature range between the boiling point of sodium and the melting point of sodium chloride—only 80° C—means that the temperature of the reaction must be carefully controlled.

## ELECTROLYSIS

Attempts by Keyes and Swann[34] to deposit titanium electrolytically from solutions of its simple salts in water or other solvents having high dielectric constants were unsuccessful, and they concluded that the metal could be plated only from highly ionized complex compounds. Porkony[35] succeeded, however, in electroplating it from a strongly alkaline solution of titanic oxide or hydroxide, employing a layer of the finely divided metal as cathode and copper or iron as the anode.

No cathode deposit was obtained on electrolyzing a weakly acid (0.1 normal) sulfate solution containing 1 mg titanium dioxide per milliliter at a current density of 1 to 2 milliamperes per square centimeter.[36] Depending upon the acidity of the solution and the metal used for anode, titanium was precipitated as a basic sulfate, was reduced to the trivalent state, or was oxidized to pertitanic acid. With a cathode of gold or silver amalgam, pertitanic acid was formed, particularly if the two electrodes were separated by a porous kaolin partition and oxygen was passed into the cathode compartment. Employing a lead cathode, the yellow color of pertitanic acid changed gradually to the violet color of titanous ions, while the anode became covered with orange-brown peroxide. Basic titanic compounds were precipitated by employing a zinc cathode at lower acidities. A portion of the original solution, treated with a small proportion of hydrofluoric acid and neutralized with ammonium hydroxide, gave, on electrolysis with a lead cathode, a coating of black titanium monoxide. The addition of sodium sulfate prevented hydrolysis as the hydrogen ion concentration of the solution was reduced, and thin films of metallic titanium were deposited on cathodes of heavy overvoltage such as lead, tin, and zinc. A similar deposit was obtained from a more acid solution after addition of hydrogen peroxide. The metal was deposited from a 5 to 10 per cent tartaric acid solution containing 0.5 to 1.0 per cent titanium dioxide at a current density of 30 milliamperes per square centimeter. According to Groves and Russell [37] titanium was deposited on mercury at a rate of 0.2 g per hour from a solution of the dichloride in 2 normal hydrochloric acid at a current density of 5 amperes per square centimeter. The metal was deposited on a copper cathode up to 1 micron in thickness from an electrolyte of sulfanilic acid and titanium hydroxide, using platinum or titanium-zinc anodes and a current density of 0.1 ampere per square decimeter at 20° C.[38]

The electrolysis of titanium tetrachloride [39] in the presence of commercial-grade hydrogen peroxide containing nitrogen and operating above 1000° C gave a very thin yellow adherent coating of titanium nitride on the iron cathode. The nitride coating was quite resistant to weathering; for instance, the action of mineral acids and atmospheric agencies at 1000° C was practically negligible. Electrodeposition of the metal was effected from solutions of complex organic salts of tetravalent titanium, having a concentration of titanium of 35 g per liter or more, to produce a continuous adherent film. Such a coating, however, retarded corrosion of the iron base to a slight extent only.

The electrolytic reduction of tetravalent titanium to the trivalent state was found to proceed to completion on electrodes of lead, copper, and bright platinum, but was incomplete on dark platinum, since the equilibrium $2Ti^{++++} + H_2 \rightleftarrows 2Ti^{+++} + 2H^+$ was set up and could be established from either side.[40] From polarigraphic studies with the dropping mercury cathode, Zeltner[41] found that titanium gave an increase of current at 0.8 volt as a result of the reduction from the quadrivalent to bivalent state. No reduction was observed in alkaline solutions. The cathode curve[42] for polarization as a function of current density for tetravalent-trivalent electrolytes employing a mercury jet electrode showed two branches corresponding to the reaction $Ti^{++++} \rightarrow Ti^{+++}$ and $Ti^{+++} \rightarrow Ti^{++}$. The electrode potentials[43] of titanium in ¼ molar solutions of the tetrachloride and trisulfate were found to be 0.23 and 0.18 volt, respectively, as measured against the standard hydrogen electrode. With the latter electrolyte, the addition of a trace of hydrofluoric acid increased the potential to 0.22 volt. Hydrochloric and sulfuric acids caused a decrease in the potential, while alkalies gave the usual slight increase. Except with solutions containing hydrofluoric acid, replacement experiments did not agree with the electromotive force measurments.

The oxide layer formed by anodic polarization on titanium metal was found to be of the anatase crystal modification.[44]

## PROPERTIES

Titanium falls in Group IV of the periodic classification of the elements, according to Mendeleeff, and heads Subgroup A, which also includes zirconium, hafnium, and thorium. Titanium forms salts which hydrolyze readily in solution, on heating or dilution, to yield a precipitate of hydrous oxide, and this property is taken advantage of in the commercial manufacture of titanium pigments.[45]

Analogous salts of the other members of the group show increasing stability in solution with higher atomic weights, and metallic properties become more pronounced in the same direction. Similarly, the amphoteric nature of the dioxides, which is very pronounced with titanium, ends with zirconium, and thorium oxide exhibits basic properties only. The metallic (basic) role of titanium is exhibited in such compounds as sulfates, chlorides, and phosphates, while the nonmetallic (acidic) characteristic appears in a long series of titanates and titanofluorides or fluotitanates.

In addition to the characteristic valence of four, divalent, trivalent, and pentavalent compounds are well known and others have been reported. Quadrivalent titanium is isomorphous[46] with silica in garnet, hornblende, enigmatite, mica, tourmaline, and olivine, with zirconium in the astrophyllite group of minerals, and with columbium in lavenite. Trivalent titanium is isomorphous with aluminum in titanaugite and with the rare earths in the titanite group.

Titanium is one of the few elements that can be made to "burn," that is, react with incandescence in an atmosphere of nitrogen.[47] The combination takes place with great readiness and begins at a temperature of around 800° C. The metal ignites in air at 1200° C, in oxygen at 610° C, and burns with incandescence. It decomposes steam at red heat, 700° to 800° C, to form the oxide and hydrogen. Titanium is attacked rapidly by concentrated sulfuric and hydrochloric acids, and slowly by dilute sulfuric acid, but concentrated and dilute nitric acids have no appreciable effect. It does not tarnish on exposure to laboratory atmosphere or to a salt spray for 30 days. With the halogens it combines directly to form the corresponding tetrahalide; the reaction takes place with fluorine at a temperature slightly above atmospheric, with chlorine at 350° C, with bromine at 360° C, and with iodine at 400° C.[48]

Titanium has an atomic weight of 47.90 and an atomic number of 22. It is a member of the first transitional series of elements, and consequently has variable valence, forms colored ions, and its compounds yield colored aqueous solutions. From repeated mass spectrum analyses, Aston[49] detected five isotypes having atomic weights in order of abundance 48, 46, 47, 50, and 49. The principal line corresponded to isotope with mass number 48, and all the others were faint. Actual proportions of the isotopes[50] were found to be 46, 10.82; 47, 10.56; 48, 100; 49, 7.50; 50, 7.27. The upper limits for other isotypes are: 42, 0.001; 43, 0.01; 44 and 45, 0.002; 51, 52, and 53, 0.0001; and 54, 0.004. From these values the atomic weight was calculated to be 47.88. Walke[51] reported that titanium became very radioactive after bombardment with deuterons, and that the emitted radiations were mainly positrons and hard gamma rays. A radioactive isotrope of half life, 41.9 minutes, was produced by alpha particle bombardment.[52] The same effect was produced by bombardment with neutrons.[53] Exposure of titanium 46 to lithium gamma radiation gave a yield of 7.4 per cent of a radioactive isotope of half life 3 hours.[54]

From X-ray photographs of diffraction in titanium tetrachloride vapor, the atomic radius of titanium[55] was found to be 1.54 Ångstrom units. The paramagnetic character[56] decreased as the valence of the associated ions increased.

The earlier literature on the specific properties of titanium metal reveals many discrepancies which evidently arose from studies of impure samples. Hunter,[57] who was probably the first to prepare and study a practically pure sample, reported it to be a metal resembling steel in appearance, having a specific gravity of 4.50, and a melting point of 1800° to 1850° C. The sample was hard and brittle at ordinary temperatures but could be forged at red heat. Titanium prepared by the iodide method, and probably of even higher purity, was ductile while cold.[58] Rods 7 mm in diameter were cold drawn to 30 microns, and after heating to 650° C were hot-rolled into sheets 0.5 mm thick, and finally into foil 10 microns thick. The cold-worked material had a tensile strength of 126,000 psi, yield strength of 100,000 psi, proportional limit of 72,000 psi, elongation of 4 per cent in 2 inches, density of 4.5, melting point of 1725° C, and hardness Rockwell A of 65. According to Gillett,[59] samples exhibited a tensile strength of 142,000 psi at 1 per cent elongation. Vacuum annealing at 925° F resulted in a tensile strength of 56,000 psi at 12 per cent elongation. More recently a number of investigators, working with large samples of the pure metal produced on a pilot-plant scale by the modified Kroll process, have studied the physical properties of titanium using standard procedures for pressing compacts, sintering, forging, and annealing.[60] The granular metal was consolidated by pressing into compacts, at a pressure of 50 tons per square inch, and sintering for 16 hours at 950° to 1000° C in a vacuum of $1 \times 10^{-4}$ mm of mercury. The compacts are ductile and malleable, are readily cold-rolled, and may be made into sheets and bars by specific fabricating methods. Annealed samples have a tensile strength of 82,000 psi, with an elongation of 28 per cent, proportional limit of 55,000 psi, hardness of 55 on the Rockwell A scale, density of 4.5, electrical resistivity of $56 \times 10^{-6}$ ohms per cubic centimeter, and a melting point of 1725° C. The cold-worked material has a tensile strength of 126,000 psi, with 4 per cent elongation, proportional limit of 72,000 psi, and a Rockwell hardness of A 65. These excellent physical properties, together with its noncorrosive characteristic, place titanium high in the list of strong, light metals. The surface of ductile titanium can be hardened by heating the sample in a controlled atmosphere containing small proportions of oxygen or nitrogen. In a specific test, 1.3 mg of oxygen per square centimeter of surface produced an outside layer having a Rockwell hardness of C 58. The hardening effect appeared to be a solution of TiO in titanium and could be modified by temperature regulation. Samples of the ingot form of titanium metal, after annealing, showed a tensile strength of 80,000 psi, yield strength of 72,000 psi, proportional limit of 38,000 psi, and a hardness of 60 on the Rockwell A scale. Hot-forged rods showed slightly higher values. The metal can be process-annealed by heating in air at 1200° F for one hour; it is readily forged within the temperature range of 1600° to 1800° F; machinability is similar to that of austenitic stainless steel; it may be cold- or hot-rolled; and it can be readily spot-welded in an inert atmosphere. Test exposures to a 2200° F flame for one-half hour indicate that titanium is no more affected than stainless steel. The metal has excellent resistance to sea water, salt spray, and humidified atmosphere. At temperatures above 375° C it can absorb large quantities of hydrogen, resulting in embrittlement. Even small amounts of absorbed nitrogen make the titanium brittle. At 500° to 600° C, an oxide coating is formed, and at higher temperatures the gas diffuses into the lattice, causing brittleness. Titanium in ingot form has a density of 4.54, a specific heat of 0.142, a coefficient of linear expansion of $8.5 \times 10^{-6}$ per degree Centigrade, and an electrical resistivity of 8 microhms per centimeter. Tests after exposure for 450 hours at 1050° C indicate that the room-temperature yield strength, ultimate tensile strength, per cent elongation, and hardness are not affected significantly. If rubbed against hard surfaces, titanium has the unusual property of producing marks or smears believed to be the result of chemical reaction.

Titanium occurs in two modifications: *alpha,* which crystallizes in the hexagonal system, is stable up to 900° C, and *beta,* having a body-centered cubic lattice, is stable above 900° C. The transition temperature was found to be influenced greatly by traces of oxygen and probably also by nitrogen. Its heat capacity varies uniformly from 6.507 calories per mole at 200° C to 8.901 calories per mole at 817° C and then increases to an extremely high value as the temperature approaches 900° C. Above this temperature the heat capacity seems to remain practically constant at 7.525 calories per mole, which value corresponds to that of the beta modification. Electrical resistance of the metal over the temperature range of 100° to 1000° C was found to be sensitive to traces of oxygen whether free or combined, and over the same temperature range the thermoelectric force, as measured against gold, was also found to be variable and sensitive to oxygen.[61] Potter [62] found that the electrical resistance does not become a linear function of temperature at higher temperatures, as is approximately true for nontransitional elements. From resistance measurements, de Boer, Burgers, and Fast [63] determined the transition temperature to be 882 plus or minus 20° C.

Titanium and zirconium were reported to be isomorphous at nearly all temperatures in both the hexagonal and body-centered cube (high temperature) form.[64] By means of resistance measurements, the solid solution having an atomic ratio of 1 to 1 was found to have a transition point minimum at 545° C. Melting-point diagram of the system showed a minimum at 1575° C, with an atomic ratio of titanium to zirconium of 2 to 1. The transition point of pure titanium was found to be 885° C. From studies of X-ray diffraction patterns Hull [65] concluded that titanium crystallized in the holohedral class of the hexagonal system with an axial ratio of 1.59. The side of the unit triangular prism measured 2.97, and its height 4.72, Ångstrom units. Two sets of these triangular prisms make up the lattice, and the atoms of one set are in the center of the prisms of the other set. Precision measurements on a specimen 99.9 per cent pure demonstrated an hexagonal, close-packed structure with $a_0$ equal to 2.951, and an axial ratio $c$, equal to 1.590 Ångstrom units.[66] The density was computed at 4.49. According to Burgers and Jacobs,[67] the beta form of titanium, which is stable above 900° C, has a body-centered cubic lattice with two atoms in the unit cell. At temperatures just above this value, the length of the sides of the cube was found to be 3.32 Ångstrom units. Hull[68] also reported a centered-cube structure, with each atom surrounded by 8 other atoms. Hydrogen dissolved in the titanium changed the structure to a nearly cubic face centered lattice.[69] The paramagnetic susceptibilities of the system titanium-hydrogen differ from those of titanium.

Bridgman[70] subjected thin discs of titanium simultaneously to hydrostatic pressure up to 50,000 kg per sq cm and shearing stress up to plastic flow. The curve of the maximum shearing stress plotted against pressure showed no break, indicating that titanium had no transition within the range studied. The electrical behavior further evidenced the absence of transition points.[71] Irregularities in the behavior of commercial titanium at temperatures above 500° C were explained on the basis of impurities. The electrical resistance decreased slightly with increased pressure.[72] The pressure coefficient of electrical resistance at 30° C gave an abnormal direction of curvature. Compressibility of the metal was observed to increase greatly with the temperature. Clausing and Moubis[73] observed that between 70° and 273° A the change in electrical resistance with temperature was normal. At 0° C the specific resistance and temperature coefficient were found to be $8 \times 10^{-5}$ and 0.00469, respectively. Measurements were later extended to 2.2° A, and these showed that titanium gives normal temperature-resistance curves at the extremely low temperatures employed.[74] In magnetic fields up to 100,000 gausses, the increase of electrical resistance[75] was proportional to the square of the strength and, in stronger fields, followed a linear law. The electrical resistance of a single crystal of titanium containing 0.2 per cent zirconium and 0.03 per cent lead decreased greatly at 1.15° A, and the character of the curve obtained was interpreted as indicating true supraconductivity which could not be attributed to impurities.[76]

Mixter[77] burned pure ground titanium in powdered sodium peroxide and determined the heat of combustion to be 215,600 calories, but as a result of later work the value was revised to 218,500 calories. There was evidence of slow oxidation of the powder in air, even at room temperature. Specific heat curves for both titanium and titanium carbide, over the temperature range of 51° to 298° K, were normal.[78] From these data the entropies of titanium and titanium carbide were calculated to be 7.24 and 5.8, respectively, at 98.1° C.

The coefficient of expansion was found to increase from $5 \times 10^{-6}$ per degree at $-150°$ C to $12 \times 10^{-6}$ per degree at 650° C, indicating the absence of polymorphic transformation between the ranges of temperature studied.[79]

Kroll[80] investigated the effect of alloying elements on the deformability of titanium prepared by the reduction of the oxides, as determined by hot rolling. Oxygen showed the greatest tendency toward brittleness, and copper, sulfur, and all gases were harmful. The most ductile alloys were those containing nickel, iron, cobalt, molybdenum, tungsten, and tantalum. All additions increased the hardness. Titanium dissolved up to 10 per cent in molten lead and tin, and 12 per cent in molten zinc. With the first two elements it was mutually soluble in the molten state, but it was completely insoluble in the solid state.

Tests of explosibility[81] of dusts showed titanium to be one of the most explosive substances.

Molten titanium reacts with certain of the refractory oxides.[82] In vacuo, the molten metal reacts vigorously with aluminum oxide, actively with beryllium oxide, and slightly with thorium oxide. Of these highly refractory materials, only thorium oxide showed any promise as a crucible material for melting titanium.

Titanium has a higher melting point than steel but above 800° F the strength of the metal drops rapidly, and above 1300° F it absorbs excessive oxygen and nitrogen which causes embrittlement. The physical properties of high purity titanium are given in Table 6–1.

The atomic mass of titanium 48 is $47.96405 \pm 0.00019$.[83] Experimental data on the potential of the titanium III–titanium II couple support the theoretical value of $-2$ volts.[84] The oxidation reduction potential of the titanium II–titanium III system in molten potassium chloride and sodium chloride is $-1.807 \pm 0.008$ volts.[85] Trivalent titanium is bound in a more stable complex than the divalent. The standard oxidation-reduction potential of titanium II–titanium III in fused lithium chloride-potassium chloride eutectic at infinite dilution is $-1.550 \pm 0.005$ volts.[86]

The surface tension of titanium at the melting point is 1427 dynes per centimeter.[87] The temperature coefficient of surface tension is 1.075. A 5 per cent contraction was observed during the beta-alpha transformation of purified titanium.[88] This phenomenon is related to the shear mechanism occurring along one preferential crystallographic plane of the beta phase of titanium. According to Bibly and Parr[89] the martensitic transformation, beta to alpha, in pure titanium is isothermal.

## TABLE 6–1

### Physical Properties of High Purity Titanium Metal

| | |
|---|---|
| Atomic volume, cc/gram atom | 10.6 |
| Density at 20° C | 4.507 |
| Hardness, Vickers | 80–100 |
| Melting point, ° C | 1660 ± 10 |
| Boiling point, ° C | 3260 |
| Linear coefficient of expansion 20 to 300° C, micro inches per ° C | 8.2 |
| Latent heat of fusion, cal per gram | 100 |
| Latent heat of transformation, cal per gram | 14 |
| Heat of combustion to $TiO_2$, cal per gram | 4700 |
| Electrical conductivity, per cent of copper | 3.6 |
| Superconductivity, ° K | Below 1.73 |
| Crystal structure | |
|   Below 882° C | Closed packed hexagonal alpha |
|   Above 882° C | Body centered cubic beta |
| Specific heat, cal/gram/° C | 0.13 |
| Lattice constants | |
|   Alpha | a = 2.9504 |
| | c = 4.6833 |
|   Beta | a = 3.3065 |

The solubility of hydrogen in liquid titanium can be expressed by the equation, log S = 2370/T + 1.325.[90] With pressure and temperature factors the equation becomes, log S = 2370/T + 0.626 + 0.5 log p $P_h$, where $P_h$ is the pressure of hydrogen in millimeters of mercury.

Titanium is evaporated by electron bombardment in a magnetic field.[91] A suitable apparatus has a cooled copper terminal through which a titanium wire is fed into the atomization zone.

According to the mechanism of the titanium-liquid oxygen explosion reaction, the impact of a titanium surface immersed in liquid oxygen generates sufficient heat to gasify a pocket of oxygen.[92] In addition the impact tends to compress the oxygen at the local impact sites. A rapid reaction occurs at the fresh surface formed by the impact. A fresh titanium surface formed by the rupture of a test specimen reacts in gaseous oxygen under 100 psi pressure at temperatures as low as −250° F. In contact with liquid oxygen, titanium can be ignited by impact.[93] Oxidized titanium ignites at 500 pounds per square inch oxygen. Ignition of titanium at high oxygen pressures is promoted by pressures of 80 atmospheres, breakdown of the passive protective oxide film by abrasion or other means, increased temperature, fine state of subdivision, and excitation by an electric spark.[94]

The oxidation rate of titanium increases markedly in the presence of carbon monoxide, carbon dioxide, and water vapor.[95] A two-way diffusion of oxygen and titanium ions through the scale occurs during the oxidation of titanium in steam.[96] Intrusive diffusion of titanium ions occurs at lower temperatures in air.

The corrosive action of acids on titanium surfaces can be inhibited or considerably retarded by adding titanium IV ion to the solution.[97] A study of the corrosion, pyrophoric reactions, and stress corrosion cracking of pure titanium resulting from storage in fuming nitric acid showed that the tendency toward ignition sensitivity depends on a dark coating of finely divided titanium metal on the surface.[98] Titanium shows good resistance to acids and to chloride salts,[99] and it is not corroded by chlorine containing less than 0.1 per cent of water.[100]

A miniature getter pump containing 70 mg of titanium evaporated by electric heating provides a pumping rate of several liters per second with pressures of $10^{-9}$ to $10^{-10}$ ton.[101] A charge transfer complex of iodine and butyl phosphate with lambda maximum of 5000 Ångstrom units is used as a boundary lubricant for titanium.[102]

The resistance of titanium to corrosive materials is probably the result of a protective film of stable oxide or absorbed oxygen, or specifically to sea water to a hydrochloride layer. Air facilitates the formation and repair of this film. Consequently the presence of air at the metal surface is frequently an important factor in corrosion behavior. Studies by Jordan and Fischer [103] showed that titanium in many ways has superior corrosion resistance to stainless steel. Consequently it has many applications in the construction of chemical plant equipment.[104]

Crude chlorine gas from the electrolysis of sodium chloride solutions is purified by means of an electrostatic precipitator with titanium or titanium clad electrodes.[105] Titanium is sufficiently resistant to chlorine, organic chlorides, and ionized chlorine gas to be used in the process.

# 7

# Oxides and Hydroxides of Titanium

**OXIDES**

Titanium forms four well-defined oxides: the monoxide, TiO, which is feebly basic; the sesquioxide, $Ti_2O_3$, which has decided basic properties; the dioxide or titanic acid, $TiO_2$, which is amphoteric; and the trioxide or pertitanic acid, $TiO_3$, which exhibits acidic properties only. The titanosic oxide, $TiO \cdot Ti_2O_3$, $Ti_3O_4$, or $Ti(TiO_2)_2$, does not seem to be definitely established, and others corresponding to the chemical formulas $Ti_2O_5$, $Ti_3O_5$, and $Ti_7O_{12}$ have been reported, but evidence of their chemical individuality is not unequivocal. Hydroxides and hydrous oxides corresponding to most of these have been described.

**Dioxide.** From a practical standpoint, the dioxide is the most important member of the group. It is generally prepared by hydrolytic precipitation from titanic salt solutions (sulfate or chloride) by dilution, by heating, or by addition of an alkaline agent. Such products are in the hydrous form but the closely held water may be driven off by calcination at elevated temperatures.

The compound may also be obtained by the direct oxidaton of titanium metal or the lower oxides, by heating the anhydrous tetrachloride with oxygen or air at high temperature, by decomposition of salts, and by electrolysis.

Bohm[1] noted that precipitated hydrous titanium oxide glowed brilliantly for a short time on heating to dull redness, and that the extent of the change and the temperature at which it took place depended upon the method of preparation of the sample. By Debye-Scherrer X-ray spectrograms, the glow was shown to be the transformation from the amorphous to crystalline state. Wohler[2] found, however, that the titanic acid precipitated from both hot and cold solutions showed this luminescence at 500° C, and he suggested that the phenomenon was due to a type of surface fusion or sintering. The effect appeared only with rapid rise of temperature.

After ignition at 1000° C, the compound became practically insoluble in hot sulfuric and hydrochloric acids; however, if the temperature of ignition does not exceed 700° C, solution in concentrated sulfuric acid is accomplished without great difficulty. Hydrofluoric acid is more effective, and fusion with potassium acid sulfate or alkali metal bases converts the calcined product into a readily soluble form.[3] It may be dissolved more conveniently in a hot mixture of sulfuric acid and an alkali metal or ammonium sulfate. A suspension of the solid compound in dilute sulfuric or hydrochloric acid assumed a pale lilac or light blue color on addition of metallic zinc, indicating the formation of titanous ion.[4] Titanium dioxide may be volatilized[5] by heating with boron fluoride. It has been reported to react with liquid hydrogen chloride at low temperatures and with the gas at 200° C to form a yellow basic dichloride, $Ti(OH)_2Cl_2$.[6] In the presence of water it forms the trihydrate.

The oxide exists in three crystal modifications,[7] anatase, brookite, and rutile, all of which have been prepared synthetically. Experiments show that rutile, the most stable modification, is formed at the highest temperature, brookite at a considerably lower value, and anatase at a point still lower on the scale. The findings are also in agreement with the known occurrences of the three species in nature. Heating a single crystal of anatase between 800° and 1000° C yielded an aggregate of rutile, but the transformation was too rapid to follow any certain direction.[8] Brookite was converted into rutile in a comparatively ordered fashion by heating at 700° C, and in a partially ordered fashion by heating at 800° C. Pamfilov and Ivancheva[9] were not able to prepare titanium dioxide of the brookite structure from aqueous media. The product precipitated from solutions of its salts by thermal treatment in the laboratory had only rutile or anatase structure. Normally, sulfate solutions yield anatase directly, while halides and nitrates give rutile. Under special conditions, however, the results may be reversed. For instance, by seeding sulfate solutions with rutile nuclei all the titanium content may be precipitated in this form. Furthermore, solutions prepared by dissolving in sulfuric acid hydrous oxide precipitated from aqueous titanium tetrachloride have yielded rutile on hydrolysis. Titanium tetrachloride and other solutions, which normally yield precipitates of the rutile form, may be hydrolyzed in the presence of phosphoric or sulfuric acid so as to give anatase. Concentration and temperature of hydrolysis are also factors.

According to Bunting,[10] the stable form of titanium dioxide above 400° C is rutile with a melting point of 1852° C. It formed no compound with silica, but an eutectic mixture containing 89.5 per cent silica by weight was obtained at 1540° C. With alumina, a compound corresponding to the chemical formula $TiO_2 \cdot Al_2O_3$ and having a melting point of 1860° C was produced, and at 1850° C this compound formed an eutectic with alumina at 62 per cent aluminum oxide. Titanic oxide was readily crystallized from fused borax.[11]

Specific gravity varies from approximately 3.84 for anatase to 4.26 for rutile. It has the highest index of refraction of any white inorganic crystalline material, and ranges from 2.55 for anatase to 2.71 for rutile.

Baraleff [12] reported the molecular volume of titanium dioxide as 18.0. The heat of formation at constant volume,[13] as determined in a bomb calorimeter by using paraffin oil, was found to be 218.1 kg cal, and the density at 21° C was 3.863. At 20° C and constant pressure, the heat of formation of the rutile modification was determined as 218.7 kg cal.[14] The heat of formation of the oxide, under constant pressure at 19° C derived from the heat of combustion, determined calorimetrically, was reported to be 225.3 kg cal.[15] Entropy at 298.1° K was found to be 12.45 eu.[16]

Buttner and Engle [17] found that the dielectric constant of powdered titanium dioxide presented no anomaly within the temperature range of −180° to 0° C. The value decreased, however, with increasing temperature over the range of 20° to 300° C, by the same proportion for all wave lengths. From 20° to 120° C this change was 8 per cent. At higher temperatures some reduction occurred, and slightly reduced samples showed a strong dispersion, with a minimum in temperature coefficient for wave lengths below 60 m or above 350 m.[18]

From studies of three samples of rutile over a frequency range from 60 to 106 cycles and over a temperature range from 30° to 600° C, the dielectric constant [19] was found to be approximately 100. The value fell off slightly in the infra-red region, and it had a temperature coefficient of $-8.2 \times 10^{-4}$ per degree centigrade. The dielectric losses were low, particularly at higher frequencies. In the higher temperature region an exponential relationship was observed between conductivity and the reciprocal of the absolute temperature.

The surface area [20] of finely divided titanium dioxide, as measured by the absorption of vapors of relatively volatile liquids, agreed well with values obtained from the absorption of nitrogen. Volumes of 13.8 square meters per gram were obtained for rutile. Foex [21] measured the electric conductivity of compressed cylinders of oxides of titanium under various conditions. According to Earle,[22] titanium dioxide is an electrical semiconductor in which the current carriers are actually free electrons. Shomate [23] measured the heat capacities of rutile and anatase in the temperature range 52° to 298.16° K, and computed the molal entropies of rutile as 12.01 and of anatase as 11.93 at the highest temperature studied, 298.16° K.

A new system of triplets, B, has been found more recently in the band spectrum of titanium dioxide, in addition to the A, C, and D systems.[24]

Cathode rays acting on recently fused titanium dioxide excited three blue bands at 0.625, 0.550, and 0.425 micron, while the aged material exhibited only faint orange-red bands. Ultraviolet light had no effect, but under flame excitation a grayish-blue luminescence was visible at 425° C, which changed to red at 677° C, and finally to yellow at still higher temperatures. On heating to incandescence in an oxyhydrogen flame with hydrogen in excess, a reddish luminescence reached a maximum at 985° C, although with an excess of oxygen a modified blue glow was observed throughout all the visible region and at all temperatures below 1200° C.[25]

A composition of titanium in concentration of $10^{-5}$ to $10^{-1}$ with alumina, heated at 1300° C for one hour, cooled quickly and excited in a Braum tube by cathode rays under 4000 volts, exhibited a faint violet luminescence.[26] Ions[27] of titanium were emitted on heating the oxide to a white heat on a tungsten filament.

St. Pierre[28] determined the melting point of titanium dioxide to be 1840 ± 10° C. Oxygen is lost on heating, and the residue is nonstoichiometric $TiO_{2-x}$.[29] The x value depends on the temperature and on the pressure of the oxygen. Titanium dioxide and titanous oxide melt at 1870 and 1820° C, respectively. The melting-point apparatus uses a sun mirror which develops temperatures up to 3000° C in any atmosphere.

Rutile and anatase crystals take up cations with radius of 0.54 to 0.96 micron, and the anions fluorine and oxygen provided the cation-anion ratio of the added materials is 1 to 2 and the neutrality is preserved.[30] An example is the fusion of 5 g of anatase mixed with 0.2 g of zinc fluoride at 800° C for 0.5 hour and then at 1000° C for another 0.5 hour. This gives a yellowish-white pigment of the rutile structure. The combination of 5 g anatase, 0.2 g of magnesium fluoride and 0.2 g of nickel fluoride gives a canary-yellow product with the rutile structure.

Below 715° C the rate of transformation of brookite to rutile is extremely slow with little or no induction time.[31] The data are well fitted by the first order equation. The energy of activation is 60 kcal per mole, and the frequency factor is of the order of $10^{13}$ hours. A broad and small exothermic peak at 750° C is indicated by differential thermal analysis. Heat of transformation is $-100 \pm 75$ calories per mole. Except for a small energy of activation, the kinetics and mechanism of brookite-rutile transformation are similar to the anatase-rutile transformation. The kinetic results are explained qualitatively in terms of the order-disorder theory to diffusionless transformation in solids.

Titanium dioxide of stoichiometric composition has a sound lattice with n' = 1.9999 ± 0.0009 molecules per unit cell.[32] The structure of the grayish and deeper colored samples is imperfect, caused by missing oxygen ions. Up to 0.034 ions per unit cell or 0.85 per cent of all the oxygen positions are vacant. Crystal parameters of rutile as determined by Legrand and Delville [33] are a = 4.584 ± 0.002 and c = 2.953 ± 0.001 Ångstrom units. The titanium-oxygen distance is 1.975 Ångstroms, and the oxygen-oxygen distances are 2.54, 2.77, and 2.95 Ångstroms. For anatase a = 4.584 ± 0.002 and c = 2.953 ± 0.001 Ångstrom units. The titanium-oxygen distance is 1.937 Ångstroms, and the oxygen-oxygen distances are 2.41, 3.03, and 3.87 Ångstroms. More precise determination of atomic parameters of rutile titanium dioxide gave values for a = 4.594 ± 0.003, c = 2.959 ± 0.002, and x = 0.306 ± 0.001. From crystal analysis of anatase a = 3.785 ± 0.001 and c = 9.514 ± 0.006 Ångstrom units.[34] The oxygen parameter is 0.2066 ± 0.009. For rutile a = 4.5929 ± 0.0005 and c = 2.9519 ± 0.0003 Ångstrom units. The oxygen parameter is 0.3056 ± 0.0006. There are four titanium-oxygen distances of 1.946 ± 0.003, and two of 1.984 ± 0.004 Ångstroms in rutile. In anatase there are four titanium-oxygen distances of 1.937 ± 0.003 and two of 1.964 ± 0.009 Ångstroms. These values are correlated with color differences.

Absorption bands observed near 0.5 and 0.66 micron prove that oxygen deficient rutile contains anion vacancies with localized electrons.[35] Optically smooth and parallel sample plates are cut from rutile single crystals prepared by the Verneuil method. Oxygen nonstoichiometry is obtained by annealing the samples in a hydrogen atmosphere. Hurlen [36] observed that some of the data on the defect structure of rutile is inconclusive; some indicate a defect structure of the anion vacancy type, but most favor a cation interstitial defect structure. The available data are best explained by assuming the defect structure to be a combination of a stoichiometric and a nonstoichiometric one, but involving titanium interstitial ions.

Titanium hydroxide freshly prepared and dried to free flowing has a titanium to water ratio of 1 to 1.29.[37] Further drying yields the 1 to 1 meta form. Beyond this stage drying is slow. The hydroxide is amorphous. Differential thermal analysis of 1 to 1.2 and 1 to 0.7 hydrates shows dehydration with a peak temperature of 160° C, and conversion from anatase to rutile from 600 to more than 700° C. The exotherm due to crystallization after dehydration is obscured by the dehydration peak in the wetter samples. The hydroxide shows a moderately strong 3125 cm$^{-1}$ infra-red band, indicating that the amorphous state is polymeric. According to Wilska [38] the initial hydrous titanium dioxide obtained from both sulfate and chloride solutions, even after drying at 110° C, is always amorphous. At elevated temperatures the first crystallization from samples obtained from sulfate solutions is anatase. Hydrolysis by direct boiling of chloride solutions produces rutile as the initial crystalline product. Other methods of hydrolysis give either anatase or mixtures of the two crystal forms. All products from both sulfate and chloride solutions are finally converted to rutile at 700 to 920° C. No brookite is formed at any step.

A titanium dioxide compound of very high purity is prepared for the production of rutile single crystal boules.[39] Such starting material must be free from elements having ionic radii incompatible with the rutile crystal lattice, that is, those with ionic radii outside the range of 0.60 to 0.75 Ångstrom units. The material must also be free from elements that react with titanium or titanium dioxide to form chemical compounds. Ammonium titanium sulfate may be employed to prepare titanium dioxide of this purity. It is obtained by the reaction of titanium tetrachloride with ammonium sulfate. The crystals obtained after purification by recrystallization are dried and decomposed at 900° C for 1 to 2 hours to obtain the titanium dioxide starting material. This titanium dioxide has an open structure and a fluffy appearance.

Pure titanium dioxide prepared by hydrolyzing iso- and normal and secondary butyl titanate in acetic acid, dilute ammonium hydroxide and tartaric acid, yields in all cases comparable products having a surface area of approximately 300 sg meters per gram after drying at 400° F.[40] On calcination at higher temperatures, surface area and pore volume decrease substantially. The rapid degradation of pure titanium dioxide with increasing temperature can be arrested by adding small amounts of other metal oxides and silica. Dorin [41] carried out an electron microscopic investigation of the sintering of titanium dioxide crystals. Samples of the dioxide were heat treated and sintered at 800, 1100, and 1400° C for periods of time from 1 hour to 3 days.

The characteristics of titanium dioxide suspensions are determined by sedimentation and by nephelometric methods of analysis.[42] Kurylenko [43] reviewed the mineralogical, crystallographic, and optical characteristics of the oxides. The effects of oxide impurities in rutile was studied. Isotope exchange with oxygen 18 indicates the existence of anion vacancies in titanium dioxide.[44] Far infrared dielectric measurements show that rutile exhibits strong modes near 190 cm$^{-1}$ which are connected with its large value of low frequency dielectric constant. These are quite similar to the low frequency modes in strontium titanate.[45]

Reduction of titanium dioxide with hydrogen at higher temperatures gives mixtures of $Ti_3O_5$ and $Ti_2O_3$.[46] The structure of anosavite is analogous to that found in titanium dioxide.

Powdered titanium dioxide is fused and dropped from a vibrating hopper through an oxygen-hydrogen flame onto the top of a refractory rod surrounded by a muffle.[47] The melt accumulates as a black boule which on heating in nitrogen or argon at 500 to 1500° C yields a lustrous, beautifully colored gem.

**Sesquioxide.** Lower oxides of titanium have usually been prepared by reduction of the dioxide. Reduction with hydrogen at 1200° to 1500° C yielded primarily the sesquioxide, but at 1500° C some monoxide was found in most tests.[48] A sample of the dioxide, heated in a stream of hydrogen for 20 minutes at 1000° C, lost only 4.64 per cent in weight as compared with a theoretical loss of 10 per cent for complete conversion to the sesquioxide.[49] According to Nasu,[50] within the temperature range of 750° to 1000° C, the equilibrium can be expressed as $2TiO_2 + H_2 \rightarrow Ti_2O_3 + H_2O$.

Amorphous hydrous titanic oxide was converted to the sesquioxide by heating at 800° to 1000° C [51] with a carbonaceous reducing agent. Crystallization was promoted by the addition of 0.5 per cent zinc oxide or ammonium fluoride. A similar reduction was effected by calcining a finely ground mixture of the hydrous oxide with 2 to 20 per cent of a carbonaceous material such as sawdust, sugar, or charcoal.[52] Solid solutions of titanium sesquioxide in the dioxide were prepared by heating the dioxide at 800° to 1100° C in a nonoxidizing atmosphere with a reducing agent such as a titanous compound, aluminum, chromium, iron, or vanadium.[53]

According to Billy,[54] reduction by hydrogen, aluminum, and carbon failed to give well-defined lower oxides, but by reaction with an excess of the metal itself the sesquioxide was formed at 700° C and the monoxide at 1500° C.

Titanium dioxide was found to be quite stable at its melting point, about 1800° C, but dissociated into the sesquioxide at 2230° C. The heat evolved in the reaction $1/2Ti_2O_3 + 1/2O \rightarrow TiO_2$ was determined as 99,000 calories. In the presence of carbon monoxide, carbon effected reduction to the sesquioxide at 870° C, but by employing a higher temperature or by melting it with carbon, a mixture of sesquioxide and carbide was formed. By heating the dioxide in nitrogen, incomplete conversion to the sesquioxide took place, and this was also true of reduction with hydrogen. This lower oxide melted at 1900° C and tended to lower the melting point of titanium dioxide.[55] Ruff [56] obtained the sesquioxide, along with other lower oxides, by heating the dioxide in an electric vacuum furnace. The original material lost oxygen rapidly. It has also been prepared by the reaction of titanium tetrachloride and formaldehyde.[57]

Electrolysis,[58] under various conditions, of titanium dioxide dissolved in fused salts and mixtures (calcium chloride, calcium chloride and sodium chloride, cryolite, sodium pyrophosphate and sodium chloride), and of potassium titanofluoride ($K_2TiF_6$), dissolved in mixtures of fused sodium chloride and potassium chloride, gave in every case lower oxides of titanium instead of the metal.

The Debye X-ray method of following the reaction indicated that at 1000° to 1500° C reduction of titanium dioxide by a mixture of hydrogen and nitrogen gave $Ti_2O_3$ and TiO.[59] Above 1700° C some nitride was formed. At 1900° C the dioxide was completely reduced in 3 hours, with carbon in nitrogen, to the carbide and nitride which formed a solid solution.

The specific heats of TiO, $Ti_2O_3$, $Ti_3O_5$, and TiN were measured between 52° and 298° K, and the molal entropies were computed as 8.31, 18.83, 30.92, and 7.20, respectively.[60] Naylor[61] measured the heat contents above 298.16° K and developed the specific heat equations. TiO, $Ti_2O_3$, $Ti_3O_5$ exhibited transition points.

According to Breger[62] the unit cube edges of the compounds titanium oxide, nitride, and carbide decrease with the atomic weight of the nonmetal owing to decreasing radius, increasing ionization of titanium atoms, and increasing electron gas density.

The sesquioxide is a hard, violet-colored solid, crystallizing in the hexagonal system. It is readily oxidized to the dioxide, is slightly soluble in the latter compound, and imparts a bluish color. It is soluble in dilute sulfuric and hydrochloric acids, but is oxidized by nitric acid. From thermoelectrical data, the lattice energy[63] was calculated to be 3569.51 kilogram calories.

Titanium dioxide in a melt of sodium chloride is reduced with a magnesium-zinc alloy to produce the sesquioxide.[64] This mixture is heated at 850 to 880° C for 1.5 to 2 hours and then to 950° C after which it is cooled, washed with water to remove the sodium chloride, and then with dilute hydrochloric acid to remove the titanium monoxide byproduct. Powdered titanium dioxide with two times the theoretical quantity of ash-free powdered coke is heated first at 700 to 750° C under a hydrocarbon gas containing nitric oxide, carbon monoxide, or oxygen, then heated at 900 to 950° C under carbon dioxide and finally at 1000° C under hydrogen to give titanium sesquioxide suitable for semiconductors.[65]

**Monoxide.** Titanium monoxide has been prepared by the reduction of the dioxide with various agents. Employing hydrogen, small yields were obtained at 1500° C; reduction with charcoal at 1100° to 1500° C gave a solution of the monoxide and titanium carbide,[66] varying from 33 to 60 per cent of the latter. Billy[67] obtained the compound by reducing the dioxide with magnesium metal. A mixture of titanium dioxide and titanium metal, pressed into pieces 4 by 40 millimeters in size and heated in an exhausted atmosphere at 1550° to 1750° C for 2 hours, gave a sintered stick of golden yellow color. Chemical analysis corresponded to the composition of the monoxide, and X-ray examination confirmed the finding.[68]

The compound dissolved in dilute sulfuric acid with the evolution of hydrogen and it was insoluble in nitric acid, although oxidized to the dioxide. On heating the material in air, the finely powdered product became dark blue and crystalline after 1 hour at 400° C, gray-green after 1 hour at 600° C, and was converted rapidly to the dioxide above 800° C. Its density was 4.93, melting point was 1750° C, and specific resistance $2.49 \times 10^3$ ohms per centimeter. Hardness was about four times that of glass. It had a structure of sodium chloride type [69] with $a = 4.25$ Ångstrom units. The data indicated that it was not ionized. Reduction of titanium dioxide with calcium proceeded smoothly to the formation of the metal, but reduction with magnesium stops at TiO after the intermediate formation of $Ti_3O_4$.[70] Titanium monoxide of a deep chestnut color, on heating at 150° to 250° C in the presence of oxygen, became deep violet as a result of the formation of $Ti_2O_3$; at 250° to 350° C it changed to the deep-blue $Ti_3O_5$; above 350° C it reverted to the dioxide. Reduction of $TiO_2$ with hydrogen gave $Ti_3O_5$ as the final product. Calcium and magnesium, molded into pellet form with titanium dioxide, produced TiO as the end product.[71] Reduction started at 500° to 575° C.

A product reported to be titanosic oxide, $Ti_3O_4$, was prepared by heating a mixture of the dioxide and microcosmic salt in a loosely covered crucible in air. Cubic crystals of this composition were obtained by the high-temperature reduction of titanium dioxide. The chemical constitution was regarded to be $TiO \cdot Ti_2O_3$.

A blue oxide corresponding to the formula $Ti_3O_5$ was obtained by reducing the dioxide with hydrogen or carbon monoxide at 700° to 1100° C.[72] With the former gas, complete conversion was effected.

Titanium monoxide suitable for use as a starting material in the electrolytic production of titanium or as a semiconductor is prepared by reducing pigment-grade titanium dioxide with lampblack or petroleum coke in a two-stage process.[73] The mixture is first calcined at 1000 to 1200° C to produce the trioxide, which is in turn reduced at 1700 to 1750° C. An excess of carbon is avoided. Calcium carbide effectively reduces titanium dioxide dissolved in chlorides of lithium, calcium, strontium, or barium to the monoxide or carbide.[74] A homogenized mixture of powdered titanium hydride with an equivalent amount of the dioxide is calcined in vacuo above 1400°C in a corundum crucible to form pure titanium monoxide having a light brown color.[75]

Reactions as well as X-ray examinations show that titanium monoxide is not a true bivalent oxide but a metal type oxide that behaves as a solid solution of oxygen in the face centered cubic lattice of beta titanium.[76]

The trioxide behaves as a true trivalent oxide. Titanium monoxide is similar to the nitride and carbide with which it forms mixed crystals.

The pentoxide, $Ti_3O_5$, is prepared by reduction of the dioxide with metallic titanium and a mixture of hydrogen and steam in a high frequency, high vacuum apparatus at 1200 to 1600° C. In a similar manner $Ti_5O_9$ is formed.[77]

Calculations from measured heats of combustion for titanium, the monoxide, the trioxide, and the pentoxide gave values for $-\Delta H f°$ in kcal per mole: TiO 123.9 ± 0.2, $Ti_2O_3$ 362.9 ± 0.4, $Ti_3O_5$ 586.9 ± 0.6, $TiO_2$ 225.5 ± 0.2. Values of $\Delta F f°$ are TiO 116.9 ± 0.2, $Ti_2O_3$ 342.3 ± 0.4, $Ti_3O_5$ 553.1 ± 0.6, $TiO_2$ 212.3 ± 0.2.[78]

**Peroxides.** Acid solutions of titanic salts acquire a bright yellow color on addition of hydrogen peroxide, and the properties become quite different from those of the original compound. Since the coloration develops only on the addition of peroxides or percompounds, the product is thought to have a peroxide structure. Sulfate solutions give an immediate precipitate with ammonium hydroxide, but from similar solutions containing hydrogen peroxide the precipitate is insignificant and becomes appreciable only after destruction of this agent. It follows that the introduction of active oxygen into the molecule of titanium oxide promotes formation of complex ions.[79] The coloration produced by adding hydrogen peroxide to solutions containing from 4 to 80 milligrams of titanium per liter proved stable over a period of 2 years.[80]

Titanium may be precipitated by hydrogen peroxide in alkali or ammoniacal solutions as peroxide hydrate, but it redissolves in an excess of the cold liquor. From such solutions, alcohol throws down a pertitanate, as, $K_4TiO_8 \cdot 6H_2O$. On adding hydrogen peroxide slowly to an ice-cold solution of titanium tetrachloride in ethyl acetate, an orange coloration first appeared, and later a white precipitate formed which contained titanium, chlorine, and hydrogen peroxide in the approximate ratio of 1 to 1 to 1. The product was very soluble in water, yielding a clear orange-colored solution. The apparent loss of three atoms of chlorine per molecule of tetrachloride can be attributed to oxidation to free chlorine by the hydrogen peroxide on partial decomposition of the ethyl acetate, with the formation of ethyl chloride and titanium acetate. Both solution and precipitate were thought to be hydrogen peroxide solvates of a monochlorotitanium compound.[81] Viscosity measurements indicated complex formation on adding hydrogen peroxide to aqueous titanium tetrachloride, and study of the movement of the ions in an electric field showed that the titanium was part of the anion.[82] Schwarz and Sexauer[83] prepared the peroxide of titanium in aqueous solution and extracted the uncombined water with ethyl alcohol at 0° C. Combined water and oxygen were then determined, and the ratio of titanium dioxide, oxygen, and water was found to be 1 to 1 to 2. From these data they assigned the formula $TiO_2 \cdot O \cdot 2H_2O$.

On treating freshly prepared titanic acid with neutral hydrogen peroxide, or by treating a solution of titanium sulfate with this reagent, and then adding ammonium hydroxide, a product was obtained which, after drying over concentrated sulfuric acid, yielded a yellow, horny mass of hydrated titanium peroxide or pertitanic acid, $TiO_3 \cdot H_2O$. The trioxide may also be obtained by dropping titanium tetrachloride into aqueous alcohol and adding a large excess of hydrogen peroxide. On adding ammonium hydroxide or carbonate, or potassium carbonate to the solution, a bright yellow precipitate corresponding to $TiO_3 \cdot 3H_2O$ formed after a time. At 0° C the trioxide, in the presence of potassium and alcohol, yielded crystals of potassium peroxide hypertitanate, $K_2O_4 \cdot TiO_3 \cdot K_2O_2 \cdot 10H_2O$, and with sodium hydroxide a compound of the formula $(Na_2O_2)_4 \cdot Ti_2O_7 \cdot 10H_2O$ was obtained.

Work by Billy [84] indicated that the pertitanates are complex compounds with hydrogen peroxide and that they correspond to the chemical formula $Ti_2O_5$ rather than $TiO_3$, as generally assumed. Hydrates of pertitanic salts may be obtained by hydrolytic precipitation from dilute aqueous solutions. Analysis of the salt obtained at 10° to 20° C corresponded to the chemical formula $Ti_2O_5$, but products thrown down at around 0° C gave a ratio of titanium to oxygen of 2 to 5.27, which was far from the suggested formula $TiO_3$.[85] McKinney and Madson [86] suggested that titanium pentoxide may be a molecular compound of $TiO_2 \cdot TiO_3$, but they also pointed out the possibility of a ring structure such as

$$
\begin{array}{cc}
\begin{array}{c}
\phantom{Ti}O \\
\phantom{Ti}\parallel \\
Ti-O \\
\mid \phantom{Ti-}\mid \\
\phantom{Ti-}O \\
\mid \phantom{Ti-}\mid \\
Ti-O \\
\phantom{Ti}\backslash\backslash \\
\phantom{Ti}O
\end{array}
&
\begin{array}{c}
O=Ti-O \\
\mid \\
O \\
\mid \\
O=Ti-O
\end{array}
\end{array}
$$

By substituting peroxide groups for the oxygen atoms attached to the titanium atoms, a large number of different formulas can be postulated. Compounds of this type would exhibit isomerism and polymerism. In explaining the structure of such compounds, Hakomori [87] proposed induction valence which is produced by electrostatic induction and is not measurable by integral numbers.

Freshly prepared and well-washed titanium hydroxide dissolved in dilute aqueous hydrogen peroxide to give a clear yellow solution which exhibited colloidal properties. A suspension of 0.0035 mole titanium hydroxide in 50 ml water treated with 0.4 ml of 30 per cent hydrogen peroxide (also 0.0035 mole) cleared after one hour at room temperature and after a few minutes on warming. Larger proportions of hydrogen peroxide hastened dissolution, and smaller amounts gave solutions with pronounced opalescence. Addition of electrolytes effected almost complete precipitation, but higher concentrations of hydrogen peroxide reduced the recovery. Concentrated solutions solidified to gels on cooling.[88] However, titanic acid is quite soluble in hydrogen peroxide acidified with sulfuric acid and yields true solutions, as determined by the ultramicroscope.[89]

Ozone was evolved from peroxide sulfate compounds of titanium. The formula suggested for pertitanium sulfuric acid, a member of this group, is $HSO_4$-$TiO_2$-O-O-$SO_3H$.[90]

Potentiometric and calorimetric measurements indicated the formation of $TiO_3$ and $TiO_2 \cdot H_2O_2$.[91]

## COLLOIDAL COMPOUNDS

Titanic acid is generally considered to exist in two modifications, the alpha or ortho corresponding to the chemical formula $Ti(OH)_4$, and the beta or meta having the composition $TiO(OH)_2$. The former may be obtained as a voluminous white precipitate by adding aqueous ammonia or alkali metal hydroxides or carbonates to solutions of titanic chloride or sulfate, in the cold. It may also be prepared by decomposing potassium titanate with hydrochloric acid. The precipitate should be washed and dried in the cold, for if the temperature is raised, some meta variety is formed. As long as it remains fully hydrated, the ortho modification is soluble in dilute sulfuric and hydrochloric acids, and in the stronger organic acids, but loses water on heating and forms complex, less soluble products. It passes slowly into the meta form if allowed to remain in contact with water at room temperature, and more rapidly on heating. The ortho acid gives an orange-red coloration with tannin.

Metatitanic acid is precipitated directly from hot solutions of titanic salts by adding alkalies or by boiling; by the action of nitric acid on titanium metal, by drying the ortho acid in vacuum, or by heating it at 140° C. It is a white powder, insoluble in water, very slightly soluble in dilute acids, and soluble in hot concentrated sulfuric acid.

From X-ray studies, Levi [92] concluded that both varieties were completely amorphous. The greater part of the water was found to be loosely bound by capillary or osmotic forces.[93] The preparations were represented as penetrated by canals of radii of the order of 100 Ångstrom units. The remainder of the water was considered as being bound in the interior of the titanic acid. Dehydration by the Van Bemmelen, the Bolte, and the thermal methods proceeded without interruption or breaks, showing that the hydroxides are not crystal hydrates of definite structure. The compounds were represented as holding water by adsorption (capillary forces). On heating to incandescence, the hydrous oxide became crystalline and was a mixture of anatase and brookite. At 1100° C, both changed to rutile.[94]

These differ primarily in the size and complexity of their ultimate particles as a result of the amphoteric nature of the compound. As a result of relatively greater surface, the orthotitanic acid precipitated from cold solutions has a greater absorptive capacity for dyes than metatitanic acid obtained from hot solutions. In general, the effectiveness of both varieties decreases with dehydration. Samples prepared from alkaline solutions absorbed basic but not acid dyes, while the opposite was true of products obtained from slightly acid solutions. Titanic acid precipitated from chloride solutions by calcium carbonate, however, absorbed both acid and basic dyes but had a low capacity. This difference in the behavior of samples obtained from acid and alkaline solutions may be attributed to the amphoteric nature of the oxide, resulting in the former case in the precipitation of a highly basic titanium salt, and in the latter of an alkaline titanate. Another possible explanation is that the positively and negatively charged colloidal particles are present as a result of hydrogen and hydroxyl ions absorbed from the medium. By the same reasoning, titanic acid obtained by precipitation with calcium carbonate is neutral and should have no acid or basic ions.[95]

No alpha acid prepared at 25° C was detected, in the colloidal state, in hydrochloric acid solutions stronger than 0.25 normal, while with beta acid precipitated at 100° C the colloidal condition persisted until the concentration of the hydrochloric acid reached 1.5 normal.[96] The mode of precipitation and the temperature of drying exerted considerable influence on the solubility.

From X-ray, phase rule, and potentiometric titration studies, Weiser and Milligan[97] concluded that the alleged ortho and meta acids do not exist, and that the different preparations which can be obtained are not isomers but differ in size and extent of coalescence of the primary particles into secondary aggregates. Titania gel was found to be hydrous titanium dioxide. This work indicated that, in general, gelatinous precipitates of the oxides are not polymerized compounds or condensation products resulting from the splitting off of water from hypothetical metallic hydroxides, but, rather, that they consist of agglomerates of extremely minute crystals of oxides or simple hydroxides which hold large amounts of water by adsorption or capillary forces.

According to the same workers, hydrous titanic oxide gives X-ray diffraction patterns of the amorphous type and relatively sharp ring electron diffraction patterns.[98] This was interpreted as meaning that the oxide was crystalline, but that the primary crystal size was so small that the X-ray diffraction pattern consisted of broad bands. After aging at room temperature, gels of titanium dioxide gave weak X-ray patterns corresponding to anatase.[99]

Many compounds of titanium in solution have a tendency to assume the colloidal form [100] under a variety of conditions, and this has made studies of the general chemistry of the element more difficult. This property, objectionable to theoretical chemists, has been mastered by industrial research workers and is applied in the production of titanium pigments. Careful control of the formation of colloidal hydrous titanium dioxide and the conditions which exist during its subsequent coagulation are regarded as prime determining factors in the manufacture of titanium dioxide of highest pigmentary properties.

These compounds are unique in that they can exist in the colloidal state in solutions containing large proportions of electrolytes. For instance, dispersed systems, containing 60 g colloidal titanium dioxide in a liter of a solution containing 400 g sulfuric acid, may be prepared readily, and under most favorable conditions stable aqueous systems containing more than 600 g colloidal titanic acid per liter can be obtained. Such solutions are almost as fluid as water and can be prepared commercially at a cost comparable to that of titanium dioxide pigment. Hydrolytically precipitated titanium dioxide serves as a convenient starting material. The washed pulp is treated with an alkali to neutralize the adsorbed sulfuric acid and washed until free of sulfates. Two grams concentrated hydrochloric acid is then added for each 100 g of titanium dioxide, and on agitation the thick slurry becomes quite fluid. The peptization process is readily reversible, and coagulation is effected by the addition of a small proportion of a sulfate or other polyvalent anion. Colloidal titanium dioxide dispersed in acid media acquires a positive charge, although in alkaline systems the charge may be reversed.

Colloidal solutions may be prepared in any degree of dispersion by peptizing titanic acid or hydrous oxide obtained by hydrolytic precipitation.[101] The washed pulp is first freed from sulfuric acid by treatment with ammonium or alkali metal hydroxide, followed by additional washing, and is then agitated with a small proportion of hydrochloric acid or chlorides of titanium or silicon to effect dispersion. Alternatively, both these stages may be carried out simultaneously by employing barium chloride which converts the sulfuric acid to an insoluble salt and liberates hydrochloric acid as a peptizing agent. In an example, 1 kg of titanium oxide, or metatitanic acid precipitated by the hydrolysis of sulfate solution, was washed and treated with ammonium hydroxide to neutralize the residual sulfuric acid. The ammonium sulfate was then washed out and 20 g concentrated hydrochloric acid was added. Peptization was immediately effected and the system became completely fluid. The product was dried and dehydrated for pigment use.

Stable solutions containing colloidal titanic oxide may be prepared by adding aqueous titanium sulfate to hot water and cooling the mixture before any appreciable precipitation has taken place.[102] Such dispersions may be coagulated by adding concentrated hydrochloric acid.

Colloidal titanium oxide in a coagulated, paste-like form was peptized by adding ammonium hydroxide or an organic base, such as methylamine, with agitation, until the product was slightly alkaline to litmus. On introduction of a large quantity of distilled water, a pseudo solution resulted, which appeared opaque in reflected light but transparent in thin layers.[103] Similar solutions were prepared by subjecting finely divided titanium dioxide, at moderate heat, alternately to the action of dilute acid and dilute alkaline solutions. After each treatment the product was washed with distilled water, and the process was continued until transformation to the colloidal state was complete.[104]

By reducing a 1 per cent solution of palladium chloride, together with 15 per cent aqueous titanium tetrachloride which had previously been almost neutralized with sodium acetate, and heating the system to boiling, a solution consisting of colloidal titanium dioxide with adsorbed palladium was obtained.[105] Muller[106] found that colloidal solutions of titanium oxide would be produced and stabilized by an aqueous suspension of cherry gum. Mixtures were separated by this procedure.

From a study of the peptization of titanium hydroxide and of the resulting solution, Parravano and Caglioti[107] found that the particles were charged positively, but that the sign of the charge could be changed by alkaline reagents such as sodium hydroxide, ammonium hydroxide, and sodium phosphate, or by precipitation of the colloid by the tartaric acid method. Among organic compounds, fuchsin showed an extremely high flocculating power. Electrolytes were not only adsorbed at the surface of the titanic oxide micelle, but were probably held within the particles in the frozen state, and, consequently, sometimes hindered the polymorphic transformations during calcination of the flocculated products. Gels prepared by hydrolysis were extremely lyophilic. Hydrochloric acid and titanium tetrachloride were employed as peptizing agents.

Colloidal titanium dioxide was found to behave as a typical positively charged hydrosol.[108] Ethyl and methyl alcohols sensitized this sol toward a majority of coagulating agents. Dispersed titanic oxide precipitated from aqueous tetrachloride migrated to the anode under a potential difference,[109] and such particles were deposited electrophoretically from a suspension in an aqueous solution of a soap or an alkali salt containing an amphoteric metal in the acid radical, such as potassium zincate or sodium hexametaphosphate.[110] The deposited particles were enclosed in a film of fatty acid or metallic oxide so that their original form was maintained and they were not plastic in the dry condition.

Studies of dialysis coefficients indicated that titanic acid is capable of forming molecularly dispersed acids, and that these are unstable and that they alter rapidly to colloidally dispersed hydrous oxide.[111] Dispersions of saponified titanic acid esters stiffened to the consistency of jelly after standing for 6 to 22 hours. According to Joseph and Mahta,[112] chlorides of potassium, magnesium, and aluminum, dissolved in a mixture of water and methyl, ethyl, or propyl alcohol, increased the viscosity and rate of coagulation of dialyzed titanium dioxide solution beyond that observed with aqueous electrolytes. The periodic nature of the time-viscosity curves disappeared with increased purification of the colloidal material. The autocatalytic nature of the coagulation process disappeared on progressive dialysis.

Colloidal titanium oxide obtained by heating dilute tetrachloride solutions in contact with air acted as a strong protective colloid.[113]

Titanium hydroxide sols, prepared by dropping the tetrachloride into water at 18° C, could not be completely freed of hydrochloric acid by dialysis. Such a solution, containing 15.2 g titanic oxide per liter, became gradually more viscous on prolonged dialysis and coagulated to a gelatinous mass at a hydrogen ion concentration corresponding to pH of 4.1. The particles adsorbed appreciable amounts of anions from solutions of sodium and potassium salts. Adsorption of hydroxyl ions was relatively small, and their high coagulating power was due to the removal of hydrogen ions by neutralization. The sols adsorbed more anions than cations, resulting in liberation of hydroxyl ions. Titanic hydroxide acted more basic than acidic, and the stability of the sol precipitated at 18° C was attributed to the adsorption of tetravalent titanium ions.[114]

Hydrosol of titanium dioxide, prepared by peptization of suspensions of the hydrolysis product of tetrachloride solutions, was treated with various compounds by Thomas and Stewart.[115] Addition of potassium salts resulted in an increase in the pH, the degree of which decreased in the following order of anions: citrate, maleate, oxalate, succinate, propionate, tartrate, lactate, sulfate, and nitrate. The decrease in hydrogen ion concentration was ascribed to the substitution of anions for hydroxyl groups of the hydrosol. Aging at room temperature resulted in a slight increase in acidity, and boiling accelerated this effect as a result of the formation of oxo complexes of the type shown below.

$$(H_2O)_4 - Ti \begin{matrix} O \\ \diagup \diagdown \\ \diagdown \diagup \\ O \end{matrix} Ti - (H_2O)_4$$

Potassium salts of citric, lactic, and tartaric acids, potassium hydroxide, and ammonium hydroxide formed anionic micelles because these ions displaced sufficient water groups to form negatively charged particles. Oxalic, tartaric, malic, and sulfuric acids peptized hydrous titanic oxide to anionic micelles, while nitric, hydrochloric, acetic, formic, and propionic acids yielded positively charged particles. Succinic acid had no peptizing effect. These data indicate that the ability of the acid to disperse titanium dioxide depends upon the strength of the acid and the capability of the anion to displace the water groups.

Vasilev and Deshalit [116] coagulated titanium dioxide sols with sodium sulfate and then titrated the products potentiometrically with barium and calcium chlorides. On precipitation of the sulfate ions the material became peptized and at the same time chloride ions were again adsorbed. By this method the same sample could be alternately peptized and coagulated many times. Measurements carried out on such solutions showed that the activity value of the hydrogen ions increased almost linearly with increasing sol concentrations.[117]

Negative titanium dioxide sols showed exchange adsorption of cations and replacement of hydrogen ions. On the other hand, positive sols did not dissociate on dilution to yield hydrogen ions and corresponding anions, and the charge on the colloidal particles remained unchanged.[118] Gold sols assumed a purplish color on addition of titanium compounds.[119]

**Gels.** Patrick [120] prepared titanium dioxide gels by drying the hydrogel obtained from colloidal titanic acid solutions in a current of air at 75° to 120° C, and then raising the temperature slowly to 300° to 400° C. By washing the hydrogel with water at 21° to 80° C before drying, a gel was obtained whose apparent density was less affected by strong heating.[121] Hard, tough gels having a density below 0.5, after being subjected to a temperature of 871° C, were prepared by heating the hydrogel slowly to 79° to 163° C under conditions which did not permit dehydration, followed by slow cooling.[122] This heat treatment was carried out by passing heated gas saturated with water vapor over the hydrogel immersed in water, by circulating heated air repeatedly over the material in perforated trays in a drying chamber, or by soaking the washed hydrogel in strong sulfuric acid which reacted with the water to produce the desired temperature rise, 50° to 135° C. The hydrogel could be washed before or after the treatment, or the dried gel could be washed. A similar product, prepared by heating a dilute sulfate solution of titanium with 20 per cent aqueous ammonia, was washed by decantation, coagulated with acetic acid, again washed to remove soluble salts, and dried at 100° to 110° C. The porosity of titania and alumina gels was found to be approximately equal, and the higher activity of the latter was ascribed to smaller pores.[123]

Stowener[124] neutralized an aqueous solution of titanium tetrachloride at a temperature below atmospheric to obtain a precipitate which was washed and dried to yield a hard, absorbent gel. A product of improved properties was obtained from hydrogels which had been previously purified by treatment with strong sulfuric, nitric, or hydrochloric acids.[125] Similarly, Barclay[126] mixed equal quantities of 10 per cent solutions of titanium sulfate and aluminum sulfate at −2° C and added slowly, with stirring, dilute aqueous ammonia until the alkalinity of the system was 0.005 normal. The precipitate was washed and dried to yield a hard, stable, and highly porous composite gel.

Vibrant jellies[127] were obtained on neutralizing a solution of sodium titanate in 33 per cent hydrochloric acid by the dropwise addition of aqueous alkali carbonate (sodium carbonate).

Only those inorganic gels which contain hydrogen, replaceable by silver, are considered to be true hydroxides, and since the corresponding compounds of titanium do not react to form silver salts, Krause[128] classified them as hydrated oxides. Passerini[129] found that the water in titania gel gave the same infrared absorption bands as pure water, and concluded that the water was absorbed physically by a phenomenon similar to capillarity. Dehydration curves of hydroxy gels of $H_2TiO_3$ and $H_2Ti_2O_5$ passed through phases of changing composition and through singular points corresponding to "daltonites."[130] Auger[131] observed that titanium dioxide gel was only slightly soluble in basic solutions, and that it varied from 20 mg per liter in 10 per cent sodium hydroxide to 1200 mg per liter in 40 per cent potassium hydroxide. It was insoluble in a saturated sodium carbonate solution. Bicarbonates seemed to form a double carbonate which hydrolyzed very readily.

The sorptive power of titania gel for sulfur dioxide, as measured by contact for 5 hours at 0° C under a pressure of 1 atmosphere, varied according to the method of dehydration and activation.[132] Concentration of titanium dioxide, the precipitant added, and the method of coagulation were of minor importance. Sulfate ions were more effective than chloride, and the presence of a small proportion of ferric oxide increased the activity to a moderate degree. Low temperature of dehydration decreased the sorptive power. The water content was the most important factor, and gels, regardless of dehydrating conditions, had a maximum sorptive power corresponding to an optimum proportion of water. At 16° C, under a pressure of 1 atmosphere, the gel adsorbed 8.5 per cent of its weight of ammonia. It was found to be an excellent adsorbent of many organic compounds, and, at room temperature and 1 atmosphere pressure, absorbed 32 per cent of its weight of benzene, 30.4 per cent of acetone, 55.4 per cent of chloroform, 43.2 per cent of carbon disulfide, and 27.3 per cent of ether.[133]

Higuchi [134] studied the sorption and desorption of sulfur dioxide within the temperature range of −40° and 40° C and explained the hysteresis phenomena on the basis of the capillary adsorption theory. Sorption and desorption of gases and vapors by titania gel at higher pressures exhibited hysteresis, but a low-pressure, reversible equilibrium occurred.[135] Similar studies,[136] at temperatures from −22.5° to 50° C, revealed that Polanyi's adsorption equation was more applicable over the entire range than Patrick's formula which held at lower temperatures. The process proved to be not strictly reversible. Sorption and desorption isotherms of methyl alcohol, ethyl alcohol, normal and iso-propyl alcohols, monochlorobenzene, toluene, hexane, heptane, acetone, and ethyl acetate by titanium dioxide gel were measured at 0°, 10°, 20°, and 30° C, and as in the previous work, the isotherms showed a hysteresis loop in the range of high relative pressures.[137] This characteristic indicated two phenomena, namely, adsorption of the vapors on the surface of the gel and capillary condensation of the sorptives in pores of the gel. Acetone acted somewhat anomalously. The permanence of the hysteresis loop [138] in the water vapor-titania system through 30 adsorption-desorption experiments was established by measurements on a McBain-Bakr balance at 30° C. The results supported the cavity concept for permanent hysteresis effect with rigid gels.

Titania gels break down on standing, to give granular precipitates.[139] An approximate value of 22,000 calories for the energy of activation was determined, as compared with 17,000 calories for silicic acid gel. Acidic mixtures set more rapidly with increases in the pH, the titania content, or the temperature. Gels set at a higher temperature were liquefied by rapid cooling.

Further work by Klosky [140] indicated that titania gel was a poorer adsorbent and catalyst support than silica gel.

A coprecipitated titania-silica gel [141] was prepared by passing a stream of nitrogen through titanium tetrachloride at 99° C, and then into a violently agitated dilute solution of sodium silicate, at such a rate of flow that all the titanium tetrachloride was absorbed before reaching the surface of the solution. The mixed precipitate was separated, washed, and dried.

Titania gels are prepared by heating at 85° C sols made by hydrolyzing titanium tetrachloride containing chlorides of calcium, magnesium, aluminum, iron, or manganese.[142] The addition of salts to the sols increases the adsorption of the gel in the decreasing order aluminum chloride, calcium chloride, magnesium chloride, and manganous chloride.

# 8

# The Chemistry of Titanium Salts

## SULFATES

Titanium forms di-, sesqui-, and mono-sulfates, as well as a number of basic, double, and complex compounds, and hydrates.

**Disulfate.** The disulfate, $Ti(SO_4)_2$, can be obtained by the action of hot sulfuric acid on the dioxide and may be crystallized from fuming sulfuric acid, but it decomposes on contact with even traces of moisture. Preparation by oxidation of the sesquisalt with nitric acid has been reported. Von Bichowsky [1] added concentrated sulfuric acid to hydrous titanic oxide slowly to avoid local overheating, and heated the mixture carefully just below the boiling point until sulfation was complete. The product was reported to be the disulfate.

Tetravalent compounds are known, containing a much lower proportion of sulfuric acid than that corresponding to the normal salt. For example, sodium hydroxide or other alkalies may be added to solutions of normal sulfates in such proportions as to reduce the ratio of sulfuric acid to titanium dioxide to 1 to 1 by weight, or even less, without causing a precipitate. Such acid-poor solutions may also be produced directly from the ore or the precipitated oxide. Titanium oxytrisulfate, $TiO_2(SO_4)_3 \cdot 3H_2O$, has been obtained by heating a solution of the dioxide in sulfuric acid in a closed tube at 120° C. On raising the temperature to 225° C, however, it converted to anhydrous titanyl sulfate. There is experimental evidence to indicate that the titanyl salt is the only sulfate of tetravalent titanium stable enough to persist under ordinary conditions, and it is this compound which forms on reacting ilmenite ore with sulfuric acid in the first stage of the commercial process for pigment manufacture. Such solutions give the dihydrate ($TiOSO_4 \cdot 2H_2O$) on crystallization, and they are readily hydrolyzed by heating or dilution to yield oxide of the same type as the more acid solutions corresponding to the normal salt.

A number of hydrates have been reported. The pentahydrate was obtained by boiling titanic acid with alcoholic sulfuric acid, followed by evaporation of the solution.

Crystalline titanyl sulfate having two molecules of water has been obtained from the product formed by heating hydrous titanic oxide with 2.5 parts concentrated sulfuric acid or with an excess of dilute acid at 60° to 120° C.[2] The crystals were washed with dilute sulfuric acid, in which they are only slightly soluble, although solutions containing up to 400 g titanium dioxide equivalent per liter were prepared by dissolving the salt in water. The crystalline dihydrate was also prepared by heating a solution containing 1 to 15 per cent titanium oxide and 35 to 40 per cent free and combined sulfuric acid.[3] The crystals were washed with alcohol to remove adhering liquor. Kirkham and Spence[4] obtained a fine granular form of titanyl sulfate by heating a solution of specific gravity 1.50 to 1.75, containing 4 to 12 moles of sulfur trioxide to each mole of titanium oxide at 90° to 135° C, with constant agitation. The dihydrate was prepared in 94 to 98 per cent yields by dissolving commercial titanium dioxide in hot 50 to 60 per cent sulfuric acid at the ratio of 4.5 to 3.5, and further heating the syrupy mass at 95° to 100° C for 3 to 4 hours, with stirring, until precipitation was completed.[5] The crystals were filtered, washed with acetone or alcohol, and dried at 70° to 80° C.

Water-soluble titanic and titanyl sulfates in solid form were produced by spraying saturated solutions in a finely divided state into a current of air at a temperature of 250° to 350° C.[6]

Solutions of low content of free sulfuric acid for use in tanning and mordanting were obtained from regular titanyl or titanic sulfate liquors such as are prepared from ilmenite by adding ortho or meta titanic acid at not over 60° C to react with the free acid.[7] Concentration of the neutralized solution was effected by evaporation under reduced pressure.

The titanium content may be separated from a sulfate solution containing iron and other impurities by adding a sulfate of an alkali metal or ammonium which forms a relatively insoluble double sulfate with the titanium.[8] Such compounds are quite stable to heat and may be prepared at elevated temperatures to obtain higher concentrations, and consequently higher yields, by crystallization on cooling. In an example, the solution obtained by dissolving 1000 parts of a reaction product of ilmenite and concentrated sulfuric acid containing 17 per cent titanium dioxide in 2500 parts of water was heated quickly with 400 parts of potassium sulfate at 95° C. The solution was filtered and cooled to 0° to 5° C, at which temperature the double sulfate crystallized with a yield of 65 to 75 per cent of the titanium. Most of the remainder was recovered by evaporation or hydrolysis. Wohler and Flick[9] separated sulfates of titanium and iron, as obtained from ilmenite, by selective dissociation with heat. Ferric sulfate decomposed at 715° C and titanic sulfate at 635° C, so that if the mixed compounds were heated at a temperature between these limits in a stream of sulfur trioxide, the latter was converted to the insoluble oxide without altering the former.

A basic titanium sulfate, insoluble in water but soluble in acids and containing 0.25 to 0.50 mole sulfur trioxide per mole of titanium dioxide, was precipitated from a titanic sulfate solution under controlled but incomplete neutralization with calcium carbonate at temperatures below 50° C.[10] Any iron present was held in the ferrous state. McCoy[11] converted basic titanium sulfate into a water-soluble form by treating it with fuming sulfuric acid.

From a study of the system, titanic oxide, sulfur trioxide, and water at 150° C, Sagawa[12] found that the solid phase consisted of these components in the ratio of 6:5:4, 1:1:1, 1:1:0, and 2:5:5. Reinders and Kies[13] investigated the solubility of hydrous titanium dioxide and of hydrous titanium sulfate in sulfuric acid-water mixtures at 25° C. A triple point was observed at 38.1 per cent sulfuric acid and 3.5 per cent titanium dioxide. The solubility of the sulfate in water or dilute acid was high, decreased to 0.05 per cent in 65 per cent sulfuric acid, and again increased in more concentrated acid. These solutions, particularly in water and dilute acid, were metastable but actually remained clear for months.

A number of double sulfates are known. Typical of the group are titanic potassium sulfate, titanic sodium sulfate, and titanic calcium sulfate.

According to Miyamoto,[14] hydrogen reacts with titanic sulfate $Ti(SO_4)_2 \cdot 3H_2O$, in a silent electric discharge to yield hydrogen sulfide. In dilute solutions the compound gave a color reaction with ascorbic acid.[15] At pH of 3.0, a yellow coloration developed which gradually changed to an intense reddish brown at pH of 4.6 and disappeared at pH of 5.2. In alkaline solutions the color was pale rose.

A study of the titanium dioxide-sulfur trioxide-water system shows that solutions of different concentrations are saturated with $TiO_2 \cdot xH_2O$, $TiOSO_4 \cdot H_2O$, $TiOSO_4 \cdot 2H_2O$, $TiOSO_4 \cdot H_2SO_4 \cdot 2H_2O$, $TiOSO_4 \cdot H_2SO_4 \cdot H_2O$, and $Ti(SO_4)_2$.[16] The highest solubility at 25 to 75° C is exhibited by the hydrates of $TiOSO_4$. For all of the compounds, solubility increases with temperature. The solutions stable at low temperatures hydrolyze at 100 to 105° C. The three titanium sulfates, $TiOSO_4 \cdot 2H_2O$, $TiOSO_4 \cdot 4H_2O$, and $TiOSO_4$, are obtained in crystalline form from titanium sulfate solutions containing different concentrations of sulfuric acid.[17] To prepare the solutions, titanium dioxide is heated with 200 weight per cent of sulfuric acid. At a concentration of 6 to 7 per cent titanium dioxide and 43 to 44 per cent sulfur trioxide $TiOSO_4 \cdot 2H_2O$ forms. Titanyl sulfate tetrahydrate forms at a concentration of titanium dioxide of 8 to 9 per cent and of sulfur trioxide of 50 to 52 per cent. Titanyl sulfate separates at a concentration of titanium dioxide of 9 to 10 per cent and a sulfur trioxide concentration of 61 per cent.

The solubility isotherm shows that there are five crystalline regions in the system of titanium dioxide, sulfur trioxide, and water at 100° C, $TiO_2 \cdot H_2O$, $TiOSO_4 \cdot H_2O$, $TiOSO_4 \cdot H_2SO_4 \cdot 2H_2O$, $TiOSO_4 \cdot H_2O$, and $Ti(SO_4)_2$.[18] Crystals of $TiOSO_4 \cdot H_2O$ are formed at 55 per cent sulfur trioxide and crystals of $TiOSO_4 \cdot H_2SO_4 \cdot 2H_2O$ at 64 to 69 per cent sulfur trioxide. Anhydrous $TiOSO_4$ is formed at 150° C. At 200 to 250° C, $TiOSO_4$ and $Ti(SO_4)_2$ form, while at 275 to 300° C only anhydrous titanium sulfate forms. In a study of the decomposition of the sulfates, titanyl dihydrate, monohydrate, and the hydrous salt were heated at a rate of two degrees per minute and the weight loss-temperature curves were observed by a thermal balance.[19] Dehydration of the dihydrate and monohydrate occurred at 140 and 320° C and at 240 to 400° C respectively. Sulfur trioxide was evolved from each sample at approximately 500° C. The decomposition compound produced at 500 to 800° C was anatase, while the product formed by heating at 900° C for 30 minutes was rutile. According to Pamfilov and Khudyakova,[20] of all of the titanium sulfates described in the literature only titanyl sulfate dihydrate can be isolated with certainty from solutions of titanium dioxide in sulfuric acid. That titanium dioxide in the hydrolysis product is chemically bound by sulfuric acid follows from the lowering of the boiling temperature as compared with solutions of sulfuric acid of the same concentration.

The precipitation of titanium from sulfate solutions with sodium hydroxide commences at a pH of 1.5 for the quadrivalent form and at a pH of 4 for the trivalent state.[21] At the boiling point the pH values are shifted to greater acidity. Precipitation of titanium IV from dilute sulfate solutions by the addition of sodium hydroxide takes place in three stages.[22] The first stage is the coprecipitation of sulfuric acid with $H_2TiO_3$; next is the neutralization of the sulfuric acid in the precipitate by sodium hydroxide until free $H_2TiO_3$ is formed; and finally the coprecipitation of sodium hydroxide with $H_2TiO_3$.

Crystalline ammonium titanyl sulfate, $(NH_4)_2SO_4 \cdot TiOSO_4 \cdot 2H_2O$, is obtained by adding ammonium sulfate solution to a titanium sulfate solution.[23] More than 90 per cent of the titanium dioxide is precipitated from a solution containing 92 gpl titanium dioxide and 220 gpl sulfate ion by adding 2.5 mole proportions of ammonium sulfate. The purified salt is soluble in water but the water solution hydrolyzes rather rapidly. Sodium titanyl sulfate is obtained by the reaction of titanium tetrachloride with a solution of sodium bisulfate.[24] Water soluble double sulfates of titanium and ammonium are formed by heating titanium dioxide with ammonium sulfate at 400 to 500° C. The reaction begins at

350° C and at 550° C thermal dissociation of the double sulfate starts yielding titanium sulfate and gaseous ammonia.[25]

In hydrochloric and sulfuric acid solutions the dominating cation at pH less than 1.3 is $TiO^{++}$.[26] At a pH of 1.4 to 1.6 the dominating cation is $TiO(OH)^{+}$. In 10 to 11 normal hydrochloric acid solutions $TiCl_6^{--}$ forms. The ions were absorbed from the solution by ion exchange resins.

Differential thermal analysis and X-ray diffraction studies show that $TiOSO_4 \cdot H_2O$ dehydrates at 240° C to titanyl sulfate.[27]. Titanyl sulfate dihydrate dehydrates completely at 225° C to give an amorphous material which on heating to 500° C forms $TiOSO_4$. Titanyl sulfate decomposes at 625° C to titanium dioxide of the anatase crystal form and sulfuric acid.

**Sesquisulfate.** Titanous sulfate $Ti_2(SO_4)_3$ may be prepared readily by the electrolytic reduction of the tetravalent salt in aqueous solution, and if the introduction of impurities is not objectionable, metallic zinc or iron, sodium thiosulfate, or other reducing agents may be employed. To a crude titanic sulfate solution obtained from ilmenite, McKinney[28] added a stabilizing agent (citric, tartaric, or oxalic acid, or their soluble salts) and treated the product with calcium hydroxide to remove excess sulfuric acid, followed by hydrogen sulfide to precipitate the iron. The filtered liquor was then electrolyzed to reduce the titanium to the trivalent state. The stabilizing agent prevented hydrolysis during the initial purification steps. Such a solution is used in the discharge of coloring agents employed in dying textiles. Spence and Craig[29] found that more stable solutions can be obtained by reducing aqueous titanic sulfate or chloride with aluminum than by dissolving the lower oxide directly in acid. Iron-free titanous sulfates, pure enough for use in refined analysis, were prepared by the electrolytic reduction of potassium titanium oxalate,[30] $K_2TiO(C_2O_4)_2 \cdot 2H_2O$.

Titanous sulfate is a bluish, crystalline solid, soluble in water. The compound does not hydrolyze in solution under normal conditions, but it is slowly oxidized by air to the tetravalent state, and at a more rapid rate in the presence of a small amount of an added copper compound which acts as a catalyst.[31] It is a strong reducing agent and precipitates metallic copper from moderately concentrated solutions of cupric salts.[32]

The sesquisulfate forms double salts with sodium, potassium, and ammonium sulfates, which may be crystallized from solution as hydrates, $3Ti_2(SO_4)_3 \cdot (NH_4)_2SO_4 \cdot 18H_2O$ and $Ti_2(SO_4)_3 \cdot Na_2SO_4 \cdot 5H_2O$. It forms alums with rubidium and cesium; cesium titanium alum, $Cs_2SO_4 \cdot Ti_2(SO_4)_3 \cdot 24H_2O$, was made by Piccini[33] in 1895. Hydrous titanic oxide was dissolved in dilute sulfuric acid, the proper proportion of cesium sulfate was added, and the solution was electrolyzed to reduce the titanium to the trivalent state. A deep violet color was developed and crystals of alum separated from the solution. Rubidium titanium alum,[34] $Rb_2SO_4 \cdot Ti_2(SO_4)_3 \cdot 24H_2O$, was prepared by the same method.

**Monosulfate.** Titanium monosulfate, corresponding to the chemical formula $TiSO_4$, has been prepared by dissolving metallic titanium in dilute sulfuric acid, and the salt has been obtained in crystalline form by evaporating the solution. It has also been prepared by dissolving the monoxide in dilute sulfuric acid.

Ehrlich[35] found that oxygen dissolved in metallic titanium up to the ratio of 1 Ti to 0.45 O. The lattice constants and the mole volume changed only slightly.

## CHLORIDES

Titanium forms well-defined di,- tri,- and tetra-chlorides, as well as a number of basic and complex salts and hydrates.

**Tetrachloride.** According to Vigoroux and Arrivant,[36] titanium tetrachloride was first prepared by George in 1825 and later by Dumas, Wohler, Merz, Demarcay, and Moissan. The more common methods comprise the direct action of chlorine gas on a great variety of titaniferous materials, including the relatively pure metal, alloys, carbide, cyanonitride, the dioxide, and ores, at elevated temperature. However, employing oxygen containing raw materials such as ilmenite, rutile, or the processed oxide, a carbonaceous reducing agent is added to the charge to combine with the oxygen. Mixtures of chlorine and gaseous hydrocarbons (coal gas) have also been employed in the absence of solid carbon.

The compound has also been obtained by the reaction between titanic oxide and carbon tetrachloride in a sealed tube at elevated temperature, and by the action of carbon tetrachloride, chloroform, and sulfur chloride on heated titanic oxide.

As a rule the metal and its alloys dissolve in hydrochloric acid, but because of the reducing conditions, lower chlorides are formed largely.

Titanic chloride obtained by the chlorination of titaniferous materials usually has a yellowish to reddish color which has been attributed to vanadium oxychloride, ferric chloride, and dissolved chlorine. Such products may be purified by redistillation after treatment with sodium amalgam[37] or metallic gold, silver, mercury, iron, carbon, copper, bronze, bismuth, tin, cadmium,[38] iron, antimony, zinc,[39] lower valent titanium compounds[40] which may be formed in place by reducing a portion of the tetrachloride, as with hydrogen, polymerizable organic compounds such as rubber and drying oils,[41] crystalline ferrous sulfate,[42] an active sulfide,[43] or a heavy metal soap.[44] The crude tetrachloride may also be purified by bringing the vapor in contact with an absorbent such as silica gel, activated alumina, or bentonite.[45]

Titanium tetrachloride dissolves in water to form clear solutions which are hydrolyzed readily to yield a series of basic chlorides, $TiCl_3(OH)$, $TiCl_2(OH)_2$, $TiCl(OH)_3$, and finally $Ti(OH)_4$ or hydrous titanic oxide. Similar solutions may be obtained from another approach by dissolving orthotitanic acid in hydrochloric acid, or by adding an alkaline earth chloride (barium chloride, calcium chloride) to sulfuric acid solutions of titanium. The insoluble alkaline earth sulfate formed may be removed by filtration or settling. Stannic chloride was found to exert a considerable protective action [46] against flocculation of titanium tetrachloride solutions.

Anhydrous titanium tetrachloride is a colorless liquid which fumes strongly in moist air. Thorpe,[47] in 1880, determined the specific gravity at 0° C to be 1.7604, and the boiling point as 136.41° C. Specific gravity decreased to 1.74 at 10.1° C, and to 1.52 at the boiling point. Later, Sagawa [48] found the specific gravity at 0° C, and the molecular weight to be 1.76139 and 107.83, respectively, and the viscosity coefficient 0.007921 at 20° C and 0.007458 at 25° C. The fluidity [49] from −15° to 50° C was found to be a linear function of the specific volume. The solidification temperature has been reported by various workers from −23° to −30° C. From freezing point data of dilute solutions, Nasu [50] calculated the heat of fusion as 12.90 calories per gram. Arii [51] measured the vapor tension from 25° C to boiling and determined from the pressure temperature curves that the boiling point should be 136° C, while the experimental equation gave 135.8° C. The molecular heat of vaporization, as calculated by the Claussius-Clapyron equation, was 8960 calories at 25° C and 8620 calories at the boiling point. The Trouton constant was found to be 21.07, showing that the tetrachloride is a normal liquid. Specific heat of the liquid between 13° to 99° C was found to be 0.19, and that of the vapor from 163° to 271° C was 0.13. Heat of solution was determined as 58.5 calories. Spencer [52] determined the empirical heat capacity equation of gaseous titanium tetrachloride.

According to Bhagavantam,[53] molecules of titanium tetrachloride show a lack of optical symmetry. Diffraction patterns [54] obtained, on allowing an electron beam to penetrate vapors of the tetrachloride at right angles, were characteristic of a single molecule of the gas or vapor. The atomic distance was found to be 3.61 Ångstrom units, and the atomic form tetrahedral. Lister and Sutton [55] investigated the structure of $TiCl_4$ and $TiBr_4$ by electron diffraction of the vapors. Where the configuration would have an important effect upon the mode of scattering, it was found that the assumption of a tetrahedral molecule was entirely satisfactory. The bond lengths in Ångstrom units were Ti-Br 2.31 and Ti-Cl 2.18. Index of refraction measurements also showed the tetrachloride molecule to be a regular tetrahedron, with no dipole moment, and an electron polarization of 43.24 in carbon tetrachloride.[56] The atomic radius of titanium was determined as 1.51 Ångstrom units. The compound was found to be diamagnetic, with a specific susceptibility at 35° C of $-0.287 \times 10^{-6}$. The paramagnetism of other compounds of titanium was explained as being due to the unsymmetrical nature of valency links, and that of titanium as being due to two pairs of valency electrons being in different orbits.[57] According to Siertsema,[58] titanium tetrachloride is the only diamagnetic substance which shows negative magnetic rotation. Its parachor was reported to be 262.5, and the atomic parachor for titanium 45.3. Viscosity[59] varied from 34.03 at 13° C to 26.65 at 75° C. Delwaulle and François[60] observed that the fourth line of the Roman spectra of titanium tetrachloride is a doublet, and that the chlorobromides exhibit three new frequencies.

Titanium tetrachloride does not conduct electricity. It is miscible with organic solvents, and the molecular conductivity[61] of the compound in methyl and ethyl alcohols resembles that of true salts and increases with dilution. It also shows a high conductivity in acetonitrile.[62] Urbain and Schol[63] found that a potential sufficient to cause an arc to burst out in the midst of titanium tetrachloride dielectric caused an intense liberation of chlorine.

The approximate free energy of the reaction with water was calculated as $-16.6$ calories.[64] Roth and Becker[65] determined the heat of formation of titanium dioxide (rutile), at constant pressure at 20° C, to be $218.7 \pm 0.3$ kilogram calories, and from this value calculated the heat of formation of the tetrachloride to be 185 kilogram calories. Entropy calculations[66] from spectrographic data and from the third law of thermodynamics gave values of 84.4 and 81.5, respectively.

The thermal diagram of the system titanium tetrachloride-hydrochloric acid showed the existence of the compounds $TiCl_4 \cdot 6HCl$ and $TiCl_4 \cdot 2HCl$, melting congruently at $-86°$ and $-30.8°$ C, respectively.[67] Titanium, antimony, tin, arsenic, and phosphorus formed halides with readily interchangeable halogen atoms so that, on bringing them together, reciprocal conversion took place and the original compound could not be isolated in the solid or gaseous form.[68]

A solution of titanium tetrachloride in ethyl alcohol formed with anhydrous hydrogen peroxide an insoluble white addition product which was considered a coordination compound of the tetrachloride and hydrogen peroxide.[69] It slowly assumed a red color on standing. Decomposition was accompanied by the liberation of oxygen, hydrogen peroxide, and water.

A series of chlorosulfonate chlorides of titanium has been reported.[70] The monochlorosulfonate trichloride was obtained by the action of chlorosulfonic acid on titanium tetrachloride, and the dichlorosulfonate dichloride in low yields was formed by the action of sulfur trioxide in great excess. All were yellow crystalline compounds unstable in the presence of water. In 1879 Clausnizer [71] prepared a yellow amorphous solid, corresponding to the chemical formula $ClSO_2 \cdot O \cdot TiCl_3$, by adding sulfur monochloride to titanium chloride. Thermal analysis showed that titanium tetrachloride and $SO_2Cl_2$ reacted to form $TiCl_4 \cdot 2SO_2Cl_2$.[72] Titanium tetrachloride reacted with phosphorus trichloride to give small nearly octahedral grains of $TiPCl_7$, and with phosphorus tribromide to form red acicular crystals of $TiP_2Br_{10}$.[73] Gnezda [74] prepared a number of complex compounds, some of which formed colloidal solutions in water. Qualitative evidence indicated the formation of a compound between manganese difluoride and titanium tetrachloride.

A reaction product corresponding to the formula $TiNCl \cdot TiCl_4$ was obtained by passing the vaporized tetrachloride in a current of nitrogen through a silent electric discharge at 4500 volts.[75] One per cent titanium nitride was formed. With traces of oxygen present, the product contained this component but no nitrogen, and appeared to be a mixture of oxychlorides. On adding a solution of the tetrachloride in chloroform to NOCl in the same solvent, a material corresponding to the chemical formula $TiCl_4 \cdot 2NOCl$ was formed.[76]

At low temperature the tetrachloride reacted with liquid hydrogen sulfide to form a compound corresponding to the formula $2TiCl_4 \cdot 2H_2S$.[77]

The solubilities of tantalum chloride ($TaCl_5$) and columbium chloride ($CbCl_5$) in titanium tetrachloride at 25° C were found to be 14.3 and 0.24 per cent by weight; at 50° C the values were 27.1 and 0.58 per cent; and at 100° C they were 33.9 and 1.32 per cent, respectively.[78]

The liquid-vapor curves of titanium tetrachloride with carbon tetrachloride, silicon tetrachloride, and tin tetrachloride, showed no maxima and minima.[79] A study of the freezing-point curves revealed that the system titanium tetrachloride-silicon tetrachloride yielded an eutectic mixture containing almost 100 per cent of the latter, which solidified at −66° C. Carbon tetrachloride and antimony pentachloride yielded eutectics which contained 40 and 60 per cent titanium tetrachloride and solidified at −66° and −50° C, respectively. Stannic chloride formed a solid solution.[80]

The heat of vaporization of titanium tetrachloride changes from 7.050 at 57.7° C to 8.96 kcal per mole at 135.2° C.[81] Entropy is constant within this range. As determined from X-ray diffraction data the lattice constants of the tetrachloride are: a 9.70, b 6.48, c 9.75 Ångstrom units and beta 102° 40′.[82]

Titanium oxychloride is formed by the action of dry ozone on boiling titanium tetrachloride in a flask provided with a reflux condenser.[83] The oxybromide is prepared by a similar process. Hildreth [84] produced the oxychloride by the reaction of the tetrachloride with nitrogen dioxide, nitrogen tetroxide, or mixtures below 450° C. Below 300° C the product is amorphous or noncrystalline. This reaction can be carried out in the vapor or liquid phase, or in carbon tetrachloride solution. The product may be calcined to anatase or rutile according to the treatment and the temperature employed. Titanium oxychloride, $TiOCl_2$, is obtained in quantitative yields by the oxidation of the tetrachloride with arsenic trioxide, antimony trioxide, mercurous oxide, or bismuth trioxide.[85] The pale yellow, moisture-sensitive product is washed with dry pentane and vacuum dried. Reaction of the trichloride with oxygen yields the oxychloride.[86] The titanium trichloride is suspended in a chlorinated hydrocarbon and the reaction is carried out at 30 to 40° C.

The compound, $K_2TiCl_6$, is prepared by treating potassium chloride with an excess titanium tetrachloride in molten antimony trichloride in a sealed tube at 90° C for 2 hours.[87] Excess titanium tetrachloride along with the antimony trichloride is extracted from the products with carbon disulfide.

**Trichloride.** Billy and Brasseur [88] found that titanium trichloride was best prepared by the action of powdered antimony (obtained by reducing solutions in hydrochloric acid with zinc) on the tetrachloride at 340° C. Any unreacted tetrachloride was removed from the product by washing with carbon tetrachloride, and antimony trichloride was washed out with ether. All operations were carried out in an atmosphere of carbon dioxide, with absolute exclusion of air. Sodium amalgam reduced the tetrachloride to dichloride, but some trichloride was formed by reaction between these components. At slightly elevated temperature the titanic chloride was reduced to the trichloride by aluminum, magnesium, zinc, hydrogen, arsenic, and tin. Aluminum chloride catalyzed the reaction, and in the presence of this agent the reduction was quantitative at 200° C.[89] By carrying out the reduction with metallic silver in a sealed tube at 180° C, the trichloride was obtained along with silver chloride.[90]

Anhydrous titanium trichloride was obtained by reducing the tetrachloride with hydrogen at 600° C.[91] The product was a crystalline solid, black in reflected light and violet in transmitted light. By further heating the reactants at 660° C in an atmosphere of hydrogen, a mixture of di- and tetra-chlorides was obtained.[92] According to Georges and Stahler,[93] at 1200° C there was 94 per cent conversion to the violet crystalline trichloride.

Bock and Moser [94] prepared brown and violet varieties of titanium trichloride; the former could be converted into the latter form but the reverse was not true. Heat of solution of the brown type was higher than that of the violet. As a result of this work the compound was considered to be polymorphous. At 400° C the violet type was produced by the action of a silent electric discharge on a mixture of titanium tetrachloride and hydrogen.[95]

Crystalline titanium trichloride of the chemical composition $TiCl_3 \cdot 6H_2O$ was prepared by electrolytic reduction of a 25 per cent solution of titanic chloride, followed by evaporation under reduced pressure at 60° to 70° C to a specific gravity of 1.5, and cooling.[96] Such solutions can be obtained by dissolving the tetrachloride in water, or by dissolving titanium hydroxide in hydrochloric acid and reducing with zinc, iron, or similar agents, but obviously these materials would introduce impurities. Metallic titanium in the presence of hydrochloric acid reduced the aqueous tetrachloride almost wholly to the trivalent state.[97] Dilute sulfuric acid catalyzed the reaction.

Trivalent titanium halides are formed by the reaction of finely divided titanium carbide, TiC, with dry halide acids in a molten alkali metal or alkaline earth halide both at 250 to 400° C under an inert atmosphere.[98] In an example, dry hydrogen chloride was passed through a molten mixture of 45 parts of lithium chloride, 55 parts of potassium chloride and 50 parts of titanium carbide by weight in a graphite crucible under argon. After 4 hours at 450° C the melt contained 3 per cent by weight of the trichloride. Lower halides are prepared by the vapor phase reduction of the tetrachloride with titanium carbide.[99] For effective reaction, the carbide must have a particle size below 0.074 mm. The temperature of the reaction should be lower than the temperature of disproportion of the lower halide but high enough to vaporize the higher halide, preferably from 300 to 1100° C. The lower halide is liberated from the reaction mass in a fused salt, which consists mainly of an alkali metal halide. Thus a shuttle containing 22 g of titanium carbide, TiC, of high purity and 0.074 mm particle size was placed in a quartz crucible. The apparatus was evacuated and heated to 550° C, after which titanium tetrachloride was distilled into the reaction vessel at a pressure of 1 atmosphere. After heating at 750° C for 4 hours, the reaction mass was cooled and leached with 1 normal hydrochloric acid. An aqueous solution of pure titanium trichloride resulted. By the reaction of solid titanium carbide with the gaseous tetrachloride at 1000 to 1500° C followed by the gradual cooling of the reaction products to below 500° C, the trichloride is obtained.[100] The solid trichloride produced is suitable for disproportionation at an elevated temperature to titanium metal and the tetrachloride.

The reaction of carbonyls and carbonyl halides of metals of groups VIB and VIIIB at 30 to 80° C with the tetrachloride in the presence of saturated liquid aliphatic hydrocarbons gives the trichloride.[101] After separation from the hydrocarbons the reaction mixture is useful for the reduction of organic nitro compounds to form amines, and also as a catalyst for the polymerization of olefins. The trichloride is also prepared by the reduction of the tetrachloride in the presence of an alloy of aluminum and titanium.[102] This alloy acts as a catalyst and simultaneously as an acceptor for the hydrogen chloride formed. The temperature of the reaction must be maintained at 800° C, because below 750° C the rate of the reaction is too low. Above 850° C the formation of the dichloride predominates.

Zinc amalgam reduction of the tetrachloride to the trichloride proceeds at room temperature.[103] Reduction with titanium and with aluminum proceed at the boiling point of the tetrachloride. Aluminum reacts only in the presence of aluminum chloride. The action of the aluminum chloride is recognized as activation or cleaning of the passive surface of the aluminum powder. Hydrogen reduction starts at 500° C and proceeds rapidly at 1000° C. Above 1200° C the dichloride is produced. Reduction of the tetrachloride with elemental antimony produces the trichloride.[104] The antimony trichloride formed is separated from the titanium compound by distillation with kerosene and reduction with zinc.

Reduction of the tetrachloride with an excess of hydrogen at 750 to 1700° C yields titanium trichloride.[105] After reaction the gases are cooled rapidly to a temperature between the freezing point of titanium tetrachloride and 200° C. The apparatus consists of a reaction pipe made of Hastelloy C, stainless steel, or molybdenum. Surrounding the pipe is an inert atmosphere and a cooling chamber where the titanium trichloride is separated from the unreacted gases which are drawn off and recycled.

In a continuous process the trichloride is obtained by passing dry hydrogen saturated with titanium tetrachloride through a stainless steel tube heated at 850 to 900° C.[106] From the tube the products pass directly into a cooling chamber. An electric arc at 35,000 volts above a surface of titanium tetrachloride surmounted by an atmosphere of hydrogen yields brown, solid titanium trichloride.[107] The tetrachloride reacts with titanium to produce the trichloride which is simultaneously sublimed under a pressure of 200 to 700 mm.[108] The reaction of the gas phase reduction of the tetrachloride with hydrogen to the trichloride appears to be first order with respect to the tetrachloride and hydrogen, giving an overall empirical second-order rate expression.[109] An activation energy of 41 kcal per mole is indicated. Traces of mercury vapor catalyze the reaction. Lower titanium chloride melts are produced by the reduction of the tetrachloride with titanium scrap in a fused salt bath.[110] Titanium trichloride prepared by the reduction of the tetrachloride with titanium may be purified simultaneously by sublimation at a pressure of 200 to 700 mm.[111]

The heat of formation of crystalline titanium trichloride at 25° C is −172.4 ± 0.8 kcal per mole.[112]

Titanium trichloride is oxidized by carbon tetrachloride by a route involving carbonium ion.[113] A mixture of 3 to 4 millimoles of titanium trichloride, 5 millimoles of aluminum chloride, 5 ml of carbon tetrachloride and butyl bromide was stirred at 25 to 30° C. Irrespective of the amount of butyl bromide added, titration after 3 hours contact showed 14 per cent and after 3 days contact showed 50 per cent conversion to the tetrachloride. Hexachloroethane was identified by gas chromatography.

Titanium trichloride formed by heating the tetrachloride at 1000° C is used in the preparation of a catalyst for the polymerization of propylene at atmospheric temperature and pressure to give solid polypropylene.[114] The dark purple trichloride on exposure to air and water decomposes and may ignite spontaneously.

Vaporized and preheated titanium tetrachloride and glacial acetic acid are introduced simultaneously in a reaction zone at 140° C to yield a yellow chloride acetate anhydrous powder.[115] Hydrogen chloride vapor formed as a by-product of the reaction is vented from the top of the zone. The $TiCl_2(C_2H_3O_2)_2$ powder is easily handled and is stable stored under anhydrous conditions. If heated to 850° C in the presence of air, the dichloride diacetate decomposes to produce rutile titanium dioxide of pigment grade.

Titanium trichloride oxidizes slowly in the air at ordinary temperatures, and at 440° C decomposes into the tetrachloride and dichloride. It is soluble in water to form stable solutions which act as strong reducing agents. The rate of oxidation in aqueous solutions,[116] as determined from the velocity of oxygen absorbed at 30° C, was a linear function of the oxygen partial pressure and varied inversely as the concentration of free hydrochloric acid in the solution. Hydrolysis and complex formation largely influenced the rate. The trichloride reacts with conjugated double bonds in organic compounds to give colored products,[117] although only ethylenic linkages next to ( = C = O ) groups are reduced by the reagent in ammoniacal solutions. It was particularly effective in reducing flavanones and flavones.[118] At low temperatures it combined with ammonia to give $TiCl_3 \cdot 6NH_3$, which reacted violently with water.[119] The latter product gave off ammonia at 300° C and yielded a black product, probably $TiCl_3 \cdot 2NH_3$, which reacted so violently that it caught fire on exposure to moist air.

Solutions of titanium trichloride and tribromide gave absorption maxima at the wave length of 7440 Ångstrom units.[120] From a study of specific rotation of titanous ion near its absorption region, the existence of a doublet structure was found by Bose.[121]

**Dichloride.** Titanium dichloride may be produced by reducing the tetrachloride in a current of hydrogen at 700° C, although at higher temperatures, 800° to 900° C, the metal is obtained.[122] Three reactions are involved in the reduction, $TiCl_4 + H_2 \rightleftarrows TiCl_2 + 2HCl$, above 600° C; $TiCl_2 + TiCl_4 \rightleftarrows 2TiCl_3$, at 400° C in a current of hydrogen; and $2TiCl_3 + 2HCl \rightleftarrows 2TiCl_4 + H_2$.[123] Schmidt[124] concluded that the dichloride could not be obtained as such by this method, because of the side reactions, but that it was always mixed with other products of the reaction, $2TiCl_2 \rightleftarrows Ti + TiCl_4$ and $TiCl_2 + TiCl_4 \rightleftarrows 2TiCl_3$. At 600° C the tetrachloride was completely decomposed, but there was 82.6 per cent loss of titanium dichloride by sublimation. At 800° C only a 0.4 per cent yield was obtained.

Anhydrous titanium dichloride, prepared by the thermal decomposition of the trichloride, decomposed slowly on heating in vacuum at 475° C, but was not volatile at 300° to 600° C. It reacted with ammonia at low temperature, giving gray $TiCl_2 \cdot 4NH_3$, which decomposed in air and dissolved in water with the evolution of hydrogen.[125] Titanium monoxide dissolved in hydrochloric acid with the evolution of hydrogen, but gave only 10 to 18 per cent yield of the dichloride.[126] Solutions of the dichloride were unstable and readily changed over to the trichloride.

Titanium dichloride may also be prepared by reduction of the tetrachloride with sodium amalgam, and by dissolving the metal in hydrochloric acid.

Titanium dichloride is formed by the reaction of scrap titanium metal with the tetrachloride at 1020° C in the absence of air.[127] The dichloride is withdrawn from the bottom into a receiver heated at 400° C in argon to remove unreacted tetrachloride. A mixture of the dichloride and the tetrachloride is manufactured by the reaction of unrefined titanium metal with hydrogen chloride, titanium tetrachloride, or tetrabromide in the vapor form.[128] These lower halides are dissolved and electrolyzed in molten salt baths to yield pure, ductile, compact cathode deposits of titanium. The dichloride is formed by passing the tetrachloride through a fluidized bed of 150 to 200 mesh aluminum heated to 475 to 500° C.[129] The aluminum is fluidized with argon. A coarse, free-flowing, black-powder form of the dichloride collects in the bed. Above 700° C titanium reacts with the gaseous tetrachloride to form the dichloride.[130] Gaseous or liquid tetrachloride is introduced continuously in another air-free vessel, where it reacts at 300° C with the dichloride to form the crystalline trichloride.

Highly reactive low valence titanium chlorides in the powdered form with large surface areas are obtained by heating the tetrachloride in anhydrous tetrahydronaphthalene under reflux at 80 to 150° C.[131] Hydrogen chloride is passed over titanium carbide or a crude alloy of titanium and copper heated to 1000° C to form gaseous lower chlorides of titanium.[132] These products are shock cooled to below 500° C and then reheated to 600° C in the presence of titanium tetrachloride to form the trichloride. Subchlorides are produced at low temperatures and atmospheric pressure from the tetrachloride, sodium and sodium chloride.[133] In an example, 24 parts of sodium were dispersed in 207 parts of finely divided sodium chloride in a stirred reactor under an inert gas. To this mixture was added slowly 96 parts of the tetrachloride at 290 to 330° C to produce light gray dichloride.

Polarigraphic studies showed that the species present in a solution of titanium III and titanium IV ions in hydrochloric acid are $Ti(OH)Cl^{++}$, $Ti(OH)Cl_2^+$, $Ti(OH)_2Cl_2$ and $Ti^{+++}$.[134]

The dichloride is a black solid, unstable in air. It dissolves in water, with the liberation of hydrogen, and gives a black precipitate with ammonia, which is oxidized slowly in air.

The magnetic susceptibilities of titanium dichloride, dibromide, and diiodide show deviations from pure ionic magnetism.[135] These deviations are attributed to the type of atom binding between the metallic ions.

## FLUORIDES

In addition to the characteristic di-, tri-, and tetra-valent salts, titanium forms a number of complex and basic fluorides.

The normal tetrafluoride may be obtained by the action of fluorine or anhydrous hydrofluoric acid on metallic titanium at elevated temperatures. Anhydrous hydrofluoric acid and antimony tetrafluoride react separately with the tetrachloride to form this compound. The product obtained by dissolving titanium dioxide in aqueous hydrofluoric acid is probably fluotitanic acid, although evaporation of such solutions at gentle heat results in a syrupy liquid which gives crystals of the tetrafluoride.

On heating titanium dioxide compounds with hydrofluoric acid or ammonium fluoride, a compound is formed which by further heating evolves titanium tetrafluoride or titanium diaminotetrafluoride, depending on the agent used. These reactions are involved in a process for pigment manufacture.[136]

**Fluotitanates.** Potassium fluotitanate, $K_2TiF_6$, can be prepared by the action of potassium hydrogen fluoride on a solution of titanium dioxide in an excess of concentrated hydrofluoric acid. The hydrated salt, $K_2TiF_6 \cdot H_2O$, may be crystallized from the solution obtained, either by adding potassium hydroxide to aqueous hydrogen titanofluoride or by fusing titanium dioxide in potassium carbonate and dissolving the melt in dilute hydrofluoric acid. The analogous sodium salt is prepared in a similar manner. Ammonium fluotitanate may be crystallized from the solution obtained by adding ammonium hydroxide to aqueous titanium tetrafluoride in a proportion just short of that required to yield a precipitate.

Fluotitanates of the alkaline earth and certain of the heavy metals have been prepared and crystallized in the hydrated form.

Titanium tetrafluoride is a white solid at ordinary temperatures; it has a specific gravity of 2.80 at 20° C and a boiling point of 284° C. It dissolves in water to form clear solutions from which the dihydrate, $TiF_4 \cdot 2H_2O$, may be crystallized. Titanium tetrafluoride differs from the other tetrahalides in that it does not hydrolyze in aqueous solution on heating or on dilution. It absorbs ammonia, forming at low temperatures titanium triaminotetrafluoride, $TiF_4 \cdot 4NH_3$, and at 120° C the diaminotetrafluoride, $TiF_4 \cdot 2HN_3$, which are decomposed by hot sulfuric acid to give the dioxide.

According to Natta,[137] the tetrafluorides of titanium, silicon, and carbon give excellent X-ray photographs which show that these compounds crystallize in the cubic or tetragonal system.

From an investigation of ammonium hexafluorides of titanium, Hartmann[138] found that titanium tetrafluoride begins to distill off above 150° C. By electrolytic reduction of a solution of normal titanium fluoride containing ammonium fluoride, Piccini[139] obtained a violet crystalline precipitate of $TiF_4 \cdot 3NH_4F$ that was slightly soluble in water. The whole of the titanic acid content was precipitated from solutions of ammonium titanofluorides, $TiF_4 \cdot 2NH_4F$, by adding an excess of ammonia.[140] Quinine titanium fluoride separated on mixing alcoholic quinine with a hydrofluoric acid solution of titanium hydroxide, and the strychnine salt was prepared in a similar manner by mixing a solution of strychnine in hydrofluoric acid with $H_2TiF_6$ in the proportion of 2 to 1.[141]

Oxyfluopertitanates, derivatives of titanium trioxide, are prepared by adding alkali or alkaline earth metal fluorides to solutions of titanium tetrafluoride after oxidation with hydrogen peroxide.

Titanium trifluoride is obtained as an insoluble violet powder by igniting potassium titanofluoride in a current of hydrogen and leaching the product with hot water, or by reducing potassium titanofluoride in solution with zinc and hydrochloric acid or sodium amalgam.

Fluotitanites have been obtained as violet precipitates by the reaction of metal fluorides with titanium trifluoride. For example, a violet precipitate of ammonium pentafluotitanite, $(NH_4)_2TiF_5$, is formed by adding a concentrated solution of ammonium fluoride to one of titanium trifluoride. The analogous potassium salt, $K_2TiF_5$, is obtained as a deep violet precipitate by adding potassium hydrofluoride to the liquor obtained by treating titanium trifluoride with water. It is sparingly soluble in water but dissolves readily in dilute acids and yields a precipitate with ammonia which is readily oxidized to titanic acid.

Sodium titanium fluoride, $Na_2TiF_6$, and potassium titanium fluoride, $K_2TiF_6$, are prepared by the reaction of frozen titanium tetrachloride and the corresponding aqueous alkali fluoride.[142] These compounds are relatively insoluble in water so that they may be filtered from the solution and washed. The yield of sodium titanium fluoride is 84 per cent of the theoretical. Yields of the potassium salt are higher due to the lower solubility. These products are useful as electrolytes in the production of metallic titanium. To prepare potassium titanium double fluoride $K_2TiF_6$, titanium dioxide is first calcined with potassium carbonate and carbon at 1800° F for 1 hour.[143] After cooling, the sintered product is crushed to pass 200 mesh and dispersed along with calcium fluoride in a dilute solution of potassium chloride. After vigorous stirring, enough sulfuric acid is added to reduce the pH of the solution to 3 to 5 and the mixture is boiled to precipitate calcium sulfate. The calcium sulfate is separated by filtration and the filtrate is concentrated by evaporation. Crystals of potassium titanium fluoride separate from the solution. Calcium fluoride is added to a sulfate solution of ilmenite at 100° C to convert the sulfate to the fluoride.[144] Ferric iron is then reduced to the ferrous condition with a limited amount of scrap iron. To the filtered solution is added a mixture of hydrogen fluoride, sodium fluoride, or potassium fluoride and potassium chloride at 60° C to form potassium titanium fluoride. Sodium fluotitanate is obtained at 80 per cent yield by heating an equal molar ratio of titanium dioxide and sodium fluosilicate at 600° C for 30 minutes.[145] From the reaction products, sodium fluotitanate is extracted with hot water.

Alkali metal fluotitanates in which the titanium has a valence of two or three are produced by the reaction of titanium carbide with an alkali metal fluotitanate.[146] A graphite crucible loaded with 3 moles of anhydrous potassium titanium fluoride and one mole of −100 mesh titanium carbide was placed in a reactor which was then sealed, evacuated, supplied with a continuous argon atmosphere. After heating at 850° C for 8 hours the crucible was cooled rapidly to room temperature. The product, a deep reddish brown salt cake containing 24.7 per cent titanium, indicated 98.8 per cent conversion to the trivalent fluotitanate.

## BROMIDES

**Tetrabromide.** Titanium tetrabromide is obtained as an amber-colored, hygroscopic, crystalline solid by the action of bromine vapor on the heated metal or carbide or on a mixture of titanium dioxide and carbon at red heat. It may also be produced by passing anhydrous hydrobromic acid into the tetrachloride at temperatures below the boiling point. Young [147] heated an intimate mixture of titanium dioxide and sugar charcoal at 300° C in a stream of carbon dioxide to remove all moisture and air, then bubbled the carbon dioxide through liquid bromine and raised the temperature to 600° C. The compound had a specific gravity of 2.6, a melting point of 39° C, and a boiling point of 230° C. It hydrolyzed readily in aqueous solutions to oxybromides, and finally to titanic acid and hydrobromic acid. Basic salts corresponding to the formulas $Ti(OH)_2Br_2$ and $Ti(OH)_3Br$ have been reported.

Titanium tetrabromide is formed from a gaseous mixture of the tetrachloride, bromine, and hydrogen at 300 to 500° C.[148] For example, 587 g per hour of titanium tetrachloride, 1043 g per hour of bromine and 350 l per hour of hydrogen were passed through a tube 2000 mm long and 60 mm in diameter. The first half was filled with 4 mm activated carbon particles at 450° C and the remainder with 8.8 mm ceramic rings at 250° C. Following the reaction space was a condenser at 160° C which removed titanium tetrabromide in 87 per cent yield, and then a second condenser at 20° C for removal of titanium tetrachloride. With 3 to 5 mm pumice in the first half of the tube a yield of 80 per cent titanium tetrabromide was obtained.

A thermodynamic study of the titanium-bromine system shows that the reaction of titanium with bromine occurs at 20° C with the production of the tetrabromide.[149] The reaction between titanium and titanium tetrabromide yields titanium dibromide first. If the tetrabromide is present in excess, solid titanium dibromide reacts with gaseous titanium tetrabromide to form a tribromide. On the other hand, if titanium is present in excess, solid titanium monobromide is formed above 350° C.

Fritsch [150] observed a large negative magnetic rotatory power of titanium tetrabromide paralleling that of the tetrachloride. The sample prepared from titanium tetrachloride and hydrobromic acid was a yellow, crystalline solid having a melting point of 39.5° C. Seki [151] determined the vapor pressure near the melting point, and from the results determined the thermodynamic properties of the compound.

Ammonium bromotitanate is obtained by mixing ammonium bromide with a solution of titanium tetrabromide in hydrobromic acid. It is a dark-red, crystalline solid, less stable than the corresponding chlorotitanate. The mixed halides, titanium dibromide-dichloride and monobromide-trichloride, are obtained by the action of bromine on titanium dichloride and trichloride, respectively, in regulated proportions.

Tribromide. Titanium tribromide has been obtained by the electrolytic reduction of the aqueous tetrabromide, and by the reduction of the tetrabromide with hydrogen. The compound was reported to be unstable.

According to Young,[152] titanium tetrabromide gave the dibromide at 400° C in vacuum.

## IODIDES

Tetraiodide. Titanium tetraiodide can be prepared by passing iodine vapor over heated titanium metal, by passing dry hydriodic acid into the tetrachloride, and by passing a mixture of titanium tetrachloride, iodine vapor, and hydrogen through a red-hot tube. The compound melts at 150° C and distills without decomposition at a temperature slightly above 360° C. It fumes in air and dissolves in water to form solutions which are readily hydrolyzed to basic salts and finally to titanic oxide and hydriodic acid. Fast[153] prepared the tetraiodide by the reaction, at room temperature, between iodine and a large excess of metallic titanium. At 180° C pure tetraiodide distilled over and was condensed in a second bulb. However, by heating the compound and metallic titanium for 1 hour, the diiodide was formed. At 550° C the latter compound had an appreciable vapor pressure and partially reconverted to the tetraiodide, according to the reaction $2TiI_2 \rightarrow Ti + TiI_4$.

Titanium tetraiodide is prepared by heating titanium metal with a solution of iodine in carbon disulfide or benzene in an atmosphere of nitrogen.[154] After the reaction the solvent is removed from the tetraiodide by distillation. Free-flowing crystalline titanium tetraiodide is produced by combining the tetrachloride and hydrogen iodide in an inert solvent such as a saturated hydrocarbon, aromatic hydrocarbon or, halogenated hydrocarbon.[155] Thus, 13.8 ml of titanium tetrachloride was dissolved in 100 ml carbon tetrachloride in a flask maintained at −25° C, and hydrogen iodide vapor was bubbled through the flask for 2 hours until hydrogen chloride fumes were no longer evolved. The contents were warmed to room temperature and the carbon tetrachloride was filtered from the tetraiodide crystals. These crystals were washed with carbon tetrachloride, and the excess was removed by evaporation under vacuum. Conversion to the tetraiodide was essentially complete. Titanium tetrachloride is used as the solvent and as the reactant with hydrogen iodide gas to produce titanium tetraiodide.[156] The exothermic reaction may be carried out in batches or continuously in a column in which liquid titanium tetrachloride flows down and hydrogen iodide flows upward. A temperature between 20 and 30° C is maintained. Crystals of titanium tetraiodide form if the solution is cooled below 15° C. Fractional distillation may be used to separate the titanium tetrachloride which boils at 136° C from the tetraiodide which boils at 200° C.

Iodine reacts vigorously with metallic titanium to form the tetraiodide at temperatures above 130° C.[157] The vapor pressure in millimeters of mercury of liquid titanium tetraiodide is described by log P = $-(3.06/T)10^3 + 7.57$. In the presence of excess titanium, the triiodide, diiodide, and monoiodide are formed. These compounds evolve the tetraiodide at higher temperatures.

A study of the reaction of crude titanium with iodine to form the tetraiodide and its subsequent decomposition to produce the pure metal shows that the reaction is rapid at first but slows down soon because of the passivity of the crude material.[158] The calculated total heat of the reaction is 148,822 calories, the isobar-isothermal potential of the reaction is 114,850 calories per mole per degree, and the integration of the Gibbs-Helmholtz equation is 130 to 135 calories per mole per degree. An alloy of 80 per cent nickel and 20 per cent chromium is the best material for the reactor.

Titanium dioxide is heated at 1300 to 1800° C in the presence of petroleum coke or charcoal to form titanium carbide, which is then subjected to the action of iodine vapor at 1100° C, preferably in an electrically heated, graphite-lined furnace to produce the tetraiodide.[159] The product is cooled, condensed, and fractionally distilled at 150 to 379° C to effect purification. The formation of the tetraiodide in the reaction of titanium carbide with iodine is facilitated by adding 1 to 5 per cent of the tetrachloride to the iodine used.[160] This addition results in the same rate of formation of the tetraiodide at 1000° C as to 1100° C without it.

From the vapor pressure curve for liquid titanium tetraiodide, a number of physical constants were determined.[161] The boiling point was reported to be 377.2° C.

**Triiodide.** Titanium triiodide, as the corresponding bromide, can be prepared by the electrolytic reduction of a solution of hydrous titanic oxide in hydriodic acid, and may be crystallized as a violet hexahydrate, $TiI_3 \cdot 6H_2O$. It is also obtained by the reaction between the diiodide and tetraiodide, $TiI_2 + TiI_4 \rightarrow 2TiI_3$.

By heating the triiodide in a high vacuum above 350° C the reaction may be reversed to yield the original products. The compound dissolves slowly in water to form stable solutions, and is readily oxidized on exposure to air with the liberation of hydriodic acid.

Titanium diiodide can be prepared by heating the tetraiodide in a current of hydrogen and mercury vapor, or by heating a mixture of the tetraiodide and titanium metal at 700° C in a reaction bulb and collecting the vapor in a second bulb at 44° C. The compound is a black, hygroscopic solid, which readily oxidizes on exposure to the air, with the liberation of hydriodic acid. It reacts violently with water, with the liberation of hydrogen.

Periodic acid and its salts react with salts of titanium to form complexes.[162]

## COMPOUNDS WITH NITROGEN

*Nitrates.* Nitrates are probably formed by dissolving hydrous titanic oxide or titanates (potassium titanate, barium titanates) in nitric acid. On evaporation, however, the solutions yield a basic salt, $5TiO_2 \cdot N_2O_5 \cdot 6H_2O$. Such solutions are readily hydrolyzed on heating or dilution to yield titanic oxide and nitric acid. The solvolysis reaction of an excess of the tetrachloride with anhydrous nitric acid, cooled with ice, yields $TiOClNO_3$, which on contact with additional nitric acid forms a white, finely divided precipitate consisting of 20 per cent titanium dioxide and 80 per cent nitric acid along with water.[163] This product can be obtained directly by the reaction of the tetrachloride with excess nitric acid.

*Nitride.* Titanium has a strong affinity for nitrogen and burns in an atmosphere of the gas at 800°C with incandescence, forming the hard, bronze-colored mononitride, TiN. This compound may also be formed by heating titanic oxide very strongly in the electric furnace in the presence of nitrogen, or by heating the dioxide for 6 hours at 1400° to 1500° C in a stream of ammonia gas. It has a specific gravity of 5.18. It is hard enough to scratch rubies and slowly cut diamond. Friederich[164] reported that titanium nitride was stable at its melting point, 3200° A, and that the electrical resistance at ordinary temperatures was 1.8 ohms per square millimeter.

The nitride of quadrivalent titanium, $Ti_3N_4$, may be obtained by the successive action of liquid ammonia and potassamide on the compound $TiBr_4 \cdot 8NH_3$, or by heating the compound $TiCl_4 \cdot 4NH_3$. It is also reported to have been prepared by burning titanium metal in an atmosphere of nitrogen. The compound is a copper-colored, hard, crystalline solid. According to Ruff and Eisen,[165] however, titanium nitride of formula $Ti_3N_4$ was found to consist of impure titanium mononitride with the oxide and TiNCl, and by strong heating in a stream of ammonia this product was converted to pure titanium nitride, TiN.

Bosch[166] prepared nitrides of titanium by heating a mixture of a titanic compound, carbon, and alkali metal salt, such as sodium sulfate or potassium carbonate, in a current of nitrogen at comparatively low temperatures. The alkali metal compounds acted as nitrogen carriers. As an example, a mixture of 80 parts titanic acid, 20 parts wood charcoal, and 2 parts sodium sulfate was heated in contact with nitrogen at 1240° C. Titanium nitride, as well as other nitrides or cyanonitrides, were formed.

By passing vaporized titanium tetrachloride in a current of nitrogen through a silent electric discharge at 4500 volts, about 1 per cent titanium nitride was formed, along with a compound corresponding to the formula TiNCl·TiCl$_4$.[167] The compound was also formed by heating a charge of 40 parts titanium dioxide, 10 parts sodium carbonate, 10 parts lampblack, and 40 parts iron filings at 1000° C in the presence of nitrogen.[168] A hydrocarbon gas or vapor, e.g., benzene, gasoline, or kerosene, may be used to replace part of or all the lampblack. A crystalline [169] structure was developed by heating the regularly prepared powder in an electric arc. The material had a specific gravity of 5.18; was harder than quartz; had high mechanical strength and was resistant to sudden changes in temperature. Purification [170] was effected by heating the initial product in air at 700° C to burn out any carbon and then treating with sulfuric or hydrochloric acid and ferric chloride to dissolve silicides, phosphates, and sulfides without decomposing the nitride to form ammonia.

Rossi [171] proposed the use of titanium nitride in fertilizer as a source of nitrogen. Von Bichowsky [172] produced titanium nitride or cyanonitride for use as fertilizer or in the manufacture of ammonia by heating ilmenite or the purified oxide with carbon and sodium carbonate in proportions less than required for complete conversion to titanate at 1100° C in an atmosphere of nitrogen. The copper-colored crystals found in interstices of the lining bricks of a blast furnace corresponded to approximately 6 parts TiN and 1 part TiC.[173]

Aagaard and Espenschied [174] calcined an intimate mixture of 10 parts by weight of titanium pyrophosphate and 4 parts of carbon black in nitrogen at 1500° C for 2 hours to produce titanium nitride, TiN, of 1 to 5 microns particle size. This product containing 79 per cent titanium and 21 per cent nitrogen is useful in powder metallurgy, as an abrasive, in the production of cutting tool alloys, and in heat-resistant alloys suitable for use in the manufacture of gas turbine blades. Espenschied [175] prepared titanium nitride, TiN, by heating a dried mixture of titanium sulfate or chloride with finely divided carbon at 1350 to 1500° C in nitrogen. Thus, titanium nitride of 1 to 10 micron particle size was prepared by mixing an aqueous slurry of 134 parts of basic titanium sulfate containing 10 per cent sulfuric acid and 39 parts of lampblack, drying at 200° C for 6 hours, grinding the solids to a powder, and calcining in nitrogen at 1285° C for 3 hours.

A mixture of 10 parts of titanium tetrachloride vapor and 1 part of nitrogen is passed over sponge titanium in a quartz tube at 700° C to produce the nitride as a yellow-brown porous powder.[176] Titanium nitride can be isolated from the products formed by heating a mixture of the tetrachloride and ammonia in an electric arc above 2500° C.[177] The nitride is prepared in a plasma jet from the tetrachloride and nitrogen.[178]

The compound Ti$_2$AlN is prepared by heating a pressed mixture of titanium nitride, titanium, and aluminum at 850° C.[179]

**Cyanonitride.** Titanium cyanonitride may be prepared by heating to whiteness a mixture of titanium dioxide and charcoal in a stream of dry nitrogen gas; by heating potassium ferrocyanide and titanic oxide in a closed tube to a temperature above the melting point of nickel; or by fusing potassium cyanide in contact with vapors of titanium tetrachloride. It is also formed in blast furnaces smelting titaniferous iron ores, and may be produced by heating a mixture of titanium dioxide and carbon in air at high temperature. The compound is a hard, copper-colored, crystalline solid, often appearing iridescent, having a specific gravity of 5.28, and possessing a metallic luster. In fact it was mistaken for a metal by early investigators. It is attacked by nitric and hydrofluoric acids, and the ignited material decomposes in a current of steam, yielding hydrocyanic acid, ammonia, hydrogen, and titanium dioxide. Chlorine reacts with the cyanonitride at red heat to form the tetrachloride. On a commercial scale,[180] cyanides of titanium, TiCN$_2$ or TiC$_2$N$_4$, can be made by heating a mixture of titanic oxide with an excess of carbon in contact with nitrogen at 1800° C under a total pressure of 2 atmospheres. The partial pressure of carbon monoxide evolved was kept below 200 mm of mercury. Carbon may be supplied by coke, coal tar, or pitch. Ammonia may be obtained by treating the cyanide with steam.

In 1849 Wohler[181] concluded that the copper-colored crystals so often observed in blast furnaces were not simple bodies but cyanide and nitride of titanium. Dry chlorine reacted with the crystals at elevated temperatures to form titanium tetrachloride. On passing steam over the crystals heated to redness in a porcelain tube, ammonia, hydrocyanic acid, and hydrogen were set free, leaving titanium dioxide of the anatase crystal form. Later Rudge and Arnall[182] reported these copper-colored crystals to be titanium nitride in admixture with carbon. Graphite and yellow crystals of titanium nitride were produced by passing chlorine over this blast-furnace product. Titanium oxychloride, 3TiCl$_4$·TiO$_2$, as yellow crystals, was obtained by passing chlorine over the titanium nitride, with the introduction of moisture.

Von Bichowsky[183] found that nitric acid reacted with titanium nitride to form the dioxide and nitrogen gas. Sulfuric acid and sodium nitrate in aqueous solution reacted violently and evolved copious brown fumes of oxides of nitrogen. By adding the nitrate solution slowly and keeping the system cool, however, titanium sulfate was formed and a part of the nitrogen was converted to ammonia. Thus nitric acid alone, or in the presence of water and sulfuric acid, can completely oxidize titanium nitride to titanium oxides or sulfates and the nitric acid is at the same time reduced to nitrogen, nitrogen oxides, or ammonia. Higher recovery of combined nitrogen was obtained from the cyanonitride, $Ti_5CN_4$.

For example, 1 part sulfuric acid, 1 part sodium nitrate, and 0.1 part phosphoric acid were added to 1 part crude titanium cyanonitride suspended in 20 parts of water and the mixture was heated at 80° C for 24 hours. More water was added to complete the precipitation of titanium oxide, and ammonium sulfate was recovered from the liquor. The titanium dioxide obtained was soft, fine grained, and of high purity.

In a similar manner, pure titanic acid was obtained by the action of sulfuric acid and sodium nitrate on the nitride, $Ti_2N_2$ at 50° to 60° C.[184] The yield of ammonium compounds as by-products in the decomposition of titanium nitride or cyanonitride with nitric acid in the presence of sulfuric acid was increased by the action of a phosphate compound,[185] which also catalyzed the oxidation reaction.[186]

Ammonia was also prepared from titanium nitride and cyanonitride by heating the material with an acid or a salt which splits off acid under nonoxidizing conditions, or by heating these compounds with steam in the presence of an agent which serves as an oxygen carrier.[187]

Guignard[188] treated titanium nitrogen compounds with water vapor at a temperature of 360° to 500° C and removed the ammonia or cyanide compounds from the reaction zones as fast as formed. By heating titanium nitride ($Ti_3N_2$) with a halogen oxyacid (HOCl) and sulfuric or phosphoric acids, titanium dioxide and the corresponding ammonium salt were formed.[189]

In 1862 Riley[190] prepared hydrous titanium oxide by oxidizing the nitride in a mixture of nitric and hydrochloric acids. The crude material was washed and fused with potassium bisulfate, the melt was dissolved in water, and the solution was boiled to effect precipitation. The hydrous oxide obtained could not be washed free of sulfuric acid, but none was detected in the calcined product. Farup[191] developed a process of preparing titanic oxide from titanium-nitrogen compounds in which the nitrogen was also recovered in usable form. By heating such compounds with concentrated sulfuric acid, the nitrogen was transformed to ammonium sulfate. At higher temperatures corresponding to superatmospheric pressure, comparatively dilute sulfuric acid could be employed to advantage. Titanium sulfate formed dissolved in the solution and was immediately hydrolyzed, almost quantitatively, and the sulfuric acid liberated again took part in the reaction so that only a negligible amount of acid was required over that necessary to combine with the nitrogen and possible impurities.

In an example, titanium nitride, prepared by reduction of ilmenite in an electric furnace in the presence of nitrogen, was treated with dilute acid to remove most of the iron and some of the silicates. One part of this product, containing 54 per cent titanium dioxide, 10 per cent nitrogen, and 5 per cent iron, was mixed with 1.33 parts by weight of 50 per cent sulfuric acid and heated in an autoclave for 3 hours at a pressure of 25 atmospheres. At the end of this period the reaction product was filtered and washed. The precipitate, which consisted of 85 per cent titanium dioxide, 2 per cent nitrogen, 0.5 per cent iron, and 5 per cent sulfuric acid, represented practically all the titanium of the original material, while the solution contained the remainder of the nitrogen in the form of ammonium sulfate, together with some iron sulfate and free sulfuric acid. Titanium pigments and other titanium compounds were prepared from the residue by further treatment, according to conventional methods. If necessary, the titanium content was dissolved in strong sulfuric acid and reprecipitated.

A process for recovery of titanium dioxide from ores, combined with fixation of atmospheric nitrogen, was developed by Andreu and Paquet.[192] Ilmenite, rutile, or titaniferous iron ore was mixed with charcoal and sodium carbonate and heated in an atmosphere of nitrogen up to 1850° F. This product was then treated with steam at a temperature not exceeding 1800° F to liberate ammonia, leaving a residue which was washed to obtain a titanate. From this, titanic oxide was recovered by hydrolysis or by treatment with dilute acid.

Alternatively, if it was not required to recover titanium dioxide, the ore was mixed with carbon and heated in nitrogen to effect the fixation. The product was then treated with steam to split off ammonia, and the residue was used again.

In a related process,[193] ilmenite was mixed with a reducing agent and heated in an atmosphere of nitrogen in an electric furnace at 1200° to 1400° C to form titanium nitride and metallic iron. If the charge included carbon, titanium cyanonitride was also formed. The iron was removed by magnetic separation or dissolved selectively by dilute acid, and the residue was heated with strong sulfuric acid to form titanium and ammonium sulfates. These compounds were dissolved in water and the solution was boiled to cause hydrolytic precipitation of titanium oxide, which also contained a small proportion of sulfuric acid. The filtrate was either evaporated to cause the ammonium sulfate to crystallize or heated with calcium hydroxide to liberate ammonia. After washing, the precipitate was treated with barium chloride and calcined to give a pigment consisting of titanium dioxide and barium sulfate. Alternatively, the product obtained after treatment with strong sulfuric acid was held at 400° C to precipitate titanium dioxide, which was then washed free of iron.

## PHOSPHATES

Titanic phosphates precipitated from aqueous solution by reaction between soluble phosphates and salts of tetravalent titanium are more or less basic and thus of variable composition, and, in general, more uniform products may be obtained by fusing the dioxide with alkali metal phosphates or phosphoric acid. Titanyl metaphosphate $TiO(PO_3)_2$ may be obtained by dissolving titanic oxide in fused sodium or potassium metaphosphate and slowly cooling the melt. With a large excess of potassium metaphosphate, octahedral crystals of $TiO_2 \cdot P_2O_5$ were formed. Octahedral crystals, assumed to have the composition $TiO_2 \cdot P_2O_5$, were prepared by dissolving hydrous titanic oxide in orthophosphoric acid at temperatures around that at which dehydration began.

On adding a large excess of ammonium hydrogen phosphate to a solution of titanic chloride at 100° C, a precipitate formed slowly,[194] which, after washing with hot 5 per cent aqueous ammonium nitrate and water was calcined, first in a closed and then in an open crucible. Analysis showed a weight ratio of titanium dioxide to titanium dioxide plus phosphorus pentoxide in the product of 0.5234. Hydrated basic titanic phosphate [195] may be precipitated in a granular, readily filterable and washable form, practically free from iron and other impurities, by adding gradually, with constant stirring, solutions of phosphoric acid or soluble phosphates to a solution of titanic salt (sulfate or chloride), under carefully controlled conditions of concentration, free acidity, and temperature. Concentration of the titanium sulfate should not exceed 5 per cent titanium dioxide, and the precipitation temperature should be 50° C or higher. To hold in solution undesirable impurities, such as iron and chromium, it was necessary to have free acid present. The presence of chlorides favored the formation of a granular product. Composition of the precipitate corresponded to the formula $3TiO_2 \cdot P_2O_5 \cdot 6H_2O$.

For example, a solution of titanium, containing 1 to 5 per cent titanium dioxide and free acid equivalent to 8 to 15 per cent sulfur trioxide, was subjected to reducing conditions to convert the iron component to the ferrous condition and heated to 80° to 90° C. A solution of sodium phosphate was added gradually in a proportion sufficient to provide slightly more than 1 mole phosphorus pentoxide for each mole of titanium dioxide. Under these conditions practically the whole of the titanium was precipitated in the desired form, and little excess phosphoric acid was necessary. The precipitate was dried, calcined, and ground to give a white pigment of excellent hiding power.

In a similar process, nelsonite ore was heated with 80 per cent sulfuric acid [196] at 175° C, and the reaction product was dissolved in 1.2 parts of water. After separation of the insoluble residue, the solution was treated with scrap iron to reduce all iron salts to the ferrous state, and then cooled to remove ferrous sulfate by crystallization. The supernatant liquor was then boiled to effect precipitation of basic hydrous titanium sulfate-phosphate, which, after washing, was calcined at 815° C to yield a product containing 72 per cent titanium dioxide, 26 per cent phosphorus pentoxide, and 2 per cent impurities having satisfactory pigment properties.

Titanic salts that are hydrolyzable only with difficulty, for use in dyeing, mordanting, tanning, and lake manufacture, were prepared by combining the hydrated oxide with phosphoric acid and an organic acid, oxalic, lactic, tartaric, in a proportion of 1 or more moles for each mole of the oxide.[197] Alternatively, such compounds were obtained by treating titanic phosphate with organic acids, and also by the reaction between acid-soluble titanium salts of organic acids with phosphoric acid. By all methods the compound could be recovered in crystalline form by evaporating the solution.

**Pyrophosphate.** Barnes[198] prepared a soluble double pyrophosphate of titanium and sodium by double decomposition from solutions of sodium pyrophosphate and basic titanium sulfate. According to Levi and Peyronel,[199] the pyrophosphate has a cubic form with 4 molecules in the cubic cell. The lattice constant was found to be 7.80 Ångstrom units, and the value of $d$ was calculated as 3.106 Ångstrom units.

**Titanous Phosphate.** The blue or purplish phosphate of trivalent titanium may be produced by the reaction between a soluble phosphate and a titanous salt (sulfate, chloride) in aqueous solutions. It is soluble in water, and the solutions are relatively stable toward hydrolysis.

Titanium phosphides, $Ti_2P$ and $TiP$, are obtained by passing phosphine over powdered titanium at 800 to 850° C.[200] The phosphine is freed of moisture and oxygen, and mixed with argon. At 800° C $Ti_2P$ is produced but at 850° C $TiP$ is obtained.

## SULFIDES

A number of sulfides of titanium, in addition to the characteristic di-, sesqui-, and mono-compounds, have been described in the scientific literature.

**Disulfide.** The disulfide can be obtained by heating a mixture of rutile, sodium carbonate, and sulfur, by the vapor phase reaction between titanium tetrachloride and hydrogen sulfide, by the action of hydrogen sulfide on titanium tetrafluoride, and by passing carbon disulfide over red-hot titanic oxide. It is stable in air at ordinary temperatures, but burns at elevated temperatures to titanium dioxide and sulfur dioxide. The heated compound is readily decomposed in a current of steam to yield hydrogen sulfide, sulfur, hydrogen, and titanic oxide. It dissolves in hydrochloric and sulfuric acids, with the evolution of hydrogen sulfide. In appearance it resembles mosaic gold.

Titanium disulfide is a better lubricant than graphite or molybdenum sulfide.[201] The compound is made by heating powdered titanium with a slight excess of sulfur. To avoid rapid increase in temperature to 2000° C, at which temperature the disulfide is unstable, 20 to 60 weight per cent of chlorides of sodium, potassium, calcium and magnesium that boil at 1100° C is added to the mixture. This maintains the temperature of the exothermic reaction at the boiling point of the salts.

Primary sulfides of titanium and an alkaline earth element in equimolecular amounts are sintered at 800 to 1400° C in an inert atmosphere such as nitrogen or helium to give ternary metal sulfides useful as pigments and as opacifiers in ceramics, glazes, enamels, and glassware.[202] Instead of using the preformed sulfides, they may be made in situ by passing hydrogen sulfide over a mixture of oxides or hydroxides of titanium and of an alkaline earth metal heated to at least 500° C. After the evolution of sulfur and of water vapor ceases, the hydrogen sulfide is replaced by a stream of an inert gas, and the temperature is maintained at 800 to 1400° C for 10 minutes to 6 hours. In an example, 4.19 parts of an intimate mixture of equimolecular parts of titanic sulfide and barium sulfide are heated in an atmosphere of hydrogen sulfide up to 1100° C in 2 hours. The mixture is held at this temperature for 35 minutes and cooled to give $BaTiS_3$ as a granular brownish black solid containing no unreacted barium sulfide or titanium sulfide. By the same process $SrTiS_3$ and $CaTiS_3$ are prepared.

**Sesquisulfide.** Titanium sesquisulfide, $Ti_2S_3$, is formed by igniting the disulfide in a current of an indifferent gas or by passing a mixture of moist hydrogen sulfide and carbon disulfide over titanium dioxide heated to bright redness. It is a grayish black, crystalline solid, stable in air, insoluble in dilute acids, but soluble in concentrated nitric and sulfuric acids with decomposition.

**Monosulfide.** The monosulfide, TiS, can be obtained by heating the higher sulfides, di- and sesqui-, in a current of hydrogen or passing titanium tetrachloride over a heated tungsten filament. It is a reddish solid of metallic appearance, stable in air at ordinary temperatures, but glows on heating and yields sulfur and titanium dioxide. It is not affected by water or dilute acids, but dissolves in concentrated sulfuric acid.

Picon[203] prepared four sulfides of titanium, $Ti_3S_4$, $Ti_4S_5$, $Ti_3S_5$, and $Ti_2S_3$, under nonaqueous conditions at high temperatures, and studied their properties. The first of the group, $Ti_3S_4$, was obtained by heating titanic oxide with graphite in an atmosphere of dry hydrogen sulfide at

1650° C. This, in turn, yielded $Ti_4S_5$ on heating in vacuum beginning at 1300° C and at 1400° C in an atmosphere of dry hydrogen sulfide gave $Ti_2S_3$. By heating $Ti_3S_4$ at 800° C for 1 hour in dry hydrogen sulfide, $Ti_3S_5$ was formed. Dry hydrochloric acid attacked the sulfides, $Ti_3S_5$, $Ti_2S_3$, $Ti_3S_4$, and $Ti_4S_5$, at 220°, 200°, 250°, and 315° C, respectively. Chlorine attacked all at 175° C. Aqueous hydrochloric acid gave no more than traces of hydrogen sulfide with any of the compounds at room temperature, and at 100° C only $Ti_3S_5$ showed rapid liberation of this gas. Above 800° C hydrogen sulfide, or a mixture of sulfur and hydrogen, slowly transformed the lower sulfides to $Ti_3S_5$. Water reacted slowly at 250° C to form hydrogen sulfide, titanium oxide, and basic sulfates. Sulfur dioxide combined with the sulfides, $Ti_2S_5$, $Ti_2S_3$, $Ti_3S_4$, and $Ti_4S_5$ at 185°, 205°, 210°, and 225° C, respectively, but the reaction became appreciable only at 300°, 400°, 400°, and 400° C, and, as before, sulfur, titanium oxide, and basic sulfates were formed. At 1000° C, anhydrous ammonia yielded a mixture of sulfides and titanium nitride. Concentrated nitric and hot sulfuric acids attacked all the sulfides; the latter yielded $Ti_2(SO_4)_3$. The speed of the reaction with hot sodium hydroxide solutions increased with the sulfur content of the compound. Aqueous potassium permanganate slowly converted the sulfides to potassium sulfate and titanium oxide.

The dioxide reacted with zinc sulfide and carbon to form a sulfide of titanium which volatilized only at high temperatures.[204] Sulfur reacted directly with titanium metal at high temperatures.

Alkali metal and alkaline earth thiotitanates are prepared by heating the corresponding titanate in an atmosphere of carbon disulfide.[205] Typical examples are $Li_2TiS_3$ and $BaTiS_3$. These compounds are useful as lacquer pigments. Barium titanium sulfide, $BaTiS_2$, is synthesized by the reaction of carbon disulfide with the oxide at elevated temperature.[206] Barium titanium sulfide is hexagonal, but its sulfur content decreases with increasing temperature of preparation.

The structure of $Ti_5Se_8$ has an unusual unit cell, monoclinic with centered faces.[207]

**Aluminides.** Titanium aluminides, TiAl and $TiAl_3$, are prepared from titanium and aluminum, and from titanium iodide and aluminum by the powder metallurgy method.[208]

## CARBIDES

Moissan,[209] one of the first to prepare titanium carbide, fused titanic oxide and silicon carbide together in an electric furnace. Other methods involve heating titanium dioxide with carbon at 1700° to 2000° C and fusing the dioxide with calcium carbide in the electric furnace. Ridgway [210] heated a charge of titania and carbon in a closed electric resistor furnace having a graphite core and a steel casing. The reaction took place at 1600° to 1800° C, and as it proceeded an enlarged ingot of the carbide was progressively formed. An atmosphere free from oxygen and nitrogen, such as carbon dioxide from the reactant, was maintained throughout the reaction and until the product cooled.

Similarly, compressed masses of a mixture of carbon and titanium-iron alloy were heated to 1800° C in an inert atmosphere to effect conversion to the carbide, and the crude product was purified by treatment in turn with hydrochloric and hydrofluoric acids, followed by removal of loose graphite after each treatment by a flotation process.[211] Parker[212] obtained carbides by heating finely divided titanium or titanium-iron alloy with a carburizing mixture of charcoal, manganese dioxide, potassium hydroxide, and ammonium chloride in a hermetically sealed box at 1800° C.

The carbide may also be made directly from ores.[213] Rutile was heated with the calculated amount of carbon to a high temperature, under reducing conditions, to form a plastic mass which was finally sintered. Fluxing materials were added to the charge if required. Carbides prepared in this manner contained impurities from the ores and unreacted charge. Graphite was removed by heating at a temperature below the melting point of the hard alloy to form the corresponding carbide, which was readily leached out by treatment with acid.[214] Crystalline titanium carbide[215] and metallic iron may be obtained by heating an equal mixture of ilmenite and coke in a carbon container in an electric resistance furnace. Iron was recovered as a by-product. According to Meerson,[216] titanium carbide forms at a much lower temperature in the presence of tungsten. Ballhausen[217] pressed into shape an intimate mixture of titanium dioxide with carbon black, coke, or graphite, and heated it by resistance offered to the passage of an electric current to form the metal carbide.

The reaction produced by electrically heating a carbon filament in titanium tetrachloride vapors was followed by observing the change in resistance of the wire and by X-ray studies.[218] Titanium carbide (TiC) was formed, but to produce the pure compound the reaction product was heated either in a hydrocarbon atmosphere or in a vacuum. The final product was in the form of a hollow tube.

Titanium carbide is a crystalline solid of specific gravity 4.25, melting at 3410° C. Water and hydrochloric or sulfuric acids have no action even at 600° C, but it is attacked by aqua regia and burns in oxygen at red heat. Mott[219] estimated the boiling point to be 4300° C. It crystallizes in the cubic class of the isometric system, sodium-chloride type, with an $a$ value of 4.3 Ångstrom units.[220] The compound forms a complete series of mixed crystals with vanadium carbide.[221]

Titanium carbide prepared by heating a mixture of titanium dioxide and carbon at 1200° to 3000° C had a cubic lattice with a parameter of 4.26 to 4.31 Ångstrom units.[222] A sample, prepared by heating the dioxide and carbon in an electric furnace in an atmosphere of hydrogen at 1700° C, was a gray powder of metallic appearance with melting point about 3430° C. It was soluble in aqua regia. The specific resistance at room temperature was found to be 1.8 to 2.5 ohms, and at the melting point 7.0 ohms.[223]

Carbonization of titanium was studied at 1900° C by heating the dioxide with carbon in an atmosphere of carbon monoxide and hydrogen.[224] Maximum saturation of the carbide by carbon (19.5 per cent carbon) occurred in the beginning of the process, and the value for bound carbon gradually dropped to 17 per cent. Treatment in the presence of nitrogen, such as would be possible under factory conditions, led to the formation of the nitride as a result of carbon displacement. For the preparation of pure titanium carbide, vacuum carbonization was necessary.

Naylor[225] measured the heat content of titanium carbide and nitride from 373° to 1735° K, calculated the entropies, and developed specific heat and heat capacity equations.

Higher carbides[226] corresponding to the chemical formulas $TiC_2$ and $TiC_4$ were prepared by heating titanium oxide with carbon under a total pressure of 2 atmospheres or more, while the partial pressure of the evolved carbon monoxide was maintained below 200 mm of mercury.

Kinzie and Hake[227] reported a friable, amorphous, reactive compound of the chemical formula $Ti_2OC_2$, prepared by heating a mixture of rutile and carbon enclosed in an insulated mixture of coke, sand, and sawdust, in an electric resistance furnace without fusion.

A hard intermetallic, crystalline compound, corresponding to the chemical formula $TiWC_2$, was formed by heating a mixture of tungsten, titanium, and carbon for a prolonged period in a bath of molten nickel above 2000° C in a graphite crucible.[228] On cooling, the compound was separated from the nickel by digestion, first with aqua regia and finally with hydrofluoric acid. These crystals had a hardness of 9.8 on Mohs' scale.

Kovalskii and Umanskii[229] made a study of the ternary titanium carbide-tungsten carbide-columbium carbide system.

A mixture of 1 mole of titanium dioxide with 2 to 3 moles of calcium carbide is pressed into tablets and heated in an oven under a vacuum or an inert gas at a temperature between 1200° C and the melting point of the charge to produce titanium carbide.[230] The reaction mixture is boiled with concentrated hydrochloric acid. Titanium carbide containing 44.8 per cent titanium and 52.7 per cent carbon was obtained in 99 per cent yield. In the synthesis of titanium carbide from titanium dioxide and carbon black, titanium trioxide and titanium monoxide solid solution are produced as intermediate compounds.[231] Metallic titanium appears temporarily during the process but is dissolved in the carbide-rich solid solution at a higher temperature in the presence of enough carbon. The carbide is produced by the explosive reaction of titanium powder and acetylene.[232]

## BORIDE

Sixty mesh boric oxide and 100 mesh titanium dioxide blended in a 5.2 to 1 ratio are added slowly to 5 parts of agitated sodium at 600° to 750° C to produce titanium diboride.[233] The mixture is leached with liquid ammonia, methyl alcohol, and an aqueous solution of ammonium hydroxide and ammonium chloride to give a product of 81.9 per cent purity. This product is purified by calcination at 1400° to 1490° C under argon at a pressure of $1 \times 10^{-4}$ mm to yield 96.8 per cent titanium diboride. Boric oxide may also be added to molten sodium, leached and then blended with titanium dioxide to give titanium diboride at 50 per cent yield with a particle size of 10 to 50 microns. The diboride obtained by heating briquets of titanium dioxide, boron carbide, and carbon at 1400 to 1500° C at a pressure of $10^{-4}$ mm of argon for 2 to 3 hours contained 0.70 to 0.87 per cent carbon, and the titanium to boron ratio was not stoichiometric.[234] Heating for 3 hours at 1700° C reduced the carbon content to 0.26 per cent and gave a titanium to boron ratio of 0.5. The impurities in the original agents, silicon, iron, calcium, aluminum, and copper, were readily eliminated in the course of the reaction. Magnesium and cobalt were eliminated more slowly.

## SILICIDE

Finely powdered titanium hydride and silicon are sintered in vacuum at 1315° C to produce titanium silicide, $TiSi_2$.[235] Sintering at 1650 to 1800° C gives $Ti_5Si_3$.

## FERRO- AND FERRICYANIDES

The X-ray diffraction patterns of titanium ferrocyanide and titanium ferricyanide so closely resemble those of the corresponding compounds of iron, copper, cadmium, indium, aluminum, and zinc, with respect to the position and relative intensities of the diffraction lines, that it is concluded that these compounds form an isomorphous series having the same face-centered cubic symmetry.[236]

## HYDRIDES

Archibald and Alexander [237] heated granular magnesium and titanium dioxide in an air-free atmosphere to produce metallic titanium, and then admitted hydrogen gas to the reaction zone to form the hydride. Klauber [238] obtained a good yield of titanium hydride by the discharge (spark) electrolysis of 0.1 to 0.2 normal sulfuric acid at a current density of 0.2 to 0.4 amperes and a potential of 240 volts, employing titanium electrodes, after a modification of the process developed by Paneth. The hydride was a colorless and odorless gas which burned in air with a colorless flame and deposited titanium dioxide. One gram of titanium absorbed 407 ml of hydrogen [239] at room temperature, and 46 ml at 1100 C, and at the lower temperature the hydride was formed.[240] The general behavior of the titanium-hydrogen system [241] was found to be analogous to that of the palladium-hydrogen system.

Titanium hydride is prepared by passing hydrogen over titanium above 33° C, the temperature at which the reaction begins, followed by rapid cooling.[242] Sponge and ingot titanium absorb 42.96 and 38.14 liters of hydrogen per 100 g of metal, respectively. Mixtures of 9 parts of hydrogen to 1 part of titanium tetrachloride saturated with mercury were irradiated with light at 2537 Ångstrom units to obtain low yields of $TiClH_3$, $TiCl_2H_2$ and $TiCl_4$.[243]

Titanium hydride is used in welding.[244] It is decomposed by heat in the zone of the weld to evolve nascent hydrogen and titanium as the molten metal which serves to facilitate the weld and improve its strength.

## TITANATES

Titanium forms an extensive series of titanates with the alkali, alkaline earth, and heavy base metals in which it appears as the acidic constituent. The more common members of the group, sodium, potassium, calcium, barium, magnesium, zinc, iron, nickel, cobalt, and lead, are usually prepared by heating the corresponding oxide, hydroxide, or carbonate, or in some cases neutral salts, with an intimate mixture of titanic oxide at elevated temperatures to bring about a direct reaction. With some of the stronger bases, however, the reaction may be effected in aqueous media. For example, barium and potassium titanate have been prepared by heating a suspension of hydrous titanium dioxide in a concentrated solution of the corresponding hydroxide at the atmospheric boiling point.

Titanates of the alkaline earth and heavy metals are very stable compounds, insoluble in water, but are attacked slowly by acids. The alkali metal titanates, which are insoluble in water but soluble in dilute acids, are employed as intermediaries in preparing other compounds of titanium from the ores. Potassium metatitanate, $K_2TiO_3$, may be formed by fusing titanium dioxide with potassium carbonate, and the tetrahydrate, $K_2TiO_5 \cdot 4H_2O$, by boiling a suspension of titanic acid in aqueous potassium hydroxide. The hydrated salt is very soluble in water. Sodium metatitanate and orthotitanate can be prepared by fusing titanium dioxide with chemically equivalent proportions of sodium carbonate. The meso- and para-titanates have been reported. According to Washburn and Bunting,[245] the binary system, sodium oxide-titanium dioxide, forms compounds of the chemical composition $Na_2TiO_3$, melting at 1030° C, $Na_2Ti_2O_5$, melting at 985° C, and $Na_2Ti_3O_7$, melting at 1128° C.

Sodium and potassium titanates, soluble in organic and dilute mineral acids, were prepared by heating hydrous titanic oxide with 1.6 to 2.0 parts of the corresponding hydroxide or carbonate at 100° to 200° C.[246] The white, crystalline, slightly hygroscopic solids were used for making other compounds of titanium. Similar compounds were obtained by Rockstroh[247] by heating dried titanic oxide with an alkali metal hydroxide or carbonate at 150° to 220° C. Titanates of the type $Na_2Ti_5O_{11}$[248] were produced by heating rutile or bauxite residues with alkali metal compounds such as sodium carbonate, sodium sulfate, potassium carbonate, and potassium hydroxide at 800° to 1000° C, in the ratio of 2 moles alkali metal oxide to 5 moles titanium dioxide. Coggeshall[249] heated rutile with anhydrous alkali metal carbonates at a dull red heat. Water-soluble compounds (titanates) were obtained by igniting oxide with sodium carbonate in a directly heated rotary tube furnace.[250]

The mono- and di-titanates of lanthanum, praseodymium, neodymium, samarium, and yttrium have been prepared by fusing the corresponding oxide with titanium dioxide in stoichiometric proportions.[251] The samarium and yttrium compounds, $Sm_2Ti_2O_7$ and $Y_2Ti_2O_7$, have a cubic crystal structure. Results generally paralleled those obtained with the corresponding silicates.

Well-defined alkali metal titanates are obtained in better than 95 per cent yields by sintering for 1 to 4 hours at 500 to 850° C, a titanium ore, high titanium slag or bauxite residue with a stoichiometric amount of alkali in the presence of 0.1 to 10 per cent of a catalyst.[252] Suitable catalysts are a fluoride, bromide, or iodide together with aluminum oxide, silicon dioxide, or magnesium oxide. Thus, 100 parts of rutile containing 95 per cent titanium dioxide, 133 parts anhydrous sodium carbonate, and 2 parts of sodium bromide were sintered for 2 hours at 825° C to give $Na_2TiO_3$. In another example $Na_2Ti_2O_7$ was prepared by sintering at 850° C for 2 to 4 hours a mixture of 100 parts of ilmenite containing 50 per cent titanium dioxide, 23 parts of anhydrous sodium carbonate, and 5 parts of aluminum oxide. Sintering of alkali metal hydroxides with titanium dioxide to form titanates takes place at a lower temperature and in a shorter cycle in vacuo at a final pressure of 1 mm.[253]

Fibrous alkali metal titanates are synthesized hydrothermally by the reaction of titanium dioxide with sodium or potassium hydroxide or carbonate.[254] Potassium compounds give $K_2Ti_6O_{13}$, while sodium compounds give $Na_2Ti_6O_{13}$, $Na_2Ti_5O_{11}$, $Na_2Ti_4O_{11}$, or $Na_2Ti_3O_7$ depending on the conditions. The fibers are 0.5 to 1.0 micron thick and 1 to 2 mm long. Sodium and potassium titanates of fibrous form are prepared by hydrothermal reaction of alkali solutions and titanic hydroxide.[255] The products are white cotton-like fibers insoluble in water and alkali solutions, with a softening point of 1100° C and a melting point of 1300° C. Thickness of the fibers is less than 1 micron and the length from 0.01 to 20 mm. These fibers are resistant to weak acids.[256]

Gier and Salzberg[257] produced asbestos-like alkali metal titanates having the general formula $M_2Ti_6O_{13}$ in which M represents potassium, sodium, rubidium, or cesium under conditions resembling those in nature. The products are formed by the reaction between titanium dioxide and the corresponding alkali metal hydroxide or carbonate at 600 to 700° C in the presence of water under a pressure of more than 200 atmospheres for a period of 30 minutes to 3 hours. They possess a fibrous structure in which one dimension of the crystal is larger than the other dimensions by a factor varying from 5 to 1000. Water of crystallization is not present. The fibrous material is insoluble in water, boiling 50 per cent sodium hydroxide solution, sulfuric acid, and concentrated hydrochloric acid.

Crystalline fibrous potassium titanate of the formula $K_2Ti_6O_{13}$ has high heat-insulating properties.[258] The fibers have an average diameter of 1 micron, a length of 0.5 to 1 millimeter, a melting point of 2500° F, a surface area of 11 square meters per gram, a density of 3.2 g per cubic centimeter and a hardness of 4 on the Mohs scale. At 47 and 90 per cent relative humidity the moisture pick up is 0.34 and 1.7 per cent, respectively. The specific heat is 0.22 British thermal units per degree Fahrenheit per pound. Tensile strength is 24,000 pounds per square inch; elongation at break is 0.59 per cent; and modulus in tension is $4 \times 10^6$ pounds per square inch. These fibers scatter infrared radiation because of their dimensions and high index of refraction.

Barium titanate was prepared as an intermediary step in the conversion of anatase to rutile by boiling a suspension of acid-free hydrous titanic oxide in an equivalent amount of barium hydroxide solution.[259]

Pulp obtained from thermal hydrolysis of a sulfate solution of ilmenite was treated with dilute aqueous sodium hydroxide to neutralize the residual sulfuric acid, and the salts formed by the reaction were removed by further washing with water. A portion of this product, as a thick paste containing 80 g titanium dioxide, was suspended in a solution of

315 g barium hydroxide octahydrate, Ba(OH)$_2 \cdot$8H$_2$O, in 2 liters of water and the mixture was boiled at atmosphere pressure for 1 to 2 hours to complete the reaction.

A solid solution of 7BaTiO$_3 \cdot$SrTiO$_5$ showed at room temperature a tetragonal structure, but changed at 80° C to the cubic type.[260] Around the transition point the two structures coexisted over an interval of several degrees. Barium titanate shows several specific heat anomalies.[261] At the Curie point, 120° C, there is a maximum in the heat capacity as a result of disappearance of the spontaneous polarization. Another anomaly occurs at 5° C and coincides with the dielectric constant maximum.

Materials used as additives in compounding ceramic bodies with special electrical properties include barium titanate, barium-strontium titanate, calcium titanate, magnesium titanate, and strontium titanate. More recently a new field of high dielectric ceramics has been developed, based on the compounds of titanium dioxide. The most frequently used material is barium titanate which finds applications in many high dielectric ceramic products as well as in uses where its piezoelectric properties are important.

Capacitors of high dielectric constant combined with a low-power factor are produced from barium titanate fired at high temperature, followed by reduction and reoxidation.[262]

Barium titanate at room temperature exists in a different form from the normal tetragonal with varying axial ratios in different samples.[263] This produces square pattern texture. Barium titanate is an example of a substance in which the crystal structure is not fixed at a given temperature. This is reflected in its ferroelectric properties. Tetragonal barium titanate irradiated at 35° C with thermal and fast neutrons in the ratio of 7 to 1, to 20 to 1 transforms to the cubic form.[264] The drop in the dielectric constant with time of barium titanate insulations is minimized if barium carbonate is calcined with titanium dioxide in the presence of an oxygen containing compound of bismuth.[265] Typical compounds are the trioxide, pentoxide, carbonate, nitrate, and stannate. The addition of tin compounds appears to have a synergistic effect in combination with bismuth, specifically bismuth oxide and stannic oxide. These wet-milled ingredients were dried and calcined at 1100° C for 1 hour. The product was ground to pass 200 mesh, mixed with an aqueous solution of polyvinyl alcohol, pressed into disks, and fired at a maximum temperature of 1375° C for 0.75 hour. This vitrified product had a dielectric constant at 25° C of 1900, a power factor of 1.2 per cent, and the aging at the end of 1000 hours was 2 per cent. Stoichiometric amounts of barium carbonate and titanium dioxide give best results.

At 400° C the reaction of titanium dioxide with barium carbonate begins.[266] With increasing temperature three modifications of barium titanate are formed, none of which is purely cubic at room temperature. Pure monocrystals of barium titanate are prepared by heating a mixture of barium carbonate 1.5, titanium dioxide 1.0, and barium chloride 3.0 moles in a crucible to 1200° C and cooling the melt at a rate of 20 degrees per hour.[267] The crystals are freed from the cooled mass by boiling for a long time in distilled water. Crucibles of the type required are prepared by calcining a shaped mixture of 95 parts titanium dioxide with 5 parts of barium oxide or titanate in 40 to 50 parts of a 1 per cent aqueous solution of ammonium humate in an oxidizing atmosphere at 1300 to 1350° C. High purity barium metatitanate is prepared from crude raw materials by a two-stage process.[268] Rutile ore or other form of titanium dioxide and barium carbonate in a 2 to 5 mixture are fed into an oil fired rotary kiln at 480° C. After 6 hours the material leaves the kiln at 1150° C with a composition of barium oxide 63, titanium dioxide 34, and impurities 3 per cent. A 46.6 g portion of this crude product is added to 100 ml of hydrochloric acid at 95° C. After stirring for 20 minutes, 100 ml of water at 95° C are added and agitation is continued for 10 minutes more. After standing for several hours the slurry is filtered and washed with water. The filtrate and washings are added to a solution of oxalic acid at 65° C to precipitate barium titanium oxalate. This precipitate is separated by filtration, washed until free from chlorides, dried at 100° C, and calcined at 1050° C for 5 hours to give barium titanate of high purity.

A coating of barium titanate on titanium and its alloys improves the antigalling and impact properties.[269] The glass-forming region of the $BaO$-$TiO_2$-$SiO_2$ system includes only three compounds which melt below 1500° C and which do not devitrify during cooling to room temperature in 18 hours.[270] Glasses of this system are unique in comparison to most optical glasses in that they have a high deformation temperature, exceptional chemical durability, and very low hygroscopicity. These glasses show 80 to 90 per cent transmittance to 3 micron and 50 to 60 per cent to 4.5 micron light. The n values of the glasses are 1.63 to 1.88 and the v values are 25.3 to 49.9.

Hedvall[271] obtained the metatitanate $CaTiO_3$, but no orthotitanate, by igniting at a high temperature an intimate mixture of titanium dioxide and calcium oxide. However, a study of the melting-point diagram[272] of mixtures of these oxides showed maxima at the composition corresponding to $CaO \cdot TiO_2$, $2CaO \cdot TiO_2$, and $3CaO \cdot TiO_2$, and eutectics containing 12 per cent calcium oxide at 1420° C and 52 per cent at 1780° C. In a similar study,[273] various amounts of these components were heated for 10 hours at 1350° C. With equimolecular amounts of calcium oxide and titanium oxide, the compound $CaO \cdot TiO_2$ formed, but with calcium oxide in excess it formed a solid solution with the titanates. The crystal structure of the product, synthetic perovskite, was found to belong to the regular system, and the lattice constant was 7.61 Ångstrom units. According to Parga-Pondal and Bergt,[274] if calcium oxide was present in the system in equivalent proportion or in excess, the metatitanate formed at temperatures up to 1300° C, but at 1400° C a compound corresponding to the formula $3CaO \cdot TiO_2$ was formed. Melting corresponding raw mixtures in the oxyacetylene flame indicated the existence of titanates rich in lime. In the presence of silica there was no evidence of the formation of double compounds between calcium silicate and calcium titanate.

A solution containing titanium oxalate obtained from titanium tetrachloride and oxalic acid is mixed with a calcium chloride solution at 40 to 45° C to precipitate calcium titanium oxalate.[275] The precipitate is filtered, and calcined to burn off all of the carbon, leaving calcium titanate of high purity. For example, 1000 parts of titanium tetrachloride was dissolved in 2000 parts of water at 30 to 35° C over a period of 1 hour. With 373 ml of this solution was mixed 520 g crystalline oxalic acid dissolved in 2000 ml of water at 45° C. On addition of a solution of 156 g of crystalline calcium chloride dihydrate in 600 ml of water, calcium titanium oxalate was precipitated. Preferably 520 g of glacial acetic acid was added to the solution. The precipitate was filtered, washed with water, dried, and calcined at 1050° C for 2 hours to decompose the oxalate. If the precipitation was carried out in the presence of acetic acid a yield of 98 per cent was obtained. Without the acetic acid, the yield was only 82 per cent. The particle size of the product was 0.1 to 1.0 micron.

Calcium titanate, $3CaO \cdot 2TiO_2$, undergoes dimorphic transformation at 765° C.[276] At the melting point in a reducing atmosphere, the dititanate is converted to $3CaO \cdot Ti_2O_3$, which exhibits cubic symmetry with a = 7.724 Ångstrom units. The extent of the reaction can be observed microscopically by following the disappearance of birefringence. Pure free flowing, finely divided calcium titanate suitable for use as a dielectric material or for the preparation of single crystals is prepared by calcining a mixture of calcium and titanium sulfates at 1300 to 1500° C.[277] Hot sulfuric acid solutions of titanium and calcium sulfates are mixed and heated with stirring until a dense crystalline precipitate of $CaTi(SO_4)_3$ is formed. The precipitate is filtered and washed with cold water until free from sulfates, after which it is air dried and calcined at 1400° C for 2 hours to form calcium titanate of high purity.

Monocrystalline calcium titanate formed by heating the finely powdered material in an oxy-hydrogen flame at 1850 to 1950° C followed by slow cooling because of its high index of refraction and low reciprocal dispersion is useful in making gems, lenses, and prisms.[278]

The formation of zirconium titanate by heating zirconium oxide and rutile takes place at a lower temperature and at a faster rate in the presence of lithium fluoride, calcium oxide, or sodium oxide.[279]

Rare earth titanates such as $Ln_2Ti_2O_7$ are prepared by direct combination of the oxides.[280] Weighed samples of the rare earth oxide and titanium dioxide are mixed in stoichiometric amounts in an agate ball mill under acetone. After drying, the mixture is fired at 1050° C for 14 to 15 hours, furnace cooled, and ball milled again. This milled product is pressed in 1 inch diameter pellets which are refired at 1200 to 1350° C for another 14 to 15 hours.

A compound having the formula $Fe_2TiO_4$ and possessing the same structure as magnetic iron oxide was obtained by heating in vacuum a mixture of one mole of titanium dioxide and two moles of ferrous oxide, or by reducing a mixture of titanium dioxide and ferric oxide with hydrogen at 900° C.[281] Heating a mixture of titanium dioxide and ferrous oxide gave $FeTiO_3$, which formed two series of solid solutions, one limited by $2Fe_2O_3 \cdot FeTiO_3$ and the other by $2FeTiO_3 \cdot Fe_2O_3$. The latter compound was paramagnetic. Slightly ferromagnetic $Fe_4(TiO_4)_3$ was also prepared.

A study of the system $Bi_2O_3$ and $TiO_2$ resulted in the formation of $Bi_{24}TiO_{38}$, $Bi_4Ti_{13}O_{12}$, and $Bi_2Ti_3O_9$.[282] Further studies of the system $Bi_2O_3$, $TiO_2$, PbO revealed the compounds $Bi_4PbTi_4O_{15}$ and $Bi_2PbTiO_{12}$.

The addition of ammonium hydroxide to a solution of gadolinium or disprosium oxides in cold concentrated hydrochloric acid containing 15 per cent titanium trichloride gave the compounds $Gd_2TiO_5$ or $Dy_2TiO_5$.[283]

A compound having the empirical formula $BaO \cdot TiO_2 \cdot B_2O_3$ is made by heating at 700 to 1300° C a mixture of titanium dioxide and barium metaborate or a mixture of barium and boron compounds capable of being converted into barium metaborate.[284] Preferably a mixture of barium sulfide, titanium dioxide, and boric acid in a 1 to 1 to 2 mole ratio is used as the starting material. The fused product is milled to obtain the desired particle size for pigment use.

A titanite spinel [285] of the formula $MgO \cdot Ti_2O_3$ has been made by heating magnesium with titanium dioxide, metallic titanium with magnesium oxide, or titanic oxide with magnesium oxide under reducing conditions. The product was a violet to black crystalline solid. Tanaka [286] produced magnesium ortho-, meta-, and di-titanates by high temperature reactions between magnesium oxide and titanium dioxide. Calcium oxide and titanium dioxide yielded only one addition product, $CaTiO_3$, but formed solid solutions. The compound was sensitive to light.

Cole and Nelson [287] followed the reaction between mixtures of hydrous titanium dioxide and zinc oxide, ranging from 33 to 59 per cent of the former on the dry basis at temperatures between 400° and 1000° C, by determining the size of the unit cell of the product. Zinc orthotitanate was the only compound formed, but solid solutions of titanium dioxide in this compound were found to extend to the composition represented by the formula $Zn_2TiO_4 \cdot 5TiO_2$. These products were not stable at high temperatures and began to dissociate at 775° C into titanium dioxide and another solid solution that was poorer in this component. The primary reaction was initiated at 430° C, and the formation of zinc orthotitanate was practically complete at 800° C after 3 to 6 hours. A product corresponding to the composition of zinc metatitanate was found to be in reality a solid solution of titanium dioxide in zinc orthotitanate. The unit cell of the orthotitanate was determined by X-ray measurements to be 8.46 Ångstrom units, and the specific gravity of the pure compound was 5.12.

Cobalt and nickel metatitanates, corresponding to the chemical formulas $CoTiO_3$ and $NiTiO_3$, have been reported formed by fusing the corresponding chloride with titanium dioxide.

From X-ray data, Cole and Espenschied [288] found that the reaction between titanium dioxide and litharge, in proportions from 25 to 89 per cent of the latter component, yielded only one compound, lead metatitanate. Combination started at 360° C and was completed at 400° C in 4 hours. The samples for study were prepared by calcining an intimate mixture of litharge with hydrolytically precipitated titanic oxide pulp freed from sulfuric acid. Crystals were obtained by dissolving the product in sodium tungstate at 850° C and cooling the melt. Such crystals were also obtained from a system rich in litharge. For instance, at 900° C a charge of 2 moles lead oxide to 1 mole titanium dioxide gave a mixture of crystalline lead titanate and molten lead oxide. Samples of lead titanate held for 24 hours at 400°, 575°, 700°, 775°, 860°, and 1000° C, and cooled in air, gave identical diffraction patterns, and a hard-sintered sample, held at 1200° C for 2 hours, yielded the same pattern. No difference was observed in the products obtained, employing excess of either ingredient or of the crystalline material. The titanate crystallized from melts as orthorhombic holohedral pyramids. Nicholson [289] likewise made X-ray studies of the products obtained by calcining mixtures of titanium dioxide and litharge over a wide range of proportions at 700° C, and was able to detect only three crystalline phases, tetragonal lead monoxide, anatase, and lead titanate ($PbTiO_3$). Lines corresponding to the titanate were obtained consistently from the calcination products of other lead and titanium compounds.

In a study of speeds of reaction, rhombic, yellow, lead monoxide and tetragonal, red, lead monoxide were mixed in equimolecular amounts with commercial titanium dioxide and heated in glass ampoules at 400° to 800° C.[290] Velocity of formation of lead metatitanate from the two oxides in air, nitrogen, and carbon dioxide did not vary greatly but was more pronounced at different temperature intervals. The yellow rhombic oxide reacted more rapidly below the conversion temperature, 488.4° C, and the tetragonal form more rapidly at higher temperatures.

Ferrous and ferric metatitanates, corresponding to the minerals ilmenite and arizonite, have been prepared synthetically from the appropriate oxides.

A study of the system $Fe_2O_3$-$FeO$-$TiO_2$ at high temperatures revealed that only rutile, iron glance, and pseudobrookite, $Fe_2TiO_3$, separated from the melt.[291]

A mixture of at least one mole of calcined titanium dioxide with one mole of reactive lead oxide free of metallic lead and having a particle size less than 0.5 micron is heated in a neutral or oxidizing atmosphere at 600 to 750° C to form lead titanate, $PbTiO_3$.[292] Heating is continued until all of the lead oxide is combined with the titanium dioxide and the resulting lead titanate has a particle size below 0.5 micron.

To obtain titanates as finely crystalline powders of high purity, a solution of titanium tetrachloride is treated with hydrogen peroxide, cooled to 10 to 15° C, then treated with ammonia and finally with a nitrate or a chloride of the metal other than titanium.[293]

# 9

# Organic Compounds of Titanium

Many organic compounds of titanium have been reported in the scientific literature, but few of these are of commercial value. Titanium oxalate [1] soluble in water is obtained by treating titanium sulfate or a double alkali metal sulfate with oxalic acid or an alkali oxalate. The reactants may be intimately mixed in a solid or moist state or may be dried and heated together. In another modification,[2] titanium sulfate solution was added to a solution of a carbonate of an alkali metal or ammonium until a pH of 6 was reached, and this intermediary titanium carbonate compound was washed and gradually added to an aqueous solution of oxalic acid containing a small proportion of octyl alcohol which served to slow down the reaction. The precipitated titanium oxalate was filtered and washed. Basic titanic oxalate [3] may be obtained from solutions of tetravalent titanium in mineral acids by adding oxalate ions under regulated conditions of neutralization, dilution, and temperature. For instance, such a compound was precipitated by adding 0.15 to 0.30 mole oxalic acid or an equivalent amount of an alkali or alkaline earth metal oxalate to a solution containing 0.6 mole sulfuric acid and 1 mole of titanium dioxide. Such a solution was strongly basic. The temperature was held below 50° C, and the solution was diluted to effect precipitation. In the presence of chloride ions, precipitation may be effected from a more acid solution, thus yielding a purer product.

Titanium compounds soluble in organic and in dilute inorganic acids may be prepared by heating metatitanic acid or basic sulfate with an alkali below the fusing point, 150° C. By treating the reaction product with appropriate acids, double salts, such as titanium sodium tartrate, lactate, and sulfate, and titanium potassium oxalate, were obtained.[4]

The complexity of titanium oxalates is proved by the ease with which they may be crystallized from solution and by their resistance to hydrolysis on dilution and heating, in contrast to other salts. Titanium sulfate gives an immediate precipitate with ammonium hydroxide which almost completely redissolves on addition of hydrogen peroxide. Thus it appears that the introduction of active oxygen into the titanic oxide molecule promotes the formation of complex ions. Pechard [5] reported that titanic acid prepared by the action of sodium carbonate on aqueous titanium tetrachloride dissolved readily in a warm solution of potassium acid oxalate, and on cooling a precipitate of crystalline titanium potassium oxalate formed.

In preparing double formates and double acetates in the form of water-soluble crystals, Stahler [6] added a saturated solution of an alkali, alkaline earth, or ammonium formate or acetate in excess to the concentrated solution of a salt of trivalent titanium (sulfate, chloride). The crystalline salt was separated from the liquor, with exclusion of air.

Fenton [7] obtained a bulky chocolate-colored precipitate on adding a warm aqueous solution of a quadrivalent titanium salt to a solution of dihydroxymaleic acid. Titanium tetrachloride reacts with high molecular weight monobasic organic acids, such as stearic and oleic, at 70° to 110° C, under anhydrous conditions, to form soaps which may be used as driers in paints and varnishes.[8] At the temperatures employed, the hydrochloric acid was eliminated as formed. Tartaric acid combined with titanium tetrachloride to form a complex [9] in which the two constituents were present in equimolecular amounts, while the hydroxide and sodium tartrate formed a similar product in which the mole ratio was 2 to 3.

A tervalent titanium compound, prepared by reducing alcoholic $(C_2H_5)_4$-$TiO_4$ with sodium, had a dark blue color and dissolved in dilute hydrochloric and sulfuric acids to form the corresponding titanous salt.[10] Solutions obtained in this manner were entirely free of iron and other metallic or acidic impurities.

Titanium phthalate [11] can be made by adding titanium sulfate solution, adjusted to slight acidity, to an aqueous solution of sodium phthalate prepared by neutralizing phthalic acid or anhydride with sodium hydroxide in stoichiometric proportions. One mole of titanic sulfate reacted with 2 moles of sodium phthalate. A dense white precipitate formed which usually contained more or less titanium hydroxide, depending on the degree of hydrolysis. On drying at 105° C, the washed precipitate became gelatinous and had a tendency to form semivitreous or horny aggregates resembling silica gel. The acidity of the titanium sulfate solution was an important factor. For instance, if highly acid solutions were employed, precipitation was delayed and the yield and composition of the product were frequently unsatisfactory. On the other hand, a practically neutral sulfate solution gave a precipitate containing an appreciable proportion of titanium oxide with the result that the phthalate content was considerably less than normal. Basic titanium phthalate was produced by adding

½ mole proportion of aqueous sodium phthalate to a solution of titanyl sulfate.[12]

An aqueous solution of a titanium salt was added to a solution of a carbonate of an alkali metal or ammonium until a pH of 6 was obtained, and the precipitate, after thorough washing, was added to a water solution of oxalic acid containing a small proportion of octyl alcohol to obtain a basic titanium oxalate corresponding to the chemical formula $(TiO)_2(C_2O_4)_2$.[13]

In concentrated sulfuric acid, titanium dioxide gives complex compounds of various colors with hydroxy aromatic compounds such as phenol, salicylic acid, and metol.[14] These colored compounds have been suggested as a basis for the colorimetric determination of titanium. Titanium tetrachloride reacted with phenols in the presence of ammonia or organic basis in anhydrous media to form alcoholates and phenylates.[15] Employing higher alcohols, a lower alcoholate was made first and this compound was reacted with the higher alcohol. For example, titanium tetrachloride was mixed with ethyl alcohol, and ammonia dissolved in ethyl alcohol was added. Typical compounds of this class are titanium tetraethylate and tetraphenylate. Titanium tetrachloride added slowly to dry methyl alcohol gave the solid dimethyloxydichlorotitanium, and triethoxychlorotitanium was prepared by refluxing tetramethoxytitanium and acetyl chloride at 70° to 80° C for 2.5 hours. Addition of ethyl alcohol gave a solid product containing ethyl alcohol of crystallization. Refluxing triethoxychlorotitanium with acetyl chloride, or titanium tetrachloride with ethyl alcohol, for 24 hours at 80° to 100° C gave diethoxydichlorotitanium, a pale yellow solid. Crystals containing methyl and ethyl alcohols were obtained by adding the corresponding reagent. Diisopropoxydichlorotitanium was prepared by refluxing titanium tetrachloride in isopropyl alcohol for 9 hours. Diisobutoxydichlorotitanium was prepared in a similar manner. These dichloro compounds were readily hydrolyzed in air and dissolved instantly in water, yielding clear solutions.[16]

McCleary[17] produced a water-soluble compound by heating titanium tetrachloride with a tertiary alcohol (butyl) which was used to prepare a slightly opalescent water solution containing more than 100 g per liter titanium dioxide. $Ti(OPh)_4 \cdot PhOH$, obtained by heating the anhydrous tetrachloride with phenol, was an orange red crystalline powder soluble in organic solvents, but which decomposed slowly in water.[18] The dark red crystalline $TiCl_2(C_{10}H_7O_2)_2$, obtained by the reaction of the tetrachloride with 2-naphthol in dry carbon disulfide, was decomposed by moist air.

Titanium gluconate[19] may be prepared by the double decomposition reaction between an alkaline earth gluconate and aqueous titanic sulfate. After filtering off the alkaline earth sulfate formed, the gluconate was obtained as a greenish-yellow, syrupy liquid, and the solid compound was precipitated by adding an organic compound miscible with water. The purity of the titanium gluconate was found to depend in a large measure upon the relative proportions of titanium dioxide and sulfuric acid in the original solution. In general, a molecular ratio of 1 to 1 gave best results. In an actual operation, powdered calcium gluconate was added, with agitation, to a solution of titanyl sulfate. Reaction occurred at once, with the precipitation of calcium sulfate and the formation of a greenish-yellow, viscous liquid. Water was added to maintain the required fluidity, and, after the calcium sulfate had settled, the supernatant liquor was separated by decantation and filtration. Titanium gluconate was then precipitated from the solution by the addition of methyl alcohol. The washed and dried product was a white powder containing 28.7 per cent titanium dioxide and consequently represented almost pure gluconate.

Titanium alcoholates reduce aldehydes and ethers.[20] The reduction of titanium tetrachloride by alcohol is accompanied by autoxidation and formation of hydrogen peroxide. Reduction of mandelic or lactic acid was free from such complications, and in daylight or mercury arc proceeded with a speed proportional to the concentration of the reducing agent and of the tetrachloride.[21]

Organic titanic compounds may be prepared by treating titanium tetrachloride with aromatic hydroxyl compounds, such as phenol, in the presence of ammonia or an amine.[22] The reaction may be carried out in an inert solvent, such as benzene, and it may be initiated in the absence of and completed in the presence of the organic base. Titanium tetrachloride reacted with phenol to form trichlorotitanium phenolate,[23] $TiCl_3OC_6H_5$, which readily hydrolyzed to give titanium hydroxide, phenol, and hydrochloric acid. The tetrachloride also reacted with para chlorophenol to form $TiCl_3(OC_6H_4Cl)_2$.[24] A precipitate of $(C_2H_5)_2COTiCl_3$ [25] was formed by treating benzophenone in anhydrous benzene with titanium tetrachloride. Triphenylcatechol titanates have been separated into their optical isomers by the use of cinchonine.[26]

Demarcay[27] observed that titanium chloride combined with oxygenated ethers as well as with sulfides and sulfhydrates of alcohol radicals. Methyl ether combined with anhydrous titanium tetrachloride in molecular proportions at temperatures above 150° C to give a yellowish, solid complex having the composition corresponding to the formula, $TiCl_4CH_3 \cdot O \cdot CH_3$.[28] Titanic chloride and benzoic chloride combined directly to form a double chloride corresponding to the formula, $TiCl_4 \cdot C_6H_5COCl$.[29] Addition of titanium tetrachloride in chloroform to solutions of diphenyl ketone, heliotropin, and methyl phthalate in the same solvent gave precipitates of crystalline addition compounds which were unstable in moist air.[30] The oxonium salt, $TiCl_4 \cdot 2(C_2H_5)_2O$ formed by the reaction of titanium tetrachloride and anhydrous ethyl ether, was very hygroscopic and was readily decomposed by water.[31]

Complex compounds are produced by the reaction of isovitamin C, or its first reversible oxidation product, with an alkali, alkaline earth, or ammonium salt, and titanium tetrachloride in a medium consisting of one or more mono- or poly-hydroxy alcohols or ketones.[32] A basic compound of quadrivalent titanium for use in treating paper or leather was formed by the reaction in aqueous solution of titanium tetrachloride with a sulfamate of a metal which yields an insoluble chloride, such as lead.[33] 5 Chloro-2-furyl ethyl ketone may be prepared by the Friedel and Craft reaction of 2-nitrofuran, propionyl chloride, and titanium tetrachloride.[34] Titanium tetrachloride reacted with organic compounds of nitrogen to form complexes in which the ratio of the components was not generally accountable on the basis of a constant coordination number of 6 for titanium.[35] Para nitro- and 2-4 dinitrophenyl-hydrazones were reduced by titanium tetrachloride to $p$-$C_6H_5(NH_2)_2$ and 2-4 $(H_2N)_2C_6H_3NH \cdot NH_2$, respectively.[36] Titanium halides form crystalline binary complexes with nitrilles [37] and with nitro compounds.

Organo metallic compounds [38] may be prepared by treating titanium tetrachloride with eight mole proportions of a primary, secondary, or tertiary amine in an anhydrous solvent. Gilman and Jones [39] prepared compounds of this type from titanium tetrachloride, and studied their reactivity.

The reaction of isobutene with titanium tetrachloride at low temperature is induced by traces of moisture but is not affected by dry hydrochloric acid.[40] Methyl maltoside heptaacetate with ortho ester structure, on treatment with titanium tetrachloride, gives the normal acetochloromaltose and not the third acetochloromaltose from which it was prepared,[41] and ethyl tetraacetyl-$\beta$-d-glucosido-glucose is similarly converted to aceto chloroglucose.[42]

Rothrock [43] produced plastic molding and casting compositions by polymerization of esters of unsaturated alcohols with inorganic acids of titanium, such as methyl methacrylate titanate and tetramethallyl titanate. Interpolymers of tetramethallyl titanate and vinyl acetate give white enamel-like surfaces that are resistant to wear.

The esterification of linseed-oil fatty acids with glyceryl titanate yielded oils of good drying properties but dark in color.[44] By heating butyl titanate with glycerol, a white solid glycerol titanate formed which was insoluble in water and in the common organic solvents. A reddish, viscous liquid, which in thin layers dried in air to give transparent, glassy, brittle films, was formed by refluxing phenyl titanate with formaldehyde in butyl alcohol.

Organic esters of titanium, especially polymeric butyl titanate, are used in the formation of paints, thin layers of which covered finally with a sheet of aluminum metal can withstand a temperature of 700° C for an extended period of time.[45] The butyl titanate polymerizes at 200 to 250° C, and the polymers contain 65.4 per cent titanium dioxide. Alkyl titanates are used as fabric treatments and paint vehicles.[46]

Potassium titanium oxalate is obtained by dissolving potassium titanate in a stoichiometric amount of oxalic acid.[47] In an example, 5.5 g of titanium dioxide was placed in a stainless steel apparatus of 200 liter capacity provided with a water jacket. To this was added a solution of 10.7 kg of technical potassium hydroxide equivalent to 9.2 kg of the pure compound in 24 liters of water. Steam was admitted to heat the reaction mixture to dryness with occasional stirring. The dry mixture was then calcined for 5 hours at 500° C. After cooling, the potassium titanate formed was added to a solution of 20.7 kg technical oxalic acid in 80 liters of water in a 150-liter enameled container. The mixture was filtered, thickened, and crystallized by cooling. The mother liquor was again concentrated by evaporation and crystallized. On filtering and drying, 24.2 kg of potassium titanium oxalate was obtained.

Extremely large numbers and varieties of organic complexes of titanium have been prepared.

Herman and Nelson[48] prepared compounds characterized by the titanium to carbon bond. Phenyl magnesium bromide and barium titanate yield a phenyl titanium derivative which is too instable to be isolated. Phenyl lithium and isopropyl titanate yield relatively insoluble crystalline $C_6H_5Ti(iso\ C_3H_7O)_3 \cdot LiOC_3H_7 \cdot LiBr \cdot (C_2H_5)_2O$ which with titanium tetrachloride yields $C_6H_5Ti(iso\ C_3H_7O)_3$ having a melting point of 88 to 90° C. This product with diphenyl ketone gives triphenylcarbinol. The compound $C_6H_5Ti(iso\ C_3H_7O)_3$ catalyzes the polymerization of styrene.

In an example, 1 mole of lithium phenyl in 750 ml of ethyl ether is added to 1 mole of titanium propoxide in 600 ml of ethyl ether maintained at 8 to 15° C to give a slurry containing $C_6H_5Ti(OC_3H_7)_3 \cdot LiOC_3H_7 \cdot LiBr \cdot (C_2H_5)_2O$. A 10 per cent excess, 0.276 mole, of titanium tetrachloride in 1 liter of ethyl ether is then added in 1.5 hours with stirring and cooling. The clear supernatant liquor is decanted from the lithium chloride and lithium bromide into a nitrogen filled distilling apparatus. The ethyl ether is stripped in vacuum at $-40$ to $-20°$ C leaving a dark gray crystalline residue. This residue is purified by dissolving it in ethyl ether followed by filtering and stripping hot of the ethyl ether at 100 to 200 mm nitrogen pressure to obtain a white crystalline precipitate which is washed with cold ethyl ether and dried. The purified product is $C_6H_5Ti(OC_3H_7)_3$, a white crystalline solid melting at 88 to 90° C. It decomposes rapidly with oxygen and water.

Compounds of this type may be represented by the general formula $R_nTiX_{n-4}$. In general the stability of the compounds and of the titanium-carbon bond increases as R increases in electronegativity in the order: butyl, methyl, ethinyl, p-anisyl, phenyl, naphthenyl, indenyl. The stability decreases as n increases from 1 to 4, but decreases in the following order as X is represented by butoxide, methoxide, chlorine, fluorine. Such compounds formed in ethyl ether as the solvent are more stable than those formed in benzene or petroleum ether. Addition of amines such as ethylenediamine and pyridine increases the stability of the titanium to carbon bond but undesirable side reactions occur.

## PHYSIOLOGICAL EFFECTS

A number of investigators have demonstrated experimentally that titanium compounds are not toxic.[49] The noninjurious effect of the dioxide was demonstrated by Heaton,[50] who within a short time ate a pound of commercial pigment, titanium dioxide-barium sulfate composite, mixed with a little glucose. Passage through the alimentary canal was observed by X-ray photographs and a fluorescent screen, and complete elimination required 24 hours. No harmful effect was felt or detected, and on the contrary it apparently cured an ailment which had baffled physicians for years.

Titanic acid, along with silica and similar substances, has been detected in the lungs of patients suffering with silicosis and other lung diseases, although there was no indication that the titanium compound was responsible for the disease.[51] The dioxide, fed to various animals for a period of 6 months, proved harmless.[52] Daily ingestion by sheep of 2 to 3 grams showed no ill effects after 3 months, and complete elimination in the excreta suggested its application as a reference substance for digestion studies.[53] After feeding the compound to guinea pigs in considerable amounts, no ill effect was noted, only traces remained in the intestinal tract, and none was detected in other organs.[54]

Small doses of the double titanium and sodium tartrate increased the respiration rate in frogs, mice, and rats, and larger amounts caused paralysis and death.[55] Carteret[56] ascribed the toxicity of preparations of this type to the effect of the acid radical. Alkaline titanium citrates, lactates, and stearates were studied. Experiments in vitro showed that titanium trichloride markedly inhibited blood phosphate activity.[57]

The titanium complexes of ascorbic and dehydroascorbic acids have been successfully employed in the chemotherapy of cancer,[58] but they occasionally show toxic action. Pick[59] reported the use of the disulfide and the salicylate of titanium in the treatment of diseases of bacterial origin. These compounds were especially effective in lupus. Intermuscular injections of an olive-oil suspension of hydrous titanic oxide or titanocitrate into rats having Jensen sarcoma reduced the rate of mortality and checked the growths.[60] From a study of erythropoietic action of inorganic elements fed to anemic rats daily, along with 0.5 mg of iron, Beard and Myers [61] found that recovery occurred in 2 to 3 weeks with the addition of 0.1 mg of titanium, as compared with 4 to 6 weeks with iron alone.

Diseased skin areas treated with titanium dioxide ointment healed more quickly than untreated areas.[62] No irritation resulted on applying the dioxide to the skin of normal rabbits for 15 days. An ointment containing this agent was used during the first World War for the treatment of erythema caused by exposure to mustard gas.[63]

From the results of extensive studies, Kahone [64] concluded that the use of titanium dioxide and its derivatives in pharmacy and cosmetics was justified by the innocuousness of these compounds.

Growth and fermentation were accelerated by ascorbic acid complexes of titanium.[65] According to Pick,[66] all soluble organic and inorganic compounds of titanium check the putrefaction of animal and vegetable proteins. Twenty grams of putrescent liquid was deodorized by addition of 0.1 g titanium sulfate.

Hanzlick and Tarr [67] found titanium tetrachloride to be a severe skin irritant. Application to the skin of dogs produced hyperemia, swelling and edema, ulceration and necrosis. Similar changes, together with vesication, took place on human skin.

Titanium compounds have been reported to act as catalysts in the oxidation reactions of vegetable cells, and there is evidence that the element is essential in the formation of soils from rocks.[68]

Yasue [69] proved the insecticidal value of titanium dioxide powder against tribolium ferrugineau.

In acute inhalatory intoxication of white rats with titanium tetrachloride a large amount of titanium was found immediately in the blood.[70] The main route of elimination was by the intestines but much was also found in the urine. Titanium remained relatively long in the animal. On repeated application, hydrochloric acid and titanium trihydroxychloride from the hydrolysis of the tetrachloride showed the greatest toxicity.[71] The titanium trihydroxychloride dust caused markedly greater irritation and oxygen demand than hydrogen chloride gas. Based on these results, the concentration of the products of tetrachloride hydrolysis in the air should not exceed 0.001 mg per liter. Titanium dioxide inhibited the growth of microorganisms.[72]

# 10

# Chemical Analysis of Titanium and Its Compounds

The voluminous literature on the analytical chemistry of titanium includes processes for its separation and determination, gravimetrically, volumetrically, colorimetrically and by instrumental methods. Gravimetric estimations are generally based on the hydrolysis of dilute, weakly acid solutions by long boiling, or precipitation by special reagents such as "cupferron," the ammonium salt of nitrosophenylhydroxylamine, and tannic acid. Hydrolysis of quadrivalent salt solutions usually gives erroneous results because of the interference of aluminum, phosphoric acid, iron, chromium, and other substances which normally accompany titanium, particularly in ores. Titanic acid, precipitated by the slow process of thermal hydrolysis, frequently adheres to the surface of the glass vessel so tenaciously as to make complete removal practically impossible, and during the long period of boiling there is some solvent action on the glass, with the result that contamination inevitably occurs. Although this was the first method employed for the determination of titanium, in view of the many disadvantages it is now practically obsolete.

Cupferron is accepted as an excellent reagent for the precipitation of titanium, although ferric iron, copper, zirconium, thorium, tin, vanadium, and uranium are also precipitated quantitatively under favorable conditions. In addition, certain other elements, particularly silicon, tungsten, cerium, silver, lead, mercury, and bismuth, are known to be partially precipitated or more or less occluded with the titanium. Fortunately most of these may be separated by well-known methods before adding the cupferron, and besides, since comparatively few of these elements are commonly associated with titanium, the possibilities of serious interferences are not so great as might be supposed. The degree of acidity of solutions from which total precipitation may be effected varies widely with the different elements, and with titanium, iron, and zirconium it is unusually high. This property offers a means of at least partial separation. Iron is usually present in crude titaniferous materials in appreciable proportions, and its removal may be accomplished readily by precipitation with hydrogen sulfide from an ammoniacal solution containing also ammonium salts of certain hydroxy organic acids, particularly citric and tartaric, followed by filtration. Under these conditions titanium has the special property of remaining in solution. Solutions for analyses may contain almost any acid, and sulfuric in amounts up to 40 per cent by volume exhibited no solvent action on the precipitate, although such high proportions have not been found necessary in any of the actual separations. Nitric acid in high concentrations must be avoided because of its oxidizing action on the reagent.

In operation, the test solution, containing the equivalent of 20 ml concentrated sulfuric acid and 0.2 g titanic acid, is brought to a volume of 400 ml, and the beaker is placed in a bath of cracked ice. A recently filtered 4 per cent solution of cupferron is added, drop by drop, with constant stirring, in considerable excess. The titanium comes down as a canary-yellow, flocculent precipitate which coagulates almost immediately and may be filtered without the aid of suction. After careful washing with ice-cold dilute hydrochloric acid, the precipitate is dried in an oven and ignited at high temperature to produce titanium dioxide, which is weighed as such. The operation is carried out in the cold because the cupferron appears to decompose even below the temperature of boiling water, forming an oily material which solidifies to a resinous mass on cooling.[1]

Volumetric methods are generally employed if large numbers of samples are analyzed regularly, because of the simplicity and speed of the operation. Such analyses involve reduction of the titanium by suitable means, usually a special type of Jones reductor, followed by reoxidation with a standard oxidizing agent. Ferric solutions are usually employed, for with this reagent the iron, which so often accompanies titanium in the sample and is reduced along with it, is not reoxidized and thus causes no interference. The presence of other impurities often found in titanium ores, however, causes erroneous results, and of these vanadium is the worst offender. The end point is readily detected by aid of an alkali thiocyanate which is added directly to the main solution and which gives a strong, brown color in the presence of ferric ions.

An oversize Jones reductor which has proved satisfactory consists of a dispensing burette 2 inches in diameter and 21 inches long, connected through a glass stopcock to a delivery tube approximately 0.5 by 3.5 inches. A plug of glass wool is fitted in the bottom of the burette, and on top of this is placed a 7-inch column of 10-mesh and 20-mesh zinc

(500 g of each), and above this a 5-inch column, approximately 500 g of sticks about 0.5 by 2.5 inches in size. Before charging into the reductor vessel, the zinc is amalgamated by slowly pouring a solution of 22 g mercuric chloride in 200 ml concentrated hydrochloric acid onto 500 g of the metal in granular or stick form, covered with water. After thorough mixing, the treated product is washed with hot water until free from chloride.

The delivery tube of the reductor vessel is led into a 1-liter flask through one opening of a three-hole rubber stopper. A second hole serves as an inlet for carbon dioxide or a nonoxidizing gas, and the third opening as outlet for the gas.

The standard oxidizing solution is prepared from ferric ammonium sulfate, 30 g of the crystalline salt being dissolved in 300 ml of water to which has been added 10 ml of concentrated sulfuric acid. Potassium permanganate solution is then introduced, drop by drop, until a faint pink color persists, to insure oxidation of any ferrous iron, and the volume is brought to 1 liter. The solution is then standardized against potassium permanganate solution of known strength. Sodium oxalate, supplied by the U.S. Bureau of Standards, serves as a splendid primary standard. A 50-ml portion of the ferric solution is made up to 100 ml in 5 per cent sulfuric acid, passed through the reductor as used for titanium, and titrated with the potassium permanganate solution of predetermined strength. Each milliliter is equivalent to approximately 0.005 g titanium dioxide, and may be adjusted to this exact value if desired. A saturated solution of ammonium thiocyanate is used as indicator.

Before each use, the reductor is drained to the top of the granular zinc, and 200 ml of hot 5 per cent sulfuric acid is added. This solution is drained slowly, the zinc is washed in this manner several times with boiling water, and drained again to the top of the layer of granular zinc. The hot solution of the sample to be analyzed, containing not more than 0.25 g titanium dioxide equivalent in 125 to 150 ml of 10 to 15 per cent sulfuric acid, is washed into the reductor. It is covered with a watch glass and the carbon dioxide is turned on so that a slow stream flows through the reducing flask to prevent oxidation by the air. After 15 to 20 minutes, with the gas still flowing continuously, the reductor is drained slowly to the top of the column of granular zinc and boiling 5 per cent sulfuric acid is added in amount just sufficient to cover the layer of stick zinc. The reductor is again drained to the granular-stick contact and washed with boiling water in the same manner until the total volume is almost 1 liter. The receiving flask is removed, 7 ml of the indicator solution is added, and the titanium is titrated rapidly with the standard ferric ammonium sulfate solution to a faint brown color, which persists for several minutes.

In carrying out the titration, greater accuracy is obtained by adding nearly all the standard solution required before agitation, and then mixing with a gentle rotary motion until the final end point. When not in use, the reductor should be filled with water above the column of zinc and covered with a watch glass.

Titanium dioxide pigments in general, and similar manufactured products, may be dissolved directly in a hot mixture of sulfuric acid and an alkali metal or ammonium sulfate, although more refractory materials, such as ores, may require fusion with potassium bisulfate, sodium carbonate, or a similar compound. A 0.25 to 0.50 g sample of the pigment is weighed into a 250 ml beaker, and 20 to 30 ml concentrated sulfuric acid and 10 g of ammonium sulfate are added. The beaker is covered with a watch glass and the contents are carefully boiled over a flame until solution is effected. A slight residue of undissolved siliceous material may be disregarded. After cooling, the product is diluted with water to a volume of 125 ml to 150 ml; if a white precipitate forms, indicating barium sulfates, it is filtered off and washed with hot 5 per cent sulfuric acid and finally with hot water until free from titanium. The washings are added to the main solution and the combined product is brought to the desired concentration before being transferred to the reductor. Other composite pigments are treated in a similar manner, although it may not be necessary to remove the insoluble material.

Ores may require a fusion treatment to convert them to soluble form. To illustrate, 0.5 g finely ground ilmenite is mixed with 10 g potassium bisulfate and heated in a covered crucible at a red heat for 1 hour. The melt is cooled, dissolved in dilute (10 per cent) sulfuric acid, and adjusted to the concentration specified under the section on pigment analysis. The insoluble residue is usually small and need not be removed from the solution.

From the following chemical equations involved in the volumetric analysis of titanium, chemical equivalents may be calculated readily.

$$2Ti^{++++} + Zn \rightarrow 2Ti^{+++} + Zn^{++}$$
$$3Ti^{++++} + Al \rightarrow 3Ti^{+++} + Al^{+++}$$
$$MnO_4^- + 5Ti^{+++} + 8H^+ \rightarrow Mn^{++} + 5Ti^{++++} + 4H_2O$$
$$2MnO_4^- + 5C_2O_4^{--} + 16H^+ \rightarrow 2Mn^{++} + 10CO_2 + 8H_2O$$
$$MnO_4^- + 5Fe^{++} + 8H^+ \rightarrow 5Fe^{+++} + Mn^{++} + 4H_2O$$
$$Ti^{+++} + Fe^{+++} \rightarrow Ti^{++++} + Fe^{++}$$

A simplified method of analysis employs zinc or other liquid amalgam as the reducing agent in place of the Jones reductor, although the hazards of using mercury still exist. To effect solution the finely ground sample is heated with a mixture of sulfuric acid and ammonium or sodium sulfate in the usual manner. Water is added to take up the cooled melt. The solution is transferred to a separatory funnel along with the liquid amalgam. Carbon dioxide is passed in to displace the air, after which the funnel is closed with the stopper and shaken until the reduction of the titanium IV to titanium III is complete. The amalgam is drawn out through the cock at the bottom of the separatory funnel and the solution is titrated in the funnel with standard ferric chloride or ferric sulfate solution employing potassium thiocyanate as the indicator.

According to a method developed by Rahm [2] the complete analysis of titanium in pigments and ores is carried out in a 500 ml Erlenmeyer flask. The sample is dissolved by heating with a mixture of sulfuric acid and ammonium sulfate, reduced with aluminum metal, and titrated with standard ferric chloride solution. No special equipment is required. The wide-mouthed Erlenmeyer flask is fitted with a one-hole rubber stopper carrying a glass delivery tube which extends into a beaker of saturated sodium bicarbonate solution. A convenient concentration of ferric chloride solution is such that one milliliter is equivalent to 0.005 g of titanium dioxide or 0.0626 molar. This solution is prepared by dissolving 16.9 g of ferric chloride hexahydrate in 800 ml of distilled water containing 15 ml of 18 molar sulfuric acid and diluting with water to 1 liter. It is standardized against the National Bureau of Standards titanium dioxide sample 154.

In carrying out the analysis 0.1000 to 0.2000 g of pure titanium dioxide or 0.5000 g of composite pigment is transferred to a 500 ml Erlenmeyer flask, and 25 ml of 18 molar sulfuric acid and 25 g of ammonium sulfate are added. The flask is heated over the flame of a Meker burner and swished occasionally to facilitate the solution of the sample. After cooling the hot, golden yellow colored solution to room temperature, 130 ml of distilled water and 20 ml of 12 normal hydrochloric acid are added. The solution is reheated to boiling and 1 g of aluminum metal is added. Without delay the rubber stopper containing the delivery tube is inserted in the neck of the flask so that the discharge end of the tube extends into the sodium bicarbonate solution. After all of the aluminum has dissolved and evolution of hydrogen has ceased, the flask is cooled in a bath to below 60° C. The delivery tube is removed, 2 ml of 45 per cent potassium thiocyanate solution is added as an indicator, and the titanium III is titrated with standard ferric chloride solution to the appearance of a light orange end point.

Titanium ore may be heated with potassium bisulfate to effect solution. A larger amount of aluminum is required to effect reduction since the iron is reduced before the titanium.

Guerreiro [3] developed a method for the analysis of titanium ores which includes the separation of the titanium as the soluble complex ion,

$TiO_2(SO_4)_2^-$. A 0.5 g sample of the ore, mixed with 2.0 g of fused potassium pyrosulfate and 3 drops of concentrated sulfuric acid, is heated in a platinum crucible for 10 minutes with a small flame and then completely fused. After cooling, the melt is taken up in hot dilute sulfuric acid solution to prevent hydrolysis. The silica residue is separated by filtration, fused with potassium pyrosulfate, dissolved, and added to the main solution. Water is added to bring the total volume to 250 ml. To a 25 ml aliquot is added bromine water followed by boiling to remove the excess. After cooling, hydrogen peroxide and ammonium hydroxide in excess are added. The color of the solution disappears due to the formation of the titanium complex, $TiO_2(SO_4)_2^-$. Ferric hydroxide and aluminum hydroxide precipitate. If the last heating causes the precipitation of titanium hydroxide, more hydrogen peroxide is added to dissolve it. After removing the iron and aluminum hydroxide by filtration, the ammoniacal solution is made acid with hydrochloric or sulfuric acid. Then sodium thiosulfate or sulfurous acid is added, after which the solution is boiled to effect hydrolysis. After cooling to 40° C, ammonium hydroxide is added to precipitate the remaining titanium hydroxide which flocculates and settles rapidly. The precipitate is filtered, washed, calcined, and weighed as titanium dioxide.

Titanium may be determined colorimetrically by comparing the intensity of the yellow color produced on adding hydrogen peroxide to a test solution with that of a solution of known titanium content similarly treated. This test is very delicate and 0.0003 g titanic oxide in 100 ml of solution gives a noticeable color under favorable conditions. Solutions deficient in acid, however, do not develop the maximum intensity of color, and it is necessary to have approximately 9.0 g sulfuric acid in each 100 ml.

Methods of analysis given here are based on the assumption that no interfering elements are present. Detailed accounts of procedures for the separation of such elements are too numerous to be presented here.

In the hydrolysis of sulfate solutions of ilmenite, the active or potential acid concentration is a critical factor and must be controlled carefully. It constitutes the sum of the free acid and the sulfate ions associated with the titanium, expressed as sulfuric acid, and represents the free sulfuric acid that will be present in the hydrolysis liquor after precipitation of the titanium as hydrous oxide. Sulfate combined with the iron, for example, is not released during the hydrolysis process, and for this reason is not included in this classification. The important active or potential acid concentration may be determined directly by titration with a standard alkali.[4] An accurately weighed sample of approximately 1.5 g of the ilmenite solution is washed into a 400 ml beaker containing 300 ml of a 1 per cent solution of barium chloride and titrated, at room temperature, with a 0.25 normal sodium hydroxide solution employing methyl orange as indicator. The barium chloride serves the purpose of transforming the acidity from sulfuric to hydrochloric acid, which gives a sharper end point with the methyl orange. Just before the end point is reached, the barium sulfate coagulates and settles rapidly.

Chemical analysis alone cannot distinguish between blended and coalesced titanium dioxide-barium sulfate composite pigments; the two may be differentiated, however, by "dyeing" the pigment with methyl violet and examining it microscopically.[5] Mechanically mixed products show small violet particles of titanium dioxide and large white particles of barium sulfate, while the coalesced material exhibits gray particles of uniform size. Identification can also be established by appropriate fractionation by levigation. The various fractions from precipitated pigments have about the same composition, while blends give residues rich in the filler.

The presence of titanium dioxide in paint films may be readily detected colorimetrically by a reagent prepared by adding a few crystals of potassium iodide, followed by an equal quantity of phenazone, to 1 ml of 2 normal hydrochloric acid.[6] One drop of this reagent is placed on the paint film and allowed to remain for 1 hour. In the presence of titanium a deep golden-brown color develops.

Newer type automatic polarigraphic analyzers and polarigraphic cells are used for the continuous determination of titanium III and iron III in titanium dioxide pigment manufacture.[7] A radioactive tracer is added to titanium sulfate solution for control of the continuous hydrolysis and washing process.[8] The tracing component is not separated from the solution by crystallization of the ferrous sulfate or by settling. A preferred tracer solution is prepared from lanthanum nitrate containing a known amount of ytterbium 90.

Up to 1000 parts per million of 19 elements present in titanium metal, titanium oxide, and titanium tetrachloride are determined by spectrochemical analysis.[9] All of the products are converted to the dioxide. Excitation by a spark-ignited unidirectional arc gives optimum precision and line to background relation.

Anatase and rutile are usually determined quantitatively by X-ray diffraction, electron diffraction, or infra red spectroscopy, although indirect methods of identification have been employed.

According to Lamprecht,[10] rutile and anatase forms of titanium dioxide can best be distinguished by X-ray diffraction. Other physical measurements are not reliable because of small differences in physical properties between the oxides. A simple iodimetric method based on the difference in photochemical activity of anatase and rutile is proposed. Quantitative analysis of anatase-rutile mixtures may be carried out with an X-ray diffractometer.[11] The intensity of scattering for each component of an allotropic mixture is essentially proportional to its concentration. Consequently the ratio of anatase to rutile in titanium dioxide is obtained by measuring the reflectance for Copper K alpha radiation at the Bragg angles of 12.68 degrees and 13.73 degrees, respectively. Rutile and anatase forms of titanium dioxide are determined in a mixture by diffuse reflectance measurements between 360 and 420 millimicrons.[12]

A method for distinguishing rutile from anatase is based on the great difference in the tendency to chalk of these two crystal forms of titanium dioxide, which is related to their difference in reflecting violet and ultraviolet light.[13] A Phillips HPW lamp, 125 watts, screwed into a mat reflector, hangs vertically on a support, which is attached to a wooden stand holding the camera together with a small metal frame for the pigment samples to be photographed. The frame with three microscopic slides rests on the lower wall of the wooden stand which is adjusted at an inclination of 45 degrees. From the lamp to the center of the frame is 30 cm. A well with a hole 1 cm in diameter is placed parallel to the slides. The inner wall of the wooden stand is painted dull black. A ring is inserted between the lens and camera for a sharp picture. The lens is adjusted to infinity. To provide a smooth surface to be photographed and to eliminate any shadow effect, the pigment samples are placed under the slides. The pigment samples and dye are mixed with water to form a homogeneous paste. Too much water can cause nonuniform reflections. The pictures are taken immediately to prevent drying. Three slides are photographed at the same time, one of rutile and one of anatase, whose tendency to chalking are known, and the test sample. A simple densiometer is used to express the density of the three samples on the developed negative. Identification of the unknown sample is made from the densities of the negatives.

An effective method of identification using the spectrophotometer is based on the fact that the anatase modification reflects light more strongly than does rutile.[14] A calibration graph is used for the measurements which shows the relation between the remittance of a standard white plate and the wave length. This remittance is known only of the visible region of the spectrum, 400 to 750 millimicrons, but since the calibration curve has a linear course at the short wave side, the values for wave lengths in the ultraviolet are determined by linear extrapolation. The curves of a number of titanium dioxide samples show clearly an anatase and a rutile group. Differences are caused by different after-treatment of the pigments. These curves confirm the fact that no direct relation exists between ultraviolet reflection and resistance to chalking. Pigments dispersed in methyl cellulose showed a stronger ultraviolet reflection than pigments of the same type dispersed in alkyd resin compositions. Two commercial anatase and three rutile titanium dioxide pigments tested with the photographic method showed a direct relationship between crystal type and density of the negative, but failed to give a correct evaluation of the chalking resistance of different types.[15] An anatase and a rutile titanium dioxide used as standards differed greatly in reflectance at wave lengths of 350 to 450 millimicrons.

The amount of aluminum in alumina modified titanium dioxide pigments is determined colorimetrically by using the free acid form of a stilbazo dye reagent which at pH 5.4 forms a stable pink complex with the aluminum ion.[16]

# 11

# Commercial Production of Titanium Metal

The production of titanium is a very precise operation since the hot metal combines with oxygen, nitrogen, and moisture of the air, with carbon and most construction metals. It decomposes any present refractory material absorbing the oxygen. These contaminants render the metal so hard and brittle as to be useless. If once picked up there is no practical method for the removal of these impurities. This problem is solved by carrying out the reduction reaction in a mild steel vessel under an atmosphere of helium or argon at around 800° C. At this temperature alloying of the titanium with the iron is at a low level, but even so, a layer of the metal in contact with the vessel contains too much iron to meet specifications and must be discarded.

In operation, magnesium ingots cleaned by pickling in dilute acid to remove surface oxidation followed by rinsing in water and drying are charged into a cylindrical, flat-bottomed mild steel pot.[1] After welding the cover in place the closed reactor is tested for leaks. If there is no leak, all the air is removed from the pot by evacuating twice followed by releasing the vacuum each time with helium or argon. Inert gas pressure is left on the vessel until it is ready to use. The magnesium charged pot is placed in a vertical cylindrical furnace, heated by electricity or burning fuel, and equipped for temperature regulation. As soon as the magnesium begins to melt, purified titanium tetrachloride is fed in at controlled rates while the inert gas pressure is maintained to prevent any inward air leakage. Since the reduction reaction is highly exothermic the titanium tetrachloride is fed in at such a rate that the heat generated in the vessel is dissipated through its walls. External heat is no longer needed to maintain the vessel temperature between the required 750 and 1000° C.

Although the actual reaction mechanism is probably complex, the net reduction is represented by the equation, $TiCl_4 + 2\,Mg \rightarrow Ti + 2\,MgCl_2 +$ heat, which shows that 4 pounds of titanium tetrachloride react with 1 pound of magnesium to produce 1 pound of titanium and 4 pounds of magnesium chloride, approximately. In practice excess magnesium is employed to prevent side reactions so that the magnesium efficiency does not exceed 90 per cent. The by-product magnesium chloride is tapped from the pot in molten form from time to time during the reaction period and is drained as completely as possible after the reaction is completed in such a manner that no air enters the furnace.

As soon as it has cooled sufficiently to lose the reddish color, the reactor is removed from the furnace and cooled to room temperature. A small positive pressure of inert gas is maintained at all times. The vessel is opened in a dry room, 0.006 pound of water per 1000 cubic feet of air at 70° F, after grinding away the bead weld using a swinging grinding wheel. If exposed to ordinary air, the hygroscopic magnesium chloride product picks up moisture. During the subsequent heating in the purification step any moisture absorbed contributes its oxygen to the titanium.

The half filled pot, containing 250 pounds of titanium in the sponge form, 100 pounds of magnesium chloride, and 35 pounds of magnesium metal, is chucked in a lathe, and the mixed products are turned out in chips up to 0.5 inch in size. These turnings are received in a steel basket in which they are later distilled. Care is taken to leave a half-inch layer of reaction mass on the side walls and bottom of the steel vessel which is contaminated with iron from the reactor.

The basket of chips is placed in a clean condenser which is immediately sealed. After evacuating to a pressure of 1/20,000 of an atmosphere, the retort is placed in an electrically heated vacuum furnace so arranged that the pressure in the furnace outside the condenser is approximately the same as the pressure inside. If this were not done, the retort would collapse at the temperature employed during distillation, 930° C. This distillation vessel is 30 inches in diameter by 8 feet high with half inch steel walls. In operation the lower 5 feet of the condenser extend into the furnace being supported by a flange placed at this height. The upper portion of the retort is water jacketed and thus serves as a condenser for the impurities distilled from the hot chips. Heavy layers of magnesium metal and magnesium chloride vaporized from the chips condense on the cold still surfaces of the upper area. Distillation is continued under vacuum for 30 hours, after which the vacuum in the return is released with inert gas simultaneously with the release of the vacuum in the furnace with air. After the water and vacuum connections are broken, the condenser is removed from the furnace and cooled. Cooling water is circulated through the jacket, and a small pressure of inert gas is maintained. Then the retort is opened and the titanium sponge is chipped from the basket, passed through a crusher, and screened to obtain pieces of the required size.

The sponge must be melted into an ingot before it can be rolled, forged, or extruded. This is accomplished by first pressing it into bars which serve as consumable electrodes in the arc melting of the titanium in a water cooled copper crucible the size and shape of the desired ingot.

The process is also shown in the flow sheet in Fig. 11–1.

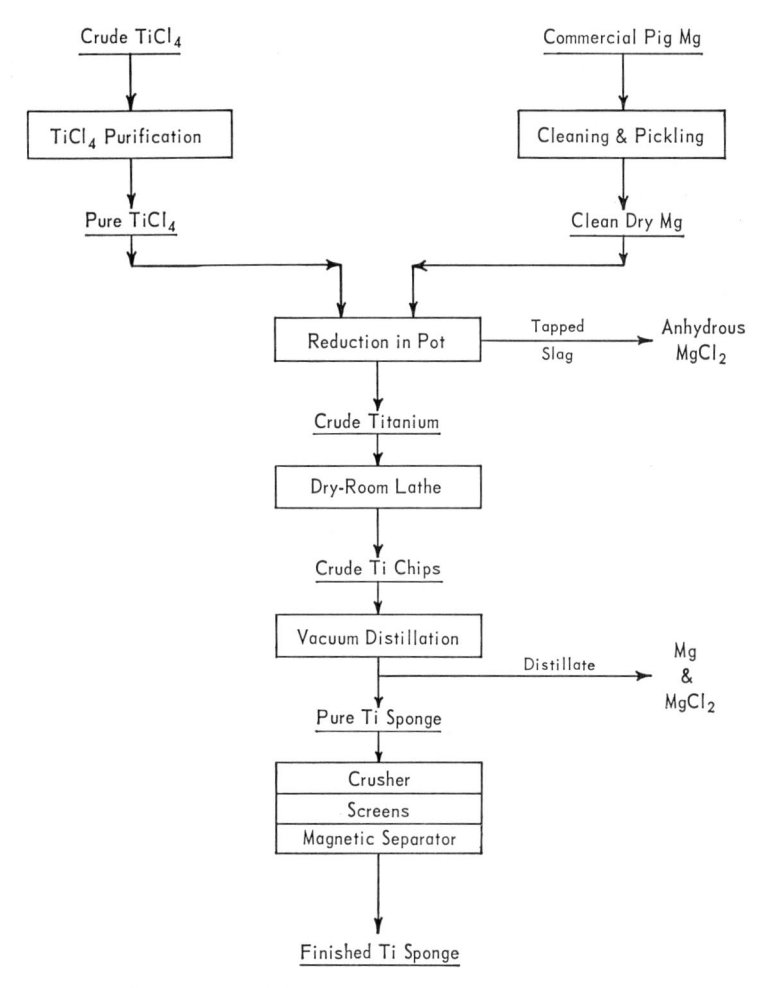

**Fig. 11–1.** Flow sheet of the process for producing titanium sponge. (Reproduced from Bureau of Mines Information Circular 7791.)

## REDUCTION OF TITANIUM TETRACHLORIDE WITH MAGNESIUM

The design and operation of the plant at Boulder City, Nevada, having a capacity of 1500 pounds of titanium a day is reported in detail.[2] Crude titanium tetrachloride is purified by refluxing with hydrogen sulfide followed by distillation. After placing a weighed amount of clean magnesium in a steel pot the lid is welded in place. The air is exhausted and the pot is back filled with helium. Next the pot is placed in a furnace and heated until the magnesium begins to melt. The titanium tetrachloride is then added at a predetermined rate and the outside pot temperature is held at 750 to 800° C. Since the reaction is highly exothermic, the rate of addition of the tetrachloride is regulated to control the temperature. The magnesium chloride produced is removed by tapping several times during the batch operation. After completion of the reaction the pot is removed from the furnace and cooled. The pot is opened inside a room with very dry atmosphere and the products are removed by placing the pot on a large lathe and cutting the reaction mass out by turning. The titanium chips are collected in a basket and heated in a vacuum furnace under 40 microns absolute pressure at 910° C or higher to volatilize the magnesium and magnesium chloride which are collected in a water cooled retort.

Systematic elimination of detrimental impurities introduced with the raw materials, and improved handling and processing techniques insure consistent production of high quality titanium by reduction of the tetrachloride with magnesium.[3] These improvements include stripping of dissolved gases from the tetrachloride with helium or argon, preliminary dehydration of the crude titanium sponge, and careful regulation of feed rate of the titanium tetrachloride.

Kingsbury[4] reduced titanium tetrachloride with magnesium to form a molten reduced titanium chloride-magnesium chloride intermediate product which was cooled and pelletized. The pelleted mixture was dropped into molten magnesium to form titanium metal in pelletized form, which was heated at 800 to 1000° C in vacuo to distill off the impurities. Cold malleable titanium is manufactured by chemically treating titanium tetrachloride with fused magnesium at atmospheric pressure with helium as a protective covering.[5] The reaction temperature is near the boiling point of magnesium. The helium must be free from nitrogen and oxygen as these are absorbed by the titanium at high temperatures. Magnesium chloride resulting from the reaction is melted and decanted while the helium blankets the titanium. The titanium is then fused under high vacuum, and residual magnesium chloride and unconverted magnesium are driven off. Holst and Proft[6] prepared titanium sponge by the reduction of the tetrachloride with magnesium in an evacuated reactor.

The titanium tetrachloride is added in the gaseous state and the magnesium chloride is distilled off as formed.

Crude titanium sponge suitable for anode material in the electrolytic refining process is produced by passing the tetrachloride through molten magnesium chloride on which a layer of magnesium is maintained.[7] The magnesium reduces the tetrachloride to metal sponge which may fall to the bottom of the bath or collect on the cover of the reactor which is made of steel in the form of an inverted torus.

To effect continuous reduction, titanium tetrachloride is fed into a closed, tall reactor from an internal perforated annular trough around its top so that the liquid flows down the walls into a pool at the bottom.[8] Powdered magnesium is dropped down the central axis through an upper zone of titanium tetrachloride vapor heated at 800 to 900° C. In falling, the particles react with the vapor and the products of titanium and magnesium chloride are quenched in the bottom pool. A screw conveyor then carries these particles up an inclined tube where the titanium tetrachloride drains off. Titanium tetrachloride is reduced continuously in a closed reactor by magnesium or sodium dissolved in a fused bath of sodium, lithium, and calcium chlorides.[9] The reactor space above the bath is divided vertically into two chambers, one of which is used to supply the reducing metal.

Crystalline titanium of very high purity is produced by admitting vapors of titanium tetrachloride and magnesium into the top of a columnar steel reaction chamber at 800 to 950° C.[10] The titanium deposits on a network of thin titanium ribbons, while the magnesium chloride and excess magnesium are drawn downward as vapors into a water-cooled condenser. Massive and compact titanium is deposited in crystalline form in high yield if mixed vapors of titanium tetrachloride and magnesium impinge and react on the surfaces of titanium ribbons stretched in an evacuated chamber at 900 to 950° C.[11]

Increasing the partial pressure of inert gas, argon, decreased the yield, the particle size, and the iron and chlorine content of titanium sponge as well as the coefficient of utilization of magnesium.[12] Increasing the rate of addition of the titanium tetrachloride feed to the reactor had a similar effect. With a baffle installed in the center of the reactor, the effect of argon partial pressure and of the titanium tetrachloride feed rate were decreased. The yield and particle size of titanium sponge were improved.

Pure titanium is produced by the reduction of the tetrachloride vapor at 750 to 1200° C and 0.01 to 300 mm pressure without an inert gas atmosphere.[13] The product, being mostly a fine loose powder, separates on formation from the by-product salt and excess reducing agent, and collects on baffles in the externally heated steel chamber. Reduction with magnesium is carried out at 800 to 900° C. According to an improvement by Conklin [14] the retort is charged with magnesium and purged with argon, after which the addition of titanium tetrachloride is started. After at least 10 per cent of the titanium tetrachloride is added, magnesium chloride is drained off at a rate equivalent to the addition of the tetrachloride. Finally the product is vacuum distilled. To reduce the titanium tetrachloride with magnesium a stream of molten magnesium chloride is maintained at 755 to 800° C between an upper and a lower reservoir.[15] Molten magnesium and titanium tetrachloride are added to the lower portion of the current. A suspension of titanium and magnesium chloride is withdrawn from the lower portion of the lower reservoir.

Pure titanium is produced by the reduction of the tetrachloride with magnesium followed by electrolytic refining.[16] Crude impure titanium from the reduction process is melted and cast in 3-inch ingots, which become the cathode in an electrolytic cell. The bath of the cell is fused sodium chloride at 850° C containing 3 per cent titanium dichloride under an inert atmosphere. Takeuchi [17] injected titanium tetrachloride and magnesium both in the vapor phase into a reaction tower maintained at 800 to 1100° C containing pieces of titanium to deposit high purity titanium metal in the crystalline form on the metal packing. The by-product magnesium chloride condenses in a cooler zone of the tower. In a two-stage process the tetrachloride is reduced with magnesium or sodium.[18] Titanium dichloride and titanium trichloride produced at 200° C in the first stage form a free flowing mixture with the excess reducing metal. In the second stage this mixture is heated to 800 to 1200° C to complete the reduction to titanium which begins at 145 to 300° C, and to agglomerate the metal product. Schmidt and Stoddard [19] reduced titanium tetrachloride with magnesium in a bath of molten magnesium chloride. The titanium metal produced is periodically consolidated to a compact mass which is expelled from the bottom of the reactor. Contamination of the titanium with iron is prevented if the reduction of the tetrachloride at 750 to 1100° C is carried out in a vessel coated with iron oxide (mill scale).[20]

Ishizuka [21] produced titanium by the reduction of the tetrachloride with magnesium in an autoclave at 1100 to 1500° C and a pressure of 5 to 20 atmospheres. In a continuous process the tetrachloride is reduced with magnesium in a moving mass of molten chlorides to avoid adhesion of the metal to the walls of the reactor.[22] Movement of the mass is effected by a jet of titanium tetrachloride vapor.

Before carrying out the reduction reaction a small amount of titanium tetrachloride is added to molten magnesium at 650° C to purify the metal.[23] Similarly a small amount of molten magnesium is added to titanium tetrachloride at an elevated temperature to purify the tetrachloride. High purity sponge titanium is obtained by decreasing the interdiffusion of titanium and the sheet iron liner of the reactor by coating the inner walls with 1 to 5 g of carbon per square foot and heat treating under vacuum to remove the volatile constituents of the coating.[24]

Titanium tetrachloride and magnesium are mixed at the upper part of a vacuum furnace and brought in contact with the molybdenum furnace walls maintained at 750 to 1100° C to deposit titanium.[25] Magnesium chloride vapor formed condenses at the lower temperature part of the furnace and is removed at the bottom. Titanium metal is produced by burning particulate magnesium metal in titanium tetrachloride vapor at a temperature above the melting point of the magnesium chloride produced.[26] In a two-stage process titanium tetrachloride is reduced with titanium particles in a molten magnesium chloride bath to form subchlorides.[27] This product is then heated with magnesium at 720 to 1300° C to reduce the subchloride to the metal in particulate form.

Titanium tetrachloride is reduced with magnesium in a molten salt bath so that continuous billet compacted metal is extracted.[28] An inverted cylinder filled with fused magnesium chloride or sodium chloride is heated to a temperature at which magnesium or sodium evaporates but not their chloride.[29] Excess titanium tetrachloride is allowed to react with the magnesium or sodium to produce titanium, which is deposited and sintered into rod shape in the cylinder. As the first step the tetrachloride is reduced with magnesium at 650° C to give titanium trichloride crystals containing some titanium dichloride.[30] These crystals are heated in an eutectic mixture of sodium chloride and potassium chloride at 680° C with addition of titanium sponge.[31] A yield of titanium of 46 per cent is obtained. Fletcher [32] produced titanium by heating magnesium in a closed reaction vessel by positive ion bombardment while continuously directing a stream of titanium tetrachloride through a jet orifice onto the heated magnesium. The metal is prepared by heating the tetrachloride with magnesium in a protective atmosphere or in vacuo. By controlled mixing of the reactants a compound of titanium and magnesium, $Mg_2Ti$, forms first and decomposes at a higher temperature. A mixture of titanium tetrachloride and magnesium powder is forced through a jet with an electric current passing between the pipe forming the jet and another electrode.[33] The heat of discharge causes the tetrachloride and magnesium to react to produce titanium. The melt of magnesium chloride is allowed to flow out and the sponge titanium is purified by distillation in vacuo. Titanium tetrachloride is reduced with magnesium so that a stream of magnesium chloride at 755 to 800° C is drawn off at the top of the reactor, and a mixture of titanium and magnesium chloride is drawn from the lower portion of the mass.[34]

According to an improved process the retort is charged with magnesium and purged with argon before the addition of titanium tetrachloride is started.[35] After at least 10 per cent of the titanium tetrachloride is added, magnesium chloride is drained at a rate equivalent to the addition of titanium tetrachloride. After removal of 90 per cent of the magnesium chloride, the reaction is stopped, and the remaining magnesium and magnesium chloride are removed by vacuum distillation. According to a refinement of the process, the tetrachloride is reduced with magnesium in a reaction zone out of contact with the walls.[36] These walls are externally cooled to a temperature below the melting point of the by-products. Continuous withdrawal of the solidified reaction product is possible. A magnesium rod fed continuously through an atmosphere of titanium tetrachloride limits the reduction of titanium to a zone maintained out of contact with the reactor walls.[37] The reaction zone is maintained at 750 to 1400° C. By this process the titanium sponge and the magnesium are not contaminated with iron through contact with the reactor walls. Sponge is obtained by the continuous reduction of the tetrachloride with magnesium at 750 to 1450° C.[38] The rate of withdrawal of cooled ingot from the reaction zone is adjusted to equal the rate of metal production so that 90 per cent of the by-product salt is removed. Magnesium particles in the form of a train react with titanium tetrachloride vapor to produce titanium sponge.[39] The magnesium train is laid down on a heat-controlled surface which has a pile-forming zone, preheat zone, ignition zone, smoldering zone, draining zone and sponge removal zone. Titanium tetrachloride vapor is used as a propellant to spray stoichiometric quantities of liquid magnesium into a reaction chamber maintained at 1400 to 1600° F containing an inert gas.[40]

A stream of argon is passed over boiling magnesium to carry magnesium vapor into the reaction vessel, where it is mixed with titanium tetrachloride vapor. This process permits a convenient separating of titanium metal from the salt.[41] Magnesium is heated in an iron vessel in vacuo at 140° C and titanium tetrachloride is introduced slowly with heating up to 800 to 950° C.[42] The product is ground in a ball mill, washed with water, and dried to give titanium of 99.8 per cent purity. Vapors of titanium tetrachloride and magnesium are injected simultaneously through separate nozzles into the top of a tall, tubular, evacuated reactor containing below the nozzles a network of titanium wire at 800 to 1100° C through which the products pass.[43] Titanium crystals collect in the network. Magnesium chloride and excess magnesium are deposited in a lower condenser. In the continuous production of titanium a horizontal or slightly inclined iron pipe with a screw conveyor set in a furnace is filled with molten magnesium at 700 to 950° C.[44] Titanium tetrachloride is passed in under pressure. Molten magnesium chloride and molten magnesium are separated from the solid titanium by screening. Titanium tetrachloride is reduced with magnesium and the magnesium chloride produced is removed continuously from the reaction zone to increase the efficiency of the reaction and to increase the furnace capacity.[45] The final residue of titanium, magnesium and magnesium chloride is distilled in vacuo to yield purified titanium. The temperature is maintained at 900 to 980° C in the reaction zone. Jordan [46] reduced titanium tetrachloride with gaseous magnesium, sodium, or potassium.

Titanium tetrachloride is reduced to metallic sponge with molten magnesium supported on the surface of a vortex formed by molten magnesium chloride contained in an argon filled chamber and rotated with a stirrer.[47] The reduced sponge does not come in contact with the walls of the chamber or accumulate on them, since the sponge is continuously removed from the bottom of the chamber together with some magnesium chloride. Gaseous titanium tetrachloride in small quantities is added at low pressure and at low temperature to wash the remaining air from the vacuum reactor.[48] The temperature is then raised to 700° C and titanium tetrachloride is introduced to react with the calcium, magnesium, or sodium to produce titanium. This procedure avoids the use of expensive inert gas.

Increased yield of titanium is obtained by adding the titanium tetrachloride to the magnesium in stages.[49] Titanium is reduced from its chloride by magnesium produced in place in an air tight electrolytic cell held at 650 to 750° C.[50] To standardize the process, the titanium tetrachloride is fed into the reactor to obtain titanium with the aid of a shaped program disk, kinematically and electrically connected to a regulator valve installed in the feed line.[51] Blue [52] developed a crucible for reducing titanium tetrachloride with magnesium under controlled rate of reaction and temperature conditions. Argon can be substituted for helium as the protective gas.[53] However, argon tends to cause clogging of the vent and feed pipes, and condensation of the mixed chlorides on the surface. Gillemot [54] carried out the reduction of the tetrachloride in an inclined reaction chamber.

Titanium metal is produced by the reduction of the tetrachloride with a 2 to 20 per cent solution of magnesium in zinc at 750 to 1100° C.[55] The product is distilled to remove the zinc, magnesium, and magnesium chloride, leaving titanium of high purity. High yields of crystalline titanium are obtained by feeding a molten reducing agent onto the surface of a molten bath of a lower halide of titanium dissolved in a fused salt.[56] The process is conducted so as to form a thin crust of sintered titanium fines over the bath. A layer of fused salt in which the reducing agent is concentrated collects on the upper side of the sintered layer. Briquets of lower chlorides of titanium with magnesium are charged into a molten salt in a sealed reactor and heated to the reaction temperature to produce titanium which is scooped out of the melt.[57]

Strict adherence to the optimum conditions of time of heating, starting temperature and pressure, pouring off time, magnesium chloride content, and feeding schedule for titanium tetrachloride, together with automatic regulation of the inlet valves for titanium tetrachloride improve and standardize the process of producing titanium metal.[58] Production is also increased substantially. Automation is applied to the batch reduction of titanium without forced cooling.[59] The equipment was used successfully for two years.[60]

To avoid the formation of lower chlorides and to remove heat more effectively from the reaction zone, the lid of the reactor is made in the form of a hollow vessel, the bottom of which extends into the high temperature area.[61] Molten magnesium or sodium is forced through a jet with gaseous titanium tetrachloride into a reaction vessel to produce titanium.[62] To prevent corrosion and clogging, the jet nozzle is protected by a sheath of an inert gas. The difficulty of removing titanium sponge produced by the reduction of the tetrachloride with magnesium is overcome by inserting a thin cylindrical liner of titanium in the reactor.[63] Winter and Tully[64] removed by-product magnesium chloride by passing the products from the reduction reaction through a slightly inclined conduit at atmospheric pressure under controlled temperature to accomplish successive melting, drainage and vaporization.

The magnesium and magnesium chloride may be removed from titanium by dissolving them in methyl alcohol at its boiling point.[65] These materials are also removed from the sponge by heating the mixture in a retort at 900° C at $10^{-4}$ to $10^{-5}$ mm pressure.[66] Several pieces of titanium sponge are placed in the cooler part of the retort as a getter. The magnesium chloride and calcium chloride contained in titanium powder are extracted with ethyl alcohol or ethylene glycol.[67] This avoids contamination of the metal by oxygen and hydrogen produced during extraction with aqueous solvents. Titanium sponge free from hydrogen and oxygen compounds is prepared by leaching the reaction product with dilute mineral acids to which a complexing agent such as oxalic acid or acetic acid has been added.[68] With the complexing compound present, the bivalent titanium compounds do not evolve hydrogen. The addition of 0.01 to 8 per cent of a titanium complexing agent to the oxidizing acid solution used for leaching the crude sponge improves its effectiveness.[69] Effective complexing agents are citric, tartaric, lactic, malic, and other organic acids. The evolution of hydrogen during the leaching of sponge containing magnesium or sodium with dilute acids is suppressed by adding to the acid solution 0.25 to 5 per cent of a sugar, aldose, ketose, keto acid, furan derivative, or pyran derivative.[70] A crude titanium metal product is leached with a water solution containing 2 to 10 per cent mineral acid, 0.5 to 6 per cent ferric compound, and 2 to 10 per cent nitrate compound to remove magnesium and magnesium chloride.[71] Magnesium chloride may be dissolved effectively from the reduction product with acetone, an unsymmetrical ketone, diacetone alcohol or isopropyl alcohol.[72]

From a study of the thermodynamics of the reduction of titanium tetrachloride with metals, Sergeev[73] concluded that reduction with magnesium will continue to be the leading method. However reduction with sodium offers some advantages.

Titanium dioxide is reduced to titanium of over 95 per cent purity with sodium-calcium sludge.[74] Sodium is distilled and recovered as a by-product. Titanium tetrachloride is reduced to the metal with sodium in liquid ammonia.[75]

Subchlorides of titanium in a molten solution of alkali or alkaline earth metal chlorides are reduced by sodium vapor at a temperature above the melting point of the salt mixture formed.[76] For example, 70 kg of $Na_2TiCl_4$ is heated in a vacuum furnace to 810° C under argon. The pressure is slowly lowered to evaporate 2.4 kg of sodium. A mixture of sodium chloride and titanium crystals up to 10 mm across remains. Crystalline titanium metal of needle form and high purity is produced by adding molten sodium dropwise through titanium tetrachloride vapor at 400 to 750° C to form titanium dichloride as an intermediate.[77] The temperature is then raised to 805 to 875° C, and more sodium is added to reduce the dichloride to metal. Titanium, lower titanium chlorides, sodium, and sodium chloride from a continuous reactor are fed into an inclined tube where they are maintained at 850 to 950° C.[78] In the tube the titanium particles adhere to form a sintered mass. The molten sodium chloride acts as a lubricant to move the mass through the tube. Vent holes are provided in the tube to prevent pressure build up. In a cyclic process, the sodium chloride by-product is electrolyzed to produce chlorine and sodium.[79] The chlorine is used to produce more titanium tetrachloride and the sodium is employed to reduce it.

## REDUCTION OF TITANIUM TETRACHLORIDE WITH SODIUM

Advantages of the sodium reduction process for producing titanium are lower cost and greater efficiency of the reductant, ease of handling in the molten state, and lower temperature for the reduction of the tetrachloride.[80] The disadvantages include greater heat of reaction and the necessity of leaching the entire quantity of sodium chloride produced from the titanium metal. In the magnesium process 90 per cent of the magnesium chloride is tapped molten from the reactor. The reduction of the tetrachloride by sodium takes place in two stages, the first of which produces at 800 to 1000° C a mixture of titanium dichloride and trichloride in sodium chloride.[81] In the second stage, sodium reduction of the subchlorides is a heterogeneous reaction between a solution of sodium in sodium chloride and another solution of the subchlorides in sodium chloride at 800 to 900° C. This is a primary electrolytic reaction between sodium ions produced on the reactor walls and soluble titanium ions attracted to the cathodic site. Titanium tetrachloride vapor and sodium vapor introduced into an evacuated reaction chamber react in a flame.[82] The metal is deposited on a curved surface at a temperature below the melting point of titanium. The sodium chloride formed is condensed within the chamber.

The flame generating nozzle in which the titanium tetrachloride vapor reacts with sodium vapor leads into a tube closed at the nozzle end but provided with openings at the far end where the titanium is deposited.[83] An electric coil around this tube preheats the sodium. Gaseous or liquid titanium tetrachloride and gas or liquid sodium are sprayed into a closed, upright, cylindrical chamber having water cooled metal walls.[84] The reaction products are deposited on the side walls in the form of a ring. By starting near the base of the reaction chamber and gradually raising the spray nozzles, the deposit is extended up the tube. The salt by-product separates on the inner surface as a liquid film which drips into the cooling chamber at the bottom. The reduction reaction is started by electric arc ignition and is regulated by cooling the walls of the reactor.[85] By this process titanium with a particle size of 13 to 100 mm is obtained.

Sponge titanium is produced in a two-stage process by first reducing the tetrachloride to a subhalide by the metal itself and then reducing the subchloride by sodium, thus avoiding melting of the metal.[86] Similarly in the first reactor, titanium tetrachloride is introduced below the surface of molten sodium chloride, while liquid sodium is sprayed onto the surface of the bath.[87] The molten sodium chloride bath is then transferred to a second stage reactor where liquid sodium is sprayed onto its surface to produce titanium.

Titanium is produced by introducing streams of titanium tetrachloride and molten sodium in a closed reactor at 1000° C.[88] The 5 per cent balance of the tetrachloride is added after all of the sodium has been added. Similarly a stream of liquid titanium tetrachloride is injected into molten sodium in a closed reactor with an argon atmosphere externally heated at 800 to 1100° C.[89] Titanium sponge of good quality is produced by reduction of the tetrachloride with a slight deficiency of sodium, leaching the crushed product with hydrochloric acid in two stages and washing with water.[90] Reduction of the tetrachloride takes place on the surface of the growing sponge at the tetrachloride-sodium vapor interface by forming an intermediate phase of subchlorides.[91] The sponge is formed by agglomeration of metallic titanium and subsequent fusion of separate grains. Utilization of the sodium is 98 to 100 per cent of the stoichiometric amount. Reduction is carried out in an inert atmosphere between 200° C and the melting point of the sodium chloride formed while maintaining a fluidized bed in the reactor.[92] The product is then heated to 600 to 780° C in a separate vessel.

Gullett[93] added titanium tetrachloride and sodium to a reactor in small increments to effect the reaction at 600° C. The reaction product was added to molten sodium chloride at 850° C in an electrolytic cell to obtain titanium. Titanium of low hydrogen content is produced by reducing the tetrachloride with a deficiency of sodium followed by leaching with dilute hydrochloric acid.[94] Finely divided titanium metal, lower chlorides, or a mixture is prepared in a continuous operation at 105 to 205° C under pressure of 2 pounds per square inch by the reaction of molten sodium with vaporized titanium tetrachloride in an agitated bed of finely divided reaction products.[95] Films of sodium approaching colloidal dimensions spread over the inert solids. This is known as high surface sodium.

Titanium sponge is produced continuously by the reduction of the tetrachloride with sodium dissolved in molten salt mixtures.[96] The titanium metal formed sinks to the bottom of the reactor and is drawn off as produced. A suitable mixture is lithium chloride and potassium chloride. Coarse crystalline metal is produced from the reduction of the tetrachloride with molten sodium injected near the bottom of a vigorously agitated fused sodium chloride bath in a closed reaction vessel.[97] The area in the vessel above the bath contains the tetrachloride to be reduced. A mixture of liquid titanium tetrachloride and solid particles of sodium in a closed reactor is heated above the melting point of sodium to initiate the reduction reaction.[98] The heat of the reaction is more than sufficient to melt the titanium. The cooled product of titanium metal and sodium chloride is crushed and passed through a mill fitted with a one-sixteenth inch sieve plate.[99] The final product retained on 100 mesh contains 73.7 per cent titanium.

Titanium tetrachloride and sodium are added to a fluidized bed in such proportions that the particulate titanium-salt mixture contains 0.3 per cent excess titanium tetrachloride.[100] The volume of the reactor bed is maintained constant by continuously withdrawing the reaction products. The metal is produced by feeding the tetrachloride vapor and sodium vapor as fuel into a modified blow torch and impinging the flame on a pool of molten metal.[101]

To prevent the formation of masses too large for discharge during the deposition of sponge on added titanium pellets in a continuous process, an excess of the tetrachloride is added to the reactor at 600° C along with the sodium.[102] Pure titanium having particles large enough to be stable in air at room temperature is produced by heating the initial reduction product on a moving plate type belt of stainless steel in a furnace at 850 to 950° C and draining most of the sodium chloride through the belt.[103] The sintered titanium sponge containing some sodium chloride is discharged into a cooler. In a two-stage process the amount of the sodium added in the first stage is less than that required for complete reduction.[104] In the second stage more sodium is added to complete the reduction and a temperature high enough to sinter the titanium powder is employed. Similarly, about half of the total amount of molten sodium required to produce titanium is allowed to react with gaseous titanium tetrachloride. Excess molten sodium is then introduced below the surface of the sodium chloride bath to reduce the titanium trichloride and dichloride to the metal.[105]

By carrying out the reduction of titanium tetrachloride with sodium in a centrifuge the products are clearly separated according to their densities.[106]

A chloride, bromide, or iodide of titanium is reduced to metal powder in a fused salt bath by an alkali metal produced in situ by electrolysis of the bath in a compartment cell.[107] Titanium tetrachloride is reduced to titanium in a cell containing molten sodium produced in place by electrolysis of the fused salt.[108] Large crystals of titanium several centimeters in size are produced by reducing the subchloride with sodium vapor at 800 to 850° C.[109] The sodium chloride produced forms a molten bath in which the crystals collect.

Finely divided $Na_2TiCl_4$ and molten sodium are fed into a sealed reaction vessel equipped with a blade stirrer maintained at 500 to 600° C to produce titanium sponge in a continued process.[110] The granulated titanium mixed with sodium chloride is screened out of the reaction mixture and leached with hydrochloric acid. By careful control of the stoichiometric relation between titanium tetrachloride and sodium, titanium sponge is produced which is not contaminated with sodium or by the presence of lower chlorides.[111]

The process of producing titanium by the interaction of the chloride with sodium vapor in a closed vessel is improved by using a lining of compressed, spongy titanium metal.[112] This lining prevents contamination of the product with the material of the reaction vessel. To prevent clogging of the nozzles during the reduction of titanium tetrachloride with sodium vapor droplets of either substance are interposed between the streams of the reactants.[113]

Titanium tetrachloride is reduced with sodium at temperatures between 97.5 and 200° C in ball mills and in high surface sodium reactors.[114] The reaction is carried out in two stages. Yamaguchi[115] found that titanium tetrachloride is reduced to titanium in liquid sodium at 105° C. The black pyrophoric titanium powder reacts with water to form anatase.

Subhalides of titanium prepared from the tetrachloride in a molten solution of alkali or alkaline earth metal salts are reduced by sodium vapor.[116]

Titanium is produced by slowly pouring molten potassium fluotitanate into molten sodium at 850° C.[117] After all of the fluotitanate is added, the temperature is raised to 1100 to 1300° C for 15 minutes. The reacted mixture is then drained from the vessel, and cooled after which the salt is dissolved from the titanium pellets.

An apparatus for the production of titanium consists of a reaction chamber with a retort.[118] The space between the walls of the retort and the reaction chamber is filled with liquid sodium or magnesium. As required the liquid reducing agent is forced by pressure into the reaction chamber containing titanium tetrachloride. An apparatus for the continuous process is described by Uebayashi.[119]

The reaction products from the reduction of titanium chlorides with sodium are cooled from the bottom to solidify the sodium chloride and titanium, while allowing the excess sodium to collect at the top, where it is removed before leaching the salt.[120] By-product chloride salt and residual sodium or magnesium are removed from titanium sponge by passing an inert gas through the mass at a high temperature in a closed apparatus followed by cooling and leaching in water.[121] Titanium is separated from sodium chloride by washing with a dilute mineral acid and with water on a filter.[122] The washed titanium is removed from the filter, centrifuged, and dried.

Titanium metal of low hydrogen content may be obtained if the reduction by sodium is carried out in the presence of an excess titanium tetrachloride and if there is present during the leaching step a small quantity of an oxidizing agent such as hydrogen peroxide, chlorine, nitric acid, or iron III ion.[123] Low hydrogen content is also obtained by carrying out the reaction between titanium tetrachloride and sodium so that subchlorides of titanium are present in the product and leaching the crushed material with a water solution of hydrochloric acid and a fomate or oxalate, or with a solution of hydrochloric acid and sodium phosphite.[124]

The reaction products from the reduction of titanium tetrachloride with sodium are leached with liquid ammonia to dissolve the sodium and sodium chloride.[125] The metal is vacuum-dried to remove traces of ammonia, melted in an inert atmosphere, and cast in shapes.

To prevent particles of sodium embedded in sodium chloride from adhering to the vessel lining during the reduction of titanium tetrachloride with sodium, a rotating steel drum is provided in the reactor.[126] A closed metallic container is half filled with powdered sodium chloride, filled with argon, and molten sodium and titanium tetrachloride are added to the vigorously stirred mixture.[127] The sodium reduces the titanium tetrachloride with the formation of solid sodium chloride and titanium.

The bed remains in a powdery state and is removed in portions at intervals without interrupting the process. The cooled mixture is leached with water.

Titanium tetrachloride is reduced with sodium in a molten bath of lower chloride of titanium and sodium chloride in a vessel in which the reaction zone is isolated from the main salt bath.[128] By this improvement the costly process of removing the titanium metal from the reaction vessel is eliminated. To consume the excess heat which is generated in the reaction between titanium tetrachloride and sodium above 800° C, the rate is controlled so that part of the sodium is maintained in a state of reflux.[129] The sodium is fed to the reaction mixture after condensing on the conical lid of the vessel.

Below 800° C the activation energy of the reduction of titanium tetrachloride with sodium is 3.6 to 5.7 kcal per mole and the process is autocatalytic, with solid titanium, apparently, being the catalyst.[130] Above 800° C where sodium vapor is the principal reductant, the activation energy is 13.0 to 19.1 kcal per mole.

Titanium salts such as $TiCl_3$, $TiCl_2$ and $Ti_2(SO_4)_3$ are reduced with sodium in liquid ammonia to produce metallic titanium.[131]

## REDUCTION OF TITANIUM TETRACHLORIDE WITH SODIUM OR MAGNESIUM

Titanium metal is produced by the reduction of the tetrachloride with sodium or magnesium which are produced in the reactor by the electrolysis of sodium chloride or magnesium chloride.[132] Chlorine is liberated at the carbon anodes. Molten lead, zinc, or tin serves as the cathode. Hood[133] produced titanium by the reaction of the tetrachloride with sodium or magnesium in the vapor phase at 2000° C. The metal droplets formed during the reaction are collected to form continuously cast ingots. Heat for vaporizing the reactants and melting the product is supplied by the exothermic reaction. Alkali or alkaline earth metals heated to 700° C and titanium tetrachloride are jetted together from concentric nozzles into a chamber where they are ignited spontaneously.[134] The impure titanium is collected below the molten salts and vacuum distilled. A pure form of metal is produced by the reduction of the halides in alkali chloride or bromide fused bath with an alkali or alkaline earth agent.[135] The excess reducing agent is brought near zero by the addition of titanium tetrachloride or tetrabromide. Finally excess titanium halide is reduced with magnesium which is insoluble in the fused bath.

McKinney[136] produced titanium in a continuous process by reducing the tetrachloride with an alkali or alkaline earth metal and recovering the metal through a sealing pool of the salt formed. The tetrachloride is treated in vacuo in the vapor phase with an alkali or alkaline earth metal to form sponge.[137] A chamber containing titanium tetrachloride is connected to the reaction crucible by an open tube without valves to avoid plugging. The tetrachloride contained is cooled to keep it from vaporizing during evacuation of the system. It is then warmed to introduce titanium tetrachloride vapor into the crucible containing the metal. The temperature of the container and of the crucible is adjusted so that the vapor pressure of the tetrachloride is always greater than that of the metal. Titanium tetrachloride is reduced with sodium or magnesium first to the dichloride, and the fused mass is syphoned to a second stage reactor for a more gradual reduction of the dichloride.[138] This slower reduction produces larger crystals of titanium. Reduction of the tetrachloride to the metal is effected with sodium or magnesium in a vessel which acts as the anode.[139] The salt formed serves as melt for refining which is carried out in the same vessel.

A dispersion of titanium chloride or oxide with an excess of alkali or alkaline earth metal is impinged on a solid surface with a force ten times gravity in an inert atmosphere at elevated temperature to produce titanium metal which is left on the solid surface.[140] To regulate the temperature during the reduction the reactor is cooled by refluxed zinc chloride.[141] Titanium metal or its alloys are produced from the chlorides dissolved in a fused alkali or alkaline earth halide by the gradual addition of sodium, potassium, or magnesium to the surface of the bath.[142] The metal is obtained by partially reducing titanium tetrachloride with a molten alkali or alkaline earth to a mixture of a lower chloride of titanium and of a chloride of the reducing metal.[143] After cooling, this mixture is pelletized in an inert atmosphere and conveyed through the molten reducing metal to complete the reaction.

Titanium is separated from an alkali or alkaline earth chloride by melting the contaminated sponge in an inert atmosphere at 1725° C under a pressure of 5 to 10 atmospheres.[144] Under these conditions the fused salt does not evaporate at the melting point of the titanium. The two molten phases form two layers and may be separated by drawing them off separately. The purity and ductility of titanium produced by reduction of the tetrachloride with sodium or magnesium are improved by heat treatment of the reaction products at 1000 to 1400° C.[145] Metal sponge produced by reduction of the tetrachloride with sodium or magnesium is allowed to drain at high temperature to separate most of the molten by-product salt, cooled in an inert atmosphere, and crushed.[146] The particles are heated by induction or internal resistance to reduce the surface area to permit leaching with dilute acid without embrittlement.

## REDUCTION OF TITANIUM TETRACHLORIDE WITH SODIUM AMALGAM AND OTHER AMALGAMS

A continuous process for reducing titanium tetrachloride with 0.5 to 1.5 per cent sodium amalgam is effected at 100 to 170° C in an atmosphere of argon.[147] The titanium, sodium chloride, and sodium amalgam are removed from the retort by a screw conveyor as soft lumps of finely divided solids. These impurities are removed from the sponge by distillation. By the reaction of the tetrachloride with sodium amalgam at a temperature of 500 to 1200° C high purity titanium is produced continuously in high yield.[148] The tetrachloride vapor is fed upward while the amalgam vapor is fed downward into the vessel. In a two-stage process titanium is prepared by the reaction of the tetrachloride with a mixture of potassium and lithium amalgams.[149] Sodium amalgam or sodium is used to reduce the tetrachloride to a metal sponge suitable for the production of ductile titanium.[150] Pure, crystalline titanium is obtained by sodium or potassium amalgam reduction of the tetrachloride.[151] Titanium is not soluble in mercury and does not form a stable compound with it. The amalgam is maintained at 200° C.

While liquid titanium tetrachloride flows downward through a hollow electrode to a liquid bath of mercury with sodium or calcium in a closed chamber having an argon atmosphere, an arc is maintained between the electrode and the bath to produce titanium metal.[152] The titanium is carried as a slurry with excess tetrachloride draining into a collecting tank. Titanium tetrachloride is reduced to the metal by magnesium amalgam at 1200° F and a pressure of 40 atmospheres.[153] Alkali and alkaline earth metal amalgams in general are effective reducing agents.[154]

A mixed product of titanium, sodium chloride, and mercury obtained by the reduction of the tetrachloride with sodium amalgam is compressed below the melting point of sodium chloride to force out a large part of this impurity.[155]

Titanium tetrachloride dissolved in liquid ammonia is reduced with sodium amalgam at room temperature to form titanium amalgam.[156] This product amalgam is removed from the system and stirred to release titanium metal on the surface of the mercury. A mixture of titanium tetrachloride, sodium bromide, and liquid ammonia is heated with sodium amalgam at 80° C to effect reduction to titanium powder.[157]

## ELECTROLYSIS OF TITANIUM TETRACHLORIDE IN A FUSED SALT BATH

Porous, coarsely crystalline, and ductile titanium is deposited on the inner surface of a basket cathode having a series of perforations facing graphite anodes.[158] By controlling the current and the flow rate of titanium tetrachloride gas into the electrolytic bath, the titanium is reduced without diffusing out of the cathode. The bath maintained at 825 to 950° C is sodium chloride or an eutectic of sodium chloride with lithium chloride, magnesium chloride, strontium chloride, or barium chloride. Heat is supplied by graphite electrodes in the bath. Titanium is produced electrolytically from titanium tetrachloride in a stainless steel vat, lined with alumina and heated from the outside.[159] A porous cup in the center containing the cathode is surrounded by a ring of anodes. The electrolyte is a mixture of sodium chloride and strontium chloride at 560° C containing 6.25 per cent of dissolved titanium dichloride or trichloride made by reducing the tetrachloride with titanium or produced in place. Added titanium tetrachloride dissolves in the electrolyte. Liquid titanium tetrachloride is fed through a hollow electrode into an electric arc between the electrode and a liquid bath of mercury, sodium, calcium, or an alloy below to form the metal.[160] The titanium metal and the excess tetrachloride drain into a collecting tank. In the electrolytic production of titanium by reduction of the tetrachloride in a divided cell, the output is increased by dissolving metallic titanium in the catholyte.[161]

Molten alkali or alkaline earth metal chlorides are saturated with titanium tetrachloride and electrolyzed to form titanium trichloride and dichloride at the cathode.[162] The cathode is temporarily removed from the bath and placed in gaseous titanium tetrachloride, which is reduced to the metal and more lower chlorides. The cathode is then returned to the bath to allow the lower chlorides to dissolve. To produce titanium the bath is then electrolyzed with a nonremovable cathode. Pure titanium is produced as a cathode deposit by electrolysis of a chloride in a fused mixture of alkali or alkaline earth metal chlorides with anode rods of titanium carbide or sulfide.[163] The chlorine evolved at the anode reacts with the carbide or sulfide to replenish the titanium chloride in the electrolyte. The metal is deposited electrolytically from fused alkali chlorides by a contact cathode operating in an atmosphere of titanium tetrachloride and hydrogen.[164] According to a process by Wolski,[165] the tetrachloride is fed to the fused electrolyte through a hollow graphite anode. The chlorine evolved at the anode stops foreign materials from entering the melt.

Titanium crystal intergrowths are obtained by electrolyzing a molten alkali metal halide bath containing titanium ion.[166] The crystalline titanium is attached to a titanium plate by a layer of solid salt interspersed with the crystals. By confining the electrolytic reduction of the tetrachloride to the interior of a porous basket cathode, the electrolyte between the anode and the cathode is kept free of lower chlorides of titanium, thus eliminating side reactions.[167] High purity titanium is prepared continuously from the tetrachloride in an electrolysis cell containing separated cathode and anode compartments connected by a bipolar electrode of molten titanium alloy.[168] The cathode is withdrawn from the bath of sodium chloride, magnesium chloride and titanium dichloride as the titanium accumulates. A bath of $Na_2TiCl_4$ in a sodium chloride-potassium chloride eutectic is electrolyzed at 575° C with nickel or tungsten rod anodes to produce titanium.[169] A glass diaphragm encloses the cathode chamber to prevent mixing of the anolyte and the catholyte. Titanium tetrachloride is fed into a mixture of sodium chloride and sodium fluoride held at 900° C in a graphite crucible through a graphite tube to form a double fluoride.[170] Electrolysis of this mixture at 700° C gives titanium metal of high quality. By feeding titanium tetrachloride gas into the cathode region of a sodium chloride-lithium chloride bath at 900° C soft titanium is produced by electrolysis.[171] Chlorine gas is used as the protective atmosphere. A stainless steel basket filled with sponge titanium is placed between the anode and cathode in a fused bath containing sodium chloride, potassium chloride, magnesium chloride, and titanium dichloride.[172] Titanium III formed at the anode is reduced to titanium II at the added sponge to avoid the decrease of titanium deposition caused by the titanous ions.

A fused salt type electrolytic cell for the production of titanium is provided with a fixed anode, a movable cathode, a heater placed above the cathode, and an exchange chamber for the cathode.[173] To prevent corrosion of the refractory lined nickel dome for the removal of chlorine during the electrowinning of titanium, the back electromotive force of the dome is maintained at a constant 2.2 volts by use of a bias current.[174] With a fused bath of lithium chloride and potassium chloride titanium may be produced electrolytically at 550° C without the use of a protective gas.[175] At this temperature titanium is not attacked by the ambient atmosphere. Howard[176] recovered titanium by electrolysis of a fused salt electrolyte by a process which effectively disintegrates the deposit and separates the metal particles. Titanium trichloride formed by the reaction of the tetrachloride with titanium is electrolyzed in a bath of potassium chloride and lithium chloride at 550° C to produce the metal.[177] Wainer[178] electrolyzed the molten product of the reaction of titanium tetrachloride with titanium monosulfide-titanium dichloride and sulfur monochloride. A carbon anode and a nickel cathode are employed.

In the electrolytic production of titanium from the tetrachloride an electrolyte carrier having a low melting point and containing at least 70 per cent aluminum chloride is used.[179] If mercury is used as the cathode the lower chlorides of titanium are formed in the cell and the process is continuous. Snow and McCord[180] produced titanium directly from the tetrachloride by electrolysis in a fused salt bath. In the electrolysis, anode polarization is prevented by the presence of fluoride salts comprising from 0.02 to 0.10 combined fluorine.[181] This depolarizing salt can be added to the fused bath before or during electrolysis. High purity, firmly adhering titanium is deposited at the cathode of an electrolytic cell containing fused alkali or alkaline earth metal chlorides or fluorides into which are simultaneously passed gaseous tetrachloride and hydrogen.[182]

Titanium tetrachloride is passed through the perforated cathodes into a fused mixture of 18 per cent potassium chloride, 2 per cent calcium fluoride, and 80 per cent magnesium chloride, and electrolyzed at 6 volts and 800 amperes.[183] A graphite anode is used at the center of the cell. The titanium metal is drawn from the bottom of the vessel and purified by heating in vacuum. Electrolysis of titanium tetrachloride in an equal mixture of sodium chloride and potassium chloride at 675° C yields the metal.[184] A current of 60 amperes passes through the cell. Titanium tetrachloride is introduced under pressure. A zinc cathode is employed for electrolyzing titanium tetrachloride dissolved in alkali or alkaline earth metal chlorides. The separation of the titanium from the zinc is accomplished by electric arc melting or vacuum distillation.[185] The fused salt cell is provided with means to convey the purified titanium to storage without exposure to the air.[186] Titanium is produced as pure coarse crystals by electrolysis in fused salt baths in two stages.[187] In the first cell the tetrachloride is reduced with sodium at the cathode, the fine metal is allowed to settle and only the lower concentrated portion is admitted to the second cell.

The metal is produced in a continuous process by electrolysis of fused salts in a graphite lined furnace heated with silicon carbide bars.[188] In the process the titanium is deposited on a slowly rotating metallic cathode made of blocks of titanium carbide. A scraper removes the metal from the cathode.

Measurements of electrode potentials at 600° C in melts as used in the electrolytic deposition of titanium gave for Na, NaCl 3.22; Mg, $MgCl_2$ 2.84; and Ti, $TiCl_2$ 2.04 volts.[189]

## ELECTROLYSIS OF LOWER CHLORIDES OF TITANIUM IN FUSED SALTS

Titanium metal is produced by the continuous electrolysis of the dichloride or trichloride in a fused salt electrolyte.[190] The lower chlorides are formed by bubbling the tetrachloride through molten titanium immersed in the electrolyte. By a similar process titanium trichloride is prepared in the electrolytic cell by bubbling the tetrachloride through molten lead.[191] A ternary mixture of titanium trichloride in the eutectic mixture of 60 mole per cent lithium chloride and 40 mole per cent potassium chloride is electrolyzed at 550° C with tungsten electrodes to produce titanium.[192] The applied potential is 1.2 volts at a current density of 6 amperes per square centimeter. Titanium metal is produced by the electrolysis of titanium dichloride in a fused salt mixture of sodium chloride and strontium chloride in a compartmented cell containing a porous diaphragm.[193] The tetrachloride is fed into the electrolyte through a hollow tantalum cathode and is reduced to the dichloride by electrolytic action. By electrolysis of the dichloride dissolved in a fused mixture of 73 per cent strontium chloride and 27 per cent sodium chloride at 750° C the metal is obtained.[194] The surface of the electrolyte is protected by an atmosphere of argon or helium. A graphite anode and two tantalum cathodes are provided. Titanium tetrachloride introduced into the cell near the cathode is reduced to yield the dichloride and trichloride in the ratio of five to one. A mixture of titanium dichloride and trichloride manufactured by the reaction of titanium sponge with hydrochloric acid or titanium tetrachloride in the vapor form is electrolyzed in a molten salt bath to yield metal of high purity.[195] Titanium is produced by electrolysis with an inert anode and a titanium cathode in a two phase fused salt system.[196] The lighter anolite is a mixture of sodium chloride and magnesium chloride while the heavier catholite is a mixture of sodium chloride, titanium dichloride and titanium tetrachloride. A temperature range of 450 to 700° C is employed. The electrolysis of a solution of the dichloride and trichloride in a fused mixture of calcium chloride, sodium chloride and potassium chloride at 600° C with a silicon anode and a titanium plate cathode gives pure sponge titanium.[197] Silicon tetrachloride is produced at the anode. A mixture of the trichloride in fused lithium chloride and potassium chloride is electrolyzed at 500 to 700° C in a graphite crucible.[198] The cell employs a molybdenum cathode and a graphite anode. On raising the cathode to obtain the titanium, it is flushed with an inert gas to prevent peeling off of the deposit.

Reduced titanium chlorides dissolved in fused alkali and alkaline earth chlorides can be electrolyzed in diaphragmed cells under an inert gas atmosphere to yield ductile titanium as adherent, crystalline deposits.[199] The reduced chlorides are produced in situ by adding the tetrachloride through a hollow cathode at a controlled feed to current ratio. Titanium dichloride formed by the reaction of the tetrachloride with spongy titanium at 1020° C is electrolyzed in a bath of potassium chloride and sodium chloride at 750° C to yield the metal of 99.3 per cent purity.[200] Titanium tetrachloride is reduced with hydrogen on a hot silica surface to the trichloride for the electrolytic production of the metal.[201] The electrolyte consists of an eutectic of lithium chloride and potassium chloride. According to the process the preparation and the electrolysis of the titanium tetrachloride are carried out in a single unit, thus making the process semi-continuous. For electrolysis, titanium trichloride formed by heating the tetrachloride with the metal at 400 to 500° C is dissolved in fused potassium chloride or mixed chlorides of lithium, sodium, potassium, or calcium to a concentration of 20 to 30 per cent.[202] This solution is electrolyzed at 700 to 850° C to obtain coarsely crystalline deposits of the metal. The cathode current density is 0.1 to 0.5 amperes per square inch at 2.0 to 2.4 volts.

Melts of titanium dichloride in sodium chloride are desirable as a starting material for the production of titanium metal or as an electrolyte for refining.[203] Sodium reduction of the tetrachloride with an 18 hour period of holding produces the dichloride and trichloride in the molar ratio of 8 to 1. Reduction with scrap metal in molten sodium chloride produces melts containing the dichloride and trichloride in the ratio of 4 to 1. Fused mixtures of titanium dichloride and sodium chloride for use in the second stage of sponge production from the tetrachloride or for electrolytic production of the metal are produced having a low concentration of the undesired trichloride.[204] Reduction of the tetrachloride is effected with molten sodium in a closed evacuated reactor at 650° C in contact with a pile of titanium fragments in the molten sodium chloride product.

The reaction of the trichloride with titanium metal in molten sodium chloride at 800 to 950° C is surface controlled and the heat of activation is 8.29 kcal per mole of the trichloride.[205] An active site or reaction nucleus is created by adsorbing a titanium trichloride molecule at the metal surface. Reaction of the second titanium tetrachloride species forms an activated complex that decomposes into three titanium dichloride molecules. The trichloride and impure sponge metal are added to a fused mixture of sodium chloride and potassium chloride at 900° C to obtain a product of 99.8 per cent purity.[206] Both the dichloride and trichloride are produced by striking an arc between electrodes immersed in liquid titanium tetrachloride.[207] These compounds are disproportionated at 500° C to yield chlorine and titanium of high purity.

Titanium is produced by the electrolysis of the dichloride in a fused mixture of 73 per cent strontium chloride and 27 per cent sodium chloride at 750° C.[208] The surface of the electrolyte is protected with argon or helium. The tetrachloride introduced in the cell is reduced to the dichloride and trichloride. These products yield titanium at the deposition cathode. A simplified cell design permits efficient electrolysis of the dichloride and the trichloride to metallic titanium and chlorine.[209] The electrodes are placed so that chlorine gas does not approach the cathode.

## ELECTROLYSIS OF FLUOTITANATES

Titanium is electrolytically produced from a fused salt bath containing sodium fluotitanate or potassium fluotitanate, alkali metal chlorides, bromides or iodides, and titanium tetrachloride, tetrabromide or tetraiodide at 600 to 850° C.[210] The tetrachloride is introduced in the gaseous state. The voltage is so regulated that the fluotitanate does not decompose. Drossbach[211] electrolyzed potassium fluotitanate at 300° C in sintered alumina apparatus. According to Wurn, Gravel, and Potvin[212] the formation of trivalent titanium double fluorides of sodium and potassium always precedes the production of metallic titanium during electrolysis of a molten bath of sodium chloride and potassium chloride containing sodium fluotitanate and potassium fluotitanate. These trivalent compounds have the formulas $K_2NaTiF_6$, $K_3TiF_6$, and $Na_3TiF_6$. The metal is electrodeposited from a fused mixture of potassium fluotitanate, potassium fluoride and calcium fluoride employing an anode of compacted titanium dioxide, charcoal and pitch.[213] In the reactions titanium tetrafluoride is formed and decomposes to titanium and fluorine. A fused mixture of lithium chloride, potassium chloride, potassium fluotitanate and sodium fluoride is electrolyzed at 450° C with a current density of 200 amperes per square decimeters at 3 to 5 volts.[214] In a step process a fused bath of potassium fluotitanate and sodium chloride is electrolyzed first at a low voltage and current density and finally at much higher values.[215]

Molten potassium fluotitanate containing sodium and magnesium fluoride to raise the melting point is electrolyzed at 760° C with a graphite anode to produce a continuous titanium plate free from impurities.[216] The bath is agitated. Electrolysis of $NaTiCl_4$ in a bath of sodium chloride-lithium chloride eutectic gives titanium powder at higher than 90 per cent yield.[217] The particle size of the titanium metal deposited from sodium or potassium fluotitanate is greatly increased by maintaining the total electrolysis bath concentration of aluminum, chromium, iron, and vanadium below 0.1 per cent of the titanium.[218] Sodium and potassium fluotitanate used in the electrolysis may be readily purified by recrystallization from water. The metal is produced by electrolysis of an anhydrous fused salt bath consisting of an alkali metal fluotitanate and one or more halide salts of an alkali or alkaline earth metal.[219] Potassium fluotitanate prepared by the reaction of potassium fluoride and sodium fluoride with titanium sulfate solution is mixed with sodium chloride and electrolyzed at 700° C in an inert atmosphere.[220] From 5 to 8 volts are applied at a current density of 200 to 500 amperes per square decimeter.

Titanium granules are deposited on an iron cathode by electrolysis at 650 to 780° C of granular titanium monoxide added to molten mixture of

7 to 20 per cent purified sodium or potassium fluotitanate and 80 to 93 per cent sodium chloride or potassium chloride.[221] Electrolytic deposition on base metals is made from a molten salt bath containing a major amount of an alkali or alkaline earth metal halide together with potassium fluotitanate at 750 to 850° C in an atmosphere of argon.[222] Electrolysis of potassium fluotitanate in molten sodium chloride at 760° C gives good yields.[223] The current density does not appreciably affect the quality of the product. Iron is eliminated by using graphite electrodes connected with nickel screws. The formation of titanium carbide is eliminated by coating the electrodes with nickel.

Titanium tetrachloride is fed into a fused mixture of sodium fluoride and sodium chloride at 900° C to form a double fluoride which is electrolyzed at 700° C to give the metal.[224] Electrolysis at 750 to 850° C in a mixture containing 85 per cent sodium chloride and 15 per cent potassium fluotitanate yields 91 per cent recovery of titanium from the carbide and 62 per cent recovery from the cyanonitride.[225] High purity metal is electrodeposited from fused double salts as $Na_2TiCl_4$.[226] The double salt is produced by reducing the tetrachloride with an alkali metal or its amalgam in a nonpolar solvent (benzene or toluene) at 100° C.

## ELECTROLYSIS OF TITANIUM DIOXIDE IN A FUSED SALT BATH

Titanium dioxide dissolved in calcium fluoride is electrolyzed in a graphite cell at 1900° C with a graphite anode and a titanium tip cathode cooled with water to produce block titanium somewhat contaminated with carbon.[227] Electrodes of titanium dioxide and carbon are used to maintain an electric arc in vacuo at 3000° C to produce metal of high purity.[228] Electrolysis of a fused bath of calcium fluoride, magnesium fluoride, and titanium dioxide at 1000 to 1400° C gave a deposit of pure titanium.[229] A catholite composed of sodium chloride, calcium chloride, and titanium dichloride, and an anolite of sodium chloride, calcium chloride, and titanium dioxide, separated by a diaphragm of porous alundum, are electrolyzed to produce the metal.[230] Spongy titanium is deposited at the cathode by the electrolysis of a fused bath of calcium chloride, calcium oxide, and titanium dioxide at 600 to 800 amperes per square decimeter and a temperature of 1150 to 1250° C.[231]

The metal is produced from electrolytes consisting of 5 to 10 per cent titanium dioxide dissolved in molten alkali metal phosphates, borates, or fluorides, or of mixtures of alkali metal fluotitanates.[232] Electrolysis at 840 to 950° C with a current density of 1 to 3 amperes per square inch yields titanium in the neighborhood of the molybdenum cathode. The metal is recovered from the cooled electrolyte by leaching with water or ammonium citrate solution.

A melt of titanium dioxide, calcium chloride, calcium oxide, and hydrogen chloride is electrolyzed to produce titanium suitable for electrorefining.[233] Titanium dioxide is reduced by electrolysis in a fused bath of calcium chloride to give the monoxide.[234] The cathode is a sheet of molybdenum. Titanium monoxide obtained in this manner is used for the production of the metal in a similar process. A fused bath of magnesium chloride and sodium chloride is electrolyzed at 700° C with a titanium dioxide-coke anode and an iron cathode to produce titanium tetrachloride and magnesium metal.[235]

The solubility in mole per cent at 1000, 1100 and 1200° C of titanium dioxide in lithium fluoride is 0.13, 0.23, and 0.55; in sodium fluoride it is —, 14.65 and 18.72; and in potassium fluoride it is 26.17, 37.94 and 50.40.[236]

## ELECTROLYSIS OF THE CARBIDE, NITRIDE, AND MONOXIDE

A titanium carbide anode is electrolyzed in a bath of calcium fluoride and magnesium fluoride at 1670° C to liberate molten metal at the cathode.[237] The metal is electrodeposited continuously from a titanium carbide anode in a fused bath of potassium chloride and lithium chloride.[238] To remove the titanium deposit the cathode is continuously rotated. Electrolysis of a carbide anode in a bath of fused alkali or alkaline earth metal chlorides gives titanium in a yield of 75 per cent.[239] The anode is produced by heating a shaped mass of titanium dioxide and coal at high pressure followed by reduction in a stream of hydrogen. Electrolysis is carried out with an anode of titanium carbide containing small amounts of the nitride and lower oxides in a bath of fused calcium chloride, calcium fluoride, magnesium fluoride, or strontium fluoride, or mixtures having a melting point below that of titanium, 1660° C, and a boiling point higher than that of titanium.[240] The liberated molten metal settles to the bottom of the cell where it solidifies and acts as the cathode. A current passing through the electrolyte maintains the temperature. In a similar process a fused bath of sodium chloride is used.[241]

Wainer [242] produced titanium electrolytically from a fused alkali or alkaline earth halide bath of a solid solution of titanium carbide and monoxide. Coarse crystalline metal is recovered from its carbide by electrolysis in a fused bath of halide salts at 800 to 900° C.[243] The cathode is an axial rod of titanium, nickel, tungsten, or molybdenum in an upright cylindrical cell. A melt of the carbide suspended in alkali, alkaline earth or aluminum chloride is electrolyzed with a titanium carbide anode at 500 to 600° C to produce the metal.[244] The anode disintegrates to replenish the melt as it is decomposed by chlorine evolved during electrolysis. Mixtures of the carbide and monoxide pressed into pellets are fused in an electric arc in an atmosphere of argon.[245] The silvery buttons containing 89.5 per cent titanium are especially suited as a cell feed material in the electrolytic production of the pure metal.

Titanium nitride is electrolyzed in a fused salt bath of potassium fluotitanate and sodium chloride at 850° C to produce the metal.[246] A steel, cone-shaped cathode is employed to recover the nitrogen. A fused bath of calcium chloride containing 5 per cent titanium monoxide electrolyzed in an atmosphere of argon at 900° C and a current density of 150 to 450 amperes per square decimeter gives the metal at a current efficiency of 33 to 51 per cent and a yield of 88 per cent.[247] Fluoride, sodium, potassium and lithium ions are very detrimental, practically eliminating any possibility of obtaining titanium.

## ELECTROLYSIS OF ALLOYS

Molten titanium alloys are used as the anode in fused salt electrolysis to deposit crystals of the metal.[248] The electrolyte of potassium fluoride, potassium chloride, and potassium fluotitanate is maintained at 980 to 1050° C. A typical alloy contains 60 per cent nickel and 80 per cent titanium. Electrolysis using anodes of ferrotitanium or ferrosilicon titanium in a fused bath of sodium chloride, ammonium hydrogen fluoride and titanium dichloride at 870° C gives a deposit of high purity metal.[249] A crude alloy of titanium and iron produced by the reduction of an ore with carbon or aluminum is made the anode in a fused bath of calcium fluoride.[250] Electrolysis at 1200 to 1800° C yields pure metal. Anodes of alloys of titanium with copper, iron, silver, and tin are refined in molten electrolytes of calcium chloride, sodium chloride, and titanium dichloride at 800° C.[251] High purity titanium is deposited at the cathode. A titanium-magnesium alloy is produced electrolytically and dissociated into the free metals at 1000° C.[252] Titanium tetrachloride is introduced into the cathode compartment of the molten bath of potassium magnesium chloride and calcium chloride during electrolysis at 680° C.

## REDUCTION OF TITANIUM TETRACHLORIDE WITH HYDROGEN

Titanium tetrachloride vapor is effectively reduced by hydrogen if the reactants are heated together above the titanium boiling point of 3535° C.[253] At this temperature the reaction is calculated to be exothermic, evolving 105 kcal per mole. Hydrogen chloride is 90 per cent associated. A mixture of the tetrachloride vapor and hydrogen passed through an electric discharge between tungsten electrodes yields metal of high purity.[254] Titanium halides are blown in a plasma flame alone or with hydrogen to produce metal of high purity.[255] In a process by Ferraro [256] the tetrachloride is reduced with hydrogen in a cyclic process. The reduction at 1000° K yields gaseous titanium trichloride and hydrogen chloride, which are separated by fractionation. The liquefied trichloride is heated to 550° K to form the tetrachloride and the solid dichloride. Finally the solid dichloride heated above 2000° K disproportionates into the trichloride and titanium metal. Lower chlorides produced by the reduction of the tetrachloride with hydrogen in a mixture of sodium chloride and potassium chloride is treated with magnesium or sodium at 750 to 850° C to produce sponge.[257]

Titanium produced by hydrogen reduction of the tetrachloride reacts readily at 800 to 1200° C with nickel and at 1050 to 1350° C with graphite of the reaction vessel.[258] The carbide is a product in the graphite reactor. Aluminum also reacts with the gas mixture.

Large single crystals of titanium are prepared from the hydride.[259] Stoichiometric titanium hydride, $TiH_2$, prepared from sponge metal and hydrogen is powdered to 4 microns and pressed. The molded plate is evacuated at 500° C, sintered at 1250° C, and left overnight. Large crystals of the metal are obtained. In the presence of hydrogen, the tetrachloride is thermally decomposed at the surface of an aluminum or copper object to be metallized.[260] Dry cracked ammonia is used as the raw material for the combustion of the chlorine formed. Vapors of the tetrachloride are reduced continuously in a mixing jet at 200° C by hydrogen preheated to 2500° C in an electric arc to form subchlorides as an intermediate product.[261] To manufacture titanium trichloride, a gaseous mixture of excess tetrachloride and hydrogen is passed continuously into a heated reaction chamber.[262] From this chamber the reaction mixture flows out rapidly and is cooled to obtain a dispersion of the trichloride and tetrachloride.

## REDUCTION OF TITANIUM DIOXIDE

A mixture of titanium dioxide and calcium hydride compressed at 250 kg per square centimeter and heated in vacuo in a molybdenum crucible begins to react at 500° C.[263] At 900° C the reduction is complete in 5 minutes. In a two-step process the dioxide is reduced to a lower oxide by the hydrogen evolved by heating to 1000° C a mixture of calcium hydride, carbon, and aluminum.[264] Then calcium vapor reacts with the lower oxide to give powdery titanium. The reduction of the dioxide with calcium hydride yields a titanium-hydrogen product of variable composition.[265] This product separated from the other components by washing with 10 per cent hydrochloric acid is especially suitable for the preparation of the metal powder. From an energy point of view reduction of the dioxide with hydrogen obtained as a result of dissociation of calcium hydride is less likely than reduction by calcium.[266] Consequently calcium hydride as a reductor for titanium dioxide does not present any energy advantage over direct reduction with calcium. Calcium reacts with the dioxide in the liquid as well as in the vapor phase.

Titanium dioxide with a stoichiometric amount of carbon as the anode is reduced in a high erosion arc to the metal.[267] The product gases are rapidly cooled to prevent the reverse reaction. A mixture of the dioxide and glycerin subjected to radiation in a cobalt bomb equivalent to 10,000,000 r/g yields titanium metal, glycerin acid, and other products.[268] An intimate mixture of the dioxide and a carbide of an alkaline earth metal or aluminum is heated at 800 to 1200° C to produce a mixture of titanium, carbon, and an oxide of the alkaline earth element.[269] The product is ground, concentrated, and leached with dilute acids to obtain a titanium rich residue which is electrolyzed in a molten salt bath. The dioxide is treated in a hydrofluoric acid solution, heated first at 250 to 400° F to drive off the water, and later at 400 to 600° F to sublime the titanium tetrafluoride which is condensed in another vessel.[270] The tetrafluoride thus produced is treated with silicon at 500 to 800° F to form silicon tetrafluoride vapor and a residue of powdered titanium. Reduction is effected by heating the dioxide at 800 to 1200° C with calcium carbide in excess. The dioxide is reduced to the monoxide by titanium followed by reduction of the monoxide to the metal by titanium carbide.[271]

Titanium and its alloys are produced by the reduction of the dioxide.[272] A mixed crystal of 5 to 10 per cent carbon, 7 to 14 per cent oxygen, and the remainder titanium is heated to at least 2500° C until the carbon and oxygen form a gaseous compound. The required temperature depends on the composition of the mixed crystal. Titanium is prepared by the reduction of the dioxide at a pressure of 1000 atmospheres and a temperature of 400° C.[273]

By addition of an excess powdered aluminum or magnesium and liquid oxygen to produce more heat, the Thermite process may be employed to reduce the dioxide.[274] The reaction may be carried out in a magnesium oxide crucible. Reduction of the dioxide to lower oxides is effected by reaction with an excess magnesium, with up to 30 per cent magnesium chloride flux at 700 to 1000° C in a closed steel reactor having an atmosphere of hydrogen.[275] This product is used as anode material in the electrolytic production of the pure metal in fused salts. The dioxide mixed with granular magnesium and magnesium chloride is heated in an inert atmosphere at 1000° C to produce titanium in 90 per cent yield.[276] The product is leached with hydrochloric acid, mixed with calcium and calcium chloride, and heated to 1000° C to complete the reduction reaction. Reduction of the dioxide with magnesium and calcium in stages yields a low oxygen product.[277] The dioxide is first reduced with magnesium at 1000° C, leached in 5 per cent hydrochloric acid, followed by evaporation of the residual magnesium at 1250 to 1400° C. This product is reduced with calcium at 1000° C, leached with 5 per cent hydrochloric acid, and degassed at 840° C.

Pure titanium dioxide is electrolyzed in a molten bath of anhydrous calcium chloride at 800° C.[278] Molten zinc serves as the cathode. After the titanium content of the cathode metal reaches 2.5 per cent the alloy is drawn from the furnace and distilled to remove the zinc which is recycled. The electrolysis of the dioxide in molten cryolite at 950° C results in the deposition of an alloy with aluminum containing 7 to 30 per cent titanium.[279]

A titanium sulfide obtained by heating rutile with carbon disulfide and hydrogen sulfide or with sodium carbonate and sulfur reacts with sodium or potassium at 800 to 100° C in an inert atmosphere to produe titanium.[280] After cooling to 500° C, sulfur is added to form liquid sodium and potassium polysulfides which are drained from the titanium.

Thermite reductions of slags from ilmenite smelting produce a crude product containing 65 per cent titanium.[281] Ilmenite is reduced in molten aluminum to produce an iron alloy containing 41 per cent titanium.[282] The reaction vessel is made of magnesium oxide or aluminum oxide. In a two stage process the dioxide is reduced partially with magnesium in a calcium chloride bath and finally with calcium to produce titanium metal.[283]

## IODIDE METHOD OF PRODUCING TITANIUM

The rate of decomposition of titanium tetraiodide on a heated filament is independent of the filament surface area but directly proportional to the vapor concentration.[284] Decomposition rates vary with filament temperature according to an Arrhenius type reaction. A preliminary determination of activation energy is 28.6 kcal per mole. Results indicate that the reaction is first order. The mass of the filament increases linearly with time. The process is modified by film boiling titanium tetraiodide on a heated filament resulting in similar deposition on the filament and liberation of iodine in the vapor state.[285]

Titanium tetraiodide is decomposed in an electric arc to produce the metal and iodine.[286] One of the electrodes may be hollow for introduction of the tetraiodide. Titanium is produced by the thermal decomposition of the tetraiodide or the tetrachloride by the use of high temperature energy from an electric arc.[287] The thermal deposition of the gaseous tetraiodide on an indirectly heated Vycor or quartz finger in a modified De Boer bulb gives good yields.[288]

Dunn [289] produced titanium by the decomposition of the tetraiodide in a dissociating furnace free of contaminating gases. The walls of the furnace are kept at a temperature below that at which the tetraiodide decomposes, while metallic particles are introduced and inductively heated to 1200° C to provide a surface on which the titanium deposits. An inclined furnace permits continuous operation. In a cyclic process the tetraiodide is formed in one chamber and decomposed to titanium and iodine is another.[290] Crude titanium is packed in a cylinder of thin titanium plate and heated in vacuo in an induction furnace to 1350 to 1500° C.[291] Then titanium tetraiodide purified by sublimation and iodine are introduced into the furnace. Metal of 99.99 per cent purity is deposited on both surfaces of the titanium plate after first forming and then decomposing the tetraiodide.

In the production of the metal in an inclined rotary furnace, the tetraiodide is prevented from adhering to the surface by cooling the walls with water circulating through the induction coils and by introducing into the furnace particles of titanium or tungsten as centers of accumulation.[292]

Loonan [293] produced metallic titanium from the carbide and iodine. Iodine vapor is brought in contact with titanium carbide at 1100° C to produce titanium tetraiodide. The gaseous tetraiodide then reacts with titanium metal at 600 to 1000° C to yield a stable mixture of the tetraiodide and triiodide. This gaseous mixture is decomposed on a heated surface at 1100 to 1700° C to liberate titanium metal and iodine which is reused to produce more iodides.

Electrolysis of a solution of the tetraiodide in a fused mixture of lithium, potassium, and sodium iodides at 300 to 375° C gives pure titanium.[294] A crude electric furnace product containing titanium, titanium carbide, and titanium monoxide reacts with the tetraiodide at 1025° C to form the diiodide, which is condensed to a liquid at 625 to 1025° C.[295] Sheathed electric heating elements at 800 to 1500° C immersed in this liquid effects disproportionation of the diiodide to the metal which deposits on the sheaths and titanium tetraiodide which passes off as vapor.

## PURIFICATION OF TITANIUM BY ELECTROLYSIS

To remove oxygen, the impure titanium metal is made the anode in an electrolyte of fused magnesium chloride.[296] Pure titanium is deposited at the cathode. An anode of impure titanium is decomposed electrolytically in an electrolyte of fused alkali or alkaline earth halide salt and redeposited in the pure form.[297] The bath is purified to a high degree before electrolysis. Refinement of the metal is effected by electrolysis in a fused bath of lower titanium bromide and sodium bromide.[298] A current density of 0.2 ampere per square centimeter of the iron rod cathode are employed. With a soluble anode of scrap metal and a sodium chloride electrolyte containing 1.5 to 4 per cent lower chlorides up to 1000° C, pure coarsely crystalline titanium deposits.[299] The form and size of the dendritic crystals, but not their purity, are strongly influenced by the current density and by the concentration of titanium in the melt. Titanium III appears to migrate more readily toward the cathode than titanium II. Pure titanium is deposited at the cathode of a cell containing an impure metal anode and a fused potassium chloride-fluotitanate electrolyte at 1900° C.[300]

Impure titanium mixed with 1.5 times its weight of petroleum coke is melted in a graphite crucible in an induction furnace to form billets which are made the anode in an electrolyte of sodium chloride at 780° C.[301] Coarse crystals of pure titanium are deposited on the nickel cathode. Molten titanium scrap is used as the anode in a carbon container in a fused bath of potassium fluotitanate, sodium chloride, and potassium chloride.[302] Pure titanium is deposited at the tungsten cathode. Sponge containing impurities is made the anode in a refining process.[303] Electrolysis is carried out in a bath of sodium chloride, potassium chloride, and magnesium chloride at 600° C with a current density of 50 amperes per square decimeter. A titanium plate serves as the cathode.

Titanium can be electrorefined almost completely from binary alloys with molybdenum, tin, and zinc.[304] Refining from aluminum and chromium is only slightly less effective. Flexible titanium is obtained by electrolyzing an oxygen absorbed brittle product with a carbon anode in a bath of calcium chloride at 1000° C.[305] A part of the calcium chloride may be replaced by magnesium chloride or sodium chloride. Titanium metal more amenable to arc melting is purified by an electrolytic process.[306] The magnesium chloride is removed from the sponge by leaching with water. Then the impure titanium is made the anode in an electrolyte consisting of 0.1 to 10 per cent hydrochloric acid. The magnesium is removed from the impure metal and deposited on the cathode. Oxygen can be removed from titanium to obtain a malleable product by electrolysis in a fused bath of sodium chloride and potassium fluotitanate in a graphite crucible.[307] The metal sinks to the bottom and serves as the anode. The pure metal is plated at an inert cathode with a molten alkali or alkaline earth metal chloride electrolyte and an impure titanium anode.[308] For successful operation of the process it is necessary to introduce chlorine or titanium tetrachloride as a depolarizer through the cathode or adjacent to it.

Crude alloys of titanium are refined to the metal of high purity by electrolysis in a fused chloride or bromide salt of an alkali or alkaline earth metal which also contains a small proportion of free sodium and of a soluble salt in which the titanium has a valence of less than four.[309] Titanium having high proportions of copper, iron, nitrogen, and other impurities is refined by electrolysis at 1500 to 1650° F in a sodium chloride bath containing titanium dichloride and trichloride with a current density up to 2000 amperes per square foot of cathode surface.[310] The electrolyte contains some free sodium. To insure the dissolving of titanium only from an anode containing nobler metals, it is necessary to add to the electrolyte of fused alkali or alkaline earth metal chlorides 0.25 to 5 per cent of a lower chloride of titanium and 0.1 to 2 per cent of dissolved alkali or alkaline earth metal.[311]

In refining titanium the anode and cathode chambers are connected at the cell bottom through a bipolar molten lead electrode.[312] The electrolyte is a sodium chloride melt containing 1 to 3 per cent dissolved titanium in the form of a lower chloride.

## NON-ELECTROLYTIC METHODS OF PURIFYING TITANIUM

Titanium may be purified by passing a stream of argon saturated with titanium tetrachloride at 20° C over the impure material at 1050° C.[313] The reaction product, mostly titanium dichloride, is condensed in a cooler zone of the tube. Metallic titanium free from impurities remains in the tube. The titanium formed by decomposing the lower chlorides above 750° C consists of small flakes. Oxygen and nitrogen are removed from the metal by heating at 500 to 1800° C in the presence of germanium just sufficient to form germanium oxide and germanium nitride, which are sublimed at temperatures of 710 and 650° C, respectively.[314] Gallium and indium are also effective. The oxygen content is reduced by treatment with calcium in an atmosphere of argon.[315] For convenience the calcium may be dissolved in molten calcium chloride. Quin[316] treated impure titanium containing oxygen or nitrogen with molten zinc, cadmium, or lead to form a liquid alloy and a solid residue containing the impurities. The liquid alloy is filtered and distilled to remove the solvent metal, leaving the purified titanium as a residue. Oxides are removed from titanium by heating with calcium at 1000° C in fused calcium chloride.[317]

Wainer[318] removed oxygen from titanium scrap by heating the metal with carbon in a fused bath of sodium or potassium fluotitanate at 950° C. Titanium sponge contaminated with salt is crushed and suspended in molten sodium chloride and magnesium chloride.[319] The suspension is then placed in a high frequency electromagnetic field to heat the particles to 1600 to 1720° C, thus decreasing their surface area twentyfold so that they may be effectively leached in dilute acid. Crushed titanium sponge is freed from other reaction products and excess reductant by passing it in an inert atmosphere through a high frequency electromagnetic field to volatilize the impurities and fuse the metal particles to reduce the surface area. The specific surface area is decreased from 0.45 square meter per gram in the feed to 0.035 in the product. This material is readily washed with 5 per cent nitric acid without contamination.[320]

## ELECTROPLATING TITANIUM FROM WATER SOLUTIONS

Electrodeposition of titanium from water solutions takes place mostly through depolarization of the titanium ions by the formation of an alloy with the cathode metal.[321] X-ray examination shows that the deposit has a hexagonal lattice similar to metallurgical titanium. Electrolysis of aqueous solutions of complexes of the tetrachloride with pyridine or with phosphorus oxychloride gives deposits of titanium.[322] A 10 per cent solution of titanous sulfate in ammonium bisulfate at 150 to 200° C is electrolyzed to give titanium coatings resistant to acids.[323] Carbon and platinum anodes are most effective. Shiny, compact titanium is deposited at room temperature from an electrolyte containing 150 to 300 g of sodium hydroxide per liter and 20 g per liter titanium with 4 to 5 ml per liter of glycerol and 0.3 to 0.5 ml of glucose per liter.[324] Optimum current density with stainless steel anodes is 20 amperes per square decimeter. Titanium powder is deposited electrochemically from aqueous ammonium fluotitanate solutions.[325]

Electrolysis of an aqueous solution of a compound of titanium II or titanium III such as the dichloride, trifluoride, trihydroxide, or silicofluoride containing formaldehyde and pyrogallol gives metallic titanium.[326] The electrolysis of titanium sulfate solution solidified with sulfuric acid with a mercury anode gives titanium of 99.8 per cent purity at a current efficiency of 21 per cent.[327] A clean mercury surface is provided by removing the film of titanium continuously. The metal is deposited from an electrolyte consisting of an aqueous solution of titanium oxalate, malonate, or succinate.[328] A titanium plate anode and a copper plate cathode are employed at a potential of 10 volts and a current density of 20 amperes per square decimeter (cathode).

An ingot of impure titanium as the anode is refined electrolytically in a bath of aqueous ammonium hydroxide containing titanium tetrabromide and sodium bromide at 25° C.[329] A current density of 1 ampere per square decimeter of cathode surface is employed. The cathode is a plate of titanium metal. Titanium is plated on a cathode of iron, copper, or brass employing a platinum anode from a solution of 100 g titanium hydroxide, 40 g hydrogen chloride, and 100 g of ammonium chloride in water to a total volume of 1 liter.[330] A temperature of 30 to 50° C is employed and the pH is maintained at 4 to 5. The current density is 3 to 4 amperes per square decimeter. Thick deposits of titanium are electrodeposited from solutions of divalent and trivalent titanium ions.[331] To prepare the solution titanium hydroxide is dissolved in sulfuric acid and reduced electrolytically. Concentrated aqua ammonia is added to obtain a blue precipitate of the dihydroxide and trihydroxide which is dissolved in hydrochloric acid. This solution, after the addition of glue, is electrolyzed with a platinum anode and a lead or zinc cathode. Titanium is electroplated from a bath of titanium hydroxide 100 g, 50 Be′ hydrochloric acid 250 g, orthoboric acid 100 g, ammonium fluoride 50 g, and glue 2 g in water to make 1 liter of solution with a platinum or carbon anode.[332] A current density of 2 to 3 amperes per square decimeter at 2 to 3 volts at 20 to 50° C gives a uniform plated surface without precipitation of the hydroxide. Sato[333] deposited titanium on a copper cathode from an aqueous solution employing platinum as the anode. An electrolyte containing 70 g potassium hydroxide, 2.5 ml of gluconic acid, 3 g of mandelic acid, 10 g of titanium hydroxide, and water to 100 ml electrolyzed for 4 hours at a current density of 27 amperes per square decimeter gave 6.8 g of titanium. The coating was not thick enough to protect the copper.[334]

Parts to be electroplated with titanium are anodized in a nonaqueous bath of hydrogen fluoride and ethylene glycol, washed with acetone after a nickel strike is applied.[335] The current efficiency of the anodic oxidation of titanium in aqueous electrolytes is only 67 to 68 per cent due to oxygen evolution at the metal surface.[336] In nonaqueous solutions the value is larger than the theoretical. Electrolysis of a titanium tetrachloride-ammonia complex salt solution at 20° C with a graphite anode and a mercury cathode gave the dichloride and the trichloride along with ammonia complexes.[337] Copper plating of titanium proceeds in three stages: activation in a solution of hydrofluoric acid, copper strike from a cyanide plus Rochelle salt or pyrophosphate electrolyte, and copper plating either in conventional sulfuric acid or pyrophosphate electrolyte.[338]

## ELECTROPLATING FROM NONAQUEOUS SOLUTIONS

A liquid ammonia solution of titanium tetrachloride, sodium bromide, and zinc cyanide was electrolyzed with a graphite anode and a titanium cathode to produce a titanium alloy containing 36 per cent of zinc.[339] The alloy was decomposed by heating in vacuo at 800° C. Metallic titanium may be electroplated from solutions of the lower chlorides in liquid ammonia.[340] The deposit is often pyrophoric. Von Bichowsky[341] electrodeposited titanium from a solution of the tetrachloride and copper chloride in methyl alcohol. To preclude the formation of lower valent forms of titanium, $(NH_4)_2S_2O_8$ is added to the catholite.

Base metals are clad with a firm, adherent deposit of titanium in a fused salt bath having an anode of titanium carbide or a solid solution of titanium carbide and titanium monoxide [342] at a cell bath temperature of 900 to 1000° C and a cathode current density of 25 to 200 amperes per square decimeter. The bath contains at least 5 per cent of $K_2TiF_6$. Titanium dispersed in fused salts can be plated on metals, especially copper and iron.[343] For example, a copper sheet placed near a titanium sheet in a bath of fused sodium chloride or potassium chloride received a coating of titanium. The thickness of the layer increased with temperature and time to a maximum of 0.007 inch. This coherent coat protected the base metal from corrosion. The mechanism is considered to involve the formation of a titanium pyrosol first, then a deposition of these particles on the other metal surface, forming a titanium rich alloy. Base metals are similarly clad with a firm adherent deposit of titanium employing a cell bath temperature of 900 to 1000° C and a cathode current density of 25 to 200 amperes per square decimeter.[344] Electrolytic deposits of titanium from a metal fluoride bath up to 40 mils thick diffuse into different base metals to form an impermeable layer with high resistance to sea water and some chemicals.[345]

Decorative and nongalling coatings resistant to corrosion and abrasion are applied on titanium and its alloys by anodizing in acid or basic aqueous solutions of many compounds at 1300 to 2000 volts.[346] Suitable solutions are boric acid, citrates, sulfuric acid, or potassium hydroxide. Color may be produced by heating in argon containing a small proportion of nitrogen. Above 1625° F the titanium recrystallizes and decorative colored spangles are formed.

## CLADDING METALS WITH TITANIUM AT HIGH TEMPERATURES

Steel is coated with titanium by heating in a closed vessel with a mixture of granular titanium, magnesium chloride, and magnesium at a temperature above the melting point of the salt at a pressure below 100 microns.[347] Titanium is plated on a metal surface by the decomposition of a gaseous heat decomposable compound.[348] In an example, dicyclopentadienyltitanium was passed over an iron dish previously heated to 500 to 900° F in an atmosphere to deposit the coating. Steel panels were cladded with titanium by electrophoresis but only some of the titanium could be fused into the base metal by heating.[349] Mild steel plates are firmly bonded to titanium sheets by interposing a thin layer of chromium, cobalt, molybdenum, or silver.[350] The assembly is hot rolled with 50 per cent reduction in thickness. An explosive for bonding titanium to stainless steel is very fine pentaerythritol tetranitrate 20, red lead oxide 70, and butyl rubber resin binder 10 per cent.[351] The process is well adapted to tubes. A tube of titanium is placed inside the stainless steel tube and bonded by the explosion which is propagated lengthwise.

## SPECIAL METHODS OF PRODUCING TITANIUM METAL

The metal is formed by the disproportionation of titanium trichloride at high temperatures.[352] Large, pure, purple crystals of the trichloride are prepared by passing the tetrachloride over powdered titanium monoxide heated in a kiln to 1000 to 1500° C followed by cooling the reacted vapor in another tube below 500° C to effect condensation. This form of the trichloride is resistant to water vapor. The tetrachloride in the vapor is reduced to lower chlorides or to the metal in the form of aluminum alloys by passing the vapor over a moving layer of solid aluminum particles.[353] The reaction zone is above the melting point of aluminum, 800 to 1000° C. Titanium slag is treated with gaseous titanium tetrachloride at 700 to 750° C at atmospheric pressure to form the liquid dichloride which is removed from the slag.[354] It disproportionates to titanium and the tetrachloride in the presence of a titanium sheath, heated at 1100° C, immersed in the pool.

Powdered titanium metal is prepared by the dehydrogenation of the finely divided hydride in vacuo in thin layers at 600° C at a pressure of 0.001 torricelli.[355] Sulfides, halides, and oxides of titanium are dissolved in a flux or slag-forming agent, and firmly comminuted aluminum, magnesium, sodium, or calcium is blown into the liquid bath to effect reduction to the metal.[356] Titanium sulfide with a composition between $TiS_2$ and $Ti_2S_3$ is reduced with hydrogen, calcium, calcium hydride, aluminum, and magnesium.[357] However, only magnesium gives a titanium metal of high purity. Samples analyzed from 98 to 99.5 per cent pure. Von Zeppelin[358] reduced the tetrachloride with a binary or ternary alloy such as cadmium-magnesium-zinc.

Coherent titanium metal is produced at 550° C by allowing liquid tetrachloride or tetrabromide to react with liquid lithium on a bath of fused chlorides or bromides of potassium and lithium in a closed reactor.[359] After completion of the reaction the sponge is raked out of the bath and purified by vacuum distillation. Titanium is codeposited with copper or other metal from the dioxide in a fused salt electrolyte to increase the size of the crystals and to decrease the amount of salts carried by the deposited metal.[360] Titanium is recovered from the alloy. Silicon reduces the tetrafluoride at 500 to 800° F to form powdered titanium and silicon tetrafluoride.[361] First titanium dioxide is added to hydrofluoric acid at 300° F to form titanium tetrafluoride which is dried, purified by sublimation, and mixed with the silicon. Titanium chips are ground to powder, cleaned, and amalgamated with mercury, after which the mercury is distilled off leaving the pure metal.[362]

Surface hardening of titanium and titanium base alloys is effected by heating boric acid fused above 700° C to boric oxide without scaling.[363] By this treatment the tensile strength of the titanium is increased, but the ductility is decreased.[364] A higher surface hardness is achieved by spark treatment. Carbon is added by the spark treatment in a hydrocarbon liquid. If the sparking is performed in air, nitrogen and oxygen are added. Nitrogen only is supplied by sparking in the gas while oxygen is supplied by sparking in water. Both titanium and its alloys are surface hardened by heating in an argon atmosphere containing ethane or other hydrocarbon.[365]

## MELTING AND FORMING TITANIUM

The ability of molten titanium to dissolve the materials of the container poses the biggest problem in melting the metal. Contamination with carbon that results from induction melting is eliminated by using arc melting techniques in which the melt is made in a water cooled copper crucible with a consumable titanium electrode. Molten titanium does not wet the chilled copper, and the rapid removal of heat solidifies the molten metal at the walls. Other electrodes such as carbon and tungsten introduce some impurities.

Melting may be carried out under a vacuum or an inert atmosphere of argon or helium. Vacuum melting is preferred because this procedure eliminates more hydrogen from the metal. The sponge is heated by an electric arc established between the metal in the crucible and the electrode. It is consumed in the melting process and serves as the source of the titanium being added to the crucible.

Consumable electrode arc melting of titanium sponge in high vacuum is greatly affected by the extinction of the arc every half cycle.[366] This problem is overcome by incorporating into the electrode material a small amount of an element having a lower ionization potential than that of the titanium metal. All of the alkali metals except sodium are effective. Ingots are formed by melting sponge in a water cooled copper crucible with a tungsten tipped negative electrode.[367] The copper crucible and the titanium charge serve as the positive electrode. Up to 900 amperes are employed. Alternating current may be used with titanium electrodes.

The consumable electrode skull casting technique is reported to be the best method of producing titanium castings.[368] A water cooled crucible is charged with a previously produced skull and scrap metal and evacuated. After striking the arc the current is raised to 1000 amperes per inch of diameter of the crucible. The electrode is advanced to compensate for consumption. Consumable electrodes are made by joining the briquets with metallic titanium or titanium hydride cement.[369] Powders of these cements are placed between the briquets and heated at 800° C in vacuum to effect cementation.

In the arc melting in a vacuum or inert atmosphere, contamination with tungsten or carbon from the nonconsumable electrodes is avoided or minimized by employing electrodes of titanium carbide cemented with small proportions of carbon and iron.[370] A nonconsumable electrode of titanium oxide and nitride is used in the arc melting of sponge in an atmosphere of argon.[371] The oxide and nitride may be a coating on the titanium.

Large titanium castings are made by consumable electrode arc melting and casting in vacuum.[372] The melt is contained in a water-cooled copper crucible. An automatic, continuous-casting arc furnace employing a nonconsumable electrode and a direct current arc gives titanium of higher ductility, formability, and toughness than does the induction furnace.[373] A crucible lined with a refractory of thorium oxide 85, hafnium oxide 8, and ytterium oxide 2 per cent is suitable for melting and purifying titanium.[374] In processing, the lining is heated to 3270° F and evacuated. Scavengers are added to the titanium metal at 3100° F through a vacuum trap.

The use of forms of purest graphite powder with phenolic resin binder makes possible the production of titanium castings with slight impurities and good surface qualities.[375] Phenolic resin is superior to other binders, including sodium silicate and tar. Consolidated shapes are formed by compressing titanium sponge at 30 tons per square inch.[376] The sides of the shapes may be fused by shielded arc welding.[377] Pressure butt welding of titanium bar is effected at 500 to 900° C. The tensile properties of any weld are improved by heating for 30 minutes at 950° C.

Titanium is chemically milled at a controlled rate of 50 to 175 mils per hour, leaving a smooth surface by water containing hydrofluoric acid and chromic oxide or its alkali metal salt.[378] Parts or all of the surface of a piece of titanium may be machined electrochemically to fixed depth.[379] Local areas may be masked to prevent the etching. To prevent contamination of the metal by hydrogen, and resulting embrittlement, the bath contains an oxidizing agent such as chromic acid in addition to the halogen acid (hydrofluoric).

Alloying elements in titanium fall into two groups: those that strengthen and stabilize the alpha or room temperature form and those that strengthen the beta or high temperature form. Alloying agents of the first group are oxygen, nitrogen, carbon, and aluminum, while iron, manganese, chromium, molybdenum, and vanadium fall in the latter group. Alloying metals are used to raise the tensile strength and hardness of the titanium, but they also reduce the ductility. Consequently, the highest strength alloys are relatively brittle. Successful alloys have been developed with manganese, chromium, molybdenum, tin, aluminum, and iron. The alloys are manufactured commercially by adding the alloying metal to the titanium during melting of the sponge.

On September 14, 1948, E. I. du Pont de Nemours and Company announced the commercial production of titanium metal in sponge and ingot forms of 99.5 per cent purity.[380] Initial output was 100 pounds daily and the initial price was $5 a pound.

With government subsidy this plant was greatly expanded and many other companies entered the field. In general, the government agreed to buy the titanium metal produced at $5 a pound until the plant was paid for. After building up large stockpiles the government discontinued purchasing the metal, and industrial uses did not materialize as anticipated. Consequently production from the 1957 peak was cut back drastically. Many of the plants were closed. In 1964 the Titanium Metals Corporation of America, Henderson, Nevada, owned jointly by the National Lead Company and the Ludlum Steel Corporation, accounted for 50 per cent of the United States production. Reactive Metals, Inc., Miles, Ohio, owned jointly by the National Distillers and Chemical Corporation and the United States Steel Corporation, produced 25 per cent of the total. The remaining one fourth was divided between Crucible Steel Company, 10 per cent; Republic Steel Company, 8 per cent; Harvey Aluminum Company, 6 per cent and Oregon Metallurgical Corporation, 1 per cent.

The total United States production of titanium sponge in 1963 was 7865 tons. Sponge processed amounted to 8865 tons, 15 per cent of which was imported from Japan. Mill product shipments came to 6112 tons with a value of $72,000,000. The price of titanium sponge averaged $1.32 a pound in 1963 and that of the finished milled products $5.90 a pound.

Of the total consumption of titanium metal in the United States in 1964 missiles and space vehicles accounted for 21 per cent. Thirty-nine per cent went into frames, and 29 per cent went into jet engines of military aircraft. Civilian aircraft took 8 per cent of the total, mostly in the frames. Seven per cent found application in industrial uses, chiefly corrosion-resistant equipment for chemical plants, for example, pipes, heat exchangers, tank linings, pumps, and valves. Hydrospace accounted for 1 per cent.

## USES FOR TITANIUM METAL

Titanium has many actual uses and the potential uses are very great, based for the most part on its properties as a silvery white, light, corrosion-resistant, tough, strong metal. Among the metals available for construction there is a gap between aluminum and steel. Aluminum, with a desirably low density of 2.7, is easily formed and machined, but it has relatively low strength and is not resistant to corrosion. Iron, at the other extreme, can be alloyed to give high strength and resistance to corrosion, but its greater density, 7.87, is a decided disadvantage where weight is an important factor. Titanium, with a density of 4.5, coupled with its strength, ductility, high melting point, and noncorrosive characteristic, is the present outstanding candidate to fill this gap. It combines the properties of stainless steel with those of the strong aluminum alloys and possesses certain definite advantages over both. One outstanding advantage is its high proportional limit which is comparable to that of heat-treated steels and aluminum bronze, while its density is only a little over half of these materials. As a result, wrought titanium is in a class by itself so far as the weight of a section having a given proportional limit is concerned. Titanium is a preferred structural material in aircraft design where a minimum weight combined with a continued high stress is important. Its resistance to corrosion is an added advantage in airships for use over the sea or along the coast. A use of great importance is for making reciprocating mechanical parts in jet engines where heat and pressure are great. The metal seems almost ideal for ocean-going vessels because of its outstanding properties of lightness, strength, and great resistance to corrosion. As the cost of production is decreased, titanium will be used extensively for structural purposes. Titanium seems well suited for textile machinery where a considerable saving of power can be effected by using such a light, strong metal for high-speed spindles, spools, warp beams, and other moving parts. It does not stain the threads as do aluminum and magnesium alloys.

The surface-hardening property of titanium gives it a definite advantage over the really light metals in the construction of parts subject to frictional wear. It seems suited for automobile pistons, because, in addition to the characteristic properties of lightness and strength, it has a coefficient of expansion a little less than that of cast iron, which is ordinarily used for cylinders. The high heat conductivity suggests its use for handles for aluminum pans and cooking utensils. It has been proposed for many sports uses, such as tennis rackets and fishing rods, where its excellent physical and working properties would be utilized. Combination of stainlessness, high proportional limit, and low modulus makes it an ideal material for springs, and its use should make possible the construction of greatly improved spring balances and watch springs. Its properties also recommend it for use in tool mountings where a certain amount of give is desirable to prevent breakage, and in making pen points and styluses. Rubbing titanium metal against a hard surface often produces a smear which is difficult to remove, and this characteristic is employed for a variety of purposes, including the production of very stable high electrical resistance glass, simply by marking the surface with a titanium point.[381] Such smears can be used to coat materials with a metallic film and to etch glass without the use of hydrofluoric acid. Herenguel[382] investigated the use of titanium powder as a paint pigment.

The really large-scale use of titanium seems to depend only on its availability in suitable form at a price in line with the common metals such as iron and aluminum.

Recently Sikorsky replaced steel with titanium in the rotor hubs of Marine Corps helicopters and added 1000 pounds to the lifting power. The two-thousand-miles-an-hour A 11 jet plane, announced early in 1964, is constructed largely of titanium. To the supersonic planes titanium contributes resistance to heat as well as light weight. North American's XB-70 A uses 12,000 pounds of titanium or about 9.5 per cent of the airframe weight. Of this amount 6000 pounds is in the sheet form, 3000 pounds in plate, and 3000 pounds in extrusions, bars, and forgings. The Corvair TXF is expected to use 4000 pounds of titanium.

On a strength-to-weight basis titanium is four times as expensive as stainless steel, but the price is expected to come down 25 per cent within a few years. The airframe industry and its suppliers have learned to form titanium parts, large and small, from sheet metal. Generally these parts, referred to as details, are hot formed under pressure. At the temperature employed the metal becomes malleable and ductile so that it can take complex shapes.

Titanium can be machined with the same equipment and in the same way as stainless steel. However, fusion welding requires an inert atmosphere of helium or argon to prevent contamination by oxygen, nitrogen, and hydrogen, with consequent brittleness at the weld.

Cost of fabrication of titanium parts for airplanes is somewhat higher than for aluminum alloys or stainless steel, but the big obstacle is the cost of the basic metal. Titanium costs $10 a pound for airplane components, steel $1.80, and aluminum $1.10.

# 12

# Leaching, Roasting, and Smelting of Ores

**RESIDUES**

The first processed titanium pigments were obtained as residues from ilmenite after leaching out a large part of the iron and other soluble constituents. Naturally, silica and other insoluble components, as well as some of the soluble compounds, remained with the titanium dioxide and exerted important and unfavorable influences on the properties of the final product, particularly on the color.

Farup [1] subjected ilmenite to an oxidizing roast at 500° C to obtain an unsintered reddish-yellow product which disintegrated on addition to water and formed a fine suspension. Small quantities of electrolytes (sodium chloride) were then added to effect rapid coagulation and settling of the exceedingly fine-grained powder, which was then filtered and dried. The product was applicable for paint pigments and polishing powder.

In the next step forward, ilmenite was fused with an alkali metal compound; the melt so obtained was subjected to an oxidizing roasting process and heated with sulfuric acid to dissolve the soluble salts.[2] Sulfur dioxide was introduced to reduce the iron to the ferrous condition, and the suspension was then filtered and washed. The insoluble residue of crude titanic oxide was purified by repeated treatment with sulfuric acid, and the ferrous sulfate liquor was evaporated for recovery of sulfuric acid. Barton [3] heated ilmenite with sodium sulfide to form ferrous sulfide, and electrolyzed the product in aqueous sodium chloride so that the chloride liberated effected decomposition and solution of the sulfides. The residue of relatively pure titanium dioxide was washed and dried for use. In a similar manner the ore was heated with sodium sulfide or a mixture of sodium sulfate and coal below the fusion temperature (625° to 825° C) to form sodium ferrosulfide and sodium titanate.[4] The fritted product was then treated with dilute acid of such strength as to dissolve the iron compounds but not the titanium, and the residue was washed and calcined. Instead of employing dilute hydrochloric acid, the iron could be extracted by passing sulfur dioxide into a water suspension of the fritted product.[5] By holding the melt of ferrosodium sulfide and sodium titanate in the fused state, a gravity separation into a titanium-rich upper layer and an iron-rich lower layer took place. After cooling, the titanate portion was separated mechanically and treated with dilute acid to remove the iron and other impurities, leaving relatively pure titanic oxide.[6] A residue of 90 to 99 per cent purity was obtained from the product of fusing ilmenite with such alkalies as sodium hydroxide or carbonate.[7] The melt was lixiviated to remove the excess alkali, and the residue was digested with hydrochloric acid of such concentration as to combine only with the iron and any remaining alkali. At this stage cold water was added to promote settling of the titanium oxide and to hold other compounds in solution.

Ilmenite was fused in a reducing medium with an alkali in such proportions as to form metatitanates.[8] The reduced iron was separated, and the slag was treated with an acid to dissolve the alkali oxide. A residue of titanium dioxide was obtained. Ilmenite suspended in a fused mixture of $NaCl$-$FeCl_3$ reacted above 550° C with a preferential attack on the iron content.[9] The beneficiated titanium dioxide was recovered in the same particle size as that of the original ore. A mineral containing titanium, aluminum, and iron was ground to 100 mesh and converted to soluble compounds by fusing with ammonium sulfate under nonoxidizing conditions.[10] The solution was boiled to precipitate the titanium as hydrous oxide. Ammonia was recovered and reused. An artificial rutile [11] was obtained by heating titaniferous iron ores with coke in a proportion insufficient to reduce the iron completely, to form a fused mass which separated into a lower layer of molten iron and an upper slag layer consisting primarily of titanium dioxide. A mixture of titanite or sphene, with sodium chloride, was subjected at 900° to 1000° C to superheated steam.[12] The melt was treated with a solution of hydrochloric acid obtained in the first stage, and the solid residue was used as a pigment or was converted to pure titanium dioxide in the conventional manner.

According to von Bichowsky,[13] titanium dioxide free from iron may be prepared by subjecting ilmenite to a reducing roast, leaching the impurities as far as possible with dilute mineral acid, and heating the dried suboxide with sulfur or sodium sulfide under nonoxidizing conditions to convert the last trace of iron to the sulfide, in which form it may be extracted readily with hydrochloric acid. The purified residue was then heated in air to convert it to titanic oxide of pigment grade.

Jebsen [14] crushed the melt obtained on fusing titaniferous ores with alkaline agents and subjected it to an oxidizing roasting process. This product was heated with sulfuric acid, and sulfur dioxide was introduced into the solution to reduce the iron to the ferrous condition. The resulting ferrous sulfate liquor containing a small amount of titanium sulfate was employed in a subsequent roast. By repeated acid treatments the ferrous sulfate was extracted, leaving pure titanic acid. According to a later development, solution of the reaction product obtained by heating ilmenite or other titaniferous ore with sulfuric acid was carried out in the presence of a sulfurous reducing agent such as pyrite, which was added before or during the reaction.[15]

In a method developed by Booge [16] for producing a practically iron-free titanium dioxide concentrate, ilmenite or similar ore was heated in air at 500° to 1000° C and the resulting oxidized product was subjected to the action of gaseous hydrochloric acid at 600° to 800° C to convert the iron to ferric chloride. This compound, which was volatile, distilled off at the temperature employed. Crude hydrous titanium oxide containing iron and other foreign substances was first treated with concentrated sulfuric acid or a bisulfate at an elevated temperature, and the reacted mass was disintegrated and ground with sodium chloride.[17] This mixture was heated at 300° to 500° C with periodic stirring to produce a porous sintered mass which was cooled and leached according to the counter-current principle. The residue of purified titanic oxide was then filtered, washed, and dried. Any titanium salts in the filtrate were precipitated hydrolytically by boiling, and then added to the main product.

By carbon reduction of Wyoming titaniferous iron ores, followed by acid leaching, residues containing 2.1 to 2.8 times the titanium dioxide content of the original ore were obtained.[18] Reduction in the presence of sodium carbonate, followed by wet magnetic separation, gave a magnetic fraction containing 90 to 95 per cent iron and a nonmagnetic fraction containing 40 to 70 per cent titanium dioxide. Ravenstad and Moklebust [19] heated ilmenite or titanomagnetite with a reducing gas at 700° to 900° C, or with carbon at 1000° C, in a rotary furnace to obtain metallic iron in a finely divided form which was reoxidized by air and steam in the presence of carbon dioxide at 50° to 100° C. The iron oxides were removed by magnetic or electroflotation methods, leaving a concentrate containing 68 per cent titanium dioxide. Brassert [20] produced titanium-free steel and a highly concentrated titanium slag from ores containing ilmenite. The magnetic iron oxide was selectively reduced to sponge iron, and the product was heated in a melting furnace with lime to form a slag with the titanium oxide and gangue of the ore. Carbon was then added to the bath to reduce sufficient iron oxide of the ilmenite in order to effect a concentration of titanium in the slag.

Enriched iron ores containing titanium, vanadium, and chromium were heated with charcoal at 500° to 800° C in the presence of a sodium compound to reduce the magnetite to sponge iron and convert the chromium and vanadium compounds to the corresponding sodium chromate and vanadate.[21] Ilmenite remained unchanged. The vanadates and chromates were dissolved and the iron was removed by magnetic separation. Dunn[22] extracted phosphorus and other impurities from ilmenite of the Virginia and Norway type, to be used in the manufacture of pigment, by heating the unground ore with 5 to 10 per cent sulfuric acid at 82° to 93° C. Such treatment did not dissolve the titanium content of the ore. According to an interesting procedure, a mixture of ilmenite and pyrite was subjected to the action of moisture and air to effect a preliminary oxidation of the sulfides, and the product was heated with added sulfuric or hydrochloric acid to dissolve selectively the iron content.[23] Sulfur dioxide was employed by Stahl[24] to extract titanium from ores and slimes.

Iron and other impurities were separated from titaniferous material by electrolyzing an alkaline solution or suspension (for example, a suspension of ilmenite) at 90° to 120° C under a pressure of hydrogen or air of 20 atmospheres, employing an insoluble anode of carbon, nickel, or chromium steel.[25] The iron was deposited in a metallic state.

In a modified process, the ilmenite was reduced with coal at 1000° to 1100° C, and the sponge iron was leached with ferric chloride solution.[26] Mixtures of 2 parts concentrate (44.7 per cent titanium dioxide and 37.7 per cent iron), and 1 part brown coal at temperatures above 1000° C gave best results. By treating the reduced product at room temperature with 15 parts of solution containing 20 per cent ferric chloride and 3 per cent hydrochloric acid, 98 per cent of the iron was removed. Moore[27] heated finely crushed hematitic ilmenite with 34 per cent pulverized charcoal in a covered crucible for 7 hours at 900° to 1000° C to produce a product containing 24.6 per cent titanium dioxide and 55.2 per cent iron of which 50.4 per cent was in the metallic state and 11.1 per cent carbon. Magnetic concentration was unsatisfactory, so leaching with ferric chloride was employed. Carbon was removed by ignition. A sponge product made by reducing ilmenite ore with charcoal gave a yield of metallic iron of 79.5 per cent, and leaching with ferric chloride solution extracted 96 per cent of the iron, leaving a concentrate containing 53.6 per cent titanium dioxide.[28] Small quantities of copper, nickel, lead, and cobalt were separated in concentrated form. Electrodeposition of the iron gave a product 99.96 per cent pure with a current efficiency of 94.3 per cent.

Titanium dioxide may be separated from ores (ilmenite) by concomitant reduction and oxidizing volatilization.[29] An air blast heated at 750° to 850° C was passed over thin layers of a mixture of powdered ore and reducing fuel which were moved forward by air jets. Bancroft [30] subjected titaniferous iron ores to a reducing roast in the presence of fluorspar to produce metallic iron without reducing the titanium dioxide.

Tikkanen [30a] extracted 98 per cent of the iron from 40 mesh ilmenite after roasting at 800 to 850° C with 10 to 20 per cent sulfuric acid at 100 to 120° C in an autoclave. A titanium dioxide residue is recovered from ilmenite by fusion with sodium hydroxide followed by leaching with water.[31] The sodium hydroxide may be recovered from the leach liquor for reuse. In an example, 90 g of sodium hydroxide pellets and 23 g of powdered ilmenite containing 59 per cent titanium dioxide and 35 per cent ferric oxide were heated at 750° C in a nickel crucible. As soon as gases ceased to come off, the furnace was cooled to 680° C, and the undissolved solids were allowed to settle. The liquid mixture separated from the undissolved ferric oxide contained 8.5 g titanium dioxide, 0.87 g ferric oxide, and 71.7 g of sodium hydroxide per liter. This product was leached with four portions of boiling water to yield a residue containing 8.12 g titanium dioxide, 0.86 g of ferric oxide, and 2.6 g of sodium hydroxide. Granular ilmenite is converted to rutile by heating at 650 to 750° C in an atmosphere of carbon monoxide or sulfur for 1 to 4 hours, followed by leaching the product with dilute sulfuric acid.[32] If a considerable proportion of sulfur is added to the ore, the mixture is heated in a pressure vessel with water and compressed air or oxygen at 120° C and a pressure of 60 psi. The residue is washed, dried, and treated by conventional ore dressing methods to remove contaminating minerals as chromite and quartz. Both leaching methods may be used in series, the pressure digestion to remove iron sulfides, and the acid to remove iron oxides and hydrates. Magnetic separation to remove chromite leaves rutile of 88 per cent purity. Silica and zirconium sulfate are the chief impurities.

Most of the iron is removed from ilmenite as sulfide by a reduction roast with sodium sulfate followed by leaching with dilute sulfuric acid.[33] In an example, 100 parts of ilmenite containing 53.1 per cent titanium dioxide, 35.1 per cent ferrous oxide, 1.28 per cent silica, 0.58 per cent alumina, 0.67 per cent calcium oxide and magnesium oxide, and 0.20 per cent vanadium pentoxide was mixed with 107 parts of sodium sulfate and 24 parts of anthracite (81.9 per cent fixed carbon and 10.8 per cent ash) and heated at 1100° C in a rotary furnace. The sintered product was leached with 10 times its volume of water and 15 times its volume of 4 to 5 per cent sulfuric acid to give a residue containing 71.5 per cent titanium dioxide, 7.0 per cent titanium trioxide, 3.9 per cent ferrous oxide, 0.05 per cent vanadium pentoxide, 10.4 per cent sodium oxide, and 1.0 per cent silica. One kilogram of the concentrate blended with 0.4 kg of anthracite was chlorinated at 650° C to produce titanium tetrachloride.

A mixture of one part ilmenite (51.4 per cent titanium dioxide and 41.0 per cent ferrous oxide), 2 parts of sodium sulfate, and 0.6 part of coke was fused at 950 to 1000° C, cooled, crushed to 4 mm particle size, leached with eight times the weight of water at 80° C for 15 minutes, and vacuum filtered.[34] The filtrate contained 22.1 gpl ferrous sulfide, 32.4 gpl sodium oxide, 15.2 gpl sodium sulfide, and 1.5 gpl sodium sulfate. Sulfuric acid, 70 to 80 gpl, was added to the filter cake in a porcelain reactor to a pH of 3. The acid suspension was filtered, repulped with 8 per cent sulfuric acid, boiled for 1 hour, and again filtered. This residue was dried and heated to 600 to 700° C to obtain a final product containing 97.5 per cent titanium dioxide at a recovery of 92 per cent.

A mixture of 300 parts of ilmenite, containing 53.1 per cent titanium dioxide, 35.1 per cent ferrous oxide, 1.28 per cent silica, and 0.58 per cent alumina with 321 parts of sodium sulfate and 72 parts of coke was roasted at 1050° C for 3 hours.[35] This reaction product was leached with water to recover sodium sulfide and sodium hydroxide and dried to give a material composed of 51.5 per cent titanium dioxide and 27.1 per cent sodium oxide. The dried titanate was decomposed by reaction under pressure with 10 per cent sulfuric acid for 30 minutes at 180° C. After washing and drying, the residue contained 92.6 per cent titanium dioxide.

Titanium dioxide concentrates are produced by heating ilmenite with a reducing agent to yield metallic iron under conditions of temperature which prevent slag formation.[36] The addition of boric oxide, calcium fluoride, sodium hydroxide, sodium chloride, or a mixture of boric oxide and a phosphate to the ore prior to reduction facilitates the electrochemical separation of the metallic iron. Dilute hydrochloric acid, aqueous ferric chloride, or dilute sulfuric acid, returned as mother liquor from the hydrolysis step, can be used to dissolve the iron at temperatures above 70° C. In a step process the ferric oxide of ilmenite is reduced first with iron powder to ferrous oxide; the ferrous oxide is reduced with coke to metallic iron; and the metallic iron combines with excess coke to form iron carbide, $Fe_3C$.[37] Reduction of ferrous oxide takes place at 900 to 1000° C; metallic iron is liberated at 100 to 1150° C; and the carbide is formed between 1150 and 1200° C. In an example, a 100 kg batch of Malayan ilmenite ground to 250 mesh was fused with 5.4 kg of iron powder, 9.8 kg of carbon, and 25 kg of sodium carbonate. The well-mixed powder batch was fed to a rotary furnace and heated for 30 minutes at 1100° C. At this time the temperature was raised to 1250° C and held at this value for 30 minutes to yield a completely fluid mass. The slag was poured into water to obtain a friable mass containing 73 per cent titanium dioxide and 0.8 per cent ferric oxide. This product was washed with hot 10 per cent sulfuric acid to obtain a material containing 95.4 per cent titanium dioxide with silica as the principal impurity. From this product pigment-grade titanium dioxide was produced. A titanium ore is sintered with 20 to 40 per cent of sodium salts and reduced with coke at 1200 to 1400° C to form pig iron and a titanium slag.[38] After cooling, the slag is crushed and extracted with dilute acid to remove impurities. In an example, 1000 parts of ore containing 40.44 per cent titanium dioxide, 31.34 per cent ferrous oxide, 21.07 per cent ferric oxide, and 1.73 per cent silica was sintered with 132 parts of sodium carbonate and reduced at 1350° C with charcoal. The molten pig iron was separated, and the slag was heated at 1200° C for 1 hour. By this treatment a product containing 68.5 per cent titanium dioxide, 1.81 per cent ferrous oxide, 0.36 per cent ferric oxide, 5.72 per cent silica, 2.13 per cent alumina, 6.01 per cent magnesium oxide, 1.11 per cent manganese oxide, and 13.48 per cent sodium oxide was obtained. An additional step of extracting the crushed slag with dilute sulfuric acid gave raw titanium white containing 86.07 per cent titanium dioxide, 1.63 per cent ferrous oxide, and 0.85 per cent ferric oxide. A mixture of 150 parts powdered ilmenite containing 53.1 per cent titanium dioxide with 128 parts of gypsum and 38 parts of coke breeze was heated for 4 hours at 1200° C, leached with 10 per cent sulfuric acid, and washed with water to obtain 93 parts of a concentrate containing 81.5 per cent titanium dioxide, 11.2 per cent calcium oxide, 1.5 per cent ferrous oxide, 1.3 per cent silica, and 0.4 per cent vanadium.[39] One kilogram of this product was mixed with 300 g of coke breeze and chlorinated at 700° C with 3.5 kg of chlorine to form 2.2 kg of crude titanium tetrachloride. After distillation the content of impurities was reduced to 0.001 per cent silica, 0.005 per cent vanadium, and only traces of iron.

Ilmenite is beneficiated to produce a heavy-grade rutile for use in porcelain enamel.[40] The crushed ilmenite is blended with calcium carbonate or sodium carbonate and carbon and sintered at 500 to 1000° C for several hours. In the sintering process the titanium is converted to titanates and the chromium is converted to chromate. The sinter is crushed, screened, and roasted at 600° C to remove the excess carbon. After roasting, it is broken up and leached with a dilute sodium peroxide solution to remove the soluble chromate compounds, after which it is digested in hydrochloric acid at temperatures slightly below the boiling point for 1 to 2 days to dissolve the iron compounds as chlorides and to decompose the titanates to form hydrated titanium dioxide. The fraction through 325 mesh is washed free of iron compounds and calcined to produce a heavy-grade rutile suitable for the manufacture of ceramic enamels. In an example, 100 g of ilmenite ground to pass 325 mesh was intimately mixed with 40 g of calcium carbonate and 8 g of carbon black and heated in a covered carbon crucible for 15 hours at 900° C. The resulting sinter was broken up and leached in water at 85 to 90° C for 4 hours. After this treatment the excess carbon was removed by decantation. The dewatered cake was dried and digested in concentrated hydrochloric acid at 90° C for 1 day. This digestion product was filtered and wet screened to separate the fine hydrated rutile from the iron chloride in the digestion liquor and from the coarser silica and sulfur impurities. After washing with hydrochloric acid solutions, the hydrated rutile was calcined for 1 hour at 800° C to obtain a product containing less than 0.0015 per cent chromium expressed as chromic oxide.

The iron oxides of ilmenite are reduced by subjecting the finely divided ore to the action of coke oven gas, hydrogen, carbon monoxide, or a mixture of these at 1800 to 2200° F in a fluidized bed.[41] After cooling, the product is leached with 20 to 40 per cent nitric acid to dissolve the metallic iron, filtered, and washed. The leach solution is hydrolyzed at 400 to 500° F under pressure of air or oxygen to regenerate the nitric acid which is recycled, and to precipitate ferric oxide. After separation and drying, the ferric oxide is reduced by hydrogen in a fluidized bed to iron powder. For example, ilmenite ground to 200 mesh was reduced at 2000° F by hydrogen in 1 hour. Then 400 g of the product was leached for 30 minutes with 2 liters of a solution containing 300 gpl nitric acid in a closed vessel containing oxygen at 100 psi. The leach residue contained 70.5 per cent titanium dioxide and only 10 per cent of the original iron. The solution containing 79 gpl iron was heated for 30 minutes in an autoclave at 450° F and 50 psi oxygen pressure to precipitate 89 per cent of the iron by hydrolysis as ferric oxide. After filtering and washing, the ferric oxide was reduced by hydrogen at 1500° F to pure iron powder. The filtrate contained 291 gpl nitrate ion.

Ilmenite of different types is reduced with hydrogen or coke to metallic iron at less than slagging temperature.[42] The products are dependent on the type of ilmenite, temperature, time, and the reducing agent used. Hydrogen is the most effective reducing agent at lower temperatures.

Sodium titanate $Na_2TiO_3$ is decomposed by reaction with carbon dioxide at 650 to 800° C to form sodium carbonate and titanium dioxide.[43] The products are treated with water to dissolve the sodium carbonate, and the suspension is filtered to recover the titanium dioxide. Sodium titanate is recovered from rutile ore by sintering with sodium carbonate or potassium carbonate in the presence of a small proportion of an inorganic halide for removal of the chromium.[44] A mixture of 100 parts of ore of such fineness that 90 per cent passes a 325 mesh screen, 75 parts of finely divided sodium carbonate or potassium carbonate, and 3 to 5 parts of sodium fluoride, aluminum fluoride, sodium chloride, cryolite, or sodium nitrate is sintered for 120 to 150 minutes at 1600° F, washed with water counter-currently, and dried. The product is suitable as a component of porcelain enamel frit.

A mixture of ilmenite concentrate sodium sulfate, and carbon in the ratio of 1 to 2 to 0.6 is heated at 1000° C to produce a melt of iron sulfide and sodium titanate.[45] The melt is digested with water and acidified with sulfuric acid to obtain a product containing 98.8 per cent titanium dioxide. This method consumes much less sulfuric acid than does the typical direct digestion of ilmenite with sulfuric acid. Ilmenite and sodium sulfate in the ratio of 1 to 2 are heated with coke at 1200 to 1250° C for 5 hours to obtain a product from which practically all of the titanium dioxide, almost iron free, can be obtained by extraction with dilute sulfuric acid.[46]

A titanium dioxide concentrate is obtained from red mud from bauxite by reduction followed by leaching.[47] The red mud is reduced by hydrogen or other reducing gas at 700 to 900° C and cooled in the absence of air. This reduced product is leached with 10 per cent sulfuric acid to dissolve the iron. The hydrogen liberated in the reaction is used to reduce more red mud residue. From an original mud containing 11.4 per cent titanium dioxide, nearly all of the titanium is recovered in the form of a concentrate containing 64.8 per cent titanium dioxide.

A titanium dioxide concentrate suitable for chlorination is obtained in a single stage process by heating ilmenite, coke, and sulfuric acid in an autoclave at 100 to 200 psi.[48] The amount of coke is from 20 to 35 per cent of the titanium dioxide in the ore. From 25 to 100 per cent of 15 to 50 per cent sulfuric acid in excess of that required to react with the iron, calcium, and magnesium in the ore is employed. Before treatment the ore is ground to 200 mesh. The filtered, washed, and dried product contains 50 to 70 per cent titanium dioxide, 0.5 to 10.0 per cent iron, 25 to 40 per cent carbon, and no more than 2 per cent total sulfate. This product may be readily chlorinated at 400 to 700° C. In an example, to 1000 parts of ilmenite through 200 mesh particle size containing 59.5 per cent titanium dioxide, 9.5 per cent iron, and 25.2 per cent ferric oxide was added 1100 parts of 50 per cent sulfuric acid and 210 parts of pulverized coke. The mixture was heated in an autoclave at 500 psi for 30 minutes, after which the product was removed, filtered, washed, and dried to obtain a concentrate containing 69.1 per cent titanium dioxide, 1.2 per cent iron, 0.1 per cent calcium oxide, and 0.01 per cent magnesium oxide. The iron removed from the ore was 96.2 per cent, and the loss of titanium was 2.0 per cent. This concentrate was chlorinated to give water white titanium tetrachloride.

Extraction of ilmenite with 20 per cent hydrochloric acid at 270° C leaves a residue containing 97 per cent titanium dioxide and 0.04 per cent chromium suitable for pigment manufacture.[49] The original ilmenite contains 2.0 per cent chromic oxide.

A practically iron-free titanium dioxide concentrate is prepared in a single-stage operation by reacting under pressure ilmenite, a carbonaceous reducing agent, sulfuric acid, and a small amount of an organic flotation agent.[50] Pressures of 500 to 2000 psi are employed. The final product has a particle size of 0.2 to 1 micron. Coke equal to from 2 to 4 per cent of the ilmenite reduces the iron III to iron II. From 0.01 to 0.1 per cent by weight of the flotation agent (oleic acid) aids in dissolving the iron without dissolving the titanium.

For example, 1000 parts of minus 200 mesh ilmenite concentrate (46 per cent titanium dioxide, 40 per cent ferrous oxide, and 4.5 per cent ferric oxide), 0.5 part of oleic acid, 21,160 parts of 50 per cent sulfuric acid, and 20 parts of carbon were first mixed to form a slurry. The slurry was heated in an autoclave at 500 psi for 30 minutes. During the reaction the slurry was agitated to improve the contact. The reacted slurry was removed from the autoclave, filtered, and washed to remove the iron and other impurities. After drying, the concentrate contained 92.7 per cent titanium dioxide and 0.4 per cent iron. Ninety-nine per cent of the iron was removed with a loss of only 1 per cent of the titanium. The processed product possessed the properties of white titanium dioxide pigment.

Leaching with sodium hydroxide solution at a pressure of 9 to 11 atmospheres and 190° C increased the titanium dioxide content of an ore from 50 to 72 per cent by decreasing the silica from 28 to 4 per cent.[51]

## FUSED PRODUCTS SOLUBLE IN ACIDS

The initial products obtained as residues from ilmenite were fairly pure titanium dioxide with relatively high tinting strengths, but were decidedly yellow, were expensive to manufacture, and fell far short of the original conception of a high-grade white pigment.

The next improvements were processes in which the titanium content of the raw material (ilmenite) was converted to soluble form, dissolved, and reprecipitated from the dilute solution after removal of suspended solids. This allowed a more complete separation of iron, silica, and other undesirables, and thus gave titanium dioxide of higher purity and improved pigment properties, particularly color. The available ores of titanium are highly refractory. In the pioneer period, fusion with sodium hydroxide, sodium hydrogen sulfate, potassium carbonate, sodium sulfide, and similar reagents was resorted to, and the resulting melts were dissolved in dilute acids to obtain the desired solution.

In an early process, ores of titanium were fused with alkali metal sulfides[52] and the melt was boiled with dilute sulfuric acid to dissolve only the iron. The residue was washed and heated with stronger sulfuric acid to form titanium sulfate, which was dissolved in water. Any iron present was reduced to the ferrous condition, and the solution was clarified and boiled to precipitate the titanium as hydrous oxide. To break down rutile or other not easily soluble titanium ore, Ryan and Knoff [53] fused the finely divided material with an alkali metal compound such as potassium carbonate or sodium acid sulfate in proportions sufficient to convert the titanium to acid-soluble form, i.e., about 5 parts of titanium oxide to 1 of sodium oxide. The crude product was digested with 93 per cent sulfuric acid at 200° to 220° C until the titanium was converted into soluble sulfates. Rossi [54] fused rutile and ilmenite with sodium acid sulfate and dissolved the melt in boiling water. By a similar method [55] the ore was heated with a mixture of sodium bisulfate and sulfuric acid below the fusion point (120° to 300° C) until a solid cake was obtained. The cooled mass was crushed and dissolved in water. Iron may be removed in a preliminary stage by heating the ore with carbon at 900° to 1050° C and leaching the spongy metal with ferric chloride.

Barton and Kinzie [56] heated an intimate mixture of ore, alkali metal sulfate, and sulfuric acid at 200° to 350° C (below the fusing point) to form water-soluble sulfates. Readily soluble titanium compounds [57] were prepared by heating ilmenite with sulfuric acid in a solution of potassium sulfate. According to another modification, finely ground rutile was heated with sodium hydroxide at 280° to 650° C without fusion, and the product was leached with water and dilute sulfuric acid. The residue of crude titanium dioxide was redissolved in strong sulfuric acid and reprecipitated to improve its purity.[58] A product consisting of calcium titanate and magnetite was obtained by roasting titaniferous iron ore [59] with calcium oxide in equimolecular amounts, and the iron was removed by magnetic separation. By an alternate method, the iron oxides were reduced to the metal and separated magnetically as before. Lubowsky [60] calcined finely ground rutile with magnesium oxide at a temperature of 1400° to 1500° C to form magnesium titanate. This was cooled, ground, and heated at 90° C with concentrated sulfuric acid to form soluble sulfates. The slurry was diluted with water and cooled to 0° C to effect separation of magnesium sulfate by crystallization.

On roasting finely ground ilmenite with a mixture of calcium oxide and sodium carbonate at 900° C, a mixture of calcium titanate and sodium ferrate was formed.[61] The soda was extracted from the product by leaching with water, and over 95 per cent of the iron dissolved in 10 per cent sulfuric acid. The residue of calcium titanate was heated at 165° C with concentrated sulfuric acid to produce calcium sulfate and soluble titanium sulfate. On fusing crude titaniferous material containing alumina and silica with lead monoxide, a mixture of titanates and silicates resulted from which the aluminum oxide crystallized on cooling.[62]

In a combined fusion and reduction process,[63] a mixture of ilmenite, sodium carbonate, and anthracite coal dust was heated at 800° C in a roasting furnace. The product was discharged in water, and the solution was carbonated to effect separation of the sodium carbonate. After extracting the iron with dilute sulfuric acid, the residue was filtered, dried, and heated at 150° to 180° C with strong sulfuric acid to convert the titanium to soluble salts. The temperature was then raised to 200° to 225° C, and the sulfur dioxide liberated by the reaction between sulfuric acid and carbon reduced the titanium to the trivalent state. In a similar process developed by the same workers, the ore was heated with sodium carbonate, a reducing agent (coke), and a flux at 1000° C. The product was then washed with water and treated with sulfuric acid to form titanium sulfate.[64] Raffin[65] heated ilmenite with sodium carbonate and coal to reduce the iron to the metal, and, after leaching it with dilute sulfuric acid, treated the residue with stronger acid to form soluble titanium sulfate. In the presence of an alkali chloride, total reduction of the iron oxides was effected at 700° to 800° C.[66] By fusing titanium ore with barium carbonate and coal, a slag composed essentially of barium and titanium oxides and metallic iron was formed. After removing the iron by magnetic separation, the product was mixed with sulfuric acid and heated to form titanium sulfate and barium sulfate.[67] Similarly, ilmenite or titaniferous iron ore was heated with barium sulfate and coke (a small amount of sodium sulfate could be included in the charge), and the product was leached with water to dissolve barium hydroxide and barium sulfide. The residue was then separated and treated with dilute acid to dissolve the iron sulfide, and the crude titanium dioxide remaining was treated with concentrated sulfuric acid to convert it to soluble form.[68]

Buckman[69] heated a mixture of ilmenite, sulfur, barium sulfate, and coke in a blast furnace and dissolved the melt in strong hydrochloric acid. A product composed primarily of sodium titanate and metallic iron, obtained by smelting titaniferous iron ore with coke and sodium hydroxide, was lixiviated with hot water and treated with 50 to 60 per cent sulfuric acid to convert the titanium content to soluble form.[70] A similar method applicable to sphene has been described by Tatarskii.[71] One part of the finely ground concentrate was mixed with 4 parts of sodium sulfite and 0.8 part powdered coke and heated for 1 hour at 850° C. The cooled melt was digested with water, and the residue was washed with sulfurous acid solution and calcined at 800° to 850° C. This crude titanium dioxide product was fused with potassium acid sulfate (1 to 12) at 430° C for 5 hours and poured into water to obtain a solution of titanium sulfate. Yamamoto[72] fused a mixture of titanium ore, an acid sulfate, and carbon, and extracted the melt with water. By treating the product with dilute mineral acid and sulfur dioxide, a titanate was obtained as residue.

By another approach Bachman [73] heated ground ilmenite with an excess of coking coal in a retort to obtain a product consisting primarily of metallic iron, titanium oxide, and carbon. After leaching the iron with dilute sulfuric acid, the residue was treated with concentrated sulfuric acid to yield a solution of titanic and titanous sulfates. In a similar process [74] ilmenite or titaniferous iron ore was subjected to a reducing roast at 900° C. The product was lixiviated with dilute acid to remove the sponge iron, and the titaniferous residue was then heated with concentrated sulfuric acid and sodium bisulfate. Whittemore [75] heated titaniferous iron ore at 950° C with 25 to 33 per cent carbon and dissolved the resulting sponge iron in dilute acid. The residue, as before, was separated and heated with concentrated sulfuric acid to form soluble titanium sulfate. Magnetic concentrates from titaniferous iron ores were heated with coke, oil, or tar at 900° to 1100° C to reduce the iron component to the metallic state, and this was then dissolved from the friable product with dilute hydrochloric or sulfuric acid.[76] The titanium residue was dried for use or dissolved in strong sulfuric acid. Farup [77] heated ilmenite with a carbonaceous reducing agent at a temperature high enough to reduce the iron component to the metallic state but below that required to reduce the titanium compounds, and heated the product with concentrated sulfuric acid of such concentration as to dissolve the titanium compounds without attacking the iron. According to Fitzgerald and Bennie,[78] the iron oxides of ilmenite were completely reduced to metallic iron by heating the ore at 1000° C with 7.3 to 10.6 per cent carbon. The sintered mass was crushed and the iron was removed by magnetic separation, followed by treatment with 10 per cent sulfuric acid. Von Bichowsky [79] reduced the titanium content of ilmenite to an amorphous suboxide by heating the ore with carbonaceous material, leached out the impurities (iron) with dilute hydrochloric acid, and heated the residual suboxide of titanium with sodium sulfate under reducing conditions so as to convert any remaining iron compounds to sulfides. The latter were leached out with hydrochloric acid and the purified residue was oxidized to titanium dioxide of pigment grade.

In a related process, ilmenite was heated with coal until the titanium was reduced to the trivalent state.[80] The product was treated with sulfuric acid in such proportions as to form a solution of titanium sulfate, and this was diluted and boiled to precipitate basic titanium sulfate. By heating a mixture of titania and zircon with carbon under reducing conditions at 900° to 1000° C without sintering, and cooling in a nonoxidizing atmosphere, dititanosotitanic carboxide [81] was formed, leaving the zircon unattacked. This carboxide may be used in preparing other compounds of titanium. A method has been described by which the ferric oxide of oxidized ilmenite was reduced with hydrogen at high temperatures.[82]

The iron was separated, leaving a high-grade titanium dioxide concentrate which was used in pigment making.

To dissolve selectively the iron content, ilmenite or a similar ore was heated with 40 per cent sulfuric acid at 170° to 180° C for 6 hours in the presence of a reducing agent such as tervalent titanium compounds, copper and sulfur dioxide.[83] Carbon and cupric iodide were found to catalyze the reaction. After most of the iron had been extracted, the residue was dissolved by heating with stronger sulfuric acid, and the solution was hydrolyzed to produce iron-free titanium dioxide.

Farup[84] effected a selective solution of the iron content of ilmenite by heating the pulverized ore with three times its weight of 20 per cent hydrochloric acid at 70° to 80° C for 12 hours. The titanium dioxide residue was then converted to soluble form by digestion with concentrated sulfuric acid. In an improved process[85] the finely ground ilmenite was leached with hydrochloric acid of specific gravity 1.5 for 2 or 3 days at 50° to 60° C, after which the temperature was raised to 85° to 95° C to precipitate any dissolved titanium, leaving a solution of ferric and ferrous chlorides. The crude titanium dioxide residue was purified by dissolution in strong sulfuric acid and reprecipitation.

Titanium minerals which are practically insoluble in acids are converted to a soluble form by roasting with iron minerals and coke.[86] The mixture ground to 325 mesh is sintered at 2100 to 2500° F for 10 to 30 minutes in a nonoxidizing or slightly reducing atmosphere containing carbon dioxide. On withdrawal from the furnace the sinter is immediately quenched in water. Coke is added to the original charge to maintain the iron in the lower valent form. The ratio of titanium dioxide to iron in the product should be between 1 to 1 and 4 to 1. If the titanium ore contains sufficient iron, no iron mineral need be added. Minerals of titanium mixed with an alkali are sintered by heating below the melting point in the presence of a catalyst to obtain a product suitable for the manufacture of pigments.[87] The catalyst consists of at least one metal fluoride, bromide, or iodide, or oxide of a metal of groups 2, 3, or 4, preferably alumina, silica, or magnesia. In an example, 100 parts of ilmenite containing 50 per cent titanium dioxide, 23 per cent sodium carbonate, and 5 parts of alumina were heated for 2 to 4 hours at 500 to 850° C until all of the combined carbon dioxide was removed. Actual time and temperature depended on the type of ore used. The chromium, vanadium, and niobium were converted to chromates, vanadates, and niobates, which were easily removed from the sintered product by washing with water.

A concentrate of titanium dioxide is obtained from low-grade titaniferous ore by a process which includes roasting with Glauber's salt and anthracite.[88] To illustrate, an ore containing 36.3 per cent titanium dioxide, 48.2 per cent ferrous oxide, 2.86 per cent silica, 0.2 per cent calcium oxide, 2.10 per cent magnesium oxide, and 5.66 per cent alumina was roasted with Glauber's salt and anthracite at 1100° C for 3 hours under reducing conditions. After immersing the products in water to wash out the alkalies, the slurry was added to a solution prepared by dissolving ilmenite, containing 51.0 per cent titanium dioxide, 41.5 per cent ferrous oxide, 0.6 per cent silica, and 2.35 per cent alumina, in 95 per cent sulfuric acid and boiled with agitation. As soon as the generation of hydrogen sulfide ceased, titanium dioxide was precipitated. The washed and dried product contained 87.5 per cent titanium dioxide, 0.38 per cent ferrous oxide, 0.19 per cent silicon dioxide, and 0.06 per cent calcium. Evans and Gray [89] roasted a mixture of 50 to 80 per cent rutile ore, 10 to 40 per cent soda ash, and 5 to 40 per cent chalk at 800 to 1100° C for 2 hours. A dry porous mass resulting from the digestion of this reaction mixture with 94 per cent sulfuric acid was dissolved in water or dilute sulfuric acid. The clarified solution represented a recovery of 90 to 95 per cent of the titanium dioxide in the original ore. A dried, ground ilmenite ore containing at least 60 per cent titanium dioxide is mixed with an alkali metal compound, sodium carbonate, and heated at 900° C for 2 hours in an oxidizing atmosphere to give a pale yellow titanate product.[90] The reaction mixture may also contain calcium or magnesium compounds to improve the free-flowing properties of the mixture. Titanium dioxide pigment can be made from the titanate product by the usual sulfate process.

The iron content of ilmenite is converted to the sulfate by sulfur and oxygen treatment in a fluidized bed, after which it is leached with a mineral acid.[91] Iron compounds, acid, and sulfur can be recovered.

Improved pigments are produced by removing phosphates of chromium, iron, and vanadium present as impurities in washed titania hydrolyzate before calcination.[92] The titanium dioxide is converted to an insoluble metal titanate and the phosphates to soluble derivatives by reaction with the alkali compound at elevated temperature. After washing out the phosphates, the titanate is reacted with concentrated sulfuric acid to form titanium sulfate, which is dissolved in water and filtered. The purified solution is hydrolyzed to produce hydrous titanium dioxide which is filtered, washed, calcined, and finished as usual.

## SLAG

In a process developed by Campbell,[93] titaniferous magnetite was crushed and screened; fines were sintered with sodium carbonate, and the coarse ore and sinter were smelted in a blast furnace. The pig iron containing most of the vanadium was converted to steel in a Bessemer converter. The slag was ground and leached with sulfuric acid, and titanium dioxide was precipitated from the purified solution by thermal hydrolysis and processed in the usual manner to produce pigment. Recovery of titanium was 80 per cent. The sodium carbonate was recovered and reused.

Iwase [94] developed a novel process of separating titanium dioxide directly from titaniferous iron ores, which involved reduction with carbon. To this sponge-iron-containing product were added as fluxes from 25 to 40 parts of calcium oxide and 35 to 55 parts of manganese oxide for each 100 parts of titanium dioxide, and the mixture was then smelted. The silica content of the charge was so regulated that the slag which was produced consisted of an eutectic mixture of titanate and pyrophanite, with a resulting content of titanium dioxide higher than 35 per cent and of manganese oxide higher than 15 per cent, and having a correspondingly low melting point. By reducing titaniferous magnetite with solid carbon, with reducing gases, or with a mixture of the two at 900° to 1000° C, and then smelting the residue, pure iron as one product and a slag rich in titanium, vanadium, and chromium were obtained.[95]

Heating ilmenite ores at 1000° C before smelting improves their flowability and reducibility and lowers the melting point of the slag.[96] The cause of disintegration of titanium-alumina slag is the oxidation of the lower oxides of titanium to the dioxide with resultant increase in volume.[97] Titania slags are made with controlled alkali addition so that no titanium-alkali complexes are formed.[98] The slag retains the viscosity necessary to separate the pig iron formed by the high temperature reduction. Molten slag is cooled gradually to the transformation point of titanium dioxide to produce an easily digestible product.[99] This results from a volume change at the transformation point.

A concentrate containing 60 per cent titanium dioxide is obtained from waste slag produced in the manufacture of iron-titanium alloys from ilmenite by sintering the slag with soda ash.[100] The slag containing 18 to 28 per cent titanium dioxide, 40 to 60 per cent alumina, 10 to 15 per cent calcium oxide, 0.3 to 15 per cent ferrous oxide, and 2.9 per cent silica with traces of magnesia is heated with the soda ash at 100 to 1200° C for 4 hours. A mole ratio of sodium oxide to alumina plus ferrous oxide of 1.05 is employed. Leaching is carried out at a concentration of 240 g of sinter in 1 liter of sodium carbonate solution containing 40 gpl sodium oxide. In addition to the high titanium dioxide residue, a solution of sodium aluminate of high purity containing 100 gpl alumina suitable for the production of aluminum is obtained as a by-product.

Ilmenite or titaniferous iron ores are smelted with coke and a small proportion of limestone and sodium oxide flux at 1300 to 1500° C to produce a high titania slag readily soluble in sulfuric acid.[101] Appreciable quantities of alumina and silica normally produce viscous slags, which are difficult to separate from the iron and to remove from the furnace unless fluxing agents in proportions greater than normal are used. With large amounts of fluxing agents the slags have a lower titanium dioxide content. According to this process, titaniferous iron ore containing alumina and silica is mixed with calcium carbonate (limestone) and sodium oxide as flux along with carbonaceous reducing agent. The mixture is heated to form molten iron metal and a high titania slag which separate by gravity. Before leaching, the slag contains 1 to 5 per cent ferrous oxide and from 1 to 5 per cent sodium oxide or potassium oxide. The slag is leached with dilute sulfuric acid at 25 to 80° C to form a digestible titanium concentrate.

In an example, 100 parts of ilmenite containing 43.7 per cent titanium dioxide, 36.8 per cent ferrous oxide, 7.1 per cent ferric oxide, 2.2 per cent calcium oxide, 5.2 per cent silica, 2.6 per cent alumina, and 2.4 per cent magnesium oxide was mixed with 7.6 parts of limestone, 1.7 parts of soda ash, and 11.0 parts of coke. This mixture was charged into an arc furnace and heated at 1400° C for 1 hour, after which the molten iron and slag were separately tapped from the furnace. The fluid and free-flowing slag was cooled and ground to a fineness that 95 per cent passed through a 325 mesh screen. Analysis of the slag was titanium dioxide 62.7, ferrous oxide 2.7, sodium oxide 1.4, calcium oxide 7.9, silica 9.5, alumina 11.6, and magnesium oxide 4.2 per cent. The ground slag was leached in 10 per cent sulfuric acid for 1 hour at 75° C with agitation to give a concentrate containing 76.2 per cent titanium dioxide, 2.9 per cent ferrous oxide, 0.4 per cent sodium oxide, 8.5 per cent calcium oxide, 1.1 per cent silica, 6.8 per cent alumina, and 4.1 per cent magnesium oxide. The leached slag was digested in 89 per cent sulfuric acid to give a titanium dioxide recovery of 97 per cent.

Ilmenite is smelted in a low shaft blast furnace to produce pig iron and a titanium-rich slag.[102] An alkaline flux, such as sodium oxide or sodium carbonate, is preferred. Temperatures employed are 600 to 1200° C during a 1 to 8 hour initial period; 1800 to 2200° C for 10 to 45 minutes; and 1300 to 1800° C during a 30 to 120 minute period, during which separation of the slag from the cast iron takes place. A slag containing 90 per cent titanium dioxide was obtained from ilmenite containing 55 per cent titanium dioxide and 33.4 per cent iron. A mixture of ground titaniferous ore, coking coal, and a small proportion of soda ash is heated in a coking oven until all volatile matter is driven off.[103] Metallic iron and titanium oxide are deposited in the pores of the coke. The metallized coke, together with some calcium carbonate and magnesium carbonate, is then smelted in a cupola to obtain molten iron and a slag rich in titanium dioxide.

Iron sand or titanium-bearing iron ores are smelted to obtain metallic iron and a slag rich in titanium dioxide.[104] For example, a mixture of

100 parts of iron sand containing 52.2 per cent iron, 11.0 per cent titanium dioxide, 0.46 per cent vanadium oxide, and 1.08 per cent silica, 30 parts of charcoal, and 27 parts of sodium carbonate was melted in a crucible furnace at 1370° C. A slag containing 40.3 per cent titanium dioxide and pig iron containing 4.1 per cent carbon, 0.17 per cent silica, and 0.4 per cent vanadium were obtained. Twenty parts of slag was further heated in an electric furnace with 6 parts of charcoal to obtain a crude mixture of titanium carbide, titanium trioxide, and titanium monoxide. The volatilized sodium oxide is absorbed in a solution of carbon dioxide to precipitate sodium carbonate. This crude product was treated with chlorine in an electric furnace at 480° C in the presence of oxygen to produce titanium tetrachloride which is then decomposed to titanium dioxide and chlorine.

One part of Tawahus magnetite containing 18.8 per cent titanium dioxide, 57.5 per cent ferric oxide, 11.8 per cent silica, 7.5 per cent alumina, 0.5 per cent vanadium pentoxide, 0.5 per cent magnesium oxide, and 0.1 per cent calcium oxide, ground to 20 mesh was mixed with 0.4 part sodium carbonate and 0.18 part of 20 mesh coke, and roasted at 1080° C to form metallic iron and a high titania slag.[105] After wet grinding, the iron was separated by magnetic or gravitational means. The slag was roasted in air at 850° C. After leaching with sodium hydroxide to recover the vanadium, the slag was further leached with dilute sulfuric acid to remove sodium oxide and silica. After leaching, the slag was digested in concentrated sulfuric acid for pigment manufacture.

For the selective reduction of 100 parts of titanium ore with 50 to 150 parts of carbon, silica or silica-containing iron ores are added as a flux.[106] The amount of carbon is such that the ferric oxide is completely reduced to iron, and a large proportion of the titanium dioxide is reduced to the mono- and trioxides. By this process a slag is obtained containing all of the titanium compounds but only a small part of the iron. To illustrate, 68 kg of a molten titanium ore at 1450° C containing 17.15 per cent titanium dioxide and 69.51 per cent ferric oxide was mixed with 28.1 kg of silica and 12.7 kg of coke. The slag obtained from this melting process contained 45.8 per cent titanium dioxide, 1.6 per cent iron, 41.1 per cent silica, 16.3 per cent alumina, and 1.9 per cent magnesium oxide. This slag may be chlorinated or digested with sulfuric acid for pigment manufacture.

In the smelting of titaniferous iron ores a deficiency rather than a preponderance of basic oxide is necessary for producing a manageable slag rich in titanium dioxide.[107] With lime as the fluxing agent a substantial deficiency of calcium oxide below the $CaTiO_3$ level is recommended. In the process of smelting ilmenite with carbon to produce cast pig iron and a titanium-rich slag, magnesium oxide is not reduced until the ferrous oxide content drops below 1 per cent.[108] The reduction of magnesium oxide converts $MgTi_2O_5$, which is readily digested in sulfuric acid, to undesirable $MgTi_2O_4$, which is digested with difficulty. This reaction can be carried out at 1200 to 1350° C, which produces a sinter, or preferably at 1350 to 1600° C, at which temperature both the titanium dioxide-rich slag and the iron are liquids. A more fusible titanium dioxide slag is obtained by fluxing with calcium oxide and magnesium oxide.

Readily digestible titanium concentrates are obtained from titaniferous iron ores containing compounds of aluminum and silicon by a two-step process.[109] The first step is a smelting operation at 1250 to 1400° C in which most of the iron oxide is reduced to metal, while the silica and alumina are collected in a freely flowing slag. In the second step, the silica and alumina compounds are removed by leaching the slag with 5 to 15 per cent sulfuric acid at 25 to 80° C. A slag containing readily soluble silica and alumina compounds is obtained by smelting the ore with sodium carbonate and coke in such amounts that the ratio of sodium oxide to titanium dioxide, alumina, and silica in the slag is 0.30 to 0.40 if the titanium content of the ore is between 5 and 22 per cent, and 0.24 to 0.40 if the titanium content is from 22 to 50 per cent. The ratio of sodium oxide to titanium dioxide in the leached slag must be 0.16 to 0.24 to obtain a titanium recovery of 85 to 94 per cent on treatment with 70 per cent sulfuric acid.

Ilmenite is heated with iron powder at 900 to 1000° C to reduce the ferric oxide to ferrous, and then with coal or coke at 1100 to 1200° C to reduce the ferrous oxide to metallic iron, and add 4.2 per cent carbon to provide fluid cast iron.[110] The mixture is then heated at 1250° C with sodium carbonate to produce a fluid titaniferous slag. The three additions to the ore may be made together. An ilmenite containing 54.2 per cent titanium dioxide, 24.0 per cent ferrous oxide, and 15.4 per cent ferric oxide required 5.4 parts of iron powder, 9.8 parts of carbon, and 25 parts of sodium carbonate per 100 parts of ore. The slag was calculated as $Na_2Ti_3O_7$ with 5 per cent extra sodium carbonate to combine with the impurities. It contained 73 per cent titanium dioxide and 0.8 per cent iron as ferric oxide. By treating the crushed slag with dilute sulfuric acid, washing with water, and drying, a concentrate containing 95.4 per cent titanium dioxide and 4.2 per cent impurities was obtained.

A process for producing soluble titanate slag from low-grade titaniferous ores involves two main steps.[111] The ore is sintered at 1050° C with carbon and soda ash to reduce the iron to metal powder retaining the titanium dioxide in the slag. After wet grinding, the powdered iron is separated from the slag by magnetic or gravity methods. The slag may be decomposed with sulfuric acid to produce titanium dioxide pigments. Titaniferous iron ores are smelted with a solid carbonaceous reducing agent and limestone, sodium carbonate, or sodium hydroxide as the fluxing agent at 1250 to 1700° C until the molten iron formed settles out from the slag.[112] One mole of fluxing agent is added for each mole of acid material other than titanium, and 5 to 12 moles for each mole of titanium dioxide. After separating the iron, the slag is rapidly cooled below 900° C by quenching with water or with a blast of air, or by spreading it in a thin layer on a cold surface to avoid formation of slowly soluble rutile form of titanium dioxide. Iron is removed from low-grade ilmenite by smelting to produce a high-grade slag that can be further processed into titanium pigment.[113] This method is limited to ilmenite ores in which the titanium dioxide is diluted with iron oxides which are reduced readily by carbon. Two campaigns with a 2500-pound furnace yielded 12.2 tons of slag having an average analysis of 67 per cent titanium dioxide, 10 per cent calcium oxide, 7.2 per cent magnesium oxide, 4.5 per cent silica, 4.8 per cent alumina, 6.2 per cent ferrous oxide, and 1.2 per cent sulfur. Recoveries and power required are such that the process is economically feasible.

Ilmenite and titaniferous iron ores may be smelted with silica only as the flux to yield metallic iron and a titanium silicate slag.[114] From 0.5 to 1.5 parts of silica are added for each part of titanium dioxide after heating to at least 1450° C. These slags retain their fluidity and workability in the presence of the carbonaceous reducing agent in the charge. Slags produced by this method are readily soluble in 60 per cent sulfuric acid. Working with low-grade ores containing alumina, it is possible to adjust the conditions to reduce the titanium to the divalent state. This slag may be chlorinated directly to obtain a 100 per cent yield of the tetrachloride. A titanium-rich slag is obtained by smelting ilmenite without the addition of fluxes.[115] A reducing agent is added to the original charge in an amount sufficient to reduce more than 10 per cent of the ferrous oxide. The remaining ferrous oxide is reduced after the charge melts completely by adding the reducing agent directly to the melt.

Titaniferous magnetite from Iron Mountain, Wyoming, is roasted with sodium carbonate and leached to remove the vanadium, after which it is smelted to produce a high titanium dioxide slag and pig iron.[116] The titanium mineral of the ore is finely disseminated in the magnetite. Surface samples analyzing 49.3 to 52.3 per cent iron, 21.8 to 24.2 per cent titanium dioxide, 0.61 to 0.75 per cent vanadium pentoxide, 4.4 to 5.6 per cent alumina, 1.9 to 2.6 per cent magnesium oxide, 0.9 to 1.1 per cent silica, 0.02 to 0.05 per cent sulfur, and 0.02 to 0.05 per cent phosphorus were roasted at 950° C with 15 per cent sodium carbonate. The product was leached cyclically with water to remove 90 per cent of the vanadium as red cake analyzing more than 80 per cent vanadium pentoxide. After leaching the residue was smelted with coke in a graphite lined arc furnace to recover 98 per cent of the iron as pig iron containing less than 0.1 per cent vanadium and titanium. Practically all of the titanium dioxide was recovered in the slag, which was amenable to digestion with sulfuric acid or to chlorination for the production of titanium dioxide pigments.

A slag containing 64.2 per cent titanium dioxide was purified to yield a product containing 96.7 per cent titanium dioxide, 0.5 per cent silica, 0.7 per cent ferrous oxide, 0.6 per cent alumina, 0.1 per cent calcium oxide, 0.7 per cent magnesium oxide, 0.6 per cent manganese oxide, 0.1 per cent sulfur, and a trace of phosphorus.[117] To the 60 mesh slag was added an equivalent amount of 13 per cent sodium hydroxide solution, and the mixture was heated with an excess of calcium hydroxide at 100 to 130° C for 1 hour. After cooling, the suspension was filtered and washed with water. The residue was subjected to gravity concentration, washed with dilute sulfuric acid and water, and dried. Finely ground ilmenite or titaniferous magnetite, mixed with a sulfur or phosphorus compound in such proportions that the sulfur or phosphorus is 0.3 to 5 per cent based on the iron, is heated with coke at 1450° C to reduce at least 90 per cent of the iron content to the metal.[118] After grinding the sinter to 325 mesh under nonoxidizing conditions, the iron is extracted by magnetic separators. A preliminary oxidation of ferrous oxide to ferric aids the reduction. This operation may be combined with an alkali extraction of vanadium with a favorable effect on the reduction step, since alkali salts act as catalysts to lower the temperature required for reduction of the iron oxide. For example, a slag fraction containing 82.3 per cent titanium dioxide, 8.2 per cent total iron, and 0.11 per cent phosphorus, and a metal fraction containing 88.2 per cent total iron, 86.4 per cent metallic iron, 2.6 per cent titanium dioxide, and 0.91 per cent phosphorus were produced. A titaniferous mineral containing 40 per cent iron and 6 per cent titanium dioxide is mixed with 20 per cent of its weight of coke and heated at 975° C to effect reaction.[119] The product after wet magnetic separation yields an impure metal containing 62 per cent iron used in steel making, and dross containing 65 per cent titanium dioxide used for pigment manufacture. The reduction fusion of ilmenite in the presence of an alkali flux yielded a high titania slag low in iron.[120] With calcium carbonate as the flux, the slag contained 83.5 per cent titanium dioxide. Barium carbonate flux produced a slag containing 60 per cent titanium dioxide. With sodium carbonate as the fluxing agent the slag contained 74 per cent titanium dioxide. The iron content was from 1 to 2 per cent. Much higher working temperatures were required with calcium carbonate and barium carbonate than with sodium carbonate.

A digestible titanium concentrate is obtained by smelting titaniferous iron ore with a carbonaceous reducing agent, silica, alumina, and sodium oxide or potassium oxide at 1300 to 1700° C until the ferrous oxide content of the slag is reduced to between 2 and 20 per cent.[121] The amount of silica, alumina, and sodium oxide plus potassium oxide in the mixture is sufficient to produce a slag in which the sum of silica, alumina, and sodium oxide plus 0.66 potassium oxide is from 0.75 to 2.2 parts for each part of titanium dioxide. Preferably the silica is from 40 to 55 per cent, the alumina is from 20 to 35 per cent, and sodium oxide plus 0.66 potassium oxide is from 20 to 30 per cent of the total of these constituents. After separation and cooling, the slag is ground to 325 mesh and leached with 10 per cent sulfuric acid for 30 minutes at 75° C to remove the silica, alumina, and sodium and potassium compounds. The titanium dioxide-rich residue is well suited for digestion with concentrated sulfuric acid in the pigment manufacturing process. From 50 to 80 per cent of the titanium content in concentrates prepared by the reduction of ilmenite followed by magnetic separation of the iron is in the reduced titanium III form.[122] This component may be oxidized by pelletizing the material with dilute sulfuric acid and heating the pellets in air first at 100° C and later at 250 to 400° C. The solubility of the titanium dioxide in the slags in sulfuric acid increases with the degree of reduction. Slags produced with calcium carbonate flux contain calcium titanate, titanium dioxide, calcium oxide, and lower oxides of titanium. The solubility of the oxide increases with its degree of reduction.

In the production of high titania slags the titanium dioxide may undergo reduction to the composition $TiO_{1.67}$ without affecting its recovery from the slag.[123] The solubility of the oxide increases with the degree of reduction. Mixed titanates derived from $3TiO_2 \cdot Na_2O$ are present in the sodium slags. The maximum solubility of titanium in the sodium slags is 80 per cent, depending on the condition of formation. Calcium titanate, titanium dioxide, calcium oxide, and lower titanium oxides are present in the calcium slags. Here too the extent to which titanium goes into solution depends on conditions during slag formation, and it increases with the degree of reduction of the free oxide. The solubility limit is 50 per cent. Among the three fluxes investigated, carbonates of barium, calcium, and sodium, the latter gave the lowest temperature of slag formation, 1200° C; the widest temperature range may be utilized, 1200 to 1380° C; and the degree of solubility of titanium is highest, 80 per cent. The primary constituents of high titanium slag are anisovite, titanium dioxide, solid solutions of $2MgO \cdot TiO_2$, $Ti_2O_3$, and $(FeMg\ Mn)O \cdot TiO_2\text{-}Ti_2O_3$.[124] The last two completely miscible solutions were identified for the first time. In addition to these constituents a commercial slag contained ilmenite, $Mg_2TiO_4$, $(Mg\ Fe)_2TiO_2$, $MgTi_2O_3$, $CaTiO_3$, $MgAl_2O_4$, and $Mg_2SiO_4$. The viscosity of the molten slag as a function of the $Ti_2O_3$ to $TiO_2$ ratio at a constant ferrous oxide concentration first decreases to a minimum and then increases sharply as the ratio increases.[125] Increasing the ferrous oxide concentration reduces the viscosity, but the effect is complex and related to the titanium trioxide-dioxide ratio. The melting point of the slag at constant ferrous oxide concentration is a complex function of this ratio, having alternately minimum and maximum values.

Ilmenite ore from Malaya contains 2.0 per cent monazite, 0.5 per cent zircon, and 0.5 per cent cassiterite.[126] In high-titania slags from this ore zirconium and hafnium oxides constitute 0.5 to 2.2 per cent and rare earth oxides 0.2 to 1.0 per cent. The main component in these slags is the solid solution of titanium dioxide, titanium suboxide, alumina, manganese oxide, and rare earth oxides in $(Mg Fe)O \cdot 2TiO_2$. Rutile makes up about 5 per cent and ilmenite-hematite perhaps several per cent.

Titanium III in slag is oxidized to the tetravalent form at 40 to 90° C by exposure to air in the presence of water and an oxidation catalyst.[127] Effective catalysts are 0.1 to 2.0 per cent sulfuric or hydrochloric acid, or metallic salts of these acids. In an example, a concentrate containing 36 per cent titanium dioxide and 44 per cent titanous oxide was wetted with dilute sulfuric acid. One part of acid was added to 20 parts of concentrate. The temperature of the 12 mm layer increased to 84° C during 15 minutes, and the content of titanium III oxide decreased from 55 to 18.5 per cent of the total titanium oxides. Similarly an ore concentrate containing 80.5 per cent titanium oxides, 78 per cent of which was trivalent, was ground to 2 to 3 mm and wetted with dilute hydrochloric acid in a proportion of 1 part of acid to 20 parts of concentrate. The temperature of the 20 mm layer increased to 84° C, and the content of titanium III salts decreased from 78 to 25.5 per cent.

A review of the field of high-titania slag is given by Takei.[128]

The concentrate obtained from Tatar bauxite containing titanium 22.26, ferrous oxide 5.96, ferric oxide 60.40, alumina 5.18, and silica 1.26 per cent was briquetted with wood charcoal and heated for 3 hours at the rate of 10 to 15 degrees a minute to 1500° C to produce titania rich slag and iron.[129] The proportion of carbon calculated on the basis of the amount required to reduce ferrous oxide to iron varied from 100 to 40 per cent. As the carbon content decreased over this range, the yield of slag increased from 46.0 to 86.7 per cent; the yield of iron decreased from 46.0 to 13.3 per cent; the titanium dioxide content of the slag decreased from 77.09 to 52.49 per cent; and the content of ferrous oxide and of iron increased from 14.35 to 25.86 per cent, to lower the melting point of the slag to 1420° C. As the carbon content in the original mix increased, the loss of titanium dioxide and silica with the metal was reduced from 5.5 to 1.0 and from 7.5 to 1.0 per cent, respectively. Titanium carbide was not detected in any of the slags. The addition of sodium carbonate to the briquetted mixture reduced the melting point of the charge to 1100° C. Analysis of the slags showed titanium dioxide 36.7 to 54.27 and ferrous oxide 1.21 to 2.71 per cent. Ilmenite ore may be mixed with 10 to 25 per cent red sludge from bauxite before smelting to produce iron and titanium dioxide-rich slag.[130]

## ELECTRIC FURNACE SLAG

Large-scale electric furnace smelting of ilmenite-magnetite ore using coal or coke as the reducing agent to produce a slag rich in titanium dioxide and refined iron began in 1952. The methods employed are outlined by Knoerr.[131] Ensio [132] conducted the smelting of titaniferous ores in an electric arc furnace in such a way that a low sulfur, high carbon iron and a titanium-rich slag with less than 2 per cent iron are obtained. For this purpose the ilmenite is mixed intimately with such an amount of carbonaceous material that its carbon content together with that of the carbon to be burned by the electrodes is sufficient only to reduce all of the iron present. The carbonaceous material consists of log back, wood chips, and splinters; because of its low density, it serves as an insulator against the heat of the furnace. Thus, 100 pounds of ilmenite ore containing 55.8 per cent iron oxide, 36.7 per cent titanium dioxide, and 0.04 per cent sulfur was mixed with 105 pounds of wood paste and 8.3 pounds of coke (17.1 pounds of fixed carbon per 100 pounds of charge). This charge was smelted in an electric arc furnace at 1650° C to give a titaniferous product containing 0.7 per cent ferrous oxide and 85 per cent titanium dioxide equivalent of which 50 per cent was in the form of trivalent titanium. The iron contained only 1.07 per cent of sulfur.

Beneficiated ilmenite ore is smelted in a continuous three-phase, 250 kva, open-top furnace employing coal or coke as the reductant to produce metallic iron and a titania-rich slag.[133] Minus one-fourth inch ore was smelted by the standard method. Ore of minus 14 mesh size was smelted in briquetted form by the open bath method, and in the unagglomerated form by the cold dry-top method. This method consisted in working a foot-deep bed of ore, coke, and wet-paper mill bark over a molten bath. Slag averaging 82.5 per cent titanium dioxide was produced over an 87 hour period. The maximum concentration was 85.0 per cent titanium dioxide. Such a slag is attractive for the production of titanium tetrachloride. Low phosphorus, highly purified pig iron and high titanium dioxide slag are obtained by the electric furnace smelting of Japanese iron sands.[134] Vanadium is recovered in the process.

A high titanium dioxide slag was produced by smelting ilmenite with magnesia flux in an electric furnace.[135] An Indian ilmenite (57 per cent titanium dioxide, 25.2 per cent ferrous oxide, 1.38 per cent silica, and 2.8 per cent alumina) was heated with coke and magnesia in an electric furnace at 1650° C for 5 to 7 hours. The amount of coke was 1.4 to 1.8

moles for each mole of ferrous oxide. The content of ferrous oxide reached a minimum of 2.0 per cent at a ratio of magnesium oxide to titanium dioxide in the slag of 0.2. Recovery of titanium dioxide in the slag was from 80 to 95 per cent. Ishizuka [136] fused in the electric furnace a mixture of 100 parts iron sand containing 33.3 per cent titanium dioxide, 25 parts of charcoal, 10 parts of limestone, and 1 part of fluorite to produce a titanium concentrate. The fusion product was run through the bottom and the titanium dioxide concentrate containing iron collected on top. Chlorine was passed through a mixture of 4 kg concentrate and 1 kg of charcoal powder heated at 500° C for 6 hours at a rate of 2 liters a minute to obtain 900 ml of condensate. Fractionation of the condensate yielded 800 ml of titanium tetrachloride, 100 ml of silicon tetrachloride, 25 g of vanadium concentrate, and 50 g of ferric chloride.

Slags containing 80 to 83 per cent titanium dioxide can be consistently produced by smelting ilmenite without the use of flux, provided the reducing agent consists of the proper proportions of wood chips and coke.[137] A magnesite lining proved more satisfactory for continuous dry-top electric smelting operation than a carbon lining. The smelting operation is very sensitive to small changes in operating temperature if the slag contains 85 per cent or more titanium dioxide. Additional energy input and higher tapping temperatures are required to produce slags containing the higher percentages of titanium dioxide. The addition of manganese ore to the smelting charge has a definite beneficial effect in displacing iron from the slag. Slags containing 80 to 85 per cent titanium dioxide chlorinate readily and produce an acceptable grade of titanium tetrachloride. Best results are obtained with slags that have been ignited at 800° C. Ilmenite concentrates can be chlorinated directly, but the large amount of ferric chloride produced is an economic problem. A final product containing 94 per cent titanium dioxide can be produced by chlorinating the slag directly without the presence of carbon, followed by water leaching to remove iron and manganese chlorides.

Smelting of ilmenite in the electric furnace to produce metallic iron and titanium-rich slag is facilitated by the addition of sufficient carbon to form 20 per cent ferrous oxide in the initial melt, followed by gradual reduction in the liquid phase.[138] Smelting without flux, the titanium monoxide and titanium trioxide content of the slag increased with decreasing ferrous oxide. The titanium monoxide increased with the titanium trioxide content. From 2 to 5 per cent of ferrous oxide had to be left in the slag to prevent solidification in the furnace. Smelting without flux yielded a slag with a lower amount of impurities. Ilmenite is reduced in a resistance furnace at 1800 to 2500° C with only enough carbon to reduce all of the iron to metal and the titanium to the monoxide.[139] The iron forms globules. A dense, nonmagnetic coarsely crystalline titanium product containing oxycarbide which can be chlorinated exothermically is obtained.

Heating ilmenite ores before smelting in the electric furnace improves their flowability and reducibility and lowers the melting point of the slag.[140] Raw ilmenite ore is heated in a kiln at 1000° C for 0.5 hour. The sulfur content is reduced from 0.3 to 0.017 per cent in the ore and from 0.45 to 0.2 per cent in the smelted product. A mixture of 50 to 60 per cent high titania slag and 50 to 40 per cent coke is melted in the furnace by an electric arc to form a refractory lining containing 60 per cent titanium dioxide and 0.5 per cent titanium carbide.[141] An arc furnace lined in this manner can be operated at 1600 to 1800° C to produce slags containing up to 90 per cent titanium dioxide from ilmenite. Miller and Hatch[142] worked out a charge pattern for a three phase arc furnace with such a manner that six portholes between adjacent electrodes of different phase connection receive about 50 per cent of the central charge, representing 85 to 95 per cent of the total charge while the remaining 50 per cent is introduced through 12 lateral ports. That other than the central charge is introduced through outer charge ports to maintain the sloping side banks for protection of the furnace walls. In an example, a furnace with a power output of 18,000 kw melted an intimate mixture of 308 tons of ilmenite, containing 37.3 per cent titanium dioxide, 26.3 per cent ferrous oxide, and 30.0 per cent ferric oxide, and 46.2 tons of coal in 24 hours. Ninety-eight per cent of the charge was fed through the center ports and the remainder through the outer ports. The slag tapped at 1600° C contained 70.3 to 71.5 per cent titanium dioxide and 12.1 to 13.0 per cent ferrous oxide, while the iron tapped at 1615° C contained 1.77 to 1.81 per cent carbon and 0.16 per cent sulfur.

# 13

# Titanium Sulfate Solutions for Pigment Manufacture

The many methods in the literature for the production of titanium dioxide pigments fall in three main groups: hydrolysis of aqueous solution of the sulfate, chloride, nitrate, chloroacetate, and other salts of titanium IV at elevated temperatures; the vapor phase oxidation of anhydrous titanium tetrachloride, tetrabromide, or tetraiodide with oxygen or air at high temperatures; and precipitation from purified water solutions of titanium tetrafluoride or other compounds by the addition of ammonium hydroxide or other base. Methods based on the hydrolysis of sulfate solutions are by far the most important commercially, although the vapor phase oxidation of the tetrachloride is gaining in favor. Only these two methods are employed commercially. Methods based on the hydrolysis of aqueous solutions of titanium tetrachloride have been developed through the pilot plant stage. A commercial plant based on the fluoride process proved unsuccessful.

Nitrate solutions are prepared by indirect means, such as dissolution of orthotitanic acid, sodium titanate, or barium titanate in nitric acid.

Hydrous titanium dioxide of high purity may be precipitated satisfactorily by thermal hydrolysis from sulfate and from chloride solutions containing large proportions of iron as well as smaller amounts of other impurities. However, the iron must be in the ferrous state of oxidation which is stable toward hydrolysis.

Fluorides of titanium, on the other hand, are stable, and, unlike the sulfates and chlorides, their solutions do not hydrolyze on boiling or dilution. The titanium is usually precipitated by adding an alkaline agent (ammonium hydroxide) after carefully purifying the solution to remove the heavy metals which would also be thrown down as hydroxides and so injure the properties of the pigment. Since extremely small amounts

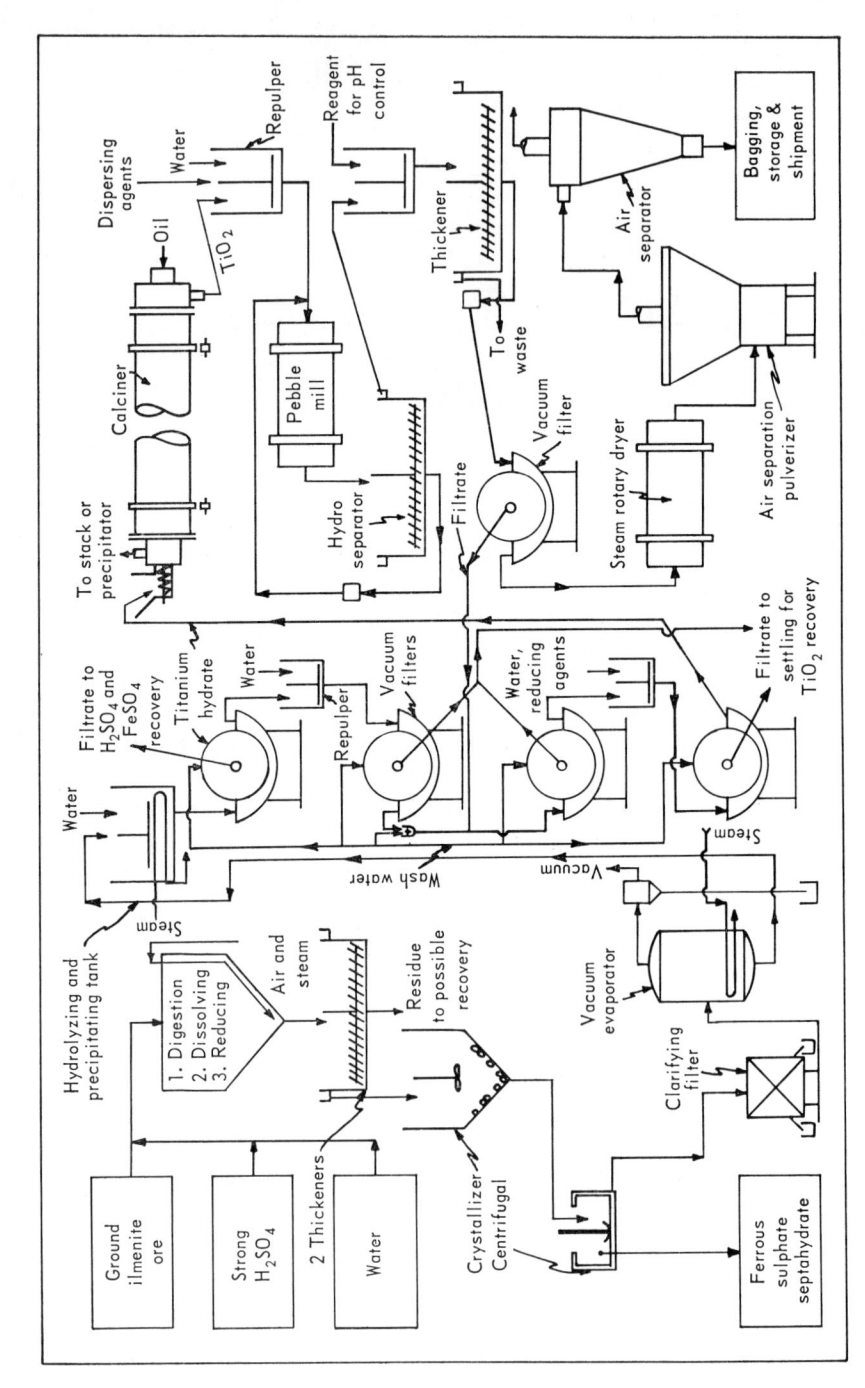

**Fig. 13-1.** Flow sheet of the sulfate process for the manufacture of titanium dioxide pigments.

of compounds of iron, vanadium, chromium, and manganese that are normally present in the ore would seriously injure the color and other properties of the calcined product, previous removal must be practically complete.

A unique departure from the conventional processes based on the precipitation of titanium dioxide by the thermal hydrolysis of aqueous solutions, known as thermal splitting, is particularly applicable to titanium tetrachloride prepared by the action of chlorine on rutile ore or titanium dioxide concentrate. By this process anhydrous titanium tetrachloride is converted directly to the dioxide with the liberation of elemental chlorine by heating the vapor in admixture with oxygen or air at high temperatures either indirectly or in a flame. This process has the advantage that chlorine rather than hydrochloric acid is recovered, but on the other hand the tetrachloride must be previously purified to a very high degree, since compounds of iron, vanadium, chromium, and other heavy metals normally picked up from the ore are precipitated with the titanium and seriously impair the color of the pigment.

Regardless of the type of solution employed, the fundamental steps of the hydrolysis process include decomposition of the raw material (usually ilmenite ore) to obtain a soluble salt; dissolution of the reaction product in water; reduction of the iron component to the ferrous condition; clarification and filtration of the solution; adjustment of the absolute and relative concentration of the components, which may involve crystallization and vacuum concentration; precipitation of the titanium component as hydrous oxide by boiling the solution to effect hydrolysis after proper seeding; filtering and washing the product; treating the pulp with conditioning agents; calcination to convert the amorphous hydrous oxide to crystalline form having the desired pigmentary properties; and, finally, grinding, classifying, and pulverizing the calcined product.

A widely used commercial process for the production of titanium dioxide pigments is based on the thermal hydrolysis of sulfate solutions prepared by dissolving ilmenite in sulfuric acid. Nuclei are formed in place in the solution to be hydrolyzed.

As originally employed, the process gives pigments of the anatase crystal modification. Other methods employ the same essential steps, although the actual procedure of nucleating the solution and of carrying out the hydrolysis may vary.

A flow sheet of the process based on the hydrolysis of sulfate solutions to produce the anatase type pigments is shown in Fig. 13–1. To produce pigments of the rutile crystal form by this process rutile inducing nuclei are added in the hydrolysis step or a rutile promoter is added to the hydrous titanium dioxide just before calcination. These nuclei or promoters are suspensions of very small particles of rutile, usually in hydrochloric acid.

Flow sheets of the process are given by Forbath [1] and by Coates [2]; and have been published by the National Lead Company,[3] the U.S. Bureau of Mines,[4] and the Glidden Company.[5]

## DIRECT REACTION OF ILMENITE OR SLAG WITH SULFURIC ACID

The fusion, reduction, and combination processes were slow, cumbersome, and expensive of operation, but fortunately later investigations revealed that the titanium content of ilmenite, the most available ore, and electric furnace slag can be converted economically to water-soluble form by heating the finely ground material directly with concentrated sulfuric acid.

Weintraub [6] found that finely divided ilmenite can be decomposed quantitatively with concentrated sulfuric acid in a proportion equal to that theoretically required to give normal salts of the base constituents of the ore. Pulverized ore was mixed with approximately 2 parts by weight of 93 per cent sulfuric acid and heated in a cast-iron vessel. At 100° C the exothermic reaction started and the ilmenite was converted into a solid mass composed largely of soluble titanium and iron sulfates. Farup [7] mixed finely ground ilmenite with a relatively large proportion of concentrated sulfuric acid (60° Bé), and heated the slurry gently to start the reaction which proceeded vigorously with a great evolution of heat. The grayish solid mass obtained was then roasted at 900° C to decompose the sulfates and distill off the sulfuric acid, and the product was ground to a yellowish pigment. A later development involved the separation of most of the iron from the reaction product.[8] Finely divided ilmenite was mixed with one or two parts of concentrated sulfuric acid and heated gently to initiate the reaction, which then proceeded vigorously with evolution of much heat. The solid sulfate mass obtained was then heated in air at about 600° C, and at this temperature sulfates of titanium were converted to insoluble oxides while the ferrous sulfate remained unchanged. The cooled mass was then leached with water, whereby practically all the iron was dissolved, leaving the titanium oxide together with the gangue. This crude product was further refined by heating with sodium chloride at 700° C and extracting the impurities from the mass with water.

Rossi and Barton [9] digested ilmenite with 95 per cent sulfuric acid, in a proportion by weight of 2.5 to 2.6 times the titanium dioxide content, to obtain a soluble product that was leached in about three times its volume of water. Blumenfeld [10] found that titanium-bearing ores (ilmenite) were decomposed by heating with 70 to 90 per cent sulfuric acid at 130° C, and finally to 220° C. The reaction product was dissolved in water in such proportions that the final solution contained 15 per cent titanium dioxide. In a similar procedure [11] the ore was decomposed as before by heating with one or two parts by weight of 80 per cent sulfuric acid, and the reaction product was treated with three parts of water or less. A soluble product was obtained by heating ilmenite with 70 to 90 per cent sulfuric acid at 150° to 180° C for 1.5 hours or for some time after the mass had assumed the form of a dry powder. After cooling to 50° C, a limited quantity of water was added in small portions to dissolve practically all the titanium sulfate and only 15 to 20 per cent of the iron sulfate. Sponge iron was introduced to reduce any ferric salts to the ferrous state. After separating the solution, the iron sulfate left in the residue was leached out and treated to recover sulfuric acid.[12] Similarly, finely ground ilmenite was heated with 96 per cent sulfuric acid at 160° C until the reaction was complete.[13] The product was agitated with a limited amount of water to dissolve the titanium component together with some ferrous and ferric compounds. Finely divided metallic iron was introduced to convert the latter component to the ferrous state. After filtration, the ferrous sulfate was largely removed by crystallization.

A further improvement [14] was the production of a solution of titanyl sulfate, which gave titanium dioxide of the same chemical and physical properties as the normal sulfate solution. Finely ground ilmenite was mixed with concentrated sulfuric acid so that on the 100 per cent basis the ratio of acid to titanium dioxide was 1.2 to 1.8. The mixture was gradually heated to 120° C to initiate the reaction, and the paste obtained was further heated to 140° to 250° C to produce a solid product. The temperature, however, was held below the constantly increasing boiling point. A solution containing 150 to 250 g per liter titanium dioxide was obtained by leaching the reaction cake.

Jebsen [15] mixed pulverized ilmenite with 94 per cent sulfuric acid in amount less than that required to form normal sulfates of the base-forming elements present and obtained a slimy product which on heating to 100° C reacted vigorously, yielding a more or less solid cake. A ratio of acid to ore of 1.44, instead of the theoretical 1.9 required to decompose the particular sample of ilmenite and transform its constituents into normal sulfates, gave good results. The solution of titanium and iron sulfates obtained by leaching the cooled product on the counter-current principle was treated with metallic iron to reduce the ferric compounds to the ferrous condition. In preparing such a water-soluble compound by the thermal reaction between finely divided titaniferous material and concentrated sulfuric acid, Schmidt [16] employed proportions so as to get a solid reaction mass on heating, and added 0.5 to 10 per cent (based on titanium dioxide) of a crystalline seeding agent such as hydrated or anhydrous titanyl sulfate or the products of a previous operation. Ilmenite or artificially prepared hydrous or dehydrated titanium oxide served as the starting material. If the more refractory rutile ore was used, it was first fused with an alkali to form the corresponding titanate. For example, a mixture of 1000 parts of pulverized rutile and 1250 parts of a 20 per cent solution of sodium carbonate was evaporated to dryness and roasted at 950° C to obtain sodium titanate. On heating 600 parts of this material with 1140 parts of concentrated sulfuric acid and seeding the product with crystalline titanyl sulfate, a crystalline, porous, readily soluble mass was obtained. By another approach, pulverized ilmenite was impregnated with sulfuric acid and heated to 300° C in such a manner as not to introduce water.[17] The concentration and proportion of acid was such as to react with the titanium and other bases to give a semidry product. The ore in pulverized form may also be decomposed with sulfuric acid in steps.[18] In this process the ilmenite is first heated with less acid than that required for reacting with all the titanium, the product is lixiviated, and the insoluble residue is again treated with another portion of acid. The hydrous titanium oxide obtained from the solution by thermal hydrolysis is further purified by treating it in the same manner as the original material.

The product obtained by reacting finely ground ilmenite concentrate with twice its weight of sulfuric acid (specific gravity 1.72) at 140° C was leached with water at 97° C for 8 hours to obtain a solution of titanium and iron sulfate.[19] After reducing any ferric ions by introducing metallic iron or by electrolysis, the solution was cooled to crystallize the ferrous sulfate. According to the method of Monk and Ross,[20] titanium-bearing ores (ilmenite) were agitated with a small proportion of sulfuric acid of specific gravity 1.55 and heated to about 175° C until the mass thickened. More acid was added, either continuously or intermittently as the reaction proceeded, to convert the titanium to the normal sulfate. The initial addition of acid may be sufficient to form ferrous sulfate and titanyl sulfate. Wrigley and Spence[21] heated a mixture of ilmenite and sulfuric acid to the reaction temperature by injecting steam directly into the mass. By this method the strength of the acid may be regulated. The solution of iron and titanium sulfates obtained by extracting the digestion cake with water was treated with metallic iron to convert all ferric salts to the ferrous condition and hydrated ferrous sulfate was separated by crystallization on cooling. A proportion of other acids, e.g., hydrochloric and oxalic, may be added to the solution at this stage. Kingsbury and Grave[22] added phosphoric acid or soluble phosphates to the normal ilmenite-sulfuric acid mixture in such a proportion that the content of this agent expressed as phosphorus pentoxide was 0.1 to 2 per cent of the weight of the ore. This produced a better dispersion of the finely ground ilmenite in the acid so that on initiating the reaction by the conventional methods a higher recovery of titanium was obtained.

Kramer [23] treated titanium containing residues such as mud or slime from clarification underflow with strong sulfuric acid (oleum) in amount considerably in excess of that theoretically required to produce normal sulfates of the titanium and other metallic constituents, and utilized the resulting mixture in acid attack of a fresh charge of ore. In general such mixtures were employed as would yield an acid concentration of approximately 96 per cent at a temperature of 150° C or above, whereby the acid-titanium dioxide ratio obtained was many times that normally employed in attacking ilmenite. Yields of titanium of 96 to 98 per cent were obtained. In an example, 315 pounds of mud suspension containing 300 to 500 g per liter solid was placed in a pan, and 2000 pounds of oleum was added with agitation. The mixture became quite hot and reaction took place. Acid strength was adjusted by adding recovered hydrolysis liquor containing 25 per cent sulfuric acid, and the temperature was raised to 215° C by the introduction of steam, during which time almost complete conversion to soluble sulfate of the titanium dioxide present in the mud was accomplished. This liquor was used for diluting ilmenite-oleum mixtures to set off the reaction or in dissolving the reaction cake. Alternatively, the mud acid reaction product was mixed with cold acid and ground ilmenite, and the reaction was set off by injection of steam or addition of water to raise the temperature. Mud was mixed with strong acid, above 85 per cent, and the mixture was added to an attack vessel with a suspension of ilmenite and cold acid. The reaction was initiated by conventional methods. The insoluble portion of the reaction product of ilmenite and sulfuric acid left after leaching out the soluble sulfate was concentrated by a mechanical flotation process and the concentrate of unattacked ore was again treated with sulfuric acid.[24] Moran [25] mixed such residues with dilute sulfuric acid from a previous hydrolysis reaction to form a free-flowing slurry, and added this product to a regular digestion mixture. To get more complete reaction [26] the residue was dried and heated with hydrogen or illuminating gas at 800° to 1300° C to reduce the iron to the ferrous state. The reduced product was mixed with regular ilmenite ore and treated with concentrated sulfuric acid in the usual manner. Bousquet and Brooks [27] heated the residue with an excess sulfuric acid at 200° to 300° C to effect reactions, and further heated the solid product at the same temperature to expel the excess acid.

Shtandel [28] treated titaniferous magnetite containing 39.1 per cent titanium dioxide and 49.4 per cent iron with sulfuric acid solution of 10 to 60 per cent concentration at various temperatures up to the boiling point. In all the experiments the ratio of acid to ore was 1.5 times the theoretical. The highest extraction, 38.6 per cent titanium dioxide and 45.5 per cent iron based on the original ore, was obtained with 60 per cent acid at the boiling point. On refluxing ilmenite or blast furnace slag from titaniferous iron ores with 40 to 60 per cent sulfuric acid for 6 to 8 hours, 90 to 95 per cent of the titanium was dissolved. By employing stronger acid, however, a water-soluble precipitate containing much titanium was obtained.[29] Under optimum conditions, 83 per cent of the titanium dioxide content of such slag was extracted with sulfuric acid of specific gravity 1.84 in 2.5 hours.[30] The sample was ground to 200 to 250 mesh and heated with 1.7 parts by weight of the acid to 180° to 200° C. From a study of the reaction between ilmenite from Korea and strong sulfuric acid, Matsubara[31] concluded that the ore must be finely ground and that too much excess acid reduces the recovery of titanium dioxide in the later hydrolysis step.

A mixture of soluble sulfates of titanium and iron was obtained by gradually adding oleum to damp pulverized ilmenite with constant stirring.[32] The reaction product was dissolved in water and diluted to obtain a solution of the desired concentration.

According to Grachev,[33] the titanium and iron components of titanomagnetite concentrates were dissolved in 40 to 50 per cent sulfuric acid two to three times faster if the iron was kept reduced electrolytically to the ferrous condition as dissolved. Recovery of titanium of 70 to 100 per cent was obtained at 94° to 97° C in the presence of 10 to 25 per cent excess acid.

Ilmenite or slag is digested in tanks made of concrete or heavy steel plates lined first with lead followed by an acid resistant brick, 30 feet high by 12 feet in diameter with a 6 inch discharge at the bottom.[34] A typical charge consists of 50,000 pounds of 66° Bé sulfuric acid, 40,000 pounds of ilmenite, and 175 pounds of antimony trisulfide. After heating the mixture with steam, water is introduced to decrease the acid concentration to 85 per cent and initiate the exothermic reaction. After cooling, dilute sulfuric acid and water are added to dissolve the solid cake. The solution after clarification and filtration is cooled to remove ferrous sulfate by crystallization as the hydrate. At this stage the solution contains 140 to 150 gpl titanium dioxide, 30 gpl iron, and 65 to 70 gpl free sulfuric acid. After concentration by vacuum evaporation the solution contains 200 gpl titanium dioxide, 60 to 65 gpl iron, and 85 to 95 gpl free sulfuric acid.

To prepare rutile hydrolysis seed, titanium orthohydrate is first precipitated from titanium sulfate solution with sodium carbonate and washed repeatedly to remove sulfate ion to a very low residual value. The titanium hydroxide is peptized in hydrochloric acid solution to a titanium dioxide concentration of 45 gpl. Citric acid is added to prevent agglomeration on standing. Inorganic multibasic acids hinder the conversion to rutile during calcination, but the organic acids are decomposed before they can do any harm. The stabilized orthotitanic acid in hydrochloric acid is then dehydrated carefully to form a rutile seed suspension having the most desirable porperties. By the loss of one molecule of water the orthotitanic acid is converted to stable metatitanic acid. For hydrolysis 1 per cent of seed is sufficient. A yield of 95 per cent is obtained after boiling the solution for 4 hours.

The pigment particle size is determined by the agglomeration of crystallites in the hydrate having a size of 0.01 to 0.02 micron. During calcination these anatase crystallites convert to rutile aggregates having a particle size of 0.2 to 0.3 micron. Presumably the crystallites formed in the hydrolysis step are a molecular compound containing two titanium hydroxide groups tied in closely with a sulfate group. The sulfuric acid content is 10 per cent. During calcination the condensation can go in two ways depending on the way the cleavage happens. With the rutile seed the cleavage results in the simplest type of molecule, $Ti_2O_4$. If another type of splitting results, anatase is obtained, consisting of a more complex and less stable molecule of the general composition, $Ti_4O_8$.

To prepare a sulfate solution suitable for hydrolysis to yield pigment, ilmenite ore containing 48 per cent titanium dioxide is ground to pass 200 mesh and heated with sulfuric acid.[35] At 165° C an exothermic reaction is initiated which causes a rise of temperature to 200° C with violent evolution of steam. Duration of the reaction is from 1 to 2 minutes. The product is then heated to dryness at 220° C, after which it is cooled and water is added to effect solution. Filtration and washing with hot 5 per cent sulfuric acid gives a recovery of 90 to 92 per cent of the titanium dioxide in the solution. Ilmenite concentrates from wehrlites containing 30 to 44 per cent titanium dioxide are decomposed by reaction with 85 per cent sulfuric acid at 180° C for 2 hours.[36] Consumption of sulfuric acid is 4.6 kg per kilogram of titanium dioxide in the ore. The solution is used for the manufacture of titanium dioxide pigment. The optimum conditions for digesting Taiwan ilmenite in sulfuric acid are: uniform ore particle size of 200 mesh or finer, 75 per cent sulfuric acid concentration, and two parts of acid (100 per cent basis) to one part of ilmenite.[37] At 140 to 150° C the reaction time required is 7 hours.

Titanium dioxide pigment containing less than 0.5 per cent chromic oxide is manufactured from an ore containing more than 0.5 per cent of this impurity.[38] In an example, 200 g of ilmenite ore, ground to pass a 200 mesh screen, was placed in an iron pot and 393 g of 83 per cent sulfuric acid was added. The iron pot was placed in an oil bath to maintain the temperature of the mixture at 160° C for 1 hour. After cooling, the reaction product was added to cold water to obtain a solution containing 83.5 per cent of the titanium dioxide and only 8.6 per cent of the chromic oxide in the original ore. The titanium hydroxide precipitated from the clarified solution by hydrolysis contained 0.18 per cent chromic oxide, although the starting ore contained 1.75 per cent. This ilmenite ore contained titanium dioxide 48.7, chromic oxide 0.85, ferrous oxide 30.4, ferric oxide 15.3, manganese oxide 2.0, vanadium pentoxide 0.22, phosphorus pentoxide 0.035, and silicon dioxide 0.3 per cent.

Pulverized ilmenite is first digested with 35 per cent hydrochloric acid to remove 90 per cent of the iron. The residue is then digested with 95 per cent sulfuric acid.[39] This process is advantageous if the iron content of the ore is high and the cost of hydrochloric acid is lower than that of sulfuric acid.

A mixture of ilmenite with sulfuric acid is fed into a vertical furnace against a current of air at 500° C to effect reaction.[40] The soluble titanium sulfate is cooled to 150 to 250° C by the evaporation of water. The reaction of sulfuric acid with ilmenite or titanium slag is regulated by the length of the reaction trough and the rate of transport of the reaction mixture.[41] At the end of the trough the granular mass drops into a receptacle where the reaction terminates at a higher temperature. A slag containing 98 per cent titanium monoxide is digested with concentrated sulfuric acid and extracted with hot water.[42] After adding sugar, ammonium hydroxide, and sodium sulfide, the extract is heated to give a precipitate which is removed from the solution by filtration. The titanium remains in the filtrate.

The digestion reaction of titanium ores or slag with sulfuric acid is caused to proceed uniformly without periods of violent effervescence by charging the reactants slowly in large tanks so that the first charge forms a solid cake before other portions are added.[43] With 200 mesh material the ratio of sulfuric acid to titanium dioxide is 1.7 to 2.4. The sulfuric acid concentration at the start is 91 to 98 per cent, but water is added to dilute it to 86 to 91 per cent and to raise the temperature to 100 to 175° C for rapid reaction. For example, 17 per cent of a charge of 20 tons of ore containing 48.3 per cent titanium dioxide and 33 tons of 93 per cent sulfuric acid reacted with 455 pounds of water and 455 pounds of steam in 13 minutes to form a cake at 150° C. The remainder of the charge was then added over a 45-minute period at a rate of 2360 pounds of ore-acid mixture and 69 pounds of water per minute. A maximum temperature of 203° C was reached. Digestion of ilmenite or slag with sulfuric acid is carried out in a batch type operation in which the heat of reaction and heat of dilution are absorbed by the reactants so that only minor amounts of sulfur trioxide, sulfur dioxide, and sulfuric acid mists pass out of the reaction exhaust.[44] For example, 2000 g of 325 mesh ilmenite is mixed with 2929 g of 95.6 per cent sulfuric acid. The slurry in the feed tank is heated to 97 to 160° C with injected steam. After reaction the temperature rose to 202° C. The temperature was then lowered by passing secondary heating zone gases through a by-pass, thereby decreasing the volume of gases passing through the primary heating zone.

In a process for decomposing titanium ore, 5000 g of circulating 60 per cent sulfuric acid, preheated to 130° C is mixed with 1000 g of ilmenite and 1600 g of concentrated sulfuric acid, and stirred for 2 hours.[45] In a few minutes the temperature rises to the boiling point of the mixture, about 165° C. At the end of the reaction the temperature is 155 to 165° C. The resulting acid-crystal mixture is cooled to 60° C and filtered. Approximately 2700 g of sulfates and the amount of circulation acid introduced are recovered. The filter cake is washed with water to remove part of the acid if required, and extracted with 0.8 part of water at 60° C. During the extraction iron III is reduced with scrap iron until a small proportion of titanium III is formed. After the usual processing, the liquor contains 250 to 260 gpl titanium dioxide, 40 gpl iron, and 500 gpl active sulfuric acid.

Espenschied [46] dissolved ilmenite directly in sulfuric acid without the formation of a solid product. Titaniferous iron ores and concentrates are reacted with sulfuric acid at an acid-to-ore ratio slightly in excess of the stoichiometric proportions under refluxing conditions to maintain the materials in the fluid state. In carrying out the process sulfuric acid of approximately 70 per cent concentration is added to finely ground ilmenite or concentrate with stirring, after which the mixture is heated to boiling and refluxed for 4 to 6 hours. Under these conditions the reaction mass remains liquid during the entire dissolution period. The amount of acid must be calculated from the known contents of titanium, iron, and other acid-consuming elements in the original ore. Any excess acid should not exceed 10 per cent of the amount needed to react with the titanium content to form titanium sulfate. In an example, 200 g of ilmenite ore analyzing 45 per cent titanium dioxide and 36 per cent iron oxide was ground so that 90 per cent passed through a 325 mesh screen. The ground ore was added to 543 g of 70 per cent sulfuric acid. This amount of acid was sufficient to combine with all the titanium and iron present plus an excess amounting to 8.5 per cent of that required to combine with the titanium. The mixture was agitated, rapidly heated to boiling, and maintained at the boiling temperature with agitation in a vessel equipped with a reflux condenser for 5.5 hours. As the reaction proceeded, the temperature of the boiling mixture decreased. The digestion mass remained in a fluid state throughout the entire reaction. After dissolution was completed, the specific gravity of the solution was adjusted by dilution with water. Clarification and filtration were carried out in the normal manner. The clarified solution had a specific gravity of 1.525 at 60° C and contained 96 per cent of the titanium in the original ore.

## DIGESTION WITH WASTE ACID

Titanium is extracted from low-grade titaniferous magnetite with waste sulfuric acid from petroleum refineries.[47] The ores contain about 7 per cent titanium dioxide. More than 80 per cent of the titanium is extracted by the 17.4 per cent sulfuric acid solution. Higher concentrations of sulfuric acid result in lower yields of titanium. A 2-hour treatment with 1.5 to 2 equivalents excess acid at 180 to 210° C gave optimum extraction. Heating the solution for 3 hours at 85 to 90° C at a pH of 2.01 after reduction of the high ferric iron content with iron filings precipitated hydrolytically 97 per cent of the titanium dioxide. Waste acid liquor from the hydrolysis of titanium sulfate solution is reused to digest ilmenite ore until the ratio of the concentration of vanadium pentoxide to titanium dioxide reaches the value of 0.07 to 0.1.[48] At this stage the vanadium is separated. The solution is treated with scrap iron to convert the ferric iron to the ferrous state, after which it is cooled to crystallize out most of the iron as ferrous sulfate. After separation of the crystals, the solution is hydrolyzed to produce hydrous titanium dioxide and sulfuric acid at about 25 per cent concentration. Waste acid from the hydrolysis of sulfate solutions is reused for the digestion of ilmenite ore or slag.[49] Heat liberated in the exothermic reaction between the ore and concentrated sulfuric acid is utilized for the evaporation of the dilute hydrolysis acid. By judicious use of hydrolysis acid and new concentrated sulfuric acid it is possible to effect savings of 0.9 ton of acid per ton of titaniferous slag. Incompletely decomposed portions of the titanium ore after treatment with hydrolysis acid can be filtered off and retreated to give a total recovery of 97 per cent as compared with the one-step recovery of 93 per cent.

## TREATING THE ORE TO IMPROVE DIGESTION AND PREVENT FROTHING

The efficiency of digestion in sulfuric acid of ilmenite from inland deposits is improved by preliminary washing of the ore in caustic solutions of 0.01 to 20 per cent concentration to remove organic contaminants that hinder the settling of the mud residue from the sulfate solution.[50] The grain size of the ore should be 200 to 230 mesh. Preferably it is treated as a 50 to 60 per cent slurry with 0.5 to 5.0 per cent caustic based on the weight of ore for 1 to 30 minutes before drying or other treatment. If the ore is chlorinated this treatment reduces the chlorine consumption up to 50 per cent. In the treatment of ilmenite from Trail Ridge, Florida, an aqueous slurry of 60 per cent solids in 0.1 per cent sodium hydroxide solution is agitated for three consecutive periods of 3 minutes each in scrubbers. After a 30 per cent dilution with water, the slurry is classified. The residue is washed, dried, and treated electrostatically and magnetically to give a 60 to 80 per cent titanium dioxide concentrate. This concentrate is ground until 90 per cent is finer than 325 mesh, digested with sulfuric acid, and dissolved in water. The solution, clarified with copper sulfate and ferrous sulfide, contains 90 per cent of the titanium dioxide in the overflow. Without the sodium hydroxide treatment, the recovery is 85 per cent, the capacity of the settling tank is only 70 per cent as high and the volume of the underflow mud is 1.55 times as large.

Ilmenite flotation concentrates produced with the use of fuel oil–oleic acid mixtures are treated with 0.05 to 0.2 per cent of amines to prevent excessive frothing during subsequent sulfuric acid digestion.[51] In the most effective amines two of the groups are alkyl containing 12 to 20 carbon atoms, while the third is hydrogen or a saturated alkanoic acyl group containing up to 20 carbon atoms. Suitable treating agents are dioctadecylamine, dioctadecylstearamine, and dioctal decylstearamide sulfate. In the digestion of flotation concentrates of titaniferous iron ores with 88 per cent sulfuric acid, violent unpredictable foaming is prevented by adding to the ore 0.05 to 0.2 per cent by weight of a monoglyceride of a saturated fatty acid containing 10 to 20 carbon atoms.[52] The organic materials which cause foaming include fuel oil and asphalt which are used to decrease wind losses in ore piles, as well as stearic acid, oleic acid, soap, and other agents used in the flotation. Effective antifoaming agents are glycerol monostearate, glycerol monopalmitate, glycerol monomyristate, or glycerol monolaurate. To remove fuel oil and oleic acid used in the flotation of ilmenite the concentrate is slurried in water, acidified to a pH of 1.5 to 2.5, and agitated with 0.5 to 2.0 pounds per ton of sodium hexametaphosphate.[53]

## ORES OTHER THAN ILMENITE

In a method of producing titanium dioxide pigment from natural rutile,[54] a finely ground mixture of the ore and magnesium oxide was first heated to form the titanate and this product was dissolved in aqueous sulfuric acid. The resulting solution was filtered and cooled to 0° C to crystallize the magnesium sulfate, and the filtrate was heated to effect hydrolytic precipitation of metatitanic acid. This material was filtered, washed, and calcined according to accepted methods, and the sulfuric acid was recovered from the waste liquor.

Ryan and Knoff [55] fused finely ground rutile ore with an alkali metal compound, such as sodium hydrogen sulfate or potassium carbonate, in the proportion required to produce the corresponding acid-soluble titanate. This required the equivalent of 1 part sodium oxide to 5 parts of titanium dioxide. The reaction product was digested with 93 per cent sulfuric acid at 200° to 220° C to convert the titanium component to soluble sulfates. These were dissolved in water, and the solution was clarified and run slowly into boiling water containing a small amount of oxalic acid to effect precipitation of hydrous titanium oxide.

Titanium dioxide in the form of rutile having excellent pigment properties is produced from rutile ore by a process including the hydrolysis of a sulfate solution.[56] Pulverized rutile ore is sintered for 1 to 2 hours at 700 to 900° C with such an amount of an alkali metal hydroxide or carbonate that the titanate obtained contains 60 to 120 parts of alkali metal per 100 parts of rutile. After cooling, the titanate is treated with sulfuric acid for 0.5 hour and the resulting sulfate is dissolved in water. After seeding with 5 per cent of a separately prepared nucleating suspension, the solution is boiled to hydrolytically precipitate the titanium. The precipitate is filtered, washed, and calcined at 700 to 900° C to give pure titanium dioxide in the form of rutile having excellent pigment properties. If ilmenite is employed as the raw material, the process yields pigment of the anatase crystal form.

A process employing siliceous ores of the sphene class was reported by Alessandroni,[57] who found that if the finely ground material was heated with hydrochloric acid of more than 15 per cent strength, the titanium component was dissolved and held in solution along with the calcium and minor metallic constituents such as iron, magnesium, aluminum, and vanadium, leaving a residue primarily of silica. In operation the ore was heated with 10 to 20 per cent excess acid under a reflux at 40° to 50° C for several hours to avoid loss of hydrogen chloride and inhibit hydrolysis until an appreciable amount of titanium had gone into solution. During the next 24 hours the temperature was raised gradually to the boiling point, and heating was continued until the optimum yield of titanium was obtained. The system was cooled, and the solution was separated from the siliceous residue by decantation and filtration and heated with sulfuric acid to convert the chloride to sulfates. For reason of economy the hydrogen chloride liberated was recovered for future use. By controlling the temperature, concentrations, and method of mixing, anhydrite of pigment grade could be produced as one product. The suspension was filtered and the sulfate solution was concentrated by evaporation under reduced pressure to 10 to 20 per cent titanium dioxide and subjected to thermal hydrolysis by conventional methods to produce relatively pure titanium dioxide.

Composite pigments containing calcium sulfate or calcium sulfate and silica as extenders were obtained by leaving one or both of these components suspended in the titanium sulfate solution and carrying out the hydrolysis in their presence. As in other methods, any iron in the solution was reduced to the ferrous state.

The decomposition of perovskite with sulfuric acid is carried out in the presence of additives to accelerate the solidification of the titanyl sulfate.[58] The resulting decomposition product is heated to 300 to 400° C to drive off the excess sulfuric acid. Perovskite is decomposed by sulfuric acid with the formation of a double sulfate of titanium and calcium.[59] The product is leached under conditions to produce a coarsely crystalline precipitate of gypsum. Titanyl sulfate is separated by a salting out process.

Perovskite and sphene concentrates are decomposed by sulfuric acid or a mixture of sulfuric acid and ammonium sulfate in a continuous process in a paddle mixer type of apparatus.[60] The concentrate should be ground to the extent that only 5 per cent is retained on a 200 mesh screen. If the ore is fused with a mixture of sulfuric acid and ammonium sulfate, the resulting melt contains a coarse crystalline double sulfate, $(NH_4)_2$-$Ti(SO_4)_3$. These melts have a medium viscosity, do not set and are easily conveyed by small paddles. Melts consisting of coarse crystalline double sulfates of titanium with calcium and rare earth elements lower the capacity of the apparatus as a result of the increased viscosity. Heat transfer is a determining factor in the production capacity of the paddle mixer.

Perovskite[61] containing 40.8 per cent titanium dioxide, 30.7 per cent calcium oxide, and 5.2 per cent ferric oxide was converted to water soluble salts by heating the finely ground ore with twice its weight of 93 per cent sulfuric acid at 150° to 170° C for 1 hour, and then at 150° C for 1 hour more. Recovery of the titanium dioxide was 90 to 93 per cent. The mineral titanite[62] was likewise converted to a dry, water-soluble solid mass by digesting with 1.5 parts of 80 per cent sulfuric acid at 140° to 145° C for 10 to 11 hours. The product was agitated with water at 97° C for 7 hours, and metallic iron was added to the solution to reduce the ferric component to the ferrous state.

According to a process for producing titanium pigments from sphene[63] (titanite), the material ground to 200 mesh was mixed with 1.6 parts of hydrochloric acid as a 5 to 12 per cent solution and boiled under reflux to extract the soluble nonhydrolyzable constituents such as oxides of alkali metals and iron. Employing acid of this concentration, any titanium that dissolved was immediately hydrolyzed at this temperature and precipitated as hydrous oxide. Such a treatment required 72 hours for completion and yielded a solution containing calcium, magnesium, iron and manganese and other constituents which form nonhydrolyzable salts, and a residue consisting essentially of the titanium dioxide, silica, and alumina of the original ore. The undissolved portion was then removed from the liquor by decantation or filtration, washed, and dried. A partial mechanical separation was effected at this stage by selective dispersion in water with such agents as chlorides of aluminum, titanium, cerium, and zirconium which dispersed the titanium oxide to a greater degree and held it in aqueous suspension for a longer period of time than the impurities. The titanium-containing residue was agitated for 1 or 2 hours with water containing 0.5 to 5 per cent aluminum chloride (based on the solids) to form a uniform free-flowing slurry. Stirring was then discontinued and the mixture was allowed to stand until the silica and other impurities settled; this took place fairly rapidly. The suspension of hydrous titanium oxide was separated by decantation or elutriation, coagulated with magnesium sulfate, and filtered. In either operation the concentrate was dissolved in strong sulfuric acid to form solutions from which pure oxide was precipitated by thermal hydrolysis.

In an example, 150 pounds of finely ground sphene containing 32 per cent titanium dioxide was added to 180 gallons of 12 per cent hydrochloric acid at room temperature and boiled under reflux for 48 hours. The residue was separated from the supernatant liquor by filtration, washed, and dried. Analysis showed 46.2 per cent silica, 5.7 per cent ferric oxide, and 48 per cent titania which represented 98.8 per cent of the titanium in the original ore. One half the residue was heated with an equivalent amount of concentrated sulfuric acid, and the reaction product was leached with water to obtain a solution of titanium sulfate. All the insoluble material, largely silica, was separated, and the solution was treated with zinc dust to reduce the iron salts to the ferrous condition and a small proportion of the titanium to the trivalent state. It was then subjected to thermal hydrolysis and the precipitate of hydrous titanic oxide was washed, filtered, calcined, and pulverized. The other half of the original residue was treated in the same manner, except that the silica was not separated from the solution before hydrolysis. This gave a composite pigment consisting of approximately equal parts of titanium dioxide and silica.

Anderson and Williams [64] sintered an intimate mixture of titanium silicate mineral (sphene) with limestone and sodium sulfate, leached the cooled product with water, and treated the clarified solution with carbon dioxide to precipitate titanium hydroxide.

According to a method developed for processing Japanese iron sands containing titanium and vanadium, the solution obtained by dissolving the finely divided material in sulfuric acid was first diluted with water to 10 times the volume of the original ore and treated with scrap iron to neutralize the excess acid.[65] By thermal hydrolysis 93 per cent of the titanium was precipitated from the clarified solution as hydrous oxide. Iron was recovered from the filtrate by crystallization as ferrous sulfate, and the vanadium was obtained from the evaporated liquor as the pentoxide.

Indian bauxites normally contain from 9 to 12 per cent titanium dioxide. A method for the recovery of this constituent from residual muds after the manufacture of aluminum sulfate was described by Chakravarty.[66] Such muds were first heated with dilute sulfuric acid to dissolve most of the remaining iron, then washed with water, dried, and pulverized. After this purification step, the residue was digested with strong sulfuric acid and the reaction product was dissolved in water. The solution was clarified, treated with scrap iron to reduce the ferric sulfate to the ferrous salt, and boiled to effect hydrolytic precipitation of the titanium sulfate as hydrous oxide. This product was filtered, washed, and calcined. From 100 pounds of mud was obtained 42 pounds of 98 to 99 per cent titanium dioxide suitable for white pigment, vitreous enamel, and soap making.

Swarup and Sharma [67] obtained a 75 per cent recovery of titanium dioxide of 99 per cent purity from the red mud left behind in extracting aluminum oxide from Indian bauxites. The mud was calcined and mixed with sulfuric acid of specific gravity 1.6. Steam was passed in until the specific gravity was reduced to 1.1, and then the product was filtered and washed. This titanium-rich residue was treated with sulfuric acid of specific gravity 1.6, and steam was passed in until the temperature reached 60° C. The liquor was diluted with water to a specific gravity of 1.2 and filtered. Scrap iron was introduced to reduce all ferric ions to ferrous, and the solution was hydrolyzed after adding freshly prepared titanic hydroxide nucleating agent.

According to another modification, the air-dried sludge was treated with sulfuric acid of specific gravity 1.4 at 80° to 90° C, and the titanium-rich residue was separated, washed, dried, mixed with sulfuric acid of specific gravity 1.89, and heated at 250° to 300° C in a muffle furnace.[68] The reaction product was dissolved in water, and the solution was clarified and hydrolyzed according to conventional methods. Desai and Peermahomed [69] digested bauxite sludge with 10 normal sodium hydroxide at 100° C, then heated the washed and dried product with concentrated sulfuric acid for 8 hours at 130° C. Iron filings were added to reduce the iron component and part of the titanium to the next lower state of oxidation, and the solution was filtered and heated at 100° C to effect hydrolytic precipitation of $H_2TiO_3$. The precipitate was filtered, washed with dilute sulfuric acid, water, dilute sodium hydroxide solution, and again with water, then calcined to produce a snow-white product containing 97 per cent titanium dioxide.

Solutions containing from 2 to 3 parts of titanium dioxide to 1 part of ferric oxide are prepared by the extraction of red mud with 5 to 50 per cent sulfuric acid.[70] Any residual sulfuric acid can be used for the extraction. The solution is neutralized with the final product consisting of titanium dioxide and iron, and then the titanium dioxide concentrate is precipitated with a basic reagent. For example, 10 g of a red mud containing titanium dioxide 11.4 and ferric oxide 76.5 per cent was extracted with 25 ml of 10 per cent by volume sulfuric acid for 120 hours to give a solution containing 16.3 gpl titanium dioxide and 5.7 gpl ferric oxide. After separation of the solid residue, the original acidity of the solution was restored with sulfuric acid. The extraction was separated twice under identical conditions to give a final solution containing 40.3 gpl titanium dioxide and 14.4 gpl ferric oxide.

A concentrate containing 72 to 78 per cent titanium dioxide is obtained from red earth from bauxite.[71] For example, 200 g of red earth containing titanium dioxide 38.4, ferric oxide 20.4, and sodium oxide 7.6 per cent was heated for 1 hour at 45° C with 16 ml of an acid solution containing sulfuric acid 47, alumina 7.1, and ferric oxide 17.8 gpl. The product was filtered, washed, and dried to obtain 152 g of residue containing titanium dioxide 49.8 and ferric oxide 36.6 per cent. This residue was heated for 2 hours at 150° C under a pressure of 4 atmospheres with 610 ml of a solution containing 295 gpl sulfuric acid, filtered, and washed to give 120 g of a concentrate containing 72.8 per cent titanium dioxide, 4.5 per cent alumina, 11.8 per cent ferric oxide, and 2.5 per cent silica. The acid solution was recycled. Titanium dioxide can be recovered from bauxite sludge by the formation of $RSO_4 \cdot Ti(SO_4)_2$, which can be hydrolyzed and ignited to give a pure form of the oxide suitable for pigment purpose.[72] Red mud from bauxite purification is digested to dissolve the oxides and pigment grade titanium dioxide is produced by hydrolyzing the solution in the presence of oxalic acid.[73] About 50 per cent of the alumina is recovered as aluminum sulfate, and the iron is recovered as Prussian blue. The acid can be concentrated and reused.

Titanium dioxide pigment prepared by the usual procedures from Indian bauxite sludge is of inferior quality as a result of the large amount of alumina present. To obtain good quality pigment, the sludge is first upgraded by treatment with 5 normal hydrochloric acid for 2.5 hours at 80 to 90° C.[74] A slight excess of concentrated sulfuric acid is then added. The solid reaction product is powdered and heated to 250 to 300° C for 4.5 hours. This cured product is extracted with hot water, and potassium sulfate is added to the hot solution. On cooling about two-thirds of the total amount of dissolved alumina crystallizes out as potassium alum. The remaining solution of titanium salts now containing but little alumina is hydrolyzed by the usual method to yield titanium dioxide of high quality. Attempts to upgrade red muds containing 1.4 per cent titanium dioxide by chemical treatment such as extraction with hydrochloric acid or sodium hydroxide, or reduction with carbon or methane were economically successful.[75] Extraction with a mixture of hydrochloric and perchloric acids gave recoveries of 60 to 70 per cent of the titanium dioxide and a hydrolysis of 80 per cent purity.

## DIGESTION OF TITANIA SLAG WITH SULFURIC ACID

Forbath[76] described the method of production of titanium dioxide pigment from Sorel slag, rather than from ilmenite, with the aid of a pictured flow sheet of the Varennes, Quebec, plant of the Canadian Titanium Pigments Company. The slag is digested in sulfuric acid and processed in the usual way except that crystallization of the ferrous sulfate, filtration, and concentration of the solution are eliminated.

High titania slag made by electric arc smelting and containing 60 to 70 per cent titanium dioxide is used for pigment manufacture.[77] This slag is digested with sulfuric acid, and the product is dissolved in water to form a solution of titanyl sulfate. On hydrolysis in the usual manner, the solution yields a product which is processed to pigment grade titanium dioxide. According to Irkov and Reznichenko[78] slags from the smelting of titanomagnetites dissolve readily in sulfuric acid, and from the leach liquor more than 90 per cent of the titanium dioxide is recovered. The consumption of acid is 10 to 12 per cent greater than in the manufacture of titanium dioxide from ilmenite as the raw material. By leaching the slag with hydrochloric acid, concentrates containing 69 per cent of titanium dioxide are obtained. Leaching with hydrochloric acid removes the main content of aluminum. Chlorination of the concentrates gives titanium tetrachloride with 99 per cent yield. For chlorination roasting, the fluo-solids procedure is most suitable.

Titanium dioxide pigment is produced from slag containing from 65 to 90 per cent titanium dioxide together with 1 to 16 per cent ferrous oxide by the sulfate process.[79] To control the rate of the digestion reaction the slag is ground so that all of the material passes through a 200 mesh screen, but not more than 30 per cent of the particles have an average diameter smaller than 8 microns, and at least 90 per cent of the particles are smaller than 44 microns in average diameter, that is, through 325 mesh. Sulfuric acid of 92 to 98 per cent concentration is employed. The slag-acid mixture is heated with steam to 160 to 190° C before reaction takes place. Heat is liberated to carry the reaction to completion. The solid digestion cake is blown with air to hasten cooling and to produce a porous mass. Water is added to dissolve the soluble sulfates and produce a slurry having a density of 1.47 to 1.55. The addition of water is regulated to hold the temperature of the solution below 75° C to avoid premature hydrolysis. Ordinarily all of the soluble materials dissolve in 4 hours, but it may be necessary to agitate the slurry for periods up to 16 hours to insure complete solution.

The aqueous slurry is clarified by the addition of 0.03 to 0.06 per cent base on the weight of the slurry of animal glue and by the addition of antimony oxide in 10 per cent aqueous sodium sulfide solution. After the solids have flocculated, the slurry is filtered to yield a clear solution of a density of 1.4 to 1.5. This step may be carried out in a continuous process. The liquor contains much less ferrous sulfate than liquor obtained from ilmenite, with the result that it contains 9.5 to 15.5 per cent titanium as titanium dioxide. Concentration by evaporation is not necessary if the solution contains 13 per cent or more titanium dioxide. A separately prepared nucleating agent is employed, and the solution is boiled to precipitate the titanium as hydrous oxide. The washed product is calcined to either anatase or rutile pigment.

In an example, a slag containing 67 per cent titanium dioxide was ground so that all of it passed through a 200 mesh screen, 75.8 per cent was greater than 10 microns in diameter, and 54.4 per cent was larger than 20 microns in diameter. Sulfuric acid of 98 per cent concentration equivalent to the basic components except titanium and 70 per cent of the equivalent to convert the titanium to titanyl sulfate was added to the digestion and the mass was heated to 60° C. The ground slag was gradually added with stirring to insure thorough wetting of the particles, after which the mixture was heated with steam to 180° C. At this temperature a vigorous exothermic reaction took place. After a reaction period of 4 hours in which time the temperature reached 215° C air was blown through the cake. Water was then run into the digester, care being taken to keep the temperature of the solution below 75° C. The slurry was agitated for 4 hours. A 10 per cent animal glue solution was then added to the slurry and a solution of antimony trioxide containing 0.004 per cent antimony trioxide based on the weight of the slurry was added in a 10 per cent aqueous sodium sulfide solution. After flocculation was complete a clear liquor having a density of 1.48 was obtained. Recovery of titanium dioxide was 71.9 per cent. The clarified titanium sulfate solution was treated with 3 per cent, based on the titanium dioxide, of a sodium titanate nucleating agent, the liquor having been heated to 50° C prior to the addition. The mixture was heated to boiling and boiled for 5 hours to effect hydrolysis. The precipitated hydrolyzate was washed, treated with 2.5 per cent rutile seed, and calcined at 850° C for 4 hours.

In actual production, to recover 1 metric ton of titanium dioxide pigment from slag requires 2.51 metric tons of sulfuric acid, while the recovery of 1 metric ton of titanium dioxide pigment from 42 per cent titanium dioxide ilmenite requires 3.75 metric tons of sulfuric acid.[80] Grinding the slag so that 90 per cent passes through a 250 mesh screen and maintaining a ratio of sulfuric acid to titanium dioxide of 1.9 to 2.1 give optimum recovery of 95 to 96 per cent for slags containing 75 to 85 per cent titanium dioxide. Chemical composition of the slags was titanium dioxide 65 to 85, ferrous oxide 0.5 to 10, silica 4 to 8, alumina 7 to 14, magnesium oxide 7 to 8, manganese oxide 1 to 5, and calcium oxide 1 to 5 per cent. High purity, white titanium dioxide is obtained from high-titania slag containing large amounts of chromium and silicon.[81] One kilogram of high-silica slag containing 62 per cent titanium dioxide, pulverized to 100 mesh, is mixed with 1 kg of powdered calcium fluoride and transferred to a lead pan. Three kilograms of concentrated sulfuric acid are added and the mixture is heated for 2 hours. After cooling, the product is leached several times with 3 kg of denatured alcohol. The acid of the filtrate liquor is neutralized and dehydrated, and 5000 ml of concentrated aqueous ammonia is added to completely precipitate titanous hydroxide. After filtration and washing, the hydroxide is calcined to obtain 573 g of high purity, white titanium dioxide.

A study of the digestion of high titania slags indicates that optimum conditions are two parts of 92 per cent sulfuric acid to one part of slag having a particle size of 40 to 60 microns.[82] Initial temperature is at least 115° C with a reaction temperature of 180° C. Time of the reaction is from 2 to 3 hours. The sulfuric acid consumption, fineness of grinding, and yield (96 per cent) are more advantageous in comparison with ilmenite as the raw material, but the decomposition cannot be carried out in troughs in which the time of delay is 20 to 30 minutes. The liquor obtained by dissolving the decomposed mass contains only a small amount of titanium III, 200 to 220 gpl titanium dioxide and 30 gpl ferrous oxide equivalent so that in comparison with solutions from ilmenite, the reduction of iron III, evaporating in vacuo, and separation of ferrous sulfate before hydrolysis can be eliminated.

The recovery of titanium dioxide from slags containing 75 to 85 per cent titanium dioxide and from ilmenite containing 43 per cent titanium dioxide with the sulfate process were studied in 5 and 10 cubic meter reactors charged with 1 to 2 tons of the powdered solid and 91 to 92 per cent sulfuric acid at a solid to liquid ratio of 1 to 1.9.[83] After solidification, the reaction mass was treated with steam at 175° C. The temperature rose to 185° C by the heat of the reaction. After 4 to 6 hours of leaching, the solution obtained from the slag product contained 206 to 232 gpl titanium dioxide, had a density of 1.56 to 1.58, had an iron to titanium dioxide ratio of 0.04 to 0.09, and contained 17 gpl titanium III. Under the same conditions the solution obtained from the ilmenite product contained 120 gpl titanium dioxide after 7.5 hours leaching and the iron to titanium dioxide ratio was 1. Thus without reducing the recovery of titanium dioxide, substitution of the slag for ilmenite eliminated the formation of ferrous sulfate. This eliminated the need for crystallization, filtration, and concentration and decreased the amount of sulfuric acid required. The anatase content of high titania slags readily dissolves in sulfuric acid, but rutile is dissolved with difficulty.[84] The anatase-rutile transformation is irreversible. In the reduced state much titanium dioxide dissolves in $(MgFe)O \cdot 2TiO_2$, but the solid solubility decreases rapidly with oxidation. Rutile is precipitated at high temperatures while anatase is precipitated at low temperatures. The prevention of high temperature oxidation is very important in obtaining good yields of titanium dioxide. Rapid quenching is also effective with slags containing 82 to 74 per cent titanium dioxide. The addition of 6 to 8 per cent magnesium oxide is most effective.

In the digestion of titanium-containing slags with sulfuric acid, the titanium III in the slag undergoes an oxidation-reduction reaction with the liberation of hydrogen sulfide and sulfur dioxide.[85] This reaction with the evolution of the gas is prevented by adding to the slag-sulfuric acid mixture 2 to 10 per cent of the amount of iron III as hematite or pigment grade ferric oxide, theoretically required to oxidize the trivalent titanium. Trivalent titanium in slags or concentrates is oxidized to the tetravalent state that can be digested in concentrated sulfuric acid.[86] The titaniferous material is agglomerated and oxidized partially in the presence of dilute sulfuric acid with air at temperatures below 100° C, after which oxidation is completed by air at gradually increasing temperatures within the range of 250 to 400° C. In the process the temperature and the amount of air are controlled in such a way that autooxidation with subsequent glowing of the material is avoided. By beginning the oxidation at a temperature at which spontaneous ignition does not take place and subsequently increasing the temperature as the oxidation goes on, it is possible to carry out the complete oxidation in a relatively short time. A material oxidized in this way consists of a mixture of anatase and a compound corresponding to iron-magnesium-dititanate. In an example, a concentrate having an average particle size less than 5 microns and containing the equivalent of 80 per cent titanium dioxide with 55 per cent of the total titanium in the trivalent state was rolled to pellets 8 to 10 mm in diameter. Dilute, 1 to 20, sulfuric acid was added in the pelleting process. The pellets were treated with air so that the temperature rose to 83° C after 6 minutes. After 10 minutes the content of trivalent titanium in the pellets was reduced to 28 per cent of the total. The pellets were then placed in an 8 cm deep layer on a stationary grate in a furnace. A mixture of combustible gas and air was introduced at the bottom of the furnace and drawn through the layer of pellets by induced draft. The temperature of the introduced gases was held at 290° C.

The trivalent titanium of slag is oxidized to the tetravalent state by blowing air through the porous sulfuric acid digestion mass maintained at an elevated temperature.[87] The process is applicable both to the porous cake obtained in the batch digestion and to the granular mass obtained in continuous digestion. The length of the oxidation treatment for a porous cake produced from a slag of any specific reduced titanium content varies inversely with the temperature maintained. For example, a porous cake which can have its reduced titanium oxidized to the desired value in 2.5 hours at 170° C can be oxidized in 30 minutes at 200° C. Heat may be supplied by a heating jacket surrounding the digester, by electrical resistance space heaters applied to the interior surface of the digester, or by the introduction of steam into the porous reaction cake along with the air. Temperatures employed may be from 150 to 260° C, although the preferred range is from 170 to 200° C. Trivalent titanium is not precipitated in the hydrolysis step and is lost in the pigment process unless previously oxidized to the tetravalent state.

High-titania slags are subjected to an oxidation roast before digestion with sulfuric acid to produce solutions for hydrolysis.[88] These slags produced in the electric furnace contain 60 to 80 per cent titanium oxides with the remainder primarily iron, calcium, magnesium, aluminum, and silicon compounds derived largely from the reducing agent and flux. From 10 to 20 per cent of the titanium is in the reduced form. Before digestion the slag is ground so that 90 to 95 per cent passes through a 325 mesh screen. If this untreated slag is used in the digestion step an oxidation-reduction reaction takes place between the lower valent titanium compounds and the sulfuric acid with the result that some of the sulfuric acid is reduced to sulfur dioxide, sulfur, or hydrogen sulfide. This results not only in the loss of sulfuric acid but also in a fume nuisance. Furthermore, the lower valent titanium compounds are not hydrolytically precipitated.

If the slag is subjected to a preliminary oxidation roast it may be handled in much the same manner as ilmenite ore. Such an oxidation roast may be carried out in a rotary calciner, a Wedge- or Henshoff-type furnace, a fluidized bed, or a stationary or a rabbled hearth furnace. The finely ground slag is heated in air at 400 to 500° C for 1 hour until all of the lower valent titanium has been oxidized to the dioxide and a small part of the ferrous oxide has been reduced to ferric oxide. In an example, slag containing 72 per cent titanium dioxide, 14.1 per cent titanous oxide, 8.17 per cent total iron and 0.06 per cent metallic iron, and ground to a fineness that 94.1 per cent passes through a 400 mesh screen, was roasted in air in a rotary calciner at 500° C for 0.75 hour. All of the titanium oxides were oxidized to the dioxide and a small part of the iron was oxidized to the ferric state. A sample of the slag roasted at 400° C for 0.75

hour contained 2 per cent titanous oxide, and none of the iron was oxidized to the ferric state. The oxidized slag contained 72 per cent titanium dioxide, 8.0 per cent total iron, 0.8 per cent titanium trioxide, and 0.1 per cent ferric oxide. A batch of 380 pounds of this slag, 72 per cent of which was finer than 400 mesh, was added with agitation to 45.5 gallons of 98 per cent sulfuric acid, after which 5.4 gallons of water were introduced. Steam was introduced and the reaction proceeded vigorously to completion.

The trivalent titanium in slag is oxidized to the tetravalent state by blowing a mixture of air and superheated steam through the porous sulfuric acid digestion cake maintained at an elevated temperature.[89] Slags from the electric furnace smelting of ilmenite contain up to 30 per cent of the titanium in the lower valent form. By mixing superheated steam with the air the elevated temperature is maintained so that the oxidation process proceeds rapidly. The temperature of the cake is maintained at 170 to 200° C so that oxidation is rapid. A cake that is oxidized at 170° C in 2.5 hours may be oxidized in 0.5 hour at 200° C. Steam is delivered at 100 psi pressure. In a typical reaction 98.2 per cent of the titanium content of the slag was converted to the sulfate. After the violent digestion reaction ceased the porous cake was allowed to bake for 3 hours. At this stage 94.6 per cent of the titanium was converted to the sulfate. The solution obtained by leaching the porous cake with water contained 313 gpl titanium and 35.4 gpl reduced titanium expressed as titanium III.

Wet slag containing reduced titanium compounds is dried and ground continuously while inhibiting oxidation of the lower valent compounds to products insoluble in sulfuric acid.[90] These electric furnace slags contain from 3 to 8 per cent moisture depending on the weather conditions. Efficient reaction of the slag with sulfuric acid requires that it is ground to 325 mesh. Before the slag is ground it must be dry, for any moisture would dilute the sulfuric acid. Furthermore in the grinding operation wet slag is converted to putty-like masses that contain a large proportion of oversize particles which cannot be separated for further grinding as is usual for dry materials. The dried product contains about 0.1 per cent moisture. If the drying and grinding are carried out in air at elevated temperatures a part of the reduced titanium is oxidized to a rutile form which does not react with the sulfuric acid. The rate at which the oxygen of the air converts the reduced titanium compounds to acid insoluble form is a function of both the temperature and the particle size. So after the coarse particles are dried at 175° C or below they are cooled to below 120° C before grinding. In an example, a 2 to 10 mesh slag was dried in a direct fired rotary kiln at 160° C to a moisture content of 0.1 per cent and passed through a hold up bin where the temperature was reduced from

140 to 70° C before it was ground in a ball mill. Only 2.2 per cent of the titanium content was changed to the acid insoluble form.

Finely ground mixtures of high titanium slag containing titanium in a lower valent form and ilmenite are digested in concentrated sulfuric acid.[91] The ilmenite supplies ferric oxide to oxidize the reduced titanium to the tetravalent form and prevent losses of sulfuric acid by conversion to sulfur dioxide or hydrogen sulfide. The ilmenite can be partially replaced by ferric oxide, barium peroxide, or pyrolusite ore. A mixture of ilmenite containing more than 60 per cent titanium dioxide with more than half of the iron in the ferric condition and a slag containing at least 60 per cent titanium dioxide equivalent with some of the titanium in the reduced form is digested with concentrated sulfuric acid to form a solid mass containing soluble titanium sulfate.[92] In general 85 to 95 per cent sulfuric acid is used. The cooled mass is dissolved in water to produce a solution which is hydrolyzed in the conventional manner. The washed hydrolyzate on calcination at 900 to 1050° C yields high-grade titanium dioxide pigment. The lower valent titanium compounds in the slag eliminate the addition of scrap iron to the solution to reduce the iron III to iron II.

## POROUS DIGESTION PRODUCT

The next steps forward were methods of carrying out the reaction between finely divided ilmenite or slag and concentrated sulfuric acid to produce a porous, water-permeable, solid product which could be dissolved readily in the reaction tank without the necessity of first digging out and breaking up the cake. Several procedures have been described, but in general this effect is obtained by blowing air through the reacting mass as it solidifies.

Washburn[93] reacted finely ground ilmenite with concentrated sulfuric acid (72 to 92 per cent), under such conditions as to produce a porous, water-permeable solid to facilitate its subsequent solution. The acid-ore mixture was heated by blowing in hot air or live steam from 80° to 120° C, at which temperature a small amount of water was added to initiate the reaction. A local dilution of the acid took place, producing a localized overheating. At this region the exothermic reaction immediately commenced and energetically propagated throughout the mass with copious evolution of sulfur trioxide fumes. As a rule one injection of water was enough, although more injections were necessary in some cases. During the reaction the mass was kept in an agitated condition by injection of air or other gas to obtain the porous liquid-permeable product. The air stream was continued until the mass solidified to ensure a highly uniform and cellular cake. Instead of heating the mixture in the reaction vessel, the components may be preheated before bringing together. The reaction, which is exothermic, usually liberates enough heat to carry it to completion. With certain ores, notably Norwegian ilmenite, the mixing and stirring of the finely ground material with sulfuric acid may cause the reaction to take place without preliminary heating. In any case the reaction may be controlled by adjusting the volume, temperature, and period of application of the gas. The porous cake is dissolved in the reaction vessel without further treatment.

In an example, 10 tons of pulverized ilmenite was mixed in a cylindrical, conical-bottom tank with 10 to 24 tons of sulfuric acid (72 to 92 per cent), and steam was injected into the mixture through a nozzle at the bottom until the temperature reached 80° to 120° C, usually 100° C. The mass thickened, and at this stage from 2 to 12 quarts of water was introduced through a pipe at the bottom of the vessel to initiate the reaction which rapidly spread throughout the mass. During the reaction, and while the product was solidifying, a vigorous stream of air was injected through a nozzle at the bottom of the tank. After the mass had solidified and cooled to the desired temperature, the air supply was shut off and water or dilute sulfuric acid was introduced from the bottom through the nozzle formerly used for air. To the resulting solution, consisting of a mixture of titanium and iron sulfates, scrap iron was introduced in a lead basket to reduce all the ferric component to ferrous, and a small proportion of titanic sulfate to titanous condition. The residue, consisting essentially of unreacted ore and silica, became suspended in the solution during the process.

According to Cauwenberg,[94] a porous cake may be obtained by adding a small proportion of a carbohydrate or other carbonizable material to the ilmenite-sulfuric-acid mixture before initiating the reaction. The normal reaction is exothermic, and the large amount of steam and other gases liberated within the mass tends to make it porous. The degree of porosity attained, however, depends upon the viscosity of the mass, which in turn determines the extent to which the evolved steam is retained as the product solidifies. Optimum viscosity may be developed by adding the proper proportion of the organic material.

To illustrate, ilmenite ground to 200 mesh was mixed with twice its weight of 66° Bé sulfuric acid and the suspension was heated rapidly to a temperature of 120° to 130° C, and 0.6 per cent dextrin in dry form, based on the ore, was added. The reaction between sulfuric acid and dextrin resulted in a voluminus production of finely divided carbon which greatly increased the viscosity of the mass, and the heat liberated initiated the reaction between the ore and the acid, which then proceeded without external application of heat. Steam was rapidly evolved within the mass, and with the increased viscosity a large proportion of it was retained so that a liquid permeable, solid cake resulted. This porous reaction product was more readily dissolved in water to yield a solution of titanium and iron sulfates. To attain this end, Coffelt [95] first heated ilmenite with sulfuric acid of 76 per cent strength to obtain a semiliquid mixture. A further quantity of acid of at least 95 per cent strength was then added, and the reaction was completed at an elevated temperature to produce a porous solid cake.

The addition of 2 to 10 per cent peat moss to ilmenite facilitates attack by concentrated sulfuric acid and yields a porous digestion mass which can be dissolved easily.[96] The peat moss reduces the ferric iron to the ferrous state and eliminates the use of metallic iron as a reducing agent.

## PRESSURE DIGESTION

Buckman [97] treated titanium ores such as ilmenite and rutile, which had been pulverized and preferably air floated, with sulfuric acid at high temperatures and pressures to obtain water-soluble products. Pressures up to 200 pounds per square inch were employed. After the reaction had gone to completion, water was added to dissolve the titanium and iron sulfates formed. According to a later process, finely ground ilmenite was heated under pressure at 180° C with 20 to 40 per cent sulfuric acid to form a solid mass which was leached with water containing enough trivalent titanium to reduce the ferric iron to the ferrous state.[98]

Titanium ores containing iron in the ferric state, such as arizonite,[99] were subjected at 550° to 650° C to the controlled action of a reducing gas to convert the ferric compound to the ferrous condition without the formation of metallic iron. The treated ore was then digested with sulfuric acid under pressures corresponding to temperatures of 135° to 180° C to form soluble salts of the titanium, iron, and other basic constituents.

Powdered slag from the smelting of iron is heated with 35 per cent sulfuric acid at 300° C under 3 atmospheres pressure for 1.5 hours.[100] The residue containing 2 per cent iron, 0.07 per cent vanadium, 1 per cent titanium, and 0.5 per cent aluminum is boiled for 5 minutes with nitric acid, potassium perchlorate, perchloric acid, or sodium bismuthate to form sulfates of vanadium, titanium, and iron of the highest valence. A 40 per cent solution of sodium hydroxide is added to convert the sulfates to hydroxides and the aluminum hydroxide further to sodium aluminate. The mixture is filtered, and the residual ferric hydroxide and titanium hydroxide are stirred with 15 per cent hydrochloric acid and boiled for 10 minutes with small amounts of oxalic acid and glycerol catalyst. The iron forms a complex oxalate salt while the titanium is precipitated as titanyl hydroxide. After washing and drying, the titanyl hydroxide is calcined to obtain titanium dioxide of 99.8 per cent purity in a yield of 93.7 per cent.

## CONTINUOUS PROCESSES OF DIGESTION

A continuous process for decomposing titanium ores with sulfuric acid was developed by Booge, Krchma, and McKinney.[101] Previous intermittent methods present inherent difficulties, so that such digestions are usually carried out in relatively small or extremely large batches. In the latter case large cylindrical tanks with cone-shaped bottoms are used, and the charge is agitated by injection of steam or air. Ilmenite can be mixed with sulfuric acid of different concentrations at ordinary temperatures without any apparent combination, but on heating gently the reaction becomes so violent, because of its exothermic nature, that it can scarcely be controlled. On feeding the ground ore into heated acid, moreover, the product solidifies before complete decomposition is accomplished; if a large excess of acid is used to overcome this difficulty, the process becomes uneconomical. To overcome these difficulties, a mixture of comminuted ilmenite and strong sulfuric acid was fed continuously into a primary zone where the reaction was initiated by adding a small amount of water, steam, or dilute acid, or by heat. Once started, the reaction proceeded indefinitely as fresh charges were added, particularly if the acid component consisted of oleum and dilute acid liquor. This partially reacted mass, in the form of a semifluid paste, was moved continuously into a second zone held at an elevated temperature where combination proceeded to completion and solidification was effected. The product was subjected to attrition as it was moved forward, and it was discharged as a granular solid.

In a typical operation, oleum was mixed with 55° Bé sulfuric acid in such proportions that a 91 per cent acid at 70° C resulted. One and one-half parts of this product were mixed with one part finely ground ilmenite (54 per cent titanium dioxide) and run into the first chamber, where the reaction was initiated by heat. It then overflowed into a second zone where the temperature was 180° C. In another example, the ore (61 per cent titanium dioxide) and 93 per cent sulfuric acid were added continuously to the initial zone and moved forward as before. The operation could be carried out in a rotary kiln, a conveyor arrangement, or a series of pans. According to a similar process, applicable both to ilmenite and titanium-rich residues or muds obtained as by-products in various steps of the commercial manufacture of pigments (from ilmenite), the attack was carried out in such a manner that the reaction mixture was always in the dry or lumpy state, thereby ensuring the elimination of all operating difficulties arising from a change in state during the operation.[102] Finely ground ilmenite or mud was mixed on a conveyor with solid iron and titanium sulfates from a previous operation, and enough sulfuric acid was added to complete the reaction. The proportion of ingredients of the mix were adjusted so as to obtain a solid granular product before the titaniferous material had combined with the acid, and this granular material was then fed continuously into a revolving, gas-fired kiln in which the reaction took place. Temperature of the reacting zone was held at 180° C. The reacted mass discharged from the kiln was crushed and ground, and a considerable proportion, around 20 per cent, was returned to the conveyor to be mixed with more ilmenite or mud and acid. An advantage of the process is that dilute acid can be used to effect the decomposition. If a comparatively weak sulfuric acid is used, however, it is necessary to lower the rate of solid feed to the kiln and to increase the amount of decomposed product recirculated in order to have sufficient acid for complete reaction after the water has been driven off in the feed end of the kiln. In such cases temperatures above 180° C may be employed to effect faster evaporation.

In an example, the material used was a mixture of 114 parts of wet cake from filtering ilmenite solution, which contained 49.5 per cent solids, and 23.1 per cent titanium dioxide on a wet basis, and of a dry material containing 44.2 per cent insoluble titanium dioxide and 23.8 per cent iron as oxides. This was slurried with 135.4 parts of sulfuric acid as fresh acid and the suspension was divided into three equal batches. The first portion was heated at 180° C and then crushed and added to the second. This mixture was baked, after which a portion was again baked with the third batch of the original slurry. The final product was a dry mass which gave an over-all recovery of soluble titanium dioxide of 56.6 per cent.

An improved process consisted of acting on ilmenite with sulfuric acid in a continuous manner under such conditions that the ore-acid mixture was broken up into films which were later subjected to attrition.[103] In this manner any solid reaction product which englobed the particles of untreated ore were broken away so that fresh surfaces were exposed continuously to the action of the acid, and the inert bodies (balls) upon whose surfaces the films were deposited acted as heat stabilizing agents in that they disseminated the heat of reaction and prevented local overheating. On the other hand, their residual heat initiated the reaction of the incoming mix.

The most convenient apparatus for carrying out the attack was found to be a tubular ball mill which could be operated continuously. Ilmenite and sulfuric acid in reactive proportions were fed in at one end and the rotation of the mill was so regulated that the reaction product emerged at the discharge end as a solid, disintegrated mixture of soluble sulfates of the base constituents of the original ore.

In a typical operation, ilmenite containing 53 per cent titanium dioxide and 33.3 per cent iron in the form of both ferrous and ferric oxides was employed. The oleum was 104 per cent sulfuric acid, and the amount used in the feed was 1.33 times the weight of titanium dioxide in the ore. This mixture was fed continuously from the agitated tank to the ball mill at a rate of one-half pound a minute, and at the same time steam was admitted to initiate the reaction. Decomposition started almost immediately and the steam was shut off. The reaction was kept going by the addition of dilute acid liquor recovered from the hydrolysis of a titanium sulfate solution which contained 23 to 25 per cent sulfuric acid and 8 g per liter titanium expressed as the dioxide. The rate of addition of the liquor was held constant at 6.4 pounds per hour, and with this dilution the average concentration of the acid available for reaction was 82.4 per cent. A composite sample of the solid ball mill discharge gave, as soluble constituents, 18.9 per cent titanium dioxide, 11.0 per cent iron, and 15.3 per cent sulfuric acid in excess of that required to form soluble titanic and ferric salts. Recovery of titanium was 88.5 per cent.

Employing a pugmill attack, pulverized ilmenite and oleum in predetermined proportions were continuously added to an agitated tank and the mixture was allowed to flow into the feed end of a rotary screw conveyor.[104] Simultaneously with the acid-ore mixture, water or dilute waste acid was fed into the conveyor. At once heat was developed, and the exothermic reaction between the acid and ore was initiated. The mass gradually thickened as the reaction proceeded, and it was moved forward by adjacent and cooperatively acting helicoidal parallel screws rotated at different speeds and adapted to prevent accumulation of reacted material. The temperature was easily controlled by varying the speed of rotation of the paddle shafts and by varying the amount of diluent. Best results were obtained by operating under such conditions that the mass became dry in the first half of the conveyor length. Instead of using oleum and an aqueous diluent, 90 per cent sulfuric acid could also be added at the start, but in this case the reaction had to be initiated by maintaining the feed end of the apparatus at an elevated temperature.

To illustrate, 3340 pounds of finely ground ilmenite containing 53 per cent titanium dioxide and 5000 pounds of oleum (104.5 per cent sulfuric acid) were fed per hour into a mixing tank. This overflowed into the conveyor, where the reaction was started by adding water, and a switch was then made to dilute waste acid (24 per cent sulfuric acid) which was then admitted at the rate of 1800 pounds per hour. The mass was continuously moved forward by the conveyor and discharged at the opposite end as a solid reacted mass.

Similarly, on treating finely ground ilmenite with 80 per cent sulfuric acid at 65° C in a rotary kneading machine heated by an oil jacket, a vigorous reaction took place with the formation of soluble sulfates of the base components of the ore.[105] As soon as the reaction had subsided, the temperature was raised gradually to 250° C, and kneading was continued until the initial pasty product solidified to a dry crumbly mass consisting primarily of titanium and iron sulfates. Ninety per cent sulfation was readily effected by this process. The sulfates formed were dissolved by leaching the cake with water or dilute sulfuric acid.

In carrying out the continuous reaction of ilmenite with sulfuric acid on an endless conveyor, Moran and Nelson [106] found that a layer of a mixture of paper pulp and asbestos protected the steel belt employed and prevented the digestion reaction product from adhering, so that it could be discharged continuously from the surface of the protective layer.

Titanium ore or slag is digested continuously with sulfuric acid in a rotary kiln having a rigid longitudinal member fitted with projecting finns.[107] The interior surface of the kiln to which the ore and acid are charged is not only scraped to remove the plastic cake before it sets, but the solidifying and solid reaction mass is maintained in a disintegrated condition. A slurry of 1.5 parts by weight of 93 per cent sulfuric acid to 1 part of slag was charged into a kiln. Temperatures of 300 to 375° C were maintained in the heating chamber by the combustion of gas which maintained a temperature of 200° C throughout the entire rotating mass inside the kiln. The reaction product was discharged to another kiln, where it was maintained at a baking temperature of 200° C for 1.5 to 2 hours. Recovery of the titanium was 95 per cent.

In the continuous extraction of titanium from ilmenite the amount of sulfuric acid needed is added in two portions of different concentrations at different temperatures.[108] From 20 to 50 per cent of the sulfuric acid is added at 50 to 70 per cent concentration, obtained by evaporation of waste acid, and the remainder is added in the form of 98 per cent sulfuric acid. The ilmenite is treated with the first portion at 100 to 150° C. On addition of the second portion, the temperature is increased to 160 to 200° C. After heating the mixture in a rotating drum at 200 to 300° C for 1 to 1.5 hours, more than 90 per cent of the titanium may be extracted with water.

Ilmenite or slag ground to 200 mesh is digested with concentrated sulfuric acid in a continuous process in the presence of foaming agents as N-(cis-9-octadecenyl) trimethylenediamine.[109] In an example, 650 g of ilmenite, containing 50 per cent titanium dioxide, 30.2 per cent ferrous oxide, and 10 per cent ferric oxide was mixed with 5711 ml of 93 per cent sulfuric acid and 16.3 g of N-(cis-9-octadecenyl) trimethylenediamine. The slurry was pumped into another tank at the rate of 83 g of ore per minute and 8.5 ml of water was added per minute. Here the water set off a violent foaming reaction and the slurry flowed vigorously toward a horizontal reactor equipped with a screw conveyor. The temperature was 100 to 140° C at the entrance and 170 to 195° C at the exit. Small beads of digestion cake formed and were discharged. The cooled cake was soaked in water for 4 hours at 55° C to obtain a recovery of the titanium of the ore of 92.2 per cent.

Ilmenite ore is heated with 45 to 75 per cent excess sulfuric acid of 60 to 65 per cent concentration at 150 to 175° C to form ferrous sulfate and titanyl sulfate which precipitate from the acid solution.[110] The excess acid is separated from the solid sulfates and reused. An autoclave may be employed or the process may be carried out continuously. In an example, 5000 g of 60 per cent sulfuric acid preheated to 130° C was stirred in a vessel equipped with a high speed stirrer with 1000 g of ilmenite and 1600 g of concentrated sulfuric acid, and stirred for 2 hours. Within a few minutes the temperature reached the boiling point of 165° C. At the end of the reaction, during which 50 to 100 g of water evaporated, the temperature was between 155 and 165° C. The acid crystal mixture was cooled to 60° C and filtered to obtain 2700 g of solid sulfates filter cake, which was washed with water. The total filtrate, 5200 g containing 59 per cent sulfuric acid, 1.0 per cent of iron, and 0.4 per cent titanium dioxide was heated again to 130° C to effect concentration to 60 per cent sulfuric acid and recycled. The filter cake was dissolved in water at 60° C; the iron III was reduced to iron II with scrap iron until a small amount of titanium III was formed. The liquor was settled, filtered, and crystallized. Each liter of the solution contained 250 to 260 g of titanium dioxide, 40 g of iron, and altogether 500 g of sulfuric acid. This solution was hydrolyzed to produce titanium dioxide, which was processed in the usual manner to yield pigment.

Ilmenite, high titanium dioxide slag, and other titanium-containing materials are digested with circulating sulfuric acid at 160 to 180° C in a fluidized furnace.[111]

## DISSOLUTION OF THE DIGESTION PRODUCT

Regardless of the method employed, the reaction product obtained by digesting finely divided ilmenite or slag with concentrated sulfuric acid is leached with water or dilute sulfuric acid to obtain a solution consisting primarily of titanic, ferrous, and ferric sulfates, with minor proportions of sulfates of vanadium, chromium, manganese, magnesium and other metals, and of phosphates, depending upon the original composition of the ore. If the digestion is carried out so as to produce a porous cake, dissolution may be effected directly in the cylindrical, conical-bottom reaction tank without breaking up the mass by mechanical means. In such cases the leaching liquor is introduced at the bottom of the porous mass. Obviously the temperature should be held relatively low (well below the boiling point) to avoid hydrolysis, but at the same time it should be maintained as high as practical to speed the rate of solution. To avoid possible loss of titanium, resulting from hydrolysis while dissolving the soluble sulfates from the reaction mass, Farup [112] employed dilute solutions of sulfuric acid as the solvent instead of water. This method was also followed by Washburn [113] and others. In general, it has been found desirable to employ dilute sulfuric acid for the initial leaching, followed by water in the later stages. Dangers of hydrolysis may be minimized by adding the solvent liquor in relatively small portions so as not to cause excessive dilution of the dissolved titanium sulfate, particularly in the initial stages.

Dissolution of the sulfate cake may be accelerated by maintaining a low concentration of trivalent titanium [114] during the operation, and this practice is usually followed. Agitation of the solution is necessary, since it contains suspended particles of the reaction product as well as some unreacted ilmenite which would tend to settle out in the tank and resist the solvent.

For reasons of economy the concentration of the final solution should be as high as is consistent with efficient operation, although the upper limit for practical handling is determined by the viscosity and the tendency toward crystallization. This value in general is around 8 to 10 per cent titanium dioxide. Such liquors always carry a small proportion of unreacted ore, siliceous residue, and gangue material in suspension, in more or less colloidal condition.

## REDUCTION OF THE SOLUTION

All the ferric component of the final ilmenite solution is reduced to the ferrous state, and to prevent later reoxidation during processing the treatment is continued until a small proportion of the titanium is converted to the trivalent condition. Ferrous sulfate is more stable in solution than is the ferric salt, and it thus minimizes the precipitation of iron compounds along with the titanium oxide during the hydrolysis and washing steps. Furthermore the intrained ferrous salts are more easily removed from the hydrous titanic oxide pulp by washing than are ferric salts, since the latter have a strong tendency to change over to insoluble basic salts or oxides as the acid concentration of the liquor is reduced.

In the earlier days of the industry, reduction was effected in the cathode compartment of an electrolytic cell, the anode compartment of which contained dilute sulfuric acid. Electrolysis was continued until all the ferric ions and a small part of the titanic ions were reduced to the divalent and trivalent states, respectively.[115] Such solutions were also treated in a cell containing an unglazed porcelain compartment filled with 30 per cent sulfuric acid in which the lead anode was mounted. The ilmenite liquor was circulated through the cathode compartment until all iron compounds were reduced to the ferrous condition and enough titanous salts were produced to prevent reoxidation of the iron salts during subsequent treatment.[116] Later work showed that good yields of titanous irons could be obtained without the use of a diaphragm in the electrolytic reduction of such a solution.[117]

Weintraub[118] reduced the iron component of ilmenite liquors to the ferrous condition by introducing metallic zinc, sulfurous acid, or sodium thiosulfate, although metallic iron has been reported to give best results.[119] Employing metallic iron or its alloys[120] as reducing agents, greater efficiency was obtained at concentrations of 93 to 130 gpl titanium dioxide and at temperatures around 60° C. Krchma[121] reduced such solutions by introducing two metals in discrete form, such as tinned iron, which are between hydrogen and calcium in the electromotive series. For reasons of economy, iron is usually employed in large-scale operations. Supported in lead baskets, it may be introduced into the hot solution to effect reduction.

Rau and Swartz[122] employed sponge iron, produced by the reduction of ilmenite or titaniferous magnetite, to reduce the sulfate solutions for hydrolysis. The agent was added in small increments with agitation, and the temperature of the solution was held just below 60° C. Use of the titanium-bearing sponge instead of scrap iron resulted in an increase in the titanium to iron ratio in the solution and conserved sulfuric acid.

Ryan and Cauwenberg[123] circulated the ilmenite solution through or over the reducing agent at such a rate as to avoid overheating or local change in acidity factor. More stable solutions were obtained by this method.

Instead of treating the entire body of ilmenite solution with metallic iron until a small amount of trivalent titanium was formed, a small proportion was removed and its constituents were reduced as far as practicable. This solution, rich in titanous salts, was added back to the main batch of liquor, in slight excess of that required to convert all ferric iron to the ferrous state, so as to furnish a low concentration of titanous ions to the final mixture.[124] This operation may be carried out in a cyclic process. The required amount of water and trivalent titanium solution were added to the reaction product of ilmenite with sulfuric acid to dissolve the mass and reduce the ferric iron to the ferrous state. The reduced solution was then run into a settling tank where it was separated from insoluble matter by decantation, and the clear overflow was led to storage tanks for future use. The residue was withdrawn from the bottom of the tank and reslurried in another tank with dilute titanium sulfate solution. Muds from this tank, which contained much less titanium than those from the first settling operation, were discharged to waste. If economic considerations warrant, however, more titanium can be recovered by employing a third settling stage. Wash liquor obtained as a clear overflow from the second tank contained 70 g titanium dioxide, 50 g ferrous iron, and 250 g total sulfuric acid per liter. This solution was treated with tin scrap at 30° to 50° C to reduce the titanium to the trivalent state. An efficiency of 90 to 95 per cent was obtained, and 0.38 pound of tin scrap was used for each pound equivalent of titanium dioxide reduced. After reduction the titanous solution was added to the reaction vessel to reduce the ferric iron and act as a solvent for fresh digestion cake. The proportion used was such that the final solution contained a slight excess of the trivalent titanium to accelerate dissolution and to compensate for air oxidation in subsequent operations. The method was also applicable to scrap iron as the reducing agent, but wash liquors of higher concentration (75 to 80 per cent) were required to obtain satisfactory efficiency. Mechanism of the reduction may be expressed by the following equations:

$$Fe + 2Fe^{+++} \rightarrow 3Fe^{++}$$
$$Fe + 2Ti^{++++} \rightarrow Fe^{++} + 2Ti^{+++}$$

Reduction may also be effected by adding titanous sulfate in crystalline form.[125] This method has the advantage in that it is practically instantaneous and does not introduce foreign materials into the solution. Any other soluble titanous compound whose acid radical does not interfere with subsequent treatment may be employed.

Brooks [126] treated ilmenite solution with sulfur dioxide in the presence of activated carbon to effect reduction of the ferric component to the ferrous condition. In a specific application, to 100 parts by weight of a sulfate solution of ilmenite was added 0.18 part activated carbon made from wood or vegetable products, and 100 parts sulfur dioxide gas was then passed through at a temperature of 30° C for 2 hours. Ninety per cent of the ferric sulfate was reduced to the ferrous condition. Employing the same conditions without the carbon (gas was passed through at the same temperature for the same time) only 3.1 per cent of the total ferric iron was reduced. Scrubbing towers may be used to advantage.

The proportion of the titanium component that should be converted to the trivalent state has been reported as high as 5 per cent,[127] although much lower values, 1 or 2 per cent, have given satisfactory results. Obviously it is only necessary to have a high enough proportion of this agent to maintain reducing conditions throughout the processing operation. Large excess should be avoided since it is not precipitated in the hydrolysis step, and, furthermore, reoxidized titanous sulfate tends to yield discolored oxide.

## CLARIFICATION OF THE SOLUTION

At this stage filtration by conventional methods would be unsatisfactory, for the solution is strongly acid and contains finely divided suspended materials that are at least in part in the colloidal condition. Since ordinary filter materials are either attacked by strong acid or clogged by the mud, the liquor is generally subjected to a primary clarification process. The colloidal material is first coagulated by the conventional method of introducing a second colloid of opposite electrical charge, such as glue or a metallic sulfide, after which the liquor is passed through a series of settling tanks and finally passed over a specially designed plate or rotary filter coated with an acid-resisting material such as diatomaceous earth (precoat), nitrated cloth, glass cloth, vinyl-resin cloth, or perforated rubber sheets. Instead of relying on the force of gravity, centrifuging or reduced pressure may be employed to speed up the operation.

According to Goldschmidt,[128] the suspended solids in sulfate solutions of ilmenite are electropositive colloids that settle rapidly on neutralization of this charge by precipitated sulfur, which is electronegative. The colloidal sulfur may be precipitated in the solution by the oxidation of hydrogen sulfide by a ferric salt or by sulfur dioxide. Rapid settling of the suspended material may also be effected by the addition of organic colloids having an electrical charge of opposite sign. Glue, albumen, and casein have been found to be particularly effective. On mixing a small proportion of the agent into the suspension, the solids are rapidly coagulated and settled, leaving a clear supernatant liquor which is decanted or filtered or both. In an example, 2 or 3 decigrams of noncoagulated albumen were added per liter of sulfate solution having a specific gravity of 56° Bé, with thorough mixing. After standing for 24 hours, the clear supernatant liquor was decanted and the muddy residue was filtered.[129] Rapid settlement was also effected after treating the slime products with gelatin [130] in a similar manner.

Frequently glue and similar organic colloids fail to remove the suspended materials as completely as desired, and at least part of the coagulating agent remains in the solution. Glue, for example, may cause frothing in a later step, and on standing it is also subject to decomposition or putrefaction. In an improved process, glue, gelatin, alginates, albuminoids, protein, or dextrin were added to hasten the settling of the suspended solids as before, and the partially clarified solution was then treated with a mutually coagulating organic colloid, particularly tannins, which removed any added material as well as any original colloid remaining in suspension in the solution.[131]

As an example, a concentrated solution containing 1 kg of glue was added to 1 cubic meter of liquor obtained by dissolving ilmenite in sulfuric acid. After primary settling, the solution which carried in suspension 0.04 per cent inorganic impurities and practically all the glue added was treated with 2.5 kg of tannic acid as a concentrated aqueous solution. The tannic acid coagulated the glue to form flocs which settled rapidly and carried down with them the remaining suspended material. Such solutions have also been coagulated and clarified by adding gelatin or glue of animal origin, together with a light, solid material such as sawdust or asbestos,[132] or by treatment with an extract of flaxseed or its husks.[133]

Solutions of titanium which are difficult to filter may be clarified by mixing in colloidal metallic sulfides having an electric charge opposite in sign to that of the suspended matter to be removed. Heat may be applied to reduce the viscosity if necessary. After settling, the supernatant liquor is separated by decantation or filtration or both. In an example, a sulfate solution of ilmenite, of specific gravity 50° Bé and containing 200 gpl titanium dioxide, was brought to 50° C and 0.5 g arsenious oxide, as a soluble salt dissolved in water, was added per liter. Solid ferrous sulfide sufficient completely to precipitate the arsenic as sulfide was then introduced. Flocs immediately formed and the suspension could be filtered immediately, or, after standing, the supernatant liquor could be decanted from the residue. The proper amount of dissolved arsenic may be introduced into the solution by adding arsenious oxide to the ilmenite-sulfuric acid mixture before digestion.[134] To effect coagulation, a sulfide was added to the solution to precipitate the arsenic as sulfide.[135] Zhukova[136] treated such solutions with copper sulfate and sodium sulfide to effect clarification. The copper sulfide formed, which was colloidal in nature, neutralized the charge on the suspended solids so that coagulation and settling were effected rapidly. In this process the composition of the solution was important, since titanous ions reduce copper sulfate to metallic copper. Concentrated solutions may be clarified by a two-stage process consisting first of sedimentation, followed by centrifugal separation at 40° to 90° C.[137]

## PURIFICATION OF THE SOLUTION

By the generally employed method of manufacture, some ilmenite ores, for example those occurring in Virginia, tend to yield pigments sensitive to light and of unsatisfactory color. These undesirable properties are the result of impurities, such as chromium and manganese compounds, derived from the ore, and are either dissolved or colloidally dispersed in the sulfate solution so that they are precipitated along with the titanium dioxide in the hydrolysis step. Very small proportions of these impurities are sufficient to render the pigment photosensitive and off color. These injurious manganese and chromium compounds may be effectively removed from the solution by adsorption on gypsum or strontium sulfate, just after the crystallization step.[138] Hydrous titanium oxide precipitated after this treatment yielded a pigment of good color and high stability to light.

To illustrate, a regular sulfate solution of Virginia ilmenite, after the crystallization stage, was mixed with 9 per cent of its weight of ground gypsum and agitated for 12 hours at room temperature, after which it was filtered to remove the gypsum together with the adsorbed impurities. The filtrate having been processed in the usual manner, the finished pigment had excellent color and was not photosensitive. These undesirable impurities were also removed from solutions of this type by treatment with nonsiliceous adsorbents such as activated carbon and cotton fibers.[139] The impurities adsorbed on the added agents were removed by filtration.

Bachman[140] treated clarified solutions with potassium ferricyanide or potassium ferrocyanide to precipitate undesirable impurities, and then boiled the diluted filtrate to effect hydrolytic precipitation of the titanic oxide of improved purity.

Sulfate solutions stable to premature hydrolysis[141] may be produced from ores having a high iron to titanium dioxide ratio by adding dried precipitated titanic oxide before the sulfuric acid attack. In an example, the ratio was adjusted to 0.72. Such solutions, after removal of ferrous sulfate by crystallization, may be recycled to the dissolving step to increase the iron to titania ratio.[142]

The acidity of sulfate solutions of ilmenite may be reduced without disturbing their crystalloidal properties by adding ferrous carbonate[143] or metallic sodium or potassium[144] at a temperature not higher than 60° C. The same results were obtained by treating the solution with a neutral, soluble, organic extracting agent such as a ketone, alcohol, ester, or cyclohexanone.[145] The mixture was agitated and allowed to settle, after which the upper layer, rich in acid, was drawn off for recovery of the organic agent by vacuum distillation. By another modification, hydrogen peroxide or barium peroxide was added to the solution to oxidize the ferrous iron to the ferric state and thereby tie up a larger quantity of sulfuric acid.[146]

Grachev [147] found it feasible to remove iron and aluminum from ilmenite solutions by electrolysis, employing a ceramic diaphragm and lead electrodes. Lead compounds were removed by adding barium or strontium hydroxide to the solution.[148] The precipitate of barium sulfate, which adsorbed the lead component, was removed from the solution by settling or filtration.

By gradually diluting impure titanium salt solutions with water, impurities such as vanadium, columbium, and tantalum were precipitated progressively.[149]

Air is bubbled through the solution obtained by dissolving in water the product from the digestion of ilmenite in sulfuric acid to remove the residual hydrogen sulfide before adding antimony trioxide to effect the clarification.[150] After dissolving the antimony trioxide in sulfuric acid solution, sodium sulfide is added to precipitate antimony sulfide in the colloidal form, which neutralizes the charges on the particles of suspended silica and unreacted ore to cause flocculation. Antimony sulfide precipitated by residual hydrogen sulfide is not effective in coagulating the suspended material so that this amount of antimony is lost. For example, 22.5 tons of ilmenite ore, 38.25 tons of 93 per cent sulfuric acid, and 2.06 tons of water were reacted in the usual way to produce a cake of soluble sulfates. The cake was dissolved in 52.8 tons of water, and the iron was reduced to the ferrous state by immersing scrap iron in the solution until the analysis showed 3 gpl of trivalent titanium calculated as the dioxide. Air was then bubbled into the bottom of the tank at a rate of 125 cu ft per minute for 120 minutes to remove the hydrogen sulfide. Fifteen pounds of antimony trioxide were added to and dissolved in the solution. It was then pumped to a clarification tank and 14.4 cu ft of 11 gpl sodium sulfide solution were added. The antimony sulfide precipitate settled to the bottom of the tank and carried with it the suspended solids in the solution leaving a clear supernatant liquor. If the residual hydrogen sulfide was not removed 60 lb of antimony trioxide were required to effect clarification.

## FILTRATION

Instead of employing the time-consuming clarification process for ilmenite liquors, Weise and Raspe [151] filtered such solutions directly through nitrated cellulose cloth. Trivalent titanium compounds should not be present, however, since they cause a slow denitration of the fabric, particularly at elevated temperatures. In the operation of the process, the sulfate solution was first reduced with scrap iron to a point just short of the formation of trivalent titanium and was then filtered through the nitrated fabric, after which the reduction was further carried to the desired stage by electrolysis or by adding titanous salts directly. For example, 100 kg of ilmenite in a finely divided state was intimately mixed with 200 kg of 80 per cent sulfuric acid and heated at 150° C for 1 hour. Water was added to the reaction product in such proportions as to obtain a solution of 1.45 specific gravity. Metallic iron was then introduced, with stirring, at 70° C until ferric ions just disappeared. At this point the reduction was interrupted, and after cooling to 40° C the solution was passed through a filter press, the cloths of which consisted of nitrated cotton. Luttger [152] recommended Vinyon cloth as coating for continuous filters. Macerated paper pulp, deposited from a water suspension on perforated sheet lead, has been used successfully.[153] In a continuous process, a drum, coated with a porous filter medium such as diatomaceous earth, was revolved at 700 rpm in the solution to be filtered. At each revolution of the drum a cleaning element or blade removed the accumulated slimes, together with a thin layer of the filter material.[154]

Filtration of such solutions, whether first clarified or not, was facilitated by stirring diatomaceous earth or a similar filter aid into the hot acid-bearing liquor.[155] Lewis [156] found that the filtration rate was increased by adding 1 gallon of sulfonated oil to 10,000 to 20,000 gallons of the crude solution.

## REMOVAL OF FERROUS SULFATE BY CRYSTALLIZATION

Sulfate solutions of ilmenite contain practically all the iron of the original ore. After reducing this component to the ferrous state, the solutions, on cooling to room temperature, still carry an appreciable amount of iron, depending upon the initial concentration. This large proportion of ferrous salt may be a source of trouble in subsequent operations. Sulfuric acid, liberated on hydrolysis, reduces the solubility of the ferrous sulfate, and crystals may become mixed with the titanium dioxide. On the other hand, dilution for overcoming this factor would increase the cost of recovering the sulfuric acid. These disadvantages can be avoided if the iron content is first reduced by cooling the solution to subatmospheric temperatures, 5° to 15° C. This temperature may be attained conveniently by a process of forced evaporation. Heptahydrated ferrous sulfate crystallizes out without forming a double salt with the titanium, and the crystals may be separated by decantation, filtration, centrifuging, or a combination of these, with practically no loss of titanium salts. The mother liquor may then be concentrated to relatively high titanium content without excessive thickening, and the sulfuric acid liberated on hydrolysis may be recovered more economically.[157] Furthermore, in certain methods of hydrolysis, particularly that of Blumenfeld, the ratio of titanium to iron must be adjusted to within specific limits for best results. On the other hand, solutions from which too large a proportion of iron has been removed may give pigments of unsatisfactory properties. For this type of hydrolysis crystallization is generally carried to such a stage that the content of ferrous sulfate is approximately 0.9 that of the titanium dioxide. By employing other methods of hydrolysis applicable to lower total concentration, however, it may not be necessary to remove any of the ferrous sulfate, as, for example, in the precipitation of composite pigments.

## SOLVENT EXTRACTION

In a modification of the usual sulfate process, titanium sulfate and iron sulfate are extracted from the sulfate solution of ilmenite with mono- or dialkyl substituted orthophosphoric acid.[158] Especially effective extractants are di (2-ethylhexyl) orthophosphoric acid, monododecyl orthophosphoric acid, and monoheptadecyl orthophosphoric acid. Although the organic agents may be used directly to extract the iron and titanium ions from the aqueous phase, it is advantageous to employ them as solutions in organic solvents. The most effective solvents are kerosene, isopropyl ether, and Stoddard solvent, although aromatic hydrocarbons, petroleum derivatives, and others are effective. In actual operation the extraction step can be carried out using mixer-settlers, countercurrent extractors, or other units in single or multiple stages. After separation of the organic phase the iron present is removed by scrubbing with hydrochloric acid. This operation is carried out using conventional liquid-liquid scrubbing and stripping techniques. Stripping of the titanium component from the iron-free organic phase is obtained by adding an aqueous solution of an alkali metal hydroxide or carbonate, or an acidic ammonium or alkali metal fluoride such as ammonium bifluoride or potassium bifluoride. Hydrated titanium oxide is precipitated from the basic stripping solution while the double ammonium titanium fluoride precipitates from the acid fluoride solution if its solubility in the aqueous phase is exceeded.

As an illustration of the process, a 100 g sample of ilmenite containing 35.5 per cent titanium and 24.0 per cent iron was heated with 50 ml of 90 per cent sulfuric acid to effect reaction. The product was leached with 200 ml of water and the leach liquor was separated from the residual solids by filtration. This solution was agitated in a liquid-liquid extractor for 30 minutes with a 30 weight per cent solution of di (2-ethylhexyl) orthophosphoric acid dissolved in kerosene. The titanium and iron bearing organic extract was scrubbed with 5 molar hydrochloric acid to strip the iron, after which the two phases were separated. To this titanium-rich, iron-free phase was then added 150 ml of a saturated solution of ammonium carbonate. The resulting aqueous and organic phases were separated. Hydrous titanium dioxide was precipitated from the alkaline solution by gentle heating. After separation by filtration, the precipitate was washed and calcined for 4 hours at 850° C to give white titanium dioxide possessing the rutile crystal structure. Titanium is removed from sulfate solutions by ion exchange or by solvent extraction.[159] Suitable agents are alkyl phosphates and dibutyl pyrophosphate.

Cation exchange resins in the H-form absorb titanium from solutions up to 1 normal containing 0.1197 mg of titanium per liter.[160] The amount absorbed decreases as the acidity of the solution increases. Anion exchange resins absorb titanium from 6 to 12 normal hydrochloric acid, and the amount absorbed increases with the acidity. Apparently in dilute hydrochloric acid titanium is in the form of $TiO^{++}$; in concentrated solutions it is in the form of a complex.

In the absorption of titanium IV by ion exchange resins at pH below 1.3 the dominating action is $TiO^{++}$. At pH of 1.4 to 1.6 the dominating cation is $TiO(OH)^+$.[161] In a hydrochloric acid solution less than 4 normal the electrochemical $TiOCl_2$ is adsorbed by anion exchange resin AN-2f. In 10 to 11 normal hydrochloric acid solution $TiCl_6^{--}$ forms. These ions are poorly adsorbed because of the competing effect of the chloride ion. In sulfuric acid solutions adsorption increases with the acid concentration above 0.1 molar.

Acidic titanium sulfate solution such as that obtained by extracting ilmenite with sulfuric acid, is passed through a column of the H form of a cation exchange resin containing sulfonic acid groups, for example, Dowex 50, Permutit H, Wofatit P, K or KS, Zeo Rex, or Nalcite HGR.[162] The first portion of the effluent contains only sulfuric acid which can be concentrated and recycled to the extraction step. The second portion containing titanium sulfate practically free from impurities such as iron, aluminum, and manganese is used for the manufacture of titanium dioxide pigments. After the resin has become ineffective, it is regenerated by treatment with hydrochloric acid.

Ellis[163] leached 200 mesh or finer ilmenite with four times the equivalent of 9 to 10 molar hydrochloric acid at 60 to 75° C to dissolve the iron and titanium compounds. The iron is extracted from the solution with a solution of tributyl phosphate after which the titanium is precipitated as hydrate by neutralization with a base. Alternatively the titanium compounds may be selectively extracted with mono- or dialkyl-substituted orthophosphoric acid solution. Suitable alkyl groups are mono-dodecyl, monoheptadecyl and di (2-ethylhexyl). The organic extractants may be used directly but are generally employed as 1.5 to 3.5 molar solutions in toluene, benzene, or kerosene. Titanium is stripped from the organic with an aqueous solution of an alkali, ammonium carbonate, hydrofluoric acid, or acid fluorides. The used butyl phosphate solution in an organic solvent is stripped from the iron by extraction with water for reuse.

## CONCENTRATION OF THE SOLUTION

Concentration of the crystallized mother liquor is effected by evaporation under reduced pressure at a corresponding low temperature to prevent premature hydrolysis at this stage.[164] However, solutions for the manufacture of the relatively pure titanium dioxide by some methods, and of coalesced composite pigments in general, are normally not crystallized or concentrated.

# 14

# Hydrolysis of Titanium Sulfate Solutions to Produce Pigment

Several versions of the mechanism of the thermal hydrolysis of titanium salt solutions have been presented. It does not seem to have been definitely established.[1]

In any case, the crystalloidal particles grow in size until the colloidal state is reached, and finally precipitation begins. As a result the hydrolytic equilibrium is disturbed, a new quantity of colloidal titanium oxide is formed by decomposition of more of the crystalloid salt in solution, and this, in turn, is coagulated. The process continues until the greater part of the titanium is precipitated in the form of a relatively insoluble hydrous oxide.

According to Mironov[2] the formation of titanyl hydroxide proceeds according to the scheme: $Ti^{++++} + 2\,OH^- \rightarrow TiO^{++} + H_2O$ and $TiO^{++} + 2\,OH^- \rightarrow TiO(OH)_2$.

Rigid control of the formation of colloidal hydrous titanic oxide and of the conditions during its subsequent coagulation is of primary importance in the manufacture of titanium dioxide of highest pigmentary properties. This has been mastered, at least in part, by extensive research.

The hydrous oxide precipitate obtained commercially by thermal hydrolysis of sulfuric acid solutions of titanium contains sulfuric acid held so firmly that it cannot be removed by washing with water, but can be driven off at temperatures in the neighborhood of 700° C. Experimental evidence indicates that this residual acid is not chemically combined but adsorbed on the hydrous oxides.

Factors controlling the hydrolysis reaction are quality of the seed, proportion of the seed, concentration of the titanium sulfate solution, the acid factor of the solution, and the rate of heating.[3] For anatase, boiling up to 6 hours is necessary to complete the hydrolysis. Three hours boiling is sufficient for rutile.

## ANATASE AND RUTILE MODIFICATION OF TITANIUM DIOXIDE

Normally, sulfate solutions on hydrolysis yield directly titanic oxide of the anatase crystal form, while halides and nitrates give rutile. Under special conditions, however, the results may be reversed. For example, by seeding sulfate solutions with rutile nuclei all the titanium content may be precipitated in this form. Furthermore, solutions prepared by dissolving in sulfuric acid the hydrous oxide originally precipitated from aqueous titanium tetrachloride have been reported to yield directly an oxide having the rutile structure.

On the other hand, titanium tetrachloride and other solutions which normally yield oxides of the rutile crystal modification may be hydrolyzed in the presence of phosphoric and sulfuric acids to give anatase.[4] For example, on hydrolysis of a 20 per cent solution of titanium tetrachloride, after addition of 5 per cent or more sulfuric acid based on the titanium dioxide, anatase rather than rutile was obtained. On the other hand, the hydrolysis product of sulfate solutions to which sodium chloride had been added was anatase rather than rutile. These results were confirmed by Parravano and Caglioti,[5] who found that hydrochloric acid solutions on hydrolysis yielded titanium dioxide which showed the rutile structure initially, but, if phosphate or sulfate radicals were present, anatase was formed first but converted to rutile on further heating.

According to Weiser and Milligan,[6] hydrous titanium dioxide freshly precipitated from sulfate solutions at room temperature gave no X-ray diffraction lines or bands, although the anatase structure was developed after aging under water for several months at room temperature or on heating to 184° C. The product precipitated at 100° C gave X-radiograms of anatase directly. Hydrous oxide formed by the hydrolysis of boiling solutions of titanium tetrachloride or tetranitrate had the structure of rutile, while that precipitated from sulfate solutions under the same conditions gave the X-radiograms of anatase. Thermal dehydration isobars and X-ray analysis of the different products indicated that the water was given off continuously and that no definite hydrate of titanium dioxide has been prepared. Products obtained by igniting at 1000° C the hydrous oxide precipitated in the cold from sulfate solutions showed the rutile structure.

The transition from anatase to rutile structure [7] has been reported to take place at 800° to 1000° C, depending on the conditions of precipitation from titanium sulfate solutions and upon the nature of the impurities. According to the same workers,[8] the appearance of a yellow color on calcination is due to the presence of impurities and has no connection with the transformation from anatase to rutile structure. Sufficiently pure samples remained white or became slightly grayish. According to Parravano and Caglioti,[9] stable anatase was precipitated from sulfate solutions with high acidity and high titanium dioxide content, while at low titanium dioxide concentration and relatively low acid ratio the product obtained was a mixture of anatase and rutile at temperatures up to 830° C, but it was converted completely to rutile after 1 hour at 850° C. Hydrolysis under ordinary conditions of a sulfate solution containing 8.1 per cent titanium dioxide and having an acidity factor of 1.28 yielded a precipitate which retained the anatase structure at 900° C, but if a wet hydrolysis product which would itself form rutile on ignition was added as a seeding agent, the titanium dioxide precipitated showed the rutile structure after calcination at 900° C for 1 hour.

Thus the nature of the calcined material was strongly influenced, not only by the initial concentration of the solution, but also by the temperature, and, finally, by the rate of hydrolysis. For example, a solution containing 5.12 per cent titanium dioxide at acidity factor 1.07 hydrolyzed on long standing to give a product that was anatase after calcination at 950° C, but higher temperature hydrolysis gave a precipitate which converted to rutile by the time it reached 950° C.

Oxides and hydroxides of titanium may be transformed readily to the rutile modification by heating in the presence of small proportions of a number of agents such as zinc oxide, lithium chloride, aluminum sulfate, magnesium sulfate, barium sulfate,[10] antimony oxide,[11] or mixtures of these. Zinc oxide, which is representative of the group, forms solid solution with the titanium dioxide and dissociates at higher temperatures. At the temperature of dissociation, rutile is separated and the zinc oxide reacts with more anatase so that the process is continuous and large amounts of rutile may be converted by small amounts of the agent. Good results may be obtained with as low as 0.5 per cent zinc oxide based on total solids, and 2.0 per cent was very effective. Conversion was inhibited by the presence of phosphates and compounds of the alkali metals.

Titanium oxide of the anatase structure may be redissolved either directly or indirectly to form halide, nitrate, or chloroacetate solutions, and then reprecipitated from such solutions in the rutile form. Though these methods are indirect, they may be carried out in a cyclic process so that the reagents are recovered and used over and over again. For instance, hydrous anatase pulp may be heated with aqueous barium hydroxide to form the titanate, which is readily soluble in nitric or hydrochloric acid.

Sulfate solutions may be converted to chloride solutions, which yield titanium dioxide of the rutile crystal form directly by adding barium chloride or other compounds which yield insoluble sulfates.

Pamfilov and Peltikhin [12] noted that titanium dioxide with the rutile crystal structure is superior to anatase as a pigment, particularly with respect to covering power and from the point of view of chalking resistance. Rutile structure is obtained directly by the hydrolysis of boiling titanium tetrachloride solution followed by calculation of the precipitate for 3 hours at 800° C. The anatase produced by hydrolysis of sulfate solutions is transformed to rutile by firing at a temperature higher than 800° C.

## PRECIPITATION WITH ALKALINE AGENTS

In the very early stages of the industry, the titanium component was selectively precipitated from solution as hydroxide or hydrous oxide by adding controlled amounts of alkaline reagents, but this gave products which were of poor color and undesirable in other respects. Thermal hydrolysis of dilute sulfuric acid solutions containing 3 to 5 per cent titanium dioxide gave pigments of better quality, but the method was slow and uneconomical. Practically pure titanium dioxide was obtained by a process based on the method of producing coalesced composite pigments. A small proportion of anhydrite was suspended in the original solutions, and after hydrolysis the calcium sulfate was removed from the precipitate by washing. This was effective because of the much greater solubility of calcium sulfate than of the titanic oxide in water. Although pigments of good quality were obtained, the process was slow and costly.

The next real advancement was the development of a process by which titanium oxide was precipitated by heating more concentrated solutions (80 g to 350 gpl titanium dioxide), in which the content of sulfuric acid was so regulated that the titanium was present not as a normal salt, but rather as one having a sulfate content between that corresponding to the titanyl and normal salts. Such concentrated acid-poor solutions yielded pigments of good quality, but complete hydrolysis required prolonged boiling under normal conditions or heating for much shorter periods in an autoclave at higher temperatures corresponding to superatmospheric pressure. The high concentrations had the additional advantage of permitting economical recovery of the sulfuric acid from the hydrolysis liquor.

To hasten precipitation under atmospheric conditions, various nucleating agents or seed were added to the hydrolysis solution, and in most cases these consisted essentially of a suspension of colloidal titanium oxide. Further developments involved a slow admixture of concentrated, acid-poor ilmenite solution with hot water in such a manner that seeding agents were formed directly within the main body of the hydrolysis mixture.

Other specialized processes have been reported, and among the more important of these are methods of producing rutile directly from sulfate solutions.

## DILUTE SOLUTIONS

One of the earliest commercial methods for the hydrolytic precipitation of titanium oxide of high purity was developed by Weintraub.[13] Finely ground ilmenite was decomposed quantitatively with concentrated (93 per cent) sulfuric acid, employing no excess over the theoretical amount required to form normal salts with the base constituents. Heat to initiate the reaction was supplied by live steam. The reaction product was dissolved in water. All the iron component was reduced to the ferrous condition by introducing metallic zinc, sulfurous acid, or sodium thiosulfate. And the solution was clarified, diluted, and boiled to effect hydrolytic decomposition. The precipitate was filtered, washed, and calcined to obtain titanium dioxide of high purity. Alternatively, a product of somewhat less purity was obtained by throwing the solution into a large quantity of boiling water. Weintraub was primarily concerned with preparing pure titanic oxide for arc lights and other electrical uses, and was not interested in or failed to recognize the possibility of this compound as a white pigment.

Rossi and Barton [14] developed one of the earliest methods of producing relatively pure titanium oxide of pigment grade which involved the hydrolytic decomposition of very dilute sulfate solutions containing iron and other impurities. The reaction product of ilmenite with 95 per cent sulfuric acid in a proportion 2.5 times the titanic oxide content was dissolved in three times its volume of water, and all iron was reduced to the ferrous state by electrolytic means. This solution was clarified, diluted to 0.5 to 3 per cent titanium dioxide, and boiled for 15 to 30 minutes to effect hydrolysis. The precipitate was washed and dried at 100° C to obtain a product containing 70 to 90 per cent titanium dioxide, 5 to 10 per cent sulfur trioxide, and 5 to 20 per cent closely held water. It was then calcined to drive off the remaining water and acid and effect crystallization.

## BASIC SOLUTIONS

To facilitate hydrolysis, Barton [15] treated solutions of this type with calcium or barium hydroxide to reduce the sulfuric acid-titanium dioxide ration to a value less than that corresponding to the normal sulfate, but short of that required to yield a precipitate of titanium. Such solutions were referred to as "basic." The insoluble residue was separated; the liquor was diluted to 3 per cent titanium dioxide equivalent, and electrolyzed without a diaphragm to reduce and maintain all the iron in the ferrous state. The solution was heated at 90° to 100° C by live steam or by external sources to effect hydrolytic precipitation of the titanium, and recoveries of 85 to 90 per cent were obtained after heating for 2.5 to 5 hours. Under these conditions ferrous iron remained in solution. The precipitate was washed and calcined at 750° C to give a practically pure titanium dioxide pigment. In addition to the above-normal agents, part of the excess sulfuric acid was neutralized with calcium, barium, strontium, or lead carbonates, sodium hydroxide or carbonate, or benzidine in preparing solutions of this type.[16] Resulting precipitates of insoluble sulfates were filtered off, and the solution was subjected to thermal hydrolysis as before.

Hydrous oxide may also be precipitated by heating more concentrated solutions containing less than 2 moles of free and combined sulfuric acid per mole of titanium dioxide, and less than 55 per cent water by weight.[17] More dilute solutions may be concentrated by evaporation, or a liquid that does not enter into reaction with the constituents, such as glycerol, ethylene glycol, or aliphatic alcohols, may be added to take up the excess water. The process is also applicable to composite pigments. Weizmann and Blumenfeld [18] precipitated hydrous titanium oxide by heating sulfuric acid solutions of ilmenite to near the boiling point. The surface of the liquor was covered with a layer of oil or paraffin during the hydrolysis step to prevent evaporation. After separation of the precipitated titanium oxide, the filtrate was cooled to crystallize out ferrous sulfate, and the liquor containing sulfuric acid and some titanium sulfate was used in preparing a new batch of ilmenite solution.

## PRESSURE HYDROLYSIS

The next step forward was a method developed by Fladmark,[19] according to which hydrolysis was effected by boiling at atmospheric pressure, or by heating at higher temperatures in an autoclave, more concentrated ilmenite solutions having a content of sulfuric acid between that corresponding to the titanyl and normal salts. These contained from 80 g to 250 g per liter titanium dioxide, and from 80 g to 400 g per liter sulfur trioxide, free and combined with titanium. Typical concentrations in grams per liter of constituents are given in Table 14–1.

**TABLE 14–1**

Analysis of Solutions, Grams per Liter

| $TiO_2$ | $Fe^{++}$ | $Fe^{+++}$ | $SO_3$ | $SO_3$ Calculated as Equivalent | Shortage of $SO_3$ |
|---|---|---|---|---|---|
| 163.5 | 101.0 | 3.3 | 363 | 478.8 | 116.8 |
| 210.0 | 49.3 | 0.0 | 311 | 490.7 | 143.9 |
| 99.0 | 53.7 | 29.3 | 297 | 337.8 | 40.8 |
| 169.2 | 84.6 | 40.8 | 488 | 547.4 | 59.4 |

Ilmenite was decomposed with sulfuric acid in such proportions that the solution obtained by lixiviating the reaction product contained less acid than would be required by the titanium, iron, and other bases to form normal salts. All iron was converted to the ferrous state, and the clear, concentrated solution was hydrolyzed in a lead-lined vessel equipped with coils of lead pipe through which steam was passed for heating. This required boiling at atmospheric pressure for long periods of time, but the process was hastened by employing higher temperatures corresponding to autoclave pressures. The precipitate of hydrous oxide was washed and calcined to produce titanium dioxide of pigment grade. Sulfuric acid was recovered economically from the relatively concentrated hydrolysis liquor.

According to Specht,[20] the rate of hydrolytic decomposition of titanium sulfate and the properties of the products obtained are determined primarily by the absolute concentration of the solution, the free acid content, and the temperature employed. Best results were obtained by effecting hydrolysis under pressure of a solution of specific gravity 1.50 containing 35 per cent potential sulfuric acid (free and combined with titanium) at a temperature of 170° to 180° C. For this range of temperature the specific gravity could fluctuate between 1.55 and 1.35, and the potential or active acid between 25 and 40 per cent. However, the specific gravity of solutions poor in titanium could be brought effectively within the specified range by adding neutral salts such as alkali metal sulfates. In a similar process, solutions of titanic sulfate containing free acid were hydrolyzed by heating to a raised temperature under pressure after treatment with 10 per cent crude precipitated titanic acid.[21]

Buckman[22] also observed that the speed of thermal hydrolysis of acid-poor ilmenite solutions increased with temperature. Higher working temperatures were obtained either by employing solutions of higher concentrations or by increasing the pressure above atmospheric. Although the first alternative was of limited application, because of the stability of such solutions, the second was accomplished by heating in an autoclave and temperatures far above the normal boiling point were reached. At around 100° C the precipitates were usually slimy and dried to hornlike masses of low tinting strength, but at temperatures of 170° to 185° C solutions of the same chemical composition gave precipitates of dense titanic oxide which dried to soft powders having very good covering power. Furthermore, the yields were much greater than those obtained by boiling in vessels open to the atmosphere. In an example, a concentrated solution of titanic sulfate containing a smaller proportion of free acid was heated to 180° C in a gas-tight, glass-lined steel cylinder at the corresponding pressure for 20 minutes. The washed and calcined product gave a fine white powder of very high covering power. This method had the advantage that solutions of much higher concentrations could be hydrolyzed without precipitation of iron, and the time required was only one fifth to one tenth that required at atmospheric pressure. The method was also applicable to the production of composite pigments and to the hydrolysis of chloride solutions.

According to a related process, solutions in which the titanium was present as normal sulfate were heated under pressure to temperatures above the normal boiling point to effect hydrolytic precipitation.[23] The temperature required for decomposition depended upon the concentration of both free acid and titanic sulfate. A solution of specific gravity 1.30, containing 0.792 g titanium dioxide and 0.0149 g ferrous iron per cubic centimeter, was not completely decomposed on heating for 2 hours at 147° C, but complete precipitation was effected at 160° C in 30 minutes. A solution of the same composition, diluted with water to a specific gravity of 1.10, threw down its entire titanium content after 30 minutes at 135° C. The washed precipitate, dried at 110° C, contained 84 to 94 per cent titanium dioxide, 1 to 5 per cent sulfur trioxide, trace to 2 per cent ferric oxide, and 4 to 10 per cent water. Calcination at 550° to 600° C was sufficient to effect removal of water and acid and develop pigment properties.

So-called "acid" solutions of titanic sulfate, that is, solutions containing acid in excess of the amount required to form normal salts with the base constituents, were hydrolytically decomposed at lower temperature and pressure by adding small proportions of a solution of a basic salt having a mole ratio of sulfuric acid to titanium dioxide between 1.8 and 1.5.[24] The solution obtained by leaching the reaction product of ilmenite with 2 parts of concentrated sulfuric acid was treated with scrap iron or electrolyzed to reduce all ferric component to the ferrous state and then clarified. A large part of the ferrous sulfate was crystallized by cooling, and titanium was precipitated hydrolytically from the mother liquor by heating in an autoclave under a pressure corresponding to a temperature of 150° C. This hydrous titanium oxide was filtered, washed, and calcined to develop pigmentary properties. The ferrous sulfate was heated to produce sulfur trioxide which was led into the residual liquor to yield strong sulfuric acid for reaction with more ore.[25]

According to a method used by Oppegaard and Stopford,[26] 1000 g ilmenite ore was mixed with 510 ml of 93 per cent sulfuric acid and added to 370 ml more of the 93 per cent acid at 130° C in a cast-iron pot. The charge was heated with stirring to fumes of sulfur trioxide, and 40 ml of water was added to initiate the reaction. After the reaction had gone to completion, the solid mass was baked for 30 minutes at 200° C, cooled, and dissolved in water at 60° C. The solution was adjusted to a specific gravity of 1.50, reduced with scrap iron to a titanium oxide content of 3 gpl, and clarified. It was then cooled to 27° C to remove part of the ferrous sulfate by crystallization, and the mother liquor was concentrated by vacuum evaporation to a specific gravity of 1.60 and hydrolyzed in an autoclave by heating at a temperature corresponding to a superpressure of 20 pounds for 80 minutes. After thorough washing the pulp was treated with 0.60 per cent potassium carbonate, then calcined and finished in the usual manner.

Titanic acid unaffected by light was produced by hydrolysis, under pressure, of ilmenite solutions having all the iron component in the divalent state and a small proportion of the titanium in the trivalent condition. In an example, the presence of 0.02 to 0.04 per cent titanous ion was sufficient.[27]

Ilmenite solutions containing 2.6 to 6.0 moles of sulfuric acid per liter were hydrolyzed by heating in a closed container at 350° to 450° C, under pressures corresponding to this temperature.[28] The product was anhydrous crystalline titanium dioxide with well-developed pigment properties. Such solutions, to which sodium chloride had been added, were heated under pressure at 300° to 320° F to obtain a hydrolysis product[29] which was filtered, washed, and calcined.

At any given temperature the rate of hydrolysis was found to increase with pressure.[30] For example, under a pressure of 3 to 11 atmospheres, reaction began at 80° C and was complete at 100° C.

## ADDED NUCLEI

To hasten the rate of thermal hydrolysis of sulfate solutions at atmospheric pressure and at the same time obtain products of pigment grade, various nucleating or seeding agents have been added. Although many types have been proposed, these in general have been suspensions of colloidal titanium compounds, but have varied considerably in the method of preparation and in method of admixture with the solution to be hydrolyzed. Frequently very small proportions of such agents have proved effective, and rarely has more than 5 per cent been required to give optimum results.

According to Mecklenburg,[31] whenever titanic salt solutions are subjected to hydrolysis there is a certain induction period between the beginning of the operation and the first visible precipitation during which colloidal titanium oxide is formed. These particles serve as centers of accumulation as hydrolysis proceeds, and the result is the formation of agglomerates which produce a visible precipitate as coagulation continues. On heating such solutions near the boiling point, hydrolysis occurs with the formation of a precipitate of hydrous oxide and the liberation of an equivalent amount of acid. Dilute solutions are more easily hydrolyzed, but on the other hand the economic operation of the process requires the use of higher concentrations, since otherwise the free acid obtained as a by-product would be too dilute to permit its recovery at a profit. On addition of as little as 1 per cent nucleating suspension "seed," based on the titanium component, however, hydrolysis of strong solutions can be effected readily. At temperatures below 50° to 60° C the mixture appears turbid, but above this range it passes through a clear stage; on further heating at 100° to 105° C for 3 hours, more than 90 per cent of the titanium is precipitated. It appears probable that the added titanium dioxide seed does not actually dissolve on heating, but is converted to a colloidal form which promotes the hydrolysis. Under the same conditions an unseeded solution of identical composition yields only 30 to 70 per cent of its titanium content.

Such nuclei were prepared by treating a titanium sulfate solution, such as that used for hydrolysis, with sodium hydroxide until neutral to methyl orange or bromophenol blue, and heating the resultant suspension of titanium oxide in a sodium sulfate solution at 100° C for 5 minutes. If the cured product was not to be used immediately, it was cooled at once to 60° C or below, for within this temperature range the nucleating property did not deteriorate significantly for several weeks. The hydrous titanic oxide of the cured suspension was the active nucleating agent, and a more concentrated seed was obtained by allowing the product to settle and then decanting part of the supernatant liquor. As an example, a small part of an ilmenite liquor containing 185 g to 213 g titanium dioxide and 440 g to 502 g potential sulfuric acid per liter was neutralized at 74° to 80° C with aqueous sodium hydroxide employing methyl orange as indicator, and held at 80° to 100° C for 15 to 30 minutes to develop the nucleating property to the maximum degree. This cured suspension was then added to the remainder of the original solution, and the mixture was heated just below the boiling point, 100° to 105° C, until hydrolysis was complete.

According to another modification, hydrolysis of concentrated solutions at atmospheric pressure was promoted and hastened by the addition of a small proportion of hydrous titanium oxide, still capable of homogeneous dispersion, which served as accumulation centers for the precipitated oxide.[32] Such nucleating suspensions were prepared by adding a portion of the main solution to an aqueous alkali to bring the pH to 4.0 to 4.5 and heating at 70° to 80° C for 15 to 30 minutes. Unless used immediately, the nuclei were cooled rapidly to below 60° C, for if held at 80° C they became overripe and lost activity. Ilmenite solutions seeded with 1 per cent of such agents, based on the titanium dioxide, gave a yield of 95 per cent after boiling for 2 to 3 hours at atmospheric pressure.

The chemical and physical properties of hydrous titanium oxide produced by the thermal decomposition of sulfate solutions varied according to the conditions under which the precipitation was carried out. A product which yielded pigment of good quality was obtained by carrying out the hydrolysis in the presence of added colloidal titanium dioxide prepared by adding a concentrated sulfate solution to a greater quantity of boiling water at a constant rate in 4 to 6 minutes. Regular nonseeded ilmenite solution diluted to 160 g titanium dioxide and 400 g potential sulfuric acid per liter to correspond to the concentration of a nucleated charge yielded only about 30 per cent of its titanium content after heating at 100° C for 24 hours. After adding colloidal titanium oxide nuclei to another portion of the solution, however, a 95 per cent yield was obtained in 3 hours under the same conditions.

Such nucleating agents were prepared by adding at a constant rate with agitation to a larger volume of boiling water, over a period of 4 to 6 minutes, an ilmenite liquor containing 200 g titanium dioxide and 500 g sulfuric acid per liter, free and combined with titanium. As the two solutions were mixed, colloidal titanium dioxide formed immediately and continued to increase in amount, after the mixing was completed, to a maximum value. After a short time, however, precipitation began, and the concentration of the colloid therefore decreased. The colloidal component may be determined quantitatively by passing a sample of the liquor through a filter medium to remove any precipitated oxide, treating the filtrate with an equal volume of concentrated hydrochloric acid, and boiling to effect complete coagulation. The coagulated material was then filtered, washed, dried, and weighed. As soon as the colloid phase reached a maximum value, the dispersion was stabilized by cooling rapidly to below 60° C. In this condition it remained active for some time.

Such a nucleating agent was mixed with an ilmenite solution containing 200 g titanium dioxide and 500 g potential sulfuric acid per liter in such proportions that the content of active acid of the seeded mixture was 400 gpl. Hydrolysis was carried out by heating the mixture at the boiling point for 3 hours. Proportions of colloidal titanium compounds in the solution to be hydrolyzed may be varied within wide limits. A great improvement in yield was noted with a concentration as low as 1.5 g to 2.0 gpl, and values up to 30 gpl were used. The final acidity of the hydrolysis liquor was also important, and values of 400 gpl were employed, since maximum recovery of titanium was obtained up to approximately this concentration. Although lower strengths gave good yields of titanium, the recovery of waste acid from the dilute liquors was not economical. The presence of iron in the solution to be hydrolyzed did not interfere with the reaction, but if the proportion was excessive it was reduced by crystallization as ferrous sulfate.[33]

Dahlstrom and Ryan [34] found that the precipitate obtained from titanic sulfate solutions poor in acid functioned as seeding agent for ilmenite liquors of usual concentrations, greatly increased the colloidal phase, and shortened the initial period before precipitation was evident. Furthermore the total time of hydrolysis was shortened appreciably and the product was of improved quality. In preparing nuclei of this type, a solution of titanium sulfate was partially neutralized with sodium hydroxide or other alkali to a pH between 1 and 2, and stirred until all turbidity disappeared. The clear product was then diluted to the desired concentration and heat-treated to develop the nucleating property to the maximum degree.

Alternatively, orthotitanic acid was dissolved in 1 to 5 per cent aqueous sulfuric acid containing sodium sulfate, magnesium sulfate, sodium nitrate, or a similar salt. These compounds apparently facilitated dissolution of the orthotitanic acid by forming complex compounds. As before, the liquor was heated to precipitate the nucleating agent. In both cases the hydrous titanium oxide nuclei were not true colloids but a definite microscopic precipitate which remained as such after mixing with the ilmenite solution and heating to boiling. The acidity at which these seeds were prepared was important, and values corresponding to a pH of 1 to 2 were found to give best results. Precipitation of effective nuclei was also a function of both temperature and time of curing; the higher the temperature the shorter the time. The most desirable range was found to be between 60° and 100° C, and heating was continued until 85 to 95 per cent of the total titanium had been precipitated. This required from 10 minutes to 4 hours.

The properly cured nuclei were added to regular sulfuric acid solution of ilmenite, and the charge was heated at or near the boiling point until the desired yield of hydrous titanium oxide was obtained. The precipitate was then processed in the usual manner to produce a pigment of high quality. Such nuclei were effective at both high and low acid concentrations, and in both "acid" and "basic" titanium sulfate solutions, that is, containing more or less active sulfuric acid than that corresponding to the normal salt. The proportion of the liquor used for preparation of the seeding agent was as high as 15 per cent of the total to be hydrolyzed, but in general 1 per cent proved effective.

In an example, an ilmenite solution containing 6.4 per cent titanium dioxide, 20.6 per cent potential sulfuric acid, free and combined with titanium, and 6 per cent ferrous oxide was employed. Of the titanium, 0.02 pound per gallon calculated as the dioxide was in the trivalent state. To 123 pounds or 15 per cent of this solution, 13.7 pounds of sodium carbonate, equivalent to one half of the active acid, was added with constant agitation, and the product was diluted with 60 gallons of water to reduce the acidity to 2.5 per cent. Turbidity which developed initially disappeared on stirring, and the clear product was heated at 100° C for 20 minutes to precipitate the desired nuclei. After cooling and settling, the supernatant liquor was removed by decantation and the precipitate was added to the main body of the solution, 697 pounds. The charge was boiled for 11 hours, at which time 95 per cent of the titanium was precipitated as hydrous oxide. In another case an ilmenite liquor from which part of the iron had been removed by crystallization as ferrous sulfate was employed. Analysis showed 12.9 per cent titanium dioxide, 3.5 per cent ferrous oxide, and 21.6 per cent potential sulfuric acid. As before, 0.02 pound per gallon of titanium expressed as the dioxide was in the trivalent state. Alternatively, orthotitanic acid precipitated by adding 1.44 pounds of sodium hydroxide dissolved in 10 gallons of water to 8.15 pounds of the above liquor was filtered and washed, and redissolved in an aqueous solution of 0.4 pound sulfuric acid and 2.3 pounds sodium sulfate. The final weight was 26.2 pounds. It analyzed 4 per cent titanium dioxide, 9 per cent sodium sulfate, and 1.6 per cent sulfuric acid, and had a hydrogen ion concentration corresponding to a pH of 1.5. This orthotitanic acid solution was then heated to 85° C and held at this temperature for 90 minutes. At the end of this period about 95 per cent of the titanium had been precipitated in a form suitable for use as nuclei. This product suspended in the mother liquor was added to 407 pounds of the original crystallized solution and the seeded mixture was boiled for 2 hours. The precipitate of hydrous titanium oxide, which represented a recovery of 95 per cent, was filtered, washed, and calcined in the usual manner.

According to a related method,[35] a sulfate solution of ilmenite was first freed from iron and then neutralized with aqueous ammonia. The precipitate formed was redissolved by agitation in the presence of ammonium chloride, and the system was heated to develop the nucleating property to the maximum degree. From 0.5 to 10 per cent of this seeding material was added to regular ilmenite solution to initiate and assist hydrolysis.

Titanium oxide precipitated by thermal hydrolysis from solutions containing considerably less sulfuric acid than that corresponding to the titanyl salt was observed to possess high nucleating power.[36] Such solutions were prepared by dissolving orthotitanic acid in titanyl sulfate solution or by neutralizing part of the active acid with sodium hydroxide or other alkali, although the latter method gave better results. The pH values were maintained below 2. A regular ilmenite solution containing from 100 to 140 g titanium dioxide, 200 g to 700 g sulfuric acid, and 50 g to 90 g ferrous iron per liter was treated at room temperature with a base such as sodium hydroxide or ammonium hydroxide in such quantity that titanium hydroxide precipitated, but completely redissolved in a comparatively short time at 20° to 30° C. The liquor was then adjusted to a concentration of 30 g to 60 g titanic oxide and 10 g to 30 g active sulfuric acid per liter, and heated at 80° C to precipitate the active nuclei. Regular ilmenite solution was seeded with 1 to 6 per cent of this cured suspension, calculated on the titanium dioxide, and boiled for 1 to 3 hours until 95 to 97 per cent of the titanium had been thrown down. The precipitate was filtered, washed, and calcined according to conventional methods to develop optimum pigment properties.

In an example, 4 per cent of an ilmenite solution containing 80 g titanium dioxide, 253 g sulfuric acid, and 85 g iron per liter, and having a specific gravity of 1.43, was removed from the main batch, mixed with dilute sodium hydroxide to bring the titanium dioxide and sulfuric acid to 40 g and 15 gpl, respectively. A turbidity appeared initially but vanished completely on stirring for 1 or 2 hours. This strongly basic solution was heated at 80° C for 2 hours to develop the nucleating property to the maximum degree and then added to the remaining 96 per cent of the original ilmenite liquor. The seeded mixture was boiled to precipitate the titanium component as hydrous oxide. A recovery of 96 per cent was obtained in 2.5 to 3 hours. The precipitate was washed, dried, and calcined to produce a pigment of good color and high tinting strength.

Schmidt[37] prepared a seeding material by heating at 70° C the product obtained by neutralizing iron-free titanium sulfate solution with sodium hydroxide until a precipitate formed which just redissolved on agitation. The cured nuclei were added to regular ilmenite solution, and the mixture was heated to effect hydrolytic precipitation of the hydrous oxide.

Allan and Bousquet[38] found that it was possible to hydrolyze a purified solution obtained directly from ilmenite ore without concentration or removal of iron merely by regulating the sulfate content and adding a suitable seeding agent. By this process titanium dioxide of good pigment properties was obtained from solutions containing as low as 100 gpl of this constituent by increasing the amount of stable sulfates. This was accomplished by adding more ferrous sulfate or any other soluble sulfate that would not hydrolyze along with the titanium, such as that of magnesium, sodium, zinc, or potassium. The factor of acidity was adjusted to 20 to 70 per cent if necessary by adding a basic compound which formed a soluble sulfate. Values between 40 and 50 per cent gave best results. In preparing such solutions ilmenite ore was digested with 2 parts concentrated sulfuric acid and the dry reaction product was dissolved in water or dilute acid. Metallic iron was introduced to reduce the ferric component to the ferrous condition and a small proportion of titanium to the trivalent state. The liquor was clarified by settling after coagulation of the slimes by glue or by filtration after the addition of diatomaceous earth, or a combination of the two, and was not crystallized or concentrated. Analysis showed the composition indicated in Table 14–2.

**TABLE 14–2**

Composition of Hydrolysis Solutions

| Constituent | Grams per Liter |
|---|---|
| Total titanium dioxide | 154 |
| Reduced titanium as dioxide | 5 |
| Iron | 115 |
| Total sulfuric acid | 454 |
| Free sulfuric acid | 64 |
| Active sulfuric acid | 252 |

Factor of acidity, 34 per cent. Specific gravity, 1.54.
Active acid equals titanium equivalent sulfuric acid to yield titanyl sulfate plus free acid.

$$\text{Factor of acidity} = \frac{\text{Free acid}}{\text{Titanium equivalent sulfuric acid}} \times 100$$

Titania gel corresponding to 2 per cent of the titanium content was added to the solution and the mixture was boiled for 5 or 6 hours to effect a 95 per cent recovery. The precipitate was filtered, washed, and calcined in the usual manner.

Other types of nuclei, such as colloidal titanic oxide formed in place, may be used. If it was desired to employ solutions of lower titanium content, additions of a soluble salt such as magnesium or ferrous sulfate were made to increase the total concentration, which was the important factor. The properties of the pigment obtained were improved for any given titanium concentration by increasing the concentration of soluble sulfates in the hydrolysis solution, and by-product ferrous sulfate was effective. The process permitted further economies in that titanium solution containing an extremely low free acid concentration could be employed.

Products of improved pigment properties were obtained from solutions containing high percentages of titanium, though poor in acid, by adding small amounts (0.05 to 5 per cent based on titanium dioxide) of alkali silicates, or oxides, hydroxides, or carbonates of zinc, magnesium, the alkali metal, or ammonia prior to or during the hydrolysis step.[39] The small proportion of alkaline agent introduced did not cause any appreciable decrease in the total acidity of the system but rather acted as initiator of the reaction. Acid-poor ilmenite liquors of this type were in a metastable condition, and the alkali broke down the metastability and started hydrolysis by a sudden local decrease in acidity. The initial solutions contained from 150 to 250 g titanium dioxide per liter, and up to 50 per cent less sulfuric acid bound to titanium than that required to form the normal salt. All the iron component was reduced to the ferrous state and a part was removed by crystallization as copperas. In some instances further purification was effected by treatment with hydrogen sulfide and the liquor was clarified. From 1 to 10 kg of the alkaline material (sodium carbonate) per cubic meter of solution was very effective, although as low as 0.1 kg gave noticeable improvement. The upper limit was a proportion equivalent to 10 per cent of the acid combined with titanium. Hydrolysis was readily effected by boiling the seeded liquor in an open vessel. The reaction proceeded more rapidly, however, at autoclave pressure. Heat was applied indirectly through steam coils or by injecting live steam. After hydrolysis had proceeded to completion, the precipitate was washed, treated, and calcined at 900° to 1000° C to obtain a pigment of higher tinting strength. A typical solution contained 200 g titanium dioxide per liter and enough sulfuric acid to combine with the iron and other bases, and an additional quantity of acid, 20 per cent less than that theoretically required to combine with the titanium present to form the normal salt.

Hydrolysis was further accelerated by adding an alkaline compound such as sodium carbonate or calcium hydroxide in the dry form to boiling titanium sulfate solution.[40] The results were a function of the ability of the dry material to react locally with the ilmenite solution to produce a change in the physical conditions of the titanium compounds present. It was the production of a particular colloidal state of titanic oxide that gave the accelerating effect on the hydrolysis. This was borne out by the fact that dry hydrated lime gave good results, dry calcium oxide was not quite so good, and dry barium or calcium carbonate was only fair. The most soluble and the lowest gravity compounds reacted most rapidly locally and thus exerted the greatest seeding effect. Distribution of the alkali throughout the solution was avoided. In an example, 2 liters of ilmenite solution containing 202 g titanium dioxide, 140 g iron, 630 g sulfuric acid, 135 g free acid, 383 g active acid per liter, and having a factor of acidity of 54.4 per cent, was heated to boiling and 5 g dry sodium carbonate was added. Boiling was continued, and after 8 hours a yield of 91 per cent was obtained. A solution having a ratio of titanium dioxide to sulfuric acid corresponding to titanyl sulfate has a factor of acidity of zero, and one corresponding to the normal salt has a factor of acidity of 100.

Leuchs [41] treated part of such a solution with a base as magnesium oxide or sodium hydroxide to precipitate titanium dioxide, and redissolved the separated material in the remainder of the original solution to bring the ratio of titanium dioxide to sulfuric acid to 1. After reduction of the iron to the ferrous state, the solution was heated at 100° C to effect hydrolysis of the titanium salts. Regular ilmenite liquors were hydrolyzed readily on heating at temperatures above 95° C after seeding with 1 to 6 per cent, based on titanium dioxide, of a suspension prepared by adding sodium or ammonium hydroxide to a solution containing 30 g to 60 g titanium dioxide per liter until the pH was brought to 2, and heating the product at 95° C to develop the nucleating power to the maximum degree.[42] The partially neutralized solution contained from 2 to 5 moles of titanium dioxide for each mole of sulfuric acid.

Nucleating agents suitable for accelerating the hydrolysis of regular sulfuric acid solution of ilmenite were prepared by adding an alkaline compound to a portion of the original solution to obtain a low but predetermined acidity, and heating at 75° to 100° C to develop the desired properties.[43]

From the opposite approach, titanium oxide obtained from hydrolysis of a sulfate solution was treated with a limited amount of sulfuric acid to dissolve the precipitate partially, and the resulting opaque suspension was heated at 80° to 200° C to develop its nucleating characteristics.[44]

In preparing composite nuclei, an impure titanium sulfate solution was added to an aqueous suspension of a hydroxide or carbonate of an alkaline earth metal in such amount that 96 to 98 per cent of the titanium was precipitated.[45] Under this condition impurities thrown down were redissolved on agitation. The precipitate was washed and the titanium component was redissolved in sulfuric acid to form a strongly basic solution containing the alkaline earth sulfate in suspension. The product was then heated for 1 to 2 hours at 80° to 90° C and cooled rapidly to below 60° C. This cured composition was added as a seeding agent in the thermal hydrolysis of ilmenite solutions.

The speed of precipitation of hydrous titanium oxide from sulfate solutions was increased by the presence of a washed oxide from a former hydrolysis.[46] Products obtained in this manner settled rapidly and were readily filtered and washed. The method was applicable to the production of both pure titanium dioxide and composite pigments. In an actual operation, the hydrolysis product of a sulfate solution of ilmenite was washed, filtered, and added to another portion of the original solution in such proportions that the final product was composed of 60 per cent freshly precipitated and 40 per cent of the added compound. The liquor was boiled to effect hydrolysis, and the precipitate was filtered, washed, and calcined.

According to another modification, a pigment containing 25 per cent titanium dioxide and 75 per cent barium sulfate was mixed wet with the solution in such proportions that the final precipitate contained equal parts of the two pigment constituents, and hydrolysis was effected by boiling. The composite precipitate was easily washed and on calcination gave a pigment of high hiding power.

Carpmael [47] added such solutions to a suspension of calcium carbonate in water and agitated the mixture for 6 to 8 hours. The washed precipitate of titanium dioxide and calcium sulfate was made into a paste containing 40 g to 60 g titanium dioxide, 15 g to 30 g sulfuric acid, and 35 g to 43 g sodium sulfate per liter, and heated at 80° to 90° C for 1 hour to develop the nucleating property. Ilmenite solutions containing iron, vanadium, chromium, and other impurities, after seeding with this material, were readily hydrolyzed on boiling to yield titanium dioxide of pigment quality.

A pure hydrous titanium dioxide which yielded pigment of high covering power was precipitated by thermal hydrolysis from such solutions after nucleating with minute amounts of accelerators obtained by the hydrolysis of titanic and titanyl sulfates in orthotitanic acid solutions.[48]

Better results were obtained by adding materials at intervals [49] during the hydrolysis process. The washed precipitate was treated with potassium carbonate and calcined at 900° C to develop pigmentary properties.

Allan [50] prepared gelatinous nuclei by treating a solution containing 201 g titanium dioxide, 437 g sulfuric acid, and 46 g iron per liter with aqueous sodium hydroxide (800 gpl) in an amount sufficient to react with all potential sulfuric acid except that combined with the iron. During the neutralization the temperature rose to 108° C, and on cooling the mixture set to a solid gel containing 6.6 per cent titanium dioxide. Hydrolysis of regular solutions was readily accomplished by boiling after seeding with approximately 4 gpl of this gel. In a typical operation, 1 g of the dry gel, obtained as described above, was added to 100 ml of a solution containing 201 gpl titanium dioxide, at a factor of acidity of 43, with 46.0 g of iron per liter, and the mixture was heated to boiling. A 96 per cent recovery was obtained after 2 hours. Nuclear gels of this type were also prepared from other than sulfate solutions, for example chloride, and any alkali or alkaline earth hydroxide or carbonate was satisfactory. Such gels were insoluble in hot and cold water. They could be washed free of soluble salts, although the washed and unwashed gel gave practically the same results. Thermal hydrolysis of ilmenite solutions were also accelerated by an admixture of a gel of titanium dioxide prepared by neutralizing a titanium salt solution with an alkali metal carbonate, and at the same time pigments of better color and brightness resulted.[51] This improvement was attributed to the fact that, employing carbonates as neutralizing agent, gel was produced at a much lower temperature than with caustic alkali. If the temperature rose too high, part of the titanium oxide was converted to a dehydrated form which did not dissolve or disperse in the ilmenite liquor and was carried into and discolored the calcined pigment. To 1 liter of a solution containing 271 g titanium dioxide, 52 g ferrous iron, 609 g total sulfuric acid, 186 g free sulfuric acid, and 518 g active sulfuric acid, and having a factor of activity of 56, was added a saturated sodium carbonate solution to neutralize the active acid completely. In this operation the temperature remained at all times below 60° C. This gel was used to seed hydrolysis solutions of compositions similar to that given above.

A composite silica-titania gel produced by intermixing a solution of sodium silicate and a titanium salt showed the same results.[52] Equally effective gels were formed in situ by adding a soluble silicate to the ilmenite solution directly. In either case the mixture was boiled to effect hydrolysis, and the precipitate was washed, dried, and calcined to give a pigment containing a small proportion of silica. Up to 7 per cent, it did not reduce the covering power of the pigment. The value fell off, however, with further addition, but not in proportion to the dilution. In an example, 2000 ml of a solution containing per liter 191 g titanium dioxide, 139 g ferrous iron, 615 g total sulfuric acid, 136 g free sulfuric acid, and 370 g active sulfuric acid, with a factor of acidity of 58, was added with stirring to 200 ml of a sodium silicate solution containing 10 per cent silica, and the mixture was boiled for 9 hours to obtain a 91 per cent yield of titanium dioxide. The precipitate was washed, calcined, and milled according to conventional methods.

According to Pollack,[53] titanium dioxide of pigment grade may be obtained by the thermal hydrolysis of sulfate or chloride solutions in the presence of 0.1 to 10 mole per cent of a soluble tin salt or of dispersed hydrated stannic oxide. Alkali metal stannates may be added directly to the main solution. Composites may be obtained by effecting the precipitation in the presence of suspended carriers, e.g., barium sulfate, light spar, or asbestine.

Regular ilmenite solutions were hydrolyzed in a shorter time and at a lower temperature after seeding with a material prepared by taking up the residue obtained on decomposing sodium metatitanate with water in the least quantity of sulfuric acid necessary to give good extraction.[54]

In making the nucleating agent, a mixture of rutile and sodium hydroxide was roasted below the melting point and leached to remove the alkali. The washed residue was then extracted with sulfuric acid of 50 per cent strength to yield a solution containing 167 pounds titanium dioxide and 251 pounds sulfuric acid in 100 gallons. One part of this product was added to five parts of the regular ilmenite solution, and the mixture was then heated to bring about hydrolytic precipitation of the titanium.

According to a later development, the titanium oxide concentrate was added directly to the ilmenite solution without being previously dissolved in sulfuric acid.[55] Rutile was heated with sodium hydroxide and the reaction product was washed with water or dilute acid to remove the alkali remaining. This residue was added to regular ilmenite solution and the seeded mixture was boiled to effect hydrolysis. Dry powdery sodium titanate, prepared by heating a paste of rutile and concentrated aqueous sodium hydroxide at 500° to 600° C, was found to aid the hydrolytic decomposition of such solutions.[56] For example, a mixture of 100 parts rutile, 100 parts sodium hydroxide, and 33 parts water was kneaded, dried, and heated at 600° to 650° C for 30 minutes. The reaction product was purified by leaching with dilute sulfuric acid. From 0.05 to 0.20 part of the crude or 0.01 to 0.10 part of the purified product was added to the ilmenite solution for each part of titanium dioxide present. The mixture was boiled to hydrolyze the titanium sulfate and the precipitate was washed and calcined at 800° to 1000° C to yield pigment.

Active nuclei were obtained by heating a solution or suspension of titanic oxide in the presence of organic acids, e,g., oxalic, tartaric, or salts of these, in proportions sufficient to prevent the formation of particles of more than ultramicroscopic dimensions but insufficient to prevent hydrolysis entirely. These agents served the further purpose of stabilizing the nuclei. It is well known that the hydrolysis of mineral acid solutions of titanium may be inhibited by the addition of organic acids or their salts, e.g., oxalic, citric, and tartaric acids, and sodium citrate. In the presence of appreciable proportions of these agents, very slight or no precipitate forms even on prolonged boiling at relatively high dilutions. In a specific operation, a 2 to 3 per cent sulfate solution of titanium was treated with 3 moles of sodium chloride and one half mole of oxalic acid per mole of titanium dioxide, and then with sodium carbonate until 90 to 95 per cent of the titanium had been precipitated.[57] This product was washed and boiled with water to give a colloidal solution for use as seed.

According to Cauwenberg,[58] the hydrolysis of titanium sulfate solutions may be initiated and maintained by the addition of organic compounds which liberate ammonia (hexamethylene tetramine, acetaldehyde ammonia, and acetamide) before or during hydrolysis. Alternatively, a small part of the solution may be activated by boiling with such agents, and then returned to the main body, whereby the dispersed particles of titanium compounds act as nuclei for further precipitation. Best results were obtained by adding the seed to the ilmenite solution at 85° C and raising the temperature to boiling within a short time.

By precipitating titanium hydroxide from sulfate solutions in the presence of a finely dispersed salt of titanium soluble in dilute mineral acids, a finely divided product of pigment grade was obtained.[59] Such compounds are titanium phosphate, the double sulfate of titanium and potassium, double fluoride of titanium and potassium, silicates of potassium, and salts of titanium and zirconium. These may be formed in solution. One liter of a solution containing 100 g titanium dioxide, 50 g iron as ferrous sulfate, and 230 g sulfuric acid was stirred with 80 ml of phosphoric acid containing 220 g phosphorus pentoxide per liter to give a voluminous precipitate which disappeared on stirring for 20 hours. The product, probably a solution of titanium phosphate, gave a fine precipitate of titanium oxide on boiling. Similarly stable titanium hydroxide compounds were obtained by effecting the precipitation in the presence of phosphoric acid or soluble phosphates.[60] After adding a relatively small proportion of solid basic titanium sulfate to sulfuric acid solutions of ilmenite, hydrolysis was rapidly effected by heating.[61]

Allan[62] reported that in the known processes by which hydrolysis of titanium salt solutions (ilmenite) are initiated or accelerated by seeding agents such as colloidal titanic compounds and calcium sulfate, pigments of improved physical properties were obtained by adding the nucleating agent in a number of portions during the precipitation.

## YIELD SEED

Sodium titanate prepared in the reaction between titanium sulfate hydrolyzate and sodium hydroxide in an aqueous medium at 85 to 95° C is an extremely effective nucleating agent for the hydrolysis of titanium sulfate solutions.[63] As shown by X-ray and diffraction analysis, the sodium titanate has no definite pattern of crystalline structure. The composition of the hydrous oxide precipitated hydrolytically from sulfate solution is reported to be $10TiO_2 \cdot 10H_2O \cdot xSO_3$. If this material is heated with sodium hydroxide solution for 1 to 6 hours, a water insoluble product is obtained which, after removal of excess alkali, contains 80 to 86 per cent titanium dioxide and from 20 to 14 per cent sodium oxide depending on the time of boiling. These analyses correspond to the compounds $Na_2Ti_5O_{11}$ and $Na_2Ti_3O_7$, respectively. This nucleating agent added to titanium sulfate solution at elevated temperatures is very rapidly transformed to colloidal titanium dioxide. Thus the entire amount of titanium dioxide nuclei formed is present in the solution at the beginning of the hydrolysis and is available at the time the titanium sulfate has become sufficiently hydrolyzed so that titanium dioxide is formed. This freshly prepared titanium dioxide adheres to the individual nuclei and is subsequently precipitated in uniform size. Such nuclei are very stable, are of uniform characteristics, and remain unchanged after storage for several months at room temperature. The hydrolysis treatment is carried out by mixing an aqueous dispersion of the nucleating agent at 20 per cent solids content with the titanium sulfate liquor at 50° C. After nucleating, the mixture is heated to boiling and boiled for 1 to 5 hours. The amount of nucleating agent may vary from 0.5 to 6 per cent, although 3 per cent based on the titanium dioxide present in the titanium sulfate solution gives optimum results.

In an example, filter cake of hydrous titanium dioxide obtained by the hydrolysis of sulfate solution and containing 100 g of titanium dioxide at 30 per cent solids was added to 150 g of sodium hydroxide with stirring. The mixture was heated for 2 hours at 85 to 90° C and constant volume to form the titanate which was diluted with water, filtered, and washed free of sulfates and excess alkali. An 88 g portion of the aqueous slurry of the nucleating agent containing 13.5 g (3 per cent) of titanium dioxide was diluted with 120 ml of water. This nuclei mixture was heated to 90° C and to it was added 2660 g of titanium sulfate liquor also at 90° C containing the equivalent of 450 g of titanium dioxide. The liquor was added at a constant rate over a period of 8 minutes after which the mixture was heated to boiling and boiled for 3 hours to effect hydrolysis. The washed product calcined with a rutile conversion seed gave a rutile pigment having very good color characteristics and a tinting strength of 1620.

A seeding material for use in the hydrolysis of titanium sulfate solution is prepared by heating hydrous titanium dioxide with sodium hydroxide at temperatures which are inversely proportional to the alkali content of the mixture.[64] If the alkali content at curing is low, a relatively high temperature must be employed, while a lower temperature must be used if the alkali content of the mixture is high. After the aqueous slurry has been heated for a period of time sufficient to alter the hydrous titanium dioxide, it is removed from the filtrate and washed to remove excess alkali. The washed product may be added directly to a sulfate solution of titanium, or it may be stored as a slurry or filter cake. If the sodium hydroxide content at curing is from 20 to 23 per cent on the titanium dioxide basis, curing is carried out at 90° C. With 25 to 35 per cent alkali, the curing temperature is from 50 to 75° C. With an alkali content at curing from 36 to 41 per cent on a titanium dioxide basis, optimum curing temperatures are as low as 30 to 40° C. In other words, the temperature used for curing the hydrate in the presence of alkali is sufficiently high to modify the hydrate so that it will form a colloidal dispersion in sulfuric acid, but not so high as to convert the hydrate to a titanate which is crystalline. To hydrolyze a sulfate solution containing the equivalent of 140 to 240 gpl titanium dioxide, the addition of 1 per cent of seed, based on the titanium dioxide content, gives a recovery of 96 to 98 per cent after boiling the solution for 1 to 3 hours.

For example, to 50 g of sodium hydroxide as a 50 per cent aqueous solution was added 333 g of the product containing 30 per cent titanium dioxide obtained by the hydrolysis of titanium sulfate solution. The total weight of the mixture was 433 g of which 42 g was unreacted sodium hydroxide; the remaining 8 g of sodium hydroxide was neutralized by the residual sulfuric acid in the hydrate. This mixture was heated for a period of time sufficient to remove practically all of the combined water and increase the sodium hydroxide content of the system at curing to 22 per cent. Curing was then carried out by heating the mixture at a temperature of 90° C for 1 hour to produce a dry yield seed. The product dispersed readily in sulfuric acid and analysis showed that 30 per cent of the titania was in the colloidal form. To 1000 ml of sulfate solution containing 260 g of titanium dioxide was added, at 60° C, 17 g or 1 per cent, on a titanium dioxide basis, of the seeding material as an 18 per cent slurry. The mixture was adjusted with water to 200 gpl titanium dioxide, heated to the boiling point, 110° C, and boiled for 2 hours. Recovery of titanium dioxide was 95 per cent.

Orthotitanic acid for the production of hydrolysis nuclei or rutile conversion seed is prepared from an iron containing titanium solution obtained by the reaction of ilmenite with sulfuric acid by adding aqueous sodium carbonate in such a proportion that the pH is between 3.5 and 4.5.[65] By this process a rapid filtering, iron-free orthotitanic acid is produced.

## NUCLEI FORMED IN PLACE

Another advance involves a slow admixture of the ilmenite liquor with hot water or dilute solutions in such a manner that the required nucleating agent is formed in place. Such processes have the advantage of simpler and more rapid operation and lower cost. Furthermore, the hydrolysis is carried out at high concentrations, thus permitting economical recovery of sulfuric acid from the waste liquor. This method has the added advantage of being particularly applicable to large-scale commercial operation.

Blumenfeld [66] effected the precipitation of hydrous titanic oxide of high purity by bringing together a sulfate solution of ilmenite and hot water, and retarding the rate of mixing so that the resulting liquor first became gradually turbid before the hydrolysis product was precipitated. Hydrolytic decomposition of such solutions by boiling alone gave low yields, particularly if the sulfuric acid content, free and combined with titanium, was as high as 100 gpl. The products were filtered with difficulty. Recovery of titanium by hydrolytic decompositions of solutions of this type, as well as properties of the product such as filterability, grain size, and color after calcination, depended not only on the final state of the system but upon certain transformations which the constituents of the solution underwent before and during hydrolysis, and upon the manner in which the final state was arrived at. By this method, however, solutions obtained by dissolving the reaction product produced by heating ilmenite with concentrated acid, and containing from 100 g to 300 g titanium dioxide per liter, were readily hydrolyzed to yield precipitates having good pigmentary properties. The concentrations of titanium dioxide and sulfuric acid were observed to be the important variables, and the mole ratio should be 1 of the former to from 1 to 2.5 of the latter.

In carrying out the operation, a solution at a temperature above 60° C was covered with a layer of water in such a proportion as to give, on diffusing or mixing, a liquor of optimum concentration. Instead of water, dilute solutions of sulfuric or phosphoric acids, aluminum sulfate, or a titanium salt could be employed. Introduction of the second liquor was effected in such a manner that gradual mixing took place. Alternatively, the initial titanium solution was concentrated even to a pasty mass and mixed with a volume of hot water, which was more or less than that removed, so as to bring the concentration of sulfuric acid in the final charge to 300 g to 367 gpl. One of the liquids, usually the more concentrated, was at the time of bringing together at a temperature above 60° C. Mixing was carried out fairly rapidly to avoid appreciable precipitation of hydrous titanium oxide before homogeneity of the liquor had been accomplished and heat had been applied. In both cases the mixture was held at or near the boiling point and the initial volume was maintained. The yield reached 85 to 95 per cent in a few hours in the first method, and in less than 1 hour in the second method. These processes were equally valid for the precipitation of hydrous titanium oxide in the presence of suspended extender materials such as barium sulfate and calcium sulfate.

As an illustration of the method, a solution containing per liter 225 g titanium dioxide, 50 g iron, and 440 g total sulfate, expressed as sulfuric acid, was heated to 95° to 98° C and covered with a layer of cold water equal to 20 per cent of its volume. This amount of water was sufficient to reduce the total sulfuric acid concentration to 367 gpl. After mixing by gentle agitation, the liquor became gradually turbid and precipitated about 94 per cent of its titanium content on boiling for a few hours. According to another modification, a solution containing 180 g titanium dioxide, 42 g iron, and 460 g total sulfuric acid per liter was concentrated to a pasty mass of one half the original volume and introduced into twice its volume of boiling water so that the final concentration was reduced to 300 g to 310 g sulfuric acid per liter. Mixing was effected in such a manner as to avoid immediate precipitation.

According to a later improvement [67] the hydrolysis was carried out in such a manner that an adequate quantity of colloidal titanium oxide was formed in the crystalloid solution prior to any actual precipitation. The exact manner in which the colloidal phase promotes hydrolysis is somewhat controversial, but apparently the titanium in crystalloid solution must pass through the colloid state before precipitation. There is wide variation in the physical and chemical properties of titanium dioxide produced by hydrolysis, dependent upon the exact conditions under which it is carried out, but by this process titanium dioxide of constantly uniform properties may be produced rapidly and in good yields. At the same time, solutions of high concentration are employed with the result that the sulfuric acid may be economically recovered from the hydrolysis liquor. On the other hand it is not advisable to operate at final acid concentrations much above 400 g per liter because of the tendency of the stronger solutions to crystallize titanium sulfate initially, and also because of the tendency of hydrous titanium oxide to redissolve in the acid liquor.

In the presence of approximately 30 g colloidal titanium dioxide per liter, the speed of hydrolysis was increased and a yield of 95 per cent hydrous oxide of improved and uniform quality was obtained on boiling for 3 hours. Iron did not interfere with the reaction, but if much was present in solution it tended to contaminate the precipitate. To overcome this tendency, the iron component was reduced initially to 20 g to 25 g per liter by crystallization of ferrous sulfate on cooling.

Solutions of this type, as obtained by the action of sulfuric acid on ilmenite, may vary widely in composition and the ratio of titanium dioxide to sulfuric acid may be designated by "free acidity factor." To illustrate, a solution of free acidity factor of 90, as employed in the examples, contained 90 per cent more sulfuric acid than that required to form titanyl sulfate with the titanium present after the iron had been satisfied. The hydrolysis was carried out with equal success, however, when employing solutions of factor of acidity less than 90 per cent. Much more important than this ratio was total or free acidity of the hydrolyzed liquor. This will be equal to the free acid originally present, plus the acid liberated as a result of the precipitation of titanium dioxide. For practical reasons the final acidity of the hydrolysis liquor was in the neighborhood of 400 gpl.

In a specific operation, the solution employed for hydrolysis was prepared by dissolving the reaction product of ilmenite with sulfuric acid in water. Scrap iron was introduced to convert all the ferric component to the ferrous state, and the major portion of the ferrous sulfate was crystallized by cooling and then filtered off. The mother liquor, which contained 190 g to 210 g titanium dioxide, 20 g to 25 g ferrous iron, and 500 g to 550 g active sulfuric acid per liter, was concentrated by vacuum evaporation until sulfuric acid, free and combined with titanium, excluding that combined with iron and other base constituents, was 600 gpl. A 100 ml portion of this concentrated liquor heated to 100° C was added at a uniform rate in 4 minutes to 100 ml of boiling water, with constant stirring. During the first one-fourth minute of the addition a turbidity appeared, but it disappeared almost immediately and the solution regained its original appearance. Temperature was maintained at 103° C, with stirring. After 10 minutes turbidity was again observed, and a few minutes later the suspension assumed a gray color and titanic oxide began precipitating. The reaction was complete in 3 hours, with a recovery of 90 per cent.

The trend of the reaction, as determined by observation and analyses for the first 10 minutes after all the solution had been introduced, is tabulated in Table 14–3. Colloidal titanium oxide in the filtered solution was determined by adding hydrochloric acid to effect coagulation, after which it was readily separated and weighed.

### TABLE 14–3

State of Titanium at Different Periods of the Hydrolysis Process

| Minutes of Reaction | Appearance of Solution | $TiO_2$ in Crystalloid Solution | $TiO_2$ in Colloid Form | $TiO_2$ Precipitated |
|---|---|---|---|---|
| 4  | Clear                   | 104 | 21.1 | 0    |
| 5  | Clear                   | 97  | 28.2 | 0    |
| 6  | Clear                   | 92  | 33.0 | 0    |
| 7  | Very slight cloudiness  | 84  | 40.8 | 0    |
| 8  | Very slight cloudiness  | 78  | 46.8 | 0    |
| 9  | Slightly clouded        | 70  | 55.0 | 0    |
| 10 | Clouded                 | 71  | 53.6 | 0    |
| 11 | Very clouded            | 83  | 41.9 | 0    |
| 12 | White                   | 45  | 31.6 | 48.2 |
| 13 | White                   | 43  | 26.6 | 55.2 |

(Constituent (Grams per Liter))

At the beginning of the process, all the titanium was in crystalloid solution. As reaction proceeded, all nonsoluble titanium dioxide was in colloidal form during the initial stages, but as the hydrolysis proceeded further this component increased to a maximum value and dropped off as the amount of precipitation increased.

The reaction occurring during the first few minutes determined the rate and yield of hydrolysis and the quality of the product obtained. At the same time this initial reaction was controlled by regulating the temperature, the speed of mixing and other variables. At low acid concentrations, such as prevail after only a part of the solution had been added to the water, the tendency to hydrolyze was greatest.

Both the yield of titanium dioxide and the properties of the resulting pigment were affected by the final acidity of the solution, which in turn was a measure of the concentration at which hydrolysis was carried out. This factor may be illustrated by a series of precipitations carried out by adding different volumes of the concentrated solution at a uniform rate to 100 ml of boiling water, as before. The data are tabulated in Tables 14-4 and 14-5.

## TABLE 14-4

### Influence of Final Acidity

| Test No. | Volume of Solution Added (ml) | Time of Addition (minutes) | Final Acidity at End of Hydrolysis (gpl sulfuric acid) | Time of Analysis after Introduction (hours) | Iron | TiO$_2$ | H$_2$SO$_4$ | Yield TiO$_2$ (per cent) |
|---|---|---|---|---|---|---|---|---|
| 1 | 60 | 2.75 | 200 | 1.5 | 26.3 | 13.0 | 215 | 85.0 |
| 2 | 100 | 4.0 | 300 | 1.5 | 36.8 | 19.0 | 297 | 84.0 |
| 3 | 200 | 8.0 | 380 | 1.5 | 42.0 | 23.2 | 380 | 85.0 |
| 4 | 250 | 10.0 | 420 | 2.0 | 44.2 | 19.8 | 410 | 88.5 |

## TABLE 14-5

### Influence of Rate of Mixing

| Test No. | Time of Addition (minutes) | Time of Analysis after Mixing (hours) | Iron | TiO$_2$ | H$_2$SO$_4$ | Yield of TiO$_2$ (per cent) |
|---|---|---|---|---|---|---|
| 5 | 6 | 2 | 43.7 | 76.2 | 415 | 57.0 |
| 6 | 10 | 2 | 44.2 | 19.8 | 408 | 88.5 |
| 7 | 14 | 2 | 44.5 | 22.6 | 400 | 87.0 |
| 8 | 18 | 2 | 45.5 | 26.8 | 425 | 85.0 |
| 9 | 26 | 2 | 50.8 | 38.0 | 418 | 77.0 |
| 10 | 34 | 1.75 | 53.6 | 110.0 | 432 | 35.0 * |

* Unfilterable colloids.

Hydrolysis effected at the highest acid concentration not only gave the highest yield, but the resulting pigment had properties superior to those of the other samples.

Development of an adequate colloid phase in the solution prior to precipitation depended in part upon the speed at which the titanium solution was mixed with the water. In the tests shown in the table, 250 ml of the concentrated solution was added hot, with stirring, to 100 ml boiling water, as before.

The most favorable results were obtained by introducing the solution of this concentration in from 10 to 18 minutes, that is, at a rate of 4 to 6 minutes per volume of solution added to a unit volume of water. More rapid addition gave poor yields, since an adequate colloid phase was not formed. On the other hand, if introduction was carried out at slower rates, there was a gradual decrease in yield until a considerable precipitation occurred before all the solution had been added. Employing still slower rates, the titanium was not precipitated, but instead it converted to unfilterable colloid form and the yield was very low.

The combined influence of all these factors is shown graphically in Fig. 14–1, in which the concentration of titanium dioxide in the various

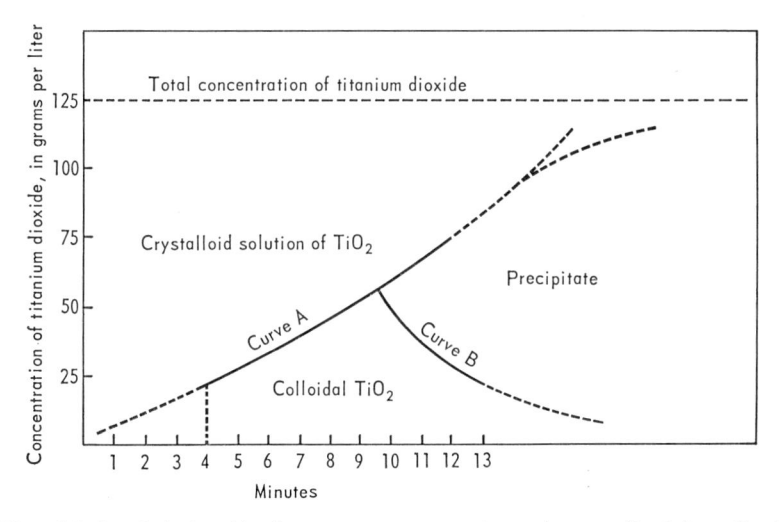

**Fig. 14–1.** Relationship between concentration of crystalloidal, colloidal, and precipitated titanium dioxide, and time of hydrolysis.

phases is plotted against the time during hydrolysis. Naturally, the total concentration of the system remains constant. At the initial state of the reaction, all the titanium dioxide was present in crystalloid solution. Curve A shows the progressive increase in the amount not in crystalloid solution. At first all the nonsoluble titanium dioxide was in colloidal form. When a point is reached, however, at the junction of curves A and B, the amount of colloid material falls off. Curve B, which shows colloidal material alone, indicates the decrease during the rest of the reaction. Obviously the difference between these two curves shows the amount of titanium dioxide precipitated. This figure shows only the first few minutes of the reaction, but as hydrolysis proceeds further, curve A rises to approach the total concentration of titanium dioxide present, and curve B drops off. Under these conditions, colloidal titanium dioxide was formed immediately on contacting the solution with water, and by the end of the period of introduction this constituent amounted to 20 gpl.

In a process employed by Olson,[68] 0.3 to 2.0 per cent of a titanium solution for hydrolysis was added quickly to 20 to 80 times its volume of water. The resulting nuclei were allowed to age and condition for 10 seconds or more, after which the remainder of the original solution was added rapidly and hydrolysis was effected at an elevated temperature.

Hydrous titanic oxide of high purity was precipitated in a very fine state of subdivision by adding a mineral acid solution of titanium to a hot dilute solution containing an organic acid or a salt thereof, such as tannic, tartaric, citric, and oxalic acids, and sodium tartrate and ammonium citrate.[69] Alternatively, the organic compound was added directly to the titanium solution, which was run slowly into hot water with constant agitation to initiate the hydrolysis. Such compounds appeared to have an effect similar to that of the inorganic accelerating agents, calcium sulfate and barium sulfate. Relatively small proportions of the soluble addition agents, much less than required for double decomposition reaction with the titanium salt, were required. More uniform results were obtained by employing hydrolysis solutions prepared from commercial uncalcined titanium dioxide containing only small proportions of impurities.

For example, hydrous titanium oxide was heated with strong sulfuric acid and the reaction product was dissolved in water. All the iron component was reduced to the ferrous constituent, and to prevent reoxidation during processing the reduction was continued until 1 gpl titanous ion was formed. Three thousand pounds of such a solution containing 7 per cent titanium dioxide was added to 1500 gallons of a 0.30 per cent solution of oxalic acid at 90° C during the course of 1 hour. The temperature was maintained constant, and agitation was continued. By the time the components were mixed, 95 per cent of the titanium had precipitated as hydrous oxide in an extremely fine state of subdivision, but in such a form that it settled and filtered rapidly. The washed product was calcined at 700° to 1000° C.

According to another variation, 32.3 pounds of oxalic acid was dissolved in 3000 pounds of the original sulfate solution containing 7 per cent titanium dioxide, and the product was added to 1500 gallons of water at 90° C in 1 hour as before. By the time the solutions were thoroughly mixed, 95 per cent of the titanium had been hydrolytically precipitated.

The hydrous oxide was filtered, washed, and calcined at 700° to 1000° C to obtain a pigment having a particle size under 0.89 micron.

According to an important method applicable in the presence of appreciable proportions of iron, a solution such as was obtained by dissolving the reaction product of ilmenite with sulfuric acid was added to a hot dilute solution containing both an organic acid or an organic acid compound and phosphoric acid or a soluble phosphate.[70] Alternatively, the reagent was dissolved in the titanium-bearing liquor and the product was added slowly to hot water. One per cent oxalic acid and 0.1 per cent phosphoric acid were sufficient to effect hydrolysis. Although iron may be present, it should be reduced to the ferrous state.

In a typical operation, 32.3 pounds of oxalic acid and 2.4 pounds of phosphoric acid were dissolved in 3680 pounds of ilmenite solutions containing 5.71 per cent titanium dioxide, 5.24 per cent ferrous oxide, and 23.81 per cent combined and 2.86 per cent free sulfuric acid, and the combined liquor was added in 1 hour to 300 cubic feet of water at 98° C. Agitation was continuous and the temperature was maintained constant. By the time mixing was complete, 95 per cent of the titanium was thrown down as a fine particle size, easily filtered hydrous oxide. A part of the phosphoric acid combined with titanium and appeared in the precipitate. The product was washed and calcined at 700° to 1000° C to yield titanium dioxide of high purity having good pigment properties.

Hydrolysis was effected by adding the titanium-bearing solution to boiling water carrying in suspension a small proportion of colloidal material, specifically silicic acid, capable of bringing the hydrous titanium dioxide to a physical state in which it was readily filtered and in which it possessed the final quality of pigment.[71] If ilmenite solutions were used, reduction was carried to the point that 5 per cent of the titanium was converted to the trivalent form and a part of the ferrous sulfate was removed by crystallization on cooling. Composites were similarly produced by effecting precipitation in the presence of fillers, such as barium sulfate, either added to or formed within the solution. The product was filtered, washed, treated with a borate, and calcined.

Thermal hydrolysis of ilmenite solutions to produce an oxide free from iron was accomplished by adding the solution slowly to an aqueous and approximately neutral dispersion at 90° C of a protective colloid such as dextrin and a salt which was relatively insoluble in water but completely soluble in the sulfuric acid liquor produced during the process.[72] Salts of this type include the fluorides and oxalates of tin, antimony, titanium, thorium, aluminum, germanium, zinc, and zirconium. According to this process, hydrolysis of concentrated titanium sulfate solutions was readily accomplished with very slight dilution of the original solution. The temporarily insoluble salt initiated the first stages of crystallization and thus formed the seed. Later the salt dissolved in the hydrolysis liquor. The function of the protective colloid was to prevent the aggregation of the seed particles of titanium dioxide already initiated by the temporarily insoluble salt, and it also acted in an auxiliary manner by furthering the maintenance in solution of the iron salts.

In an example, 200 ml of clarified and crystallized sulfate solution of ilmenite, having a specific gravity of 1.35 and containing 7.5 per cent titanium dioxide and 20 per cent sulfuric acid free and combined, was added at 30° to 80° C at a uniform rate in 15 minutes to 250 ml of a solution containing 0.3 per cent dextrin and 0.4 per cent suspended aluminum oxalate at 90° C. Hydrolysis resulting from the mixing of these solutions produced a large number of hydrous titanic oxide particles to serve as seed, and 800 ml of the original ilmenite solution was added very quickly. The system was raised rapidly to 98° C and maintained at this temperature with agitation, for 30 minutes, until the dextrin was converted to soluble sugar. This resulted in complete hydrolysis of the titanium sulfate and the precipitate was washed and calcined by conventional methods.

According to a modified procedure,[73] solutions of the same type were diluted with water containing 0.5 per cent of a colloidal polysaccharide (dextrin) and boiled to effect hydrolysis. At the same time the dextrin was converted to sugar.

Two volumes of a 10 to 20 per cent solution of titanium sulfate, and 2 volumes of a 0.75 per cent solution of dextrin, both at 80° C, were added simultaneously at a constant rate to 1 volume of water at 90° C, and the mixture was boiled to complete the hydrolysis. Ilmenite solutions containing 0.5 to 2 per cent antimony trioxide based on the titanium dioxide were hydrolyzed directly by boiling under atmospheric conditions, without seeding, to produce pigment of high quality possessing strong resistance to chalking.[74]

To improve the filtering and washing rate [75] of the hydrolysis product, 10 per cent of a sulfuric acid solution of ilmenite was set aside and the remainder, heated to 90° to 100° C, was added in 15 minutes, with constant agitation, to one fifth its volume of water or dilute acid at the same temperature. After boiling for 0.5 to 2.0 hours, the remaining 10 per cent of the solution was added, and heating was continued for 3 to 4 hours to complete hydrolysis. Wood [76] found that the filtration rate of the precipitate, as well as the ultimate pigment properties, could be improved by greatly reducing or stopping altogether the normal agitation of the liquor during the portion of the hydrolysis period that transition of the solution from a true colloid to a suspension of filterable particles is taking place. Hydrolysis was initiated by adding a relatively concentrated titanium solution to water maintained at an elevated temperature, whereby a relatively large number of nuclei were formed. Constant agitation was maintained during and after mixing until flocculation or coagulation of the hydrolyzing particles took place. The system was then allowed to stand undisturbed until it presented a cream-colored appearance, to indicate the presence of white titanic oxide flocculated into particles of the desired filterable dimensions. At this stage agitation was resumed and maintained until hydrolysis was complete. At the end of the mixing operation, the liquor appeared relatively clear, but within a few minutes a slight turbidity developed and the colloidal solution passed from black to olive green to steel gray. As this change became more rapid, agitation was suspended. The color changed rapidly through tan to the indicative cream shade. This step increased the filter rate by 50 per cent and was particularly applicable to the Blumenfeld type of hydrolysis.

In an example, 3 volumes of ilmenite solution containing the equivalent of 200 g titanium dioxide, 25 g iron as ferrous sulfate, and 600 g sulfuric acid per liter, at 100° C, was added slowly to one volume of boiling water with constant agitation. During mixing a precipitate formed, but on further addition of the solution it became dispersed so that at the end of the introduction period the solution appeared relatively clear. A few minutes after the last of the liquor had been added, a slight turbidity appeared and became more marked as the color of the colloidal solution passed successively from black to olive green to steel gray. As the color change became more rapid, agitation of the liquor was stopped entirely for a period of 10 minutes, during which time the color of the mixture changed rapidly through tan to cream. During this period the system changed from a true colloidal suspension to a suspension of flocculated titanium dioxide of filterable size. At this stage the agitation was resumed and the solution was boiled for 3 hours to complete hydrolysis.

The solution obtained from the reaction product of ilmenite with sulfuric acid is cooled as usual to crystallize ferrous sulfate and adjusted to a concentration of titanium dioxide 100, titanium trioxide 0.5, iron 36, and total sulfuric acid 265 gpl.[77] To 300 liters of the solution is added 300 liters of water at 90 to 92° C, and the temperature is held at this value for 2 hours. The hydrolysis product is separated and heated with 5 per cent sulfuric acid and 0.5 per cent titanous sulfate based on the weight of the titanium dioxide. After washing to remove the sulfuric acid, the product is calcined at 950° C for 1.5 hours to give 26.5 kg of white titanium dioxide of 99 per cent purity, possessing the anatase structure. The iron content is 0.01 per cent. Ilmenite ore from the south coast of Fukin Province containing 1 to 38 per cent titanium dioxide is pulverized and separated from the sand and most of the iron impurity.[78] One part of the concentrate is reacted with a mixture of 1 part of 88 per cent sulfuric acid with 5 parts of ammonium sulfate. The solution obtained by dissolving the reaction product in water is cooled to −2° C to crystallize the ferrous sulfate as $FeSO_4 \cdot 7H_2O$. After heating to 91° C the filtrate is poured into 4 to 5 times its volume of water at 97° C. The hydrolysis product is filtered, washed, and calcined at 830° C for 1 hour to give titanium dioxide pigment.

X-ray studies of the hydrate obtained by the hydrolysis of sulfate solutions of ilmenite after proper seeding to get satisfactory titanium dioxide pigment show an amorphous structure. Peptization to eliminate adsorbed sulfate ions changes the properties of the hydrolyzate by converting it to a sol suitable for pigment production.[79] The hydrolyzate comprises loosely formed flocs varying from 10 to 200 microns in diameter, depending on the conditions. Floc size has no effect on pigment properties. The flocs are composed of grains approximately 0.6 to 0.7 micron, the limiting dimensions found by dispersion of the flocs in various liquids such as glycerol and methyl bromide. Each grain is formed of about 1000 particles of 60 to 75 millimicrons and it is these which are separated by peptization. These particles each contain about 20 units of about 20 Ångstrom units with a reticular structure with about 10 atoms laid out in a line. The crystallites are bonded by the surface action of amorphous titanium dioxide and comprise the seeds added to the titanium sulfate solution to promote hydrolysis. However, satisfactory pigment properties depend on the presense of the 0.6 to 0.7 micron grains formed from the seed. Electron microscopic investigation of the formation of colloidal particles of titanium dioxide during hydrolysis showed the formation of initial spherical or shapeless units which in the process of aging undergo crystallization.[80] During the crystallization phase these particles break up into more numerous but crystalline units. The dimensions of the crystalline particles do not depend on the conditions of the medium of growth, and are connected only with recrystallization during the break-up stage.

A study of the various factors affecting the hydrolysis of titanium sulfate solutions was made by Sakai, Yoshikawa, Suzuki, and Kobashi.[81] A similar study was made by Ham and Kim [82] on Korean ilmenite.

## CONSTANT COMPOSITION DURING HYDROLYSIS

Farup [83] found that by effecting thermal hydrolysis in a solution of approximately constant composition, a precipitate of greater uniformity and of optimum properties was obtained. The composition, purity, and physical properties of hydrolytically precipitated titanium dioxide depend to a great extent upon the conditions under which the decomposition takes place, such as composition of the solution employed, temperature, and duration of boiling. In the commonly employed processes, great changes in the concentration of the solution take place as the precipitate is formed and an equivalent amount of acid is liberated. Thus the formation of titanic acid will take place under entirely different conditions at the beginning and at the end of the operation.

To overcome this effect, a titanium-rich solution, 50 g to 300 g titanium dioxide per liter as prepared from ilmenite, was led into a precipitation vessel and heated until practically complete hydrolysis had taken place. Four fifths of the liquor was then removed. To the remaining one fifth at precipitation temperature, fresh solution was added at such a rate as to secure a practically constant content of dissolved titanium until the vessel was filled. Heating was continued throughout the process. The supply of solution was then interrupted, and four fifths of the liquor was again removed. The operation was repeated as often as was desired. A typical solution contained 9.1 per cent titanium dioxide, 10.9 per cent ferrous oxide, 30.2 per cent sulfur trioxide, and trivalent titanium equivalent to 1.5 gpl of the dioxide. After hydrolysis had been carried to the desired stage, the product was filtered and washed.

The ilmenite solution was also added continuously to the hot mother liquor at such a rate that the composition of the solution remained constant as the titanium salts gradually hydrolyzed.

According to Saklatwalla, Dunn, and Marshall,[84] thermal hydrolysis carried out in the usual manner tends to start out at a very rapid rate which gradually falls off as the reaction proceeds so that the initially precipitated titanic oxide not only occludes more metallic impurities than that thrown down later at a slower rate but has a somewhat larger particle size. This fraction, constituting 20 to 30 per cent of the total, contaminates the entire precipitate so that the resulting pigment does not have the optimum color and tinting strength. However, by increasing the free acid content of the solution to 6 to 14 per cent, the initial rate of hydrolysis was slowed down, with the result that the quality of the first portion precipitated was greatly improved in properties. To illustrate, the reaction product of ilmenite with concentrated sulfuric acid was leached with water to extract the soluble titanium and iron sulfates. The solution containing 6 to 8 per cent titanium dioxide was treated with metallic iron to reduce the ferric component to the ferrous state, filtered, and cooled to crystallize part of the ferrous sulfate. After adjusting the free acid content to 10 per cent, the solution was heated to 180° to 190° F and diluted with 10 per cent sulfuric acid at 200° F to reduce the titanium dioxide content to 2 to 4 per cent. The adjusted solution was further heated to effect hydrolytic precipitation of the titanium. Because of the excess acid, the reaction proceeded at a much more uniform rate, and the initial fraction of poor quality which would have occurred without the excess acid was avoided. The product was washed, dried, and calcined according to conventional practice. Instead of adding acid to the solution, the proper excess was acquired by increasing the ratio of sulfuric acid to ore in the digestion step.

Titanic oxide of improved color was obtained by carrying out the thermal hydrolysis of sulfate solutions in the presence of hydrofluoric acid or a fluoride salt.[85] The clarified solution was boiled to effect hydrolytic precipitation, and water was added as the reaction proceeded to reduce the concentration of free acid and increase the yield of titanium. Pigments produced in this manner were free from iron and had a more granular structure. The decomposition of titaniferous ores and the thermal hydrolysis of the resulting solution were reported to be facilitated by adding a small proportion of calcium fluoride to the sulfuric acid used in the digestion stage.[86]

Titanic oxide obtained by hydrolyzing an ilmenite solution in the presence of reducing compounds, such as sodium sulfite, in a proportion equivalent to the ferric iron and other oxidizing agents present, did not discolor on heating to incandescence.[87] The color was also improved by digesting the hydrolysis product in the mother liquor under pressure at 135° to 140° C,[88] and by a second dissolution and hydrolytic precipitation of the hydrous oxide obtained initially from ilmenite solution.[89]

Yields from the thermal hydrolysis of sulfate solutions of titanium may be increased by adding water during the process.[90] By the usual methods involving boiling at atmospheric pressure, 90 to 93 per cent recovery is obtained in a few hours, and prolonged heating causes practically no further precipitation. The addition of water after most of the titanium sulfate has been hydrolyzed produces a sudden decrease in concentration, with a consequent acceleration of hydrolytic decomposition, so that the reaction proceeds to completion in a comparatively short time. There is some tendency at this stage to precipitate titanium oxide in a finely divided form which occludes iron compounds, but this may be effectively overcome by discontinuing the heating while the water is being added. The procedure is applicable to the production of both the relatively pure oxide and composite pigment.

For example, an ilmenite solution having less sulfuric acid than that required to form normal sulfates was reduced, clarified, filtered and cooled to crystallize ferrous sulfate, and the mother liquor was then concentrated by vacuum evaporation. Fifteen cubic meters of this solution, containing 215 g titanium dioxide, 345 g sulfuric acid, and 80 g ferrous iron per liter was boiled for 4 hours. The steam was then shut off and 1.8 cubic meters of water was added with stirring. Heat was again supplied, and after boiling for 1 hour more the recovery was 97 per cent. In another operation, to 30 cubic meters of an ilmenite solution of specific gravity 1.44, containing 90 g titanium dioxide and 260 g sulfuric acid per liter barium sulfate was added in a proportion calculated to give a product containing 25 per cent titanium dioxide and the suspension was heated by injecting live steam. After boiling for 3 hours 80 per cent of the titanium precipitated. At this stage the steam was cut off and 5 cubic meters of water was added, with stirring, causing the temperature to drop to 93° C. The system was again heated to boiling, and in 2 hours more than 98 per cent of the titanium was recovered. The water was added continuously to the hydrolysis mixture at such a rate as to maintain the concentration of sulfuric acid at a constant value approximately equal to that of the liquor just before precipitation began.

As already indicated, the hydrolysis of ilmenite solution is effected generally in the presence of a small proportion of titanous salt to hold the iron component in the stable ferrous condition. Trivalent titanium does not precipitate on boiling the solution, and, to increase the final yield, near the end of the process this component may be oxidized to the readily hydrolyzed tetravalent form by adding ferric oxide to the acid liquor.[91]

## CONTINUOUS PROCESS OF HYDROLYSIS

Saklatwalla and Dunn [92] developed an interesting process for the continuous hydrolysis of dilute titanium sulfate solution. In order to secure optimum conditions for hydrolysis, the solution was very quickly raised to the proper temperature, maintained for the exact period of time required for effecting the desired degree of precipitation and no longer, and then cooled to arrest further reaction. These conditions were readily attained by passing the solution continuously through a long hydrolyzing tube of any convenient length, depending primarily upon the velocity of flow, which, however, was sufficient to secure efficient heat transfer from the walls to the solution and to carry along the precipitated hydrous titanium dioxide. Such a reaction chamber consisted of a lead-lined pipe 1 inch in diameter with a total length of 1000 to 1500 feet.

The method was particularly applicable to ilmenite solutions containing approximately 4 per cent titanium dioxide and 6 to 10 per cent sulfuric acid in excess of that required to form normal salts with the titanium and ferrous iron present. In actual operation such a solution was supplied at room temperature and at a pressure of 35 to 45 pounds per square inch to the inlet end of the hydrolyzer, and steam, also at 35 to 45 pounds pressure, was introduced at the same end around the tube. The parallel flow provided for a maximum heat transfer at the inlet end of the tube so as to bring the solution very rapidly to the decomposition temperature, 255° to 265° F. Once attained, this temperature was maintained throughout the length of the hydrolyzer. Thus each small incremental volume of solution was immediately subjected to the maximum heating effect of the steam and was very quickly brought to the temperature required for hydrolysis. Also the velocity of flow of the solution and steam through the system favored rapid heat transfer through the walls of the tube. The time of reaction was so regulated that 85 to 90 per cent of the titanium dioxide was precipitated, since this was found to be the maximum recovery consistent with high quality of the product. One half hour was required for any one portion of the solution to travel through the hydrolyzer. The reaction product was immediately cooled on discharging so as to check any further hydrolysis, and particularly to avoid colloidal material which tended to form on slow cooling. As in other methods of procedure, the hydrous titanium dioxide was filtered, washed, and calcined.

Because of such accurately controlled conditions, the time of precipitation was shortened appreciably over that of the batch process. Although the method was applicable to various types of solutions, sufficient free sulfuric acid was present to prevent an undesirable initial precipitation of basic sulfate and to cause the hydrolysis to proceed from the start at a slower rate.

In a continuous process, leaching of the ore to remove most of the iron, digestion to convert the titanium to soluble sulfate, and hydrolysis of the resulting titanium sulfate solution are carried out under pressure at elevated temperature in circuits of steam jacketed pipes.[93] Water may be passed through the jackets for cooling as required. Each coil of the circuit is connected with a pump to develop the pressure and to maintain a velocity of the slurry of 3 to 9 feet per second to prevent the solids from settling out. Leaching is carried out at 200° C and a pressure of 300 psi, while hydrolysis is carried out at a temperature above 200° C. In the leaching stage a mixture of 200 to 325 mesh ilmenite with 45 per cent sulfuric acid is fed to a kettle equipped with a stirrer. The slurry is drawn from the kettle by a pump and fed to the expansion tank. Here air pressure is applied and the suspension is fed to the first coil of the circulating line and kept in motion by the circulating pump at 3 to 9 feet per second to prevent settling. Through by-pass pipes the preheated feed mixture can be directed to the second, third, or fourth circulating line. Sulfur dioxide gas is introduced to reduce the ferric iron to the ferrous condition. From the last line the slurry is delivered to a self-cleaning thickener and hydrocyclone to yield an intermediate solid concentrate containing 5 per cent ferric oxide and more than 80 per cent titanium dioxide. After this leaching step the concentrate is mixed with more acid until it is completely dissolved as ferrous sulfate and titanium sulfate. As in the leaching operation, digestion is carried out in circuits of steam jacketed pipes, arranged with multiple circulating pumps and bypasses to prevent "dead spots" where solids could settle out of the circulating suspension. A digestion time of 2 hours is required.

Hydrolysis, the most critical step in pigment production, follows complete dissolution of the ore. The titanium sulfate solution feed is mixed in the third line with already partially hydrolyzed solution and is directed by a pump to the first line where it is quickly heated to the desired temperature. An advantage of the process is the very rapid heating of the feed solution through the admixture of small quantities of feed to the already heated and partially precipitated solution. Another advantage is the very rapid cooling of the precipitate suspension in the last circulating line before release. At 180 to 190° F hydrolysis of the titanium sulfate is very rapid up to the point at which 30 per cent of the titanium has been precipitated as hydrous oxide. This initial reaction produces enough free sulfuric acid to slow down the rate of hydrolysis of the remaining 70 per cent. These two reactions before and after the 30 per cent point produce markedly different products. The hydrous titanium oxide produced immediately from the first reaction has large particle size, is contaminated with other metals, and gives a pigment of poor hiding power and color. Particles from the second reaction after 30 per cent yield are smaller and purer and yield a better pigment. The process avoids the initial fast reaction by injecting the feed solution of titanium sulfate directly into a circulating volume of solution that has already been partially hydrolyzed. The feed is quickly heated to reaction temperature, but the presence of free acid in the circulating solution prevents the very rapid reaction which gives a poorer product. Time for hydrolysis is 1 hour. Following the hydrolysis step the suspension is injected into a hydrolyzed mixture circulating at about 140° F to effect rapid cooling. After cooling the hydrous titanium dioxide is filtered, washed and processed to pigment as in the conventional process.

## STUDIES OF METHODS OF HYDROLYSIS

In a study of the commercial methods employed in producing titanium dioxide pigments of the anatase type, Pamfilov, Ivancheva, and Soboleva [94] compared the Mecklenburg process, in which nuclei are prepared separately and added to the solution for hydrolysis, with the method developed by Blumenfeld in which the seeding agents are formed in place within the main body of the hydrolysis solution, from the point of view of adaptability, yield, and quality of the products obtained. These two processes were selected for investigation because of their paramount practical importance and wide commercial use. Solutions employed in the greater part of the work were prepared on a laboratory scale by dissolving the reaction product of titanomagnetite concentrate with strong sulfuric acid, although some were obtained from the Leningrad plant. The concentration of titanium dioxide ranged from 120 g to 250 gpl and the iron, after reduction to the divalent condition, was brought to 25 g to 30 gpl by crystallization, as ferrous sulfate on chilling. Precipitations were carried out as nearly like plant conditions as possible. Descriptions of the methods employed are given below in detail.

**Mecklenburg.** Clarified, reduced, and crystallized sulfate solutions containing 120 g to 150 g titanium dioxide and 25 g to 30 gpl iron in the divalent state were used. In preparing the nuclei, 5 per cent of this solution was made neutral to methyl orange at 80° to 85° C by adding 20 per cent aqueous sodium hydroxide. The grayish suspension obtained was held at 80° to 85° C for 15 minutes to develop the nucleating properties to the maximum degree, and then added to the other 95 per cent of the original solution. The seeded mixture was heated at 102° to 106° C for 5 hours to complete hydrolysis, and cooled to room temperature. Turbidity was quite noticeable after 20 minutes' heating. Increase in the proportion of nuclei above 5 per cent did not increase the yield appreciably, but on the other hand gave a product of poor color. The precipitate of hydrous titanium oxide was filtered, washed, dried at 100° C, and calcined at 800° to 1000° C for 2 hours to develop the optimum pigment properties.

**Blumenfeld.** Reduced and crystallized solutions, as used before, were concentrated by vacuum evaporation to 200 g to 270 gpl titanium dioxide. The acid factor, that is, the ratio of active sulfuric acid (free and combined with titanium) to the titanium dioxide, was varied over fairly wide limits. A portion of the solution to be hydrolyzed, maintained at 94° to 96° C, was poured at a uniform rate in 16 minutes into one third its volume of water, initially at 91° C. Heat was applied so that at the end of this period the temperature of the system was 102° C. The mixture was then heated slowly to the boiling point and boiled for 5 hours. A marked precipitation occurred after 45 to 50 minutes. The hydrous titanium oxide obtained was washed, dried, and calcined as before.

Results obtained showed that, at a constant concentration of titanium dioxide in the solution, an increase in acid factor, ratio of active sulfuric acid to titanium dioxide by weight, decreased the yield. This was especially noticeable in the Blumenfeld method. Increasing the acidity of the solution and decreasing the concentration of titanium resulted in pigments of lower tinting strength. In the Mecklenburg process, employing solutions of a constant acid factor, the yield increased as the concentration of titanium decreased, but this trend was not distinct in the Blumenfeld process. The Mecklenburg method, even at acid factors greater than 4, gave yields of 96 per cent, while that of Blumenfeld gave good yields with acid factors not greater than 2.7, though best results were obtained at about 2. Color, brightness, and other physical properties were better in the Blumenfeld process.

Solutions of low acid factor but of high active acid content gave lower yields. Active acid was seldom higher than 35 per cent. By following the conditions normally employed in the Blumenfeld method of hydrolysis, good results were obtained using solutions of much lower concentration than those commonly recommended.

Work, Tuwiner, and Gloster [95] studied the hydrolysis of sulfate solutions having titanium dioxide concentrations of 1.54 to 8.97 per cent and ratios of sulfur trioxide to titanium dioxide from 1.61 to 3.75. Pigments of greater covering power were obtained at the higher concentrations of titanium and the lower acid ratio, although all unseeded solutions gave products inferior to the best commercially available pigment. The concentration of the mother liquor, after hydrolysis was complete, was found to have an important influence on the properties of the precipitate. Similar solutions [96] having ratios of titanium dioxide to sulfur dioxide between 1.0 to 1.66 and 1.0 to 0.65, and a commercial sample of ratio 1.0 to 2.26 hydrolyzed by refluxing without previous nucleation, gave products of low hiding power and poor color. However, after seeding, those liquors having ratios between 1 to 1 and 1.0 to 1.65 yielded pigments of high capacity as measured by turbidity of aqueous suspensions. Solutions more basic than 1 to 1 were autoseeded on dilution with water to an opalescence approaching turbidity, and acted accordingly on refluxing. Hixon and Plechner [97] prepared solutions of various concentrations and acidities by dissolving crystalline titanyl sulfate, $TiOSO_4 \cdot 2H_2O$, in dilute sulfuric acid, and effected hydrolysis by boiling under reflux. With increasing concentrations the proportion of hydrous titanic oxide precipitated in a given time decreased to a minimum value, then increased to a maximum and again fell off. Braun [98] found that the hydrolytic precipitation of titanic oxide from sulfate solutions was influenced by concentration, temperature, hydrogen ion concentration, manner of agitation, and presence of foreign substances both organic and inorganic.

Solutions obtained by dissolving the product formed by treating ilmenite or metatitanic acid in 70 per cent sulfuric acid at 120° C, and then at 140° to 160° C for 1.5 hours, showed no change of viscosity or composition after 6 months.[99] Very active seeding agents obtained by heat-treating a solution containing 50 g titanium dioxide and 16 g active sulfuric acid per liter had a deep brown color and blue opalescence but were transparent in transmitted light. Ilmenite solutions containing 110 g titanium dioxide and 220 g active sulfuric acid per liter, after inoculation with 1 per cent of this nucleating suspension, gave a 95 to 98 per cent recovery in 1 hour. Hydrolysis started at 96° C and was completed at 102° C. Such nuclei were more effective than the product produced by Mecklenburg.

A study of thermal hydrolysis of pure $TiOSO_4$ solutions showed that half the sulfuric acid split off rapidly and that after this the rate of hydrolysis was very slow.[100]

From studies of an analytical procedure, Kayser [101] noted that complete hydrolysis of titanium sulfate solutions took place at pH of 6.5, although other workers precipitated hydrous oxide of optimum pigment properties at a pH between 2 and 3. High concentrations of sulfuric acid were detrimental, and steps which increased the yield also tended to lower the quality of the product for pigment purposes.

Parravano and Caglioti [102] found that titanium sulfate solutions contained not only titanic and sulfate ions but also colloidal complexes of titanium dioxide and sulfuric acid. The proportions of the latter components depended upon the concentration and age of the solution and the temperature at which it had been maintained. By dialysis of such solutions, two types of gels were obtained. One was transformed by calcination at 850° C into rutile, while the other, which was a powder, remained as anatase even after exposure to 950° C for 1.5 hours. Concentration of titanium and sulfuric acid in the solution determined the two types of products.

The quantity of the metatitanic acid precipitated by thermal hydrolysis varied with the time of boiling and with the absolute and relative quantities of titanium dioxide and sulfuric acid in the solution. Its separation could be changed appreciably, however, by seeding with nucleating agents. Curves of the rate of change of the index of refraction of such solutions at various temperatures and concentrations showed a slight increase just before precipitation began. Pigmentary properties depended upon the temperature and concentration of the solution, the nature of electrolytes present, and the rate of formation of nuclei in the hydrolysis step. Precipitation conditions governed the nature and size of the micelle aggregates, and although the temperature and time of calcination controlled crystallization of the product and growth of the granules, the dimensions of the latter depended upon the size of the flocculates of the hydrolysis product. Conditions which led to precipitation of titanium dioxide of the anatase crystal form, having particles from 0.50 to 0.65 micron, were found to be the most effective for obtaining products of good pigmentary properties, while hydrolysis of solutions containing ferrous sulfate under conditions which led to the formation of rutile gave yellow products unsuited for white pigments.

Evaporation of hydrolyzing solutions of titanic sulfate showed reversibility of the reaction.[103] Resolution began as the acid value reached 12 to 17 per cent. Evaporation of such liquors changed two variables simultaneously—increased acidity of the medium, and raised the boiling point of the solution.

From a study of the equilibrium of the system titanium dioxide–sulfur trioxide-water at 100° C, Sagawa [104] concluded that the thermal hydrolysis product of sulfate solutions was titanium dioxide with a small proportion of adsorbed sulfur trioxide and water, rather than a solid solution or a compound. A precipitate containing 2 to 5 per cent sulfuric acid and 3 to 5 per cent water was obtained by the hydrolysis of acid titanium sulfate, $Ti(HSO_4)_4$, and its sulfur trioxide content increased little with higher concentrations of acid in the liquor. Such solutions prepared by dissolving titanium hydroxide in sulfuric acid in some cases yielded titanium dioxide monohydrate on hydrolysis. The precipitation took place in the metastable state, however, since the concentration of titanic sulfate in the solution was far greater than in other systems which gave titanium dioxide.

The product precipitated by boiling sulfate solutions gave the X-radiograms of anatase.[105] Thermal dehydration isobars and X-ray analyses of different products indicated that the water was given off continuously and that no definite hydrates of titanium dioxide were formed.

Sulfate solutions of ilmenite are passed through a column of steel wool to reduce the ferric ions to ferrous and hydrolyzed by several processes.[106] Such factors as the titanium dioxide concentration, free acid and time of hydrolysis affect the yield. In general better yields are obtained by the Mecklenburg process, but both this and the Blumenfeld process give white anatase on calcining the hydrolysis product at 750° C. At 920° C off-color rutile results. Fourfold dilution of the treated ilmenite solution by steam or boiling water gives poor yields but the products are white pigments of the rutile type.

## INDIRECT METHODS

Washburn and Aagaard [107] obtained practically pure titanium dioxide of pigment grade by an indirect process based on the method of producing coalesced composites by which thermal hydrolysis was effected in the presence of smaller proportions of extremely finely divided calcium sulfate which was later removed from the combined precipitate by washing with water. This was effective because of the much greater solubility of calcium sulfate than of titanium dioxide. The very large number of minute suspended particles having the anhydrite structure promoted and accelerated the hydrolytic decomposition of the titanium sulfate by breaking down the metastability of the solution by functioning as adsorptive nuclei for the compounds of titanium being precipitated. In this respect calcium sulfate acted somewhat after the nature of a catalyst; it also tended to prevent the occlusion of iron and to maintain an optimum particle size distribution.

As the reaction proceeds, sulfuric acid is one of the products of hydrolysis, and it is probable that, under the constantly changing conditions of concentration and total acidity of the solution, calcium sulfate may dissolve to some extent and precipitate to yield fresh, active particles. The anhydrite not only accelerated the hydrolytic precipitation of titanium to a much greater extent than the ordinary relatively insoluble extenders, but also yielded a pigment of excellent properties, e.g., color and opacity.

After reaction was complete, the precipitate was washed to remove the anhydrite. The solubility of calcium sulfate was much greater in salt solutions so that its removal from the hydrous titanium oxide was effected more rapidly by employing wash water containing an electrolyte, for example ammonium chloride.

To illustrate the method, a suspension of 25 pounds of hydrated lime in 32 gallons of water at 70° C was added to 164 pounds of 78 per cent sulfuric acid. The anhydrite slurry thus formed was added to 1735 pounds of an ilmenite solution containing 6.07 per cent titanium dioxide, 5.88 per cent ferrous oxide, and 6.45 per cent free sulfuric acid, and the mixture was boiled until 95 per cent of the titanium was precipitated hydrolytically. The composite product was washed free of calcium sulfate, dried, and calcined at 900° C to yield a pigment containing 99.1 per cent titanium dioxide and traces of calcium sulfate and other substances.

According to Washburn and Kingsbury,[108] the rate of hydrolysis of titanium sulfate was accelerated and a pigment of improved color and hiding power was obtained by heating the solution previously seeded with alkali-precipitated titanium hydroxide and carrying in suspension particles of calcium sulfate. Such nuclei were formed directly in the main body of the solution or in a separate portion by treating it with a base such as sodium carbonate, potassium hydroxide, calcium carbonate, ammonium hydroxide, or barium hydroxide. A proportion of the seeding compound amounting to 7 per cent of the total amount of titanium to be precipitated gave best results.

An extremely finely divided and effective calcium sulfate, having the structure of anhydrite, was obtained by adding a slurry of slaked lime or calcium carbonate to a slight excess of concentrated sulfuric acid and heating the resulting suspension near the boiling point to effect conversion of any hydrate formed.

In an example, an ilmenite solution containing 6 per cent titanium dioxide was prepared by dissolving the reaction product of the ore with concentrated sulfuric acid. All the iron was reduced to the ferrous condition, and the action was continued until a small proportion of the titanium was converted to the trivalent state. The suspended solids were coagulated and allowed to settle. Nuclear hydrous titanium oxide was precipitated by carefully pouring a solution of 6 kg to 8 kg of potassium carbonate into 26 liters of water at 80° C upon the surface of 745 kg of the clarified ilmenite liquor. The precipitate was immediately stirred into the solution, and the mixture was added to a slurry of anhydrite prepared by adding 11.3 kg of hydrated lime suspended in 120 liters of water to 93 kg of 78 per cent sulfuric acid, both at atmospheric temperature. The charge was then boiled to effect hydrolysis of the titanium sulfate, and the mixed precipitate was washed free of calcium sulfate, and calcined to yield pure titanium dioxide of pigment grade.

Composite pigments may be prepared in the same manner by employing a higher proportion of calcium sulfate, or of a mixture of calcium sulfate and barium sulfate, and washing the product less so as not to remove the extender.

According to another process of this type,[109] a sulfate solution of ilmenite containing 50 g to 250 g titanium dioxide equivalent per liter was treated at atmospheric or slightly higher temperature with a hydroxide or carbonate of an alkali or alkaline earth metal or ammonium in a proportion just short of that required to react with all the potential sulfuric acid present. The term "potential acid" is applied to the free sulfuric acid, plus that combined with the titanium, and represents the total amount of acid that would be present in the solution after complete hydrolysis of the titanium sulfate. At this stage most of the iron remained in solution, while practically all the titanium was precipitated as orthotitanic acid. Employing an alkaline earth neutralizing agent, however, the corresponding insoluble sulfate was formed. The mixed precipitate was washed and digested with sulfuric acid to dissolve the titanic acid selectively, and the resulting suspension was filtered and washed to obtain a purified titanium sulfate solution. This was concentrated by evaporation under reduced pressure, and boiled at atmospheric pressure without nucleation to effect hydrolytic decomposition. The precipitate of hydrous titanium oxide was washed and calcined at 700° to 1000° C to produce a pigment of high purity.

By carrying out the neutralization process in two steps employing an alkaline earth metal base which forms an insoluble sulfate, a purified solution of high concentration was obtained directly, thereby minimizing subsequent evaporation costs. This end was attained by adding calcium hydroxide or carbonate in the form of a slurry to the ilmenite solution in quantity short of that required to throw down the titanium. All the free sulfuric acid, and from 20 to 50 per cent of that combined as titanyl sulfate, was neutralized without precipitating the titanium. Free acid represents that part of the active acid over and above that required to form the titanyl salt with the titanium present. These relations may be expressed as

$$\text{Factor of acidity} = \frac{\text{Free acid} \times 100}{\text{Acid combined with titanium as titanyl sulfate}}$$

Such solutions were very basic and metastable. After filtration to remove the insoluble sulfates, the solution was treated with another portion of the alkali suspension, just sufficient to precipitate the titanium content. This method has the advantage of minimizing the amount of foreign matter included with the orthotitanic acid cake. Obviously, more concentrated purified solutions can be prepared from partially dehydrated cake, but this step must be carried out at low temperature to avoid decreasing the solubility of the titanium compounds.

In a specific test, to 5 liters of a solution containing 169.7 g titanium dioxide, 44.2 g iron, 522.0 g total sulfuric acid, 236.4 g free sulfuric acid and 444.4 g active sulfuric acid per liter, and a factor of acidity of 114, was added a slurry of 1500 g calcium carbonate in 10 liters of water at room temperature. The mixture was filtered and washed with 5 liters of 1 per cent sulfuric acid. The filtrate was then added to the wash liquor to get an intermediate solution containing 66.1 g titanium dioxide, 17.3 g iron as ferrous sulfate, 95.5 g total acid, and 65.1 g active acid per liter, and a factor of acidity of $-19.6$. That is, the acid content was even less than that corresponding to the titanyl sulfate. The combined solution was heated to 60° C and a slurry of 540 g calcium carbonate in 3 liters of water was added, with stirring, to produce a precipitate which contained all the titanium, mixed with calcium sulfate. This product was filtered and washed and added to 900 ml of concentrated sulfuric acid, with agitation. After cooling to room temperature, the calcium sulfate was filtered off and washed by repulping in 1 per cent sulfuric acid. The purified solution, consisting of the filtrate and washings containing 80.4 g titanium dioxide, 13.7 g iron, 208 g total sulfuric acid, 85.5 g free acid, and 184.0 g active acid per liter, and having a factor of acidity of 86.8, was concentrated by evaporation under reduced pressure to corresponding values of 210.2 g, 42.5 g, 590 g, 258 g, and 515.5 g per liter, which represented a factor of acidity of 99.7. After boiling for 4 hours at atmospheric pressure without seeding, this concentrated solution showed a yield of 91.5 per cent. The hydrous titanium oxide was filtered, washed, and calcined.

Analyses of the original, intermediate, and final solutions are shown in Table 14-6.

Booge [110] crystallized titanyl sulfate from solution at an elevated temperature and calcined the sulfate to produce the dioxide.

## TABLE 14-6

Analyses of Solutions

| Property | Original Solution | Intermediate Solution | Purified Solution | Concentrated Solution |
|---|---|---|---|---|
| Total $TiO_2$ g per liter | 169.7 | 66.1 | 80.4 | 210.2 |
| Iron | 44.2 | 17.3 | 13.7 | 42.5 |
| Total $H_2SO_4$ | 522.0 | 95.5 | 208.0 | 590.0 |
| Free $H_2SO_4$ | 236.4 | −15.9 | 85.5 | 258.0 |
| Active $H_2SO_4$ | 444.4 | 65.1 | 184.0 | 515.5 |
| Factor of acidity | 114.0 | −19.6 | 86.8 | 99.7 |

According to Lederer and Kassler,[111] a basic titanium salt free from contamination of iron may be precipitated from a sulfate solution of ilmenite by introducing hydrogen sulfide, or a substance which will liberate this gas, together with a basic compound.

A study of the hydrolysis conditions showed that commercially valuable titanium dioxide could be obtained from dilute solutions only if a considerable excess of sulfuric acid was present.[112] Zinc oxide increased the velocity of the hydrolytic reaction.

## RUTILE FROM SULFATE SOLUTIONS

The hydrolytic decomposition of titanium sulfate solutions by the more common methods, as by boiling direct, by mixing hot with hot water, by heating at atmospheric pressure in the presence of seeding materials prepared from sulfate solutions, and by heating in an autoclave, yields titanium dioxide of the anatase crystal modification. These methods have been treated in detail in previous sections.

By seeding regular sulfate solutions with rutile nuclei, however, the entire titanium dioxide content may be precipitated hydrolytically in a condition possessing the rutile crystal form.[113] For example, an alkali metal titanate was first formed by heating a mixture of 100 pounds pure titanium dioxide and 100 pounds sodium hydroxide free from sulfates at 700° C for 1 hour. It was then cooled and lixiviated with 10 per cent hydrochloric acid until neutral to methyl orange, and the product was added to 500 gallons regular sulfate solution containing 250 g titanium dioxide, 25 g ferrous iron, and 575 g sulfuric acid per liter, and the seeded charge was boiled for 4 to 5 hours to effect hydrolysis. Titanium dioxide precipitated in this manner had the rutile crystal form, and on calcination gave a pigment having one third higher tinting strength than the usual anatase product.

Any of the alkali hydroxides or any of the halogen acids may be used in the process, and the seeding crystals, once formed, will function efficiently in the sulfate solution, yielding an artificial rutile type titanium dioxide having the same crystal form as the natural material.

By hydrolyzing sulfate solutions previously seeded with a nucleating sol prepared by heat-treating a dilute solution of the tetrachloride, titanium dioxide of pigment grade was obtained, which on calcination at moderate temperatures (around 900° C) had the rutile crystal form and the correspondingly high tinting strength.[114] Pure solutions of the tetrachloride gave a product of better opacity and other pigment properties, although the presence of iron and organic impurities was not detrimental to effective nucleating properties of the heat treated sol. As in other methods, any iron salts were reduced to the ferrous state. Heat treatment of the sol developed the nucleating power to the maximum degree, and both temperature and time were regulated, since overcuring reduced the effectiveness of the product. It follows that longer periods of heating were required at lower temperatures than at higher temperatures. Best results were obtained by heating a tetrachloride solution containing the equivalent of 10 to 20 gpl titanium dioxide at 80° to 85° C for 10 to 15 minutes. The cured sol presented a slight opalescence, and the particles could not be separated by the conventional methods of filtration and washing. Degree of dispersion of the colloidal titanium oxide and its homogeneity did not seem to have a great influence on the nucleating properties, and mixtures of products having various degrees of dispersion were used as seed with satisfactory results. The method is generally applicable to solutions of all titanium salts such as sulfate, chloride, and nitrate from which the oxide may be precipitated by thermal hydrolysis, but the latter two would normally yield rutile.

Although ilmenite solutions of wide limits of concentration may be successfully hydrolyzed with such seed, better results were obtained by employing basic solutions containing 130 g to 150 gpl titanium dioxide and sulfuric acid free and combined with titanium less than that corresponding to the normal salt. The proportion of nuclei was from 9 to 12 per cent of the total titanium component; in general, the greater the basicity of the solution, and the higher the concentration of titanium, the smaller the proportion of nuclei required.

With chloride and nitrate solutions the concentration was from 140 to 160 g titanium dioxide equivalent per liter, and the proportion of nuclei was from 5 to 9 per cent.

Employing a sulfate solution containing 140 to 150 g titanium dioxide per liter and having a basicity corresponding to a proportion of titanium dioxide to active acid of 1.0 to 1.4, about 10 per cent nuclei gave the best results. If the concentration of titanium dioxide was raised to 200 gpl or more, by maintaining this degree of basicity the quantity of nuclei could be reduced. On the other hand, if the basicity was increased to a proportion of titanium dioxide to active sulfuric acid of 1.0 to 0.84, and the concentration of titanium dioxide held at 140 gpl, 7.5 per cent nuclei gave best results.

The method of mixing exerted a certain influence on the results obtained and it was found advantageous to add the solution to the nucleating sol. Carrying out the hydrolysis at the normal boiling point, 97 per cent recovery was obtained in 15 minutes to 1 hour. More rapid hydrolysis was effected at temperatures corresponding to superatmospheric pressure. The method was also applicable to continuous operation.

For example, one liter of titanium tetrachloride solution containing the equivalent of 10 g titanium dioxide was heated under reflux, with stirring, to 85° C and held at this temperature for 12 minutes to develop the nucleating property. At 65° C the opalescence began to develop. After 12 minutes at 85° C, heating was discontinued and 700 ml of a sulfate solution containing 150 g titanium dioxide, 265 g active sulfuric acid, 170 g ferrous sulfate, and trivalent titanium equivalent to 1.2 g per liter of the dioxide was added. During the addition the temperature dropped to 60° C and the system was then heated to boiling in 20 minutes. Hydrolysis was complete after boiling for 15 minutes, and a recovery of 96 per cent was obtained. The product was cooled to 80° C and the precipitate was filtered, washed, dried, calcined, and pulverized according to the conventional methods to produce a white pigment of the rutile crystal modification. It had uniform particles of 0.5 to 0.6 micron and was practically free of aggregates and of particles greater than 2 microns. Tinting strength by the Reynolds method was 1625.

Barksdale [115] prepared nuclear suspensions of the rutile type by heat-treating dilute titanic chloride solutions to which an alkaline agent such as calcium oxide, sodium hydroxide, and potassium carbonate had been added in an amount equivalent to from one fourth to one half of the potential hydrochloric acid. Acid-poor solutions obtained in this manner were diluted to the equivalent of 10 g to 30 g titanium dioxide per liter, and heated at 85° to 95° C for approximately 10 minutes to develop the nucleating properties to the maximum degree. The proper cure was characterized by the development of an opalescent appearance and was a function of both temperature and time of heating. Sulfate solution of ilmenite at 80° C was added slowly, with constant agitation, in such proportions that the titanium dioxide furnished by the nuclei was from 4 to 10 per cent of the total. The seeded mixture was heated to boiling and refluxed until the recovery exceeded 90 per cent; this in general required only ½ hour. The hydrolysis product was filtered, calcined, and milled in the usual manner to produce a pigment characterized by titanium dioxide of the rutile crystal modification and the correspondingly high tinting strength.

In a specific operation, a portion of a titanium tetrachloride solution containing the equivalent of 15 g titanium dioxide and 25 g hydrochloric acid was treated with 7 g of hydrated lime, diluted to 1 liter, heated to 85° C, and held at this temperature for 10 minutes to develop the nucleating properties. To this nuclear suspension was added in 15 minutes 1000 ml of a sulfate solution of ilmenite containing 250 g titanium dioxide, 495 g active sulfuric acid, and 190 g ferrous sulfate at 95° C; the seeded mixture was heated to boiling in 20 minutes and refluxed for 1 hour. The hydrolysis product was filtered, washed, and calcined to produce a pigment of the rutile crystal form having high tinting strength.

Similar results were obtained by employing a nucleating composition prepared by heating a titanium tetrachloride solution containing the equivalent of 5 to 30 g of the dioxide per liter, and 0.5 to 2.0 mole proportions of a nonoxidizing univalent anion such as chloride, metaborate, acetate, or formate at 80° to 90° C for 10 to 15 minutes to develop a characteristic opalescence.[116] Approximately 5 per cent of this suspension accelerated and directed the hydrolysis of concentrated sulfate solution of ilmenite to produce pigment of the rutile crystal modification possessing the correspondingly higher tinting strength and brightness. McCord and Saunders [117] precipitated titanic acid which converted directly to rutile on calcination below 1000° C by hydrolysis of sulfate solutions of ilmenite in the presence of gamma titanic acid as a nucleating agent.

By another approach, sulfate solutions nucleated or seeded with peptized oxides or hydroxides of titanium gave, on hydrolysis, a precipitate which after calcination in the usual manner consisted of particles of the rutile structure of the optimum size for pigments, with a very high tinting strength and covering power.[118] Such seeds were prepared by peptizing with a strong monobasic mineral acid, with the aid of heat, titanium oxide or hydroxide separately precipitated from a solution of a titanium salt by adding an alkaline reagent until the acidity of the solution was reduced to a value corresponding to a pH of 3 to 6. Peptization was disturbed by the presence of sulfates, and these were removed from the precipitated oxide or hydroxide by washing. In this process oxides or hydroxides of metals other than titanium, belonging to Group IV of the Periodic Table, prove effective.

Alternatively, before use in the hydrolysis the peptized seed was coagulated, for example by addition of alkali, and after being washed free from electrolytes was added in the form of a sol or gel to the sulfate solution to be hydrolyzed. The presence of phosphoric acid or a soluble phosphate during the precipitation of the seed and the later hydrolysis step was advantageous.

For producing the nuclei, a part of the sulfate solution to be hydrolyzed was removed and adjusted to an acidity not far from the neutral point by adding an alkaline agent. It was important that the neutralization be carried to a point corresponding to a pH of 3 to 6. This precipitate was washed until free from sulfates, and then peptized and converted into a sol by warming with a solution of a strong monobasic acid (hydrochloric) to develop the nucleating property. The composition was then added to the main part of the titanium sulfate liquor, after which the hydrolysis and calcination were carried out in the usual manner. Such seed proved to be extremely stable. In an actual operation, a solution was employed which was obtained by dissolving the reaction product of ilmenite with strong sulfuric acid and 1 per cent admixed phosphoric acid, and containing 200 g titanium dioxide and 470 g free sulfuric acid per liter. Thirty-two liters of this liquor was run into 120 liters of 9 per cent aqueous sodium hydroxide until a pH of 3 to 6 was obtained. The precipitate was filtered and washed free of sulfates, stirred into 38 liters of 10 per cent hydrochloric acid, and heated at 90° C until the clear solution which first formed became opalescent. This colloidal solution was added to 1000 liters of the original ilmenite liquor and the mixture was boiled for 2 hours to effect hydrolysis. A yield of 95 per cent was obtained. The precipitate was filtered, washed, calcined, and milled in the same manner as anatase titanium dioxide.

A similar nucleating composition was prepared by peptizing in hydrochloric acid the hydrous titanic oxide precipitated from sulfate solution containing 1 per cent phosphoric acid by adding sodium hydroxide solution.[119] The nuclei were used with sulfate solutions of ilmenite to yield titanium dioxide of the rutile type on hydrolysis.

Bennett [120] obtained rutile by hydrolyzing titanium sulfate solution in the presence of a nucleating agent prepared by peptizing hydrous titanic oxide hydrolytically precipitated from sulfate solution at 45° to 70° C, after adding sodium hydroxide to adjust the acidity of the solution to a pH of 2 to 7. The peptized suspension, which contained 10 to 15 gpl titanium dioxide, was added to the main solution in a proportion of 8.5 per cent based on the titanium dioxide content. The titanium content of the hydrolysis solution, after neutralization, was 55 to 75 gpl.

Similar results were obtained by employing titanium oxychloride as nucleating agent.[121]

Weber and Bennett [122] employed nuclei prepared by dispersing or peptizing in dilute nitric acid the thoroughly washed and neutralized precipitate obtained on hydrolysis of a dilute feebly acid solution of titanium sulfate. The final suspension had a pH less than 2 and possessed the properties of a colloidal solution. As an illustration of the operation of the process, a crystalloidal solution of specific gravity 1.07 containing 13 g titanium dioxide, 4.5 g iron as ferrous sulfate, and 31 g sulfur trioxide per liter was boiled to effect thermal hydrolysis. A quantity of the washed paste corresponding to 5 pounds dry titanium dioxide was neutralized with sodium hydroxide. The soluble salt formed was washed out with water and the purified product was dispersed in 20 pounds of water containing 0.85 pound nitric acid (specific gravity 1.42) to the degree that practically all passed through a filter paper but all the solid component was retained on a colloidal membrane. The pH value was below 2. This suspension was diluted to a specific gravity of 1.1 and added to 1350 pounds of an ilmenite solution of specific gravity 1.35 containing 80 g titanium dioxide, 30 g iron (all in the ferrous condition), and 172 g sulfur trioxide per liter. After thorough mixing the charge was heated to boiling and boiled for 3 hours to precipitate the titanium content as hydrous oxide. In this system 95 per cent of the titanium dioxide was contributed by the hydrolysis solution and 5 per cent by the nucleating suspension. The precipitate settled rapidly and on processing gave a pigment of improved color and tinting strength. Besides nitric acid, other monobasic acids, such as acetic and propionic, were effective.

A nucleating composition is prepared by adding an alkaline agent to a chloride solution to precipitate the titanium as hydrous oxide, and acidifying the suspension with hydrochloric acid.[123] A similar composition is prepared from the $H_4TiO_4 \cdot SO_3$ obtained as a precipitate by heating a water solution of titanyl sulfate.[124] The hydrated precipitate is treated with hot sodium hydroxide to remove excess acid and form sodium titanate which is washed with water and treated with hydrochloric acid to form a rutile seeding agent. From 1 to 5 per cent of the nucleating agent was added to titanyl sulfate solution to direct precipitation of titanium dioxide of the rutile crystal modification. Mayer[125] treated hydrous titanium dioxide, formed by heating the tetrachloride solution, with 4 to 55 millimole proportion of sulfuric acid as a stabilizer, and removed the chloride ions by dialysis. By another modification the hydrous titanium dioxide was peptized with a monobasic acid, heated with the stabilizing agent as before, and dialyzed to purify the nuclei.

A negatively charged colloidal solution of titanium dioxide, prepared by reversing the charge of the usual positive sol, was employed by Stark[126] to accelerate and direct the hydrolysis of sulfate or chloride solutions to produce pigment of the rutile form. Orthotitanic acid precipitated with ammonia from a tetrachloride solution was washed, added to 0.3 normal hydrochloric acid to get a concentration of 30 gpl titanium dioxide, and peptized by heating at 80° C for 20 minutes. The charge was changed to negative by adding 1.35 g potassium citrate for each gram of titanium dioxide in the sol. After reversal of charge, the nuclei were added in a proportion of 3 to 5 per cent, based on the titanium dioxide content, to relatively concentrated sulfate or chloride solutions, and hydrolysis was effected rapidly by heating at 95° C.

The hydrous oxide obtained by incomplete hydrolysis of sulfate solutions gave rutile on calcination.[127] A nucleated concentrated titanium sulfate solution of low acidity factor, a mole ratio of sulfuric acid to titanium dioxide of 1 or less, was boiled to effect hydrolytic precipitation of 40 to 60 per cent of the titanium dioxide. The concentration of the solution did not fall below 40 gpl titanium calculated as the dioxide. At this stage hydrolysis was interrupted and the product was filtered, washed, and calcined at 750° to 950° C. The finished pigment exhibited excellent rutile characteristics of high tinting strength and good color. Olson[128] employed a complex containing both anatase yield and rutile-inducing nuclei.

A suspension of colloidally dispersed (peptized) hydrous stannic oxide was found to induce the formation of rutile on hydrolysis of titanium sulfate solution.[129] Precipitated stannic hydrate was washed thoroughly and peptized in a 0.05 to 1.0 normal solution of a monobasic acid, and the resulting dispersion was heated at 50° to 100° C for a relatively short time to develop the nucleating properties to the maximum degree.

A nucleating agent for the hydrolysis of titanium sulfate solution to produce rutile pigment is prepared from sodium titanate by treatment with hydrochloric acid followed by heat treatment.[130] Hydrous titanium dioxide from the usual hydrolysis of titanium sulfate solution is heated with sodium hydroxide solution at the boiling point for 1 to 6 hours to obtain a water insoluble sodium titanate containing 80 to 86 per cent titanium dioxide and 20 to 14 per cent sodium oxide. This product is slurried with dilute hydrochloric acid to react with the sodium oxide, and the sodium chloride formed is removed by washing. These steps are carried out at room temperature to avoid peptization of the titanium oxide. The hydrolysis treatment is preferably carried out by mixing an aqueous dispersion of the nucleating agent at 20 per cent solids with the titanium sulfate liquor at 50° C. After nucleation, the mixture is heated to boiling and boiled for 1 to 5 hours. About 3 per cent nucleating agent based on the titanium dioxide is employed. To an aqueous pulp obtained by hydrolytic precipitation from titanium sulfate solution and containing 100 g titanium dioxide at 30 per cent solids was added 150 g of sodium hydroxide with stirring. The mixture was heated for 2 hours at 85 to 90° C. Water was added to maintain a constant volume. The titanate formed was diluted with water, filtered, and washed free of sulfates and excess sodium hydroxide. After washing, the filter cake was slurried in 2 liters of water at room temperature to which was added 75 ml of commercial hydrochloric acid. After agitation for several minutes to ensure complete reaction between the sodium hydroxide present and the hydrochloric acid, the mixture was filtered and washed with water to remove any remaining trace of sodium chloride.

To 3290 g of titanium sulfate liquor containing the equivalent of 500 g titanium dioxide at 50° C was added 78 g of a water slurry of the hydrated titanium dioxide nucleating agent containing 3 per cent titanium dioxide based on the titanium dioxide in the sulfate liquor. The mixture was heated to boiling and boiled for 5 hours. Over a period of 3.5 hours during the boiling 180 ml of water was added. A yield of 98 per cent was obtained. After the usual washing the product was calcined to rutile of good pigment properties.

A nucleating agent obtained by decomposing an alkali or alkaline earth titanate in a slight excess of a monobasic acid followed by conditioning the mixture at an elevated temperature accelerates the hydrolysis of titanium sulfate solutions to obtain an anatase hydrolyzate, which readily converts to rutile on calcination at 750 to 950° C.[131] Hydrous titanium dioxide precipitated hydrolytically from sulfate solution is treated with sodium hydroxide to neutralize the sulfuric acid present and then washed free of sulfate ion. After purifying, the titanium dioxide is boiled with a solution of barium or sodium hydroxide in equivalent proportions until the reaction is complete. Then without removal from the mother liquor, the titanate is digested with 0.3 to 0.5 normal hydrochloric acid at 80° C for 15 to 30 minutes. Temperatures of 60 to 100° C may be employed with corresponding curing times from 10 minutes to 1 hour. The resulting seed solution can be used directly, after cooling, in the conventional hydrolysis of a concentrated or dilute titanium sulfate solution, but preferably is first freed of the acid prior to use in hydrolysis. Acid removal is conveniently carried out by neutralization with alkali metal hydroxides or carbonates, ammonia, or alkaline earth hydroxides to obtain a flocculated, easily filtered suspension. Soluble chlorides are easily washed out to avoid corrosion of the plant equipment.

After adding 3 to 5 per cent of the activated seed material to the titanium sulfate liquor, the solution is heated to boiling to effect hydrolysis in the usual manner. Yields of titanium dioxide in excess of 95 per cent are obtained in a shorter time of hydrolysis. Following the hydrolysis, the raw pigment anatase titanium dioxide is subjected to the usual washing and purification treatment, following which it is calcined at 850 to 875° C to produce the improved rutile pigment.

In an example, 100 parts by weight of titanium dioxide from the hydrolysis of a titanium sulfate solution was slurried in water and neutralized with sodium carbonate. The resulting sodium sulfate was removed by filtration and washing, and the purified titanium hydroxide was digested with an equivalent amount, 214 parts by weight, of barium hydroxide. After 2 hours at the boiling temperature, the reaction was complete and the product was a suspension of barium titanate in water, containing 115 g of solid per liter. Without cooling, the suspension was added to a 20 per cent hydrochloric acid solution at 80° C. The resulting suspension analyzed 30 g titanium dioxide per liter and showed 20 g of titratable hydrochloric acid per liter. After 30 minutes the product was added to an equal volume of cold water and neutralization was effected by the addition of barium hydroxide solution. The flocculated seed material was filtered, washed to remove and recover the barium chloride, and suspended in water for use in the hydrolysis of titanium sulfate solution. Seed material containing 5 parts of titanium dioxide to 100 parts of titanium dioxide in the sulfate solution for hydrolysis was highly effective at a concentration of 170 g titanium dioxide equivalent per liter. The filtered and washed hydrolyzate, calcined at 950° C for 1 hour, converted readily to rutile and yielded a pigment having 20 per cent better hiding power than commercial anatase pigment.

Titanium sulfate solution is hydrolyzed in the presence of a small proportion of a coagulated titanium dioxide seed or accelerating agent prepared by treating a ripened, peptized sol of colloidal titanium dioxide with a soluble sulfate, or sodium hydroxide, or carbonate, in an amount sufficient to completely coagulate the sol.[132] The recovered anatase hydrolysis product is calcined at a temperature of 1000° C to obtain a rutile pigment of high quality. For example, orthotitanic acid was precipitated from titanium sulfate solution containing 145 gpl titanium dioxide, an amount of titanium III equivalent to 1.2 gpl titanium dioxide, 63 gpl ferrous iron, and 264 gpl sulfuric acid at 25° C by adding ammonium hydroxide with agitation. The orthotitanic acid was filtered and washed free of sulfuric acid, after which it was slurried in water to 20 gpl titanium dioxide. Hydrochloric acid was added to produce a 0.3 normal solution and the mixture was heated to 85° C and maintained at this temperature for 25 minutes. The resulting titanium dioxide sol was cooled to 25° C and coagulated by adding 5 g of sulfuric acid per liter of sol. After coagulation, the product was decanted, filtered, and washed until free from chloride ion. The filter cake was reslurried in water to produce a coagulated seed material having a concentration of 25 gpl titanium dioxide. This suspension was heated to 68° C and added to a titanium sulfate solution of the same composition as the starting material from which the seed were made in an amount equivalent to 5 g of titanium dioxide per 100 g of titanium dioxide in the solution to be hydrolyzed. After seeding, the solution was heated to boiling and boiled for 20 minutes. The yield was 97 per cent. The washed hydrolysis product gave rutile on calcination at 850 to 975° C.

A seeding agent for titanium sulfate solution hydrolysis is prepared by the reaction of a titanium chloride solution with sodium hydroxide so that the resulting hydrous titanium dioxide is successively maintained or ripened under both alkaline and acid conditions, and under specific alkaline and acid normalities.[133] Titanium tetrachloride solution is added to sodium hydroxide solution at room temperature in such proportions that the solution has an alkalinity of 0.4 normal sodium hydroxide. The precipitated titanium oxide is maintained in this alkaline medium for 1 to 2 minutes, and the suspension is then acidified to 0.4 normal hydrochloric acid. Either hydrochloric acid or titanium tetrachloride may be added. This suspension is heated at 75 to 90° C for 15 to 30 minutes to effect peptization and curing. If cooled to 50° C or below the sol may be stored for future use. From 1 to 5 per cent of the sol, preferably 3 per cent, based on titanium dioxide content of solution and seed is used in hydrolysis of the sulfate solution. The sol may be coagulated before use by treating it with sulfuric acid, sodium sulfate, or sodium hydroxide. Concentration of titanium tetrachloride solution is 125 gpl titanium dioxide or lower.

To illustrate, a solution of titanium tetrachloride containing 100 gpl titanium dioxide and 182.5 gpl of active hydrochloric acid, and sodium hydroxide solution containing 240 gpl sodium hydroxide are simultaneously added to a mixing vessel so that the final suspension is 0.6 normal with respect to sodium hydroxide. The suspension is then added to a relatively pure titanium tetrachloride solution of the above composition so that the final suspension contains 41 gpl titanium dioxide and is 0.5 normal with respect to hydrochloric acid. This acid suspension is heated to 90° C and held at this temperature for 15 minutes, followed by cooling and coagulation by adding sodium hydroxide solution to bring the pH to 7. Chlorides are removed from the coagulated sol by washing, after which it is ready for use. A proportion of seed, 5 per cent based on the titanium dioxide contents, is used in the hydrolysis of titanium sulfate solution containing 210 gpl titanium dioxide and an acid factor of 70. Calcination at 930° C gave a rutile pigment of good quality. From 0.2 to 0.5 per cent potassium sulfate based on the titanium dioxide is usually added before calcination.

A seeded concentrated sulfate solution of ilmenite is heated at 80 to 100° C for a relatively short time to enrich the colloidal content of the solution before carrying out the hydrolysis according to the Blumenfeld process.[134] In an example, a complex nuclei solution is prepared by blending equal amounts of: (a) a yield-inducing hydrolysis nuclei suspension, prepared by adding one volume of a titanium sulfate solution containing 150 gpl titanium dioxide to two volumes of water. Sufficient sodium hydroxide solution is added to bring the pH to 2.7 and provide a suspension containing 30 gpl titanium dioxide. The temperature is raised to 80° C and maintained at this value for 30 minutes, after which the seed is cooled, filtered, and reslurried to 75 gpl titanium dioxide, and (b) a rutile-inducing nuclei suspension prepared by adding one volume of titanium tetrachloride solution containing 80 gpl of titanium dioxide with stirring to one volume of water containing 128 g of sodium hydroxide per liter. The resulting mixture is heated at 85° C for 30 minutes, cooled, and neutralized to a pH of 7 by adding dilute sodium hydroxide solution. After washing free of chlorides the precipitate is reslurried in water to provide a suspension containing 75 gpl titanium dioxide.

Ten volumes of this complex seed suspension containing 75 gpl titanium dioxide were added to 125 volumes of titanium sulfate solution containing 200 gpl titanium dioxide, 60 gpl iron, and 500 gpl sulfuric acid at 80° C. The temperature of the solution after incorporating the seed by stirring was raised at a rate of one-third of a degree per minute. When the temperature reached 100 to 102° C and the appearance of the solution changed from black to gray, heating was interrupted and 50 volumes of water were added, after which the heating was continued with stirring. The solution was boiled for 3 hours. A 96 per cent yield of titanium dioxide, readily filterable hydrolyzate adapted to conversion to rutile on calcination below 1000° C was obtained.

A concentrated (60 to 150 gpl titanium dioxide) yield inducing and rutile inducing seed is prepared by adding 1 to 3.5 per cent hydrofluoric acid and 30 to 40 per cent hydrochloric acid, both on the titanium dioxide basis, to orthotitanic acid.[135] The seed is stabilized by heating for 5 to 30 minutes near the boiling point or by aging at room temperature for 10 to 24 hours. Concentrated titanium sulfate solution hydrolyzed in the presence of 0.5 to 1.5 per cent seed yielded hydrolyzates which after treatment with sodium carbonate 0.4, lithium carbonate 0.075, and ammonium dihydrogen phosphate 0.2 per cent form rutile on calcination at 850° C. Thus, orthotitanic acid containing 45 g of titanium dioxide was prepared and treated with a synergistic combination of 18 g hydrochloric acid and 0.5 g of hydrofluoric acid. Titanium sulfate solution seeded with this material gave, on hydrolysis, hydrous titanium dioxide which was converted to high quality rutile pigment on calcination at 840° C.

A nucleous liquid for the hydrolysis of titanium sulfate solutions to produce rutile type pigments is prepared by treating a titanate of sodium, potassium, zinc, or magnesium with a monobasic inorganic acid to produce a solution having a pH value below 1.5, diluting the solution until it has a titanium content of 5 to 80 gpl calculated as titanium dioxide, and heating at 50 to 100° C for 10 minutes to 2 hours to develop its nucleating properties while avoiding precipitation of titanium dioxide.[136] The titanate may be prepared by the reaction of hydrous or calcined titanium dioxide with an oxide, hydroxide, or carbonate of sodium potassium, zinc, or magnesium in equivalent proportions to form the metatitanate or the orthotitanate. To obtain the titanate in a form readily soluble in acids, the components are heated at 200 to 800° C for 1 to 6 hours. For the hydrolysis of titanium sulfate solution, from 1 to 15 per cent of the nucleous liquid based on the titanium dioxide is employed. Hydrolysis is carried out in the usual manner. The hydrous titanium dioxide precipitate yields rutile on calcination for 1.5 to 2 hours at 750 to 950° C.

To illustrate, hydrous titanium dioxide precipitated hydrolytically from sulfate solutions of ilmenite as such or after calcination is mixed with a 75 per cent sodium hydroxide solution to form a stiff paste in such proportions that the mole ratio of sodium oxide to titanium dioxide is 1 to 1. The paste is dried with agitation and the dry powder is calcined at 600° C for 1 hour to complete the reaction. After calcination, the sodium titanate is ground and treated with a quantity of hydrochloric acid of 10 per cent concentration amounting to 4 moles of hydrochloric acid per mole of sodium titanate. The suspension is agitated for 2 hours at room temperature, after which the temperature is raised to 50° C in the course of a further half hour. This solution, which is free from solids, is diluted with water to 10 gpl titanium dioxide and heated at 45 minutes at 85° C to develop the nuclear properties. The resulting nucleous liquid in the proportion of 10 per cent based on the titanium dioxide is added to a sulfate solution containing titanium dioxide 150, trivalent titanium as titanium dioxide 2, ferrous iron 21.6, and active sulfuric acid 327 gpl. After nucleation the solution is heated to the boiling point in 15 minutes and boiled for 30 minutes to complete the precipitation of the titanium dioxide. The precipitate is washed, calcined at 910° C for 1.5 hours, and processed to rutile pigment.

Rutile seed for the hydrolysis of titanium sulfate solution can be obtained more effectively by treating an aqueous pulp of orthotitanic acid with from 0.8 to 1.5 parts of sodium hydroxide to one part of anhydrous titanium dioxide.[137] The mixture is stirred for 1 to 6 hours at temperatures up to 100° C. In an example, 1500 ml of an aqueous slurry of 150 g hydrated lime at 30° C are added to 1060 ml of sulfate solution containing 150 g of titanium dioxide also at 30° C in 1 hour with agitation. After filtering off the calcium sulfate formed the filtrate is diluted to 15 liters. Two thousand ml of an aqueous slurry containing 30 g of hydrated lime at 30° C are added slowly to the 15 liters of solution containing the titanium and iron at 30° C over a period of 0.5 hour with agitation. The pH of the slurry is 2.7 and 90 per cent of the titanium precipitates as orthotitanic acid. The orthotitanic acid is filtered, washed free of iron and calcium, and slurried in water to 18 per cent solids. To this slurry is added 135 g of sodium hydroxide with stirring. The mixture is heated for 2 hours at 85 to 90° C with stirring at constant volume to form the titanate which is filtered and washed free of sulfates. This filter cake is slurried in 284 ml of commercial 20° Bé hydrochloric acid diluted with

1350 ml of water, and boiled at constant volume for 1 hour. A thoroughly washed aqueous pulp of hydrous titanium dioxide precipitate containing 200 g of titanium dioxide is mixed with 50 g of the neutralized and washed seed containing 6 g of titanium dioxide and with 4.1 g of potassium carbonate as a concentrated solution. The mixture is dried and calcined at 850° C for 1 hour.

A rutile inducing seed composition for use in hydrolyzing titanium sulfate solutions is prepared by heat treating a hydrochloric acid solution of freshly prepared orthotitanic acid in the presence of an ion exchange resin, IXR.[138] Thus washed metatitanic acid is dissolved in a solution of 160 g sulfuric acid per liter, and the solution is added to 160 gpl aqueous sodium carbonate to a pH of 8.5. The orthotitanic acid precipitated is filtered and washed. A mixture of 100 ml hydrate, 13 ml of 37.5 per cent hydrochloric acid and 15 ml of IXR, is heated to 70° C in 10 minutes, held at this temperature for 20 minutes, and cooled to room temperature in 10 minutes. After filtering off the ion exchange resin, the titanium dioxide is coagulated with ammonium hydroxide, filtered, and washed to yield a rutile seed having excellent hydrolysis activity which yields 99 per cent rutile after curing the hydrolysis product for 1 hour at 900° C.

Solomka[139] prepared a seeding material by adding 100 ml of sulfate solution of ilmenite containing 249 gpl titanium dioxide and 200 gpl ferrous sulfate to 375 ml of water at 96° C slowly with agitation. After heating at 96° C, 2.4 liters of solution containing 149 gpl titanium dioxide and 295 gpl ferrous sulfate was added to the hot seeding suspension. The resulting mixture was boiled for 3 hours to precipitate the titanium dioxide, which was filtered, washed, bleached, treated with 0.1 per cent phosphorus pentoxide and 0.6 per cent potassium sulfate, and calcined for 3 hours at 925° C.

Rutile-type pigment with the maximum coloring power is produced by the hydrolysis of titanium sulfate without the addition of nucleating or seeding agents followed by calcination of the hydrolyzate.[140] For example, 1000 parts by volume of a titanium sulfate solution having a density of 1.660, and containing 244 gpl titanium dioxide, 433.6 gpl sulfuric acid combined with the titanium, 39.4 gpl of iron as the sulfate, and 2.3 gpl titanium III as titanium dioxide, was heated by passing steam into it and held at the boiling point for 2.5 hours. Hydrolysis took place on the addition of 800 parts by volume of water at room temperature. The mixture was again heated and boiled for 30 minutes. After cooling to 90° C, the hydrous titanium dioxide was filtered, washed, and calcined for 2 hours at 830° C. The yield was 93 per cent based on the titanium dioxide dissolved.

Nuclei for the hydrolysis of titanium sulfate solution to produce rutile pigment are prepared by adding 24.7 liters of 20 per cent hydrochloric acid to 6.0 kg of zinc titanate and heating the mixture at 85° C for 3 minutes.[141] The mole ratio of titanium dioxide to hydrochloric acid in the seed composition is 1 to 2. A rutile inducing seed for use in the hydrolysis of titanium salt solutions prepared by heat treating a dispersion of hydrous titanium dioxide in hydrochloric acid is stabilized by the addition of sulfuric acid.[142] The mole ratio of the hydrochloric acid to titanium dioxide or oxide of a Group 4 metal is from 1 to 2.2. From 0.5 to 6.5 per cent of the multivalent ion stabilizer based on the weight of the titanium dioxide is employed. A solution containing from 20 to 90 gpl titanium dioxide, the hydrochloric acid and the sulfate ion used as stabilizer is heated at 70° C to the boiling point for 10 to 30 minutes to develop the nucleating properties after which it is cooled quickly to 60° C. A properly cured nuclei suspension has an opalescent appearance and exhibits colloidal properties.

Zinc oxide, aluminum hydroxide, magnesium oxide, rutile titanium dioxide, hydrochloric acid, zinc chloride, aluminum chloride, magnesium chloride, stannic chloride, titanium tetrachloride, formic acid, and acetic acid were introduced separately as promoters into concentrated solutions of titanium sulfate in amounts of 0.5 to 5.0 per cent based on the weight of the titanium dioxide in the solution prior to hydrolysis.[143] After hydrolysis the titanium dioxide hydrate was calcined at 850° C for 3 hours. Titanium dioxide of 99 per cent rutile crystal structure was obtained with all of the promoters except with 1 and 3 per cent of zinc chloride, 1 per cent of aluminum chloride, 0.5 per cent stannic chloride, and 1 per cent titanium tetrachloride.

A rutile inducing seed for the hydrolysis of titanium sulfate solution prepared from titanium tetrachloride solution is stabilized by an amount of sulfuric acid from 2.5 to 3.6 per cent by weight of the titanium dioxide content or from 20 to 30 millimoles of sulfuric acid per mole of titanium dioxide.[144] To illustrate, 84.4 ml of titanium tetrachloride solution containing the equivalent of 237 gpl titanium dioxide and 381 gpl hydrochloric acid, and 22.8 ml of a solution containing 17.5 gpl sulfuric acid were added to water so that the final volume was 1 liter. The concentration was then 20 gpl titanium dioxide with sulfuric acid equal to 2 per cent of the titanium dioxide content as a stabilizer. The solution was heated for 10 minutes at 90° C to develop nucleating properties and then quickly cooled to below 60° C. The resulting suspension was clear and opalescent. Such nuclear dispersions are also stabilized by the addition of monobasic carboxylic organic acids which are soluble in dilute hydrochloric acid.[145] These stabilizers disappear during calcination. In an example, titanium hydroxide is precipitated from a pure titanium sulfate solution with a sodium carbonate solution at a pH of 8. The resulting hydrate is washed until the filtrate shows no precipitate with barium chloride. After washing, the hydrate is peptized with hydrochloric acid to form a sol containing 45 gpl titanium dioxide and 27 gpl hydrochloric acid. Acetic acid equal to 2 per cent of the weight of the titanium dioxide is added, and the sol is cured by heating to 90° C in 15 minutes and holding at this temperature for 10 minutes to develop the nucleating properties. After curing the sol is immediately cooled to below 50° C. Stabilization of the nuclei is also effected by the addition of colloids, preferably gelatin or glue.[146] To prepare the nuclei, hydrous titanium dioxide gel precipitated from titanium salt solutions in monobasic acids with an alkali metal compound or ammonia, is peptized with a monohydric acid, and the liquefied mixture is poured into a fresh 1.0 to 1.5 per cent solution of gelatin. Because of this stabilization, filtration and further treatment of the gel cake can be omitted. The requirement of nuclei thus produced is reduced from 6 to 4 per cent based on the titanium dioxide content of the sulfate liquor. Rutile pigments obtained with the use of these nuclei have a tinting strength of 165 to 175 as compared with 110 to 135 for the usual pigment. Gelatinous titanium hydrate nuclei are obtained by the electrolytic hydrolysis of chloride solutions with ratios of 0.4 to 0.9 parts by weight of hydrochloric acid for each part of titanium dioxide at pH of 7.5.[147] Bone glue is added as a stabilizing agent. The temperature is raised to 85 to 95° C in 15 minutes, held at this temperature for 10 minutes, and cooled to 30° C. Graphite electrodes are employed.

An indirect method for producing rutile pigments from anatase titanium dioxide as obtained by the regular hydrolysis of sulfate solutions involved conversion to barium titanate, solution in nitric acid, and reprecipitation of the hydrous oxide.[148] Keats[149] described a cyclic operation of the process in which the nitric acid and barium hydroxide were recovered and reused. Neutralized and washed hydrous oxide of the anatase modification, as obtained commercially by the thermal hydrolysis of sulfate solutions of ilmenite, was mixed with an equivalent amount of barium hydroxide and digested near the boiling point of the solution to obtain barium titanate. This compound was treated with 5 to 10 mole proportions of nitric acid of 30 to 50 per cent strength to form titanium nitrate and barium nitrate. Since barium nitrate is only slightly soluble in the strong liquor, the greater part of the constituent was present as crystals and these were separated readily by filtration or decantation. The solution consisting largely of titanium nitrate was adjusted to a concentration of 80 g to 120 g titanium dioxide per liter and hydrolyzed by boiling, after proper seeding, to produce a rutile precipitate. The hydrous material was filtered, washed, and calcined in the same manner as anatase hydrolyzates. In carrying out the hydrolysis, solutions of concentrations above 40 gpl titanium dioxide were employed; otherwise anatase was formed. The nitric acid filtrate from the hydrolysis product was collected for reuse, and the barium nitrate crystals were decomposed by heat to give oxides of nitrogen and barium oxide, both of which were reused.

In an actual operation, a washed iron-free pulp from hydrolysis of a sulfate solution of ilmenite was slurried in water and neutralized with ammonium hydroxide. All ammonium salts formed were washed out and the sulfate-free suspension was mixed with a chemically equivalent amount of barium hydroxide in aqueous solution. The slurry was heated to boiling and maintained for 1 hour to effect formation of barium titanate. This reaction product was dewatered and the cake was treated with cold 50 per cent nitric acid, with agitation. After several hours the mixture lost its opalescence and then consisted of crystals of barium nitrate in a titanium nitrate solution. Upon filtration about 80 per cent of the titanium was recovered before washing and the remainder by washing with cold acidified, saturated barium nitrate solution. The main filtrate and washings were combined to give a solution having more than 80 g titanium dioxide per liter, and this was subjected to thermal hydrolysis after addition of the proper seeding material. Precipitated hydrous oxide was filtered, washed, and calcined according to conventional methods to produce rutile titanium dioxide of pigment grade.

The barium nitrate crystals were decomposed by heat to give oxides of nitrogen which were recovered along with the nitric acid of the hydrolysis liquor for reuse. The residue of barium oxide was added to water to regenerate the hydroxide for attacking another batch of anatase pulp.

Similarly, orthotitanic acid precipitated from regular sulfuric acid solutions of ilmenite by adding an alkaline agent, such as ammonium hydroxide, was washed thoroughly to remove any divalent anions and dissolved in nitric acid, and the resulting solution was hydrolyzed to precipitate rutile titanium dioxide which was calcined to develop pigment properties.[150]

# 15

# Filtering and Washing Hydrous Titanium Dioxide

Regardless of the type of pigment produced, the filtration and washing procedures are very similar. At the end of the hydrolysis step the system consists of hydrous titanium dioxide suspended in a strongly acid sulfuric acid liquor which also contains a large proportion of ferrous sulfate and small amounts of other dissolved salts. Added extenders may also be suspended in the liquor. Separation of the precipitate from the liquor may be effected by repeated decantation, but this would result in a dilution of the liquor to such an extent that acid recovery would probably be uneconomical and at best would be slow and cumbersome.

On the other hand, direct filtration requires special acid-resistant filters and filter media, but in commercial operation this method is usually employed. Filters of various types coated with paper pulp and diatomaceous earth, nitrated cloth, perforated rubber, glass cloth, Vinylite cloth, Dacron, or Nylon fabric or other resistant material have been employed.

Hydrous titanium oxide for pigment manufacture is precipitated normally by the thermal hydrolysis of sulfate solutions of ilmenite in which the iron content is maintained in the ferrous state. Although under these conditions there is no hydrolytic decomposition of the ferrous compounds, the freshly precipitated titanic oxide always carries entrained with it a small amount of iron in the ferrous condition, together with adsorbed or combined sulfuric acid.

Prolonged washing, consisting of a number of filtration, reslurrying, and refiltering operations, is effective in removing all the iron and other foreign salts except those adsorbed by the pulp. The adsorbed iron salts cannot be completely removed by washing with water alone, apparently because they are oxidized to the trivalent state during the operation, and ferric compounds are adsorbed to a much greater degree than ferrous compounds. At any rate it is not possible by this means to reduce the iron content of the pulp much below 0.1 per cent, or the adsorbed or combined sulfuric acid below 5 to 8 per cent. Although this proportion of iron does not render the products unsuited for industrial use, greater purity is desired to obtain pigments of improved whiteness.

If this washing operation is followed by treatment with a 0.5 to 5 per cent solution of sulfuric or hydrochloric acid at higher temperatures (50° to 60° C), pulps practically free from iron can be obtained. In an actual test 1 g hydrous titanium oxide, after washing five times with cold water, retained 50 mg ferric oxide; another sample washed four times with cold water and once with cold 1 per cent sulfuric acid contained 20 mg ferric oxide; while a third sample washed four times with cold water and once with 1 per cent sulfuric acid at 50° C retained only 0.5 mg ferric oxide.

To illustrate, the hydrous titanic oxide precipitated from a sulfuric acid solution of ilmenite was filtered from the mother liquor and washed with cold water until a very small amount of iron compounds remained. It was then made into a paste of 1 part titanium dioxide to 1½ parts of 1 per cent sulfuric acid. The mixture was heated at 50° C for 1 hour, filtered, and again washed rapidly with cold water.[1] Good results were obtained by carrying out the washing operation on filters of calcium sulfate or barium sulfate.[2] The hydrous oxide was washed alternately with cold water and hot dilute acid.

The iron content was reduced to still lower proportions (0.0005 to 0.001 per cent ferric oxide) by subjecting the pulp to a digestion treatment in the presence of added sulfuric acid and a reducing agent, such as a titanous salt, sulfur dioxide, or a soluble thiosulfate or sulfide.[3] The titanous compound may be formed in place by adding metallic zinc to the acid-pigment slurry. Precipitated hydrous titanic oxide was washed with water and suspended in a solution of 4 to 12 per cent sulfuric acid, the reducing agent was added or formed in place, and the mixture was heated at 60° to 90° C, with agitation, for 6 to 8 hours and in some instances up to 24 hours. After this treatment the pulp was filtered and washed with water to remove the remaining iron salts.

Washed, hydrolytically precipitated titanium dioxide was digested in an 80 to 120 gpl sulfuric acid solution at a temperature of 80° C to the boiling point for several hours to remove absorbed and adsorbed impurities.[4] This process was particularly effective in removing compounds of vanadium, chromium, and iron, which, if present in extremely small proportions, discolor the pigment. A small amount of titanium was also dissolved. The purified titanium oxide was separated from the acid liquor, washed with water, and calcined.

De Rohden[5] digested hydrolytically precipitated titanium dioxide with dilute hydrochloric acid in the presence of a reducing agent (hydrogen sulfide or soluble thiosulfate) to remove the iron and obtained a pigment of improved color. This bleaching effect was also obtained by digesting the hydrous oxide in an aqueous 1 per cent solution of hydrochloric acid containing a small proportion of an alkali metal salt of an oxygen acid of chlorine, such as sodium hypochlorite or potassium chlorate.[6]

Iron compounds and similar impurities may be washed from crude titanic acid after reduction with sulfur dioxide or sulfites.[7] The proportion of the agent employed should correspond not only to the trivalent iron but also to the oxidants present. Sulfur dioxide does not reduce tetravalent titanium (to the trivalent state) so there is no loss of this element in the acid liquor as titanous salts are not precipitated in the hydrolysis step. Care should be taken to prevent reoxidation of the ferrous compounds during subsequent washing. Titanic acid produced by decomposing ilmenite with alkalies was freed from iron by heating for several hours with dilute mineral acid (10 to 30 per cent sulfuric acid) in the presence of reducing agents such as titanous sulfate or sulfurous acid.[8] The acid was of such a strength as not to dissolve the titanium component.

Although the last trace of iron may be removed by digesting the washed pulp with dilute sulfuric acid in the presence of a reducing agent at an elevated temperature with agitation for 6 to 24 hours, this treatment requires a break in the washing process and acid must be added to the system. However, by mixing powdered zinc with the hydrous titanic oxide during the first repulping operation after separation from the strongly acid hydrolysis liquor, reducing conditions may be maintained throughout the washing process without the necessity of a special reduction step.[9] Residual acid is sufficient to effect the reaction. By this treatment a small proportion of titanous salts is formed which prevents oxidation of the ferrous compounds and thus permits their complete removal by the regular washing process. This operation also facilitates removal of other reducible impurities such as traces of chromium and vanadium originally present in the ore.

As an illustration, the hydrolysis product, consisting of a suspension of hydrous titanium oxide in a solution of sulfuric acid and ferrous sulfate, was filtered on a rotary vacuum or leaf filter which consisted of a series of frames, each carrying an acid-resisting covering. The cake was removed and agitated with the least amount of water to give a free-flowing slurry, and sufficient powdered zinc was added to maintain reducing conditions throughout the remainder of the washing operation and prevent oxidation of any of the iron to the ferric state. Enough closely held acid was retained by the pulp to react with the zinc. The slurry was then filtered on a rotary or leaf filter. The cake was reslurried with additional fresh water and refiltered and repulped a number of times until the impurities were removed. One addition of zinc maintained reducing conditions throughout the entire step and permitted removal of iron without interrupting the washing operation. Furthermore, no variation in the regular practice was required and no acid was added to the system. Pigments obtained from pulps treated in this manner had a high degree of whiteness and brightness. Other properties were also greatly improved.

To improve the color of the finished pulp, hydrous titanium oxide prepared from ilmenite solutions has been treated during the washing operation with a solution of a salt of thorium, zirconium, or aluminum to displace the iron present as impurity.[10] On calcination, these compounds produced white oxides in contrast to the reddish brown of ferric oxide.

The color of the pigment may be improved by subjecting the washed hydrolysis product to a bleaching treatment to reduce the compounds of vanadium, chromium, manganese, and other impurities to a minimum.[11] These hang with the pulp through ordinary washing with water. Nine parts of a slurry of 300 g hydrolytically precipitated and washed titanium dioxide per liter was mixed with 1 part by volume of 60° Bé sulfuric acid and the resulting mixture, containing 270 gpl titanium and 135 gpl sulfuric acid, was heated to 90° C and held for 1 hour. It was then cooled, filtered, washed, and calcined as usual. This pigment was noticeably better in color than a similar product in which the sulfuric acid treatment was omitted. Only 0.5 per cent of the titanium dioxide was dissolved. Chromium compounds [12] most affect the color, and the limit of sensitivity to the eye is $1.5 \times 10^{-6}$ parts of the oxide in 1 part of titanium dioxide. To overcome such discoloration, the pigment was mixed with an alkali metal carbonate, oxide, or hydroxide, or with an alkaline earth metal oxide, in a proportion slightly in excess of that required to react with all the chromium present, roasted at an incandescent heat, and washed to remove the chromate formed. If magnesium or barium oxide is used, an insoluble chromate is formed which is very slightly colored and may be left in the pigment without serious injury. The alkali chromates are soluble in water, while those of the alkaline earth metals are insoluble in water but soluble in dilute acids so that the leaching procedure depends upon the type of chromate formed. Oxidizing agents may be added before the roasting step.[13]

A process for washing hydrolysis product containing iron and copper salts was reported by Allan.[14] The hydrous titanium dioxide, after separation from the mother liquor, was washed first under reducing conditions until practically all the iron was removed. It was then treated with an oxidizing agent such as nitric acid or hydrogen peroxide to convert the copper to the soluble cupric state, in which form it was readily removed by further washing with water.

In a commercial process described by Brown [15] the hydrous titanium oxide was first separated from the hydrolysis liquor containing high proportions of sulfuric acid and ferrous sulfate by filtration on a Sweetland press, then washed to remove most of the foreign materials. It was then reslurried in water and further washed on a series of Oliver continuous filters. At the Chelyabinsk pigment plant, in the Union of Soviet Socialist Republics, microporous rubber [16] has been employed as a filtering medium. Mean diameter of pores was of the order of $10^{-4}$ mm, although the openings can be made as small as $10^{-7}$ mm, according to the method of manufacture. Glass cloth, nitrated cloth, and synthetic resin cloth have also been recommended as media for the first filtration, but, after initial separation from the strong hydrolysis liquor, further filtrations and washings may be carried out on cotton fabric media.

Ferric compounds are more completely reduced and removed, and less reducing agent is left behind if a minor portion of the titanium hydroxide slurry is acidified with sulfuric acid, a large part of the titanium is reduced to the trivalent state with aluminum or zinc, or electrolytically, and returned to the hot bulk slurry. A minimum of 0.05 gpl titanium III ion is maintained.[17] Finally the ferrous compounds are washed out. For example, 59 parts by weight of titanium hydroxide slurry containing calcium sulfate and 27 per cent solids, including 0.45 per cent ferric hydroxide based on the weight of pigment, was treated with 20 parts of 66 Bé sulfuric acid and 0.6 part of metallic aluminum. This reduced suspension was added to 22,000 parts of the original titanium hydroxide slurry and held at 55° C for 0.5 hour. For bleaching the final pigment, 0.2 pound of reducing metal per ton had been used, leaving behind 0.01 per cent iron as ferric oxide.

Titanium dioxide pigments of improved color and texture are obtained by treating hydrolytically precipitated titanium dioxide with an alkaline ammonium compound to a pH of 5 to 8, and washing the precipitate free of calcium, magnesium, and silicate ions with water as well as the ammonium sulfate formed in the neutralization reaction.[18] After neutralization and washing, the product is calcined. The tinting strength of the finished pigment can be improved by the addition of a water insoluble zinc compound after neutralization. A suspension of 600 kg of titanium hydroxide obtained by the hydrolysis of titanyl sulfate solution in 1 liter is treated with 492 ml of 15 per cent sodium carbonate solution and adjusted to a pH of 6.5, and washed twice with 7 liters of water.[19] Then 78 ml of 10 per cent phosphoric acid is added to yield a pH of 2.5, and the precipitate is filtered and washed again with 10 per cent sulfuric acid. After filtration, the washed product is calcined at 900° C to obtain 390 g of pure white titanium dioxide.

On subjecting hydrous titanium dioxide precipitated by the hydrolysis of titanium sulfate solution to the action of an alternating electric current between lead electrodes, a clear sulfuric acid solution collects on top of the cake and may be decanted.[20] In an example, 1 kg of titanium dioxide in the form of washed metatitanic acid precipitated by hydrolysis and containing 45 per cent total solids, of which 5 per cent is free or combined sulfuric acid, is placed in a glass vessel fitted with two sheet-lead electrodes having an effective surface area of 9 square inches each and placed 2 inches apart. A 60 cycle alternating current at 12 volts is passed between the electrodes. At the end of an hour the filter cake paste between the lead plates settles noticeably. The clear acid solution on top is removed, the cake is reslurried, and the treatment is repeated several times. In this way the free sulfuric acid content of the cake is reduced by one-half.

A treatment of titanium dioxide hydrate with an ion exchanger improves the pigment property directly and lowers the amount of impurities.[21] Specifically a slurry of hydrous titanium dioxide is agitated with a strongly basic ion exchanger of the NR4 type at room temperature for 3.5 hours. The pH increase from 0.4 to 0.9 indicates the splitting and desorption of sulfuric acid from the pigment. The resin is separated by a 100 mesh screen.

## NEUTRALIZATION OF RESIDUAL SULFURIC ACID

Hydrous titanium dioxide precipitated by the thermal hydrolysis of sulfate solutions retains around 5 to 10 per cent sulfuric acid even after thorough washing with water, and it may be desirable to remove or neutralize this closely held acid before calcination. Many methods have been proposed for accomplishing this end, but, in general, alkali metal or ammonium hydroxides are added to form neutral salts which may be washed out readily. The ammonium salts may be removed by volatilization at elevated temperatures. Alkaline earth metal compounds, oxides, hydroxides, or chlorides may be added to form insoluble sulfates which are retained in the precipitate. The latter method is particularly applicable to composite-pigment pulps. For example, by adding lime slurry to titanium dioxide-anhydrite pulp the residual acid will be recovered in the final pigment as calcium sulfate.

Neutralization or removal of this closely held acid is not necessary, however, for at the temperatures normally employed in calcination, 950° to 1050° C, it would be driven off along with the water content. If desirable, this sulfur trioxide can be recovered from the stack gases by electrolytic Cottrell precipitators.

Jebsen [22] neutralized the residual sulfuric acid by agitating the pulp with an excess of a hydroxide or carbonate of an alkali metal or ammonium, and removed the resulting neutral salts by washing with water. Similarly, by treating the hydrous oxide slurry with aluminum hydroxide, zinc hydroxide, sodium aluminate, or sodium zincate, the acid was converted to soluble salts which were readily washed out.[23] Monk and Ross [24] treated the washed hydrolysis product with ammonia or other volatile base, such as aliphatic amines, and then heated the mixture to a comparatively low temperature to remove the products of reaction (ammonium sulfate) by volatilization, and finally up to 1000° C to effect crystallization of the titanium dioxide. Alternatively, the hydrous oxide was dried at 100° to 700° C without decomposition before introducing the basic material. To conserve the alkaline neutralizing agent, the hydrous titanium oxide was first dried and then subjected to the action of a basic gas (ammonia) and heated in air to 500° to 1000° C to remove the sulfate compounds.[25] In a two-stage process the pulp was first treated with an alkali to combine with only a part of the residual acid, and the remainder was neutralized by kneading the mass with 5 per cent zinc oxide.[26] The product was washed, dried, and heated to incandescence to develop pigment properties.

The combined or adsorbed sulfuric acid may be transformed into neutral insoluble salts by treating the washed hydrolysis product with hydroxides, carbonates, or soluble salts of alkaline earth metals, such as calcium hydroxide and barium chloride.[27] Employing chlorides, the hydrochloric acid should be washed out before calcination. This treatment may be conveniently effected by adding a slurry of the alkaline earth base to the pigment pulp, also as a slurry, with thorough agitation, after which the product should be dewatered on a suitable filter.

# 16

# Calcining, Milling, and Processing Titanium Dioxide to Produce Pigments

## CALCINATION

The thoroughly purified and washed precipitate obtained by the thermal hydrolysis of titanium salt solutions is the amorphous hydrous oxide which contains some closely held acid. In the production of pigments, a calcination step is necessary to drive off the water and residual acid and at the same time convert the titanium dioxide to crystalline form of the required particle size. At the same time desired pigmentary properties are developed.

In 1844 Rose [1] obtained amorphous titanium dioxide as a white powder, and noted that its specific gravity varied according to the temperature to which the material had been subjected. For example, a sample calcined at 600° C had a density of 3.92, and another heated to 1000° to 1200° C was further increased to 4.25. According to Heaton,[2] calcination converted titanium oxide from the amorphous form having an index of refraction of 1.8 to the anatase crystal form having an index of refraction of 2.55 and finally to the rutile modification with a further increase in refractive index to 2.71. A mixture of titanium dioxide with barium sulfate and calcium sulfate, similarly ignited, exhibited an effective index of refraction far above the mean value of the two constituents. This was attributed to a coating of highly refractive titanium dioxide on the individual particles of the alkaline earth metal sulfate and to an intercrystallized mixture of the two materials to produce additional reflecting surfaces within the larger particles of the extender.

Amorphous titanic oxide or hydroxide, such as is obtained from sulfate solution, was converted to the cryptocrystalline modification [3] of pigment grade by calcination at 900° to 1000° C. The end point was determined by microscopic examination. Conversion was accelerated by the addition of 5 per cent zinc chloride or ammonium fluoride. During the change to cryptocrystalline form, the index of refraction was raised from 2.2 of the amorphous material to 2.55 for anatase, 2.64 for brookite, and 2.71 for rutile, with corresponding increases in tinting strength. Sulfuric acid, water, and other volatile materials were driven off. No change in particle size took place on conversion to crystalline form. Mixtures of titanic oxide and calcium sulfate, after calcination, consisted of a cryptocrystalline mass of the titanium dioxide embedded in a matrix of the extender.

In the commercial production of titanium dioxide pigments, calcination is necessary in order to develop maximum capacity and covering power. In general, these properties are dependent upon the coincidental increase in refractive index of the particles. At the higher range of temperature, development of these properties is accompanied by a partial or complete change in crystal structure from anatase to rutile. This conversion takes place more rapidly above 900° C, but at this temperature there is a danger of producing a somewhat discolored product. Below this value the last traces of sulfuric acid are expelled with difficulty. However, if the hydrous titanium oxide is treated with a small proportion of an accelerator the desired transformation can be accomplished at 850° to 900° C in a shorter time, thus helping to avoid discoloration.[4] Compounds which function as accelerators are the hydrates, sulfates, nitrates, and carbonates of potassium and sodium, and sulfates of aluminum and magnesium. Proportions of 0.5 per cent are usually sufficient. These agents not only hasten crystallization but also greatly improve tinting strength and covering power of the resulting products. Pigments processed in this manner had lower oil absorption, and their reaction, as measured by pH, was nearly or quite neutral, depending upon the degree of calcination and the nature and amount of the addition agent. If the accelerator was used in such a proportion, for example to 0.75 per cent, that some of it remained soluble in water, the calcined material was washed and dried before grinding, or this was accomplished incidental to wet milling.

Pigments prepared in this manner were characterized by clear white color, excellent brightness, covering power, and resistance to discoloration in light. These relationships are presented quantitatively in Table 16–1. All samples were calcined for the same period of time at 875° C.

Pigments of superior whiteness and opacity were obtained by treating washed titanic oxide pulp with 0.5 to 6.0 per cent of an alkali metal compound[5] before calcination. Such products were neutral or slightly alkaline in reaction, as determined by pH measurements. The added agents, which assisted the sintering action, were removed from the pigment by subsequent washing or by wet milling. This treatment was also

## TABLE 16-1

### Effect of Potassium Sulfate Additions on the Properties of the Final Pigment

| Property | None | 0.5% $K_2SO_4$ | 0.75% $K_2SO_4$ |
|---|---|---|---|
| Reaction | Very acid | Slightly acid | Very slightly acid |
| Crystal structure | Octahedrite with much rutile | Octahedrite with considerable rutile | Octahedrite with some rutile |
| Hiding power (sq cm per gram) | 105 | 124 | 147 |
| Oil absorption (parts oil per 100 parts pigment) | 36 | 32 | 31 |
| Water insoluble salts (calculated as per cent $Na_2SO_4$) | 0 | 0.24 | 0.45 |

applicable to mixtures of titanium dioxide and such extenders as barium sulfate and calcium sulfate. In an actual operation, 30 parts by weight of a pulp obtained by hydrolytic precipitation from ilmenite solutions, and containing 10 parts of titanium dioxide, was intimately mixed with 0.05 part potassium carbonate and heated at 800° to 1000° C for 2 hours. The original pulp contained sufficient residual sulfuric acid to convert the carbonate or hydroxide conditioning agents to sulfates. After calcination the product was washed to remove the potassium compounds. However, if the pigment was later ground by the wet process, all soluble compounds were extracted as an incidental phase of this operation. The improvement in color of the finished pigment was found to be a function of the alkali metal and was independent of the acid radical associated with it.

A pigment of improved brightness and tinting strength was produced by adding 0.5 to 2.5 per cent of an alkaline earth metal chloride to the hydrous titanium oxide before calcination.[6] Monk[7] added an oxide of aluminum, gallium, indium, silicon, germanium, or tin, together with an oxide of beryllium, magnesium, calcium, strontium, barium, or zinc, to hydrolytically precipitated titanium dioxide before calcination to neutralize colored impurities and yield a pigment of improved color and brightness.

Rhodes[8] calcined hydrous titanium oxide in a similar manner with 1 to 2.5 per cent of an alkali or alkaline earth metal titanate to produce a pigment of greater opacity, better miscibility, and lower oil absorption.

Untreated pigments usually have a yellow tone which is undesirable for many purposes. This discoloration may be intensified by a number of factors, such as the nature of the ore used, the conditions of manufacture, and impurities picked up during processing. The intensity of the yellowish cast generally increases to a maximum on exposure to actinic light. Although this undesirable coloration may be neutralized by blending with a pigment of similar composition having a bluish tone, the brightness or total reflectance of the product would be lowered.

White pigments having a neutral or bluish cast may be obtained without sacrificing brightness and without adversely affecting other essential properties by mixing with the hydrous oxide, before calcination, 0.001 to 0.04 per cent, based on titanium dioxide content, of one or more of the elements tungsten, molybdenum, erbium, lanthanum, tin, palladium, platinum, and thorium as convenient salts.[9] Products obtained in accordance with this procedure had a high stability to light, that is, they did not discolor on exposure to actinic rays. Working with ilmenite of the Norwegian type which had a strong tendency to give titanium dioxide possessing a relatively high degree of discoloration, a higher proportion of the conditioning agent was required.

For example, a solution of sodium tungstate containing 80 g of tungsten was added to a washed paste of 1000 kg titanium dioxide. After thorough mixing, the product was dried, calcined at 850° to 1000° C, and pulverized, to yield a final pigment having a gray-blue tone. Other essential pigmentary properties were not influenced appreciably by the treatment.

In a similar manner, pigments with a neutral white or slightly bluish cast were obtained by calcining hydrous titanic oxide in the presence of a very small quantity of a compound of tungsten or molybdenum, and subjecting the product, during or after the calcination step, to the action of a reducing gas such as hydrogen or carbon monoxide at an elevated temperature for a short period of time.[10] The effect was also produced by adding a small amount of a carbonaceous material that liberated a reducing gas on ignition. A desirable bluish tone was developed by exposing a pigment containing 0.001 per cent tungsten to the action of hydrogen at 900° C for one fifth second.

According to Heaton,[11] the yellowish color which often develops during the calcination step is not due, as originally thought, to the presence of impurities, but rather to the formation of rutile. This conversion does not take place if a small proportion of the titanium is present as a phosphate compound. A pigment of improved quality was obtained on refining crude titanium dioxide by a process involving calcination with phosphoric acid.[12] In an example, a pulp of hydrous oxide containing small amounts of iron compounds and sulfur trioxide was agitated with a mixture of sulfuric and phosphoric acids so that the final mix contained 2.5 per cent phosphorus pentoxide. After preliminary reaction the mass was ground with 1.4 parts sodium chloride for each part of titanic oxide, and the mixture was heated to 300° to 700° C. Free hydrochloric acid was given off. The ignited mass was cooled, crushed, and lixiviated, and the residue was dried and ground to obtain a white or light-colored pigment containing combined phosphorus.

Washburn [13] produced a soft white pigment by calcining hydrous titanic oxide or a composite mixture with calcium sulfate or barium sulfate in the presence of a small proportion of phosphoric acid or calcium phosphate at a temperature below the fusing point. The wet pulp was mixed with phosphoric acid in the proportion of about 15 parts phosphorus pentoxide to 100 parts titanium dioxide and calcined at 950° C for 2 hours to drive off the water and effect crystallization. Alternatively, the hydrous oxide was first dried, the phosphoric acid was added, and the mixture was calcined as before.

Titanic oxide formed by the thermal hydrolysis of sulfate solutions carries down with it most of the phosphate present so that instead of adding phosphoric acid to the pulp in a separate step it may be added in the proper proportion to the ilmenite solution before precipitation.[14] The phosphate acquired in this manner is not removed by washing with water. Phosphoric acid, phosphates, chlorides, fluorides, and organic compounds may be added to coalesced barium sulfate composite pulp before calcination as sintering agents to improve the color and other properties of the resulting pigment.[15] Alkali metal compounds and neutralizing agents may also be incorporated, in accordance with conventional practice. The color, brilliance, and covering power of alkaline earth sulfate composites were improved by the incorporation of boric acid or a borate,[16] as of calcium or barium, before calcination. Hydrous titanium oxide, as obtained by the thermal hydrolysis of ilmenite solutions, was freed from residual sulfuric acid by neutralization followed by washing. For the successful operation of the process it was necessary that the pulp be slightly alkaline rather than acid. Boric acid or a borate (calcium borate) in the proportion of 2 to 4 parts by weight to 100 parts titanium dioxide was added and the constituents were thoroughly mixed. The suspension was dewatered and ignited at 700° to 900° C, depending upon the time of exposure to develop pigmentary properties.

Allan [17] calcined titanium oxide or titanates with a small proportion of a compound of fluorine (fluotitanic acid, sodium fluoride, magnesium fluorotitanate, or hydrofluoric acid), which does not volatilize immediately at the temperature employed. Hydrofluoric acid reacted fairly rapidly to form a titanium fluoride compound in which the fluorine was fixed. In an example 1 g hydrofluoric acid as a 60 per cent solution was added to 100 g titanium dioxide of the hydrous form as a slurry. The mixture was stirred thoroughly, dewatered, and calcined at 850° C for 2 hours to produce an extremely soft white pigment having good covering power and a slightly acid reaction.

Improved properties were obtained by carrying out the calcination in an atmosphere containing vapors of zinc, antimony, cadmium, aluminum, or lead or their compounds.[18] These metallic agents were effectively introduced into the furnace atmosphere as suspensions in the fuel oil. Similar results were obtained by heating hydrous titanium oxide at 800° to 1000° C with metallic aluminum, chromium, iron, vanadium, or rhenium, or with the corresponding oxide and a trivalent titanium compound.[19] According to another modification, hydrous titanium oxide in a finely divided state of subdivision was heated at 800° to 1000° C to effect crystallization and develop pigmentary properties.[20] This operation was conveniently carried out by spraying the material directly into a furnace maintained at the proper temperature. Residual sulfuric acid was removed from the pulp by neutralization, followed by washing before the calcination step. The method proved to be equally applicable to composite products, particularly those containing alkaline earth metal sulfate extenders.

To eliminate the necessity of regular calcination, an aqueous suspension of purified titanic acid obtained by the hydrolysis of a dilute solution was autoclaved at 375° to 500° C and the corresponding pressure.[21]

By heating hydrous titanium oxide (precipitated by the thermal hydrolysis of sulfate or chloride solutions) under pressure with water near the critical temperature, in the presence of 0.01 to 0.02 equivalent of hydrochloric or sulfuric acid or an acid salt, the desired pigment properties were developed and the usual high temperature calcination step was not necessary.[22] Autoclave pressures corresponding to the vapor pressure of the solution at 250° to 375° C, or even higher, were employed.

Ancrum and Oppegaard[23] obtained pigments having a high brightness, a neutral to bluish tone, and an exceptional fastness to light by calcining the hydrous oxide with not more than 0.1 per cent antimony as a convenient compound. Although the commercial trioxide proved quite satisfactory and economical, any compound of antimony may be used. This treatment was also effective in the presence of extenders and other conditioning agents such as alkali metal salts, phosphoric acid, fluoride compounds, and borates. Furthermore, antimony compounds allowed more latitude during calcination; that is, higher temperatures could be employed without injurious effects. A washed hydrolysis product calcined with 0.01 to 0.03 per cent antimony trioxide, based on titanium dioxide, gave pigments with a neutral white to bluish cast, possessing particularly good brightness, fastness to light, and tinting strength.

Similar improvements in tinting strength and color were obtained by incorporating a small proportion of a compound of a rare earth element having atomic number between 59 and 71, or of cerium, lanthanum, or yttrium,[24] with the hydrolytically precipitated pulp before calcination. Proportions of 0.02 to 0.50 per cent gave optimum results. Phosphoric acid or phosphates in the ratio of 0.10 to 2 per cent, expressed as phosphorus pentoxide and based on the titanium dioxide, were also added to the hydrolysis solution. In a particular case, 0.10 g cerium sulfate was added to 1 liter of a solution containing the equivalent of 15 per cent titanium dioxide, 29 per cent sulfuric acid, and 14 per cent ferrous sulfate, and the resultant solution was subjected to thermal hydrolysis. Practically all the cerium and phosphate compounds came down with the titanium. The precipitate was filtered, washed, and calcined according to conventional practice.

Similarly, titanium dioxide pigments of good brightness, tone, fastness to light, and softness were produced by calcining hydrous titanium oxide obtained by the thermal decomposition of ilmenite solution after incorporation of a small proportion of a compound of columbium or tantalum.[25] From 0.01 to 2 per cent of the conditioning agent was employed, and the addition could be made at any time prior to calcination. Such compounds or minerals were mixed with the ilmenite ore before reaction, or to the solution in the form of a soluble salt before precipitation, or to the washed hydrolysis product. Frequently, if the calcination was carried out without treatment to such a degree as to obtain maximum tinting strength and hiding power, these desirable properties were accompanied by a tendency of the pigment to acquire a grayish, yellowish, or reddish tone, and in some cases to exhibit a lack of fastness to light and an excessive hardness, thus rendering the product difficult and expensive to grind. From 0.01 to 0.10 per cent columbium effectively overcame these undesirable effects.

In an example, 1000 g ilmenite was mixed with 510 ml of 93 per cent sulfuric acid, 2 ml of 85 per cent phosphoric acid, and 0.2 per cent columbite, based on the titanium dioxide content of the ore, and the charge was added to 370 ml of 93 per cent sulfuric acid, previously heated to 130° C in a cast-iron pan. The mixture was heated, with stirring, to fumes of sulfur trioxide, and 40 ml of water was added to initiate the reaction. After the reaction had gone to completion, the mass was baked for 30 minutes at 200° C, cooled, and leached with water at 60° C. The solution was adjusted to a specific gravity of 1.50 and reduced with scrap iron to a titanous oxide content of 3 gpl, after which it was clarified and cooled to 27° C to remove part of the ferrous sulfate by crystallization. The mother liquor was concentrated by vacuum evaporation to a gravity of 1.60 and hydrolyzed in an autoclave by heating for 80 minutes at a temperature corresponding to a superpressure of 20 pounds. After thorough washing, the pulp was treated with 0.6 per cent potassium carbonate, calcined at 970° C for 3 hours, and finished by the usual methods.

Titanium dioxide treated with columbium in accordance with this procedure could be subjected to a more intense calcination to develop maximum tinting strength without adversely affecting the other properties. Furthermore, the temperature range for optimum calcination was considerably increased, and values up to 1100° C could be employed.

Ryan and Cauwenberg [26] incorporated 2.5 to 10 per cent of a carbonaceous material such as sugar, dextrin, starch, or oil with hydrous titanium oxide and baked the mixture at 200° to 400° C before calcination. For convenience and economy the two steps were carried out in the same furnace. The method proved applicable to composite pulps containing alkaline earth metal sulfates. In a specific operation, a precipitate obtained by the thermal hydrolysis of a sulfate solution of ilmenite was thoroughly washed and mixed with 5 per cent, based on the dry weight, of fuel oil. The composition was then dried and baked at 250° C for 6 hours, and finally calcined at 750° to 900° C. Pigments processed in this manner were pure white and contained a minimum of impurities.

According to a related procedure, the washed pulp was treated with 2 to 20 per cent of a carbonaceous material such as charcoal, starch, sawdust, or oil, and then calcined at 750° to 900° C.[27] The carbonaceous material was burned off in the processing, leaving a pure white titanium dioxide containing a minimum of impurities and having excellent properties.

Pigments of very fine particle size [28] were said to be obtained by treating freshly precipitated titanium dioxide with gelatin, starch, or dextrin to transform part of the product into the colloidal state before calcination. According to a similar procedure, the hydrous oxide was dispersed with a peptizing agent, such as hydrochloric acid, nitric acid, or barium chloride, and coagulated.[29] The treated pulp was then washed, dried, and calcined above 840° C to obtain porous particles of titanium dioxide having a high opacity and soft texture. In incident light the pigment presented a certain number of plane faces inclined in various directions. By employing barium chloride as dispersing agent and barium carbonate as coagulating agent, the corresponding titanate was produced.

Crystalline titanium dioxide was produced by heating the hydrous material previously dried at 110° C in a compound of the same crystalline structure (magnesium sulfate) at a temperature below the melting point, 900° C.[30] By calcining a wet mixture of titanium dioxide and barium sulfate, Buckman [31] obtained a pigment in which these components were combined or heat compacted.

White titanium dioxide pigments having a controlled subordinate tint [32] were obtained by heating the regular calcined product to 900° to 1000° C and cooling in air to 750° C under carefully regulated conditions. If such cooling took place, in a few seconds a pronounced blue cast developed, while if the temperature was gradually reduced, in 5 minutes or longer a decided yellow tone resulted. By bringing about the cooling at rates between these limits, any desired tone from distinct blue to distinct yellow was obtained. Commercially, the titanium dioxide was discharged from the calciner at about 900° to 1000° C, and controlled cooling was effected easily and economically at this stage.

According to Parravano and Caglioti,[33] the pigmentary properties of titanium dioxide depend upon the sulfuric acid content of the hydrolysis product and the nature and amount of mineralizers added, and particularly upon the temperature and time of calcination. Such products crystallized at about 500° C as anatase, and although some samples were transformed into rutile by calcination at 825° C, others showed the spectrum of anatase even after calcination at 950° C for 1.5 hours. The velocity of the process of crystallization was dependent at all temperatures on the sulfur trioxide content of the hydrolysis product. Conditions which resulted in the formation of rutile did not in general produce good pigment properties. The temperature of drying the hydrous oxide before calcination was reported to have no influence on the properties of the finished pigment.[34]

Hydrous titanium oxide of the anatase modification is generally treated with a potassium compound before calcination to obtain higher tinting strength and improved color, and the rutile product may be improved by such treatment, particularly if the upper temperature range is employed. Other conditioning agents may also be included along with the alkali metal compound to bring out special and specific properties. Alkaline earth metal sulfate base composites are quite resistant to sintering, and the potassium treatment is in general not essential.

According to a commercial method,[35] the washed and treated titanium dioxide cake is fed directly from the continuous filters to a rotary, oil-fired furnace where it is completely dried and finally brought to a temperature of 1000° C at the discharge end. Calcination is one of the most critical steps in the manufacturing process. Too high a temperature or too long an exposure to a lower temperature may cause the product to become gritty and may injure the color, while undercalcination, although generally productive of good color, does not bring out the maximum tinting strength and covering power. Passage through the kiln requires considerable time—in the neighborhood of 24 hours—and control of both temperature and atmosphere at every point along the course is important.

The thoroughly washed precipitate obtained by the thermal hydrolysis of titanium sulfate solutions is the hydrous oxide consisting of tiny crystallites, 60 Ångstrom units in diameter, possessing the anatase crystal form. These crystallites are aggregated to form micelles 0.08 to 0.1 micron in diameter. In addition to the water content, the hydrous oxide contains some firmly adsorbed or combined sulfuric acid. Such tiny particles even of a high index of refraction material as titanium dioxide have very little tinting strength. The product is calcined to drive off the water and sulfuric acid, and at the same time fuse the crystallites into particles of 0.2 to 0.3 micron in diameter, half the wave length of visible light, to yield a pigment of maximum covering power. During calcination, the anatase may be converted to the rutile crystal form.

Hydrous titanium dioxide prepared from titanium sulfate by a hydrolytic reaction consists of flocculates of small anatase crystals.[36] Water and sulfur trioxide appear to be present as absorbed layers and in capillaries between and within flocculates. On heating the hydrate, water is lost at 150° C and sulfur trioxide at 650° C. Crystallite and particle size growth starts at 600° C, and the transformation of anatase to rutile takes place in the range of 700 to 950° C. According to Latty,[37] hydrolysis of titanium sulfate solutions gives a product of flocculates 0.1 to 0.8 micron in diameter which results from the aggregation of micelles 0.08 to 0.10 micron in diameter. Micelles are composed of crystallites 60 Ångstrom units in diameter. During calcination the micelles persist unchanged in size. However, the crystallites at 750° C are 200 to 400 Ångstrom units in diameter. At this temperature crystallites coalesce to form a pigment particle. In the presence of alkali, the rate of growth is accelerated.

An electron diffraction study of the transformation of titanium dioxide in thin layers by changes in temperature was carried out by Conjeaud.[38] The layers of titanium dioxide with a thickness of 500 to 1000 Ångstrom units were obtained by vacuum deposition on cold surfaces. Amorphous titanium dioxide is formed initially and is stable up to 550° C. Above this temperature, anatase and mixtures of anatase and rutile are formed. Above 800° C only rutile is stable. No brookite formation was observed. Titanium dioxide powder is compacted into cylinders with a pressure of 200 kg per square centimeter, sintered, and examined with an electron microscope for physical changes. Transformation of anatase to rutile accelerates the agglomeration.[39] The porosity increases, but does not impede the densification. In the sintering of titanium dioxide, three temperature regions are recognized. Below 900° C, densification and growth of particles do not take place. Between 900 and 1100° C, densification proceeds, but the diameter of the particles does not increase. Above 1100° C, densification does not occur, but the particle diameter increases remarkably. There are three stages in the isothermal sintering of titanium dioxide. In the first stage, the increase of particle diameter, contraction, and densification takes place to a large extent. In the second stage, little increase in particle diameter occurs, and the contraction and densification are hardly recognized. In the third stage, the particle diameter increases remarkably, but contraction and densification take place hardly at all. With sufficiently large compacting pressure, these stages cannot be distinguished from one another.

During the calcination of titanium dioxide pigments in horizontally inclined rotary kilns, a liquid fuel is introduced through a fuel line protected by cooling jackets into the gas space within the kiln at a point where the temperature is well below the maximum value.[40] By this procedure the output of the kiln may be increased by 25 to 50 per cent. Hydrous titanium dioxide is heated in a rotary kiln or spray drier up to 800° C to remove the volatile materials, disintegrated and screened if necessary, and calcined in a fluidized bed.[41] Fluidization is effected by passing a hot nonreducing gas through the bed at a velocity of 0.5 to 5.0 feet per second depending on the particle size of the titanium dioxide. A lower gas-to-pigment temperature gradient under fluidized conditions results in improved brightness. The fluid calcination of a rutile hydrate treated with 1 per cent zinc oxide is effected in 20 minutes at 850° C in an 18-inch bed. Hot gases are produced by burning 1 kg of oil with 27.8 cubic meters of air.

Anatase titanium dioxide pigments of improved hiding power, tinting strength, and dispersing properties in paint vehicles is produced by a process including double calcination.[42] Preliminary roasting of the titanium sulfate hydrolyzate is carried out in the presence of a fusible alkali metal compound until the titanium dioxide is converted to anatase crystals which are too small to possess pigment properties but which will allow a second calcination. The alkali metal salts are completely removed by washing, after which the final calcination is controlled to develop the anatase crystals fully and to prevent the formation of rutile crystals, which would result in a coarse product of low dispersability in paint vehicles and poor color.

Production of titanium dioxide is increased with a saving in fuel and dust collection by recycling 20 to 60 per cent of the hot exhaust gas back through the rotary calciner.[43] The temperature of the product is increased from 400 to 700° C in the last 100 minutes in the calciner to the discharge temperature of 800 to 1050° C. Gases from the combustion chamber enter the discharge end of the kiln at 1000 to 1300° C. The atmosphere of the calciner contains 2 to 10 per cent oxygen on the dry basis and 30 to 50 per cent water vapor.

## RUTILE FROM ANATASE

Hydrolytically precipitated titanium dioxide of the anatase crystal modification may be converted to the rutile form having desirable pigment properties by calcining at 850° to 1100° C with a small proportion, generally 0.1 to 2.0 per cent, of zinc oxide or other compounds which will yield the oxide at the temperature employed;[44] with lithium chloride, titanate, or sulfate;[45] with inorganic compounds of magnesium, such as the metatitanates, which have the crystal characteristics of the spinel, corundum, ilmenite, phenacite, or sodium chloride crystallographic group;[46] or with antimony oxide[47] or a compound which will yield the oxide at the temperature employed.

Rutile of improved pigment properties was obtained by treating the anatase hydrolysis product with a small proportion of a mixture of zinc oxide, and magnesium oxide,[48] or with a mixture of antimony trioxide with a compound of zinc, magnesium, calcium, or barium before calcination.[49]

Hydrolytically precipitated titanium dioxide which would normally yield anatase pigment was converted to rutile of improved pigment properties by direct calcination after mixing with a separately prepared compound of similar composition but which on calcination showed X-ray diffraction patterns of rutile.[50] Hydrous oxide, as obtained from sulfate solutions, was seeded with 2 to 10 per cent of a similar product obtained from chloride solution before calcination to aid and direct conversion to the rutile form. Such a mixture, calcined at 900° to 1000° C for 3 hours, showed 98 per cent rutile. By heating a mixture of 70 parts of the hydrolysis product of anatase with 30 parts of a hydrolyzate of the rutile form in an autoclave at superatmospheric pressure in a 50 gpl hydrochloric acid solution, the entire product was converted to rutile of pigment grade without subsequent calcination.[51] The product was filtered, washed, and dried. Hydrolytically precipitated anatase was also converted to rutile by calcining first at 1000° to 1300° C under reducing conditions, and then recalcining at 600° to 850° C in an oxidizing atmosphere.[52]

Booge[53] calcined an anatase hydrolysis product precipitated from sulfate solutions by the Blumenfeld process (U.S. reissue 18,854), first at a temperature below 1000° C in the presence of potassium sulfate or sodium sulfide to inhibit or prevent conversion to the rutile crystal form. The agent was removed by washing, and the product was recalcined at a higher temperature but below 1100° C with approximately 1.0 per cent zinc oxide to effect conversion to the rutile crystal form. Rutile pigments of this type showed markedly improved resistance to chalking and fading.

Peterson [54] found that by treating the hydrolytically precipitated anatase titanium dioxide with 0.5 to 0.75 per cent of a mixture of potassium sulfate and sodium sulfate and 0.25 per cent zinc sulfate, conversion to rutile could be effected by a single calcination at 975° C. Aluminum sulfate, magnesium sulfate, and barium sulfate could be substituted for the zinc sulfate.

Anatase titanium dioxide was also converted to the rutile crystal modification by a double calcination without accelerating agents.[55] Hydrous titanic oxide, as obtained from sulfate solutions, was first calcined in the usual manner with a negative catalyst (0.3 to 1.0 per cent of potassium sulfate) which retarded or inhibited rutile formation to develop pigmentary properties, and after reduction of the added agent by washing to less than 0.1 per cent, the pigment was recalcined to develop the rutile crystal structure.

From X-ray diffraction studies of titanium dioxide prepared and heated in different ways, Weiser, Milligan, and Cook [56] concluded that the rate of transformation of the more labile and less stable anatase modification into the less soluble and more stable rutile modification is opposed by two factors: retardation of the change by the adsorbed layer of ions on anatase which reduces its rate of solution, and the acceleration of the change by the ionic environment in which anatase is more soluble. Anatase formed by hydrolysis of titanic chloride and titanic nitrate solutions changes fairly rapidly into rutile at 100° C. The rate of change is decreased enormously in the presence of a large excess of alkali chloride or nitrate because of the protective action of adsorbed ions, and is speeded up in the presence of an excess of hydrochloric or nitric acid, since the solvent action of the acid predominates over the protective action of adsorbed anions. Anatase formed by hydrolysis of titanic sulfate solution is not transformed into rutile in a reasonable time even in strong sulfuric acid solution, since strong adsorption of sulfate ions is a predominating stabilizing factor. Schlossberger [57] reported that the abnormal density-temperature curve of titanium dioxide indicated that sulfur trioxide was trapped in the anatase lattice and acted as a stabilizer in the transformation to rutile.

A rutile pigment was obtained by calcining washed precipitated titanium dioxide at 1000° C with one sixth its weight of basic chloride $TiOCl_2$ and a small proportion of ferric ammonium sulfate.[58] Raw pigment anatase form of titanium dioxide is converted to the rutile modification by a disintegration treatment to reduce the particle size, followed by calcination at 750° to 1000° C in the presence of 2 to 10 parts of added rutile and 0.5 to 1.0 per cent of an alkali metal salt to each 100 parts of anatase. The added rutile acts as a conversion promoter.

Neilsen and Goldschmidt [59] eliminated the residual acid by treating the washed pulp with calcium hydroxide or calcium carbonate to form insoluble sulfates, and then heated the mixture at 900° to 1050° C to develop crystalline structure. By employing an excess of the calcium compound and prolonging the calcination, a pigment containing calcium titanate in cryptocrystalline form and an amorphous calcium sulfate was formed. A typical product contained 84.0 per cent crystalline titanium dioxide, 3.8 per cent calcium titanate, and 12.2 per cent calcium sulfate. Without an excess of the basic compound and without prolonged heating, titanium dioxide in the cryptocrystalline form was obtained. Alkaline earth metal carbonates as neutralization agents had the advantage of liberating carbon dioxide gas so as to render the product porous.[60]

Carteret and Devaux [61] obtained an acid-free product by treating the precipitated hydrous oxide with a solution of an alkaline earth metal chloride and then washing the reaction product with milk of lime or baryta to remove the hydrochloric acid. According to a combination method,[62] a part of the closely held sulfuric acid was neutralized by treatment with an alkali metal hydroxide or carbonate, and the remainder with compounds such as oxides, carbonates, and chlorides of alkaline earth metals and lead (barium chloride), which yield insoluble sulfates. Residual acid was also removed by calcining the pulp in a constantly oxidizing atmosphere in the presence of calcium chloride, ammonium chloride, or other compound capable of forming insoluble or volatile sulfates.[63]

In the sulfate process the crystallites of titanium dioxide formed at hydrolysis are strongly stabilized in the anatase phase by sulfate ions which are absorbed in the anatase structure.[64] Other polyvalent ions such as phosphate usually introduced as impurities from the ore exert even greater effect. The transformation of anatase to rutile is influenced by many factors, the more important of which are the nature and history of the anatase crystallites and the impurities present. High temperature, although effective in bringing about the transformation, causes particle growth and favors the removal of oxygen from the rutile lattice. Although the crystal can lose oxygen to the extent of 0.5 per cent without change in the unit cell size or structure, the loss of even a trace of oxygen results in a bluish color which adversely affects the brightness and color of the pigment. The conversion of anatase to rutile during calcination of the hydrate is catalyzed by the addition of rutile promoters. An effective procedure is that of inoculating or seeding the solution before hydrolysis or the hydrate before calcination with minute rutile crystallites. The effectiveness of this type of promoter is influenced by the particle size and the amount used as well as the conditions during calcination.

In the older Raymond type mill, the oil absorption of the pigment can be decreased to ranges suitable for enamels and other uses. However, the demand for pigments of improved quality has brought into use the high speed hammer mills and various fluid energy mills using high pressure steam or air. All of these means of dry particle size reduction are utilized as single operations or in combination to produce final pigments suitable for the different requirements. Regardless of the method of milling, size classification is a vital factor. Systems using intense centrifugal fields to effect separation in the fine size range are standard equipment. Most machines are provided with integrally fabricated classifiers. Research has shown that some impurities even in the concentrations below the threshold sensitivity of the spectroscope can exert an influence on the whiteness of the titanium dioxide. This has been observed in the study of single rutile crystals of large size.

Spectroscopically pure anatase converts to rutile immeasurably slowly at temperatures below 610° C.[65] Above 730° C the conversion is extremely rapid. The conversion is second order with respect to the remaining anatase and is characterized by an activation energy of 100 kilogram calories per mole. Titanium dioxide films on glass began crystallization at 250° C.[66] Above 400° C anatase forms on alkali-free glass and brookite on alkali glass. The brookite structure is induced by the presence of alkali ions. Hydrated titanium dioxide obtained by the hydrolysis of ethyl titanate at room temperature and anatase precipitated from a titanyl sulfate solution with dilute aqueous ammonia at the boiling point, heated for 30 hours at 200 to 540° C and under pressures of 200 to 1000 atmospheres showed only anatase, but at higher temperatures rutile was produced.[67] Brookite can be prepared, together with anatase, by hydrolysis of ethyl titanate or butyl titanate at 100° C under normal pressure. The crystallization is improved by heating for some weeks at 500° C. Above 600° C rutile is formed.[68] Studies of the anatase-rutile transformation by Shannon show that the rate of transformation is accelerated in a hydrogen atmosphere and retarded in vacuo. This information, interpreted in terms of the defect structure of titanium dioxide, supports other data indicating that the mechanism by which nonstoichiometric rutile is formed depends on the atmosphere in which the change occurs.

According to Knoll and Kuhnhold [69] only 0.1 per cent by weight of a stabilizing anion is necessary to prevent the transformation of anatase to normally stable rutile modification of titanium dioxide over a wide temperature range. Dilute hydrofluoric acid, sulfuric acid, hydrochloric acid, and nitric acid were added to an aqueous solution of titanium, and ammonium hydroxide was added to effect precipitation. Thermal analysis and X-ray analysis of the precipitates showed that the temperature of transformation of anatase to rutile varied from 620 to 1010° C. Nucleation to form rutile was hindered by the increasing stabilization effect of the nitrate, chloride, sulfate, and fluoride ions in that order. Hydroxide ion is as effective as nitrate ion, and it also stabilizes the brookite structure of titanium dioxide.

The phosphorus, vanadium, and chromium impurities are removed from hydrolytically precipitated titanium dioxide so that it can be calcined to high quality rutile pigment at relatively low temperature.[70] Titanium dioxide precipitated from sulfate solution of ilmenite obtained from nelsonite ordinarily contains 0.4 to 0.8 per cent phosphorus pentoxide. Phosphorus as well as boron and silicon compounds cause vitrification of the titania during calcination and inhibit the transformation from anatase to rutile. A number of impurities such as chromium, vanadium, and iron apparently enter the rutile crystal lattice during calcination, thereby causing abnormal crystal distortion and reduction of brightness of the pigment.

Sodium hydroxide is added to the washed hydrolysis product in an amount sufficient to react chemically with the phosphates and other impurities, and also with the 7 to 10 per cent of closely held sulfuric acid. To increase the rate of reaction, the slurry is heated from 60° C to the boiling point. The soluble compounds of the impurities are removed by washing. During the process the titanium dioxide may become deflocculated. Such a product is difficult to process and is unfit for calcination. Reflocculation is accompanied by the addition of sulfuric acid to produce a pH of 2 to 4, followed by boiling for 1 hour. The sulfuric acid is removed in the calcination step. From 90 to 98 per cent of the phosphorus pentoxide is removed; the chromium and vanadium are decreased to less than 20 parts per million; iron is less than 0.001 per cent of the titanium dioxide. By employing 0.5 to 3.0 per cent of a separately prepared rutile seed, calcination at 875 to 950° C effects complete conversion to rutile. The seed may be added to the hydrolysis product before or after the alkali digestion treatment.

The desired rutile form of titanium dioxide pigment is formed by heating anatase for 1 hour at 925° C if the phosphorus pentoxide content of the original titanium sulfate is reduced to less than 0.025 per cent based on the titanium dioxide before hydrolysis. There is no need of promoters or endpoint reagents.[71] If the original ore is smelted to produce pig iron, the titanium is concentrated in an acid slag of very low phosphorus content, provided the reduction of the iron compounds is sufficiently complete to produce a slag of low iron content.

Smaller particle size and larger surface area favor the transformation of anatase to rutile.[72] The rate of the transformation decreases progressively with the amount of sulfate ion with the anatase. There is an increase in activation energy with increase in the concentration of sulfate ion. Samples of anatase type titanium dioxide prepared by seven different procedures were heated at temperatures up to 1000° C to study the transformation to rutile.[73] Results were determined by X-ray diffraction patterns. The transition temperature from anatase type to rutile type differs among the samples from 400 to 1000° C. It is concluded that titanium dioxide produced from a solution containing chloride ion shows low transition temperature, while the presence of sulfate ion in the solution before hydrolysis produces titanium dioxide of high transition temperature. However, the density values indicate that the transition phenomenon is not sharp, but depends on the heating temperature and time. The transition is promoted by the presence of hydrochloric or sulfuric acid, while boric acid inhibits the transition. Lithium chloride, zinc chloride, and zinc oxide are the most effective agents for lowering the transition temperature.

A study of growing crystals of titanium dioxide on heating was carried out by electron shadow microscopy and diffraction.[74] Anatase powder was converted to rutile by heating at 1100° C for 5 hours. The rutile consisted of rods 1 to 2 microns long and 0.2 micron wide, spherical particles of 0.5 micron diameter, and smaller particles of the same size of the original anatase. After 24 hours' heating, almost all of the particles became rods elongated in the direction of the tetragonal c axis. Impurities such as lead, tin, and other metallic elements which may be occluded from the sulfuric acid used to prepare the anatase, accelerated the crystallization. Some other titanias only became sintered on heating under the same conditions if they contained less impurities than 1 part in 10,000.

The anatase form of titanium dioxide is converted to the rutile form by adding 0.1 to 5.0 per cent of boron phosphate and calcining the mixture at 870 to 960 ° C for 1 to 2.5 hours.[75] Zinc oxide, antimony trioxide, magnesium oxide, and lithium oxide or their mixtures also promote the conversion. If titanium dioxide or a compound yielding titanium dioxide on thermal decomposition is calcined with tungsten oxide or a mixture of tungsten oxide with one of several other metal oxides, crystalline titanium dioxide base solid solutions are prepared.[76] Thus, 5000 g of anatase, 0.020 g of lithium oxide (calcined from lithium carbonate), and 0.466 g of tungsten oxide are mixed well and calcined for 0.5 hour at 800° C, then pulverized and further calcined at 1000° C for 0.5 hour to obtain a white to light gray pigment. According to X-ray analysis the pigment has the rutile structure. The anatase to rutile transformation and grain growth during calcination are promoted by small additions of nickel oxide, cobalt oxide, manganese oxide, ferric oxide, and copper oxide.[77] Additions of sodium oxide and tungsten oxide retard the transformation and have no effect on grain growth. Molybdenum oxide has little effect on transformation but strongly promotes grain growth. Chromic oxide promotes slightly both effects. The method of preparation of the titanium dioxide has a strong effect on transformation. Transformation decreases with an increase in the partial pressure of oxygen in the atmosphere. Grain growth is not affected by oxygen, air, argon, or vacuum, but is promoted by hydrogen.

Anatase titanium dioxide readily forms powders of 0.1 micron particle diameter that undergo transition to rutile of 0.7 to 0.8 micron diameter at 1100° C.[78] Both anatase and rutile were compressed to 200 kg per cubic meter and then heated at 900 to 1100° C. At 900° C the anatase showed no appreciable change. The diameter of the particles increased somewhat and the density increased at 1000° C. At 1100° C the diameter increased markedly, with no change in density. The transition to rutile was promoted by the addition of 1 mole per cent of lithium chloride, zinc sulfate, or manganous chloride. Sintering of the anatase at constant temperature proceeded in three stages: an increased particle diameter and density; no external change; a large increase in diameter without change in density. If sintering takes place at high temperatures, or under high pressure, the second stage does not occur.

Rutile titanium dioxide pigment is prepared by a double calcination of an anatase precipitate obtained by the hydrolysis of sulfate solution.[79] To an aqueous pulp containing 200 g of titanium dioxide prepared by the usual hydrolysis of titanium sulfate solution, 4.2 g of potassium carbonate and 20 g of a seeding agent are added. After drying, the mixture is calcined at 850° C for 1 hour. After milling, the product is slurried in water to 15 per cent solids, and 5 ml of 20° Bé hydrochloric acid is added. The solids are filtered, washed, and calcined for 2.5 to 3 hours at 975° C. By this method from 50 to 80 per cent of the anatase is converted to rutile. Lower temperatures are used than in previous methods. Pigment grade anatase containing more than 98 per cent titanium dioxide to which 0.1 to 0.3 per cent alumina is added in the form of a soluble salt is calcined for 6 hours at 1250° C to produce rutile crystals 1 to 2 microns long.[80] The product has uniform electric properties, a dielectric constant of 100 or more, a power factor of 0.1 or less measured at 1 megacycle, and a resistivity of $5 \times 10^{12}$ ohms per centimeter.

## CALCINATION NUCLEI TO YIELD RUTILE

Anatase hydrous titanium dioxide obtained from the hydrolysis of titanium sulfate solution can be converted to rutile more effectively if a seeding material prepared by treating an alkali metal titanate with a limited quantity of hydrochloric or other monobasic acid is added before calcination.[81] In an example, hydrous titanium dioxide containing a small proportion of sulfuric acid obtained by hydrolytic precipitation from a sulfate solution and containing 100 g of titanium dioxide is diluted with water to 30 per cent solids. To this slurry is added 150 g of solid sodium hydroxide with stirring continuously. The mixture is heated for 2 hours at 85 to 90° C with continuous stirring and at constant volume. The titanate thus formed is diluted to 1.5 times with water, washed twice by decantation, filtered, and washed free of sulfates. This filter cake is slurried in 185 ml of commercial 20° Bé hydrochloric acid, diluted with 1 liter of water, and boiled at constant volume for 1 hour. In another example, the seed suspension is added without neutralization or coagulation to the hydrous titanium dioxide before calcination. Hydrolytically precipitated titanium dioxide containing 200 g of the dioxide is mixed with 4 g of potassium carbonate. The mixture is dried and heated for 1 hour at 850° C after which it is milled and slurried in water. To the slurry is added 85 g of the hydrochloric acid suspension containing 6 g of titanium dioxide. The mixture is filtered and washed. After drying, the resulting filter cake is calcined by slowly raising the temperature from 300 to 975° C in 5 hours and held at 975 to 1000° C 2 hours. A rutile product of good pigment properties results. In still another example, the seed suspension is coagulated before use.

In the preparation of seeding material for the conversion of hydrous titanium dioxide to rutile during calcination by treatment of an alkali metal titanate with hydrochloric acid or nitric acid in an amount only 20 to 50 per cent of that required to produce the tetra salt, an improved seed is obtained if from 0.1 to 5.0 per cent of the titanium is in the trivalent state.[82] Pigment of greatly improved color is obtained. The trivalent titanium can be produced in place by reduction with zinc, or it can be added in the form of a separately prepared concentrated solution of titanous chloride. To prepare the titanate, hydrous oxide precipitated hydrolytically from sulfate solution and containing a small proportion of residual sulfuric acid is treated with 1.5 parts by weight of sodium hydroxide to 1 part of anhydrous titanium dioxide, and heated for 1 to 6 hours at 80 to 100° C. Solid flake sodium hydroxide or a concentrated solution may be employed. To the sodium titanate thus formed, water is added, which may cause some hydrolysis of the titanate. The solids are washed by decantation, and then filtered and washed on the filter. Analysis of the washed solids after calcination shows 85 per cent titanium dioxide and 15 per cent sodium oxide. Prior to the formation of the titanate, the sulfuric acid in the hydrous oxide may be neutralized with sodium hydroxide and removed by washing with water. The washed solids are treated with hydrochloric acid in sufficient quantity to neutralize the sodium oxide and to convert from 20 to 50 per cent, preferably 25 per cent, of the titanium to the tetrachloride. Titanium III ion is added to or formed in the solution, after which the mixture is diluted with water and boiled for 1 hour. X-ray analysis of seed prepared by this method shows complete rutile structure. A more active seed is obtained if the alkali metal and sulfate ions are largely removed prior to the hydrochloric acid treatment.

The rutile seed suspension thus formed may be mixed with hydrous titanium dioxide obtained by hydrolytic precipitation from sulfate solutions of ilmenite at any stage prior to the final washing before calcination, or it may be flocculated by adding sodium hydroxide or sodium carbonate to a pH of 4.5 to 7.5, filtered, and washed free from chlorides before mixing with the hydrous oxide. This special rutile seed is extremely finely divided, which probably accounts for its exceptional activity. As compared with rutile seed made in the absence of trivalent titanium, the color of the final pigment is much improved. The use of 5 per cent of the seed based on the total weight of titanium dioxide induces conversion to rutile at a lower temperature or in a shorter time, thereby producing a pigment of rutile structure having better color and higher total brightness and which disperses more readily in vehicles. Smaller amounts of seed do not effect complete conversion to rutile. Calcination is carried out at 950 to 1000° C.

According to Aagaard [83] rutile seed for the conversion of hydrous titanium dioxide to rutile during calcination can be obtained more effectively by heating orthotitanic acid with sodium hydroxide to form sodium titanate followed by treatment with hydrochloric acid or other monovalent acid. This method is more rapid and less expensive, and the resultant seed is more effective. An aqueous pulp of orthotitanic acid prepared in any manner whatsoever is treated with 1.25 parts of sodium hydroxide for each part of anhydrous titanium dioxide. Either flake sodium hydroxide or a concentrated solution may be employed. The mixture is heated at 1 to 6 hours at 80 to 100° C with continuous stirring. For reasons of economy and convenience, the orthotitanic acid is prepared from a clarified, basic sulfate solution of ilmenite, from which a part of the iron has been removed by crystallization as ferrous sulfate. A typical analysis of such a solution is specific gravity 1.42, titanium dioxide 10 per cent, ferrous sulfate 10 per cent, and sulfuric acid combined with the titanium 18 per cent.

For example, 1500 ml of a slurry at 30° C containing 150 g of hydrated lime are added to 1060 ml of basic sulfate solution of ilmenite containing 150 g of titanium dioxide at 30° C over a period of 1 hour with continuous stirring. The gypsum formed is filtered out and washed with 1500 ml of cold water. The filtrate containing the titanium and iron is diluted to 15 liters and 2000 ml of an aqueous slurry containing 30 g of hydrated lime at 30° C is added over a period of 0.5 hour with good agitation. At a pH of the slurry of 2.7 about 90 per cent of the titanium is precipitated as orthotitanic acid. This is filtered and washed free of iron and calcium and adjusted with water to a calcined solids content of 18 per cent. To this slurry is added 135 g of flake sodium hydroxide and the mixture is heated at 85 to 90° C for 2 hours with continuous stirring. Water is added to replace loss by evaporation. The titanate is filtered, washed free of sulfates, slurried in 248 ml of commercial 20° Bé hydrochloric acid, diluted with 1350 ml of water, and boiled at constant volume for 1 hour. A thoroughly washed pulp of hydrous titanium dioxide obtained by hydrolytic precipitation from a titanium sulfate solution and containing 200 g of titanium dioxide is diluted with water to 20 per cent solids content and mixed with 85 g of the hydrochloric acid suspension containing 6 g of titanium dioxide. The mixture is filtered and washed free of chlorides. After adding 0.41 g of potassium carbonate, the filter cake is calcined by slowly raising the temperature from 300 to 975° C in 5 hours, and then holding at 975 to 1000° C for 4 hours. X-ray analysis shows 90 per cent rutile. The tinting strength of the hydroclassified and dry milled pigment is 1700.

A titanium dioxide sol is flocculated with hydrolytically precipitated titanium dioxide to release the hydrochloric acid for reuse.[84] The flocculated sol is employed as calcination nuclei for the production of rutile from hydrous titanium dioxide precipitated from sulfate solutions. Recovery of hydrochloric acid is from 56 to 93 per cent. In an example, 56 parts of 38 per cent hydrochloric acid were added to 417 parts of sodium titanate, which had been neutralized with hydrochloric acid, washed, and diluted with water to 18 per cent solids. This suspension contained the equivalent of 100 parts of titanium dioxide. The suspension was heated to boiling in 72 minutes. After boiling for 10 minutes there was practically 100 per cent conversion of the titanium hydrate from anatase to rutile. The cured sol was added to 700 parts of washed metatitanic acid containing 189 parts of titanium dioxide. The coagulated product was filtered and washed with 720 parts of water to remove the hydrochloric acid. Recovery of hydrochloric acid was 56 per cent. This filtrate was fortified with new hydrochloric acid and reused. The coagulated sol containing 100 parts of titanium dioxide from the sol and 18 parts of titanium dioxide from the added hydrate was added prior to bleaching to washed hydrous oxide hydrolytically precipitated from sulfate solution and containing 3145 parts of titanium dioxide. The bleached product was washed and calcined at 900° C in the presence of 0.23 per cent potassium hydroxide and 0.07 per cent alumina to effect 95 per cent conversion to rutile.

According to Tanner [85] a seeding material prepared from titanium phosphate shows a greater activity than previously prepared seeding materials for converting anatase to rutile during the calcination of hydrolytically precipitated titanium dioxide. Titanium phosphate of high purity is precipitated by the addition of phosphoric acid or a soluble phosphate salt to a solution of a titanium salt such as is obtained by the digestion of ilmenite with sulfuric acid. After a water wash the phosphate precipitate is converted to sodium titanate by heating in aqueous sodium hydroxide at 70 to 100° C for 1 to 5 hours. The sodium titanate is washed with water and treated with hydrochloric acid or other monobasic acid in sufficient amount to neutralize the sodium oxide and convert from 20 to 50 per cent of the titanium dioxide content to the tetrachloride. This hydrochloric acid slurry is diluted with water and boiled for 1 to 3 hours, during which period the rutile seed material is formed.

A pigment with rutile structure is obtained by calcining the hydrolyzate obtained from sulfuric acid solutions of ilmenite concentrate from wherlite in the presence of 9 per cent of seeding crystals prepared from titanium tetrachloride solution.[86]

## DISPERSION, HYDROSEPARATION, AND MILLING

In the manufacture of titanium dioxide pigments, the hydrolysis step tends to form a product of small particle size, but the subsequent calcination cements these into larger aggregates by compacting and sintering. To make the product suitable for most uses these aggregates must be broken down. Several methods of milling are applicable, depending upon the type of pigment and the purpose for which it is to be employed. The relatively pure oxide for incorporation into organic vehicles is generally submitted to a wet grinding and classification process. Such products for use in aqueous media, however, are normally ground in the dry state. Dry milling is in general more applicable to all grades of composite pigments, regardless of use.

Particles of 15 micron diameter are easily observed in paint films by the unaided eye, and under favorable conditions of lighting aggregates of 6 microns can be detected. These larger particles are particularly undesirable in the finished pigment, since they prevent the formation of a smooth, unbroken, and glossy paint surface. To a certain extent the aggregates can be broken down by the commonly employed grinding processes, such as passing through pebble or colloid mills, but these operations also act upon the smaller particles and reduce their size to an undesirable fineness. Screening cannot effect a separation of the coarser fraction to an appreciable extent, since the finest screen that can be used technically for this purpose, 325 mesh, has openings of approximately 40 microns.

Particles of 6 microns and larger contained in the calcined product can, however, readily be separated from the smaller fractions by an aqueous deflocculation process followed by elutriation.[87] The deflocculation must be closely controlled, and the alkalinity of the slurry should correspond to a pH between 7.2 and 10.0. If it is much above this range, the pigment will not be deflocculated to the extent that a good separation by elutriation can be obtained. Ordinarily, aqueous suspensions of the calcined product do not show any material separation of large from small particles in an elutriation process. Sodium hydroxide, sodium carbonate, sodium silicate, trisodium phosphate, and ammonium hydroxide are the most effective and commonest dispersing agents. It is interesting to note that hydrochloric acid and similar compounds, which are so powerful for dispersing hydrous titanium dioxide, are entirely useless in connection with the calcined material. A dispersion of a 1 to 7 slurry suitable for use in the elutriation process should not show any settling on standing for 10 to 30 minutes or any flocculates in the suspension. Such a stability is obtained by adding 0.11 per cent sodium hydroxide, based on total solids content, to a neutral suspension of calcined titanic oxide until a pH of 9.6 results. Table 16–2 shows optimum amounts of various agents for producing dispersions of calcined pigment having the required stability.

### TABLE 16–2

Optimum Amounts of Various Agents for Deflocculating Calcined Titanium Dioxide Pigment

| Deflocculating Agent | Per Cent Based on Weight of Pigment | Concentration of Slurry (approximately) | pH of Slurry |
|---|---|---|---|
| Sodium silicate | 2.38 | 1-7.5 | 9.6 |
| Sodium hydroxide | 0.11 | 1-6.8 | 10.0 |
| Ammonium hydroxide | 9.50 | 1-8.7 | 9.6 |
| Trisodium phosphate (dodecahydrate) | 0.71 | 1-6.9 | 9.2 |
| Sodium carbonate | 0.18 | 1-6.8 | 8.0 |

Larger or smaller amounts than those given in the table can be used effectively, provided that the pH of the suspension is kept within the workable limits of 7.2 to 10.0. Sodium hydroxide dispersed perfectly at 0.11 per cent, and half of this proportion yielded fair dispersions.

After deflocculation the suspension is caused to flow upward through an elutriation vessel at such a rate that the larger, heavier aggregates settle against the current of the liquid and fall to the bottom of the container, while the smaller particles are entrained and overflow with the liquid. A convenient type of elutriation apparatus for commercial application is the Dorr hydroseparator which in its essential parts consists of a cylindrical tank with revolving rake, central submerged feed, and overflow or launder. Other forms of equipment such as the Callow cone or simple rectangular vats may be used.

The relationship between speed of upward flow of a well-deflocculated aqueous suspension of calcined titanium dioxide and size of the largest particles that are carried over by the overflow is shown graphically by the accompanying curve.

Naturally, aggregates larger than the limiting value will fall to the bottom of the elutriation vessel. The horizontal axis shows the velocity of upward flow in centimeters per minute, and the vertical axis gives the size in microns of the largest entrained particles. Thus to obtain an overflow consisting of particles below the visible range, 6 microns, the elutri-

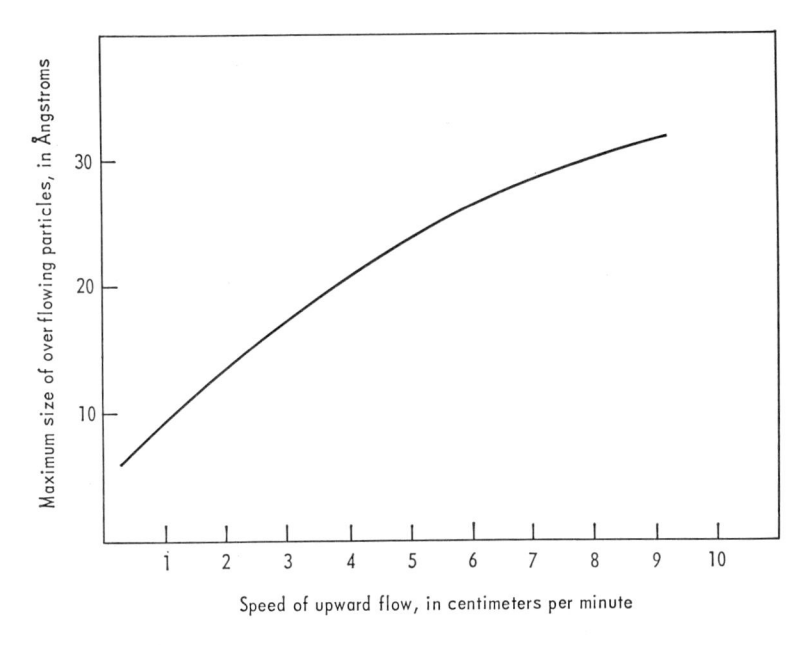

**Fig. 16–1.** Relationship between maximum size of overflowing particles and speed of upward flow.

ation speed has to be reduced to about one half centimeter per minute. These values correspond to separation at ordinary temperatures, 15° to 25° C, but at higher temperatures slightly greater speeds will produce the same operation. The sediment of larger aggregates is removed and reground in a pebble mill and again submitted to deflocculation and elutriation. On long standing the overflow will separate into a thick slurry and a supernatant layer of clear water, but for practical purposes the separation can be speeded up by acidifying the suspension to a pH above 4, or by adding magnesium sulfate or other neutral salt with polyvalent anion. The dilution of the suspension has no very material influence upon the effectiveness of the process and is determined primarily by economic considerations.

In a specific application, a slurry of 20 tons of calcined titanium dioxide in 240 tons of water was dispersed with the requisite amount of sodium hydroxide and fed continuously over a period of 24 hours into an elutriation tank of circular shape equipped with revolving paddles and having an effective settling area of 175 square feet and a depth of 9 feet. The treatment was carried out at normal room temperature of 15° to 25° C, although the capacity of the equipment would be increased by operating at somewhat higher temperature. The slurry was introduced at the center of the tank, overflow from the sides was then conducted to a smaller settling tank, and magnesium sulfate solution was introduced to effect coagulation. After this step the suspension was passed to another settling vessel of about the same dimensions as that employed for elutriation where the pigment particles settled rapidly to a pulp containing 33 per cent solids. The velocity of upward flow in the elutriation tank was adjusted to one third centimeter per minute; at this speed aggregates of 6 microns or larger settled, and the overflow carried only particles below this range. After coagulation the thick, settled slurry was filter-pressed, washed, dried, and pulverized to break up any aggregates formed during drying. Pigments produced under such conditions yielded paint films of a perfectly mirror-like appearance.

According to a similar hydroseparation process,[88] titanium dioxide was dispersed by grinding in a slightly alkaline aqueous solution of pH 9 to 12. The most effective dispersing agents proved to be alkali metal hydroxides, silicates, and phosphates, although organic compounds known as wetting agents were successfully employed within a narrow range of hydrogen ion concentration. For instance, at a pH of 7.6, phenol and sodium sulforicinate yielded marked dispersion. Partial deflocculation was obtained by the use of alkali metal carbonates and aqueous ammonia, but compounds such as sodium sulfate, barium carbonate, and calcium hydroxide were of no value and generally promoted reflocculation. For a number of carefully controlled tests on a typical pigment at pH values of 8 to 10, sodium hydroxide was found to be the most effective agent, and only 0.11 per cent, based on the titanium dioxide, was required to yield perfect dispersion of a slurry containing 1 part solids in 6.8 parts of water. As a rule titanium dioxide is treated with potassium salts before calcination, and the proportion of such compound in the final pigment was found to have a decided influence on the deflocculation. Other important factors were the dilution of the slurry and the hydrogen ion concentration of the solution. For instance, in dispersing suspensions containing 400 gpl titanium dioxide which had been calcined with 1.5 per cent potassium bisulfate, maximum fluidity occurred at pH of 10, although there was little variation between values of 9.5 and 12. But at hydrogen ion concentrations corresponding to pH values below 8.5, the degree of deflocculation was so low that the suspension would not flow through the viscosimeter, as shown in Table 16–3.

### TABLE 16–3

Relation of Viscosity to pH

| pH of Solution | Viscosity of Slurry |
|---|---|
| 8.5 | Infinite |
| 9.5 | 1.08 |
| 10.5 | 1.02 |
| 12.0 | 1.05 |

Thus the general requirements for satisfactory dispersion are that the pH should be between 9 and 12, and that soluble salts should not be present in any appreciable amount. As a rule, addition of dispersing agents beyond the optimum value caused reflocculation and injured the pigment rather than improved it. Such a suspension containing 160 g to 170 g titanium dioxide per liter was then fed into a decantation apparatus and the size of the particles carried over was regulated by the rate of flow. The overflow which contained particles up to the predetermined maximum size was collected in a vat and coagulated by adding magnesium sulfate, aluminum sulfate, or sulfuric or hydrochloric acid. After settling, the flocculated pulp was filtered, dried, and disintegrated. The coarser material (underflow) was fed back to the ball mill with a later charge of fresh pigment. By this method the grinding and separation of the pigment were carried out in a continuous operation and the smaller particles were not reduced to an undesirable fineness.

For example, in a continuous ball mill there was introduced a pulp consisting of 1000 kg titanium dioxide and 1000 kg water, in which was dissolved 1500 to 2250 g potassium hydroxide. The ground material was discharged as a slurry into a vat to which water was added at such a rate as to reduce the concentration of the dispersion to 200 gpl solids, and at this stage 2 to 3 liters more of potassium hydroxide solution containing 750 gpl of the base was added in such a manner as to obtain the maximum degree of dispersion as determined by viscosity measurements. The product was run into a continuous decantation apparatus and the overflow containing 160 to 170 gpl titanium dioxide of fine particle size was collected in a vat, coagulated with magnesium sulfate, and allowed to settle. After filtration, the solid cake was dried and disintegrated to yield a pigment of improved fineness and hiding power. The coarser material which settled to the bottom of the decantation apparatus was withdrawn in a continuous manner and returned to the ball mill.

To minimize the introduction of water-soluble compounds in the finished product and to increase its resistance to discoloration, calcined titanium dioxide was dispersed in dilute sodium hydroxide or sodium silicate solution and after hydroseparation the suspension was flocculated by adding an acid-reacting compound of a trivalent metal (aluminum sulfate) and barium hydroxide was added to neutralize the slurry.[89] The sulfate ion was precipitated as barium sulfate. Similar results were obtained by adding calcium acetate or barium acetate to precipitate the sulfate ions and flocculate the dispersed titanium dioxide.[90]

Coarser particles may also be removed from calcined titanium dioxide by a process based on elutriation of suspensions in acid media.[91] The pigment particles were dispersed by grinding in 1.3 parts of water containing 1 to 5 per cent (based on the titanium dioxide component) of an acid-reacting chloride of aluminum, zirconium, cerium, thallium, or thorium in their highest state of valence at pH between 0.5 and 5, depending upon the agent used. Of these, aluminum chloride gave the best results. This product was diluted with water to 100 g titanium dioxide per liter to yield an extremely fine dispersion in which the smaller particles remained suspended while the larger aggregates settled out rapidly. After the coarse-grained material had settled, the suspension of the fine particles was separated by decantation or elutriation and treated with a coagulating agent such as magnesium sulfate or sulfuric acid to effect coagulation and precipitation. The pulp was then filtered, washed, dried, and disintegrated, and in some cases it was subjected to an additional dry milling to reduce the oil absorption value.

Pigments treated in this manner had uniform particles below the visible range and gave paint films with smooth, glossy surfaces. The tinting strength, hiding power, oil absorption, and color were in no way impaired by this step in the processing. Such pigments were normally acidic, however, with pH values between 4 and 6, but to overcome this the pulp, before drying, was neutralized by adding alkali or alkaline earth metal hydroxides or carbonates.

To illustrate the process, a mixture of 150 kg calcined titanium dioxide pigment in 200 liters of water containing 4 to 5 kilograms of aluminum chloride was ground in a ball mill and the resulting suspension was retained in a vat for 8 to 10 hours, after which the clear supernatant liquor containing 80 gpl aluminum chloride was separated and returned for mixing with another batch of pigment. The thick sediment was then diluted to 1500 liters and agitated to reestablish the dispersion. After standing for 3 to 4 hours the upper portion of the suspension was removed, free from larger aggregates, and containing 85 to 90 per cent of the original material. A solution of 800 to 1000 g magnesium sulfate in 2000 g water was added to coagulate the particles, and the precipitated pulp was filtered, washed, dried, and disintegrated. The coarse-grained residue from the decantation step was removed and reground with a later batch of pigment.

Somewhat better results were reported employing a basic chloride instead of the normal chloride [92] as dispersing agent.

According to a novel method developed by Eckels,[93] an aqueous suspension of calcined titanium dioxide was deflocculated by adding ammonium hydroxide and the wet-milled and hydroseparated dispersion was coagulated by adding carbon dioxide. In the subsequent drying operation the ammonium carbonate was volatilized so that the final product was free from soluble salts. Pigments prepared in this manner had improved water-wetting properties and were readily dispersed in aqueous media.

Calcined titanium dioxide was separated into fractions of different particle size by gravity methods or centrifuging after dispersion in water with the aid of a basic chloride of aluminum, cerium, thallium, zirconium, or thorium.[94] To hasten deflocculation the calciner discharge was wet ground with one of these agents. The coarse particles were allowed to settle and the aqueous suspension of the fines was separated and coagulated with magnesium sulfate or similar compound having polyvalent negative ions. Alternatively, the coarser particles were removed from the suspension by centrifuging. In either case the coarse particles separated were further ground and treated as before. The coagulated slurry of the fines was filtered, washed, dried, and disintegrated according to conventional methods.

Precipitated titanic oxide was calcined to transform all particles smaller than the wave length of visible light into aggregates of distinctly larger size, and the product was then ball-milled in the presence of water to form a pigment suspension. This slurry was continuously centrifuged [95] under constant conditions of solids concentration and of effective centrifugal force to remove the coarser fraction. The finer particles were recovered from the suspension throughout the centrifuging operation by coagulation or settling, and yielded a titanium dioxide pigment of high opacity and uniform texture, while the coarse material was returned to the mill for further grinding. Pigments treated in this manner had constant particle size characteristics and thus gave paint films having smooth, glossy surfaces.

According to McKinney,[96] the wet-milling step in the manufacture of titanium dioxide pigment was improved to produce products of fine texture readily incorporable in paint and enamel vehicles and having a low sensitivity to water by grinding the discharge from the calciner in the presence of a relatively small proportion, 0.5 to 2 per cent, of one or more aluminates of the alkali metals such as sodium, lithium, or potassium.

These agents were mixed dry with the titanium dioxide or added to the slurry as a solution. Best results were obtained by carrying out the grinding operation in a closed circuit with 0.1 to 0.7 per cent of the aluminate based on the solid component.

De Rohden [97] ground calcined titanium dioxide with an equal weight of dilute ammonium hydroxide solution and added water to the residual paste to obtain a highly dispersed product. Dispersions were also produced in aqueous emulsions of organic substances containing proteins,[98] such as water casein paints and rubber latex. Up to 2 per cent alkali metal hexametaphosphate or alkali metal pyrophosphate was incorporated with the dry pigment and the mixture was added to the emulsion of the organic agent, with agitation.

Calcined rutile titanium dioxide pigment is washed with water to remove the soluble sulfate salts so that the material can be dispersed to ensure efficient wet milling and classification.[99] To effectively ball-mill the pigment, it must be dispersed in water. However, in the normal process for producing titanium dioxide pigment, the precalcined hydrate obtained by the hydrolysis of sulfate solutions is treated with various metal reagents to impart desired properties to the pigment. On calcination these additives are converted to sulfate salts which are soluble in water and inhibit good water dispersion even in the presence of dispersing agents. Potassium sulfate, sodium sulfate, and phosphorus compounds are added to prevent sintering and to insure a soft pigment. Aluminum sulfate insures good color in coating compositions. The oxides of zinc and tin are added to facilitate conversion of the crystals to the rutile modification. It is customary in producing rutile pigments also to add a seeding material to promote conversion to the rutile crystal form.

In calcining titanium dioxide pigment of the anatase modification, the temperature employed, 950 to 1000° C, is high enough to decompose the sulfate salts and volatilize the sulfate component as oxides of sulfur. However, in the production of rutile pigment the maximum permissible temperature is no higher than 950° C so that only a small proportion of the metal sulfates are decomposed. As a consequence, rutile pigments flocculate in water and cannot be effectively dispersed even with excessively large amounts of dispersing agents. The calciner discharge is leached to decrease the sulfate salts to less than 0.25 per cent based on the titanium dioxide, and then dispersed in water in the presence of 0.2 per cent dispersing agent. Leaching may be effected by simply stirring the pigment in water and washing by decantation, or by filtering the slurry and washing the cake, or by ball-milling the water slurry followed by washing by decantation, or by filtering and washing the filter cake.

Calcined titanium dioxide pigment after wet milling and hydroseparation is flocculated by the addition of 0.12 to 2.0 per cent of an organic flocculating agent such as formic acid or acetic acid.[100] Subsequent heating and drying remove all traces of the flocculating agent. Amines such as monoethanolamine are very satisfactory as dispersants. In an example, 200 parts of titanium dioxide calciner discharge was milled at 60 per cent solids with 0.1 per cent monoethanolamine for 5 hours to form a slurry, which was diluted to 12 per cent solids and hydroseparated for 2 hours. The decanted fines were flocculated with 0.15 per cent formic acid based on the titanium dioxide as a 10 per cent solution. After coagulation the mixture was filtered, washed with dilute formic acid, and dried at 110 to 130° C for 10 hours. The dried cake was ground and treated with 0.1 per cent triethanolamine as a 10 per cent solution in alcohol.

If neutral, water-soluble salts of univalent metal ions in amounts of 0.7 to 2.0 per cent, based on the weight of the titanium dioxide, are used as flocculants for hydroclassified titanium dioxide pigment in aqueous alkali suspension, redispersion in water is effected to the extent of 92 to 97 per cent.[101] A water dispersible pigment is also produced by flocculating the hydroclassified calciner discharge with 0.1 to 2.0 per cent based on the titanium dioxide of a hydrolyzable titanium salt as the tetrachloride or the tetranitrate.[102] To prepare titanium dioxide pigments capable of being dispersed in water or in an aqueous medium, the calcined material is suspended in water in the presence of 0.1 to 0.75 per cent of a dispersing agent—sodium silicate, trisodium phosphate, sodium aluminate, or tripolyphosphate.[103] After separation from the coarse material, the suspension is deflocculated by addition of 0.1 to 2.0 per cent of aluminum sulfate, based on the weight of the pigment. Then sodium carbonate, sodium hydroxide, or sodium aluminate is added to precipitate the aluminum as hydroxide on the pigment particles. After washing the pigment, 0.25 to 5.0 per cent of barium hydroxide or barium carbonate is added, and the mixture is dried and crushed. The concentration of the pigment in the aqueous slurry is from 200 to 400 gpl.

For example, 900 g of a milled rutile pigment was added during 0.5 hour to 6.75 g of sodium hexametaphosphate dissolved in 2790 ml of water with stirring. At this stage 129 ml of aluminum sulfate solution containing 70 g of aluminum oxide was added followed by 243 ml of a sodium carbonate solution containing 106 g of anhydrous sodium carbonate per liter with a pH of 7.5. After diluting to 14 liters and filtering, 1445 g of filter cake containing 50 per cent solids was obtained. This cake was mixed with 4.27 g of barium carbonate and the mixture was dried at 110° C for 16 hours. Twenty-five grams of this finished pigment was suspended in 94 ml of water at ambient temperature with stirring for 2 hours without addition of a dispersing agent. This suspension consisting of 250 g of pigment per liter contained after standing for 6 hours

185 g of pigment per liter in the suspended state and only 7 g of settled pigment.

Titanium dioxide pigment of improved electrical resistance and of improved texture is produced by flocculating the water dispersed and hydroclassified pigment with a barium salt.[104] After flocculation the pigment is dried and milled in the usual manner. The use of a barium salt minimizes the amount of water soluble salts retained by the finished pigment.

Particles larger than 4 microns are removed from suspension of calcined titanium dioxide by centrifugal hydroseparation.[105] The calcined titanium dioxide is dispersed in water by 0.1 to 1.0 per cent by weight of an alkali metal silicate, after which the suspension is agitated and within 10 minutes after cessation of agitation it is subjected to centrifugal hydroseparation.

The wet milling and hydroseparation normally involve coagulation steps with the result that the product can be redispersed with considerable difficulty, if at all, and such pigments are thus unsuited for use in aqueous media, as, for example, in latex paints and paper. Pigments for such purposes are usually ground dry and do not contain the occluded coagulating agent. This is normally accomplished on a commercial scale in mills of the ring-roll type (Raymond)[106] or in jet pulverizers (mikronizers).[107] Regardless of the type of milling, the pigment is passed through disintegrators or pulverizers of the rotary-hammer type and air-floated to insure the desired degree of fineness. The coarse material is returned for further grinding.

Composite pigments, because of their heterogeneous nature and the relatively large particle size of the extender or filler as compared with that of the titanium dioxide, are in general not amenable to the wet-milling and hydroseparation process such as is employed with the relatively pure oxide. Those products are normally subjected to dry-grinding operations as, for example, in mills of the ring-roll type or in jet pulverizers. They are likewise air-floated to such a degree that not over 0.05 per cent will be retained on a 325 mesh screen, and they are then disintegrated or pulverized.

Treating the finely divided pigment particles with a nonaqueous wetting agent more volatile than the vehicle and compatible with it facilitated dispersion.[108] The agent was removed from the mixture during the kneading operation. Composite pigments readily dispersed in organic media may be produced by milling with 0.05 to 0.9 per cent naphthenic acid or calcium naphthenate as an added step.[109] Dispersion characteristics were also improved by treating the pigment with 0.1 to 1.0 per cent of a water-soluble glycolate.[110] Berry[111] produced composites free from particles of 5 microns and larger for direct use as mix-in pigments in paints and enamels. From 0.03 to 4.5 per cent of a polar-nonpolar compound (oleic acid, naphthenic acid) was intimately mixed with a finished titanium dioxide-calcium sulfate pigment, and the treated product was dispersed in a current of air and subjected to a classification treatment to remove the coarse particles. Calcined pigment was treated with 0.1 to 5.0 per cent zinc, calcium, or magnesium resinate, abietic acid, phenolic resin, or fatty acids, pulverized in an air stream, separated and repulverized to reduce the particle size and improve the texture.[112]

The presence of water greatly increases the viscosity of paints and enamels containing the calcium sulfate composite pigment. This effect, known as bodying, may be lowered by adding alkaline compounds to the composition or it may be increased by adding acidic compounds. Aliphatic amines were found to be particularly effective in lowering this bodying tendency.[113]

## DRY MILLING

In the manufacturing process, oil absorption of the pigments is affected by the nature and amount of the conditioning agents incorporated with the hydrous oxide and by the calcination conditions themselves. Potassium carbonate or sulfate is the primary compound employed, and as the proportion is increased the oil absorption is reduced. However, the use of increased quantities of these agents is often harmful to other properties, particularly color, tinting strength, and fineness. Oil absorption of the finished pigment may be further reduced by a dry-milling process but this is effective to a limited extent, and the method is naturally more applicable to varieties having abnormally high initial oil values since in any case a minimum is reached which may again increase on further grinding. Prolonged milling may also injure the color and other pigment properties.

To determine the oil absorption value, 3 g of the finished pigment is weighed onto a glass plate and refined linseed oil just short of the estimated amount required is added from a burette. The components are thoroughly mixed by working slightly with a spatula, and more oil is added, drop by drop, until the powder is thoroughly wetted and the whole mass becomes plastic. After the end point is reached, the addition of one drop more will cause the paste to smear on the glass. Results are converted to parts of oil by weight required per 100 parts of pigment, and this number is known as oil absorption.

From a practical standpoint, oil absorption too high results in coating compositions (paints) that are too thick and have impaired gloss, while oil absorption too low results in thin paints which tend to sag or run, subsequent to application. These difficulties, resulting from variations in this property, can be corrected in general by reformulation of the paint, but this procedure is inconvenient and costly to the manufacturer and he naturally prefers a uniform product.

Pigments of lower oil absorption characteristics and increased covering power were reported to be obtained by incorporating a small proportion of sodium hydroxide, sodium nitrate, sodium sulfate, or the corresponding potassium compounds, aluminum sulfate, or magnesium sulfate in hydrolytically precipitated titanic oxide before calcination.[114] These accelerators could be formed *in situ* during the hydrolysis step and did not exceed 10 per cent of the titanium dioxide, but in general 0.5 to 1 per cent produced optimum results. Calcination was carried out at 900° C and the product was ground and pulverized according to accepted practice.

Rhodes [115] found that calcining the hydrous oxide with from 0.1 to 2.5 per cent, on the dry basis, of an alkali metal titanate (potassium titanate), which would not discolor at the temperature employed, yielded a pigment of lowered oil absorption as well as higher opacity and improved mixing properties (dispersion). Later work revealed that similar results could be obtained by employing 1 to 2.5 per cent of an alkaline earth metal titanate.[116]

According to McKinney,[117] oil absorption may be varied willfully over wide limits in the manufacturing process by treating the washed pulp with small proportions of alkali metal salts (potassium sulfate) together with phosphoric acid or phosphates so that the calcined product will have a mole ratio of alkali metal to phosphorus within the range of 1 to 1 and 1.6 to 1. For example, if a high value is desired, the hydrous oxide is calcined with from 0.2 to 0.4 per cent potassium sulfate, while on the other hand calcination with from 0.7 to 3 per cent of this addition agent gives a product having a low value. Pigments within the normal range are obtained by employing from 0.4 to 0.7 per cent potassium sulfate. The alkali metal-phosphorus ratio is held, however, between the range of 1 to 1 and 1.6 to 1. Ilmenite, the chief raw material used in pigment manufacture, normally contains phosphate compounds which are converted to the soluble form in the digestion step. During hydrolysis the phosphate constituent is mostly adsorbed by the precipitate of titanic oxide, and only a small proportion, if any, is removed in the washing operation. In calculating the required ratio, the residual phosphate is included in the total.

This treatment prevents the formation of hydrated titanic acid subsequent to calcination by decomposition of titanates and phosphates. Hydrated titanic acid has a detrimental effect upon the color of baked alkyd resin films, and it is acted upon by certain vehicles and converted to blue or gray titanous compounds that cause discoloration of the system. Furthermore, titanium dioxide pigments produced in accordance with this process have good fineness and are particularly free from coarse particles.

In an example, washed hydrolytically precipitated titanic acid was slurried to 390 gpl solids and analyzed for phosphorus pentoxide. Allowing for losses of potassium sulfate during the subsequent filtration and slight losses of phosphorus pentoxide during calcination, the necessary amount of potassium sulfate was added to produce a mole ratio of 1.17. The slurry was then dewatered in such a manner as to produce a cake containing 40 per cent titanium dioxide and 0.53 per cent potassium sulfate, based on the solid component, and the product was calcined to produce a pigment containing 0.3 per cent phosphorus pentoxide and 0.53 per cent potassium sulfate, and having a pH of 7.3. The calciner discharge was dispersed, ground, flocculated, and filtered. The dewatered cake containing 60 per cent titanium dioxide was dried at 150° to 170° C in a continuous steam-heated rotary drier and disintegrated in a mill of the hammer type to produce finished titanium dioxide pigment.

With a constant potassium content, low mole ratios (high phosphorus) resulted in low oil absorption, but as the ratio was increased by lowering the phosphorus content the oil absorption increased rapidly to a maximum and fell off, at first rapidly and then more gradually, as the mole ratio was further increased. With extremely low mole ratios resulting from high phosphate content, color and tinting strength tended to be poor.

The oil absorption of pigments processed by the conventional methods, that is, calcined at 900° to 1000° C, ground wet in a ball mill, passed through a hydroseparator to get an average particle size of 0.5 micron, dried, and pulverized, was greatly reduced by a dry pulverization process [118] in which the pigment particles were subjected to considerable pressure or momentary impact. At the same time other pigmentary qualities were likewise improved. This effect was accomplished by hand grinding a few grams in a mortar, or on an industrial scale by grinding the pigment in a dry condition in a Raymond mill or chocolate mill. In normal plant operation a grinding period of from 1 to 2 hours was required. The oil absorption of pigments subjected only to wet milling (35 to 55) was reduced to 25 or less by this additional dry-grinding process, and the higher the original value the greater was the proportionate reduction by milling. This operation had the additional advantage of increasing the hiding power of the pigment, as determined by the cryptometer, by 5 to 10 per cent, and it also increased the tinting strength up to 20 per cent.

This marked reduction in oil absorption values by dry-milling processes cannot be ascribed to actual reduction of the primary particle size since the particles are far too small to be affected by the grinding action.[119]

The effect can be explained, however, as an agglomeration of the pigment into compact particles which are not readily penetrated by the vehicle. In fact this has been demonstrated experimentally. Booge [120] attributed the reduction in oil absorption to some change in the surface or adsorption characteristics of the pigment particles. According to a method reported by Barton,[121] the pigment, after conventional wet milling, drying, and disintegration, was submitted to a high pressure without grinding or attrition to reduce the oil absorption. The size of the pigment particle was not reduced appreciably by the treatment.

From a study of the quantity of water taken up by titanium dioxide pigments at various intervals, the rate of wetting [122] was found to depend upon the size, shape, and kind of particles and upon the volume of pores, but influence of size and shape disappeared on compacting the material under pressure. The amount of oil absorbed paralleled that of water.

Finely divided pigment free from aggregates and possessing low oil absorption was produced by subjecting calcined titanium dioxide to a pulverizing type of milling followed by a disintegrating type of milling.[123] Pigments ground by the conventional wet-milling process have oil absorption of from 24 to 26, and lower values are desired in many paint and enamel formulations. Substantial reduction of this value may be obtained by a pulverizing process in which the particles are subjected to considerable pressure. Commercial ring-roll mills are preferred for this purpose, and in practice these are ordinarily equipped with an air-separation system which depends on centrifugal force to separate the coarse particles and return them to the mill for further pulverizing while allowing the fines to pass on as finished product. Such a treatment reduces the oil absorption by around 25 per cent from 24 to 26 to 17 to 20. However, this reduction is accompanied by a sacrifice in texture, and an increase in grittiness of the pigment. This decrease in texture is attributed to the compacting action of the pulverizing process so that many particles of titanium dioxide are pressed together in the form of flakelets or pellets and these persist as coarse aggregates or grit. This is a logical explanation of the unusual effect, in that vigorous pulverizing does not produce an impalpable powder but tends to coarsen the pigment and produce grittiness rather than finer texture.

Another type of dry milling is by disintegrating, such as may be accomplished in rotary-hammer mills, but this operation applied to titanium dioxide does not produce a fine subdivision of the coarse particles formed during the calcination process, nor does it reduce pigment oil absorption. However, it does break up flakes or pellets produced by compacting.

By employing these processes in series, the beneficial effects of both were imparted to the pigment. The ring-roll mill broke up hard pigment grit and reduced oil absorption, while the rotary-hammer mill broke up the relatively soft aggregates formed during the ring-roll milling. The pigment was first subjected to pulverizing milling in a ring-roll mill of 50-inch diameter equipped with an air-separator system which separated the coarse particles from the fines, returned the coarse material for further grinding, and discharged the fines. At this stage, the pigment contained an undesirable quantity of aggregates and was passed to disintegration milling in a rotary-hammer mill of 24-inch diameter. The pigment passed through the dry-milling equipment at the rate of 1600 to 2300 pounds per hour.

Sawyer [124] employed a series of grinding operations. The calcined pigment was first subjected to a grinding operation in a high-pressure, fluid-energy mill with steam at a temperature of 300° to 500° C under a pressure of 50 to 150 psi, and the product was passed through a closed-circuit, wet-grinding system equipped with a hydroseparator. The overflow was coagulated, dried, and disintegrated in a low-pressure, fluid-energy mill to break up agglomerates formed during the drying operation.

Parravano and Caglioti [125] concluded from a theoretical consideration that rutile modification of titanium dioxide should have a slightly higher oil absorption than anatase. The unit cell of anatase contains twice as many molecules as that of rutile, with the result that a unit weight of the former material has a smaller surface area.

Oleaginous coating compositions pigmented with calcium sulfate-extended titanium dioxide tend to exhibit poor gloss and poor gloss-retention characteristics. That is, they produce films having an undesirably dull, matte appearance, which for some uses is objectionable. This property was improved by adding to the pigment between 0.1 and 1 per cent benzenecarboxylic acid or an equivalent amount of one of its salts.[126] Furthermore, pigments treated in this manner possessed improved wetting, mixing, and dispersion properties in oil, varnish, and other vehicles without an increase in consistency. They had superior tint retention, higher resistance toward discoloration and yellowing, and reduced sensitivity to water. Such pigments in organic coating compositions did not settle to a hard dense cake on standing, and the aged material produced films which dried as rapidly as freshly prepared products. In an actual operation, benzenecarboxylic acid was added to a 70 per cent calcium sulfate-30 per cent titanium dioxide pigment at the rate of 15 pounds per hour while the pigment was being fed at the rate of 3000 pounds per hour into a 50-inch ring-roll mill equipped with an air separator. After treatment and pulverization in this manner, the pigment was disintegrated by passing it through a 24-inch rotary-hammer mill, also at a rate of 3000 pounds per hour.

Sutton [127] obtained better gloss characteristics by intimately mixing with the pigment a small proportion (0.5 per cent) of a trialkylphenyl or trialkylphenylalkylene ammonium hydroxide. Similarly, gloss and hardness values were improved by mixing with the calcined titanium dioxide, in aqueous suspension, a small proportion of water-insoluble basic compounds of cobalt such as the hydroxide.[128] The treated product was dried and pulverized.

Pigments of high surface-hiding power were produced by forming on the surface of the individual particles a coating of gel-like polymeric carbohydrate compound.[129] With an aqueous suspension of calcined and ground titanium dioxide was mixed 0.25 to 0.35 per cent, calculated as polyhydric carbohydrate and based on the weight of pigment, of an alkaline solution of a polymeric carbohydrate derivative such as cellulose xanthate, and dilute sulfuric acid was added to bring the pH of the system to below 7, thereby precipitating the agent in gel-like form on the surface of the titanium dioxide particles. The mixture was dewatered, and the cake was dried at a temperature not in excess of 200° C.

A pigment free from grit particles [130] was produced by heating a slurry of calcined titanium dioxide containing 0.2 to 10 per cent of an added salt (calcium chloride) or an acid or base for at least 30 minutes at a temperature above 100° C. If acid or alkali was used, the slurry was neutralized after digestion. This treatment rendered soluble certain cementing agents, and the high temperature served to dehydrate any gelatinous material, thereby preventing it from having a cementing action on the pigment during the subsequent drying operation. Such gelatinous material could result from the hydrolysis of alkali metal titanates which may have been formed during the calcination step. In an example, 10,000 parts by weight of calcined titanium dioxide pigment which had been wet ground and elutriated under alkaline conditions, coagulated with magnesium sulfate, and adjusted to a pH of 7 with sulfuric acid, was allowed to settle to slurry concentration of 400 gpl. This was diluted with water to 150 gpl and 10 per cent or 1000 parts of hydrochloric acid was added, after which the slurry was brought to boil and held with agitation for 2 hours. It was then cooled to 70° C and filtered. The cake was washed with water at the same temperature by displacement and reslurried to 150 gpl solids. After standing for 16 hours the suspension was neutralized with sodium hydroxide, filtered, and washed. This cake was dried at 175° C and disintegrated in a high-speed hammer-type pulverizer.

Micronization, fluid energy milling, improves the fineness and uniformity of particle size of titanium dioxide pigments.[131]

# 17

# Sulfuric Acid Recovery

A practical method for the recovery of sulfuric acid from dilute waste liquors obtained in the manufacture of titanium dioxide pigments was developed by McBerty.[1] In the most common process for the production of such pigments, ilmenite ore is attacked with strong sulfuric acid and the reaction product is dissolved in water. All of the ferric component is reduced to the ferrous state, and a large proportion is crystallized out as copperas on cooling the solution. The resulting sulfate liquor, containing small amounts of iron and other impurities such as chromium and vanadium, is finally hydrolyzed to precipitate the titanium as hydrous oxide. At the same time an equivalent amount of sulfuric acid is liberated, and the filtrate contains from 10 to 30 per cent sulfuric acid, appreciable amounts of ferrous sulfate, some titanium, and most of the soluble impurities originally present in the ore.

In the first step of the recovery process, this dilute waste acid liquor was concentrated to such a degree that it became saturated with ferrous sulfate at a temperature near its boiling point but short of precipitation. This operation was conveniently carried out in a vertical tower with a countercurrent of hot gases from the next step. Spent gases and water vapor from this operation were wasted into the atmosphere. Up to the end of this stage the liquor remained free flowing and contained 30 to 40 per cent sulfuric acid and 12 to 20 per cent ferrous sulfate. It was further heated until the dissolved copperas was converted to the monohydrate, but not to the point at which sulfuric acid distilled off. Ferrous sulfate, crystallized originally from the ilmenite solution, was added at this stage and likewise converted to the monohydrate. Heat required for this operation was derived from the combustion of carbonaceous fuel and the hot gases traveled countercurrent with the liquid. The most convenient form of apparatus was found to be a brick-lined rotary kiln, although the concentration may be carried out in a vacuum evaporator of the Mantius type. A considerable proportion of monohydrate crystallized out, leaving a liquor containing 65 to 80 per cent acid. If there was a demand for cheap technical-grade sulfuric acid of this concentration, the product of this stage was cooled to crystallize most of the iron compounds and other impurities. At an acid strength of 65 per cent, the precipitate of monohydrated ferrous sulfate entrained with it about 50 per cent of the chromium and 75 per cent of the vanadium present. However, if the concentration was carried to 80 per cent, it became almost impossible to separate the acid from the ferrous sulfate by physical means (filtration), and the resulting sludge was heated in a brick-lined rotary furnace below 300° C to distill off all free sulfuric acid. This temperature range was maintained as long as appreciable amounts of acid remained, and the ferrous sulfate was incompletely oxidized, since at 300° C, or above, concentrated sulfuric acid reacted with ferrous sulfate to form the ferric salt and sulfur dioxide. The acid vapors were condensed in the conventional apparatus, yielding a product of 65 to 80 per cent strength. Little decomposition took place. Distillation was carried out in an oxidizing atmosphere to further aid conversion of ferrous iron to the trivalent state and thereby minimize losses due to decomposition of iron sulfates at the relatively low temperatures. Heat for this step was provided by hot gases produced by the combustion of carbonaceous fuel, elemental sulfur, or sulfide ores in an excess of air. The hot gases were led into contact with the iron sulfate sludge, and after condensation of the sulfuric acid the exit sulfur gases were conducted to a sulfuric acid plant. The solid residue from this stage, a more or less basic ferric sulfate, was heated in another rotary kiln above 700° C by direct contact with hot gases derived from the combustion of sulfur or pyrite. A strongly oxidizing atmosphere was maintained to increase the yield of sulfur trioxide, since oxygen in the presence of ferric oxide tends to shift the equilibrium of the reaction, $2SO_2 + O_2 \rightleftarrows 2SO_3$, to the right. Gases from this step contained a considerable proportion of sulfur trioxide and variable amounts of sulfur dioxide, depending upon whether sulfur or carbonaceous fuel was used. Such gases were freed from sulfur trioxide and suspended particles of iron oxide before conversion in a contact-acid plant, otherwise the impurities would greatly reduce the effective life of the catalyst. For this reason the gases were converted to sulfuric acid in a lead-chamber plant.

The residue consisted primarily of a reddish-brown ferric oxide suitable for use as pigment or as an ore of iron. It also contained the vanadium and chromium values of the original ore, and these constituents were recovered by chemical means.

In a typical operation, a waste liquor containing 25 per cent sulfuric acid and 11 per cent ferrous sulfate was concentrated in a vertical tower, with hot countercurrent gases from the next advanced step, to 32 and

14 per cent respectively, and the discharge from this operation was further concentrated to 65 per cent sulfuric acid by heating at 153° C. At this stage the greater part of the ferrous sulfate separated as the monohydrate and was filtered off. The filtrate of 65 per cent acid, which contained only small amounts of dissolved iron and other metallic impurities, represented 65 per cent of the total free acid. The filter cake containing the other 35 per cent was charged into a direct-fired rotary kiln and free sulfuric acid was distilled off and condensed. Hot gases from the combustion of low-grade fuel oil were passed in direct contact with the sludge so that its temperature did not exceed 300° C. An oxidizing flame was maintained to prevent reduction of sulfuric anhydride to sulfur dioxide. In this operation, 95 per cent of the free acid was distilled off and recovered at a 65 per cent strength. The oxidized iron sulfate from the previous step was further heated in a direct-fired rotary kiln at 700° C in an oxidizing atmosphere, and the discharge gases, containing 6.5 per cent sulfur trioxide and 2.25 per cent sulfur dioxide, were converted in a chamber sulfuric acid plant.

Spangler [2] described a commercial method comprising the contact process for recovering the sulfuric acid values from spent liquors from titanium pigment manufacture. Such solutions normally contain 18 to 25 per cent sulfuric acid, 10 to 16 per cent ferrous sulfate, and small amounts of titanium sulfate. The free acid was first neutralized with iron oxide, which was obtained from a subsequent step of the process, and the resulting solution was evaporated to recover anhydrous iron sulfate. Both neutralization and evaporation were safely carried out in one rotary kiln heated internally by combustion of gases. The iron sulfate residue was then decomposed in a rotary roaster to produce sulfur dioxide gas of the proper concentration for subsequent conversion in a contact acid plant, and ferric oxide cinder, a part of which was returned for the neutralization of additional batches of waste liquor. After purification, the sulfur dioxide was converted to sulfuric acid of any desired strength in a modified form of the standard Chemico contact-acid plant using vanadium catalyst. This procedure, using dehydrated iron sulfate, was similar to the conventional contact process employing pyrite or sulfur, except that a heating and reducing agent was added to the raw material. Low-grade fuel or pulverized coal proved quite satisfactory, and pyrite was mixed with the sulfate residue, thus supplying part of the heat required and at the same time producing sulfur dioxide to enrich the roaster gases. A plant of this type was constructed for the National Lead Company, Titanium Division, in 1936.

Residual waste solutions containing 124 gpl sulfuric acid and 110 gpl ferrous sulfate were treated with ferrous sulfide, and the hydrogen sulfide evolved was oxidized first to sulfur and then to sulfur dioxide and used to produce sulfuric acid.[3] The liquor was added to a solution of calcium hydrosulfite to precipitate calcium sulfate and ferrous sulfide, which were separated, mixed with petroleum coke, and roasted at 300° C to convert the calcium sulfate to sulfide. The ferrous sulfide remained unchanged. These mixed sulfides slurried in water were transferred to an autoclave, hydrogen sulfide was introduced at a pressure of 5 psi, and the resulting calcium hydrosulfide solution was separated by filtration. Part of the ferrous sulfide was used to treat more waste acid liquor. The remainder was roasted to produce sulfur dioxide for the manufacture of sulfuric acid. The calcium hydrosulfide was used to treat more ferrous sulfate solution, thus completing the cycle.

In a continuous process, crystalline ferrous sulfate was heated with air to effect decomposition, and the gaseous mixture was passed into a lower temperature zone to favor oxidation of sulfur dioxide to the trioxide.[4] Additional cold air was introduced to take up the heat liberated in the oxidation reaction and hold the temperature within the effective range. In a two-step dehydration process, crystalline ferrous sulfate was first heated at 80° to 167° C, under nonoxidizing conditions, to form the monohydrate and then at 167° to 492° C, in contact with air, until the anhydrous basic sulfate was formed.[5] The anhydrous salt was heated at 700° C to yield sulfur dioxide and sulfur trioxide, free from moisture and iron oxides. The last trace of basic ferric sulfate was removed from the iron oxide by heating with charcoal.

Heinrich[6] concluded that the recovery of dilute waste sulfuric acid from the manufacture of titanium pigments could hardly be justified from a standpoint of economy, but avoidance of damage and stream pollution must also be considered. In certain localities the law would not permit the dumping of large amounts of sulfuric acid.

Titanium dioxide is more sensitive to discoloration by chromium compounds than by any other ordinarily encountered impurity,[7] and since recycled sulfuric acid tends to pick up this constituent from the ore, it should be removed before each reuse of the acid. This may be accomplished readily after concentration to 92 per cent sulfuric acid.[8] Crude acid of this strength was heated to 300° C to convert the chromium to an insoluble compound, in which form it was readily recovered by settling or by filtration. According to a similar procedure, the waste liquor was concentrated by evaporation to 60 per cent sulfuric acid to precipitate most of the ferrous sulfate, and oleum or sulfur trioxide was added to raise the acid content to 88 per cent.[9] At this concentration the metallic impurities precipitated.

Spent liquor containing more than 20 per cent sulfuric acid, together with small amounts of titanium, zirconium, and hafnium compounds, was purified by treatment with phosphates as superphosphate.[10] The resulting precipitate was separated by any convenient means. Such solutions were preliminarily decolorized [11] by treatment with active carbon or sulfur dioxide, or by precipitation, after which the oxide or carbonate of barium was added to obtain a precipitate of barium sulfate free from iron. The reaction was facilitated by adding a small proportion of nitric, hydrochloric, or acetic acid, or of one of their salts, to produce a soluble barium compound.

Waste acid liquor was purified sufficiently for use in the production of pigment-grade calcium sulfate or titanium dioxide by treating it with a soluble alkali metal chloride, sulfate, phosphate, or chlorate ranging in amount in gram atoms of alkali metal per liter of solution treated from 2 to 5 times the reciprocal of the percentage by weight of sulfuric acid.[12] The resulting precipitate was removed by settling or by filtration. Calcium sulfate and titanium dioxide pigment obtained with the use of the concentrated but unpurified waste acid differed very materially from that produced in processes employing relatively pure commercial acid. For one thing, such products were characterized by an objectionable yellowish or yellowish-brown tone which rendered them wholly unfit for many industrial uses. Furthermore, sulfuric acid costs constitute a major item of expense in titanium pigment manufacture, and it is obvious that losses occurring by reason of waste acid discard would be a serious economic factor in the process. Besides, disposal of the waste acid would constitute a serious nuisance problem.

In a typical test, 10 g of sodium sulfate was added to 1 liter of waste hydrolysis liquor which weighed 1280 g and contained 296.2 g free sulfuric acid, 16 g titanyl sulfate, and 66.7 g ferrous sulfate. Expressed as gram atoms of sodium per liter of acid solution, this was 3.26 times the reciprocal of 23.14, the percentage of sulfuric acid in the waste liquor. The mixture was agitated for 1 hour at room temperature, and the resulting precipitate, 12 g, was filtered off. Calcium sulfate prepared from the filtrate was practically equivalent in every respect to a similar product prepared from fresh acid.

Another portion of the purified acid was concentrated to 82 per cent strength, mixed with a suitable amount of fresh 98 per cent acid, and employed in attacking ilmenite by the conventional method. The resulting pigment was equivalent to that obtained under the same conditions but employing relatively pure fresh acid.

Parrish [13] neutralized the waste acid solution with gas-plant liquor, ammonium hydroxide, to form ammonium sulfate and oxides of iron. Hydrogen sulfide and carbon dioxide gases were liberated in the process. Similarly, gaseous ammonia was added to ferrous sulfate solution to precipitate ammonium sulfate and a basic iron compound.[14] The precipitate was separated and heated to 140° C to convert the iron compounds to insoluble form, and the ammonium sulfate was dissolved in water. Platonov, Zakharova, and Efros [15] recovered vanadium from the waste liquors by precipitation with ammonium hydroxide.

The residual sulfuric acid of the washed titanium dioxide pulp, as obtained commercially by hydrolysis of sulfate solutions, is driven off in the calcination step as sulfur trioxide and may be recovered from the stack gases by precipitators of the Cottrell type.

An 83 per cent sulfuric acid solution free from sludge is obtained from residual liquors from titanium dioxide manufacture.[16] The molar ratios of iron, titanium, and magnesium sulfate are adjusted so that the impurities precipitate as a complex salt. Aluminum compounds are first precipitated as alums together with most of the iron, titanium, and magnesium compounds. The filtrate is evaporated at 50 to 100° C to precipitate the remaining salts of iron, titanium, and magnesium as $(Fe,Mg)SO_4 \cdot TiOSO_4 \cdot 2(H,NH_4)SO_4$. A favorable ratio of ferrous sulfate plus magnesium sulfate to titanyl sulfate of 1 to 1 is obtained in a 33 per cent sulfuric acid solution between $-26$ and $-32°$ C, or in a 23 per cent sulfuric acid solution between $-41$ and $-47°$ C. Cooling is achieved with a coil containing cold brine, and freezing is avoided by adding an antifreeze solution. The ratio can also be controlled by the addition of suitable amounts of titanyl sulfate or hydrous titanium dioxide to the filtrate. For example, to 18 liters of acid waste containing aluminum sulfate 1.0, sulfuric acid 23.8, ferrous sulfate 10.2, titanyl sulfate 1.2, and magnesium sulfate 1.0 per cent, 162 g of ammonium sulfate was added. The solution was cooled to $-40°$ C for 4 hours. Ferrous sulfate and alums were then separated. The filtrate containing aluminum sulfate 0.05, sulfuric acid 28.3, ferrous sulfate 0.8, titanyl sulfate 1.2, and magnesium sulfate 0.5 per cent was concentrated to 75 per cent sulfuric acid and aged for 24 hours at 70° C before further filtration to remove the last of the solid salts.

The dilute sulfuric acid solution containing iron, magnesium, aluminum, and titanium sulfates obtained in the manufacture of titanium dioxide is evaporated to a free acid concentration of 75 per cent.[17] At this stage the salts are partially precipitated. Concentrated sulfuric acid is added to increase the free acid concentration to more than 85 per cent to effect precipitation of the remaining salts. The precipitation is advantageously done in two stages. For example, waste 20 per cent sulfuric acid was concentrated at 190° C to 70 per cent, mixed at 100° C with 96 per cent sulfuric acid to adjust the overall concentration to 86 per cent, and filtered. The iron, magnesium, titanium, aluminum, and vanadium sulfates in per cent in the initial solution were 6.1, 1.8, 1.5, 0.8, and 0.3, respectively, and in the final solution were 0.57, 0.06, 0.35, 0.47, and 0.07.

Part of the purified solution was further concentrated to 96 per cent sulfuric acid and recycled.

Usable products are recovered from waste acid liquor obtained in the manufacture of titanium dioxide.[18] Typical waste acid contains total sulfate calculated as sulfuric acid 300 to 400, free sulfuric acid 250 to 290, iron 30 to 50, titanium calculated as the dioxide 8 to 12, and manganese 1 to 3 gpl. Ammonia is added until the content of free sulfuric acid is reduced to 10 gpl, after which the solution is cooled to 25° C to crystallize Mohr's salt. The salt is filtered off and washed. To the mother liquor is added ammonia to pH 9. The mixture is oxidized by air at 30 to 40°C, and the ferric ammonium sulfate is filtered off, washed, and calcined at 700 to 750° C, to give a brown pigment with good color and covering power. Into an aqueous solution of Mohr's salt containing 68 g of iron per liter, ammonia is introduced until 30 per cent of the iron separates. The mother liquor is treated with ammonia to pH 9, then oxidized with air at 30 to 40° C to precipitate ferric ammonium sulfate which, after separation, washing, and calcining at 700 to 750° C, gives an iron ocher low in manganese. Very pure needle-like crystals of ferric oxide are obtained by treating the first filtrate with ammonia to pH 8, followed by oxidation with air. The combined filter liquors are evaporated to crystallize ammonium sulfate for technical purposes.

Waste liquor is concentrated to a sulfuric acid content of 60 to 65 per cent.[19] The precipitated salts are dissolved in a dilute solution of titanium sulfate low in acid and in iron at 50 to 80° C, after which the solution is cooled to precipitate ferrous sulfate heptahydrate. Part of the filtrate is recycled to dissolve more salts, and the remainder is hydrolyzed to produce titanium dioxide of pigment grade. The liquor from the hydrolysis of titanium sulfate solution is evaporated by immersion heaters with the crystallization of salts.[20] Thick acid sludge from another operation is added and the mixture is filtered to obtain 70 per cent sulfuric acid. This filtrate is evaporated to 93.5 to 95 per cent sulfuric acid with further crystallization.

According to Ozaki[21] the degree of utilization of waste liquor from titanium pigment manufacture can be raised to 90 per cent. However, vanadium accumulates as the waste liquor is recycled in the manufacturing process.

Sulfuric acid and salts of iron and aluminum are recovered from the spent liquors obtained in the manufacture of pigment from high grade slag by the action of hydrogen chloride gas.[22] The spent acid contains 10 to 30 per cent by weight of free sulfuric acid and 10 to 15 per cent ferrous sulfate, aluminum sulfate, and magnesium sulfate. Other sulfates, such as calcium and vanadium, are present in minor proportions.

The composition of the spent acid varies largely with the raw materials and the manufacturing process. Conversion from the sulfate radical in metal sulfates to free sulfuric acid is effected by an exchange reaction with hydrogen chloride. Ferrous sulfate and aluminum sulfate are easily converted to $FeCl_2 \cdot 2H_2O$ and $AlCl_3 \cdot 6H_2O$ by saturating the liquor with hydrogen chloride gas. These chlorides precipitate. Precipitation of the chlorides is 90 per cent complete with an equivalent yield of free sulfuric acid. Similar liquors are saturated with hydrogen chloride gas to precipitate fine crystals of ferrous chloride and aluminum chloride as hydrates.[23] Recovery of iron and aluminum are 74.6 and 68.1 per cent, respectively. The filtrate is evaporated at 165° C to recover sulfuric acid of 84.9 per cent concentration at 61 per cent yield. By a related process, ferrous iron in the spent liquor is first oxidized by chlorine gas, and aluminum chloride is fractionally precipitated by saturation with hydrogen chloride gas.[24] After filtration to remove the aluminum chloride crystals, the ferric ion is reduced to ferrous with scrap iron, and ferrous chloride is precipitated by adding hydrogen chloride gas to the filtrate. Ferrous sulfate and vanadyl sulfate separate as a sludge on evaporation, leaving concentrated sulfuric acid. The recoveries of aluminum, iron, and vanadium by this process are 97, 82, and 98 per cent, respectively.

Waste hydrolysis liquor is first treated with hydrogen chloride, which reacts with the ferrous sulfate to form additional sulfuric acid and ferrous chloride, which precipitates from the acid solution.[25] In the second step the ferrous chloride is reduced with hydrogen at 1000 to 1300° F to metallic iron powder and hydrogen chloride. The hydrogen chloride is recirculated to the initial step of the process. The iron powder is of a grade suitable for use in powder metallurgy processes. Preferably the spent liquor is concentrated by evaporation before treatment with the hydrogen chloride so that the sulfuric acid concentration is high enough to assure precipitation of the ferrous chloride produced. Coke oven gas, which usually contains about 50 per cent hydrogen by volume as well as some carbon monoxide, may be used as the reducing agent. In place of hydrogen chloride as the chlorinating agent and hydrogen as the reducing agent, carbonyl chloride may be used as the chlorinating agent and carbon monoxide as the reducing agent. Sulfuric acid and metal compounds are recovered in good yield from waste solutions.[26] Hydrochloric acid gas is passed into the waste solution at a temperature below 30° C to precipitate chlorides of iron, aluminum, and nickel. After separation, the precipitate is treated with an organic solvent such as ether, carbon tetrachloride, or pyridine to extract the aluminum chloride. The remaining ferrous chloride is dehydrated and baked with ferric oxide or magnetic iron oxide at 350° C in the presence of steam and air. This reaction yield is almost 100 per cent. After separating the precipitate, the solution is treated with a stoichiometric amount of hydrofluoric acid to precipitate magnesium fluoride. The recovery of sulfuric acid from the waste solution is 85 to 95 per cent.

Sulfuric acid is recovered from waste hydrolysis solution by a multi-stage process.[27] Ferrous iron is first oxidized to ferric with chlorine, and hydrogen chloride is passed into the solution. The aluminum chloride precipitate is filtered off. Iron powder and sulfur dioxide are added to the filtrate to reduce iron III to iron II and hypochlorite ion to chloride. Iron is precipitated as ferrous chloride. Sulfurous acid is oxidized to sulfuric acid by aeration, and the ferrous chloride is filtered off. Hydrochloric acid separation and sulfuric acid concentration are carried out by heating. By the concentration of the hydrochlorinated waste acid, 90 per cent sulfuric acid was recovered with a yield of 60 per cent.[28] Waste solution from the sulfate process containing 18 per cent sulfuric acid with 2.5 per cent iron and 0.6 per cent titanium dioxide is utilized for the production of ammonium sulfate by neutralization with ammonia.[29] Waste sulfuric acid containing ferrous sulfate is mixed with waste sulfuric acid containing hydrocarbons from petroleum refining and sprayed into a decomposition furnace where the hydrocarbon burns in preheated air to maintain a temperature of 2000° F.[30] Supplemental fuel may be added if necessary. The gases recovered after passing the products of combustion through scrubbing towers contain 10 to 15 per cent sulfur dioxide and are suitable for use in the manufacture of contact sulfuric acid. Iron sulfate in the waste liquor is present primarily as ferrous sulfate, but ferric sulfate or mixtures of these may be utilized. Concentrated sulfuric acid and an iron oxide suitable for blast furnace feed are recovered from the waste liquors.[31] Copperas crystals separated in the manufacturing process are converted to the monohydrate and roasted in a multibed fluidized roaster. This Dorr-Oliver process has been used by the British Titan Products Co., Ltd., to recover 70 tons of sulfuric acid a day. Waste sulfuric acid is introduced into the cold or only moderately warm zone of the roasting furnace, where it is mixed with pyrites or other sulfide ore to be roasted.[32]

In a cyclic process for recovering spent sulfuric acid, the acid solution is neutralized with ammonia and the resulting ammonium sulfate is roasted at 800° F with specially prepared zinc oxide.[33] Zinc sulfate is formed and the ammonia evolved is recycled. The zinc sulfate is then heated at 1250° C in a stream of sulfur dioxide to regenerate zinc oxide and evolve sulfur trioxide which is absorbed in dilute acid to prepare concentrated sulfuric acid in the usual way. Fresh zinc oxide is prepared by roasting pellets of activated alumina in a solution of zinc sulfate, drying, calcining at 1300° F in dry air and cooling. The cycle is repeated until the material contains 17 per cent oxide by weight. Oxides of chromium or iron can be used instead of zinc oxide, and diatomaceous earth pellets can be substituted for activated alumina.

Kingsbury and Schultz [34] recovered vanadium from the waste liquors by precipitating it along with the ferrous sulfate. Ferrous sulfate monohydrate precipitated by concentrating the waste acid solution is converted to the heptahydrate by the addition of 0.95 to 12.7 parts of water to each part of monohydrate. After aging, the system is held at 45 to 80° F for 1 hour to form solid heptahydrate and a liquid phase. Vanadium is recovered in 95 per cent yield from the separated liquid phase.

By the use of low cost chemicals, waste liquor containing 8.5 per cent sulfuric acid and 11 per cent ferrous sulfate heptahydrate is neutralized and simultaneously converted to solid end products leaving a harmless effluent which may be freely discharged.[35] In the first step the waste is neutralized with sodium hydroxide. The precipitated ferrous hydroxide is separated by filtration and heated to form ferric oxide and water. Calcium oxide is added to the filtrate to produce gypsum and sodium hydroxide. The calcium sulfate dihydrate is collected by filtration. A part of the sodium hydroxide filtrate is recycled, and the remainder is converted to sodium carbonate by treatment with carbon dioxide. The gypsum is converted to plaster of Paris. By this treatment, 1 gallon of waste liquor produces approximately 4.5 ounces of ferric oxide, 2 pounds of gypsum, and 1.11 pounds of anhydrous sodium carbonate. The sodium carbonate may be recovered in the hydrated form suitable for soap manufacture.

Schulmann [36] describes examples of waste disposal problems in the titanium dioxide pigment as well as other industries.

A study of the disposal of chemical waste by barge in an area in the New York Bight 13 miles from Scotland Lightship and 10 miles off the New Jersey coast was reported by Redfield and Walford.[37] The waste originates at the titanium pigment plant of the National Lead Company at Sayreville, New Jersey, and consists of 10 per cent ferrous sulfate and 8.5 per cent sulfuric acid dissolved in fresh water. Other substances present in minor amounts are not likely to prove harmful. The acidification of the sea water is very short lived since the acid is diluted and neutralized rapidly in the turbulent wake of the barge. Turbidity of the water due to the formation of ferrous hydroxide is the only prominent aftereffect. Biological observations revealed no evidence of damage to or exclusion of population of fish or bottom-living animals. Zooplankton introduced directly into the contaminated water of the wake are temporarily immobilized, but recover as soon as the water is diluted with clean sea water. It is concluded that the microscopic plants or animals forming the basic food supply of fishes are not being affected significantly. Concentrations of iron in the offshore area have not increased measurably since the disposal operations were started in 1948, and there has been no accumulation of iron in the water or measurable accumulation of iron on the bottom in the area of the disposal. There is no indication that accumulation of wastes being barged into the sea has altered the transparency of the offshore waters in the area of disposal. The capacity of offshore waters to receive and dispose of soluble or suspended waste without undesirable effect is very large and might properly be used more extensively. The character of the waste and selection of suitable disposal sites are major factors to be considered.

# 18

# Hydrolysis of Titanium Tetrachloride to Produce Pigment

**INTRODUCTION**

Processes for the production of pigments from titanium tetrachloride involve the high temperature oxidation (splitting) of the anhydrous vapor or the thermal hydrolysis of its aqueous solutions. Such solutions are prepared directly by the action of concentrated hydrochloric acid on selected titanium minerals and secondary products. Anhydrous titanium tetrachloride obtained by the reaction of chlorine with titanium dioxide ores or concentrates may be dissolved in water in a separate step to produce solutions of the desired concentrations.

The earliest investigators fused the ores with alkali metal compounds (sodium carbonate, potassium bisulfate) and dissolved the melt in dilute hydrochloric acid. Many titanates are soluble in hydrochloric acid, but such solutions are necessarily impure. Similar products may be prepared indirectly by dissolving orthotitanic acid at room temperature, or by adding an alkaline earth chloride to regular sulfate solutions. The resulting alkaline earth sulfate is insoluble and is readily removed by settling or filtration.

Three general methods of obtaining titanium dioxide from titanic chloride have been employed: (1) hydrolysis of aqueous solutions by heating, by dilution, or by the addition of an alkali; (2) by introducing anhydrous titanium tetrachloride in the vapor phase, along with steam, into a reaction chamber; (3) by heating the anhydrous material in the vapor phase, admixed with air or oxygen at a high temperature, either indirectly or in a direct flame.

## HYDROCHLORIC ACID ATTACK

Llewellyn [1] reported a process for decomposing titanium ores with hydrochloric acid in which the ground ore (ilmenite) was digested with the aqueous acid at 50° to 60° C for a long period of time. After cooling to room temperature, the liquor was filtered to remove any undissolved residue, and the solution containing titanic and iron chlorides was heated at 85° to 90° C to precipitate the titanic acid. A relatively low temperature was employed during the digestion step to prevent loss of dissolved titanium by hydrolysis. Dissolution was accelerated by introducing gaseous hydrogen chloride continuously or intermittently into the system to maintain the optimum concentration as the reaction proceeded.[2] According to Riley,[3] the whole of the titanium content of titaniferous iron ores may be dissolved in hydrochloric acid. Solutions containing 100 g titanic chloride per liter were prepared by dissolving the residue from bauxite after the Bayer process.[4]

By heating finely divided silicate ores of the sphene class,[5] under controlled conditions with 10 to 20 per cent excess hydrochloric acid of concentration above 15 per cent, the titanium content was converted to chloride and held in solution during the entire extraction process. The acid-ore mixture was first heated under reflux at 40° to 50° C to avoid loss of hydrogen chloride before the strength of the acid had been reduced by reaction, and to build up a concentration of titanic chloride sufficient to arrest hydrolysis as the temperature was later raised. Losses due to hydrolysis were further reduced by raising the temperature very slowly to the boiling point in 24 hours.

Ores containing iron in the ferric state, such as arizonite,[6] were first subjected, at 550° to 650° C, to the controlled action of a reducing gas to convert the ferric compounds to ferrous without the formation of metallic iron. The treated ore was then digested with dilute hydrochloric acid under a pressure corresponding to 135° to 180° C to effect dissolution.

Vigoroux and Arrivant [7] prepared the tetrachloride by the action of hydrochloric acid gas on ferrotitanium. The alloy was heated to a low redness in a porcelain tube, and dry hydrogen chloride was passed through. Reaction proceeded, with the evolution of heat. The ferric chloride formed, condensed in the cooler part of the tube, and tended to clog the passage, while the titanium tetrachloride distilled over and condensed in a receiving vessel. According to a modified procedure, the iron content was selectively dissolved from the alloy with dilute hydrochloric acid, and the residue, containing 80 to 90 per cent titanium, was separated from the oxides and treated with the dry gas as before.

Sulfate solutions of titanium, such as are obtained from ilmenite by commercial methods, were converted to the chloride by treatment with an equivalent amount of calcium, barium, or strontium chloride.[8] The precipitate of insoluble alkaline earth sulfate was removed by filtration, leaving the solution of titanic chloride.

Ilmenite is dissolved in 35 per cent hydrochloric acid at 55 to 65° C so that the final liquor contains 3 to 4 moles of hydrogen chloride per mole of titanium.[9] The iron III is reduced to iron II by electrolytic or chemical means, after which the solution is boiled to expel hydrogen chloride and precipitate crude titanium dioxide hydrate. More ferrous chloride may be added to lower the solubility of hydrogen chloride. The yield of dissolved titanium dioxide is 99 per cent. Ilmenite is dissolved in concentrated hydrochloric acid at 55 to 65° C.[10] After reducing the iron to the divalent state, gaseous hydrogen chloride is added to the cooled solution to precipitate the iron chloride, which is removed by filtration. Gaseous hydrogen chloride is driven from the solution, after which it is diluted and heated to precipitate hydrous titanium dioxide. Alternatively, the titanium content may be precipitated from the solution as $K_2TiCl_6$. This compound is decomposed at 400° C to form titanium tetrachloride and potassium chloride. Ilmenite is dissolved with more than 32 per cent hydrochloric acid with strong agitation within 2 hours.[11] For example, 600 g of ilmenite containing titanium dioxide 44.5, ferrous oxide 34.5, and ferric oxide 12.5 per cent was mixed with 2 liters of hydrochloric acid, having a density of 1.19, in a closed vessel. The vessel was rotated at 20 rpm in a water bath at 95° C. A solution containing 98.4 per cent of the titanium and 99.9 per cent of the iron was obtained. From the solution, iron was crystallized as $FeCl_2 \cdot 4\,H_2O$ by cooling to 20° C. This solution contained 142 gpl titanium dioxide, 69 gpl ferric oxide, and 56.7 gpl hydrochloric acid.

Decrease of the rotation rate to 2 rpm decreases the recovery of titanium from the ilmenite to 85.2 per cent. Increase of the temperature above 100° C leads to hydrolysis of a large part of the titanium, but the hydrolysis can be completely eliminated by phosphorus pentoxide. Thus the solution obtained contains only 23 per cent of the titanium if the reaction is carried out in an oil bath at 110° C without phosphorus pentoxide, whereas the presence of phosphorus pentoxide leads to quantitative solution of the ore. After precipitation of the titanium and most of the iron, the hydrochloric acid containing solutions are reused for further extraction. The precipitated iron chloride is heated, and the hydrochloric acid formed is passed into the extraction solution.

Rutile or other titaniferous material is heated in the presence of coke or hydrogen at 1400 to 1600° C for 1 to 15 hours before chlorination with hydrogen chloride gas.[12] The reduction may be carried out in a fluidized bed comprising the titaniferous material and coke. Inert gases such as nitrogen or carbon dioxide may be employed as the fluidizing agent. If a reducing gas, notably hydrogen, is the fluidizing gas, the presence of carbon in admixture with the titaniferous material is optional since hydrogen alone will suffice. The change that takes place in the titaniferous material is not completely understood. After rutile is heated with the coke or hydrogen it no longer shows the X-ray diffraction pattern characteristic of the rutile crystal form. In other words, the diffraction pattern of the titaniferous residue is different from that of rutile. The resulting titaniferous residue is subjected to the action of hydrogen chloride at 800 to 1000° C under anhydrous conditions to form titanium tetrachloride. Elemental hydrogen is evolved by the reaction along with the titanium tetrachloride. If ilmenite is used, ferrous chloride is formed along with the titanium tetrachloride. Ferrous chloride has a very high boiling point and solidifies at temperatures well above that of the boiling point of titanium tetrachloride so that it may be separated in a cyclone. Before it is heated with hydrogen chloride, the unreacted carbon may be separated from the reduced product by flotation in bromoform. The process may be carried out continuously. Care is exercised to remove water vapor from the titaniferous residue prior to reaction with hydrogen chloride. Usually the sweeping action of the hydrogen is adequate.

In an example, 100 g of rutile ore and 50 g of petroleum coke were heated in a static bed at 1400° C for 18 hours. The evolved carbon monoxide was drawn from the treatment zone and the residue was slurried in bromoform to float the coke from the titaniferous material.

A one inch sillimanite reactor tube was charged with 40 g of coke-free residue. Hydrogen chloride was introduced at the bottom of the tube and passed upward at a rate of 250 ml per minute for 3 hours and 10 minutes. The temperature ranged from 800 to 1000° C. Effluent gases included titanium tetrachloride and hydrogen. Eighteen grams of titanium tetrachloride were obtained by separation from the unreacted hydrogen chloride in the effluent gases.

Ilmenite ore or other titanium containing materials are dissolved by heating at 60 to 90° C with a 20 to 40 per cent solution of hydrochloric acid containing 0.1 to 10.0 per cent of fluoride ion calculated as hydrofluoric acid and based on the weight of titanium dioxide content.[13] The amount of hydrochloric acid is sufficient to form solutions from 50 per cent basic to 25 per cent acid.

Ilmenite or slag is fused with sodium hydroxide at 400 to 425° C and heated with water and hydrochloric acid.[14] The purified solution is hydrolyzed to produce titanium dioxide. A mixture of mineral rutile and a sodium or potassium compound in the ratio of 1 to 1, to 4 to 1 is roasted to form an alkali titanate without fusion.[15] After washing with water, the alkali titanate is boiled with 10 to 32 per cent by weight of hydrochloric acid or with 5 to 32 per cent by weight of sulfuric acid to effect solution. The hydrous titanium dioxide obtained by hydrolysis of the solution is washed and calcined at 500 to 1000° C. Improvement in tinting strength is obtained by ball milling before, during, or subsequent to the washing step. Ilmenite is smelted with soda ash and a reducing agent to form cast iron and a titanium-rich slag soluble in hydrochloric acid.[16] In an example, 200 parts of ilmenite containing 41.68 parts of titanium dioxide and 50.66 parts of iron oxides was mixed with 60 parts of soda ash and 18 parts of carbon, ground and melted in an electric furnace at 1450° C. From 75 to 80 parts of cast iron and 120 to 130 parts of slag were produced. One hundred parts of the slag were dissolved in 500 parts of 21° Bé hydrochloric acid at 80 to 85° C and the solution was filtered. Ferric ion was reduced to ferrous by any chemical or electrochemical means. The solution was adjusted to a concentration of 100 gpl titanium dioxide by concentration or dilution and boiled to effect hydrolysis of the titanium tetrachloride.

A potassium, ammonium, or rubidium chlorotitanate is prepared, from which an acidic solution containing primarily titanium and chloride ions is obtained.[17] The solution is suitable for the manufacture of titanium dioxide pigments or for the preparation of rutile solutions. For example, one liter of a sulfate solution of ilmenite, containing 98 gpl titanium dioxide, 340 gpl sulfuric acid and 25 gpl iron, is saturated with hydrogen chloride gas at 0° C and a stoichiometric amount of potassium chloride, 304.8 g, is added to precipitate $K_2TiCl_6$.

Formation of gaseous titanium tetrachloride, titanium trichloride, titanium dichloride, hydrogen, and nitrogen from titanium nitride and hydrochloric acid above 1300° C occurs reversibly.[18] The reaction takes place at sufficient speed that transport of titanium nitride to the cool zone of a temperature gradient is achieved.

A relatively pure titanium tetrachloride solution well suited for the production of titanium hydrolysis seed nuclei which impart easy rutile conversion characteristic to the anatase hydrolyzate is obtained from titanium sulfate solution by adding calcium chloride and barium chloride.[19] One volume of titanium sulfate solution analyzing titanium dioxide 210 and sulfuric acid 520 gpl was mixed with an equal volume of calcium chloride solution containing 500 gpl. The calcium chloride solution was added slowly to the titanium sulfate solution with agitation at a temperature below 35° C. At the end of the addition the titanium sulfate was largely converted to titanium tetrachloride, which served as a suspension medium for the calcium sulfate. After removal of the calcium sulfate by filtration, saturated aqueous barium chloride was added to the filtrate in an amount just sufficient to remove all of the remaining sulfate ion. The solid barium sulfate was filtered from the solution.

## HYDRATION OR DISSOLUTION OF THE TETRACHLORIDE

In dissolving anhydrous titanium tetrachloride in water, an initial turbidity usually occurs as a result of the separation of oxychlorides or hydrous oxides. These compounds can be redissolved, however, if additional amounts of the anhydrous tetrachloride are added. Investigations revealed that the original precipitates were caused by the great rise in temperature that accompanies hydration. If no cooling was provided, the temperature of the solvent rose to above 100° C, until a concentration corresponding to titanium dioxide 130 and hydrochloric acid 250 gpl was reached. At this point hydrochloric acid gas was evolved, and precipitation began to be apparent. If more tetrachloride was added, the temperature decreased, and the suspended material gradually dissolved to give a clear concentrated solution. The decrease in temperature in the second stage of the process occurred because the heat of evaporation of the hydrogen chloride was greater than the heat of solution of the titanium tetrachloride.

Because of losses of hydrogen chloride, the clear solutions obtained by this method contained tetravalent titanium and chlorine in an atomic ratio less than the theoretical 1 to 4, and usually about 1 to 2.7. Concentrations corresponding to 550 g titanium dioxide and 600 g hydrochloric acid per liter were readily attained.

Better results were obtained by employing dilute hydrochloric acid instead of water as the initial solvent.[20] Hydrogen chloride started boiling off at a lower temperature; the rise in temperature was less; and the degree of precipitation was decreased. Clear solutions of any practical concentration were obtained by mixing the anhydrous material in a similar manner with dilute aqueous titanium tetrachloride containing more than 150 g free and combined hydrochloric acid per liter.[21] Dissolution in this manner was accompanied by a small increase in temperature and copious evolution of hydrogen chloride so that the final product contained a lower proportion of chlorine than that corresponding to the chemical formula $TiCl_4$. The working solvent was prepared by dissolving titanium oxychloride in an excess of concentrated hydrochloric acid and adjusting the product to a concentration corresponding to 60 g titanium dioxide and 160 g hydrochloric acid per liter. Once in operation, a portion of the clear concentrated solution was diluted with water and used as the initial solvent.

The process was carried out in a continuous manner by running the anhydrous tetrachloride and aqueous solvent simultaneously into a dilute titanic chloride solution containing more than 150 g hydrochloric acid per liter. A typical product contained the equivalent of 550 g titanium dioxide and 600 g hydrochloric acid per liter, which corresponded to an atomic ratio of titanium to chlorine of 1 to 2.6.

Following these methods of preparing solutions of titanium tetrachloride, local overheating occurred, caused by the high heat of solution, with the result that some of the compound was lost by volatilization. This loss was reduced by vigorous agitation of the system and by cooling, but it could not be entirely avoided by these precautions. However, if the anhydrous material was introduced below the surface of the aqueous solvent, with constant agitation, losses resulting from evaporation were almost entirely avoided.[22] One or more submerged pipes were employed for the anhydrous material; by passing a current of dry air or inert gas along with the tetrachloride, backing up in the pipes was prevented and danger of clogging by the basic precipitate was minimized. Water, hydrochloric acid, or dilute solution of the tetrachloride served as suitable solvents in the process.

Dissolution was also carried out in a continuous manner by simultaneously introducing the anhydrous material and the aqueous solvent through submerged feed pipes into a small initial quantity of the solution from a previous run. Concentration of the titanium chloride was adjusted by regulating the rate of addition of the two components. The product was removed continuously through an overflow outlet in the vessel.

By another approach the purified titanium tetrachloride was dissolved directly in dilute sulfuric acid, and the hydrogen chloride liberated was passed into a solution of sodium hydroxide. The titanium sulfate solution obtained was subjected to thermal hydrolysis, as before. In either case the alkali metal (sodium) chloride liquor was electrolyzed to recover chlorine and the corresponding hydroxide, both of which were again used in the process. The ferric chloride was dissolved in water and reduced in solution to the ferrous salt by boiling with ferrous sulfide, according to the reaction, $2\,FeCl_3 + FeS \rightarrow 3\,FeCl_2 + S$. Sulfur was separated by filtration, and the solution was electrolyzed to recover chlorine for reuse and metallic iron as a by-product. Agnew [23] treated the effluent from the electrolytic cells with lime to precipitate the remaining iron as hydroxide, and used the separated hydroxide to neutralize more liquor prior to electrolysis so that the cyclic process resulted in no loss of iron.

If titanium tetrachloride is dissolved in hydrochloric acid solutions of 0 to 16.2 per cent concentrations at 0° C, a hydrated titanium dioxide precipitates and the concentration of titanium remaining in the solution is small. At 16.2 per cent hydrochloric acid the concentration of titanium dioxide in the solution is 0.93 per cent.[24] With a hydrochloric acid concentration of 16.2 to 36.0 per cent, the solubility increases to over 30 per cent titanium dioxide, the crystalline solid phase being $H_2Ti(OH)_3Cl_3$. These hydrated chlorides lose water on heating and are hydrolyzed endothermically at 125° C. They show amorphous to crystalline titanium dioxide at 340° C. According to Yuravskii,[24a] the intensity of attack of titanomagnetites by hydrochloric acid decreases with the content of iron and finally becomes negligible.

## PRECIPITATION OF TITANIUM DIOXIDE

**Hydrolysis of Solutions by Boiling.** Titanium dioxide may be obtained from the tetrachloride by hydrolytic precipitation effected by heating its aqueous solutions or by bringing the anhydrous material in the vapor phase into contact with steam in a reaction chamber. By regulating the conditions and employing a suitable seeding agent, titanium dioxide of pigment grade can be produced. In large-scale operation, the splitting operation has the advantage that chlorine is recovered directly rather than hydrochloric acid. Before use, however, the tetrachloride must be carefully purified, a step not so important in the methods involving hydrolysis of aqueous solutions, since most of the objectionable impurities remain dissolved in the acid hydrolysis liquor.

Thermal hydrolysis of chloride solutions in the presence of nuclear titanic acid, precipitated from a separate portion of the solution by moderate heating after reduction of the acidity to a pH of between 2 and 3 either by dilution or by neutralization with an alkali, yielded a product which on calcination gave a pigment of the rutile type having a correspondingly high tinting strength.[25] Such seeding compounds obtained at low acidity, for instance a pH of 5 to 7, had little nucleating power. On the other hand, similar products obtained from very acid solutions, for example those having a hydrogen ion concentration greater than pH 2, were very active, but stabilization required particular care. Very active nuclei were found to be less stable at high temperatures and were cooled quickly to preserve their effectiveness. To obtain the maximum degree of stability, the seed was subjected to a controlled cooling at a rate dependent upon the original acidity, and the temperature was maintained during the precipitation.

As a rule, a proportion of nucleating compound from 5 to 8 per cent of the titanium dioxide to be precipitated was required, and it was not essential that the seed dissolved in the main hydrolysis solution. Nuclei less active than normal were employed in larger proportions, and for the same results concentrated solutions required more seeding agent than dilute solution. Freshly prepared titanic chloride solutions were more reactive than older ones and thus required less nucleation.

Titanium prepared in accordance with this process was of the rutile crystal form, and the resulting pigments had higher tinting strengths than the anatase products obtained from sulfate solutions.

In a typical operation, a chloride solution containing the equivalent of 150 g titanium dioxide per liter and having an atomic ratio of titanium to chloride of about 1 to 3 was employed for hydrolysis and for preparation of nuclei. A portion of the solution was treated with aqueous sodium hydroxide to bring the pH to 2.5, heated at 80° C for 30 minutes to develop the nucleating property, and cooled quickly to below 50° C. The solid product precipitated in this manner was the active seeding agent and was usually separated and washed before use, although the original suspension in the mother liquor was effective. A quantity of the seeding material containing 7 g titanium dioxide was added to each liter of the original chloride solution and the mixture was heated at 100° C. Hydrolysis proceeded rapidly and was complete in 10 minutes. The precipitate of hydrous titanic oxide was washed, treated with 2 per cent potassium bisulfate, heated slowly to 900° to 950° C in a rotary calciner, and held at this temperature for 15 minutes.

According to another approach, anhydrous titanium tetrachloride, free from iron, copper, and vanadium, was allowed to run into 18 per cent hydrochloric acid, and the resulting oxychloride was dissolved in water to produce a solution containing the equivalent of 320 g titanium dioxide and 353 g hydrochloric acid per liter. Dilute aqueous sodium hydroxide at 40° C was added to a portion of the solution, with agitation, and during the procedure the temperature rose to 60° C. The amount and concentration of the reagents were so chosen that after mixing, the pH of the solution was 2 and the strength corresponded to 30 g titanium dioxide per liter. This liquor was then heated to 80° C, held at this temperature for 30 minutes, and cooled to below 60° C by adding cold water in such proportion as to bring the concentration to 20 g titanium dioxide equivalent per liter.

The remainder of the original solution was then mixed with concentrated hydrochloric acid to yield a liquor in which the atomic ratio of titanium to chlorine was 1 to 4, and seed was added to supply titanium dioxide equal to 8 per cent of that to be precipitated. After dilution with water to 55 g titanium dioxide and 100 g hydrochloric acid per liter, expressed as hydrolysis products, the liquor was heated to 90° C in 12 minutes and held at this temperature for 30 minutes to effect hydrolysis. The washed precipitate was reslurried in water to 250 g per liter solids, treated with potassium bisulfate in an amount equal to 1.5 per cent of the titanium dioxide, dewatered, and calcined at 950° C for 30 minutes. Both pigments were of the rutile crystal modification, and after milling and pulverizing by the conventional methods exhibited tinting strengths one fourth to one third higher than the best anatase products.

Similarly, titanium tetrachloride seeded with a sol prepared by heat-treating a diluted portion of the same solution gave a hydrolysis product which, after calcination at moderate temperature, exhibited the rutile crystal structure and the high tinting strength corresponding to this modification of titanium dioxide of pigment grade.[26] Heat treatment developed the nucleating property of the sol, and both temperature and time were important, since overcuring lessened the effectiveness of the product. Longer periods of heating at lower temperatures were required to develop the maximum nucleating efficiency. Best results were obtained by curing solutions, containing the equivalent of 10 to 20 g titanium dioxide per liter, at 80° to 90° C for 10 to 15 minutes. The resulting sol presented a slight opalescence, and the particles could not be separated by the conventional methods of filtration and washing. However, the degree of dispersion of the colloidal titanic oxide and its homogeneity did not seem to have a great influence on the nucleating property of the sol.

Tetrachloride solutions for hydrolysis contained the equivalent of 140 g to 160 g titanium dioxide per liter, and the proportion of nuclei was from 5 to 9 per cent of the total titanium dioxide. Solutions of high purity gave products of higher opacity and better pigment properties, although the presence of iron and organic impurities was not found to be detrimental to efficient nucleating properties of the sol. Any ferric iron was reduced to the ferrous state to prevent its precipitation with the titanium.

To illustrate, 1 liter of an aqueous solution of titanic chloride, containing the equivalent of 15 g of the dioxide, was transferred to a flask fitted with a mechanical stirrer and a reflux condenser, and heated to 80° to 85° C. After 10 minutes at this temperature, 1250 ml of an untreated solution at room temperature was added, containing the equivalent of 150 g titanium dioxide per liter, and the mixture was heated to boiling and refluxed for ½ hour. The precipitate was filtered, washed, calcined, and pulverized to yield a pigment of the rutile crystal form having a strength 20 to 25 per cent higher than the conventional anatase product along with excellent color and brightness.

Barksdale[27] prepared very effective nuclei by heat-treating dilute solutions of titanium tetrachloride to which an alkaline agent had been added to convert from 10 to 60 per cent of the potential or active hydrochloric acid to a salt that would not hydrolyze under the conditions subsequently employed. The efficiency of the nucleating composition, and the quality of the pigment obtained, were found to depend upon a number of correlated factors which may be more conveniently described separately.

The solutions employed were prepared by diluting commercial titanium tetrachloride with water; because of losses of hydrochloric acid gas by volatilization, the proportion of chloride was usually less than that corresponding to the chemical formula $TiCl_4$, and in some solutions was as much as 10 to 15 per cent. However, the method was also applicable to solutions produced indirectly, as by dissolving orthotitanic acid in hydrochloric acid or by adding alkaline earth metal chlorides to solutions of titanium sulfate. Iron and other polyvalent heavy metals, present as impurities, were reduced to the more stable toward hydrolysis, lower valent form, by electrolytic means or by adding zinc, scrap iron, or sulfurous acid. Concentrations expressed as hydrolysis products, according to commercial practice, between 250 g and 400 gpl titanium dioxide and 450 g to 640 gpl hydrochloric acid, gave best results, and at the same time permitted more economical operation of the process. Such solutions were used for hydrolysis and for preparation of the seeding agent.

In preparing nuclei, an alkaline compound was added to a portion of the original solution in such proportions that from 10 to 60 per cent of the potential hydrochloric acid, normally from 30 to 45 per cent, was neutralized. The composition of the alkali used was of minor importance provided that it formed a compound nonhydrolyzable under the conditions of processing, and such bases as calcium hydroxide, potassium carbonate, sodium hydroxide, aqueous ammonia, sodium sulfide, and certain organic amines fulfilled this requirement to a high degree. It is evident that the chlorides of these agents were soluble in water and that they dissolved in the solution. Calcium oxide, in addition to its low cost, gave optimum results but since it was relatively insoluble in water, it was added in the form of a slurry or in the solid state.

Concentration of nuclear solutions from 5 g to 30 gpl titanium dioxide proved effective, but best results were obtained within the range of 10 g to 20 gpl. Obviously the potential hydrochloric acid was less than that corresponding to the normal salt.

Proper curing of the nuclei was a function of both temperature and time of heating, and could be followed roughly by the degree of turbidity developed, which reached a translucent stage. In general, 10 minutes at 80° to 90° C developed the nucleating property to the maximum degree, although the same results were obtained by holding the system at a lower temperature for a longer period of time, for instance at 70° to 75° C for 1 hour. The exact change in the solution was not established, but a large part of the titanium component was converted from the crystalloidal to the colloidal condition. Partial neutralization did not render the solution of titanic chloride more susceptible to precipitation during the subsequent heat treatment, as might be expected, but on the contrary made it more stable. Such a solution, after treatment with the alkali, could be heated for a longer period of time at a given temperature without precipitation of titanic oxide. However, with both solutions precipitation would finally occur, but curing was not carried to this stage. The greatest efficiency of the nuclear composition, as measured by the properties of the pigment obtained, was found to be the combined effect of the partial neutralization and the heat treatment.

In carrying out the hydrolysis step, the manner of mixing the components was not particularly important, but for convenience the nuclear composition was usually added to the solution of titanium chloride to be precipitated. The nuclei were employed at the temperature formed or cooled to room temperature, and the hydrolysis solution could be preheated to the evolution of hydrogen chloride gas. Titanium, supplied to the hydrolysis mixture by the seeding composition of from 1 to 10 per cent of the total, exerted satisfactory nucleating properties, but from 4 to 6 per cent yielded titanium dioxide of the best pigment properties.

The efficiency of the nucleation and the quality of the resulting products were influenced by the final concentration of the hydrolysis mixture, and values from 125 g to 150 gpl titanium dioxide proved most satisfactory. After mixing the nuclear suspension with the chloride solution water was added, if necessary, to adjust the concentration to the desired value. If the hydrochloric acid content exceeded 225 gpl, hydrogen chloride gas was evolved on boiling, and arrangement for its recovery or disposal was provided. The seeded mixture was heated to boiling and maintained to bring about the hydrolytic reaction. A yield of 97 per cent or better was usually obtained after ½ hour, in some tests in 15 minutes, and rarely was it necessary to continue the boil for 1 hour.

In an actual operation, a solution prepared by diluting commercial titanium tetrachloride with water and containing 340 gpl of the dioxide and 545 gpl hydrochloric acid (expressed as hydrolysis products), was employed. To a portion of this solution containing the equivalent of 15 g titanium dioxide was added 7 g hydrated lime (the equivalent of approximately 1 mole of hydrochloric acid per mole of titanium tetrachloride), with stirring. This partially neutralized clear solution was diluted with water to 1 liter, heated to 85° C in 20 minutes, held at this temperature for 10 minutes, with agitation, and cooled rapidly to room temperature. The product was an opalescent colloidal solution of titanium dioxide. This composition was added rapidly to 1000 ml of the original solution, with constant agitation, and 500 ml water was introduced to adjust the final concentration to 144 gpl titanium dioxide. The seeded mixture was heated from room temperature to boiling in 30 minutes, and refluxed for 1 hour to precipitate 97 per cent of the titanium as hydrous oxide. This product was filtered, washed with water acidified with sulfuric acid to prevent peptization, and calcined at 800° C for 1 hour to produce a pigment of the rutile crystal form having excellent brightness and color and the very high tinting strength of 1750 by the Reynolds method.

A nucleating composition, prepared by heating a titanium tetrachloride solution containing the equivalent of 5 g to 30 g of the dioxide per liter and 0.5 to 2.0 mole proportions of a nonoxidizing univalent anion such as chloride, acetate, or formate, at 80° to 90° C for 10 to 15 minutes to develop a decided opalescence, accelerated and directed the thermal hydrolysis of relatively concentrated aqueous solutions of the tetrachloride.[28] The cation had no appreciable influence upon the hydrolysis, but only colorless compounds, for example, calcium chloride, sodium metaborate, and potassium acetate, were of practical value. Approximately 5 per cent of the nuclear suspension, based on the titanium content, was required, and the hydrolysis product after calcination at a relatively low temperature was of the rutile crystal modification and had the correspondingly high hiding power and brightness. Von Bichowsky[29] prepared a solid nucleating agent corresponding to the formula $Ti(OH)_3Cl \cdot 5H_2O$ by evaporating a solution of titanium hydroxy chloride. The product was effective in the hydrolysis of chloride and nitrate solutions.

To produce titanium tetrachloride solutions more amenable to directed hydrolysis on addition of external nucleating agents, the initial solution was subjected to heat treatment at a temperature below that at which titanium dioxide would be precipitated, and the residual colloidal material was coagulated and separated.[30] This heat treatment removed residual components which would have a nucleating influence.

Hydrolysis is effected without introducing externally prepared nuclei by adding concentrated titanium tetrachloride solution to hot water and boiling the resulting mixture to complete precipitation of the titanium dioxide.[31] The solution is added to the water at a sufficiently slow rate to insure the initial formation of colloidal titanium dioxide particles which serve as seed for the product formed later in the process. In general, the concentration of the chloride solution and ratio of solution to water are such that, after hydrolysis, the resulting liquor has a composition corresponding to constant boiling hydrochloric acid. Yields of 96 per cent are obtained in ½ hour, and the washed and calcined product gives a rutile pigment of excellent covering power and brightness. In an example, 1000 ml of a titanium tetrachloride solution containing 300 g of titanium dioxide and 480 g of hydrochloric acid, expressed as hydrolysis products, was added in 5 minutes to 1000 ml of boiling water, with constant agitation. The mixture was heated to boiling in 15 minutes and refluxed for 30 minutes to effect hydrolytic precipitation of the titanium. This precipitate was filtered, washed, and calcined, without further treatment, at 850° C to produce a pigment of the rutile crystal form possessing excellent brightness and color and having a tinting strength one third higher than that of the commercial anatase products. A similar process employs more dilute solutions.[32]

Readily filterable hydrous titanic oxide of high purity was obtained by the thermal hydrolysis of aqueous solutions of the tetrachloride containing not more than 0.1 mole of polyvalent negative coagulating ion derived from oxalic, tartaric, phosphoric, or citric acids or their alkali metal salts per mole of titanium dioxide.[33] Alternatively, the same results were obtained by adding the tetrachloride solution to hot water containing the coagulating ions. The mixture was boiled to complete the precipitation of the titanium. Since hydrous titanic acid was readily peptized in dilute aqueous hydrochloric acid, better results were obtained by employing very dilute sulfuric acid as the washing liquor.

According to a specific operation of the process, anhydrous titanium tetrachloride was dissolved in water to obtain a solution containing the equivalent of 15 per cent titanium dioxide. All iron present was reduced to the ferrous condition. To prevent reoxidation during processing, about 2 gpl of titanous ions was formed. One thousand pounds of this solution was added in 1 hour to 7000 pounds of boiling water containing 2 pounds of citric acid, and by the end of the mixing operation 95 per cent of the titanium had precipitated as hydrous oxide in a coagulated form that was easily filtered and washed. The washed pulp was calcined at 700° to 1000° C to produce a rutile pigment of good color and high tinting strength. By employing specially purified raw materials, titanium dioxide pure enough for a standard in volumetric analysis was obtained by this method.[34] A precipitate of high quality was obtained by hydrolytic decomposition of aqueous titanium tetrachloride in the presence of 25 to 30 per cent sodium sulfate.[35] At 69° to 72° C, 10 per cent solutions gave highest yields and 5 per cent solutions the poorest, although at 98° to 101° C there was no difference in recovery from solutions containing from 5 to 20 per cent titanium tetrachloride.

Darling[36] obtained finely divided titanium dioxide by passing the tetrachloride, in vapor form, into a boiling solution of glucose. During the process, glucose was converted to glutaric acid. Jenness[37] added the tetrachloride to a solution of hydrochloric acid to form a mixture of basic chlorides in the form of a dry yellow granular powder, and separated and calcined the product to obtain fluffy white titanium dioxide of pigment grade.

Processes for the manufacture of titanium dioxide pigments from siliceous minerals of the sphene class were reported by Alessandroni.[38] If the finely ground ore was heated with hydrochloric acid of concentration above 15 per cent, the titanium values were dissolved and held in solution during the extraction step, along with calcium compounds and minor constituents of the ore such as iron, magnesium, aluminum, and vanadium, leaving a residue composed essentially of silica. However, to hasten dissolution of the soluble constituents, and to obtain a maximum recovery of titanium, an excess of 10 to 20 per cent hydrochloric acid was employed. The mixture was heated under reflux at 40° to 50° C for several hours to avoid loss of hydrogen chloride before the strength of the acid had been reduced by reaction, and to build up a concentration of titanic chloride sufficient to prevent hydrolysis at the higher temperatures employed in the later stages. During the next 24 hours the temperature was raised gradually to the boiling point, and heating was continued until no more titanium dissolved. Since compounds of polyvalent elements such as iron were more readily attacked by the acid and were more stable toward hydrolysis in the lower valent condition, the reaction was carried out in the presence of 1 to 2 per cent of a reducing agent such as titanous chloride or stannous chloride. Iron, for example, was reduced before the titanium, and the added agent was not sufficient to reduce an appreciable amount of the tetravalent titanium. The final solution, containing most of the titanium of the original ore as chloride along with nonhydrolyzable chlorides of calcium, magnesium, ferrous iron, aluminum, and other metals, and some free hydrochloric acid, was cooled and separated from the siliceous residue by decantation or filtration. It was then concentrated to the desired strength by vacuum evaporation and subjected to thermal hydrolysis to precipitate the titanium as hydrous oxide. By leaving the silica suspended in the original solution, composite pigments were produced.

In an example, 1000 pounds of finely ground sphene (100 to 200 mesh), analyzing 32 per cent titanium dioxide, 29.6 per cent silica, 18.7 per cent calcium oxide, 12.9 per cent alumina, 1.43 per cent ferrous oxide, 4.29 per cent ferric oxide, 1.12 per cent magnesia, and 0.10 per cent manganese oxide, was mixed with 8300 pounds of 29.5 per cent hydrochloric acid in a vessel fitted with a reflux condenser and held initially at 40° C for 4 hours to avoid loss of hydrogen chloride. During the next 24 hours, as the acid concentration was reduced by reaction, the temperature was raised gradually to the boiling point, about 107° C. At this stage the reaction had practically ceased and the mixture was cooled and filtered. Ninety-four per cent of the titanium was recovered in the filtrate, which weighed 9100 pounds and contained 300 pounds titanic oxide equivalent, 1420 pounds free hydrochloric acid, and 369 pounds calcium chloride. This solution was next subjected to the usual reducing treatment, concentrated in a vacuum evaporator at 50° C to 11 per cent titanium dioxide, and poured into three volumes of boiling water containing an amount of oxalic acid equal to 0.05 per cent of the titanium dioxide to be precipitated. The liquor was heated to its boiling point in a tank provided with an outlet for the recovery of hydrochloric acid, and held at this temperature until 95 per cent of the titanium was precipitated as hydrous oxide. The hydrolysis product was filtered, washed, calcined, and pulverized to produce a practically pure titanium dioxide pigment of the rutile crystal form possessing excellent color and brightness and high tinting strength.

In a cyclical process, ilmenite ore was extracted with hydrochloric acid to leave a titanium dioxide concentrate which was chlorinated to form the tetrachloride.[39] The titanium tetrachloride was dissolved in water and hydrolyzed by boiling to precipitate the dioxide. Hydrochloric acid liquor formed during hydrolysis was used to extract more ilmenite ore. Leach liquor from the first step was electrolyzed to yield iron and chlorine, and the chlorine was used to form more titanium tetrachloride.

Dutt[40] prepared titanium dioxide from solutions obtained by dissolving the residue from bauxite, after the Bayer process, in hydrochloric acid. Such solutions contained up to 100 gpl titanic chloride; any considerable excess of hydrochloric acid was neutralized and the iron component was reduced to the ferrous state. Sodium acetate was added in an amount equal to one tenth of the titanium tetrachloride present, and the liquor was boiled to precipitate hydrous titanic oxide. After almost complete neutralization, the filtrate was treated with barium salts to precipitate barium vanadate.

Barton[41] treated a sulfuric acid solution of ilmenite with barium chloride, filtered off the insoluble barium sulfate formed, and heated the resulting chloride solution at 100° C for 5 to 8 hours to effect hydrolysis. The precipitate of hydrous titanic oxide was washed, dried, and calcined to produce a product of 98.2 to 99.8 per cent purity. By an opposite approach,[42] the clear solution obtained by dissolving 1 volume of titanium tetrachloride in 2 volumes of water was mixed with 0.35 volume of 95 per cent sulfuric acid and heated to distill off the hydrogen chloride. Heating was continued until fumes of sulfur trioxide were given off. During this period the temperature rose to 150° C, and most of the hydrochloric acid was recovered. The residue, consisting of a white friable mass that contained 77 to 80 per cent titanic oxide, 18 to 20 per cent sulfur trioxide, and 2 to 3 per cent chlorine, was transferred to a calcining furnace and heated to expel the remaining acids, leaving relatively pure, friable titanium dioxide of chalky appearance having a specific gravity of 3.65. Paint films pigmented with this material had a bluish cast, apparently because of its amorphous nature and very fine particle size.

A solution of titanium hydroxide in 2 molecular proportions of hydrochloric acid was boiled with sulfuric acid in the presence of kieselguhr and aluminum silicate to effect hydrolysis.[43] Similarly, aqueous titanium tetrachloride was neutralized to precipitate a gelatinous oxide, and this product was washed and redissolved in 25 per cent sulfuric acid. The solution was boiled to bring about hydrolytic reaction. Precipitation was facilitated by slight hydrogenation, which was accomplished by direct introduction of the gas or by addition of zinc.[44] The amorphous oxide was calcined to produce a crystalline pigment.

Gelatinous titanic oxide of exceptional purity was obtained by treating aqueous titanium tetrachloride with an alkali metal carbonate.[45] On boiling in a bath slightly acidified with sulfuric acid, the precipitate passed into a dense unctuous mass of pigment grade. Hydrous oxide, prepared by heating dilute tetrachloride solutions in air, acted as a strong protective colloid.[46]

The reactions of chloride solutions of titanium were carried out in an iron vessel surrounded by a casing through which water flowed.[47]

According to Atanasiu,[48] the rate of hydrolysis of titanic chloride solutions increases, with temperature and dilution, and platinum exerts a catalytic action. The reaction is bimolecular, indicating a series of consecutive reactions between the colloidal and aqueous phases. Alcohol diminishes the rate. Such solutions yield hydrolysis products which give rutile directly on calcination.[49] If precipitation is carried out in the presence of sulfate or phosphate ions, however, products are obtained which on moderate calcination show the crystal structure of anatase but are transformed at higher temperatures into rutile.

A colloidal solution effective as a nucleating agent for the hydrolysis of titanium chloride or sulfate solution is prepared from titanium tetrachloride solution concentrated by evaporation under reduced pressure.[50] For example, a solution containing 3.63 moles of titanium tetrachloride per liter and 3.63 moles of hydrochloric acid per mole of titanium tetrachloride is evaporated at a pressure of 9 inches of mercury and from 20 to 80° C to a concentration of 4.57 moles of basic titanium chloride per liter and 1.97 moles of hydrochloric acid per mole of basic titanium chloride. The solution is then diluted to 0.19 mole of basic titanium chloride per liter, heated to 85° C in 25 minutes, and held at this temperature for 15 to 20 minutes to develop the nucleating properties to the maximum degree.

Either anatase or rutile can be prepared from pure aqueous titanium tetrachloride solution.[51] If a solution containing 300 gpl titanium tetrachloride is atomized into a dry air stream at 400 to 500° C, anatase is formed while rutile is produced from the same solution by hydrolysis at 60 to 100° C. This indicates that anatase is the form first produced from pure titanium tetrachloride solution and that it can be preserved by immediate removal of the hydrochloric acid formed in the reaction. A solution containing 119 gpl of titanium tetrachloride is heated for 5 hours at 60° C with stirring to effect hydrolysis.[52] The precipitate is filtered, washed, and heated for 1 hour at 300° C to give a 94 per cent yield of the rutile form of titanium dioxide pigment. A high quality pigment of very small particle size is obtained by the hydrolysis of titanium tetrachloride solution.[53] Anhydrous titanium tetrachloride is dissolved in water at 40° C to obtain a solution containing 120 gpl titanium dioxide equivalent. About 5 per cent of the solution is neutralized partially with aqueous sodium hydroxide to a hydrochloric acid to titanium dioxide ratio of 0.25 to 0.35. This portion of the solution is heated at 70 to 80° C for 30 minutes to produce a nucleating suspension which is returned to the main bulk of the solution. After seeding, the solution is boiled for 1 hour to obtain a yield of titanium dioxide of 98.7 per cent. This precipitate is filtered and washed, 2.0 per cent by weight of potassium sulfate is added, and the mixture is calcined at 900 to 920° C for 2 hours. The titanium dioxide formed is washed with water, filtered, and dried at 500° C. It is pure white, with a light reflectance of 97 per cent, a hiding power of 35 g per square meter, and oil absorption of 44. The average particle size of 90 per cent of the product is less than 1 micron. Stability at atmospheric conditions is similar to that of anatase.

According to Fumaki and Saeki,[54] a solution containing 100 g titanium tetrachloride, 30 to 50 g ammonium sulfate, and 50 to 100 g sodium sulfate per liter boiled for 1 hour gave a 90 per cent yield of white anatase which was easy to filter and to agglomerate by heating. Use of hydrochloric, sulfuric, or nitric acid instead of the sulfate salts gave unfavorable results. Diffraction patterns of titanium dioxide produced at room temperature by adding aqueous ammonia to titanium tetrachloride showed an amorphous structure but changed to anatase patterns by heating the titanium dioxide at 400° C. Titanium dioxide produced by hydrolysis on pure water showed either the anatase or rutile structure, and presumably contained adsorbed water. Mixtures of titanium tetrachloride and water decomposed at a rate of flow of 0.15 and 5.30 ml per minute respectively (calculated as volume in the liquid state) and at 200 to 600° C commenced to produce anatase at 300° C. Mixed gases gave anatase at 800° C and rutile at 1000° C.

An aqueous titanium tetrachloride solution containing from 0.1 to 10.0 per cent of fluoride ion produced by leaching ilmenite ore is hydrolyzed by boiling after adding a nucleating and a coagulating agent.[55] The ferric ion is first reduced to the ferrous condition. At hydrolysis, the solution contains at least 250 gpl titanium dioxide and has a basicity of at least 45 per cent. A convenient nucleating agent is prepared by diluting a portion of the solution and heating it to develop the desired properties. Alternatively the titanium tetrachloride solution may be run into hot water. Such nucleating compositions contain 15 to 20 gpl titanium dioxide equivalent. The amount of nucleating agent added is from 0.5 to 5.0 per cent of the total titanium dioxide content of the hydrolysis solution. Effective coagulating agents are compounds having a multivalent negative ion such as a sulfate or phosphate. Hydrolysis yields of 95 per cent are obtained by the process. To remove the hydrofluoric acid and the coagulant, which the hydrolyzate tends to absorb, the hydrolyzate is washed with an alkaline solution such as sodium hydroxide and then with a slightly acid solution. Before calcination, a promoter to produce a rutile pigment is added, together with other conditioning agents.

The effect of the concentration of titanium tetrachloride, the acidity of the solution, and the duration of aging on the process of hydrolysis in dilute solutions were determined by the methods of dialysis and ion exchange chromatography.[56] After a few days, suddenly, the particles began to grow, the solution became cloudy and a precipitate formed. The change in the state of titanium in the solution was due to the formation of the hydroxide and not to the formation of polyions as with zirconium. The degree of adsorption of titanium by the cation exchange resin KV-2 in the H-form increased with the pH to a maximum at a pH of 1 and decreased sharply as the pH increased to 1.5. X-ray diffraction shows that titanium dioxide produced by boiling for 1 hour solutions containing 70 to 100 g titanium tetrachloride per liter was of the anatase type while solutions containing 100 to 300 gpl titanium tetrachloride gave rutile.[57] Heating at 75° C gave anatase even with solutions containing 200 to 300 gpl titanium tetrachloride. The addition of hydrochloric acid or of potassium hydroxide promoted the formation of rutile at lower titanium tetrachloride concentrations, although sulfate salts, sulfuric acid, and phosphoric acid had opposite effects. Addition of more than 5 per cent phosphoric acid gave $TiH_2(PO_4)_2 \cdot 4H_2O$. The transition point between anatase and rutile type titanium dioxide varied with the condition of hydrolysis, particularly with the temperature and the kind of compound added to the solution before hydrolysis.

Titanium dioxide pigments of the rutile crystal form are produced by the hydrolysis of titanium tetrachloride solution in a cyclical process.[58] Ilmenite ore is leached with hydrochloric acid until 95 per cent of the iron is extracted. After filtering and washing, the residue is mixed with a carbonaceous reducing agent and a binder and carbonized to form briquets which are chlorinated at 600 to 700° C. The titanium tetrachloride formed is separated from the iron chlorides, which are added to the original filtrates. This mixture is neutralized with lime. Scrap iron is added to reduce the ferric iron to the ferrous state, and the solution is electrolyzed at 2 to 20 amperes per square decimeter to yield chlorine and iron. The titanium tetrachloride is dissolved in water to form a solution of 150 to 350 gpl concentration, which is hydrolyzed until more than 90 per cent of the titanium is precipitated as hydrous oxide, which is filtered, washed, and calcined to yield rutile pigment. A part of the hydrochloric acid produced during hydrolysis is used to leach more ore and the chlorine is used for chlorination of briquets.

Chlorine and iron metal are recovered by electrolysis from ferrous chloride liquors obtained from the manufacture of titanium dioxide pigment, in which iron is extracted from the ore with hydrochloric acid.[59] In the electrolysis process the ferrous chloride solution constitutes the catholyte and calcium chloride solution the anolyte. A major portion of the dissolved iron is deposited at the cathode, while chlorine gas is liberated at the anode.

Autoclaving with water at temperatures up to 250° C effects or accelerates the hydrolysis reaction of titanium tetrachloride.[60]

**Vapor Phase Hydrolysis with Steam.** Very finely divided, soft, and uniform titanium dioxide of pigment grade was obtained by the hydrolytic reaction between vaporized anhydrous titanium tetrachloride and steam or water vapor at temperatures of 300° to 400° C.[61] The vapor phases were heated separately to the required temperature before being introduced into the reaction chamber to prevent undesirable low-temperature reactions. For example, two saturation vessels, constructed and operated after the fashion of an ordinary wash bottle, were employed to produce the vapors. One was charged with titanium tetrachloride and held at 120° C; the other was charged with water and heated to 80° C. Air was passed through to produce two gaseous streams consisting of air and titanium tetrachloride, and air and steam in the volume ratio of approximately 1 to 1. The two gaseous mixtures were preheated separately to 400° C, and introduced simultaneously into an upright cylindrical reaction vessel heated externally to 400° C, in such proportions that for each liter of reaction space 1 liter per minute of the steam and air mixture (0.5 liter steam) and 0.1 liter of the tetrachloride-air mixture (0.05 liter titanium tetrachloride vapor) entered. The vapors were directed so as to transverse the reaction vessel in a downward direction, where they reacted to form titanium dioxide and hydrogen chloride, and the products passed into a dust chamber heated at 200° to 400° C to prevent condensation or adsorption of the hydrochloric acid on the titania during the precipitation. Hydrochloric acid vapors emerging from the dust chamber were condensed and reused.

Even at decomposition temperatures as low as 300° C titanium dioxide, practically free from chlorides, was obtained in almost quantitative yield, while the material separated at 380° to 400° C possessed covering power equal to pigments produced from sulfate solutions calcined at 900° C. By this method a product suitable for direct use as pigment was obtained without the necessity of a calcination step.

According to a related process, vapors of titanium tetrachloride and water were injected into a closed collector maintained at an elevated temperature to effect hydrolysis, and the products of reaction were discharged tangentially to produce a forced circulating action within the chamber.[62] The titanic oxide was collected on a porous diaphragm while the hydrochloric acid gas and air passed through the openings.

A homogeneous finely divided titanium dioxide was obtained by the action of steam on titanium tetrachloride in admixture with a solid, inert, water-soluble salt, such as an alkali metal chloride or sulfate, at an elevated temperature.[63] One hundred parts titanium tetrachloride was mixed with 200 parts finely ground potassium sulfate to form a plastic mass which was heated to 300° to 400° C, and a current of steam was passed over the mix for 1 hour. Steam was then shut off, the temperature was raised to 800° C and maintained for 15 to 30 minutes. The mass was cooled to room temperature, leached with water to remove the potassium sulfate, and the residue of titanium dioxide was further washed, dried, and pulverized for pigment use.

Titanium dioxide having the rutile crystal form is produced by the vapor phase hydrolysis of preheated titanium tetrachloride.[64] A mixture of titanium tetrachloride vapor and an inert gas is conducted through an intake pipe, coaxially encased in a tube mantle, into the reaction zone, where it reacts with an excess of steam at 700° C introduced through the tube mantle. The titanium dioxide deposited from the gas stream is calcined for 30 to 60 minutes at 900° C.

Hudson[65] studied the hydrolysis of titanium tetrachloride at temperatures from 25 to 400° C by static methods and by flow methods. Titanium tetrachloride hydrolyzes very rapidly and homogeneously in the vapor phase at 5 to 10 cm of mercury and 25 to 100° C with $Ti(OH)_2Cl_2$ as a solid intermediate product. This is followed by slower heterogeneous liquid phase hydrolysis of the deliquescent intermediate. At 150 to 350° C titanium tetrachloride forms titanium dioxide after an induction period of a few minutes. Titanium dioxide produced by the vapor phase hydrolysis of the tetrachloride gives crystals which on heating change to globular shapes similar to those produced by liquid phase hydrolysis.[66]

## TITANIUM DIOXIDE EXTRACTION RESIDUE

Dry titanium monoxide prepared by heating a stoichiometric mixture of titanium and titanium dioxide for 10 hours at 1350° C is treated with hydrogen chloride either in a closed vessel, or by passage of a current of gas over the product to form titanium dioxide and titanium tetrachloride.[67] Reaction begins at 290° C. The titanium dioxide formed corresponds initially to a mixture of anatase and rutile, but after prolonged reaction, only rutile results. Further reaction of titanium dioxide with hydrogen chloride yields the compound, $Ti(OH)_2Cl_2$, identical with an intermediate product of titanium tetrachloride hydrolysis. At constant temperature, the rate of attack of hydrogen chloride on anatase and rutile is the same for equal concentrations of reactants. At a given temperature, the half reaction time is inversely proportional to the initial pressure. The two varieties of titanium dioxide are not simultaneously formed in the destruction of the titanium monoxide lattice. The rutile form results from the transformation of the anatase.

A process for the manufacture of titanium dioxide pigments involves leaching ilmenite with hydrochloric acid to obtain a residue consisting of essentially all of the titanium dioxide of the ore and a solution containing the major portion of the iron.[68] The titanium dioxide residue is chlorinated to obtain vapors of titanium tetrachloride, which are dissolved in water. Titanium oxide is precipitated from the solution by hydrolysis with the liberation of hydrochloric acid. The hydrous titanium oxide is washed, calcined, and finished to produce pigment. The ferric chloride solution obtained from the leaching of the ore is reduced and electrolyzed to obtain metallic iron and chlorine, which is used for the chlorination of more titanium dioxide residue from the leaching operation. Hydrochloric acid liquor from the hydrolysis step is returned to the process for subsequent leaching of fresh ore. Iron impurities are removed from ilmenite by sulfur and oxygen treatment in a fluidized bed, followed by leaching the iron as sulfate with a mineral acid.[69] Ilmenite ore from the Otanmaki mine, Finland, is leached with concentrated hydrochloric acid in three 2-hour stages.[70] A titania concentrate containing 88 per cent titanium dioxide and 3 per cent iron is obtained. Ferrous chloride is precipitated from the leach liquor by increasing the hydrochloric acid content at a low temperature. Highly concentrated hydrogen chloride required for this purpose is obtained by thermal decomposition of hydrated ferrous chloride. The regenerated leach liquor is recirculated.

Ferric chloride is volatilized from a fluidized bed of ilmenite with a stream of hydrogen chloride gas preheated to 600° C. On a pilot plant scale,[71] an artificial rutile containing 1.6 per cent total iron as ferric oxide and over 90 per cent titanium dioxide is produced from ilmenite containing 50 per cent ferric oxide. The stripping process requires a circulating hydrogen chloride gas volume of 600,000 cu ft per ton of product. Iron is removed from ilmenite or titaniferous iron ores by heating with coal or coke, then heating the sintered material with hydrogen chloride or chlorine at 750 to 1000° C and subliming off the ferric chloride.[72] The residual titanium dioxide is practically iron-free. From the sublimate, iron may be recovered by fused-salt electrolysis.

A titanium dioxide concentrate is obtained by treating ilmenite with hydrochloric acid in an autoclave under 2 to 3 atmospheres at 120 to 135° C.[73] The titanium dioxide residue is separated from the iron chloride solution by filtration, washed, and dried.

Iron is removed from titaniferous iron ores by first oxidizing the iron to the ferric state and extraction with titanium tetrachloride at 700 to 1050° C.[74] In an example, ilmenite containing 58.6 per cent titanium dioxide, 22.4 per cent ferric oxide, and 13.5 per cent ferrous oxide was treated with a mixture of titanium tetrachloride and chlorine at 1050° C to yield a residue containing 50 per cent titanium dioxide and 6.6 per cent ferric oxide.

In a similar manner, a titaniferous ore containing 74.1 per cent titanium dioxide, 4.8 per cent ferric oxide, and 20.9 per cent ferrous oxide was treated at 875° C with titanium tetrachloride and chlorine at a ratio of 2.64 to 1. A residue containing 97.8 per cent titanium dioxide and 3.2 per cent ferric oxide was obtained.

A high titania concentrate is obtained from titaniferous iron ores or slag so as to avoid the formation of uneconomical ferrous sulfate.[75] The iron oxides are selectively leached with 10 to 37 per cent hydrochloric acid at 60 to 110° C. Recovery of the titanium is 92.3 per cent. The filtrate containing ferrous chloride and ferric chloride is treated with magnesium oxide to precipitate ferrous hydroxide and ferric hydroxide which are converted to valuable ferric oxide by bubbling air through the suspension. Magnesium chloride hexahydrate is crystallized from the filtrate and heated to 500 to 600° C to regenerate magnesium oxide. Hydrogen chloride is also recovered and used for leaching more ore. Ninety-one per cent of the iron is extracted from ilmenite, with a 40 per cent excess of 20 per cent hydrochloric acid at 96 to 98° C.[76] The concentrate contains 96.1 per cent titanium dioxide.

The iron content of ilmenite or high titania slag is extracted with aqueous hydrochloric acid and the resulting titanium dioxide concentrate is employed for the manufacture of titanium dioxide pigment.[77]

The ilmenite or slag is ground to a fineness so that 70 per cent passes through a 60 mesh screen and is extracted with 10 to 37 per cent hydrochloric acid at 60 to 110° C with stirring until most of the iron has been dissolved as ferrous chloride and ferric chloride. The acid solution of iron chloride is filtered from the residue containing practically all of the titanium dioxide with minor amounts of iron and silica. This insoluble residue from the first extraction is extracted a second time with fresh hydrochloric acid. The filtrate from the second extraction is then used for the first extraction of a new batch of ore or slag. The residue from the second extraction is practically free from iron. From 0.01 to 0.5 per cent of an acid compatible, surface active agent facilitates and accelerates the reaction.

Magnesium oxide or hydroxide is added to the filtrate at 70 to 90° C to precipitate the iron as hydroxides and convert the magnesium to the chloride. After aeration to oxidize the ferrous hydroxide to ferric hydroxide, it is separated from the magnesium chloride solution by filtration decantation or centrifuging. The solution is evaporated to obtain crystalline $MgCl_2 \cdot 2H_2O$, which is fed continuously in a gas fired rotary kiln at 500 to 600° C for a retention period of 20 to 30 minutes to effect hydrolysis to magnesium oxide and hydrochloric acid. The hydrochloric acid is recovered from the kiln gases.

Most of the iron of ilmenite or titaniferous iron ores was selectively dissolved by leaching with hydrochloric acid, leaving a residue of crude titanic oxide. Such a product, obtained by digesting the finely ground ore with hot hydrochloric acid in a closed vessel, was elutriated from heavy sand, mixed with soft coal, and heated in a stream of chlorine gas at 600° to 700° C to produce the tetrachloride,[78] which, after purification by fractional distillation, was dissolved in water, and the solution was subjected to thermal hydrolysis. The precipitate was washed and calcined to yield titanium dioxide of pigment grade, while the liquor containing the liberated hydrochloric acid was employed to leach the iron from another batch of ilmenite. Hydrochloric acid effected almost quantitative extraction of iron from titaniferous magnetites.[79] On heating a mixture of 100 g of the finely divided ore and 100 ml of 20 per cent hydrochloric acid at the boiling temperature with agitation, 97.3 per cent of the iron was extracted in 6 hours and 98.2 per cent in 24 hours. The efficiency of the process increased with fineness of milling of the titanomagnetite concentrate. By a modified method,[80] ilmenite was mixed with coal and heated at 500° to 800° C to reduce all ferric iron to the ferrous form but not to the metallic state so that the ore was not appreciably changed physically. The ferrous compounds were then dissolved selectively in dilute acids leaving a residue of crude titanic oxide.

Ilmenite or titaniferous magnetite is treated with hydrochloric acid at a temperature above 175° C under pressure to dissolve the undesirable iron content before chlorination to titanium tetrachloride.[81]

A material balance showing the accumulation of soluble chlorides in the circulating liquor, the necessary bleeding off ratio and the amount of hydrochloric acid required to make up for losses was calculated from the experimental data for a recycling process for leaching iron from ilmenite.[82]

Chlorination experiments with the resulting titania concentrate showed that it reacts at rates and temperatures comparable with those of pigment grade titanium dioxide.

Takimoto[83] subjected Indian ilmenite and sands containing from 11 to 53 per cent titanium dioxide and 34 to 79 per cent ferrous oxide to reduction with various amounts of coke and alkaline earth sulfates, particularly calcium sulfate, barium sulfate, and magnesium sulfate, at 800 to 1200° C for 2 hours. The reacted mass was boiled with 15 to 30 per cent sulfuric acid or 20 per cent hydrochloric acid for 30 minutes, and the residue was treated with hydrogen chloride gas at 600° C. A final product containing a maximum of 94 per cent titanium dioxide was obtained by reducing ilmenite with calcium sulfate and carbon at 1100 to 1200° C for 2 hours, followed by treatment with hydrogen chloride gas at 600° C for 2 hours. Ilmenite ore and spent digestion liquor from paper pulp manufacture are heated in a smelter to drive off the water, oxidize the organic material, and fuse the alkali mixture.[84] The sulfur present converts the iron to iron sulfide. Water is added to dissolve the caustic and water-soluble products. To the insoluble residue, dilute hydrochloric acid is added to dissolve the iron sulfide, leaving the titanium dioxide in an active form, which is later dissolved in 12 normal hydrochloric acid to form titanium tetrachloride. Heating ilmenite concentrate with 20 per cent hydrochloric acid at 220° C and 13.3 atmospheres removes 99.5 per cent of the iron in 1 hour.[85] Less than 0.3 per cent of the titanium is lost. Iron is removed from minerals by treatment with solutions containing 0.5 to 5.0 per cent each of titanous sulfate and sodium sulfate or titanous chloride and sodium chloride at temperatures up to 100° C.[86] The titanium salts can be reduced and reused.

Kamlet[87] developed a cyclic process for the beneficiation of titanium ores and slags. Iron oxides are selectively leached from the ore by 10 to 37 per cent hydrochloric acid at 60 to 110° C. Acid compatible surface agents such as sodium hexylbenzenesulfonate speed the reaction. Recovery of titanium dioxide is 92.3 per cent with a separation of 99 per cent of the iron oxides. The filtrate from the leaching treatment containing ferrous chloride, ferric chloride, and some free hydrochloric acid is treated with magnesium oxide to precipitate ferrous hydroxide and ferric hydroxide which are converted to ferric oxide and magnetic iron oxide by bubbling air through the solution. After filtration, magnesium salts are recovered from the solution by crystallization of $MgCl_2 \cdot 6H_2O$. The crystals are separated, dehydrated, and heated to 500 to 600° C to regenerate magnesium oxide with a recovery of 92 to 95 per cent. Hydrochloric acid gas liberated is collected. From 60 to 90 per cent of the hydrochloric acid used to leach the ore is recovered. The titania concentrate is used for the manufacture of titanium pigment or metal. Most of the iron, calcium, and magnesium are removed from ilmenite or titania slag by leaching with 6 to 12 normal hydrochloric acid in a closed vessel at 150 to 250° C.[88]

Chromium is removed from ilmenite and rutile by first digesting the 10 mesh material in concentrated hydrochloric acid.[89] After filtering, washing, and redigestion in 2 per cent aqueous sodium hydroxide and 1 per cent sodium peroxide, the product is washed free of sodium ion and calcined to contain less than 0.0015 per cent chromic oxide. Oxides of iron, zirconium, manganese, tin, vanadium, and chromium are removed from ilmenite by treatment with titanium tetrachloride at 850 to 900° C and a pressure of 25 to 30 mm of mercury.[90] Silica and alumina are not attacked by the titanium tetrachloride under these conditions.

To recover crude titanium dioxide and electrolytic iron, ilmenite is first given an oxidizing roast at 700 to 900° C to form $Fe_2O_3 \cdot TiO_2$.[91] This product is reduced with a reducing gas at 900 to 1000° C to form metallic iron and titanium dioxide. The reduced product is treated with dilute hydrochloric acid containing ferric chloride or with dilute sulfuric acid containing ferric sulfate to dissolve the iron and recover the titanium dioxide as the residue. The solution is treated electrolytically to recover the iron.

Ferric chloride is formed and sublimed from ilmenite by the action of gaseous hydrogen at an elevated temperature.[92] At the temperature employed, titanium dioxide does not react with the hydrogen chloride and remains as a residue. A white titanium dioxide residue containing 0.01 per cent ferric oxide is obtained from ilmenite by an oxidation roast followed by chlorination in a fluidized bed at 1100° C.[93]

White titanium dioxide suitable for paper is obtained as a residue from ilmenite.[94] The iron oxide of the ore is reduced to metal with hydrogen at 1000 to 1100° C which is removed magnetically. Most of the titanium dioxide goes with the iron. This magnetic product is leached with hydrochloric acid to remove most of the iron, after which any remaining colored oxides are removed by chlorination at 900 to 1000° C. Treatments with hydrogen and chlorine are carried out in a fluidized bed.

A titanium dioxide concentrate is obtained by leaching slag with hydrochloric acid.[95] Molten slag from the blast furnace smelting of titaniferous sand is chilled by water to form glassy grains which are crushed to pass 150 mesh. One part of the crushed slag is kneaded with 1 to 3 parts of aqueous hydrochloric acid in an acid-resistant vessel insulated from heat. The temperature of the mixture rises to 110° C as a result of the heat of the reaction. To complete the reaction, 1 to 5 hours is needed. By this treatment, the titanium, calcium, iron, and aluminum components dissolve as chlorides, but at the high temperature employed the titanium tetrachloride rapidly hydrolyzes to metatitanic acid and hydrochloric acid. The residue of metatitanic acid and silica is filtered, washed, and calcined at 800° C. In an example, 500 g of slag containing 23.4 per cent of titanium dioxide is heated with 1.25 liters of 35 per cent hydrochloric acid. After 4 hours, the mixture is poured into 3 liters of water, stirred for 3 hours, and filtered. The residue is calcined at 800° C for 2 hours to obtain 233 g of a white powder containing titanium dioxide 44.4, silica 50.1, and ferric oxide 1.74 per cent.

High titania slag is treated with hydrogen chloride at 850 to 950° C to volatilize and remove the iron component as ferrous chloride before chlorinating the titanium dioxide to form titanium tetrachloride.[96] The slag containing 70.4 per cent TiO$_2$, 8.0 per cent iron, 0.12 per cent carbon, 1.1 per cent calcium oxide, 5.2 per cent magnesium oxide, 6.3 per cent alumina, 6.1 per cent silica, and 2.8 per cent unidentified is heated to 850 to 950° C in a closed reactor having an inlet conduct for hydrogen chloride and an outlet conduit for removing the ferrous chloride sublimate. Anhydrous hydrogen chloride is passed through the slag mass, in the absence of air, to form ferrous chloride which sublimes at the temperature employed. From 20 to 30 per cent excess hydrogen chloride is required to reduce the iron content of the slag to 0.1 per cent. Phosphorus pentoxide is also chlorinated and sublimed but the other constituents are not attacked. The sublimate of ferrous chloride is drawn off as formed, condensed at 350 to 650° C, and electrolyzed in a fused alkali metal chloride bath. Dry hydrogen gas is passed into the cell to react with the chlorine to form hydrogen chloride for reuse in the first step. The slag residue in the reactor is then mixed with an amount of carbon calculated to form carbon monoxide with all of the oxygen combined with the titanium, and dry chlorine is passed through the bed at 900 to 1000° C to form titanium tetrachloride. So long as the amount of carbon does not exceed the amount required to take care of the oxygen combined with the titanium, the other components such as calcium oxide, magnesium oxide, alumina, and silica do not react with the chlorine. The titanium tetrachloride is condensed by conventional methods.

Analysis of the residue after removal of the iron by reaction with hydrogen chloride is 81 per cent titanium dioxide and 0.12 per cent iron.

Moklebust [97] produced a titanium dioxide concentrate by reducing ilmenite with coke at 1100 to 1200° C in the presence of an alkali metal compound. Ilmenite ground to 80 mesh was mixed with 9.3 per cent sodium chloride, moistened with water, and rolled in a drum to form pellets 5 to 100 mm in diameter. The pellets were then introduced into a rotary kiln in an amount of 2 tons an hour together with 1.5 tons of coke fines. After heating the mixture gradually to 1180° C the temperature was held at this value for 4 hours. The product was cooled in the absence of air in a rotary cooler, containing 70 per cent metallized pellets, and 30 per cent excess coke and ash. From this mixture the reduced pellets were separated mechanically. The pellets containing 49 per cent titanium dioxide, 38 per cent metallic iron, and 0.3 per cent sulfur represented a yield of 93 per cent of the iron content of the ore. After fine grinding, the titanium dioxide was separated from the metallic iron particles by air classification. The concentrate containing 87 per cent titanium dioxide was suitable for the manufacture of titanium tetrachloride.

Ilmenite ore ground to 30 mesh is smelted in an electric arc furnace with coke in the ratio of 17 to 25 parts of coke to 100 parts of ore to produce a low iron slag suitable for chlorination to titanium tetrachloride.[98] Fresh charge is periodically introduced into the furnace. The highly reduced titaniferous slag is separated from the molten iron and cooled. Analysis shows 90 to 120 per cent titanium dioxide equivalent, 2 to 15 per cent iron, and 1 to 5 per cent ferrous oxide.

# 19

# Production of Titanium Tetrachloride

**CHLORINATION OF ORES AND CONCENTRATES**

**Titanium Dioxide.** According to Pamfilov and Shtandel,[1] titanium dioxide chlorinates to the tetrachloride, but carbon should be present in the charge to take up the oxygen and prevent reversal of the reaction. Up to 600° C carbon dioxide was chiefly formed, but above this temperature the monoxide was the main product. Excess chlorine reacted with the carbon monoxide to form phosgene. Ferric oxide and calcium oxide, if present, reacted with titanium tetrachloride to give titanic oxide, ferric chloride, and calcium chloride. Silicates yielded some silicon tetrachloride. Saeki and Fumaki[2] observed that the reaction between titanium dioxide and chlorine takes place readily in the presence of carbon. Anatase reacts more readily than does rutile. The main reaction below 700° C is $TiO_2 + C + 2Cl_2 \rightarrow TiCl_4(g) + CO_2$, and above 700° C is $TiO_2 + 2C + 2Cl_2 \rightarrow TiCl_4(g) + 2CO$. Phosgene produced in the reactions decomposes considerably above 500° C.

The reaction between rutile titanium dioxide and chlorine does not occur unless carbon is added, while anatase and chlorine react gradually above 850° C. Rutile reacts with chlorine above 440° C if carbon is present but slowly, but the rate of reaction between anatase and chlorine in the presence of carbon has a maximum at 540 to 650° C. Ilmenite begins to react with chlorine and carbon at 370° C. From 540° C the reaction rate increases.

Barton[3] heated a mixture of titanium dioxide and soft coal until the volatile constituents had been driven off, and treated the cinder with chlorine at 650° C to form the tetrachloride. Similarly, a mixture of crude titanic oxide and carbon was heated with chlorine at 600° C, and the reaction products were led into an expansion chamber maintained at 136° C, where the ferric chloride condensed. The vapor of titanium tetrachloride was then passed through a filter plate, also at 136° C, to a condenser.[4] According to another modification, the tetrachloride was prepared by subjecting an ignited mixture of titanium dioxide and lampblack to the action of a stream of chlorine gas.[5] A porous product, which allowed free entry of chlorine,[6] was obtained by coking at a red heat a briquetted mixture of titanic oxide and sawdust or peat. Once initiated, the heat of reaction was sufficient to maintain the chlorination temperature. In the presence of wood charcoal the reaction went to completion at 500° C.[7] The action of chlorine sufficient to cause at least 1 per cent of the titanic oxide to react per hour started at 800° C.[8] By passing a mixture of carbon monoxide and chlorine over titanium dioxide at a rate of 1 liter per hour for 3.5 hours, 37.4 per cent of the charge was chlorinated at 600° C, and 80.7 per cent at 1050° C.[9] Thus an increase in temperature greatly increased the rate. Employing lampblack instead of carbon monoxide, 67 per cent of the titanium dioxide reacted in 1.5 hours at 600° C, and 79 per cent at 1050° C.

The equilibrium gas mixture,[10] formed on treating titanic oxide and carbon with chlorine at 400° to 1000° C, contained from 30 to 50 per cent by volume of titanium tetrachloride. The other gases were carbon monoxide and carbon dioxide, and as the temperature rose the proportion of monoxide increased at the expense of the dioxide.

In a pilot-plant study of chlorination of briquets, made by heating a mixture of 3 parts rutile, 1 part charcoal, and 1.75 parts thin tar binder, unground rutile was as effective as ground and more effective than titanium dioxide pigment.[11] The chlorine efficiency was 94 per cent with 70 per cent conversion of the charge, but decreased if all the charge was consumed. The condensed liquid, after separation from ferric chloride by decantation, was clear and yellow, and contained 99 per cent titanium tetrachloride with 0.02 per cent iron. Redistillation from copper gave a colorless product.

The optimum temperature of chlorination of titanium dioxide-carbon mixtures was reported to be much lowered in the presence of manganese dioxide catalyst.[12] In an actual operation, the temperature at which the tetrachloride formed was reduced from 500° to 480° C by including 0.2 per cent manganese dioxide in the charge. Titanomagnetite concentrates gave a high yield only above 600° C, and the process was complicated by the sublimation of ferric chloride. Sphene reacted rapidly only above 800° C. Titanic oxide was attacked faster than silica, but slower than calcium oxide.

Helbig[13] mixed titanium dioxide in gel form with powdered carbon and a small amount of the oxide of a metal of the sixth or seventh group of the Periodic System, dried the product, and treated it with chlorine at a red heat. Once initiated, the reaction proceeded exothermically. Agnew and Cole[14] fed the briquetted material into the reaction chamber upon a movable framework base, and introduced chlorine above the frame and air below. Movement of the framework removed unchanged material and gangue from the reaction zone, which was maintained at approximately 600° C. To effect continuous chlorination of titanium dioxide in the presence of coke, Brallier [15] kept the temperature up to the required 700° C by causing titanium carbide or cyanonitride to react simultaneously. According to another modification, the necessary heat for the reaction was supplied by introducing silicon into the reaction zone. The silicon tetrachloride formed was separated by fractional distillation.

A small proportion of sodium hydroxide or sodium carbonate is added to the mixture of titaniferous material and coal or coke before sintering to serve as a binder and produce a product of good porosity and excellent structural strength for chlorination.[16] A sintering temperature of between 500 and 600° C is employed. From 1 to 4 per cent of sodium hydroxide is added. The sintered product is chlorinated by the usual process to produce titanium tetrachloride. Mechanically strong briquets suitable for chlorination are made from a titanium ore and a carbonaceous material with an alkali metal carbonate as a binder.[17] In an example, 28 parts of 200 mesh petroleum coke and 72 parts of 200 mesh rutile ore or 27 parts of 100 mesh petroleum coke and 100 parts of 200 mesh ilmenite are mixed in a pug mill with 3 per cent by weight of the mixture of an alkali metal carbonate to which 10 times its weight of water is added. The mixture is then formed into a cake 1 to 2 inches thick and dried in a current of air heated to 150 to 160° C until the moisture evaporates to give a compact mass.

Titanium tetrachloride is produced by chlorination at 800 to 900° C of briquets made from rutile and coke or coal with a binder of acid-treated mineral tar or pitch.[18] Briquets made with untreated tar adhere strongly to the surface of containers, mixing devices, and furnaces. In addition, such briquets deteriorate badly in the chlorination furnace. The bonding tar or pitch is treated with 2 to 5 per cent of its weight of concentrated sulfuric acid at a temperature above its melting point. Rutile or ilmenite used is ground to a particle size less than 60 mesh. The rutile, carbon, and acid-treated tar are mixed so as to produce a final composition after firing which contains rutile and carbon in the ratio of 3 to 1 to 5 to 1. After mixing, the product is fired in an open pan and then heated out of contact with air to 400 to 700° C. Titanium tetrachloride produced from such briquets contains less vanadium than usual. In an example, 31 parts of 60 mesh rutile, 10 parts of 20 mesh coke, and 5 parts of acid-treated tar were mixed, pressed into briquets, and fired at 550 to 650° C in a closed furnace. For the binder, a portion of petroleum tar having a viscosity of 19.9, Sp Engler, was mixed with 21 per cent of its weight of

96 per cent sulfuric acid and stirred until a homogeneous product was obtained. The fired briquets were chlorinated in a furnace maintained at 800 to 900° C.

A mixture of titanium dioxide with charcoal in the ratio of 3 to 1 and 15 per cent water is formed into briquets, dried, and heated for 1 hour at 800° C under a layer of charcoal.[19] The composition of the briquets is titanium monoxide and titanium carbide. Each briquet weighing 3.5 kg is placed in a vertical graphite electric furnace and heated while dried chlorine is blown through from the bottom. Up to 400° C, the gases issuing from the top of the furnace are released to the atmosphere. The temperature is then increased to 600° C and the vapors of titanium tetrachloride are condensed. After refluxing the crude titanium tetrachloride for 20 to 30 minutes with 2 per cent of copper powder to reduce the ferric chloride and vanadium oxychloride the mixture is distilled. The colorless fraction boiling above 136° C is collected to give titanium tetrachloride of high purity. Air in the distillation receiver is dried with calcium chloride. Briquets of titanium dioxide with carbon and mineral oil are dried at 100° C, calcined at 500 to 600° C, and chlorinated at 850 to 950° C to produce crude titanium tetrachloride.[20] The crude product is refluxed over copper powder, sodium hydroxide, and water, and redistilled to give pure titanium tetrachloride.

Titanium tetrachloride is produced in a continuous process by charging titanium dioxide briquets with an excess of coke at one end of the reactor and removing the spent briquets at the other end.[21] Coking coal from 75 to 135 per cent by weight of the ilmenite, rutile or slag is used in preparing the briquets which are coked at 900° C for 1 hour to remove the volatile components. The density of the briquets is 70 per cent greater than the bulk density of the original mix. Chlorine gas is passed continuously through a body of coked briquets maintained at 600 to 1000° C by the exothermic reaction. Chlorination of the titanium component of the briquets is effected without destroying the coherent structure of the briquets. The titanium-depleted briquets are withdrawn from the discharge end of the chlorinating zone. Control of the reaction temperature may be achieved by diluting the chlorine with an inert gas, by diluting the briquets mass with nonreactive materials, by spreading or contracting the reacting zone, by decreasing or increasing the size of the ore or slag particles, by varying the size of the briquets, by varying the rates of charge of the briquets and chlorine, by varying the relative proportions of preheated and cold briquets, by choice of furnace insulation, and by distribution of the chlorine in the reactor.

In an example, a mixture composed of 50 parts of slag (70 per cent $TiO_2$), 40 parts of bituminous coal, 40 parts of coke, and 7 parts of sulfite liquor and water as required, was blended, treated in a chaser mill, and briquetted on a roll press to 2 x 2 x 1.5 inch pillow-shaped blocks having a density of 100 pounds per cubic foot. The briquets were broken in half, placed in a steam drier for 2 hours, and coked at 900° C for 1 hour. Most of the volatile matter was driven off. For the chlorination, a vertical retort was filled to a depth of 2 feet with coal or spent briquets from a previous run. Above this was added a 7 foot layer of hot briquets. Air was introduced near the bottom to burn the coal to preheat the retort to 700° C. Chlorine was then introduced at the bottom of the retort at a rate of 60 pounds an hour. A reaction zone temperature of 800 to 900° C was maintained by the heat of the reaction. The less volatile calcium chloride and magnesium chloride remained in the spent briquets, which were discharged from the retort at regular intervals. Product chlorides in the vapor form together with the carbon monoxide and carbon dioxide were withdrawn from the furnace and cooled with a jet of titanium tetrachloride. The spent briquets were discharged intermittently through a star value at the lower end of the retort.

Continuous production of titanium tetrachloride is carried out in a vertical type electric resistance furnace provided with a conical residue outlet at the bottom having a water jacket and a screw conveyor.[22] A chlorine-resistant flexible bag of polyvinyl chloride is attached to the end of the outlet. Briquets of titanium dioxide and a reducing agent are fed into the furnace from the top and caused to react with chlorine, which is introduced from below at 400° C. The titanium tetrachloride produced is withdrawn continuously from the top, and the residue is collected in the bag through the conveyor and withdrawn from the end of the bag by loosening a clip. Continuous production of titanium tetrachloride from rutile is facilitated by briquetting with powdered petroleum coke or coal and a binder.[23] These calcined briquets are charged at the top of an upright chlorinator with coke. Air injected at the bottom burns some of the coal to produce heat. Chlorine is then separately injected at a slightly higher level to react with the rutile.

Titanium-zirconium concentrates are processed by the chlorination of briquets.[24] Suitable briquets made from 200 mesh concentrates with 21 to 23 per cent of 100 mesh petroleum coke are chlorinated at 900° C. The chlorides produced are fractionally condensed. Titanium tetrachloride is freed from zirconium chloride, aluminum chloride, and ferric chloride by passing the vapor through a filter of sodium chloride at 400 to 550° C or through a filter with an equimolecular amount of sodium chloride and potassium chloride at 300 to 450° C. The filter contains sodium chlorozirconate which is converted to zirconium oxide. Impurities in the condensed titanium tetrachloride are 0.003 to 0.007 per cent zirconium, less than 0.001 per cent iron, 0.004 per cent aluminum, 0.42 per cent silicon, 0.05 to 0.45 per cent vanadium, 0.01 to 0.02 per cent niobium, and

0.01 per cent chromium. The solubility of zirconium chloride in titanium tetrachloride at 25 to 80° C is 0.001 per cent, expressed as zirconium.

Pulverized titanium ores or slags are mixed with petroleum coke and aqueous hydrochloric acid, pelletized, and dried to prepare an agglomerated material for chlorination.[25] The proportion and concentration of the acid are such that the mixture contains 1 to 5 per cent hydrochloric acid and 7 to 18 per cent water. Pellets processed in this way remain porous (unglazed) during the subsequent chlorination to form titanium tetrachloride. A flowable, solid titanium ore-carbon mass is made by thoroughly mixing separately preheated finely divided coking coal to 400 to 700° F and finely divided ore to 900° F in a ratio not greater than 1 to 3.[26] The mixture is agitated while heating to not over 1000° F until the carbonaceous matter becomes sticky. Heating is continued until carbonaceous material returns to the solid form. The product is chlorinated to form titanium tetrachloride.

Rapid reaction and substantially complete utilization of chlorine are obtained by chlorination of titaniferous materials in a two stage static bed process.[27] Briquets of a mixture of the titaniferous material and a reducing agent are treated with a rapid stream of chlorine until there is a sudden increase in the chlorine content of the effluent gas. At this stage the charge is screened to remove dusty gang, and the unreacted lumps are charged to a second bed and treated with the effluent gas from the first bed after condensation of the ferric chloride and titanium tetrachloride. The chlorine in the effluent gas from the first bed reacts completely with the second bed to form additional titanium tetrachloride and ferric chloride, which are recovered separately in a second set of condensers.

Consumption of chlorine required for the chlorination of briquets made from various titaniferous materials and anthracite coal, charcoal, or carbon black in a sufficient amount to reduce the titanium dioxide, magnesium oxide, and ferrous oxide at temperatures between 250 and 900° C was measured by Takimoto.[28] The materials were products of titanium dioxide and carbon black calcined at 1800 to 1900° C for 30 to 200 minutes, containing 45 to 58 per cent titanium carbide, and 25 to 35 per cent titanium dioxide; rutile ore containing titanium dioxide 91, silica 3, and ferrous oxide 4 per cent; reduced rutile obtained by calcining the ore with anthracite at 1700 to 1800° C for 30 minutes ($TiO_2$ 8, TiC 9, $SiO_2$ 3 and FeO 3 per cent); slag obtained from two kinds of ilmenite and charcoal or coke, containing titanium dioxide 67 to 77, ferrous oxide 5 to 8, and magnesium oxide 7 to 8 per cent; titanium dioxide pigment (98 per cent $TiO_2$). The temperature required to give titanium tetrachloride in 85 to 90 per cent yield was 300, 850 to 900, 500 to 570, 700 to 800, and 500 to 530° C, respectively. Consumption of chlorine expressed as per cent of the theoretical amount required to give titanium tetrachloride in

80 per cent yield at 800° C was respectively, 170 per cent, 170 per cent, 135 per cent, and 100 per cent.

A previously carbonized mixture of spent soda base wood pulping liquor and a titaniferous material are chlorinated to form titanium tetrachloride along with chlorides of iron and other impurities.[29] An aqueous solution of the product is treated with lime slurry to precipitate hydrous oxides of titanium and of iron. The iron component is separated from the titanium dioxide by treating the mixture with sulfur dioxide. Titanium tetrachloride is formed by chlorinating rutile agglomerated with coal at a high temperature in an oven.[30] Chlorine is introduced below the briquets but near the top of the layer of coal which is combusted with air for heat. Takahashi[31] calculated thermodynamically the equilibrium constants of the chlorination reactions of alumina, silica, ferric oxide, titanium dioxide, and aluminum silicate at 600 to 1000° C.

## CHLORINATION OF RUTILE AND CONCENTRATES

Titanium dioxide concentrate to be chlorinated is suspended in a stream of chlorine together with a carbonaceous reducing agent and passed through a reaction zone at 650 to 1300° C, where the exothermic reaction takes place to maintain the temperature.[32] The product titanium tetrachloride is withdrawn concurrently. Titanium carbide requires no added reducing agent. By the chlorination of rutile in a shaft furnace the chlorine is distributed more evenly and more uniformly throughout the charge.[33]

Titanium tetrachloride is prepared by the chlorination of titanium oxide in the presence of carbon at lower temperatures than are normally used in the chlorination of titanium dioxide.[34] Intimately mixed titanous oxide and lampblack in excess of the stoichiometric amount is chlorinated in the temperature range of 300 to 500° C. Problems of construction and maintenance of the equipment within the temperature range for chlorination are less severe than in the 900 to 1100° C range required for efficient chlorination of titanium dioxide. The titanium tetrachloride produced is of very high purity.

Finely divided titanium dioxide having a particle size of less than 325 mesh is chlorinated by passing the dust through a restraining bed of coarse granular materials ranging from 20 to 120 mesh held in dynamic suspension by the chlorinating gas stream.[35] As an example, 2.1 parts per minute of titanium dioxide dust along with 0.4 parts per minute of carbon was added to the lower portion of a reaction chamber containing a 2-foot bed composed of 600 parts of titanium dioxide granules of 20 to 120 mesh size and 120 parts of carbon granules of 20 to 120 mesh. This restraining bed was suspended by an upward flowing gas stream comprising 3.8 parts per minute of chlorine and 4.7 parts per minute of carbon dioxide. The chlorine admitted was equal to the theoretical amount required to chlorinate the titanium dioxide dust. A temperature of 800° C was maintained in the restraining bed. The exit gases contained carbon dioxide 5.7, carbon monoxide 0.2 and chlorine 0.2 part. In the reaction 8.8 per cent of the titanium dioxide dust was converted to the tetrachloride, with a 92.7 per cent utilization of the chlorine. Titanium tetrachloride is prepared by first reducing rutile with coke and chlorinating the reduced product at 200 to 500° C.[36] The crude titanium tetrachloride distillate is purified by treatment with copper powder followed by distillation.

According to Patel and Kharkan [37] rutile is most efficiently chlorinated by the action of chlorine gas containing 1 part in 15 by volume of oxygen on a mixture of 10 parts of the ore ground to pass 200 mesh with 6 parts of coke ground to pass 100 mesh as the reducing agent and 0.1 part of ceric oxide, as a catalyst by weight at 525 to 600° C for 1 hour. Sulfur monochloride, hydrogen sulfide, and carbon tetrachloride promote while phosgene retards the chlorination reaction. Ferric chloride as a binder is superior to asphalt, molasses, and linseed oil in allowing more efficient chlorination of briquets.

Leucoxene containing sillimanite as an accessory mineral was mixed with carbon and chlorinated with chlorine gas under a variety of conditions.[38] The leucoxene was preferentially chlorinated up to 800° C, leaving most of the sillimanite. Manganese dioxide, lead oxide, copper oxide, ceric oxide, and calcium phosphate acted as positive catalysts for the chlorination of leucoxene only, while compounds containing sulfur acted as catalysts for both leucoxene and sillimanite. For the preferential chlorination of leucoxene, ceric oxide was the best catalyst. Incorporation of limited amounts of oxygen and water aided the chlorination of leucoxene only. The use of 50 per cent carbon based on the weight of the minerals was optimum for the chlorination of leucoxene. Ferric chloride was the best binder for briquets. Chlorination of leucoxene was most efficient by the action of chlorine containing some oxygen on a mixture of 10 parts of the ore through 200 mesh with 5 parts of carbon and 0.1 part of ceric oxide at 600° C.

## FLUIDIZED BED CHLORINATION OF ORES, RUTILE OR CONCENTRATE

The chlorination of rutile and ilmenite is accomplished by means of a fluidized bed technique.[39] A mixture of 5 parts of rutile to one part of coke or of 2.85 parts of ilmenite containing 60 per cent titanium dioxide all ground to 200 mesh is fed to a chlorination column through which chlorine preheated to 900° C flows at a rate sufficient to maintain the solids in a fluidized state. As soon as the temperature of the solids rises above 1050° C, a portion of the bed is withdrawn, passed through a heat exchanger, and returned to the chlorination chamber. Conversion of 100 per cent of the titanium dioxide in the rutile and 90.5 per cent of that in the ilmenite to titanium tetrachloride is achieved. Ninety-one per cent of the iron present in the ilmenite is converted to the chloride.

To produce titanium tetrachloride in a continuous process, rutile or ilmenite and carbon ground to 200 mesh are suspended in a current of chlorine and passed into the lower end of a reaction zone at 1250 to 1650° F.[40] Vaporized carbon tetrachloride may be substituted for the carbon and chlorine. If excessive heat losses are prevented, the reaction with carbon and chlorine is sufficiently exothermic to maintain the reaction temperature at 1200 to 1650° F, even though the materials are introduced at atmospheric temperature. The vapor of carbon tetrachloride must be preheated to 250° F to sustain the reaction at 1200° F. Air may be introduced to advantage in heating up the reactor or starting the process, but it is not desirable for normal operation. Titanium tetrachloride is recovered from the gaseous reaction product removed from the upper portion of the reaction zone.

In a continuous process, rutile ore mixed with powdered coke or anthracite is chlorinated in a fluidized state in a shaft furnace at 900° C to completion.[41] The introduction of air or oxygen with the chlorine may be necessary to supply auxiliary heat. Titanium tetrachloride, ferric chloride, and chlorides of other contaminating metals and impurities are simultaneously formed, volatilized, and removed as a gaseous mixture from the furnace. The solid materials charged into the furnace are allowed to react until they disappear or become so small that they are transported from the furnace and thereby leave a bed in which there is no substantial change in composition. On discharge from the furnace the gaseous mixture is cooled to 250° C and admitted to the base of a scrubber column connected at the lower end to a reboiler and at the upper end to a reflux condenser. The titanium tetrachloride and silicon tetrachloride which together with the entrained solids, ferric chloride and zirconium tetrachloride, from the gases are washed down the column to the reboiler. From the reboiler the titanium tetrachloride is recovered by redistillation and from the gases leaving the condenser by cooling to $-10°$ C. Continuously or intermittently the reboiler is purged to remove the solid impurities. Chlorine is passed into a quartz tube with an inner diameter of 10 cm at 600° C at a rate of 8 liters per minute, and a mixture of 4 parts of titanium dioxide with 1 part of charcoal is introduced into the tube at a rate of 17 g per minute to give 18 kg of 93 per cent titanium tetrachloride in 10 hours.[42] The residue to the amount of 560 g collected at the other end of the tube is recycled to give 100 g of titanium tetrachloride for a total yield of 98 per cent. Titanium ores mixed with carbon or other hydrogen-free reducing agent are chlorinated by the action of chlorine gas.[43] The gaseous products are cooled stepwise to separate the precipitating compounds. Iron is added to the ores to bind the oxygen as ferric oxide.

Ores and slags of titanium and zirconium are chlorinated by the fluidized bed technique without excessive corrosion of the reaction chamber, by injecting the chlorine into the bed through radially directed nozzles downwardly inclined through the reactor walls near the base.[44] By this procedure the chlorine streams contact the bed material and not the chlorinator base. Since the chlorinator outlet is at the top, the chlorine rises in streams from each nozzle through the bed, carrying bed material upward with it thus effecting fluidization. In starting the chlorination, the bed is first heated to 700 to 1200° C by burning granular fuel with air blown through the nozzles before injection of chlorine. Chlorination of 1000 lb of rutile per hour at 900° C with 200 lb of coke per hour employing nozzles of acid-resistant cast refractory material for the injection of chlorine was carried out for a period of more than 18 months without interruption.

Titanium tetrachloride is formed by the reaction at an elevated temperature of chlorine with rutile ore and coal in a fluidized condition by the rising current of chlorine through the fluidized bed. The mixture of 1 part powdered coal and 5 parts of powdered ilmenite (ground to pass a 50 mesh screen) is fed in the top of a vertical relatively narrow reaction vessel. Chlorine is introduced at the bottom to maintain the solid bed in the desired fluidized condition. The reactor is maintained at 850 to 1000° C. This temperature may be maintained by the heat of the reaction, but the initial heat may be had from external heating or by burning coke in the furnace before the reaction starts. In large units cooling may be required. The products of the reaction (titanium tetrachloride and ferric chloride primarily) leave the top of the furnace and go to a cyclone for removing the suspended solids and then to a condenser to condense the vapors.

✓It is essential that the upward flow of the chlorine through the reaction chamber is rapid enough to maintain the suspension of the solid ore and coal mixture. The height of the suspension must be great enough to provide sufficient contact of the solid reactants with the chlorine. After a bed of the desired height is obtained, the solids feed rate is reduced to that required for equilibrium. Twenty or more pounds of titanium tetrachloride per cubic foot of reaction chamber space per hour may be produced. Thus a chamber of between 50 and 60 cubic feet capacity is capable of producing over 10 tons of titanium tetrachloride a day.

In an example, a powdered mixture of 1 part of coke and 5 parts of rutile ore (through 200 mesh) is fed into a chlorine gas stream flowing at a rate of 40 feet per second to the bottom of an elongated, vertical reaction chamber previously heated to 900° C by combustion gases. The proportion of solids to chlorine was approximately the theoretical value of 100 pounds of solids per 750 cu ft of chlorine. The velocity of the chlorine stream was sufficient to maintain the solids entrained in the chlorine in a fluidized state or bed in the reaction zone. After the reaction, the products were removed from the reaction chamber, and the titanium and iron chlorides were separated from the exit gases by cooling to room temperature. The chlorination reaction produced an increase in the volume of gases due to an increase in temperature and to the formation of new volatile products (carbon monoxide and carbon dioxide) which helped to maintain the fluidized bed.[45] Recovery of the titanium was practically 100 per cent.

Titanium tetrachloride is prepared by chlorinating rutile in a fluidized bed 3 or more feet in diameter.[46] The irregularity in the density of the bed may be minimized and chlorine utilization may be improved in beds of such larger diameter by introducing the chlorine into the lower portion of the fluidized bed through a porous base or through a distributor having a number of orifices or channels in which a pressure drop of 3 to 6 lb per square inch takes place as the chlorine passes through. If the rate of chlorine introduction is from 3 to 20 times the minimum fluidizing velocity, usually less than 1 per cent of the chlorine escapes in the exit gases. Particle size of the rutile should be from 75 to 200 microns. The exact rate of flow of chlorine into the bed for optimum operation depends on the temperature of the reaction zone. With rutile of 130 micron size and a temperature of 900° C optimum efficiency of chlorination is achieved at a chlorine introduction of 100 pounds per square foot of cross section of the reaction zone. At 800° C, the rate is 60 pounds per square foot of cross section of the reaction zone.

In an example, the rutile was chlorinated in a shaft furnace consisting of an outer shell of steel lined with chlorine-resistant brick work, having an internal diameter of 2 feet 6 inches. Near the base of the furnace was a perforated plate, the perforations of which were fitted with orifices of restricted diameter and superimposed by ceramic gas-permeable barriers. The pressure drop across each of the orifices was 6 lb per square inch. These barriers permitted the uprising gas to enter the furnace but prevented dust or other solid material from passing down through the plate. Below the plate was a port through which chloride gas was admitted. At the top of the shaft was provided a star valve through which the mixture of rutile and carbon was admitted. Also on one side of the furnace near the top was provided a port from which the gases were conveyed to the condensing system. To commence the process, the chlorinator was filled with rutile to a depth of 3.5 feet and then fluidized by the admission of air, and direct gas firing to raise the temperature to 900° C. Carbon was then added to produce a bed containing 20 per cent by weight. Then chlorine was fed into the hot mass at a uniform rate of 400 pounds per hour to maintain a fluidized bed with a gas velocity of 18 cm per second. A mixture of rutile and coke was fed continuously through the star valve to maintain the bed. The rutile contained 97.4 per cent titanium dioxide and only 1.1 per cent of ferric oxide and had a particle size of 130 mesh.

In the chlorination of titaniferous material in a fluid bed reactor employing a solid or gaseous reducing agent, any heat generated in excess of that required to maintain the operating temperature within the reaction zone is absorbed by the volatilization of an added inert material, preferably titanium tetrachloride.[47] The chlorination of ilmenite and rutile in the presence of carbon is a highly exothermic reaction. In small scale operation the loss of heat is such that heat must be supplied to maintain the normal chlorination temperature of 850 to 950° C. As the size of the unit increases, a stage is reached where the chamber no longer requires auxiliary heat, but the heat generated by the reaction becomes more than sufficient to maintain the charge at the preferred reaction temperature. Overheating develops, and with it corrosion of the furnace walls, sintering of the mass and blockage of the device for distribution of the chlorine. This results in inefficient operation. Titanium tetrachloride, ferric chloride, or a mixture of these is preferable as the cooling agent since their removal at a later stage does not constitute an added operation.

For example, a shaft furnace 10 feet high and 4.8 feet inside diameter was lined with chlorine-resistant brickwork inside a steel shell which had an outside diameter of 9 feet. At the base of the shaft was a perforated plate which afforded a gas-permeable but solid-impermeable partition producing a pressure drop not less than the drop in pressure in the bed. Above the plate was a 3 foot fluidized bed consisting of rutile and coke maintained in the fluidized state by chlorine admitted through the plate at 30 to 50° C. The solid materials, 4 parts of rutile to 1 part of coke, were admitted by a screw feeder immediately above the bed. This mixture was preheated by the reaction of air admitted through the perforated plate with excess coke. The air served to fluidize the bed. In this way the temperature of the bed was raised to 870° C. At this stage the air was shut off and chlorine was admitted to the bed at a rate of 650 kg per hour. Liquid titanium tetrachloride at 20 to 40° C was admitted directly to the bed through a stainless steel pipe. The introduction of this titanium tetrachloride to the bed enabled the temperature to be maintained at 870 to 900° C. The product gases were cooled to separate the components by the usual procedures.

In chlorinating titanium-zirconium ores in a fluidized bed, the temperature is held at 870° C to prevent chlorination of the zirconium compounds.[48] This unreacted residue is periodically removed from the furnace as a concentrate containing 40 per cent zircon. The reaction is cooled by recycling a stream of liquid titanium tetrachloride to the chlorinator. A solid mixture of finely divided titaniferous ore and coke is fluidized by introducing chlorine into the reactor maintained at 800 to 1050° C through a series of helical passages in vertical bottom inlets to the reactor.[49] The passage has a helix angle at its lower boundary surface which is less than the effective angle of slide of the solids being fluidized within the vessel. A mixture of finely divided ilmenite ore, coke, and furnace ash has an effective angle of slide of 38 to 45 degrees and an effective angle of repose within about the same range. With solids of this type, inlet conduits of 0.8 inch inside diameter, a thread of ⅛ inch, and a minor helix angle of 25 degrees can be used. If the spiral inserts are removed for cleaning or replacing with the fluidizing chlorine cut off so that the bed solids settle down on the floor of the fluidizing chamber, a temporary, disposable plug is inserted in the gas inlet to retain the solids. Chlorination of titaniferous material in the presence of a carbonaceous reducing agent in a fluidized bed is most efficient if the gases entering the bed are distributed uniformly over the entire cross section of the reactor so that maximum contact between the gases and the solid particles is assured.

Titanium ore or slag is compacted with coal or coke, sintered at 600 to 1400° C, ground to pass 140 mesh, and chlorinated at 800° C to produce the tetrachloride.[50] In the electrothermic chlorination of titanium dioxide less corrosion damage results and less input energy is required if the temperature is kept at around 900° C by controlling the throughput.[51] The silicon carbide or graphite electrodes of the apparatus employed are vertically mounted and covered with quartz tubes to prevent channeling of the chlorine through the charge of titanium dioxide-carbon briquets. A shaft furnace for use in the continuous chlorination of titanium dioxide ore or concentrate includes a perforated disk to distribute the chlorine in the charge.[52] Titaniferous material is continuously chlorinated in an apparatus that has an internal diameter of 4 feet 10 inches where the static height of the bed is 5 feet.[53] This design gives a pressure drop between the wind box and the distributor tube of half the pressure drop across the bed. Less than 0.5 per cent of the chlorine gas through the apparatus is unreacted. The fluidized bed reactor for the chlorination of titanium ores consists of a porous graphite cylinder with a gas-tight corrosion resistant metal outer jacket.[54] A gap between the cylinder and the jacket is filled with soot or coal dust into which reaction gas is fed. The graphite cylinder has water-cooled electrodes for resistance heating up to 1600° C. A mixture of a powdered titanium containing iron ore is fed with coal dust and chlorine into the cylinder to form a fluidized bed. The resulting ferrous chloride is unstable and decomposes to iron droplets or vapors which leave the reactor to condense in a subsequent chamber. Titanium tetrachloride leaves the chamber as a vapor to be condensed later. A cyclone separates the coal dust from the vapors or gases.

Cracks in the ceramic lining and leaks in the iron retort used in the manufacture of titanium tetrachloride are sealed without condensation of the desired product.[55] The temperature of the contact surface between the metallic resort and the ceramic lining is lowered so that metallic chlorides present as contaminants condense in the cracks and pores to stop the leaks. To prevent deposit formation on the tubes and walls of the furnace during the chlorination of titanium dioxide ores or concentrates, titanium tetrachloride is blown into the upper part of the furnace.[56] As a result the temperature rises to only 400° C instead of the usual 700° C, with the result that no deposit formation occurs.

Titanium ore suspended in chlorine is passed through a heated, porous, graphite reaction tube surrounded by chlorine or nitrogen and housed in a steel jacket.[57] The distance between the reaction tube and the jacket is such that its temperature is too low to react with the chlorine.

Ceramic grade titanium dioxide is produced by removing iron and other colored impurities from ores or other titanium bearing material.[58] The crude material is ground, sintered, and subjected to magnetic and gravity separation for the removal of siliceous and iron impurities. This product is chlorinated at 700 to 850° C with chlorine, hydrochloric acid, phosgene, or mixtures of these gases. Air and from 0.5 to 5.0 per cent water vapor based on the gas mixture are introduced. In the chlorination step the iron content is reduced to 0.04 per cent. The water present inhibits the formation of titanium tetrachloride.

In the absence of carbon, the approximate decreasing order of reactivity of chlorine at 1000° C was reported to be iron oxide, magnesium oxide, calcium oxide, titanium dioxide, zirconium oxide, alumina, and silica.[59] More complex compounds containing these oxides, silicates, and spinel were less reactive. Furthermore, the rate of reaction depended upon the thermal history and was less for strongly calcined materials.

Ilmenite or similar raw materials were chlorinated in a continuous process,[60] by introducing the ore, carbon, and chlorine into a reaction chamber at such a rate that sufficient heat was evolved by the reaction to maintain the temperature above 600° C. Maximum efficiency was obtained, however, at 850° to 1250° C. For most purposes better results were obtained by holding the carbon content at 15 to 35 per cent of the ore. Proportions below this range were insufficient to insure complete chlorination of both iron and titanium components, while on the other hand excessive amounts made temperature control difficult. A quantity of briquets 0.25 to 0.75 inch in diameter was prepared from a mixture of 100 parts ore, 23 parts coke, and 14 parts molasses, and baked at 600° C to drive off the volatile materials. These carbonized pellets were introduced at a rate of 120 pounds an hour into a furnace having a 15-inch internal diameter and preheated to 1000° C. Chlorine was introduced at the rate of 2 to 2.5 pounds per minute, and the temperature was maintained at 850° to 1000° C throughout the run without externally heating the furnace. Vapors were withdrawn from the furnace as formed, and cooled to 40° C, with the result that 85 per cent of the ferric chloride and 25 per cent of the titanium tetrachloride condensed in this operation. The condensed chlorides were transferred to another part of the receiver, and the titanium tetrachloride was revolatilized by passing exhaust gases of the furnace over the mixture.

Titanium tetrachloride was formed by subjecting a briquetted mixture of ilmenite and coal to the action of chlorine at a temperature sufficient to volatilize the compounds formed, but not above 750° C.[61] The products were purified by redistillation. Donaldson [62] heated ilmenite or similar ores with charcoal at a temperature below 1000° C, but sufficiently high to reduce all iron compounds to the metallic state without much reduction of titanium. The iron was selectively converted to ferric chloride and volatilized by the action of chlorine at a temperature between 350° and 1000° C and the residue of titanium dioxide was chlorinated according to normal procedure. A tough granular mass having very desirable form for chlorination was obtained by drying a mixture of 20 to 25 parts powdered coke with 50 to 60 parts ilmenite and 20 parts titania in colloid suspension.[63] The suspension was prepared by agitating 100 parts of sulfuric acid-free hydrous titanic oxide (40 per cent solids) with 15 parts of concentrated hydrochloric acid. Briquettes formed by heating a mixture of titanium ore, ground to 0.5 mm, and coal, with a starch paste as binder, out of contact with air proved quite satisfactory.[64]

In a study [65] of chlorination of titanomagnetite concentrate containing 44 per cent titanium dioxide, 31.4 per cent ferrous oxide, 16.9 per cent ferric oxide, 1.84 per cent silica, 0.05 per cent chromium oxide, 0.72 per cent manganous oxide, 0.06 per cent copper, 2.76 per cent magnesium oxide, 0.15 per cent phosphorus pentoxide, and 0.16 per cent water, at temperatures up to 600° to 650° C, in a porcelain tube in the presence of solid carbon with a gas velocity of 3 liters per hour for 6 hours, a yield of 98 per cent was obtained. The concentrate was treated in the powdered form and in briquettes of various sizes. Optimum temperature of chlorination of titanium dioxide in the presence of manganese dioxide was found to be 450° to 480° C, and above 500° C without the catalyst, with a yield of about 90 per cent.

Separation of iron from ilmenite or similar ore was effected by subjecting the finely ground material (300 mesh) without previous reduction, to selective chlorination in the absence of a reducing agent at 800° C.[66] At this temperature the iron only was attacked, and the resulting ferric chloride was removed from the reaction chamber by sublimation as formed, leaving a residue of titanium dioxide together with a small amount of silica and other impurities. Chlorination of the titanium component began at 815° C, and this temperature should not be exceeded during the initial stage of the process. Best results were obtained above 800° C. The ferric chloride was deposited in cooling chambers and later decomposed by heat to regenerate chlorine for use in treating more ore. Residual titaniferous material was chlorinated in another operation, or it was used directly or was purified by conventional methods.

The amount of iron and titanium recovered as volatile chlorides from ilmenite was effectively controlled by regulating the amount of reducing agent in the charge during chlorination, and with a carbon content within an optimum range, the major portion of the iron was removed without appreciably attacking the titanium.[67] If a proportion of carbon in excess of the optimum was employed, some titanium was volatilized as the tetrachloride, and with ore mixtures containing large excesses of the reducing agent the major portions of both iron and titanium were converted to chlorides and removed from the reaction chamber by volatilization. Within the temperature range of 700° to 1150° C, changes in the temperature did not require significant changes in the carbon concentration, but below this range a slight increase in the proportion was necessary to produce best results. The optimum carbon concentration may be greatly changed, however, by mixing oxidizing gases, e.g., air and oxygen, with the chlorine. Most efficient results were obtained by chlorination above 500° C, but, in general, excessively high temperatures caused sintering of the ore and prevented proper distribution of the gases. Although reaction took place below 500° C, the rate was very slow and the heat developed was not sufficient to maintain the operating temperature. A working range of 700° to 1150° C gave most efficient chlorination, as well as maximum iron removal with minimum losses of titanium. In a series of tests, samples were prepared by mixing various proportions of finely ground ore and carbon with 12 per cent molasses (based on the ore), and the compositions were formed into briquettes ⅛ inch in diameter and baked at 400° C to remove volatile hydrocarbons. These were then chlorinated at a uniform rate of 140 liters of gas per minute per kilogram of sample for 5 minutes at 980° C, employing chlorine-air mixtures of different compositions. With no added carbon other than that derived from the molasses, and employing chlorine with no admixed air, 79 per cent of the iron initially present in the ore was removed. A similar briquetted mixture containing 4 per cent added carbon, under the same chlorination conditions, resulted in removal of 97.5 per cent of the iron with a loss of only 3.4 per cent of titanium, while treatment in the presence of 6 per cent carbon resulted in the removal of 99 per cent of the iron and 22 per cent of the titanium.

The absolute and relative amounts of iron and titanium removed were changed considerably if a mixture of equal volumes of chlorine and air was used as the chlorinating agent. Thus, with 4 per cent carbon in the charge, 85 per cent of the iron was removed without significant loss of titanium, while at a carbon concentration of 8 per cent, approximately 95 per cent of the iron and only 5 per cent of the titanium were converted to chlorides. Employing a gas mixture of 20 per cent chlorine and 80 per cent air, and briquettes containing 8 per cent added carbon, 84 per cent of the iron and practically no titanium were reacted, while 13 per cent carbon resulted in removal of 95 per cent of the iron and only 5 per cent of the titanium. Ordinarily the temperature was maintained by regulating the rate of introduction of ore, carbon, chlorine, and air, but the reaction could be cooled, if desired, by introducing a diluent gas such as carbon dioxide or nitrogen.

According to a typical operation of the process, 100 parts ilmenite ore containing 26 per cent iron and 35 per cent titanium was mixed with 7 parts of carbon and 12 parts of molasses, and the composition was formed into briquettes having an average size of ⅛ inch in diameter and baked at 400° C. Total carbon after baking was 7.6 per cent. Ten parts by weight of the briquetted mixture was heated to 980° C, and a stream of 5 parts by weight of chlorine and 1 part by weight of air was passed through the charge maintained at the initial temperature. The residue contained 93.2 per cent titanium dioxide and 0.74 per cent ferric oxide. This corresponded to a removal of 99 per cent of the original iron with a loss of only 4 per cent of the titanium.

In general, the most desirable and most economical bonding agents for forming briquettes of this type were petroleum asphalt, natural asphaltum, tar, pitch, sulfite liquor from paper making, and bituminous emulsions.

By a very similar process [68] employing chlorine without admixed air, a uniform mixture of 100 parts ilmenite ore containing 26 per cent iron and 35 per cent titanium, mixed with 6 parts of coal and 12 parts of molasses, was made up into briquettes having an average size of ½ inch and baked at 400° C to remove volatile hydrocarbons. The final product contained 7.2 per cent carbon. Ten parts by weight of the briquetted mixture was heated to 815° C, and 20 parts chlorine was passed through the charge in a continuous stream. Temperature of the system was maintained at 815° C. The unreacted residue, containing 90 per cent titanium dioxide and 0.71 per cent ferric oxide, was chlorinated to yield a relatively pure titanium tetrachloride. Removal of iron from the original ore was 98 per cent complete, while the loss of titanium was only 4 per cent. After initiating the reaction, the briquetted ore and chlorine were introduced into the reaction chamber, together with a gaseous reducing agent, at such a rate that the temperature was maintained above 600° C without adding air or a solid reducing agent other than that employed as a binder for the titanium-bearing material.[69] To prevent the upper part of the bed from cooling to cause condensation of vaporized ferric chloride, Cleveland[70] preheated the titanium-bearing material and regulated the rate of charging.

From another approach, ilmenite, chlorine, and a gaseous reducing agent, natural gas, were introduced into a reaction chamber at such a rate that the heat evolved by the chlorination maintained the temperature above 600° C.[71] At this temperature the iron and titanium chlorides were volatilized as formed.

In chlorinating ores at elevated temperatures, heat may be conveyed from the chlorination residue to fresh starting materials by a stream of gas which does not enter into the reaction.[72]

Ilmenite ore from Travancore containing 58.9 per cent titanium dioxide mixed with charcoal to the extent of 30 per cent of the ilmenite is chlorinated at various temperatures up to 800 ° C.[73] Chlorination is complete at 800° C. At 600° C 95.6 per cent of the titanium dioxide is converted to the tetrachloride. In the presence of catalysts, chlorination can be effective even at 400° C. Briquets formed from ilmenite ore with a nonreacting binder as ferric chloride, colloidal titanic acid, or dextrin are charged into a chlorinating furnace and heated to 600 to 1000° C.[74] Hydrogen chloride gas is passed over the briquets, and the iron chlorides formed are condensed and separated from the outgoing gases. After the removal of the iron salts is complete, the residue of titanium dioxide left in the furnace is subjected to the action of chlorine and a reducing gas as carbon monoxide at 700 to 1000° C to form the tetrachloride.

Ilmenite ore containing titanium dioxide 59 per cent is mixed with a large excess of charcoal and molded into briquets which are heated at 1000 to 1300° C to reduce the iron oxides to the free element.[75] The iron is then dissolved with 50 to 60 per cent sulfuric acid to leave a residue containing titanium dioxide 61.24, ferrous oxide 1.29, silica 2.19, and fixed carbon 31.35 per cent. Recovery of titanium dioxide is 95.6 per cent. Finely divided titanium ores are agglomerated with limited amounts of carbon using low temperature coke as the binder.[76] Pellets are formed.

Plant [77] prepared titanium tetrachloride by mixing a finely divided ore with coal, tumbling and calcining to produce a product for chlorination. In an example, 2000 pounds of ilmenite ore containing 31.3 per cent titanium dioxide ground to 200 mesh was mixed with 1430 pounds of a low ash, soft, coking, bituminous coal of 0.25 inch particle size and 860 pounds of a previously carbonized mixture of coal and ilmenite ore of 0.125 inch particle size. This mixture was fed into a drum rotating at 5 revolutions per minute, 8 feet in diameter and 16 feet long, and heated at 500° C for 35 minutes. At this time the heat was turned off and tumbling was continued for 10 minutes. The product was discharged, lightly crushed, and screened to give a size of 0.5 by 4 inches. Fines were recycled. The briquets were chlorinated in a shaft furnace of 16 inches inside diameter by feeding to a reaction bed at 900 to 1000° C at a rate of 125 pounds per hour. Chlorine was fed to the reactor at 220 pounds per hour. Approximately 4621 pounds of crude titanium tetrachloride was produced from 6146 pounds of briquets.

Briquets of ilmenite or high titania slag and coke are partially chlorinated with chlorine, carbon monoxide, and carbon dioxide at 800 to 900° C to remove the iron.[78] Titanium losses are low if the chlorination mixture contains more than 30 per cent carbon dioxide. The gaseous mixture of ferric chloride, carbon monoxide, and carbon dioxide from the partial chlorination step is oxidized with oxygen in a fluidized bed at 700 to 750° C. The resulting gas mixture containing 30 per cent chlorine is utilized for the chlorination of titanium dioxide. Ferric oxide obtained in the process is very pure and suitable for pigment use. Titanium ores are upgraded by selectively chlorinating the iron component.[79] Ten pounds of powdered ilmenite is mixed with powdered charcoal and 0.15 pound of powdered iron pyrite and briquetted with tar or asphalt as the binding agent. The briquets are heated at 400° C to drive off the volatile materials and then chlorinated in a furnace at 400 to 800° C. During chlorination the iron oxide in the ore is converted to ferric chloride which volatilizes. An ore treated in this manner has its titania content raised to more than 80 per cent. An ilmenite concentrate containing 44.2 per cent titanium dioxide is ground to 80 to 100 microns, compacted with petroleum coke containing 93 per cent carbon and cellulose-sulfite liquor, and chlorinated at 1600 to 2100° C in a current of chlorine passing at a rate of 7.0 liters per hour.[80] The free energies calculated for the possible reaction of chlorine in the presence of solid carbon at 1573, 1773, and 1973° K show that in the 1773 to 2373° K range, iron is chlorinated preferentially. Alumina is chlorinated more intensely than titania, but the thermodynamic probability of chlorination of even titanium carbide is higher than that of alumina. The degree of chlorination of titanium compounds in the entire temperature range varies only between 26 and 42 per cent, practically independent of the temperature. After about 90 per cent of the iron is chlorinated, the degree of chlorination of titanium increases rapidly from 30 to 80 per cent.

## FLUIDIZED BED CHLORINATION OF ILMENITE

In a single stage continuous operation, titanium tetrachloride is prepared by chlorinating a mixture of ilmenite and coke in a fluidized bed in a vertical shaft furnace maintained at a reaction temperature of 750 to 1100° C.[81] To start the process the furnace is charged with finely pulverized rutile and coke and heated with the use of a gas burner. Fluidization is maintained with a flow of air at the rate of 200 pounds per hour. At 900° C heating is stopped, the air is shut off, and chlorine is admitted at the rate of 180 pounds per hour. As the reaction begins, 120 pounds of ilmenite of 100 to 250 microns particle size and 50 pounds of coke of 50 to 500 micron particles are charged from an upper port. At another upper port titanium tetrachloride is admitted as a coolant to maintain the temperature at 900° C. Gases passed to the first cooling tower are cooled to 300° C by the addition of finely divided ferric chloride at 300 pounds per hour. In a second tower the gases are cooled to 130° C with a spray of titanium tetrachloride, and the ferric chloride is removed by means of a dust separator. Final cooling to −10° C condenses the titanium tetrachloride.

A mixture of finely divided ilmenite and coking coal are heated together with agitation to produce small composite particles suitable for chlorination in a fluidized bed.[82] Such a mixture contains 25 per cent by weight of finely divided coking coal and 75 per cent ilmenite or other ore ground to pass through a 200 mesh screen. The coking coal preheated to 400 to 700° F and the ilmenite preheated to 900° F are thoroughly mixed and heated with agitation at 1000° F until the carbonaceous matter passes through a sticky phase and finally returns to the solid form. Heating is carried out in an internally fired rotary kiln or in a fluid retort. In the fluidized bed chlorination of ilmenite at 800 to 900° C losses of unreacted chlorine and fine ore in the effluent gas and clogging of the outlet ducts with ferric chloride are minimized by passing the gas from the bed through an enlarged extension of the reactor at lower speed.[83] Here the temperature is held at 550 to 650° C so that ferrous chloride condenses on fine ore particles carried out of the bed. The condensed ferrous chloride reacts with excess chlorine to form ferric chloride. Bench scale continuous chlorination of Idaho ilmenite containing 27.9 per cent titanium and 36.5 per cent iron yielded a crude titanium tetrachloride suitable after purification for the production of titanium pigments.[84] The results of 380 hours of operation showed that 82.4 per cent of the titanium was chlorinated with 92.5 per cent utilization of chlorine.

Ilmenite mixed with petroleum coke is chlorinated at 950° C to produce ferrous chloride and titanium tetrachloride.[85] The ferrous chloride is reduced with hydrogen at 750° C to an iron powder product. Pigment grade titanium dioxide is obtained by hydrolyzing the tetrachloride continuously with steam and dry hydrogen chloride at 800 to 950° C in 5 to 9 seconds. Part of the hydrogen chloride formed in both reactions is reused. The remainder is absorbed in an aqueous solution and electrolyzed to produce chlorine and hydrogen. Ilmenite is chlorinated with a mixture of ammonia and hydrochloric acid obtained by heating ammonium chloride at 400 to 800° C and atmospheric pressure.[86] Iron is removed as ferric chloride at 500 to 600° C but as ferrous chloride at 700 to 800° C. Titanium dioxide is obtained as a residue. An apparatus for the recovery of titanium tetrachloride from ores is described by Wataribe, Fukuda, and Soko.[87]

## SELECTIVE CHLORINATION OF ILMENITE

In a cyclic process,[88] selective chlorination of ilmenite or titaniferous iron ore was carried out in two steps at different temperatures to effect a separation of iron and titanium chlorides. Finely ground ore was mixed with powdered coal or charcoal and formed into briquets with the aid of a fatty or tarry bonding material, such as oil, tar, or pitch, and subjected to a reduction roast in a furnace or rotary kiln which also served for the chlorination vessel. The briquetted material was heated, in the absence of air, at 800° C for 1 to 2 hours in a retort which was sealed, except for a small opening to allow escape of the gases driven off, and then cooled in the reducing atmosphere to 350° C. At this stage dry chlorine gas was passed through the reduced mass, and at this temperature iron only was attacked and the ferric chloride formed passed over in the vapor form and was condensed in a cooling chamber. The direction of flow of chlorine was then reversed, and the temperature was raised to 550° to 600° C. Within this range the titanium component was attacked and the resulting tetrachloride, containing traces of chlorides of iron, silicon, and vanadium, passed into a cooling tower where it was condensed. The crude material was purified by redistillation. According to Patel and Jere [89] the standard free energy change considerations show that it is beneficial to chlorinate ilmenite with chlorine in the presence of carbon and also that the iron constituent is preferably chlorinated with chlorine, titanium tetrachloride, or their mixture. The findings are corroborated from experimental work published in the literature.

Ilmenite is treated with chlorine in a unitary integrated process to form iron chloride as an intermediate product, which is converted to iron oxide and chlorine, and the released chlorine is used for the chlorination of the titanium component of the charge.[90] The ore is treated with a limited amount of chlorine to selectively chlorinate the iron component only. The ferric chloride is oxidized in the vapor state with air to form iron oxide and chlorine, and the chlorine is returned to the main reaction for chlorination of the titanium component of the ore. A mixture of beach sand ilmenite of particle size 40 to 120 mesh and powdered coal is maintained in a fluidized bed at 800 to 900° C for chlorination. Satisfactory fluidized bed suspensions are obtained using beach sand and ilmenite without grinding. The air used in the oxidation of the ferric chloride contains at least 0.0004 pound of water per cubic foot. Instantaneous chlorination of the iron component is effected and the ferric chloride is vaporized from the suspended bed. The iron oxide formed is finely divided so that it is carried along with the gas stream consisting of chlorine released and carbon dioxide and nitrogen from the original gases. A mechanical dust collector or electrical precipitator may be used to remove the iron oxide.

For example, unground, black sand ilmenite, 58.8 per cent titanium dioxide, 9.8 per cent ferrous oxide, and 25.5 per cent ferric oxide having a particle size range from 50 to 120 mesh was added to the bottom portion of a chamber at the rate of 10 parts a minute by weight along with 1 part per minute of carbonized anthracite. These materials were added to a 2-foot suspended bed of the same mixture in the chamber which was maintained in a fluidized state by passing 4.7 parts per minute of chlorine and 3.3 parts per minute of air containing 0.00089 pound of water per cubic foot in an upward gas flow through the suspension. The temperature of the bed was maintained at 850° C. Only enough chlorine to theoretically combine with the iron in the ore was admitted initially. Conversion of the iron to the chloride was 90 per cent. The reaction of the iron component with the chlorine was instantaneous and the ferric chloride was volatilized from the suspended bed. From the top portion of the bed the titanium dioxide concentrate was removed continuously by an overflow pipe to another chamber at the rate of 6.8 parts per minute. Directly above the suspension 1 part per minute of oxygen was added to oxidize the ferric chloride vapor to ferric oxide and chlorine. Conversion of the ferric chloride was 90 per cent. Removal of the suspended ferric oxide from the gas was effected by an electrical precipitator. These product gases contained 3.9 parts of chlorine, 2.3 parts of carbon dioxide, 0.3 part of hydrochloric acid, and 2.3 parts of nitrogen and were added at the bottom of the second chamber containing the titanium dioxide concentrate at a rate of 8.5 parts per minute along with 7.1 parts per minute of added chlorine to maintain a fluidized state. Carbonized anthracite was added at a rate of 1.4 parts per minute. A temperature of 850° C was maintained. The chlorine added was 5 per cent in excess of the theoretical.

A titanium dioxide concentrate is produced by the selective chlorination of ilmenite, rutile, or slag in a fluidized bed to volatilize the iron component as ferric chloride.[91] Before chlorination the iron is reduced to the ferrous state. The amount of carbonaceous reducing agent used is sufficient only to accept the oxygen combined with the iron. An amount of chlorine sufficient to react only with the iron to form ferric chloride is admitted. A bed of titanium material having a low iron content such as rutile is suspended in a reactor by passing chlorine gas up through the bed. A feed material of reduced titanium ore or slag containing iron and a carbonaceous reducing agent is added continuously to the fluidized bed. In general a fluidized bed process is carried out by continuously charging the finely divided titanium ore and coke into a reactor heated to the temperature required for the chlorination reaction to take place. Chlorine is passed upward through the bed to cause the materials to surge upwards in a turbulent fluid-like manner. The ore-carbonaceous mixture to be chlorinated is fed continuously into the top of the reactor above the upwardly surging fluid bed with which it becomes intimately mixed.

In an example, a reduced ore containing 62 per cent titanium dichloride and 31.7 per cent ferrous oxide prepared by heating ilmenite with coal or coke at 850° C was used as the feed material. In the cylindrical reactor on a perforated plate was placed a bed of rutile containing 89 per cent titanium dioxide, 3.1 per cent ferrous oxide, and 7.9 per cent other oxides. Chlorine was admitted at a rate sufficient to suspend the material in the reactor. The temperature was maintained between 790 and 810° C. To the suspended bed was added a mixture of the prereduced ore and 4 per cent carbon ground to 20 to 200 mesh at a rate adjusted to correspond to the chlorine added to react with all of the iron present in the bed. The rate of ore addition to the bed was also adjusted so that at no time during the reaction was there more than 10 per cent iron, calculated as ferrous oxide, in the suspended bed. A constant level of the bed was maintained throughout the run by continuously removing bed material at an overflow level. Ninety-four per cent of the iron was removed. The concentrate contained 90.4 per cent titanium dioxide, 3.1 per cent ferrous oxide, and 6.5 per cent other oxides.

In a continuous process the iron oxide content of ilmenite is chlorinated selectively at 800 to 950° C, after which the residual titanium dioxide is chlorinated at 950 to 1150° C by a stream of chlorine and carbon monoxide.[92] Utilization of the chlorine is practically complete. The chlorine is recovered from the condensed ferric chloride, or by burning the effluent vapor in an excess of oxygen. Solid carbon is not used in the chlorination steps.

In an example, a fluidized bed 60 cm deep and 40 cm in diameter of ilmenite containing 5 per cent ferrous oxide was fed continuously at 18 kg per hour with 40 mesh ilmenite sand preheated to 700° C and containing 37 per cent ferrous oxide and 57 per cent titanium dioxide. A gas mixture consisting of 52 liters per minute of chlorine and 58 liters per minute of carbon monoxide was passed through the charge. The whole charge was held at 900° C. Through an overflow 11 kg per hour of residue containing 95 per cent titanium dioxide and 3 per cent ferrous oxide was removed. The ferric chloride condensed from the reaction product at the rate of 11 kg per hour contained 98 per cent ferric chloride and less than 0.3 per cent titanium. In a second series ferric chloride was not condensed, but the effluent gases were treated with 90 liters per minute of oxygen. The ferric chloride was 95 per cent oxidized to ferric oxide and the resulting gas produced at the rate of 130 liters a minute contained 35 to 40 per cent carbon dioxide, 38 to 40 per cent chlorine, and 25 to 30 per cent oxygen. This gas was passed through glowing carbon at 150 liters per minute to regenerate the carbon monoxide and remove excess oxygen, after which it was recycled. The titanium dioxide-rich solid residue was heated in a second fluidized bed at 1050° C and chlorinated with a mixture of 99 liters per minute of chlorine and 106 liters per minute of carbon monoxide. Complete chlorination of the titanium dioxide was accomplished at a chlorine utilization of 95 per cent.

The standard free energy changes for the chlorination of ilmenite, rutile, and leucoxene employing carbon as the reductant and chlorine gas were calculated at different temperatures.[93] All three minerals could be chlorinated readily. Ferrous oxide and ferric oxide were preferentially chlorinated at all temperatures between 400 and 1500° C, but more efficiently between 400 and 500° C. By removing iron as ferric chloride by preferential chlorination, the titanium dioxide content of ilmenite could be upgraded. Titanium dioxide was chlorinated in preference to silica and to sillimanite in the range 400 to 1500° C but more efficiently in the range 500 to 600° C. The standard enthalpy values of the chlorination reactions of ilmenite indicated the possibility of maintaining the reaction temperature around 600 to 700° C without external heating. All the theoretical findings were verified by experimental work. The iron content of Travancore ilmenite of the composition titanium dioxide 58.78, ferric oxide 28.20, ferrous oxide 5.16, manganese oxide 2.20, alumina 2.75, phosphorus pentoxide 0.10, silica 1.38, and carbonaceous matter 0.10 per cent is chlorinated preferentially by controlling the temperature and the amount of carbon in the mixture.[94] Manganese oxide present in the one may be responsible for the ease of chlorination at lower temperatures. Ceric oxide out of iron pyrites, copper oxide, lead oxide, manganese dioxide, ceric oxide, and calcium phosphate are the best catalysts. The enhancement of the rate of chlorination by a small amount of water vapor may be due to the formation of hydrogen chloride gas and oxygen, both of which are effective. Carbon to the extent of 30 per cent by weight of the ilmenite is the optimum amount. Ferric chloride is the best binder for briquets. Efficient chlorination of ilmenite is effected by the action of chlorine containing 1 part in 15 of oxygen on the 200 mesh mineral mixed with 30 per cent of carbon and 1.0 per cent of ceric oxide at a temperature of 500° C.

The iron of ilmenite is reduced to the bivalent state and chlorinated selectively to form ferrous chloride which is separated leaving titanium dioxide concentrate suitable for chlorination.[95]

The ordinary chlorination of ilmenite results in the formation of troublesome iron chlorides. Wilska [96] by thermodynamic calculation attempted to establish conditions of temperature and pressure that favor the formation of titanium tetrachloride rather than ferric chloride. Equilibrium constants were calculated from thermodynamic data in the literature and from the possible reactions in the system. On the basis of standard free energy change calculations, ferric chloride is formed preferentially over titanium tetrachloride at all temperatures. An increase in temperature shifts the equilibrium in favor of the undesired ferric chloride. Furthermore, the pressure can have no effect on the relative equilibrium. It is concluded that unless specific catalysts are found for increasing the rate of chlorination of titanium dioxide preferentially or decreasing the rate of chlorination of ferrous oxide and ferric oxide, it is necessary to find a method of removing iron from the ore prior to chlorination. A titanium dioxide concentrate is produced from ilmenite ore by direct chlorination.[97] The optimum reaction temperature ranges from 900 to 1000° C, and optimum reaction time is 30 minutes at 1000° C.

The titanium present in ilmenite is converted to titanium tetrachloride without converting the iron present to ferric chloride.[98] Ilmenite and chlorine are separately heated to 1250 to 1450° C and maintained within this range during chlorination. Within this temperature range the titanium reacts rapidly with the chlorine but the iron compounds present in the ilmenite does not react with the chlorine. Iron practically ceases to react with chlorine at 1200° C but titanium and chlorine react at temperatures up to 1500° C. If the iron is chlorinated along with the titanium the ferric chloride tends to form a gummy solid which clogs the apparatus and it is difficult to separate from the titanium tetrachloride.

In an example, ilmenite ground to pass a 60 mesh screen and mixed with 4 per cent by weight of powdered charcoal was heated in a closed container having one chlorine inlet tube at the bottom and one gas outlet tube at the top. The ilmenite-charcoal mixture and the incoming chlorine were heated to 1300° C. Pure white fumes from the reactor were condensed to titanium tetrachloride of over 99 per cent purity. No iron was detected in the product. In another example, the ilmenite was briquetted with 10 per cent charcoal and 10 per cent molasses. A fluidized bed of ilmenite and carbon is chlorinated at 850 and 950° C with chlorine at such a velocity that the volume of the bed is doubled.[99] The temperature of the bed is controlled by a separate auxiliary fluidized bed maintained by carbon dioxide. Titanium tetrachloride and ferric chloride are continuously removed.

**Sphene.** A sphene concentrate containing 25.42 per cent titanium dioxide, 26.82 per cent silica, 28.45 per cent calcium oxide, 2.98 per cent ferric oxide, 1.32 per cent ferrous oxide, 1.29 per cent magnesium oxide, 0.13 per cent manganous oxide, 7.18 per cent phosphorus pentoxide, and 0.14 per cent water was chlorinated at temperatures between 290° and 900° C, although best results were obtained at 800° to 900° C.[100] In general, the chlorination of sphene was found to be much more difficult than of titanomagnetite. The chlorine distribution among the components of the concentrate was not favorable. Calcium chloride was formed rapidly, although the final recovery of titanium was almost complete.

**Slag Briquets.** Eight slags prepared by smelting ilmenite with commercial magnesium hydroxide were briquetted with coke and chlorinated in a laboratory apparatus.[101] At 800° C, ferrous oxide to magnesium oxide ratios of 0.7 to 1.2 gave maximum chlorine efficiency. Higher efficiencies were obtained at 900 and 1000° C. At 700° C the chlorination reaction was sluggish. The ratio of carbon dioxide to carbon monoxide in the waste gas increased as the reaction progressed. Using the carbon monoxide gas as a reducing agent for the chlorination of pigment grade titanium dioxide, the reaction was slow, but proceeded satisfactorily if carbon monoxide and a carbon monoxide-chlorine mixture were passed alternately into the charge. In the chlorination of lower oxides of titanium and lower oxide slags with carbon monoxide gas as the reducing agent, the reaction proceeded smoothly although more chlorine was required than with similar chlorinations using coke as the reducing agent. Slag briquets containing titanium dioxide 55.6, carbon 25.3, ferrous oxide 7.2, silica 4.4, magnesium oxide 3.5, and alumina 3.2 per cent are chlorinated in a vertical furnace at 700° C to produce titanium tetrachloride which is condensed.[102] In an example, 3.9 tons of briquets were chlorinated at 700° C by passing in 5.2 tons of chlorine in 24 hours. The residue in the furnace containing calcium chloride and magnesium chloride was treated by passing in air for 2 hours at 800 liters per minute for regeneration of chlorine to obtain 3.8 tons of ash. Yields of titanium tetrachloride and ash without the air treatment were 3.4 and 1.7 tons, respectively.

A mixture of titanium dioxide slag 50, bituminous coal 25, anthracite coal 25, and sulfite liquor binder 5 parts was briquetted and coked in preparation for chlorination.[103] A production of 7 tons per day was achieved with 90 per cent and higher chlorination efficiency in a 30 inch inside diameter, 15 ft high chlorinator at a hot zone temperature of 800 to 900° C. The briquet develops a sponge-like structure with absorptive capacity for the molten chlorides. From the chlorinator the gases are passed through a solid chloride condenser and then through a specially developed splash scrubber, which condenses about 40 per cent of the titanium tetrachloride vapors and is a final scavenger for about 85 per cent of the total solid chlorides. The scrubbed gas is conveniently condensed. Chlorination of up to 90 per cent is also obtained with coarse ilmenite. Coked briquets formed from 200 mesh slag produced from Idaho ilmenite, mixed with carbon and dextrin in the ratio of 100 to 35 to 6 were chlorinated at bed temperatures varying from 605 to 852° C in a vertical shaft furnace.[104] Best results were obtained with slags containing the highest percentage of total titanium and in which the titanium was in the highest state of oxidation. Low silicon, calcium, and aluminum contents of the slags enhanced their potential value for chlorination. The lowest practical chlorination temperature was found to be 480° C.

Titanium tetrachloride is produced continuously by passing finely ground agglomerates of titanium slag and carbon with chlorine gas in a parallel current flow through a fluidized bed reaction zone maintained at 700 to 950° C.[105] The agglomerates are between 8 and 200 mesh in size. Suitable bonding agents in preparing the agglomerates include alkali metal silicates, inorganic phosphates, and sodium and potassium hydroxides. The concentration of the bonding agent is from 5 to 15 per cent by weight of the solid mass. Petroleum coke is particularly suitable because of its low ash content, although ordinary coke, hard coal, and charcoal may be used. The ratio of coke to titanium slag or ore is high enough to convert all of the combined oxygen to carbon monoxide. Agglomerates can be prepared by briquetting the mixture of solids with a binder and then grinding the briquets to the desired particle size. Likewise agglomerates can be prepared by forming thin dried films of the solids with the binder and breaking up the films by passing them between rubber covered rollers.

To illustrate, solid feed agglomerates were prepared from 26.2 weight per cent coke breeze (80 per cent through 325 mesh), 65.3 weight per cent of titanium slag (80 per cent through 325 mesh), and 8.0 weight per cent of sodium silicate (silica to sodium oxide ratio of 3.22). The agglomerate particle size was 88 per cent between 35 and 60 mesh. This agglomerated material was chlorinated in a fluidized bed 3.75 inches in diameter and 7 feet high at 800° C. In the operation, 19.9 pounds per hour of chlorine and 20 pounds per hour of solid fuel were fed into the fluidized bed, and 6.6 pounds per hour of ash, 24.2 pounds per hour of condensable gases, and 67.0 cubic feet (STP) of noncondensable gases were withdrawn. The condensable gases contained 21.2 pounds per hour of titanium tetrachloride representing a titanium extraction of 91.5 per cent. A chlorine conversion of 100 per cent was attained.

Briquets formed from slag containing 68 per cent titanium dioxide and powdered coke were chlorinated under various conditions.[106] To produce the briquets the ground slag was mixed with powdered coke, compacted to the desired form, and calcined for 8 hours at 800° C in the absence of air. These briquets were subjected for 7 hours to the action of commercial anode gas of 65 per cent purity. With an increase in the carbon content of the briquets from 15 to 20 per cent the degree of chlorination after 7 hours at 800° C increased from 82 to 85 per cent to 88 to 89 per cent. At 700° C, the optimum temperature, 90.7 per cent of the titanium was chlorinated. The use of concentrated chlorine instead of anode gas resulted in an increase of 8.5 per cent in the degree of chlorination and the rate of chlorination was higher. However, the purity of the technical titanium tetrachloride was the same.

Slags produced by smelting ilmenite with magnesia flux in an electric furnace were chlorinated at 700 to 1000° C in the form of briquets with powdered coal.[107] In a specific operation, 100 parts of Indian ilmenite containing 53.1 per cent titanium dioxide and 35.1 per cent ferrous oxide, 9 to 15 parts of coke and 4 to 20 parts of magnesium oxide, as crude 65.4 per cent magnesia, were heated in an electric furnace to give various kinds of slags with varying ratios of ferrous oxide to magnesium oxide. These slags powdered to minus 150 mesh were briquetted with minus 100 mesh coke and treated with dried chlorine at 700 to 1000° C. The yield of titanium tetrachloride of 97 per cent purity was higher at higher temperatures, for example, 75 per cent at 800° C and 80 per cent at 1000° C. Slags with the ratio of ferrous oxide to magnesium oxide of 2 to 3 were chlorinated easily. The equilibrium constants of the reactions taking place in the chlorination of the slag were calculated. Briquets produced from anthracite with 97.6 per cent titanium dioxide pigment, with a reduced titanium dioxide having a ratio of titanium to oxygen of 1.0 to 1.88 and with slag obtained by smelting Malayan ilmenite with magnesium oxide are reacted with a mixture of chlorine (250 ml per minute) and carbon monoxide (100 ml per minute) at 1000° C for periods of time up to 5 hours.[108] Titanium tetrachloride in 70 per cent yield is obtained from the pigment by passing alternately through the charge a mixture of carbon monoxide and chlorine, and chlorine. Use of the mixture of carbon monoxide and chlorine is effective for the reduced titanium dioxide and for the slag, but the amount of chlorine required for the same yield of titanium tetrachloride is 30 per cent higher than in methods employing solid carbon instead of carbon monoxide as the reducing agent.

Air is blown through the residual bed from the chlorination of briquets of high titania slag and coke, or ilmenite and coke to oxidize the chlorides of calcium and magnesium, thereby recovering additional chlorine.[109] If titania slag is chlorinated at 400 to 800° C the constituents are converted to chlorides in the order, ferrous oxide, titanium dioxide, silica, and alumina, and vaporized. Magnesium and calcium chlorides remain as solids. In an example, briquets consisting of 55.6 per cent titanium dioxide, 25.3 per cent carbon, 7.2 per cent ferrous oxide, 44.0 per cent silica, 3.5 per cent magnesium oxide, and 3.2 per cent alumina were charged into the furnace and chlorine was blown in. The reaction was carried out for 24 hours at 700° C. From the products the volatile chlorides were condensed and titanium tetrachloride was separated. Then air was blown through the residual bed at the rate of 800 liters per minute for 2 hours. A total of 3.8 tons of titanium tetrachloride was obtained and the residue amounted to 1.2 tons. In a similar reaction in which air was not blown in, 3.4 tons of titanium tetrachloride were obtained and the residue weighed 1.7 tons. The chlorine liberated from the calcium and magnesium chlorides rose through a fresh charge of briquets to form additional titanium tetrachloride.

A pilot plant vertical chlorinator for the production of 5 tons of titanium tetrachloride per day was operated continuously with a briquetted charge of high titania slag, coal, and a binder.[110] Iron chloride and aluminum chloride in the product gases were collected in a multistage condenser and removed from the condenser walls by a scraper. To improve the yield of titanium tetrachloride, the slurry obtained by cooling the reaction products is fed through a spray device to the surface of the molten bath of the chlorinator.[111] Solid components are subsequently removed in dust-collecting chambers.

## CHLORINATION OF SLAG IN A FLUIDIZED BED

Titaniferous slags and ores are chlorinated successfully in a fluidized bed.[112] Titanium extraction and chlorine utilization are greater than 90 per cent for most slags and for Australian rutile. Idaho ilmenite yields only 83 per cent of its titanium.[113] Slags can be chlorinated as effectively as rutile. Magnesium causes clinkering so that the bed must be withdrawn after only 70 per cent of the charge reacts. According to Evans,

Bennett, and Groves [114] high titania slag can be chlorinated successfully if the fluidized mixture with carbon contains less than 5.0 per cent alumina, 5.0 per cent calcium oxide, 7.0 per cent magnesium oxide, and 3.0 per cent silica. Slag particles 0.0074 mm in diameter sintered during chlorination at 700° C, and the yield of titanium was only 25 to 33 per cent.[115] With 1 mm particles sintering greatly diminished and the yield of titanium increased to 84 to 87 per cent. Particles from 1 to 5 mm in diameter did not show any sintering and gave the same yield of titanium, 84 to 87 per cent. With proportions of carbon below theoretical, considerable sintering was evident.

Titanium tetrachloride is produced in a continuous process from high titania slags containing calcium and magnesium compounds by chlorination in a fluidized bed.[116] Calcium and magnesium chlorides formed fuse at the temperature of the chlorination and coat the particles of titanium dioxide to cause them to become resistant to the chlorine. To overcome this difficulty a portion of the fluidized solids are withdrawn from the reaction zone, leached with water to remove the calcium and magnesium chlorides, dried, and returned to the bed. The temperature of the reactor in the fluidized bed system is determined by a number of factors. In general the higher the ratio of ferrous oxide to titanium dioxide the lower will be the operating temperature. The rate of the reaction is also influenced by the fineness of the solid materials. Excess chlorine is admitted to prevent the formation of ferrous chloride which has a lower volatility. The total gas flow within the reactor is maintained within the range of 0.1 to 1.5 cubic feet per second to provide the flow necessary for suspension of the reactant solids in the fluidized bed.

To illustrate the process, production of 238 pounds per hour was maintained by the reaction of titaniferous slag with carbon and chlorine in a brick-lined furnace 2 feet in diameter and 12 feet high. A chlorine feed inlet was provided at the bottom and an exit for product gases at the top. A method was provided for feeding finely divided granular solid materials at the 3 foot level through an inlet on the side wall of the furnace. Any desired amount of the finely divided solid furnace bed could be withdrawn from a side wall close to the bottom of the furnace. The slag containing 72 per cent titanium dioxide, 8 per cent ferrous oxide, 9 per cent calcium oxide, 4.5 per cent magnesium oxide, and 6.5 per cent of other components, including silica and alumina, was ground so that 1 per cent was −20 to +35 mesh, 21 per cent −35 to +65 mesh, 60 per cent −65 to +100 mesh, and 18 per cent through 100 mesh. The slag was added to the furnace at a rate of 143 pounds per hour. At the same time coke containing 98 per cent carbon ground so that 90 per cent passed through a 20 mesh screen was fed to the furnace at a rate of 60 pounds per hour. The furnace bed was maintained in a fluidized condition at 850 to 900° C by an upward stream of chlorine at a rate of 0.38 foot per second or 230 pounds of chlorine per hour. This was enough chlorine to react with all of the titanium dioxide, ferrous oxide, calcium oxide, and magnesium oxide with a slight excess to ensure conversion of all ferrous oxide to ferric chloride. The gaseous chlorides were removed at the top of the furnace. To maintain the bed in a fluidized condition the amount of bed material drawn off and recirculated was controlled. The furnace bed material that was withdrawn was cooled, leached with water in a series of counter current washings to remove the calcium and magnesium chlorides. After leaching, the slag and coke mixture was dewatered, dried in a rotary kiln, and fed back into the chlorination furnace along with original slag and coke. Ferric chloride was separated from the titanium tetrachloride by cooling in a tubular cooler to lower the temperature to such a point that vaporization of liquid titanium tetrachloride condensed most of the ferric chloride. By further cooling the titanium tetrachloride was condensed. The chlorine was recovered and residual gases were allowed to escape into the atmosphere.

Slag containing titanium dioxide 83.20, silica 5.85, ferrous oxide 3.85, alumina 2.23, magnesium oxide 3.73, and manganous oxide 3.41 per cent was chlorinated in a fluidized bed with powdered charcoal added to prevent the bed from sintering.[117] Iron oxide was selectively chlorinated in a 90 per cent yield at 750 to 850° C with 10 per cent chlorine admixed with coal producer gas. Yields of other impurities were 70 per cent. After this pretreatment the titanium dioxide residue was chlorinated at 650 to 750° C with chlorine mixed with an equal volume of carbon dioxide or nitrogen. The yield of titanium tetrachloride was 80 per cent after 15 minutes.

Difficulties due to accumulation of fused calcium chloride and magnesium chloride in the fluidized bed chlorination of titanium ores or slag at 700 to 1200° C are prevented by placing the chlorine inlets two or more inches above the bottom of the bed and withdrawing periodically the coarse sticky lumps from the quiet zone below the inlets.[118]

Titanium tetrachloride is prepared by the fluidized bed chlorination of commercial titaniferous slag containing appreciable amounts of calcium oxide and magnesium oxide.[119] Such a slag containing titanium as the dioxide 76.95, titanium trioxide calculated as the dioxide 19.35, iron 8.73, silica 2.42, alumina 4.69, calcium oxide 0.24, and magnesium oxide 5.35 per cent ground to pass 20 mesh in the amount of 324 pounds was fed with ground coke at the rate of 130 pounds per hour to a fluidized bed reaction chamber maintained at 950 to 1000° C. During the same period 527 pounds of chlorine was fed into the bottom of the reactor with a recycled portion of the noncondensable exhaust gases. The product gases were quenched with sand to 700° C to condense the magnesium chloride and calcium chloride. Ferric chloride and aluminum chloride were removed in a spray condenser, and the titanium tetrachloride was finally condensed in a water-cooled condenser at the rate of 560 pounds per hour. The magnesium chloride and calcium chloride vapors may also be condensed by mixing a diluent cooling gas in an amount sufficient to constitute 65 per cent by volume of the total gaseous materials entering the reaction.

Titanium slag or ores containing small amounts of compounds of calcium, magnesium, aluminum, and silicon are heated with a carbonaceous reducing agent at 1000 to 1500° C to produce a product that can be chlorinated in a fluidized bed at temperatures between 300 to 500° C.[120] The carbonaceous reducing agent such as coke, coal, charcoal, residual fuel oils, coal gas, and natural gas is added in excess of 0.5 parts by weight to 1 part of titaniferous ore or slag. Temperatures from 1000 to 1300° C are effective if the solid carbonaceous material and slag are compressed or if gaseous reducing agents are employed. At this low temperature of chlorination 300 to 500° C, little of the aluminum and silicon present are chlorinated, and the calcium and magnesium chlorides do not cause agglomeration of the slag-coke mixture. If the slag is not subjected to the reduction treatment, temperatures of 700° C or higher are required for chlorination, with the result that the titanium tetrachloride is heavily contaminated with chlorides of other metals. Furthermore, the reaction mass becomes agglomerated by the molten non-volatile calcium and magnesium chlorides formed with the result that it is difficult to move the reacting mass through the reactor. In an example, slag containing 69.9 per cent titanium dioxide, 9.9 per cent ferrous oxide, 6.0 per cent alumina, 6.3 per cent silica, 5.2 per cent magnesium oxide, 0.9 per cent calcium oxide, 0.6 per cent vanadium pentoxide, and 0.4 per cent iron was ground in a ball mill to 50 mesh and mixed with 0.46 part powdered metallurgical coke. This mixture was charged in a graphite lined furnace. The temperature was raised to 1380° C in 5 to 9 hours, the approximate melting point of the slag where the evolution of carbon monoxide was first observed. Heating was continued for 5 hours at 1400° C. After the electric power was shut off, the furnace was allowed to cool for 2 to 3 days before the charge was removed. The solid product was ground and screened to 70 to 230 mesh. Two hundred parts of the prepared slag was suspended and preheated by a stream of nitrogen at 400° C. As soon as the temperature reached 375° C the chlorine flow was initiated and the nitrogen flow was decreased so that the gas entering the bottom of the reactor was a preheated mixture of 5.3 parts per minute of chlorine and 3.1 parts per minute of nitrogen. The temperature in the reaction zone was maintained at 375° C. Any ion chlorides present in the vapors were condensed and removed in the initial part of the condenser.

A mixture of high titania slag, petroleum coke less than 0.25 millimeter in diameter, and sulfite liquor from paper manufacture (specific gravity 1.2) is heated at 800° C for one hour to remove the moisture and leave grains suitable for chlorination in a fluidized bed.[121] Typical slag contains titanium dioxide 77.5, silica 3.3, alumina 8.7, ferric oxide 10.5, calcium oxide 0.4, magnesium oxide 4.6, and sodium oxide 0.1 per cent. During the chlorination of untreated slag in a fluidized bed, the materials agglomerate resulting in a slowing down of the reaction and in interference with the uninterrupted discharge of the reactor residue. Slag containing titanium dioxide 70.9, ferric oxide 9.2, ferrous oxide 0.6, silica 5.9, alumina 6.6, calcium oxide 1.1, and magnesium oxide 5.4 per cent is added to the charge of rutile and coke in the shaft chlorination reactor to effect agglomeration of the fine particles of rutile.[122] The slag in amounts equal to 5 to 20 per cent of the weight of the rutile prevents the charge from becoming fluidized and permits chlorine velocities as high as 120 pounds per hour per square foot of the furnace cross section.

To prepare titanium tetrachloride continuously, a mixture of finely pulverized high titania slag, carbon, and chlorine is introduced into a fluidized bed of carbon held at a high temperature.[123] Titanium tetrachloride is recovered at the top of the zone. For example, a vertical quartz tube 50 mm in diameter and 150 mm long was used as a reactor. The tube was packed with 20 mesh silica to a height of 100 mm, and powdered pitch coke was added to a height of 50 mm above the silica. After charging the tube was heated to 800° C and dry chlorine was introduced from the bottom at a rate of 1 kg per hour forming a fluidized bed of coke. At this stage a mixture of 100 parts of the pulverized slag containing titanium dioxide 76.2, ferrous oxide 5.0, silica 6.7, alumina 4.4, magnesium oxide 5.7, and manganese oxide 1.3 with 50 parts of powdered pitch coke was added from the top of the reactor at a rate of 900 g per hour. The reaction ran smoothly for 30 hours and the yield of titanium tetrachloride based on the chlorine consumption was 95 per cent.

Titaniferous compounds are chlorinated effectively in the presence of carbonaceous material in a fluidized bed at 800 to 1300° C.[124] Coke preheated while being fluidized with air is mixed with the titaniferous material and chlorine gas, and fed into the furnace at a rate sufficient to maintain the solids in a well fluidized condition. For example, to a quartz tube 4 inches in diameter and 5 feet long containing a pebble and sand support, 1 kg of calcined petroleum coke from 100 to 200 mesh in size was added. Nitrogen was used to fluidize the bed until 1250° C was reached, at which point chlorine was substituted. A chlorine velocity of 0.5 feet per second, corresponding to 42 g per minute, was maintained, while a previously prepared stoichiometric mixture of slag containing titanium dioxide 71, iron 8.4, silica 6.0, alumina 6.6, and magnesium and calcium oxides 6.5 per cent plus traces of other compounds, and coke was screw fed to the reactor at a rate of 28 g per minute. The titanium tetrachloride was collected over a period of 11 hours. Conversion efficiencies of the chlorine and titanium were 85 and 98 per cent, respectively. Agglomerates of finely divided titanium dioxide and coke breeze were chlorinated in a fluidized bed.[125] Agglomerates prepared from 26.2 per cent coke breeze, 80 per cent through 325 mesh, 65.8 per cent titaniferous slag, 80 per cent through 325 mesh, and 8 per cent sodium silicate, having a silica to sodium oxide ratio of 3.22, were chlorinated in a fluidized bed at 800° C. Recovery of titanium was 91.5 per cent. The chlorine conversion was complete. Productivity of the reactor was 39.5 pounds of titanium tetrachloride per hour per cubic foot of reactor bed.

Slag is treated with chlorine in the presence of a carbon-containing reducing agent to produce titanium tetrachloride.[126] For example, a mixture of high titania slag, petroleum coke, and chlorine are passed from top to bottom through a chamber at 1150° C. The solid particles have a size of 40 microns. In the reaction 86 per cent of the titanium dioxide present forms the tetrachloride. The yield with respect to chlorine is also 86 per cent. Nitrogen gas may be used as a carrier. High titanium slag from ilmenite is chlorinated directly to obtain the tetrachloride, which is reacted with hydrogen and oxygen from the air to form titanium dioxide and hydrochloric acid.[127] The ratio of hydrogen to air is 2 to 1, and the concentration of titanium tetrachloride in the air is 2.8 to 3.6 gpl. A temperature of the titanium tetrachloride solution of 136° C is maintained. After combustion, the product is cooled on the aluminum drum rotating at a speed of 1 rpm above the burners. By igniting the titanium dioxide at 700° C for 4 hours, a white product of 99.7 per cent purity with a particle size of 0.2 to 0.4 micron is obtained.

Slag and slag concentrate obtained in the production of ferro-titanium alloys from ilmenite are chlorinated to produce titanium tetrachloride.[128] The slag contained 18 to 28 per cent titanium dioxide. A chlorination temperature of 600° C with a 100 per cent stoichiometric excess of carbon are optimum. The chlorination efficiency of the slag is 80 to 85 per cent and of the concentrate is 90 per cent. Yield of titanium tetrachloride from the slag is 60 per cent and from the concentrate is 65 per cent. The yields are increased to 70 and 75 per cent, respectively, if the titanium tetrachloride is distilled from the ferric chloride and aluminum chloride. However, following this procedure, separation of iron chloride and aluminum chloride could not be achieved. A colorless product of high purity is obtained by the simple distillation of the crude titanium tetrachloride.

Into a fluidized bed of titanium slag or ore mixed with 10 to 50 per cent powdered carbon, a portion of the fluidizing chlorine is introduced at a level well above the bottom of the reaction furnace chamber.[129] Following this procedure, the agglomeration otherwise caused by the alkaline earth chlorides is prevented by withdrawing periodically or continuously less than 10 per cent of the charge from below the level at which the chlorine is introduced. In this lower zone, particle aggregates caused by alkaline earth chlorides accumulate.

Titanium ore or slag is mixed with powdered coke, granulated and chlorinated in a fluidized state.[130] Thus, 30 kg of slag and 12 kg of coke were finely ground and well mixed. Then 12.7 kg of 28.6 per cent aqueous sodium silicate (silica to sodium oxide ratio of 3.22) were added. The mixture was dried at 120° C for 5 hours and ground. About 67 per cent of the particles were 0.149 to 1.00 mm and 33 per cent were less than 0.149 mm. The smaller particles were recycled, and the larger were roasted by passing them through an inclined tube heated to 850° C. Of the 45.4 kg of granular product 82.8 per cent had a particle size of 0.25 to 0.5 mm. These granules were chlorinated in a reactor 9.5 cm in diameter and 210 cm high at a temperature of 800° C. The reaction of 9.07 kg per hour of granules suspended in 9.03 kg per hour of chlorine gave 3.0 kg per hour of ashes and 11.0 kg per hour of condensable gases. Condensation of the gaseous products was accomplished by bringing them in contact with cold liquid titanium tetrachloride. By this reaction 9.65 kg per hour of the tetrachloride was obtained.

The iron component of titanium slag is selectively chlorinated to produce a concentrate rich in titanium employing titanium III and titanium II as the reducing agent.[131] Ilmenite or titaniferous iron ore is smelted in a furnace at a temperature above 1500° C to produce a titanium-rich slag containing lower valent titanium compounds. At this temperature a fair separation of the iron metal from the slag is effected. With the lower valent titanium compounds present in sufficient amounts to accept all of the oxygen combined with the iron in the slag, no carbonaceous reducing agent is required in the chlorination process and no low volatile ferrous chloride is formed to slag the apparatus. If chlorine is passed through a bed of the pulverized slag at 550 to 950° C the chlorine reacts with the iron oxide to produce ferric chloride only. In a specific operation, an ore containing 35 per cent titanium dioxide and 41 per cent iron was smelted above 1550° C to convert a large part of the iron compounds to molten metal and produce a slag containing 52 per cent titanium dioxide and 15 per cent compounds of reduced titanium, 10 per cent iron oxide, and the rest gangue. A 6000 part quantity of this slag ground to 35 to 200 mesh was chlorinated at 950° C by passing chlorine through the comminuted material. Practically all of the iron oxide was converted to ferric chloride only. The residue was composed of titanium dioxide and gangue.

## METHODS OF IMPROVING THE CHLORINATION OF SLAG

Titanium phosphate is added to high titania slag before chlorination so that the calcium and magnesium compounds present as impurities are converted to harmless phosphates rather than to chlorides which are detrimental to the chlorination of the titanium.[132] The calcium and magnesium compounds in the slag are chlorinated along with the titanium dioxide to form calcium chloride and magnesium chloride which coat the titanium dioxide particles and prevent its reaction with chlorine from taking place. If the chlorination is carried out in the presence of titanium phosphate, the calcium and magnesium form phosphates which do not inhibit the chlorination of titanium dioxide. The chlorination is preferably carried out in a fluidized bed of the slag particles and pulverized carbon. Chlorine gas is added through the bottom of the bed to suspend and react with the titanium dioxide. The titanium phosphate is added to the mixture in the fluidized bed. Calcium and magnesium phosphates form, and the chlorination of the titanium dioxide proceeds in an unhampered manner. The amount of phosphate added is at least 50 per cent of the amount required to react with all of the calcium and magnesium compounds present in the slag.

In an example, a titanium slag containing 67.1 per cent titanium dioxide, 3.7 per cent ferrous oxide, 4.5 per cent magnesium oxide, 9.5 per cent calcium oxide, 8.0 per cent alumina, and 7.2 per cent silica was ground to obtain a particle size of 400 to 200 mesh. A mixture of 400 parts of this slag with 140 parts of titanium phosphate and 100 parts of carbon was placed in a vertical tube furnace. Chlorine at the rate of 96.2 parts per minute and carbon dioxide at the rate of 3.2 parts per minute were introduced at the bottom of the tube and allowed to pass upward through the charge. The slag-carbon charge was heated externally to 850° C. A velocity of the gas sufficient to suspend the solid particles was maintained thus forming a restraining bed of the mixture. The amount of titanium phosphate added was sufficient to convert all of the calcium and magnesium compounds present to phosphates. Over a 4 hour period the temperature of chlorination was held at 800 to 850° C. Recovery of titanium was 84 per cent and 90 per cent of the chlorine was utilized. Before chlorination titaniferous slag is mixed with phosphoric acid and sulfuric acid to convert the calcium and magnesium components to compounds which do not react with the chlorine.[133]

In the chlorination of titanium slags by a chlorine stream, graphite or graphite-containing carbonaceous material is used as the reducing agent.[134] For example, a mixture of 100 parts of powdered slag containing 82.2 per cent titanium dioxide with 40 parts of 100 to 250 mesh graphite was fed into a furnace from the top at a rate of 1000 g per hour, and chlorine gas to effect fluidization was introduced at a rate of 900 g per hour at the bottom. Operation for more than 10 hours gave a chlorination yield of more than 95 per cent.

In the continuous production of titanium tetrachloride from high titania slag and chlorine, the formation of high temperature condensable ferrous chloride from ferric chloride vapor in the lower temperature zone of the furnace is prevented.[135] The amount of raw materials is controlled to prevent the formation of unreacted chlorine in the reaction zone, and 0.01 to 5.0 per cent of chlorine is introduced into the low temperature zone. In a specific operation, 300 kg of pellets of 10 parts of slag containing 88 per cent titanium dioxide and 4 per cent ferrous oxide, and 3 parts of charcoal react in the lower part of the furnace at 700 to 750° C with chlorine introduced from the bottom of the charge at a rate of 0.6 liter per minute. Chlorine was supplied to the upper part of the furnace at a temperature of 500° C at a rate of 0.006 liter per minute to obtain 22 parts of titanium tetrachloride and 1 part of ferric chloride continuously from the top. A total of 2920 kg of titanium tetrachloride and 135 kg of ferric chloride was obtained from 2000 kg of pellets. In a similar process the amount of chlorine added to the reaction zone is controlled.[136]

Difficulties arising from the accumulation of fused calcium chloride and magnesium chloride during the fluidized bed chlorination of titaniferous slag are prevented by admitting the chlorine 2 or more inches above the bottom of the bed.[137] Periodically the coarse sticky material is withdrawn from the quiet zone below the chlorine inlets.

## STUDIES OF THE CHLORINATION OF SLAG

Chlorination studies of high titania slags made from ilmenites from Trail Ridge, Florida, Aiken, South Carolina, and MacIntyre, New York, and from magnetite from Iron Mountain, Wyoming, in 4 inch and in 36 inch reactors show that the performance of the slags depends on their composition and vary widely with the source of the ore.[138] All of the slags except those produced from New York ilmenite and Wyoming titaniferous magnetite can be chlorinated as effectively as rutile. From 70 to 80 per cent of the titanium is converted to the tetrachloride. These exceptional slags contain enough magnesium to cause clinkering, which necessitates withdrawal of the bed before more than 70 per cent of it can react. The iron content of the slags does not interfere with the operation of the chlorinator. Results obtained from the 4 inch and from the 36 inch units are in agreement.

Difficulties encountered during the production of titanium tetrachloride are ascribed to secondary reactions of the tetrachloride with the metal oxides present in the slag.[139] In an attempt to avoid these difficulties, the reaction of titanium tetrachloride with ferric oxide, calcium oxide, magnesium oxide, and aluminum oxide at 300 to 800° C were studied. The chlorination of ferric oxide at 600° C was complete and that of aluminum oxide at 800° C proceeded to only 5 per cent. Calcium oxide was chlorinated easier than magnesium oxide. An increase of the temperature increased the rate of chlorination. Titanium tetrachloride splits off chlorine and is transformed to titanium dioxide of the rutile form. Residual solids after the reaction had gone to completion showed the same dimensions and form of granules as the starting oxide. The amount of titanium did not decrease due to the pseudomorphism. Experiments with slags containing titanium dioxide 71.3, alumina 6.3, ferrous oxide 4.4, silica 4.8, calcium oxide 6.1, vanadium oxide 0.1, magnesium oxide 3.8, sulfate ion 0.8, manganese oxide 1.7, and miscellaneous 0.7 per cent were carried out at 400 to 800° C. All chlorides were extracted from the residue with ethyl alcohol. The degree of chlorination of magnesium oxide and of calcium oxide was the same as in the first series; that of ferrous oxide was less than that of ferric oxide; and that of alumina was twice that of free alumina. The chlorination of titania slags and titanates is remarkably hindered by the presence of calcium oxide in the material.[140]

Assuming that the chlorination velocity of titanium slag depends on the mass transfer of chlorine in the gas film, the diffusion of chlorine in the ash and the reaction velocity, each resistance is determined from the measured values of weight loss of the pelleted cylindrical solid samples of slag and coke during the chlorination under various conditions by means of a thermobalance.[141] The diffusion coefficient in the ash and other terms are determined directly from the experimental data. From a study of the chlorination of slags and other titaniferous materials, free energies were calculated for the fundamental reactions.[142] Equilibrium diagrams for the binary system vanadium oxychloride-titanium tetrachloride were determined.

## CHLORINATION OF RED MUD FROM BAUXITE

Red mud containing 6.1 per cent titanium dioxide, after heating at 800 to 930° C for 2.5 hours is chlorinated at 800° C.[143] All of the ferric oxide is transferred to ferric chloride, which is evaporated. The residue, although containing 20 per cent titanium dioxide, cannot be used for the manufacture of pure white titanium dioxide by the sulfate process since the iron cannot be removed completely. To separate the ingredients, the mud is chlorinated at 900° C to form aluminum chloride, titanium tetrachloride, ferric chloride, and silicon tetrachloride. By treating the gas with alumina, silicon tetrachloride is precipitated and an equivalent of aluminum chloride is released. Then by treating the gas with titanium dioxide, aluminum chloride is converted to titanium tetrachloride leaving a residue of alumina. The remaining gas mixture containing titanium tetrachloride and ferric chloride is treated with ferric oxide to yield titanium dioxide, which can be refined to a pure white pigment and ferric chloride by standard procedure.

Experimental enrichment of fused slags from red sludge with acid wastes from crude oil refineries at high temperature, followed by chlorination and extraction with water or cold dilute hydrochloric acid effected the transfer of 90 to 95 per cent of the original titanium content to the tetrachloride.[144] The treatment with the acid wastes inhibited the formation of compounds melting at 500 to 700° C that normally form during and interfere with the chlorination. Addition of melting point depressants was found to be undesirable. As a result of the high melting point of the slag, the fusion must be carried out in an electric furnace.

## CHLORINATION OF ORES OR SLAG IN FUSED SALT BATHS

Titanium tetrachloride is prepared by introducing chlorine into a suspension of slags containing from 43 to 73 per cent titanium dioxide in a melt of magnesium chloride or of calcium chloride.[145] At a temperature of the melt from 850 to 900° C from 89 to 98 per cent of the titanium is converted. Increasing the calcium chloride of the melt up to 80 per cent decreased the chlorination rate by 50 per cent. The magnesium chloride content of the bath had no significant effect on the reaction rate. Similarly titanium tetrachloride is prepared by the reaction of a chloride having a higher melting and boiling point as sodium chloride with titanium dioxide in the presence of sulfur trioxide at a temperature between 200 and 650° C and at a pressure at which titanium tetrachloride is a vapor and the added chloride is not.[146] The titanium tetrachloride is separated from the reaction mixture by distillation. As a specific operation, a 3 to 1 molar ratio mixture of sodium chloride and titanium dioxide is placed in a ceramic lined reaction container equipped with a heat exchange jacket, and a gas inlet and outlet. The mixture is heated to 250° C and sulfur trioxide vapor is admitted. As soon as the sulfur trioxide comes in contact with the hot solids, vapors begin to come off and emerge from the gas outlet. From these vapors anhydrous titanium tetrachloride can be condensed with an air condenser. In the reaction it seems that sulfur trioxide displaces the chlorine from the sodium chloride, which reacts with the titanium dioxide. The final reaction product in the container is sodium sulfate containing loosely bound excess sulfur trioxide. Instead of sodium chloride, other chlorides such as calcium chloride, potassium chloride, and magnesium chloride can be used. Calcium chloride is not liquefied during the reaction. Other halides as ferric chloride, ferrous chloride, stannic chloride, and aluminum chloride may be prepared by this procedure.

The chlorination of titanium dioxide mixed with powdered charcoal in a fused salt bath produces the tetrachloride.[147] In an example, 450 g of sodium chloride is fused at 900° C, and 20 g per hour titanium dioxide and 6 g per hour of powdered charcoal are added to cover the salt. Chlorine at the rate of 63 g per hour is blown into the fused salt to obtain titanium tetrachloride. Recovery is 87.9 per cent of theory.

The kinetics of the chlorination of titania slags in fused chlorides was studied indirectly by the effect of ferric chloride and aluminum chloride on the rate of chlorination of titanium dioxide with chlorine in molten carnallite mix with petroleum coke at 500 to 900° C.[148] As the temperature increased, the rate of chlorination of titanium dioxide, V, decreased. Without ferric chloride or aluminum chloride log V against $1/T$ was a linear function. The energy of activation, E, is 11,200 calories per mole. In the presence of these chlorides, V increased at 750° C with the concentration of ferric chloride and aluminum chloride.

Titania slag mixed with coke in a fused mixture of sodium chloride and potassium chloride is intensely chlorinated at 750° C with the formation of titanium dioxide and chlorides of the corresponding metals.[149] At the same time the titanium dioxide interacts with titanium tetrachloride forming melt soluble oxychloride compounds of quadrivalent titanium. In the presence of solid carbon, melt soluble compounds of trivalent titanium are also produced.

Titanium dioxide slags of different compositions are chlorinated with elemental chlorine in a molten mixture of 85 per cent potassium chloride and 6.3 per cent magnesium chloride at 900° C.[150] The progress of the reaction was followed by the amount of titanium tetrachloride formed and absorbed in sulfuric acid containing 1.0 per cent hydrogen peroxide. Slags which yielded a maximum recovery of titanium of 79 per cent by the sulfuric acid method yielded 84 to 98 per cent by chlorination. Chlorination of perovskite of African origin containing 42.9 per cent titanium dioxide and 32.2 per cent calcium oxide was partially inhibited during the first 2 hours due to the fact that calcium oxide chlorinated more rapidly than titanium dioxide. Furthermore the presence of calcium chloride in the fused electrolyte retarded chlorination of the titanium dioxide. Thus in the presence of 24, 41, and 82 per cent calcium chloride the yields of titanium tetrachloride were reduced to 92, 70, and 40 per cent, respectively. But the presence of 6.3 to 80 per cent of magnesium chloride did not affect the rate of chlorination of titanium dioxide at 850° C.

The tetrachloride is prepared by chlorinating titanium dioxide suspended in 2.5 to 10 times its weight of molten sulfur.[151] The temperature is maintained above the melting point and below the boiling point of sulfur. Chlorides of many other elements can be prepared in the same way. Titanium dioxide is chlorinated in fused carnallite at 670° C.[152] Chlorination of titanium-containing material is carried out in a melt in a reactor equipped with an electric heater.[153]

## CHLORINATION OF TITANIUM CARBIDES

According to Oreshkin,[154] gaseous chlorine combined with titanium carbide faster and at a lower temperature than with a mechanical mixture of titanium dioxide and carbon. The former reaction took place at 200° C. Carbide prepared by fusing an ore, such as rutile or ilmenite, with coke in an electric furnace gave good yields on chlorination at an incipient red heat.[155] The crude titanium tetrachloride produced was freed from vanadium by shaking with sodium amalgam and from chlorine by distillation. The slag (soluble with difficulty) formed in the hearth of aluminum furnaces employing titaniferous bauxite contained, in addition to coal and lime, 45 to 47 per cent titanium as carbide and was readily attacked by chlorine at 400° to 500° C.[156] The titanium tetrachloride formed was separated from chlorides of iron and aluminum by distillation and converted to the sulfate by heating with sulfuric acid at 160° to 200° C. Hydrolysis of aqueous solutions of the sulfate salt by boiling for 7 hours gave a recovery of 96 per cent. Brallier[157] prepared the tetrachloride by passing chlorine over the carbonitride, containing 70 to 75 per cent titanium, in a furnace consisting of a steel shell lined with carbon slabs and cooled by a water spray.

In a commercial method of manufacture the ore or concentrate, usually rutile, was first heated with coke in an electric furnace to form the carbide or cyanonitride, and this product was treated with gaseous chlorine at elevated temperatures.

Titanium tetrachloride is formed by the reaction of a titanium carbide anode with electrolytically generated chlorine.[158] In a continuous low temperature process for the manufacture of titanium tetrachloride, the carbide is first obtained by the reaction of titanium ores with carbon at 1100° C in an atmosphere of hydrogen.[159] This carbide is added from the top of the furnace containing shelves. Chlorine is supplied to each shelf and carbon dioxide is introduced at the bottom of the furnace. No external heat is necessary. The tetrachloride is prepared by feeding titanium carbide at 950 to 1400° C into a reaction zone together with a gaseous mixture comprising 95 per cent hydrogen chloride and 5 per cent by volume of chlorine.[160]

## CHLORINATION OF ALLOYS AND METALS

Favre [161] carried out the chlorination of alloys of titanium, iron, and copper, containing from 0.1 to 20 per cent of the latter element, and reported that the reaction began at 250° C. Such products were obtained by subjecting a mixture of a copper-magnesium-aluminum alloy and ilmenite to cupro-aluminothermic treatment, or by subjecting a mixture of copper and ilmenite to electrosilicothermy. A mixture of titanic and ferric chlorides was obtained by passing dry chlorine over powdered ferrotitanium at a dull red heat.[162] Industrial alloys containing up to 55 per cent titanium were employed. Titanium tetrachloride, being the more volatile, distilled past the ferric chloride in the reaction tube and condensed in a water-cooled worm. The product was colored red by ferric chloride, but because of its low solubility the greater part of this impurity was removed by filtration. Further purification by fractional distillation gave a colorless liquid which did not fume on exposure to air. According to a related method, most of the iron was dissolved from the alloy by treatment with dilute hydrochloric acid, and the residue, containing 80 to 90 per cent titanium, was chlorinated as in the previous operation. De Carli [163] treated an iron-titanium alloy with chlorine at 400° to 500° C and purified the product by fractionation after allowing it to stand over mercury or copper.

The brittle mass obtained by plunging ilmenite, heated to 1000° C, into cold water, was pulverized, mixed with half its weight of aluminum powder, heated to 500° C, and the reaction was set off by a layer of dry titanium dioxide and aluminum, which was ignited by a magnesium flame. This reduced reaction product [164] was placed in a combustion tube and heated to redness in a current of chlorine. Ferric chloride deposited in the cooler portion of the tube, while the titanic chloride was carried on to the condensing chamber and was purified by distillation. The first fraction contained some silicon tetrachloride and chlorine.

Solid titanium or titaniferous material is crushed with the exclusion of oxygen but in the presence of chlorine without external heat to form the tetrachloride.[165] The areas of fracture freshly formed react quickly with the chlorine. Other halogens may be employed. In operation, the air is replaced by chlorine and the solid material is ground under a steady stream of chlorine which carries away the titanium tetrachloride as formed. After condensation, the tetrachloride is purified by distillation. The process may also be employed to prepare germanium tetrachloride and the volatile silicon and boron compounds. Scrap titanium metal is chlorinated in the presence of sodium chloride or potassium chloride with a stoichiometric ratio of titanium to sodium chloride to chlorine of 1 to 3 to 1.5 at 300 to 600° C to form the tetrachloride.[166] The reaction takes place in the molten mass. Scrap metal is converted to the tetrachloride by reaction with chlorine at lower temperatures in the presence of lower chlorides of titanium.[167] From 2 to 3 pounds of the dichloride or trichloride are added to 100 pounds of scrap.

Ferrophosphorus containing titanium and silicon is completely chlorinated at 200 to 1200° C.[168] Chlorides of titanium and of silicon along with any phosphorus oxychloride formed during the reaction are separated as vapors from the iron and phosphorus chlorides which form a complex. Pure titanium tetrachloride is obtained from the complex mixture derived from the chlorination of ferrophosphorus.[169]

## CHLORINE WITH CARBON MONOXIDE OR PRODUCER GAS

By mixing a reducing gas with the chlorine, the two operations were carried out simultaneously and the ore was chlorinated directly without incorporation of coal or the formation of briquets. Chlorides of iron, silicon, and titanium were formed successively as the temperature was increased, and these were separated by fractional distillation. After purification the titanium tetrachloride was dissolved in cold water, and the solution was neutralized with an alkali or alkaline earth metal base. The precipitate of orthotitanic acid was filtered, washed, and redissolved in strong sulfuric acid, and this sulfate was hydrolyzed by boiling to obtain a good yield of titanium dioxide of high purity. Slight hydrogenation, brought about by introducing the purified gas or by addition of metallic zinc to the liquor, facilitated the precipitation. The product was filtered, washed, and calcined for pigment use, and the filtrate, consisting primarily of dilute sulfuric acid, was concentrated for use in dissolving more gelatinous oxide.

Dunn [170] investigated the rates of chlorination of titaniferous ores with carbon monoxide and chlorine mixtures, and with phosgene. Chlorination rates with carbon monoxide and chlorine varied more than an order of magnitude. A relation with the initial iron content was found. The rate is proportional to the weight of titanium dioxide and to the partial pressure of carbon monoxide and chlorine. The activation energy of the reaction is 20.9 kcal. Chlorination with phosgene is considerably faster than with carbon monoxide and chlorine. However, the decomposition at 600° C causes an anomaly in its behavior. The chlorination rate rises with temperature, passes through a maximum, falls to a minimum, and then rises with temperature as the carbon monoxide and chlorine reaction dominates. Hydrogen chloride has a role in the chlorination of iron from minerals.

Titanium tetrachloride is prepared by the reaction of rutile, ilmenite, or a concentrate with chlorine in the presence of carbon monoxide or carbon.[171] Liquid titanium tetrachloride is added to maintain the desired temperature. A steel vessel 10 feet high lined with refractory brick is employed as the reactor. Anhydrous tetrachloride is prepared by the reaction of titanium dioxide with carbon monoxide and chlorine or with phosgene in the presence of an alkali metal chloride or aluminum chloride at 600 to 800° C in a fluidized bed.[172] A ratio of 0.1 to 0.2 parts of chloride to 1.0 part of titanium is employed. In the presence of the chlorides, the reaction takes place at a lower temperature and the yield of the tetrachloride is increased.

The action of a chlorine-carbon monoxide gas mixture on a titanium dioxide concentrate, a high titania slag, anatase, or rutile gave high yields of the tetrachloride.[173] A 0.5 g sample of the titanium oxide was heated 200° C per hour in carbon monoxide and chlorine, each flowing at a rate of 100 cm a minute. Rutile began to be chlorinated at a higher temperature than did anatase. Rate studies at constant temperatures showed that the rate reaches a maximum at 650, 550, and 400° C for the concentrate, the slag, and the crystalline titanium dioxides. The rate increased steadily above 750° C for the concentrate and above 700° C for the slag. This maximum was attributed to the acceleration caused by carbon produced in the decomposition of carbon monoxide, which is catalyzed by iron or its oxides contained in the titanium samples. Addition of nitrogen to the reducing gas decreased the rate of chlorination. This results from the preferential chlorination of iron in the presence of nitrogen. The residue from the chlorination of the concentrate contained only rutile.

High titania slag is chlorinated by passing a mixture of chlorine and carbon monoxide through a fluidized bed of the slag.[174] The carbon monoxide may be made by passing a mixture of oxygen and carbon dioxide through hot coke. Rapid chlorination is obtained by using slag of particle size between 80 and 200 mesh. Employing carbon monoxide as the reducing agent rather than coke and at the temperature of chlorination maintained the oxides of calcium and magnesium are completely chlorinated, volatilized, and distilled off as chlorides. Practically no ash is formed within the reaction. The method also applies to ilmenite ores and rutile. In a specific operation, a gaseous mixture consisting of 60 per cent chlorine, 10 per cent carbon dioxide, and 3 per cent oxygen is conducted with a velocity of 25 cm per second through a bed of hot 110 to 180 mesh petroleum coke having a height of 1 meter. The hydrogen content of the coke is less than 1 per cent. By adjusting the carbon dioxide to oxygen ratio the temperature is held between 1000 and 1100° C. A chlorination gas is obtained consisting of 40 per cent chlorine, 50 per cent carbon monoxide, and 5 per cent hydrochloric acid. After cooling the gas to 500° C and removing the coke dust in a cyclone, it is conducted to the chlorination furnace where it passes upward through a bed of 100 to 200 mesh slag maintained in suspension in a height of 1 meter. The slag consists of 70 per cent titanium dioxide, 12 per cent ferrous oxide, 6 per cent magnesium oxide, 1 per cent calcium oxide, 6 per cent alumina, 3 per cent silica, and 2 per cent oxides of vanadium, chromium, zinc, and other metals. A reaction temperature of 1100° C is maintained. All gaseous products are conducted to a condensation system where cold liquid titanium tetrachloride is injected to lower the temperature to 300° C. At this stage the chlorides of magnesium and calcium separate in pulverant form and are removed along with the flue dust. In the second condensation zone the temperature is lowered to 150° C by admitting cooled recycled chlorine gas. Ferric chloride separates at this stage and is eliminated. The gases are then conducted into the condensation apparatus for titanium tetrachloride, where cooling is effected by cold liquid titanium tetrachloride. Also at this stage aluminum chloride and silicon chloride condense.

Australian rutile was chlorinated in the presence of carbon monoxide and carbon under different conditions.[175] The chlorination velocity in carbon monoxide is strongly affected by temperature and is proportional to the carbon monoxide concentration, but independent of the chlorine concentration. In the presence of carbon the reaction velocity is much higher. Important variables are the reactivity of the carbon and the distance between the carbon and the rutile surfaces. The reaction velocity is approximately proportional to the chlorine concentration and independent of the carbon monoxide concentration of the surrounding atmosphere. Experiments with fluidized bed chlorination of carbon-rutile mixtures indicates that the motion of the bed has little effect on the reaction velocity. At low temperatures the chlorination velocity of dense tablets is much greater than that of titanium dioxide-coke mixtures suitable for fluidization. Titanium tetrachloride is prepared from the dioxide and lower oxides with and without carbon by reaction with chlorine gas, with mixed chlorine and carbon monoxide, and with carbon tetrachloride vapor.[176] The results show that the reaction temperature generally becomes lower as the titanium constant in the oxides increases, but is exceptional in the chlorination by the chlorine and carbon monoxide mixture. Ilmenite is chlorinated with chlorine gas diluted with carbon monoxide, carbon dioxide or nitrogen.[177] Ferric chloride and titanium tetrachloride are collected and separated from one another, after which the tetrachloride is purified by distillation. In the preferential chlorination of iron in a fluidized bed of ilmenite with carbon monoxide as the reducing agent, loss of titanium dioxide is less than 1 per cent.[178] Optimum conditions are a carbon monoxide to chlorine feed ratio of 1.6 and a temperature of 900° C. About 1.5 times the theoretical amount of chlorine is needed for 97 per cent ferric oxide removal. The overall reaction rate is independent of the feed velocity above 25 feet per minute, which suggests resistance to diffusion across the film surrounding the solid particles as the controlling mass transfer step rather than mass transfer between the dense and disperse phases of fluidization. A rate equation is derived for batch and continuous operation. The reaction rate is assumed to be proportional to the carbon monoxide and chlorine partial pressures.

Oxides of titanium, aluminum, and iron are chlorinated with a mixture of chlorine and producer gas at 700 to 900° C to form the corresponding chlorides.[179] The resulting chloride vapor is purified by passing through a bed of aluminum shavings and condensed at 140 to 240° C.

## AGENTS OTHER THAN CHLORINE

With no carbon present, chloroform [180] reacted with hot powdered titanium dioxide to form the tetrachloride, carbon monoxide, and hydrochloric acid. According to Chauvenet,[181] the chlorinating power of phosgene was inferior to that of a mixture of chlorine and carbon, because of the endothermic nature of the decomposition of carbonyl chloride.

Jenness [182] heated a gaseous mixture of chlorine and di- or higher chlorides of sulfur to near the boiling point of the sulfur compound, and passed the mixed vapors over heated ores to form titanium chloride and sulfur. Similarly, on heating rutile with sulfur monochloride, the iron and vanadium components were first attacked and distilled off as chlorides, leaving almost pure titanium dioxide.[183] The latter could be converted slowly to the tetrachloride, but the mixture was difficult to purify, since the titanium compound had about the same boiling point as sulfur monochloride.

Before chlorination, the titanium-bearing silt obtained as a waste product in purifying bauxite was treated with hydrogen sulfide to convert the iron component to the nonreactive sulfide.[184] Sulfur chloride, chlorine, or carbon tetrachloride was used successfully as the chlorinating agent.

By passing the mixed vapors, obtained by chlorinating clays or similar materials in the presence of coal, into a fused mass of equimolecular amounts of aluminum chloride and alkali metal chloride containing suspended solid alkali metal chloride, the iron and aluminum fractions were removed, leaving the silicon and titanium tetrachlorides.[185]

High titania slag is chlorinated with a mixture of chlorine, carbon monoxide, carbon dioxide, and hydrogen chloride.[186] A gaseous mixture of chlorine 60, carbon dioxide 10, and oxygen 30 per cent is conducted with a velocity of 25 cm per second through a bed of 100 to 180 mesh petroleum coke, previously calcined to less than 1 per cent hydrogen, maintained in a state of suspension 1 meter in height at a temperature of 1000 to 1100° C. The temperature is controlled by adjusting the carbon dioxide-oxygen ratio of the entering gas. A chlorination gas is obtained composed of chlorine 40, carbon monoxide 50, carbon dioxide 5, and hydrogen chloride 5 per cent. After removal of the coke dust and cooling to 500° C this gas is introduced at a velocity of 15 cm per second into a chlorination furnace at a temperature of 1100° C together with 100 to 200 mesh slag containing titanium dioxide 70, ferrous oxide 12, magnesium oxide 6, calcium oxide 1, aluminum oxide 6, silicon dioxide 3, and oxides of vanadium, chromium, and zinc 2 per cent. The height of the slag bed is 1 meter. In the first condensation zone the gaseous products are cooled to 300° C by injecting liquid titanium tetrachloride. Here magnesium chloride, calcium chloride, and flue dust separate in powdered form and are eliminated. In the second zone, the temperature is lowered to 150° C by mixing with cooled, recycled chlorination gas. At this temperature the ferric chloride separates and is eliminated. The titanium tetrachloride along with the aluminum chloride and silicon tetrachloride is condensed by washing the vapors with cold liquid titanium tetrachloride. Part of the gases are recycled to the second zone. The clogging and agglomerating tendencies of the slag can be maintained if the chlorination gases are admitted to the charge with 0.1 per cent by volume of moisture. Alternatively, rutile is used as the source of titanium, and the chlorine-carbon monoxide mixture is partly replaced by phosgene.

Finely ground rutile or ilmenite and carbon are suspended in a current of chlorine and passed into a reaction zone at 1250 to 1650° F to form titanium tetrachloride. Vaporized carbon tetrachloride may be substituted for the carbon and chlorine.[187] To sustain the reaction temperature at 1200° F the vapor of carbon tetrachloride must be preheated to 250° C. Air may be introduced to advantage to heat up the reactor on starting the process, but it is not desired for normal operation. Titanium dioxide-carbon briquets are chlorinated with carbon tetrachloride at 700° C.[188] The titanium tetrachloride obtained in 93.8 per cent yield is purified by distillation. An ilmenite concentrate is treated with hydrogen chloride gas at 600 to 700° C to remove by sublimation 99 per cent of the iron as ferric chloride in 5 hours. The titanium dioxide remaining as a residue is treated with a mixture of chlorine and carbon tetrachloride at 600 to 700° C to effect complete conversion to titanium tetrachloride.[189]

## SPECIAL METHODS OF PREPARING TITANIUM TETRACHLORIDE

A bath of fused metal chloride or fluoride is electrolyzed with an anode composed of a mixture of titanium dioxide and carbon to form the corresponding halide.[190] The chlorine liberated at the anode reacts with the titanium dioxide to form the tetrachloride. Suitable fused salts are chlorides or fluorides of alkali or alkaline earth metals. Titanium tetrafluoride can be prepared by substituting a fluoride for a chloride compound. Titanium tetrachloride is prepared along with magnesium metal by electrolyzing a fused bath of 6 moles of magnesium chloride and 5 moles of sodium chloride at 700° C employing an iron cathode and a molded anode composed of 85 per cent titanium dioxide and 15 per cent graphite, coke, or charcoal with a binder of pitch, tar, or pitch oil.[191] A direct current of 200 amperes at a potential of 6 to 7 volts is applied between the electrodes to deposit metallic magnesium at the cathode and to form titanium tetrachloride at the anode. The titanium tetrachloride is withdrawn through a conduit from the anodic chamber partitioned by an insulating diaphragm with an open bottom. Purity of the titanium tetrachloride is 99.6 per cent and the current efficiency is 75 per cent.

Halides other than fluorides of metals other than alkali, alkaline earth, and rare earths can be prepared by heating an anhydrous mixture of an alkali metal double fluoride with a chloride, bromide, or iodide of calcium, magnesium, or lithium.[192] The residue can be recycled. Yields approach quantitative proportions. For example, 170 g of anhydrous lithium chloride was briquetted with 240 g of potassium fluotitanate, and the briquets were heated at 400 to 450° C to distill off the titanium tetrachloride, which was condensed. The residue consisting of 2 moles of potassium fluoride and 4 moles of lithium fluoride was leached with water to give a potassium fluoride solution which was used to make more potassium fluotitanate. An insoluble residue of lithium fluoride was converted to lithium chloride for further reaction by the action of calcium chloride. The tetrachlorides of titanium and of zirconium are prepared by heating the corresponding phosphate in a bath of a molten halide of calcium or magnesium.[193] Under these conditions the tetrachloride is removed from the reaction zone by volatilization as formed. With the use of barium or strontium halides temperatures 200° C higher are required to obtain high yields. In a specific operation, titanium phosphate is prepared from rutile and from titanium hydroxide. A titanium sulfate solution is prepared from ilmenite by the usual method and the titanium is precipitated as the hydroxide. After separating and washing the hydroxide, 85 per cent phosphoric acid is added to form a slurry. The mixture is thoroughly dried for 12 hours at 200° C and calcined for 2 hours at 900° C.

The thermal decomposition of potassium chlorotitanate yields titanium tetrachloride.[194] An acidic solution containing 100 g of titanium per liter is cooled to −20° C and saturated with hydrogen chloride. Solid potassium chloride is then added and potassium chlorotitanate is precipitated. The precipitate, after washing with ethyl acetate, is partially dried by a vacuum treatment at 14 mm of mercury, and further dried at 220° C for

2.5 hours in an atmosphere of hydrogen chloride. Decomposition of the potassium chlorotitanate at 425° C yields titanium tetrachloride and potassium chloride. Potassium fluotitanate is mixed with at least a stoichiometric amount of magnesium chloride and heated to 700° C to evolve titanium tetrachloride of high purity.[195] An inert flux such as sodium chloride may be added to form a relatively low melting mixture. Titanium tetrachloride is prepared by the action of nitrosyl chloride alone or in admixture with chlorine on a mixture of titaniferous ore and carbon at 500 to 700° C.[196] The nitrosyl-chlorine mixture is prepared by the treatment of aqueous or anhydrous hydrochloric acid with nitric acid. The nitric oxides formed in the chlorination step are converted to nitric acid and recycled. Conversion of titanium dioxide to the tetrachloride is 91.5 per cent, and 88.9 per cent of the nitric acid used in the preparation of nitrosyl chloride and chlorine is recovered. Titanum dioxide in bauxite is chlorinated with aluminum chloride to form titanium tetrachloride.[197]

Work on the laboratory scale indicates that it is possible to produce titanium tetrachloride and titanium monoxide from rutile by smelting the ore with pyrite and coke, and then processing the melt.[198] An impure iron is recovered as a by-product. Ilmenite can be treated in the same way as rutile. From 65 to 81 per cent of the titanium in rutile can be recovered as an impure titanium tetrachloride by chlorinating the melts at 150 to 250° C. By baking the mat with sulfuric acid at 200 to 250° C, 90 per cent or more of the titanium in the rutile is recovered in the form of a titanium sulfate solution suitable for the production of pigments. Chlorinating the mats followed by baking the residue with sulfuric acid give titanium recoveries greater than 90 per cent.

## COOLING AND SEPARATING THE CHLORINATION PRODUCTS

As a vaporized mixture of chlorides of iron and titanium obtained from ilmenite was cooled, the ferric chloride condensed first, but some of the solid remained suspended in the tetrachloride vapor and was carried over into other parts of the condenser system where it tended to clog the apparatus. This troublesome suspended material was removed from the vapors by spraying with cooled liquid titanium tetrachloride.[199] After removal of the iron component, the titanium tetrachloride was condensed to the liquid state. The initial condensation could be controlled to produce the required amount of liquid titanium tetrachloride to wash the suspended ferric chloride from the uncondensed vapors.[200] Following this procedure the purified tetrachloride vapor was condensed to the liquid state without clogging the apparatus. By a similar process, the vaporized mixture was washed first with cold liquid titanium tetrachloride to remove the suspended ferric chloride and then with water to condense the tetrachloride and at the same time form an aqueous solution for hydrolysis.[201]

Almost all of the condensable chlorides contained in the gases leaving the reactor furnace in which titanium tetrachloride is produced are condensed by cooling.[202] The condensates are separated mechanically in a liquid phase of titanium tetrachloride practically free of suspended matter and a sludge containing practically all of the solids, primarily ferric chloride, silicon tetrachloride, and aluminum chloride, suspended in a small proportion of titanium tetrachloride. This process has the advantage that all of the small portion of the tetrachloride remaining in the sludge is distilled. Furthermore, large amounts of titanium tetrachloride are easily transported without danger of obstruction, and only one condensation tower is needed. The condensation is effected by recycling cold titanium tetrachloride in an amount 50 to 100 times greater than the amount furnished by the chlorinating furnace. After cooling the portion of titanium tetrachloride that is not recycled to a temperature of 0 to −20° C, a second mechanical separation of solid particles can be carried out. For example, the gases from a chlorinator producing titanium tetrachloride 44, ferric chloride 5.6, aluminum chloride 4.2, carbon monoxide 13, and carbon dioxide 2.5 kg per hour at a temperature of 800° C were cooled by spraying with 3000 kg of titanium tetrachloride per hour at 25° C. Three kilograms of uncondensed titanium tetrachloride remained in the gases. The condensed chlorides were cooled from 45 to 25° C and centrifuged to give 25 kg of sludge per hour containing 10 kg of titanium chloride, which was distilled to recover all of the titanium tetrachloride. From the liquid phase, 34 kg per hour of titanium tetrachloride was recovered.

Ferrous chloride and ferric chloride are condensed from a gaseous titanium tetrachloride stream from the chlorination of ilmenite by spraying with titanium tetrachloride slurry and then washing the gas with the same solution.[203] Based on a 1 day operation, 9100 kg of rutile containing 97 per cent of titanium dioxide, 0.1 per cent iron, and 0.2 per cent vanadium was added to a continuous fluidized bed chlorinator along with 1550 kg of coke and 1500 kg of chlorine. The 25,400 kg of chlorinated gases emerging from the furnace at 900° C were colled to 165° C with a slurry of 30,000 kg of titanium tetrachloride and 35 kg of ferric chloride on an atomizing wheel turning at 14,000 rpm. After cooling, the gas stream entered a cyclone where 265 kg of ferrous chloride and ferric chloride, 50 kg of coke, and 180 kg of unreacted titanium dioxide were recovered. The gases entered a scrubber at 150° C, where they were washed with 400,000 kg of titanium tetrachloride at 110° C. From the scrubber 30,000 kg of the slurry was recycled to the primary cooler. The solids-free gases were cooled to −15° C to condense 20,500 kg of titanium tetrachloride containing only 0.002 per cent of iron. Two hundred kilograms of titanium tetrachloride was lost to the atmosphere and 3000 kg was recycled to the scrubber. Comparable results were obtained by spraying titanium tetrachloride into the chlorinator during the reaction to lower the temperature.

Gaseous titanium tetrachloride is freed from chlorides of iron and aluminum by passing it through a vessel maintained below the dewpoint of titanium tetrachloride and equipped with an agitator so that the incoming gas is violently scrubbed with a slurry containing solid particles of ferrous chloride, ferric chloride, and aluminum chloride suspended in the liquid tetrachloride.[204] Thus, a titanium slag was chlorinated to produce a gaseous mixture containing titanium tetrachloride 54, ferrous chloride 4.5, ferric chloride 6.5, and aluminum chloride 4.5 per cent with some noncondensable gases. The gas left the chlorinator at 700° C and was cooled to 125° C in a water cooled tower to precipitate 70 per cent of the ferrous chloride and aluminum chloride. From this unit the titanium tetrachloride gas containing some suspended solids was passed through an agitated scrubbing chamber maintained at 75° C containing a slurry of solid chlorides of iron and aluminum in liquid titanium tetrachloride having a solids content of 23 per cent by weight. Effluent from the scrubbing chamber was pure titanium tetrachloride. The slurry withdrawn from the agitated scrubber was sent to a conventional sludge drier to volatilize the tetrachloride which was recovered in a usual condenser. Recovery of titanium tetrachloride from the initial gas was 55 per cent and from the sludge drier was 45 per cent.

Ferric chloride in a dry form is recovered from crude gaseous titanium tetrachloride mixtures produced by the chlorination of ores or slag in a continuous process.[205] The gaseous mixture is cooled by direct contact with a spray of recycled ferric chloride-titanium tetrachloride slurry to condense the ferric chloride in apparently dry particles which settle out. To scrub out the remaining ferric chloride, the effluent gaseous mixture is sprayed with liquid titanium tetrachloride. The ferric chloride-titanium tetrachloride slurry from this step is recycled and sprayed into the hot gaseous mixture from the first step. Titanium tetrachloride containing 30 or more per cent of solid impurities as aluminum chloride, ferric chloride, or zirconium chloride is injected into a stream of air, nitrogen, chlorination gas, or titanium tetrachloride vapor at a temperature below the vaporization temperature of the impurities but 20 to 60° C above the dew point of liquid chlorides in the gas.[206] Most of the solid impurities are removed in a cyclone and the remainder by scrubbing with the liquid titanium tetrachloride to form a small amount of sludge which is recycled.

The titanium tetrachloride is condensed from the vapor and recycled in part.

Rutile mixed with powdered carbon is chlorinated in a shaft furnace at 900° C in a continuous process.[207] All the products of the reaction leave the furnace as a gaseous mixture. This gaseous mixture is cooled to 250° C and admitted at the base of a scrubber column connected at the lower end with a reboiler and at the upper end to a reflux condenser. In this unit the titanium tetrachloride and the silicon tetrachloride condense, and together with entrained solid impurities as ferric chloride and zirconium chloride are washed down the column to the reboiler. The titanium tetrachloride is recovered from the reboiler by distillation and from the gases leaving the condenser by cooling to −10° C. To remove the solid impurities the reboiler is continuously or intermittently purged.

The hot vapors from the chlorination of titanium ores are cooled by contact with liquid titanium tetrachloride in sufficient amount to condense and wash out the normally solid materials which are removed as a free flowing liquid suspension thereby avoiding clogging of the apparatus.[208] The gases produced in the chlorinator are passed through a cooling tower, where they come in intimate contact with cold liquid titanium tetrachloride in an amount more than is vaporized by the hot chlorination gases which condenses normally solid metallic chlorides but does not condense the titanium tetrachloride. The amount of liquid titanium tetrachloride used in excess of that vaporized is regulated so that there is sufficient to wash the condensed solid chlorides and dust suspended in the gases to give a free flowing suspension which is readily removed from the condenser without clogging. Titanium tetrachloride is sprayed into the cooling tower at 100° C in an example, although temperatures as low as 0° C are mentioned. The temperature of the chlorination gases is initially 800° C.

Iron, aluminum, and zirconium chlorines which are solid at normal temperatures are separated from titanium tetrachloride by a controlled cooling process.[209] The crude gaseous product from the chlorination of titanium ores contains carbon monoxide, carbon dioxide, hydrogen chloride, and chlorine, which do not condense under normal conditions; titanium tetrachloride, which is a liquid at ordinary temperatures; and ferric chloride, aluminum chloride, and zirconium chloride, which are solids under atmospheric conditions. Particles of ore and coal are suspended in the gas. Separate condensation of these chlorides is fraught with many difficulties. If cooled indirectly, iron and zirconium chlorides precipitate on the walls of the apparatus where they impair heat transfer and finally clog the pipes. These difficulties may be overcome by cooling the crude gases of the chlorination process, after they have been cooled first to a temperature of 5 to 10° C above the point of precipitation of the solid metal chlorides, by the admixture of a cooled gas. The cold gas lowers the temperature to a value below the precipitation point of the solid chlorides but at least 5 to 10° C above the condensation point of the liquid chlorides. Carbon dioxide, carbon monoxide, nitrogen or the gaseous chlorination products from which the solid chlorides have been removed may be used as the cooling gas. The permanent gases containing at least part of the liquid chlorides may be used.

In a specific operation, 100 parts of rutile ore were converted to briquets with 40 parts of coke and 10 parts of pitch. The briquets were calcined in a reducing atmosphere at 800 to 1000° C and then chlorinated at 900° C. In the process the briquets were charged into the furnace at the rate of 180 kg per hour. The crude gases consisting of 30 per cent by volume of titanium tetrachloride, 50 to 60 per cent carbon monoxide, 5 to 10 per cent carbon dioxide, 5 to 10 per cent hydrochloric acid, 1 per cent zirconium chloride, and 1 per cent ferric chloride left the furnace at 850° C and were freed from coal and ore dust in a cyclone. After leaving the cyclone at 800° C, the gases were conducted to a cooler where the temperature was lowered to 250° C. Thereafter the gas was introduced into a cooling tower at the top. Into this tower was also introduced through the annular channel 1000 kg per hour of a return gas containing titanium tetrachloride vapor at 110° C. Thereby a resultant temperature of the gas mixture of 140° C was reached. By this procedure 95 to 98 per cent of the iron and zirconium chlorides condensed. These solids were removed by the cyclone and by the electrofilter maintained at 140° C with a heat jacket. One-fourth of the gas was passed into the condenser to condense the titanium tetrachloride. The other three-fourths were cooled to 110° C and employed to cool more reaction product.

The separation of normally solid chlorides, such as ferric chloride and zirconium chloride, from normally liquid chlorides as titanium tetrachloride and silicon tetrachloride in granular form without incrustations on the walls is accomplished by precooling the chlorination gases before they enter the condenser.[210] This is accomplished by mixing with the gases a volatile liquid such as titanium tetrachloride, liquid chlorine, or liquefied chlorination gases at a temperature below 200° C which is completely vaporized. Cooling is carried to a temperature slightly above the condensation temperature of the ferric chloride.

In an example, 130 kg of briquets containing 100 kg of rutile and 30 kg of coke were chlorinated within an hour at 800 to 900° C. The product gas was titanium tetrachloride 30, carbon monoxide 50 to 60, carbon dioxide 5 to 10, and hydrogen chloride 5 to 10 per cent by volume, and about 1 per cent each based on the titanium tetrachloride of ferric chloride, zirconium tetrachloride, and vanadium oxychloride. This product gas emerged from the cyclone separator at 750 to 800° C and was chilled to 300° C by injecting 230 kg per hour of liquid titanium tetrachloride at 20° C which was vaporized in the process. In a second chilling tower 300 cubic meters per hour of stripped return gas at 20° C was added to bring the temperature to 120° C. Retention for 15 seconds in the tower separated large crystals of ferric chloride and zirconium chloride. Then passage through a Cottrell precipitator removed the remaining solids. A total of 4 kg per hour of a mixture of 50 to 60 per cent of ferric chloride and 50 to 40 per cent of zirconium chloride were recovered here. This represented more than 95 per cent of the total quantities of these chlorides originally present. From the gases 200 kg per hour of titanium tetrachloride containing most of the silicon and vanadium of the rutile ore dissolved as chlorides were recovered by cooling to 20° C.

Exit product gases and vapors from the chlorination of titaniferous ores are cooled to 400° C by contact with recycled, chilled titanium tetrachloride, ferric chloride, or chlorination gases.[211] This procedure tends to prevent plugging of the exit ducts by ferric chloride. Superheated titanium tetrachloride vapors are passed through or over a mixture of liquid titanium tetrachloride and solid ferric chloride until all of the tetrachloride is vaporized, removed as vapor, and condensed.[212] The process may be carried out continuously or in batches. On cooling the product to room temperature, titanium tetrachloride condenses to a liquid and ferric chloride to a solid. Filtering effects partial separation. The filter cake can be heated to remove the titanium tetrachloride by vaporization, or it may be purged by passing an inert gas through the cake while heating. For large-scale operation of this process a rotary evaporator is preferred. The titanium tetrachloride vapor is heated to 180 to 300° C so that it possesses sufficient heat capacity to volatilize the titanium tetrachloride liquid without the use of excessive amounts of vapor while avoiding vaporization of the ferric chloride. The tetrachloride vapor from the mixture is condensed to a relatively pure product.

Gases from the chlorination of titanium ores containing volatile chlorides are treated in a packed tower 76 cm in diameter and 2.44 meters high on a distilling pot of 4.7 cubic meters capacity containing a constant amount of titanium tetrachloride.[213] The gas mixture at 250° C is fed to the bottom of the column, and the pot is heated to vaporize 860 kg of titanium tetrachloride an hour. This condensate, 80 per cent of which is refluxed, contains less than 0.01 per cent ferric chloride. Sludge drawn from the pot contains 7 per cent ferric chloride and a maximum of 10 per cent of the total yield of titanium tetrachloride.

The gaseous mixture of titanium tetrachloride and ferric chloride is cooled to a temperature below the dew point of the ferric chloride but above the dew point of titanium tetrachloride by passing the gas stream through a bed of solid, granulated material.[214] The boiling points of the two components are 136.4 and 315° C. By proper choice of the temperature, the titanium tetrachloride can be completely purified from the other gases as well as from ferric chloride. Thus a mixture of titanium tetrachloride 13.3, ferric chloride 16.1, hydrogen chloride 17.4, carbon monoxide 37.6, silicon tetrachloride 1.45, aluminum chloride 1.05, carbon dioxide 7.5, and chlorine 5.3 per cent is passed through a column, entering at the top at a temperature of 350 to 400° C and leaving after 30 seconds at the bottom at a temperature of 180° C. This bottom gas cooled to 65° C is practically pure titanium tetrachloride.

An apparatus for the selective condensation of metallic chlorides obtained by the chlorination of titaniferous materials consists of a vertical cylinder mounted on top of an electric furnace.[215] The cylinder is separated in two compartments by a partition that does not extend to the bottom. An absorber is mounted on top of the cylinder.

Crude titanium tetrachloride from the chlorination of a concentrate with coke is conducted to a cooling chamber where a pulverized jet of crude liquid titanium tetrachloride and sand is introduced to lower the temperature to 300° C.[216] Undesired chlorides condense on the sand grains and are removed. Gas streams obtained in the chlorination of titanium ores or slag are handled in such a manner that the normally solid components are carried from the condenser in an inert liquid rather than being allowed to collect on the inside of the condenser tubes.[217] Thus, ilmenite was briquetted with coal and a binder and fed into a furnace at 900° C. Chlorine was passed into the briquets to give a gas containing titanium tetrachloride 31.7, ferric chloride 9.3, carbon dioxide 39.3, carbon monoxide 9.8, and chlorine 9.9 per cent which was passed into a condenser. During the operation, titanium tetrachloride was introduced into a reservoir at the top of the condenser and allowed to flow in a film down the inside surface of the tubes, which were cooled by passing water between the shell and the outside surfaces of the tubes. About 75 per cent of the titanium tetrachloride in the gas stream was condensed and added to that flowing down the tube wells. The ferric chloride was condensed to the solid form and removed from the condenser with the falling film. This thin slurry of ferric chloride in titanium tetrachloride was drawn off and distilled to recover the tetrachloride. Gases leaving the primary condenser were conducted to a second condenser where the remaining titanium tetrachloride was recovered.

The mixture of titanium tetrachloride with ferric chloride, magnesium chloride and calcium chloride obtained by the chlorination of slag is conducted to a condenser where a liquid titanium tetrachloride film flows along the walls to prevent the iron chlorides from clogging the apparatus.[218] By the use of cooling water outside of the condenser tubes, the temperature of the gaseous mixture is decreased to 300° C. The iron chlorides are concentrated in a high turbulent gas stream, for example, 40 to 60 feet per second. This mixture containing 50 to 80 per cent titanium tetrachloride is pumped to a separator where the titanium tetrachloride is removed and recovered by distillation.

In conventional methods for cooling the vaporized products from the chlorination of titanium ores, ferric chloride condenses directly from the gas to the solid phase, forming a hard deposit on the walls of the apparatus, which decreases heat transfer and tends to plug the apparatus.[219] This difficulty is avoided by passing the titanium tetrachloride-ferric chloride mixture through an externally cooled metal conduit, the interior of which is maintained at a temperature below the "snow point" of the hot vaporous mixture. An insulating layer of condensed ferric chloride forms on the interior of the conduit. On reaching equilibrium, sensible heat is removed from the ferric chloric vapors by condensation and then re-evaporated at such a rate that the composition of the vapors is the same on leaving as on entering the conduit. The process may also be applied to other vaporous mixtures which on cooling yield a solid as the first condensed product. Titanium tetrachloride is cooled by the evaporation of added sulfur dioxide to form crystals which are separated to give very pure tetrachloride.[220]

The mixed chlorides formed by the chlorination of titaniferous materials are separated by cooling in successive stages to effect condensation.[221] Finally, the small residual titanium tetrachloride is adsorbed on charcoal, and the hydrochloric acid is recovered last. The chlorination of titanium ores or slag in a fluidized bed with carbon proceeds at 700 to 1000° C. The gaseous chlorination product contains gases which do not condense under normal conditions such as carbon monoxide, carbon dioxide, hydrochloric acid, and chlorine; the volatile chlorides which condense to liquids at atmospheric temperatures as titanium tetrachloride, silicon tetrachloride, and vanadium chloride; metal chlorides which condense to solids under atmospheric conditions such as chlorides of iron, aluminum, zirconium, calcium, and magnesium. These different chlorides have different condensation temperatures. For example, ferrous chloride condenses at 1020° C and silicon tetrachloride condenses at 57° C. The actual condensation temperature of each of the chlorides in the gases depends upon the partial pressure of the gas in the gaseous mixture. Effective separation of the chlorides has proved very difficult. Aluminum chloride is soluble in titanium tetrachloride to the extent of 9 per cent at 80° C but of only 0.2 per cent at 20° C. In the presence of aluminum chloride the vapor pressure of ferric chloride is increased very much. It is believed that the reason for this increase is the formation of double chlorides such as $FeCl_3 \cdot AlCl_3$. Consequently titanium tetrachloride after condensation may contain up to 0.1 per cent ferric chloride, which would render the pigment worthless. After removal of the solid chlorides by cooling to 120 to 150° C the vapors are cooled to 60° C at which temperature 80 per cent of the titanium tetrachloride and 95 per cent of the aluminum chloride condense. The condensate product then contains 2 per cent aluminum chloride. It is then cooled in the next stage to 20° C, thereby separating out 15 per cent of the titanium tetrachloride together with 3 per cent of the aluminum chloride so that the condensed titanium tetrachloride contains only 0.1 per cent aluminum chloride. By further cooling in the next stage to −20° C most of the remaining titanium tetrachloride condenses, but 5 g per cubic meter of gas remains. If the gases are further cooled to −30 to −40° C more titanium tetrachloride separates as a solid, but 0.5 to 1 g per cubic meter remains uncondensed. At this stage the gas contains 3 to 10 per cent by volume of hydrochloric acid. This residual titanium tetrachloride may be removed by adsorption on dry solids as activated charcoal, silica gel, or activated alumina to 0.01 to 0.001 g per cubic meter. The titanium tetrachloride is set free by heating the active substance to 30 to 600° C. After practically complete removal of the tetrachloride the hydrochloric acid is recovered by washing with water.

The chloride pulp obtained in the production of titanium tetrachloride is placed on top of alkali metal chlorides at a temperature of 500° C to vaporize the tetrachloride which is condensed.[222] At the bottom of the melt a temperature of 700 to 800° C is maintained. At this temperature chlorides of aluminum and iron rise to the surface and keep the upper layer, on which the titanium tetrachloride is placed, stirred. If necessary additional agitation is provided by nitrogen or waste gases from the chlorination furnaces. The titanium tetrachloride obtained in this manner is free of chlorides of iron and aluminum. Titanium tetrachloride is separated from impurities, especially iron chlorides, by passing vaporized superheated tetrachloride through the mixture and condensing the combined vapors.[223] The impurities are removed in a dry, friable state. Purification by this method may be conducted as a batch or continuous process.

All of the condensable chlorides in the hot gases coming from the chlorination of titanium ores or slag are first condensed together and later separated into liquid and solid phases by mechanical means.[224] By thorough mechanical separation is obtained liquid titanium tetrachloride practically free from suspended matter, and a slurry containing practically all of the solid phases, especially ferric chloride and aluminum chloride with a relatively small proportion of titanium tetrachloride. Mechanical separation may include centrifuging, simple settling, and accelerated decantation through hydroclones. Centrifuging is preferred since it offers important industrial advantages in that it permits a rapid and continuous production. Large quantities of the mechanically separated titanium tetrachloride are used for condensation. This exceeds by many times the quantity of titanium tetrachloride contained in the hot chlorination gases. Only the titanium tetrachloride remaining in the slurry with the solids from the centrifuge must be distilled.

In an example, gases from the chlorinator at 800° C containing 44 kg of titanium tetrachloride, 5.6 kg of ferric chloride, 4.2 kg of aluminum chloride, 13 kg of carbon monoxide, and 2.5 kg of carbon dioxide per hour were led directly into the chlorination tower, where they came in intimate contact with 3000 kg per hour of titanium tetrachloride at 25° C from the centrifuge machine. The permanent gases came out of the condensation tower at 30° C and retained 3 kg per hour of uncondensed titanium tetrachloride. The metallic chlorides in the hot gases were condensed and drawn down by the heavy flow of titanium tetrachloride, forming a suspension which came out of the condenser at 45° C. After partial decantation, the suspension was passed through a centrifuge machine which collected per hour 20 kg of slurry, 10 kg of titanium tetrachloride, and 9.5 kg of ferric chloride and aluminum chloride. The liquid titanium tetrachloride freed of solids was passed through a cooler where the temperature was decreased to 25° C. About 3000 kg of the cooled titanium tetrachloride were recycled to the condensation tower, while 31 kg per hour were withdrawn from the circuit. In addition 3 kg per hour of titanium tetrachloride escaped with the permanent gases. The slurry collected from the centrifugal machine was distilled to recover 10 kg of titanium tetrachloride.

By collecting solid ferric chloride with a Cottrell precipitator during the chlorination of ilmenite followed by distillation, titanium tetrachloride containing 0.002 per cent iron is obtained.[225] The solubility of ferric chloride in the tetrachloride is 0.043 weight per cent at 110° C and 0.003 per cent at 40° C. This sets the limit of purity of titanium tetrachloride by this method. Vanadium is not removed in this way. The ordinary copper powder or hydrogen sulfide process is more satisfactory than the selective oxidation of vanadium oxychloride with oxygen for removing vanadium. Metal oxides and other suspended solid impurities are removed by passing titanium tetrachloride at 600 to 970° C through a porous refractory filter.[226] Pure titanium dioxide pigment is prepared by the vapor phase oxidation of the purified titanium tetrachloride.

A trap consisting of a column containing coarse anhydrous rock salt supported on a screen is employed to remove ferric chloride as well as other metallic chlorides from the titanium tetrachloride gas stream before it condenses.[227] The hot stream enters the column at a right angle below the screen. Directly below the screen is a chamber with a tapping port.

The entering gases are heated to 300° C or above to prevent condensation of solid chlorides. To permit free passage of the titanium tetrachloride the salt column is heated to 200° C. In the trap, ferric chloride vapors and solid sodium chloride react to form a molten mixture on the surface of the rock salt at or directly above the screen. This mixture which is fluid above 160° C drains away from the salt into the heated chamber below the screen. Additional sodium chloride is added to the top of the column as needed. Titanium tetrachloride prepared by the direct chlorination of titanium-zirconium concentrates briquetted with petroleum coke is purified in the vapor phase by passage through a vertical column packed with sodium chloride.[228] Depending on the concentrate composition, optimum conditions are a temperature of 420 to 500° C at the column bottom and a gas feed of 0.7 to 1.4 ml per cubic centimeter per minute. The titanium tetrachloride produced contains up to 0.3 per cent insoluble chlorides and less than 0.01 per cent zirconium, iron, and aluminum. Increasing the column temperature decreases the amount of zirconium tetrachloride retained by the sodium chloride and increases the insoluble chlorides content in the titanium tetrachloride product. Lower temperatures and increased gas feed can be employed with concentrates having higher iron and aluminum contents owing to the formation of low melting sodium ferric chloride and sodium aluminum chloride. The solubility of the zirconium tetrachloride in the titanium tetrachloride at 20 to 80° C is less than 0.01 per cent.

Solid materials are removed from the gases and vapors obtained in the chlorination of titanium ores in a continuous process.[229] In a separator, the gases and vapors obtained are cooled to a temperature that is above the condensation point of the titanium tetrachloride. The separated solids are fluidized with a countercurrent stream of dry inert gas as nitrogen, helium, or argon. This removes titanium tetrachloride vapors from the discharge port for solids and accelerates heat transfer to the solids. Chlorides of iron and aluminum which condense at atmospheric pressure are removed from the chlorination product of high titania slags by absorbing them on alkali chlorides.[230] The gaseous chlorination products are passed first through an apparatus which removes suspended solid particles and then through a charge of alkali metal chlorides where the undesirable, uncondensed chlorides are absorbed. The remaining titanium tetrachloride is then condensed in tubular condensers. Gaseous chlorination products entering the alkali metal chloride charge are maintained at 400 to 500° C. They leave this charge at 150 to 200° C. The fused alkali chloride and the metal chlorides which they have absorbed are maintained in a fused state and are withdrawn from the bottom of the column.

The solid chlorides, primarily iron chloride, separated from the chlorination product of titanium ores or slag is discharged into a molten mixture of ferric chloride and sodium chloride.[231] As the concentration of the ferric chloride in the melt increases some of the material is withdrawn and more sodium chloride is added. Ferric chloride separated from the chlorination product gases by condensation usually contains some occluded titanium tetrachloride and silicon tetrachloride. If discharged into the atmosphere these contaminants produce undesirable fumes and react with moisture in the air to form hydrochloric acid. These products may render the ferric chloride difficult to handle by forming corrosive hydrochloric acid and aggregates which may block the passages of the discharge system. As an illustration of the process, the hot gases from the reactor during the chlorination of ilmenite were chilled to 350° C in their passage to a primary separator, which removed any unattacked material and ash together with some iron chloride. Next the gases passed to a separator where they were cooled to 170° C by indirect heat exchange. Here the major part of the iron chloride was precipitated. The precipitated iron chloride was discharged at a rate of 70 pounds per hour through the conical bottom of the separator into approximately 350 pounds of melt of sodium chloride and ferric chloride at 250 to 280° C. Sodium chloride constituted 26 per cent of the melt. After 3 hours operation 240 pounds of melt was withdrawn. Analysis showed 82.2 per cent ferric chloride and 1.2 per cent ferrous chloride. At this stage 40 pounds of sodium chloride was added to the salt. Ferric chloride was recovered from the withdrawn melt by distillation.

Titanium tetrachloride and columbium pentachloride are recovered from ilmenite in pure form by chlorination.[232] Coke-reduced ilmenite is chlorinated at 400 to 500° C and the resulting titanium tetrachloride is freed of most impurities before condensation. A column packed with rock salt at 250° C is very effective for removing ferric chloride from titanium vapors. Most of the discoloration of the iron-free tetrachloride can be removed by passing the vapors through activated charcoal at 180° C. The small amount of remaining coloring agents can be removed by passing the vapors through the carbides of calcium or titanium which at the same time removes most of the columbium and vanadium chlorides. Titanium tetrachloride that is almost water white can be produced by using all of these column packings simultaneously. The amounts of the chief impurities found in a typical titanium tetrachloride product are iron 0.02, vanadium 0.01, silicon 0.04, columbium 0.07, and aluminum 0.05 per cent.

Vaporized chlorides from the chlorination of ores are recovered by passing the vapors through a molten mixture of ferric chloride 78 and sodium chloride 22 countercurrently.[233] Other molten mixtures are potassium chloride 42.5 and sodium chloride 57.7 per cent; barium chloride 57 and potassium chloride 43 per cent; calcium chloride 61.8 and lithium chloride 38.2 per cent; potassium fluoride 45, lithium fluoride 45, and sodium fluoride 10 per cent; and lithium bromide 75 and lithium chloride 25 per cent. The vaporized products are condensed in the anhydrous state. In an example, ilmenite ore was mixed with carbon and chlorinated. The vaporized product was passed into a scrubber countercurrently with a spray of a mixture of ferric chloride 78 and sodium chloride 22 per cent at 200° C. Considerable ferric chloride is passed through a condenser. The remaining vapors are passed through a condenser to obtain ferric chloride and titanium tetrachloride in the anhydrous state.

The vapors from the chlorination of raw materials containing titanium, zirconium, and niobium are passed through a layer of briquets pressed from coke or charcoal mixed with sodium chloride or potassium chloride to effect purification and separation of the chlorides.[234] Passing the vapors through the briquets preheated to 500° C frees niobium dichloride from chlorides of zirconium, iron, and aluminum. Subsequent passage of the vapors through the briquets preheated to 300° C frees titanium tetrachloride from chlorides of aluminum, iron, niobium, and zirconium. A slurry of liquid titanium tetrachloride and solid ferric chloride at room temperature is introduced into a moving body of gravel or sand preheated to 360° C to effect separation.[235] The titanium tetrachloride is vaporized, while the higher boiling ferric chloride remains with the moving stream of particles. Subsequently the ferric chloride is removed from the sand by screening or by chemical means. In this way the inert particles are regenerated. Kangro[236] cooled the gaseous chlorination products stepwise to separate the precipitating compounds. Iron is added to the ore to bind the oxygen as ferric oxide. Several cooling steps are employed to remove the chlorides and free chlorine impurities from crude liquid titanium tetrachloride.[237] First the crude liquid is cooled in a cooling tower to −3 to −5° C with stirring to separate the disilicon hexachloride and then to −20 to −23.5° C to separate the vanadium oxychloride by crystallization. The filtrate is introduced into another tower cooled to −27° C with vigorous stirring to crystallize the titanium tetrachloride which is separated by filtration. To illustrate, 4559 g of titanium tetrachloride of 99.92 per cent purity was obtained from 5000 g of the crude liquid of 96.47 per cent purity.

Crude, turbid titanium tetrachloride obtained by the chlorination of rutile or ilmenite is distilled and the condensate is cooled slowly so that the codistilling metal chlorides present as impurities crystallize out.[238] This permits the separation of clear titanium tetrachloride. By controlling the condensation apparatus during the chlorination of titanium ores, gases from the chlorinator are sampled by using an evacuated flask.[239] From the samples the titanium tetrachloride concentration is determined within an error of 5 per cent. Mitsui, Noda, and Nakamura[240] developed a special apparatus for the separation of titanium tetrachloride from slurries.

Conditions for the elimination of ferric chloride from titanium tetrachloride by fractional distillation was worked out based on the equilibrium diagram of the system.[241] From the vapor pressure determined with the pure titanium tetrachloride the boiling point at 760 mm of mercury and the melting point were calculated to be 135.8 and −24.3° C, respectively.

Waste gases chiefly carbon dioxide from the manufacture of titanium tetrachloride are cooled to −20 to −40° C. Any chlorine, phosgene, or titanium tetrachloride which condenses is separated.[242] The remaining gas is cooled, compressed, and treated with titanium tetrachloride to absorb traces of impurities.

The chlorine in the gaseous product obtained by the vapor phase oxidation of titanium tetrachloride is recovered by treating the exhaust gas with a inert liquid absorption agent such as carbon tetrachloride or titanium tetrachloride.[243]

## REMOVAL OF UNREACTED CHLORINE

Unreacted chlorine is recovered from the gaseous products from the chlorination of titaniferous materials, after condensation of the titanium tetrachloride, by absorption in liquid chlorinated compounds such as sulfur chloride, pentachloroethane, trichloroethane, hexachlorobutadiene, carbon tetrachloride, titanium tetrachloride, and tin chloride.[244] The absorption may take place at room temperature or at lower temperatures produced by refrigeration. Pure chlorine suitable for chlorination of additional ore is expelled from the solvent by heating. In a specific operation, 1 volume of titanium tetrachloride vapor is oxidized by 1 volume of carbon monoxide and 5 volumes of a mixture of 50 parts of oxygen and 50 parts of carbon dioxide. The gaseous product formed after separation of titanium dioxide is composed of 2 volumes of chlorine, 3.5 volumes of carbon dioxide, and 1 volume of oxygen. These gases are fed to an absorption column in which the chlorine is absorbed in titanium tetrachloride cooled to 0° C to form a solution containing 45 gpl of chlorine. The residual gases still containing 1 to 2 per cent chlorine by volume are enriched with oxygen and reused. By reaction with the recycled ferrous chloride the unreacted chlorine in the products obtained by chlorinating titanium ores is converted to ferric chloride.[245]

After condensing the normally liquid and solid chlorides from the gaseous chlorination product, the last amount of dissolved liquid chlorides is removed by washing the gas mixture with an organic liquid, which is a solvent for titanium tetrachloride and silicon tetrachloride but in which the hydrogen chloride is only sparingly soluble.[246] The resultant gases are scrubbed with water to produce an aqueous solution containing 20 per cent by weight of hydrogen chloride. Organic solvents suitable for the removal of the titanium tetrachloride and silicon tetrachloride include aliphatic hydrocarbons such as normal heptane, and halogenated hydrocarbons such as carbon tetrachloride, dichloroethane, trichloroethane, chlorinated diphenyls, and ortho dichlorobenzene. The titanium tetrachloride and silicon tetrachloride may be removed from the absorbent liquid by distillation or hydrolysis. Waste gases from the chlorination of ores are heated with 2 volumes of air saturated with steam at 200° C to precipitate titanium dioxide.[247] In an example, 5 liters of waste gas from the chlorination tower containing chiefly titanium tetrachloride is heated with two volumes of air, saturated with steam at 25° C, in an electric furnace at 200° C.

Corrosion problems encountered in processing crude titanium tetrachloride containing aluminum chloride are minimized by the addition of an amount of water stoichiometrically equivalent to the aluminum chloride followed by distillation.[248] The aluminum-water complex remains in the residue. Thus, to crude titanium tetrachloride containing 0.55 per cent aluminum chloride and 0.15 per cent vanadium oxychloride in a distilling vessel equipped with an agitator and condenser, was added an amount of water stoichiometrically equivalent to an aluminum chloride content of 0.50 per cent. After completion of the dropwise addition, the mixture was heated to volatilize the titanium tetrachloride which was condensed and recovered. The titanium tetrachloride distillate contained only 0.08 per cent aluminum chloride but was yellow because of the presence of vanadium oxychloride. Further treatment with finely divided copper followed by refluxing and redistillation gave water white titanium tetrachloride.

During the manufacture and refining of titanium tetrachloride, air is dried at the entrance of the apparatus by use of the heat from hydrolysis reaction of the tetrachloride.[249] Titanium tetrachloride is heated and distilled in a small, mild steel, vertical, tubular vessel to minimize the risk of contamination with carbon and water and of overheating.[250] Heat is supplied by low pressure steam. The rate of feed is adjusted so that the ratio of liquid to vapor at the top of the tube is 5 to 1. This insures turbulent flow at high velocity up the inner surfaces of the tubes.

# 20

# Purification of Titanium Tetrachloride

As already noted, the crude tetrachloride prepared by the chlorination of titaniferous materials usually contains free chlorine, together with small proportions of dissolved compounds of iron, silicon, vanadium, and other elements derived from the charge, and is of a yellowish or reddish color. This discoloration has been ascribed to vanadium oxychloride, ferric chloride, and uncombined chlorine. Since ferric chloride is only slightly soluble in the product, the greater part of it may be recovered, along with other suspended particles, by filtration after cooling to room temperature. The other constituents, with the exception of vanadium, may be readily separated by fractional distillation, and this impurity may be removed by a similar operation after treating the liquid with sodium amalgam [1] with gold, silver, mercury, copper, or bronze; [2] with iron; with carbon; sulfides, organic compounds, or with a reducing agent to convert the vanadium compounds to nonvolatile products.

Industrially produced titanium tetrachloride contains chlorine, hydrochloric acid, carbon monoxide, carbon dioxide, phosgene, and some organic compounds dissolved as impurities.[3] The solubility of chlorine is 11.5 weight per cent at 0° C, decreasing to zero at 135° C. Hydrogen chloride dissolves to the extent of 0.12 per cent at 20° F and 0.05 per cent at 120° F. The solubility of ferric chloride in titanium tetrachloride containing aluminum chloride increases from 0.0087 weight per cent at 70° C to 0.23 weight per cent at 127° C.[4] Over the same temperature range the solubility of aluminum chloride in titanium tetrachloride containing ferric chloride increases from 0.23 to 3.33 weight per cent. In pure titanium tetrachloride the solubility increases from 0.24 to 6.72 weight per cent. An improved method for determining the solubility of ferric chloride in titanium tetrachloride gives values at 70, 90, 105, 120, and 127° C and atmospheric pressure that agree with earlier determinations.[5] With an increase in temperature from 70 to 127° C the solubility increases more than seven-fold. Simultaneously the rate of obtaining equilibrium is increased about 10 times. The logarithmic relation of the solubility with 1/T is linear.

The solubility of calcium chloride in titanium tetrachloride is 0.0105, 0.123, 0.128, and 0.135 at 14, 39, 50, and 80° C expressed as the weight of the calcium divided by the weight of the titanium times 100.[6] Magnesium chloride is practically insoluble up to 80° C. Aluminum chloride is soluble to the extent of 0.0044 weight per cent at room temperature and 0.0170 at 78° C. The solubility of manganese chloride is 0.029 at 14° C and 0.055 at 80° C. Chromium chloride dissolves to too small an extent to be determined. Liquid-vapor equilibrium of mixtures of titanium tetrachloride with silicon tetrachloride, with vanadium oxychloride, and with trichloroacetylchloride were determined and represented graphically.[7] The liquid-vapor equilibrium of mixtures of titanium tetrachloride and vanadium oxychloride was determined in an Othmer apparatus.[8] Temperature-composition and relative volatility-composition diagrams were developed.

## COPPER

Stoddard and Pietz[9] purified titanium tetrachloride in a pilot plant by distillation with copper. One and one half per cent copper was the minimum amount that produced a colorless product. The powder used was prepared by reducing finely ground cuprous oxide with hydrogen. In a specific test, a colorless product was obtained by distillation after heating the crude tetrachloride with 1.5 per cent copper powder at 98.5° C for 15 minutes. Oleic acid also gives good results, but it offers several disadvantages.

Purification was similarly effected by a simple distillation after refluxing with small quantities of alkali metal hydroxide, powdered iron, tin, antimony, or copper and water.[10] In an example, 4 g to 10 g sodium hydroxide, 1 g to 5 g water, and 3 g to 20 g zinc powder were added to 1 liter of crude tetrachloride, the mixture was refluxed for 2 to 4 hours at atmospheric pressure, and distilled to get a colorless product containing less than 0.001 per cent vanadium. At the same time iron was reduced to less than 0.001 per cent, calculated as ferric oxide. From another approach, the vaporized material was heated at 800° to 1000° C in the presence of hydrogen to reduce the impurities to solid compounds which were readily separated.[11]

Crude titanium tetrachloride containing vanadium oxychloride, which cannot be separated by fractional distillation because of a similar boiling point, is purified at just below its boiling point by reduction with copper, hydrogen sulfide, or oils to form insoluble $VOCl_2$ which is separated by settling and decantation.[12] Additional tetrachloride may be recovered from the sludge by evaporation at 136 to 200° C in a nonoxidizing atmosphere.

Crude titanium tetrachloride is purified by passing the vapors through a column packed with pieces of metallic copper.[13] Partial purification and separation of compounds such as ferric chloride and silicon tetrachloride having boiling points sufficiently removed from that of the titanium tetrachloride can be effected by fractional distillation. Certain other chlorides, particularly those of vanadium, boil at temperatures close to the boiling point of titanium tetrachloride so that removal by fractional distillation is extremely difficult. Crude titanium tetrachloride usually contans about 0.1 per cent vanadium, and even 0.01 per cent gives a brown titanium dioxide unsuited for pigment manufacture. The crude titanium tetrachloride is distilled; the resulting vapors pass through the column packed with pieces of copper where they come in direct contact with the metal surfaces. From the column the vapors pass to a condenser. As the process proceeds a soft brown coating collects on the surface of the copper, which ultimately renders it ineffective. At this stage the flow of titanium tetrachloride vapors is discontinued and the copper is reactivated by allowing hydrochloric acid to percolate down over and through the copper until it again presents a bright surface. The hydrochloric acid wash solution contains the vanadium originally present together with some titanium and some copper compounds.

In an example, crude titanium tetrachloride containing 94.5 per cent titanium tetrachloride, 3 per cent silicon tetrachloride, and 0.28 per cent vanadium as $VOCl_3$, the remainder being ferric chloride and free chlorine, was distilled, and the resulting vapors were passed through a copper-packed, vertical steel column. The lower half of the column was packed with 200 parts of metallic copper in the form of half-inch Raschig rings. Some reflux was maintained in the column so that the temperature throughout was close to the boiling point of the tetrachloride. The vapors leaving the copper-packed column were passed through a water-cooled condenser and into a collecting vessel. After distilling off the low boiling fraction of silicon tetrachloride and chlorine, distillation was continued until only a small dark residue remained in the flask. The distilled product contained less than 0.0003 per cent vanadium. A rate of distillation was maintained so that the vapor remained in contact with the copper for 7 seconds. After acquiring the brown coating the copper packing was removed from the column and immersed in cold 20 per cent hydrochloric acid until the bright color was restored. At this stage the acid solution was decanted, and the copper was washed with water and dried. The acid solution analyzed 4.45 gpl vanadium. Purification was more easily effected by allowing 20 per cent hydrochloric acid to trickle down and over the copper column.

Crude titanium tetrachloride is purified by reaction with metallic copper which has been chemically deposited or electroplated on a ferrous metal surface at temperatures above 136° C.[14] As the reaction proceeds, undesirable impurities are converted to insoluble compounds which foul the copper surface. Periodically the copper is washed with water, which forms hydrochloric acid by reaction with the residual titanium tetrachloride until clean. The wash solution is adjusted to the proper copper content and pH and passed back through the iron pieces, where the copper is chemically redeposited on the surfaces. A tower packed with scrap iron in the form of borings, turnings, shavings, or other pieces having large surface area and adequate porosity is employed. Special tower packing shapes such as rings or saddles may be employed. The original plating solution should contain from 1 to 20 g of copper per liter and have a pH of 4. The amount of copper deposited is from 2 to 5 g per square foot of iron surface area.

In a specific operation, a vertical ceramic tube 3 by 48 inches was packed with scrap iron sheet in pieces 0.75 by 1 inch. A solution containing 13 gpl copper chloride, equivalent to 5 gpl copper with a pH of 5 was passed upward through the tube until practically all of the copper had been plated on the iron to give a coating of 3 g per square foot. Water was passed through the tower to remove chloride salts. The packing units were dried by passing hot producer gas through the tower. A crude titanium tetrachloride of dark brown color containing 0.1 per cent vanadium as the oxychloride was refluxed through the packed column. The reflux ratio was adjusted to provide adequate contact of the vapors with the copper plated pieces. The condensed tetrachloride was water white and contained less than 0.01 per cent vanadium. After the copper became ineffective the feed was discontinued; the tube was drained and washed by upward displacement of water. A hydrolysis reaction took place between the titanium tetrachloride retained on the surface of the copper, and the wash water produced hydrochloric acid which dissolved the complex precipitates formed during the purification. Nitric acid was added to the wash solution to oxidize most of the titanium III and to dissolve the metallic copper produced by the action of the titanium III. The copper content of the clarified solution was 4.6 gpl and sufficient copper chloride was added to raise the value to 5 gpl. At this stage the pH was 1.1 and sodium hydroxide was added to adjust the value to 5. This solution was passed upward through the tower to again coat the iron packing with copper.

Vanadium oxychloride and sulfur monochloride are removed from crude titanium tetrachloride by passing the vapor through gauze.[15] In an example, 3 liters of crude tetrachloride containing 2 per cent vanadium oxychloride and sulfur monochloride is heated at 136° C and the vapor is passed through 1 kg of copper gauze in 2.5 hours and condensed to obtain 2.91 liters of pure titanium tetrachloride. Zinc can be used instead of copper. Copper activated by treatment with acetone or with a lower alkyl alcohol, aldehyde, or ketone is used to remove residual impurities and color from titanium tetrachloride in the liquid or vapor phase.[16] The copper is first cleaned with a 15 per cent solution of ammonium sulfate in ammonium hydroxide of 0.88 density. Copper deposited on steel wool is a particularly useful form. After thorough washing, the copper is treated with acetone and, without contact with air, is heated in vacuum to volatilize the acetone. The titanium tetrachloride is then admitted in the liquid phase, gently agitated, and refluxed until the liquid is colorless. After impurities build up to inhibit contact, the copper is regenerated. The presence of sharp particles, sand, or quartz dust, prolongs the time before regeneration is necessary. If desired the copper can be stripped from the steel wool by the persulfate solution without affecting the wool.

The refining of titanium tetrachloride is carried out using finely divided copper powder, with intensive mixing to avoid cementation of the powder and the formation of dense aggregates.[17] This avoids the loss of copper, which is difficult to regenerate. The reaction of copper with chlorine, vanadium oxychloride, and sulfur monochloride dissolved in the tetrachloride is much quicker at 80 to 100° C. Preliminary studies indicate that some chlorine organic derivatives can be extracted from the titanium tetrachloride by copper powder. It is evident that copper powder cleans the tetrachloride simultaneously from several impurities. Titanium tetrachloride is purified by contact with at least one metal of the group below iron in the electromotive series.[18] Pieces of the metal, for example, copper tumbled through liquid or gaseous titanium tetrachloride, precipitate impurities and regenerate clean surfaces for further treatment. A practically vanadium-free product is recovered after contact with metallic copper.

Vanadium tetrachloride and vanadium oxychloride to the extent of 1.5 per cent in titanium tetrachloride are reduced to $VCl_3$ and $VOCl$, respectively, by copper, zinc, iron, or aluminum with aluminum chloride, by aluminum with 1.0 per cent iodine, and by hydrogen sulfide.[19] Sulfur and water along with carbon reduce vanadium tetrachloride only. Aluminum, aluminum-mercury alloy, magnesium, magnesium along with aluminum chloride, zinc, iron, and carbon alone are not effective. Some reduction of titanium tetrachloride to the trichloride may occur. Vanadium tetrachloride is selectively removed from the mixture by sulfur or by water and carbon. Aluminum chloride which impedes the removal by copper powder of vanadium oxychloride from titanium tetrachloride during refining can be removed by hydrolysis.[20] Gradual addition of water reduces the aluminum chloride concentration to zero. The concentration of titanium oxychloride begins to increase only after the concentration of aluminum chloride has been reduced below 0.01 per cent. Aluminum oxychloride and boehmite are the first products of hydrolysis.

## METALS OTHER THAN COPPER

Titanium tetrachloride is purified and discolorized by treatment with an especially prepared mixture of finely divided metals.[21] Impurities that boil close to the boiling point of the tetrachloride are removed by distillation after reaction with a mixture containing a free metal and an alkali metal salt. The mixture is prepared by reaction of an alkali metal with the chloride of copper, zinc, nickel, iron, or titanium at 100 to 450° C in an inert atmosphere. By introducing the reaction product directly into the titanium tetrachloride reaction chamber the process becomes continuous.

The crude tetrachloride is purified by the action at elevated temperatures of aluminum metal and aluminum chloride simultaneously followed by distillation.[22] Powdered aluminum gives the maximum surface area and should be employed. Aluminum chloride is employed in the anhydrous form to avoid the formation of undesirable oxychlorides of titanium. Crude titanium tetrachloride often contains from 0.002 to 0.01 per cent vanadium. With this type of starting material from 0.1 to 1.0 per cent aluminum powder plus 0.05 to 0.5 per cent aluminum chloride effects purification. It is preferred to reflux the mixture for 10 minutes to 2 hours before distillation. To illustrate, to 1000 g of titanium tetrachloride containing 0.01 per cent vanadium and other impurities in a glass distilling flask equipped with a reflux condenser were added 1 g of aluminum metal powder and 0.5 g of anhydrous aluminum chloride. The mixture was heated rapidly to boiling and refluxed for 1 hour. The treated tetrachloride was then distilled to give a water white product containing less than 0.0008 per cent vanadium.

Crude titanium tetrachloride is purified by refluxing with a small proportion of solid alkali metal hydroxides together with metallic sodium, magnesium, titanium, or aluminum followed by distillation.[23] A refluxing time of 2 hours is usually sufficient with the amount of purifying agents equal to 10 to 30 parts of alkali metal and 1.5 to 20 parts of powdered metal per 1000 parts of crude titanium tetrachloride by weight. With higher proportions of purifying agents a shorter refluxing time is effective. A typical crude material contains from 0.25 to 0.35 per cent vanadium, 0.04 to 0.2 per cent silica, 0.02 to 0.025 per cent aluminum, 0.01 to 0.02 per cent niobium, and 0.05 to 0.09 per cent tungsten. After purification the vanadium content is reduced to 0.0001 per cent. In an example, 2 per cent sodium hydroxide and 2 per cent magnesium metal were added to the crude titanium tetrachloride, after which it was refluxed for 1 to 2 hours and distilled. A water white distillate with only 0.0001 per cent vanadium was obtained. The residue formed in the bottom of the still comprised water-soluble vanadium compounds, the pentachloride or oxychloride, rutile, anatase, and minute amounts of other impurities.

Vanadium compounds are removed by adding a small proportion of lead amalgam, sodium-lead alloy, calcium-lead alloy, or fumed litharge-containing metallic lead followed by distillation.[24] For example, 4 g of sodium-lead alloy trimmings containing 3 per cent metallic sodium was added to 1000 g of crude titanium tetrachloride in a distilling flask at room temperature. The mixture was heated to boiling and distilled. Refluxing was not necessary. The water white distillate contained less than 0.002 per cent vanadium.

## SULFIDES

Metallic impurities, such as compounds of iron, vanadium, and manganese, were precipitated from the crude material by treatment with an active sulfide, such as hydrogen sulfide or arsenic sulfide,[25] or by passing hydrogen sulfide into the liquid in the presence of 0.05 to 0.5 per cent ferric stearate or other heavy metal soap.[26]

Vanadium and other impurities are precipitated by treating the crude product with hydrogen sulfide.[27] As a result, sulfur monochloride is formed by reaction with the free chlorine present which is soluble in titanium tetrachloride and has practically the same boiling point, 135.6° C. Chlorine is then added to convert the sulfur monochloride to sulfur dioxide which is removed by fractional distillation. Sulfur dichloride has a boiling point of 59° C. Hydrogen sulfide added to the crude titanium tetrachloride with agitation combines with the vanadium and other impurities to form solid precipitates with high boiling points. Chlorine is added by bubbling in the gas in 50 per cent excess of the stoichiometric amount to form sulfur dichloride. Although chlorine reacts with sulfur monochloride at room temperature, elevated temperatures speed up the rate.

In a specific operation, hydrogen sulfide was added to crude titanium tetrachloride in a tank with agitation at a temperature close to the boiling point until the chemical analysis showed less than 0.01 per cent vanadium.

This required 0.5 per cent by weight of hydrogen sulfide. The mixture was distilled to obtain a pure product except for 0.03 per cent sulfur monochloride. Chlorine was then bubbled into the tank 50 per cent in excess of the stoichiometric amount to convert the sulfur monochloride to sulfur dichloride. This product was fractionated to separate the titanium tetrachloride from the sulfur dichloride. The temperature at the top of the column was 110° C, and at the bottom, 136.4° C.

After treating crude titanium tetrachloride with hydrogen sulfide to effect purification, solid calcium oxide is added to increase the yield of water white product.[28] The yield may be increased from 70 to 80 per cent to 85 to 90 per cent by this method. Basic compounds such as calcium oxide, barium oxide, sodium hydroxide, calcium hydroxide, sodium carbonate, and trisodium phosphate may be employed. From 0.25 to 0.5 per cent by weight of the basic reagent is required. In an example, hydrogen sulfide was bubbled into 1000 g of crude titanium tetrachloride at a rate of 24 ml per minute for a period of 1 hour with continuous agitation. Two per cent technical grade calcium oxide was added and the mixture was agitated for 5 hours. After standing overnight at room temperature the mixture was filtered, and the filtrate was fractionated in a 4 foot column packed with Berl saddles to give a yield of 86 per cent water white tetrachloride. Purification is effected by digestion with hydrogen sulfide followed by distillation through a fractionating column to remove the low boiling constituents such as silicon tetrachloride.[29] After removal of the low boiling impurities, pure titanium tetrachloride is distilled over and condensed.

Titanium tetrachloride of very high purity is obtained in a multistage process.[30] The crude material is first distilled to remove the solid and nonvolatile materials. Hydrogen sulfide is added to the distillate to precipitate the vanadium and other remaining impurities. A second distillation separates the titanium tetrachloride from the precipitated impurities. Finally, the low boiling and high boiling impurities are removed in two separate fractional distillations. Vapors of the tetrachloride containing vanadium and other metallic impurities are passed through a fluidized bed in the presence of hydrogen sulfide or a metal sulfide.[31] The reaction temperature is held below 400° C at which temperature titanium does not form sulfides, while vanadium and other metal sulfides remain in the fluidized bed.

## INORGANIC COMPOUNDS

Espenschied[32] purified titanium tetrachloride by treatment with an alkaline earth metal hydroxide other than magnesium together with a hydroxide of an alkali metal or magnesium followed by distillation. To be effective the treating agents from the two groups must be simultaneously in contact with the crude material. The sum of the treating agents required to purify a crude material containing 0.06 per cent silica and 0.004 per cent vanadium is 1 to 1.5 per cent by weight of the crude titanium tetrachloride. From 0.5 to 2 parts of the alkaline earth hydroxide are employed for each part of alkali metal hydroxide. Refluxing is not required. In an example, 1000 g of crude titanium tetrachloride was placed in a glass distilling flask and 0.8 per cent calcium hydroxide and 0.8 per cent sodium hydroxide were added at room temperature. The mixture was heated rapidly to boiling and distilled at normal pressure out of contact with the outside atmosphere to obtain water white titanium tetrachloride containing 0.008 per cent silica and 0.0008 per cent vanadium. Purification is also effected by treating the crude material simultaneously with 1 per cent barium, calcium or strontium hydroxide and at least one of the hydroxides of sodium, potassium, magnesium, aluminum, titanium, and chromium, followed by distillation.[33] The crude tetrachloride may be purified in the liquid or vapor phase by reaction with the combination of treating agents; however, the two agents must be employed simultaneously. The quantities of the treating agents used are directly dependent on the purity of the starting material employed. A particular sample containing 0.06 per cent silica and 0.004 per cent vanadium required from 1 to 1.5 per cent by weight of the combined treating agents for purification. Effective results are obtained if from 0.5 to 2 parts of the alkaline earth metal hydroxide are added for each part of the hydroxide of the metal from the other group. The treated titanium tetrachloride is heated to boiling and distilled to obtain a pure product. No refluxing is required. In an example, 1000 g of crude tetrachloride was transferred to a glass flask and 0.5 per cent each of calcium hydroxide and hydrous titanium dioxide were added. The mixture was rapidly heated to boiling and distilled at normal pressure to yield a water white distillate containing 0.008 per cent silica and 0.0008 per cent vanadium. Meister [34] added 5 to 10 g crystalline ferrous sulfate per liter to crude titanium tetrachloride, refluxed for 1 to 6 hours, and distilled to get a product of high purity.

The crude product is purified by treatment with iodine or an iodide compound of sodium, potassium, copper I, magnesium, calcium, or aluminum, followed by distillation.[35] No refluxing is required. A sample of crude titanium tetrachloride containing 0.015 per cent vanadium and 0.01 per cent silica required for purification from 0.05 to 0.2 per cent of the treating agent by weight. In an example, 0.3 part of potassium iodide was added to 1000 parts of crude titanium tetrachloride at room temperature. The mixture was rapidly heated to boiling and distilled at normal pressure out of contact with the outside atmosphere to obtain a water white distillate containing 0.002 per cent vanadium.

Crude titanium tetrachloride is purified by treatment with titanyl chloride followed by distillation.[36] The titanyl chloride may be prepared in place by adding the titanium tetrachloride to a limited amount of water, or it may be separately prepared and added to the crude material. In a specific operation, titanium tetrachloride prepared by the chlorination of ilmenite ore was first decanted from insoluble solid matter, then subjected to simple vaporization and fractional condensation to remove the low boiling heads, primarily silica tetrachloride, phosgene, and chlorine. The heads-free product recovered was yellow in color and contained 2 per cent impurities. A batch of 86 parts by weight of this material was placed in a conventional distillation flask equipped with a reflux condenser, and 10 parts by weight of previously prepared granules of titanyl chloride was added. Evolution of hydrogen chloride was observed and the tetrachloride became slightly turbid. This mixture was refluxed for 2 hours at the boiling point and distilled to give a water white distillate. The residue in the distillation flask was a pale yellow, opaque liquid with a flocculent reddish brown solid in suspension. On cooling the residue solidified.

## TITANOUS CHLORIDE

According to Gage,[37] crude titanium tetrachloride was purified by distillation after treatment with lower chlorides of titanium. These compounds could be formed in place by reducing a portion of the original compound, as for instance with hydrogen. In an actual operation, 100 parts of crude yellow titanium tetrachloride was treated with 5 parts of the trichloride, heated under a reflux condenser at 135° C for 2 hours, and then distilled rapidly in an atmosphere of dry carbon dioxide. The condensate was a clear, colorless liquid.

Titanium tetrachloride is purified by electrolytic reduction to lower chlorides followed by recombination of the lower chlorides with chlorine from the reduction step.[38] The reduction is carried out in a bath of fused chlorides of alkali and alkaline earth metals. A conduit is provided for introducing the titanium tetrachloride into the bath adjacent to the cathode. A hollow cathode may serve this purpose. The cathode may consist of a nickel or tantalum rod or tube. Carbon is suitable for the anode. Titanium dichloride and trichloride produced near the cathode are transferred throughout the solution by diffusion and convection. Titanium tetrachloride is reformed at the anode, where it is released, condensed, and collected as a purified product. Particularly good results are obtained with a current density of 0.2 ampere per square centimeter at each electrode.

For example, an electrolyte of 7300 g of stannous chloride, 2700 g of sodium chloride, 240 g of titanium dichloride, and 1230 g of titanium trichloride was placed in a cell and heated to 700° C. Vapors of titanium tetrachloride, 4.55 g per minute, were added through the hollow cathode into the fused bath. An electric current of 1.3 faradays per mole of titanium tetrachloride was supplied to the cell at 50 amperes with an impressed potential of 1.5 volts. The titanium tetrachloride was reduced to lower chlorides which moved through the fused salt bath to the anode where titanium tetrachloride was reformed. At the surface of the fused bath adjacent to the carbon anode the tetrachloride was released and condensed. An inert atmosphere of argon was maintained above the fused bath to prevent contamination of the titanium compounds from the external atmosphere. The cell operated at a cathode current density of 0.2 ampere per square centimeter and an anode current density of 1 ampere per square centimeter.

Titanium trichloride prepared by reducing purified titanium tetrachloride with coke at 850° to 1500° C is employed to purify the crude tetrachloride.[39] For small-scale operation coke from 0.25 to 0.5 inch in diameter is preferred. The carbon particles arranged in a bed are heated by the passage of an electric current through buried electrodes. In addition to titanium trichloride and titanium tetrachloride, the product gases contain carbon tetrachloride, titanium dichloride, phosgene, and coke dust. The hot vapors are condensed by contact with cold, crude titanium tetrachloride in a spray tower, so that the mixture contains from 1 to 8 per cent by weight of titanium trichloride. This precipitates the vanadium as the oxychloride, after which the titanium tetrachloride is separated by distillation.

In an example, a vertical tube furnace of 1.5 inches internal diameter was packed for a distance of 14 inches with lumps of coke averaging 0.75 inch in diameter. Graphite electrodes were buried in the top and bottom of the coke bed, and an electric current of 35 amperes at 30 volts was passed through. The temperature of the coke rose quickly to 1300° C. Purified titanium tetrachloride vapor was introduced into the bottom of the reactor at a rate of 4.3 pounds per hour. The product gases issuing from the top of the reactor comprising 75 per cent titanium tetrachloride, 20 per cent titanium trichloride, and 5 per cent carbon tetrachloride were passed directly to the condensing tower. Crude titanium tetrachloride containing 0.4 per cent vanadium oxychloride along with small amounts of other impurities was sprayed into the top of the condensing chamber at the rate of 17 pounds per hour at 18° C. This mixture containing titanium trichloride in the amount of 4 per cent of the tetrachloride was passed directly to the still provided with a reflux condenser. The pure titanium tetrachloride distillate contained less than 0.001 per cent vanadium oxychloride. A portion of the distillate was recycled to supply titanium trichloride.

## HYDRIDES

Crude titanium tetrachloride is refluxed with 3 per cent sodium hydride or calcium hydride followed by distillation to obtain a pure water white product.[40] The fractional distillation of the crude titanium tetrachloride yields three products; the low boiling fraction consisting mostly of silicon tetrachloride and carbon tetrachloride with noncondensable gases as chlorine, hydrochloric acid, and phosgene; the predistilled titanium tetrachloride fraction; and the high boiling tails such as iron, zirconium, and chromium chlorides, and oxychlorides, which are retained as sludge in the still pot. This crude titanium tetrachloride fraction is treated with the hydride and redistilled to yield a pure, water white product. A product free from vanadium impurities and thus suitable for pigment manufacture is prepared by refluxing 863 g of the crude yellow titanium tetrachloride containing 0.1 per cent vanadium with 43 g of titanium hydride (through 200 mesh) for 3.5 hours under atmospheric pressure.[41] A water white product is separated from the reaction mixture by distillation. The titanium hydride is prepared by heating metallic titanium to 900° C and cooling in an atmosphere of hydrogen. Analysis shows 96.5 per cent titanium and 3.5 per cent hydrogen.

Purification of titanium tetrachloride is effected by treatment with iron pentacarbonyl, nickel carbonyl, or cobalt carbonyl, followed by distillation.[42] From 1 to 5 parts of the carbonyl is added to 1000 parts of the tetrachloride. As an example, to 1000 g of crude titanium tetrachloride containing 0.09 per cent vanadium, 0.004 per cent silicon, and 0.001 per cent iron was added 5 g of iron pentacarbonyl. A ruby-colored solution formed which produced a black precipitate and a slight evolution of gas. The mixture was heated slowly in a still to degas the product. Further heating distilled colorless titanium tetrachloride containing no vanadium, 0.0012 per cent silicon, and 0.0016 per cent of iron.

## ORGANIC COMPOUNDS

The vanadium in crude titanium tetrachloride is probably present as the oxychloride or tetrachloride, both of which are volatile liquids at ordinary temperature and cannot be recovered by ordinary distillation. Jenness and Annis [43] found that the addition of small amounts (0.25 to

0.50 per cent) of certain organic materials to the crude tetrachloride caused the formation of vanadium compounds which were insoluble and nonvolatile at temperatures considerably above the boiling point of the titanium tetrachloride. These complex compounds were separated by distillation or by such mechanical methods as filtering or centrifuging. Effective organic compounds polymerized in the crude tetrachloride product and reacted with the vanadium during polymerization to form nonvolatile, readily separable compounds.

Examples of this type of compounds are rubber, crude or vulcanized, balata, art gum, polymerized sulfonated oil, soy-bean oil, cottonseed stearin, Russian mineral oil, and acetylene gas. Ordinary rubber was partly polymerized, but further polymerization took place, producing a black, brittle, granular solid. The resulting compound with vanadium was separated from the liquid by distillation, filtration, or centrifuging. Russian mineral oil contained a large proportion of naphthenate compounds which were the active agents. It also appeared that titanium tetrachloride was a good polymerizing agent. The action during polymerization appeared to consist first of forming an addition compound with the crude liquor, followed by a reaction between this and the vanadium, with the liberation of the titanium tetrachloride and the formation of a finely polymerized product. The polymer contained two atomic proportions of chlorine to each of vanadium.

The extent of removal was dependent upon time, temperature of the reaction, and the per cent of agent employed. For instance, 0.10 per cent balata did not remove the vanadium after refluxing for 1 hour, but under the same conditions 0.25 per cent effected complete removal. Also, 0.5 per cent balata produced vanadium-free tetrachloride after only 15 minutes refluxing at 136° C, but contact with this proportion of agent at 20° C for 16 hours did not effect complete removal. Vanadium was recovered quantitatively from the polymerization product.

According to Pechukas,[44] crude titanium tetrachloride was purified and decolorized by treatment with carbon having absorptive or adsorptive properties, followed by distillation. Best results were obtained by using carbon black, such as lampblack, bone black, and gas black, although charcoal and petroleum coke were also found to be effective. The carbon containing the absorbed or adsorbed impurities or their reaction products was recovered and regenerated for further use. Dependent upon the initial color of the liquid and the proportion of carbon used, the time and temperature of the operation were capable of considerable variation. If the discoloration was secured by liquid contact, refluxing at atmospheric pressure (136° C) for a period of time in excess of 10 minutes, followed by distillation, was found to give a yield of purified titanium tetrachloride above 99 per cent. In treating the vapors, any convenient temperature above the boiling point of the liquid proved satisfactory.

In an example, 100 parts of crude yellow titanium tetrachloride was agitated with 4.1 parts lampblack and heated under a reflux condenser at 136° C for 15 minutes, after which the treated liquor was distilled rapidly to yield a colorless liquid.

A process developed by Meyers,[45] effective in removal of vanadium compounds, consisted of mixing an alkali metal soap with the tetrachloride, followed by evaporation. No heating was required. From 0.1 to 1.0 per cent of a sodium or potassium soap of stearic, myristic, palmitic, oleic, or lauric acid gave best results, and the vanadium content was reduced below 0.001 per cent.

Crude titanium tetrachloride containing vanadium compounds which boil at 127° C is purified by heating with a dithiophosphate ester, followed by filtration and distillation.[46] The ester reacts with the vanadium impurity to form an insoluble and nonvolatile product. By this process the vanadium content is decreased from 1 per cent by weight calculated as the pentoxide to less than 50 parts per million. For example, crude titanium tetrachloride obtained by the chlorination of ilmenite and containing 9980 parts per million of vanadium impurities calculated as the pentoxide was heated with 1 per cent dicresyl dithiophosphate at 90° C for 30 minutes in a flask equipped with a stirrer and an air-cooled condenser. After this treatment the mixture was distilled to get a product containing less than 50 parts per million of vanadium calculated as the pentoxide. By a similar process the crude tetrachloride is heated with a xanthate ester followed by filtration or distillation to remove vanadium compounds.[47] The xanthate ester reacts with the vanadium impurity to form an insoluble, nonvolatile product. In a specific operation, 2 per cent monoethyl xanthate was added to crude titanium tetrachloride and the mixture was heated under reflux at 135° C for 30 minutes followed by distillation. The vanadium content was reduced from 9980 parts per million expressed as the pentoxide to less than 50 parts per million.

Liquid chlorinated hydrocarbons of the methane series containing at least five carbon atoms are effective purification agents.[48] Polyvinyl chlorides are also effective. In an example, to 1000 parts of crude titanium tetrachloride containing 0.35 per cent vanadium at room temperature was added 1 part of chlorinated paraffin containing 42 per cent of chlorine by weight. After refluxing at 136° C for 15 minutes the mixture was distilled to give pure titanium tetrachloride containing only 0.0003 per cent vanadium. Vanadium is also removed by treatment with a carbonaceous reducing agent such as a high grade lubricating oil followed by distillation of the mixture.[49] The crude titanium tetrachloride and oil are mixed in a boiler or still. A stream of the liquid is withdrawn, superheated in another unit, and returned to the original still below the surface of the liquid mixture. The superheated vapor agitates the mixture and supplies heat to distill the tetrachloride. Superheating may be effected by heating the liquid indirectly by exchange with Dowtherm E. For example, in a closed vertical iron tank of 108 tons capacity was introduced 75 tons of titanium tetrachloride containing 0.25 per cent vanadium as the pentoxide. A pump removed 220 tons of the tetrachloride per hour and delivered it to the superheater where heat was supplied indirectly by Dowtherm E at 200° C. Vaporization did not take place until the pressure was released as the titanium tetrachloride passed through the restricted pipe into the crude liquid. After the mixture reached the boiling point, 105 pounds of high grade lubricating oil was added to the crude mixture. Refluxing was continued for 3 hours, during which time the oil formed a thioleopic complex with the vanadium. At the end of this period the titanium tetrachloride was distilled; two-thirds was collected and one-third was refluxed.

Colored metal chlorides are removed by digestion with 0.5 to 2.5 per cent of a hydrocarbon as petroleum, crude oil, paraffin, or animal or vegetable fat or oil.[50] The impurities are absorbed on the carbonized residue, which is separated by filtration. Crude titanium tetrachloride is purified by heating under reflux with animal waxes such as wool fat, beeswax, or spermaceti at a temperature up to the boiling point of the mixture, followed by distillation.[51]

Vanadium compounds are removed by heating the crude material with a mixture of a saturated or unsaturated organic compound having an iodine value less than 25 and an unsaturated compound having an iodine value of 80 or more.[52] For example, 100 parts of crude titanium tetrachloride containing 0.052 per cent vanadium by weight expressed as the pentoxide is refluxed with a mixture of 0.38 part by weight of a mineral lubricating oil having an iodine number of 12, and 0.02 part of soybean oil having an iodine number of 147. After refluxing for 8.5 minutes, vanadium was no longer detected in the vapor. The purified titanium tetrachloride was separated from the solid carbonized material by distillation. Purification is also effected by refluxing the crude material with a small proportion of a nondrying vegetable oil followed by distillation.[53] One hundred parts of the crude tetrachloride is refluxed with 1 to 3 parts of oil at 134 to 138° C for 1 to 3 hours and distilled to get a water white product containing 1 to 3 parts per million of vanadium and insignificant amounts of other impurities.

Digestion at the boiling point with 0.5 to 1.0 per cent fatty acids of 10 to 20 carbon atoms, their esters and salts effects purification of the tetrachloride.[54] An example is oleic acid. Colored impurities are removed by refluxing with 0.01 to 0.1 per cent by weight of water and 1.0

to 5.0 per cent by weight of dry powdered charcoal for 5 hours.[55] After this treatment pure colorless titanium tetrachloride suitable for pigment manufacture is recovered by distillation.[56]

## SPECIAL METHODS OF PURIFYING TITANIUM TETRACHLORIDE

Contaminating gases are removed from titanium tetrachloride by adsorption on silica gel or activated carbon.[57] The crystallized titanium tetrachloride formed by cooling the crude material to $-24$ to $-70°$ C is separated from the mother liquor by centrifuging or filtration to yield a clear, transparent liquid in 80 per cent yield.[58] Purification is also effected by the thermal decomposition of the impurities.[59] If crude titanium tetrachloride is passed from the bottom into a cylinder filled with oxygen and vapor-activated Winkler dust, the less volatile ferric chloride and aluminum chloride are absorbed first.[60] Later vanadium oxychloride and other vanadium compounds are absorbed. By the elution effect of excess titanium tetrachloride, the vanadium compounds are gradually expelled into the higher layers to form a zone with the highest concentration of vanadium. Aluminum chloride is removed from liquid titanium tetrachloride by adding water in an amount not exceeding one mole per mole of the aluminum chloride present followed by distillation.[61] Thus 49 per cent of the aluminum chloride may be removed by this method. The gases in titanium tetrachloride are eliminated by applying a vacuum and at the same time blowing a small amount of argon through the liquid.[62]

Organic matter, the most persistent impurity in titanium tetrachloride, is removed by refluxing with chlorine for 2 to 6 hours along with 1 per cent each of crystalline aluminum chloride and water as a catalyst. After reaction, the excess chlorine is removed by passing clean dry air through the boiling liquid.[63] Vanadium is best removed with clean copper after the chlorine treatment. By the chlorine-aluminum chloride-copper treatment followed by distillation, a product of 99.992 per cent purity is obtained. Redistillation in a Poidbielniak still with a 37 inch platinum-iridium heligrid at 70 to 1 reflux ratio gives a titanium tetrachloride of 99.999 mole per cent.[64] Bromine may be used instead of chlorine. In an example, titanium tetrachloride is introduced into a three-neck flask. At this stage the catalyst, aluminum chloride hexahydrate plus an equal quantity of water, is added in amount equal to 2 per cent of the tetrachloride, and the mixture is heated to boiling. Chlorine gas is introduced through a porous disc to distribute it throughout the liquid for a period of 2 to 6 hours. At the end of this reflux period all of the organic material is eliminated.

Hexachlorosiloxane is removed from titanium tetrachloride by refluxing with potassium bifluoride followed by distillation.[65] The fluorination agent is added in an amount sufficient to convert all of the hexachlorosiloxane to lower boiling hexafluorosiloxane. In an example, to a mixture of 1 kg of technical grade titanium tetrachloride and 15.9 g of hexachlorosiloxane in a distilling flask was added 30 g of potassium bifluoride. This mixture was refluxed gently for 24 hours, after which it was distilled through a vacuum jacketed column with Pyrex helices. Three fractions were taken; the first consisted of all of the material that boiled below the constant range of 133 to 134° C. Both the second and third were middle fractions boiling in the 133 to 134° C range.

The sludge obtained by the purification of titanium tetrachloride is treated with hot chlorination gases to evaporate the residual titanium tetrachloride which is recovered.[66] The solids of the sludge remain or form a suspension with the gas and are separated.

# 21

# Vapor Phase Oxidation of Titanium Tetrachloride to Produce Pigment

**THERMAL SPLITTING**

A soft finely divided titanium dioxide of excellent pigment properties is produced from vaporized titanium tetrachloride by a process in which the splitting reaction is brought about by reaction with oxygen or air at temperatures above 1000° C.[1] In carrying out the process, a saturation vessel, constructed and operated on the lines of a wash bottle, is charged with titanium tetrachloride and heated at 120° C. Nitrogen is passed through to produce a gaseous product consisting of approximately equal parts of the two constituents, and this mixture is passed simultaneously with preheated air into an upright cylindrical reaction tube heated by external means to 1100° C. The proportions and quantities of the agents are so regulated that for each liter of reaction space there is introduced, per minute, 1000 ml of the tetrachloride-nitrogen mixture (500 ml titanium tetrachloride vapor), and 1000 ml air. In passing through the reaction vessel in a downward direction, the gases react to form titanium dioxide and chlorine, and the products are discharged into a dust chamber where separation is brought about. Titanium dioxide produced in this manner possesses excellent pigmentary properties and does not require subsequent processing.

To minimize coarsening and undesirable crystallization of the particles during the splitting reaction, the gaseous mixture of tetrachloride and air is raised rapidly to the reaction temperature. The gases are preheated, but to minimize premature reaction they are not mixed before reaching the splitting chamber, which is already heated to the required temperature. The choice of splitting temperature above 1000° C, and the gas velocity, may be regulated within wide limits.

This process has the advantage over the various types of hydrolytic decomposition in that the chloride component is recovered not as hydrochloric acid but as elemental chlorine gas which is used directly for attacking more ilmenite or other titanium raw material.

The various steps of the process are shown in Fig. 21–1.

By subjecting vaporized titanium tetrachloride to thermal decomposition in the presence of oxygen, titanium dioxide of pigment grade is produced without the formation of large crystals by keeping the vapors out of contact with hot surfaces within the chamber during the splitting reaction.[2] Apparently titanic oxide formed adjacent to the hot walls of the reaction vessel, deposits on the surface, and serves as seed which grows to form large crystal agglomerates, and these reduce the average tinting strength of the product.

In carrying out this improved process, oxygen and titanium tetrachloride vapor, in the proportion of 3.3 moles of the former to 1 mole of the latter, are introduced continuously into a reaction chamber of sufficient size to permit thermal decomposition of the tetrachloride but to prevent a significant proportion from reaching the walls of the chamber. The reactants are preheated separately to the reaction temperature of 1500° to 1800° F before bringing them together in the splitting chamber, where a temperature of 1500° to 1800° F is also maintained. The collected titanium dioxide, in powder form, is further calcined to remove the last traces of chlorine. Pechukas and Atkinson[3] employed a special calcination chamber to heat titanium tetrachloride vapor with oxygen at 400° to 800° C to produce a pigment of improved properties. Schornstein[4] introduced parallel streams of the vaporized tetrachloride diluted with an inert gas, and of air mixed with 2 to 70 per cent water vapor based on the oxygen component, both preheated at 800° to 900° C, into a chamber so that an oxidation reaction took place out of contact with the walls. In a similar process, titanic chloride and air are preheated separately to 1000° to 1100° C and passed into a reaction chamber maintained at 750° C by means of a cooling jacket.[5] The proportion of chloride to oxygen is 1 to 4. During the reaction a yellow-green flame burns at the outlet of the preheating tube and a pigment smoke issues from the other side of the reaction vessel. The pigment is very voluminous and possesses excellent whiteness.

To minimize crystal growth and produce titanium dioxide of pigment grade, the chloride vapor is not allowed to come in contact with hot surfaces of the reaction chamber. This is effected by maintaining an atmosphere of nitrogen around the tetrachloride inlet.[6] A temperature of 1000° C is employed. The same result is attained by causing the titanium

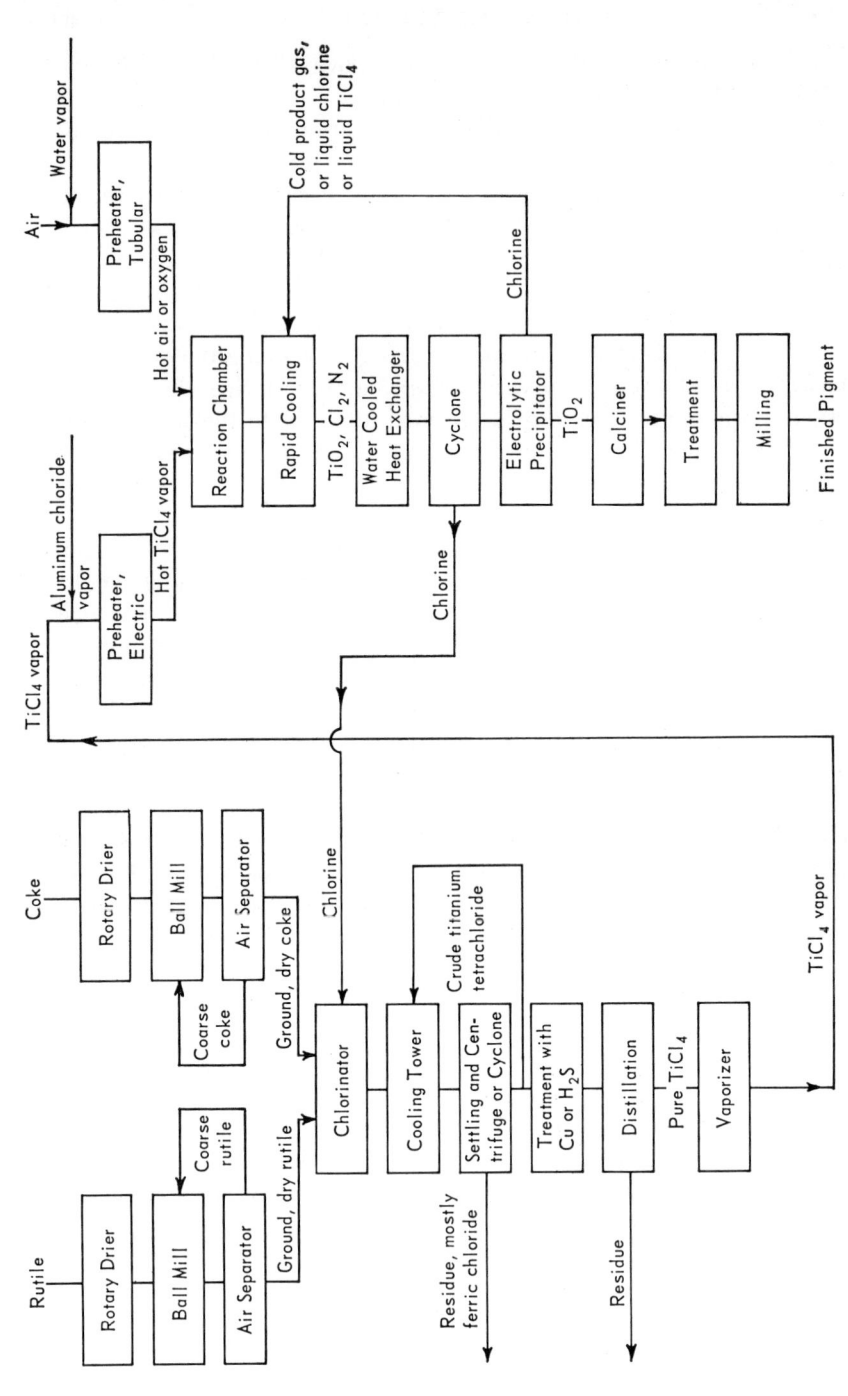

**Fig. 21–1.** Flow sheet of the chloride process for the manufacture of titanium dioxide pigment.

tetrachloride vapor to react with an oxygen countercurrent in a vertical cylindrical furnace, and withdrawing the evolved chlorine so that it surrounds the area of introduction (inlet) of the tetrachloride.[7] This procedure prevents reaction from taking place at the hot surfaces.

Shchegrov [8] calculated the free energy of the reaction of titanium tetrachloride with oxygen and the thermodynamically possible degree of conversion of the tetrachloride to the dioxide, and compared the results with experimental data. The reaction begins slowly at 500 to 600° C. Above this range the conversion increases sharply with temperature and with the increase in oxygen concentration. The products of the reaction between titanium tetrachloride vapor and dry oxygen are only titanium dioxide and chlorine. The free energy values for the reaction at 1073 and 1573° K are calculated to be $-30.28$ and $-23.06$ kilogram calories per mole, respectively.[9] Corresponding equilibrium values are $1.35 \times 10^6$ and $1.44 \times 10^3$, respectively. Oxidation begins at 700° C and accelerates sharply with increased temperature. At 1100 to 1150° C almost complete burning of the titanium tetrachloride delivered into the reaction zone is effected. Formation and growth of the titanium dioxide crystals up to 3 to 4 mm in size take place in the upper relatively cold part of the reactor. The proportion of particles of titanium dioxide smaller than 1 micron gradually increases to a maximum at 1150° C. A similar study indicates that at 1100 to 1150° C almost all of the titanium tetrachloride vapor is oxidized.[10] The oxidation probably consists of two simpler processes; oxidation of titanium tetrachloride in the vapor phase, and the oxidation of titanium tetrachloride on the surface of the solid phase. Titanium dioxide is in the gaseous phase at the moment of its formation but is immediately deposited in a crystalline form. Obtained in this manner, the titanium dioxide contains some absorbed chlorine and consists of a mixture of anatase with rutile possibly as a solid solution. After heating for 1 hour at 900° C the chlorine is driven off and the titanium dioxide is 100 per cent of the rutile crystal form.

Titanium tetrachloride before oxidation to produce titanium dioxide pigment is heated up to 2000° C by passing through electrically heated resistor elements of dense mixtures of graphite and amorphous carbon located in the heating conduit or zone.[11] For example, a vaporized mixture of titanium tetrachloride containing 1 per cent aluminum chloride at 300° C was passed at a rate of 64 pounds an hour through an electrically heated graphite resistor tube unit enclosed in a nickel shell. The resistor tubes are maintained at a temperature of 1800° C by the resistance to a 60 cycle electrical current passing through the walls of the tube. On discharge from the unit the mixture of titanium tetrachloride, aluminum chloride, and chlorine vapor at 860° C was passed directly and continuously into an associated vapor phase oxidation reactor unit for producing high grade rutile titanium dioxide pigment by reaction with air. The proportion of chlorine in the gas mixture was up to 5 per cent.

Pigments are produced by the vapor phase oxidation of the tetrachloride by a process which permits the use of metal heat exchangers for preheating the oxygen.[12] Undesirable metal compounds picked up from the heat exchangers and carried as solid particles in suspension are removed by passing the gas through a ceramic filter medium. The porous refractory filter is preferably constructed of alumina but may be of silica, boron carbide, or clay. These filters have an average pore diameter of 0.1 mm and are constructed in the form of a thimble or a flat disc. A brick-lined shell filled with silica wool filters is also satisfactory. The filter may be cleaned by passing hot concentrated hydrochloric acid through it, followed by a water wash. In a specific operation, the air preheater consisted of coiled, stainless steel tubing. The filter was a porous fused alumina thimble with average pore diameter of 0.09 mm. A silica outer container housed the filter. The metal preheater tube was heated in a gas fired furnace to 970 to 1070° C. After 5 hours service the filter was cleaned.

An organic chloride as carbon tetrachloride is added to the vaporized titanium tetrachloride to react with and remove the oxygen from the air present as a contaminant [13] before preheating to decrease the corrosion of the equipment. The most useful refractory from the standpoint of corrosion is fused silica, but it is fragile and subject to progressive deterioration at elevated temperatures. Furthermore, silica has a low heat conductivity and it is not practical at temperatures above 1000° C. Heat exchange apparatus may be designed using electrical heaters of carbon or graphite shapes. However, oxygen in the vapors derived from air dissolved in the tetrachloride or from oxychlorides present attacks the heated surfaces of the carbon units. To overcome this difficulty the oxygen is removed from the titanium tetrachloride vapor by heating with it an organic chloride to furnish a carbon content sufficient to react with the oxygen present in the vapor. Suitable organic chlorides are carbon tetrachloride and tetrachloroethane. In an example, a vaporized mixture of titanium tetrachloride containing 1 per cent by weight of aluminum chloride at 300° C was passed at a rate of 60 pounds per hour into an electrically heated carbon resistor tube unit. This titanium tetrachloride contained 120 ppm oxygen just prior to entrance of the vapor into the heating unit. An amount of tetrachloroethane vapor was added to react with the oxygen to form carbon monoxide and chlorine. Allowing for a slight excess, 11.3 ml of tetrachloroethane was added per hour. The mixed titanium tetrachloride-aluminum chloride gases were then passed through the interior of the electrically heated resistor tube maintained at 1800° C by the passage of a 60 cycle electric current through the walls of the tube where they were heated to 850° C and passed directly into an associated vapor phase cooxidation unit for the production of high-grade rutile pigment.

Preheated titanium tetrachloride vapor and water-enriched air are rapidly mixed in the reaction zone at 800 to 1400° C by charging the vapor through a restricted peripheral slotted inlet into the reaction zone in the form of a thin sheeted stream directly into the humidified air in a direction at right angles to its direction of flow.[14] Various mixing devices have been proposed. These include fuel and jet turbine types of burners; jet mixers and injectors wherein one reactant passes separately through a main conduit while the other reactant flows through auxiliary or separate tributary conduits; for example, a concentric jet type of mixer. Such mixing is not instantaneous and becomes complete at a point considerably removed from the discharge. Counter currents and eddies are set up along the internal walls of the reactor vessel which create dead spots where objectional titanium dioxide crystal growth takes place. This mixing device is constructed of silica.

Pigment is produced by admitting titanium tetrachloride vapor and oxygen to a reaction chamber at 850° to 950° C through separate feed ports in the same direction in such a way that the titanium tetrachloride stream flows within the oxygen stream.[15] The gases are preheated so that on mixing the resultant temperature before reaction is above 300° C. A molar ratio of titanium tetrachloride to oxygen of 1 to 1, to 1 to 6 is maintained. The velocity ratio of the oxygen to the titanium tetrachloride is less than one. Nonturbulent flow is maintained. By this procedure the reaction takes place out of contact with the hot surfaces including the surrounding walls of the reaction chamber and the tips of the conduits through which the gases are admitted. For example, the ratio of oxygen to titanium tetrachloride was 4 to 1 and the furnace temperature was maintained at 850 to 950° C. The titanium tetrachloride was preheated to 830° C and the oxygen to 850° C so that the temperature attained on mixing, irrespective of the heat of reaction, was 820° C. Water vapor to the extent of 2.3 per cent was introduced by bubbling the air through water at 35° C.

Vaporized titanium tetrachloride preheated to 1050° C and oxygen preheated to the same temperature are blown into a reaction chamber in which the pressure is reduced to less than 30 mm of water to produce finely divided rutile type pigment.[16] The titanium dioxide is collected immediately in a low temperature chamber, and the chlorine is drawn off. Thus, from titanium tetrachloride of 99.98 per cent purity and oxygen of 99.98 per cent purity, titanium dioxide of 99.9 per cent purity with a particle size of 0.3 to 0.5 micron and having a rutile content of 99.5 per cent was obtained in a 95.5 per cent yield. Titanium tetrachloride is introduced into the bottom of a reaction zone through a pipe, the end of which is covered by a porous cup of unglazed porcelain or aluminum carrying an orifice.[17] About 90 per cent of the titanium tetrachloride issues through the orifice, where it meets a counter-current of air or oxygen preheated to 1100° C and reacts on contact. The remaining 10 per cent diffuses through the porous cup walls and surrounds the hot surfaces with a fluid envelope containing no oxygen, thus preventing reaction at the inlet boundaries and crystal growth thereon. Titanium tetrachloride is decomposed to form the finely divided dioxide in a special burner fed with a stoichiometric mixture of oxygen with the vapor uniformly dispersed in the gas stream.[18]

Vapors of titanium tetrachloride and oxygen or air are admitted into a reaction chamber separately or mixed so that the heat of the reaction is used to heat the incoming stream without decreasing the temperature inside the reactor.[19] The titanium dioxide has a particle size of 2 to 5 microns. The oxidation reaction proceeds at 1050 to 1100° C. Wilson[20] injected an annular stream of titanium tetrachloride vapor into a central stream of oxygen. To produce pigmentary titanium dioxide the reactor is designed so that the titanium tetrachloride and oxygen injection tips are cooled to maintain their surface temperature above the vaporization temperature of the tetrachloride but below the oxidation temperature.[21] A surface temperature of 700 to 800° C is suitable to prevent buildup of objectionable scale deposits on the injection tips. Scale if allowed to form causes plugging and decreases the conversion efficiency of the reactor. Titanium dioxide of extremely fine particle size is produced by concentric impingement of an oxygen containing gas on titanium tetrachloride vapor so that contact of the oxide product with the walls of the reactor is prevented.[22] Crystal growth is further impeded by rapidly withdrawing the product from the high temperature reaction zone.

Pigmentary titanium dioxide is obtained by decomposing the tetrachloride in the vapor phase with an oxidizing gas in a reaction zone having porous walls.[23] To prevent contact of the reactants with the internal surfaces of the zone, an inert gas is diffused through the porous walls from an external source maintained under pressure. Titanium tetrachloride and air heated to 700° C are introduced separately into a reaction chamber to form a turbulent gas stream so that the oxide is precipitated in the finely divided form.[24] Simultaneously, finely divided refractory particles are directed against the walls near the gas inlets to prevent the accumulation of pigment particles. During the oxidation step, oxygen or nitrogen heated by induction or by an arc to 2000° C is added to supply auxiliary heat to achieve 950 to 1100° C in the reaction zone.[25] Additives to the inert gas to induce the formation of rutile are 0.1 to 10 parts aluminum chloride, 0.1 to 10 parts of zirconium chloride, and 0.01 to 5 parts of silicon tetrachloride per 100 parts of titanium dioxide. Titanium dioxide is added for nucleation.

Vapors of titanium tetrachloride may be superheated by entraining hot quartz sand particles in the vapors and causing the mixture to flow in a confined elongated stream as a fluidized suspension until the desired heat exchange is effected.[26] The sand is then separated from the superheated vapors. For example, the addition of 500 pounds per hour of sand at 1000° C to 250 pounds per hour of titanium tetrachloride at 138° C superheated the vapor to 800° C. The vapor phase oxidation of titanium tetrachloride containing 0.1 to 20 per cent of an organic diluent produces finely divided particles of the dioxide.[27] Air is fed into a reaction zone so that its linear speed is greater than the speed of the stream of titanium tetrachloride vapor.[28] This air stream sucks the titanium tetrachloride into the reaction zone and provides fast and intimate mixing. An inert gas may be introduced between the streams of air and vapor to prevent premature reaction, thereby eliminating crust formation on the surfaces of the feed orifices. The addition of aluminum chloride or zinc chloride increases the rutile content of the pigment to 99 per cent. Silicon tetrachloride favors anatase formation. Compounds of potassium limit the particle size growth of the titanium dioxide.

The heat transfer for the decomposition of titanium tetrachloride with oxygen is provided mainly by heating an auxiliary inert gas with an electric arc and mixing it with the reaction gases.[29] This heat transfer involves also the ionization energy as the inert gas, nitrogen, attains temperatures up to 30,000° C and reaches the reaction zone at temperatures of 5000 to 15,000° C. Reaction temperatures for the decomposition of titanium tetrachloride are 1100 to 1200° C. To obtain oxide of high purity, titanium tetrachloride is purified by distillation and then oxidized with an excess of oxygen.[30] After separating the suspended oxide, the gaseous mixture of oxygen and liberated chlorine is treated with the original chloride under pressure so that chlorine is absorbed. The chloride is treated under reduced pressure to liberate the chlorine.

In the production of pigment from titanium tetrachloride and oxygen at 800 to 1450° C, the formation of coarse dendrites on the reactor walls is prevented by passing chlorine through a porous wall to convert it to the tetrachloride.[31] A mixture of titanium tetrachloride 0.088 and oxygen containing 2.4 per cent water 0.105 mole per hour, heated to 1020° C, is passed through a porous graphite tube surrounded by a quartz jacket through which dry chlorine 0.118 mole per hour at 30° C is introduced. The tube remains free of deposits and pigment grade titanium dioxide is produced. Titanium dioxide having a rutile content of more than 40 per cent is obtained by oxidizing titanium tetrachloride in an oxygen containing gas at 1200 to 1400° C.[32] The gas burner is designed so that the gas layers are 1 mm to 1 cm in thickness.

Strong titanium dioxide granules for ceramics are obtained by passing gaseous titanium tetrachloride through titanium dioxide containing from 0.05 to 0.3 per cent water.[33] During the operation, particles of 0.2 to 0.5 micron grow to 10 microns in diameter.

## FLAME OXIDATION

Finely divided titanium dioxide is obtained by bringing vapors of the tetrachloride into the direct flame of a combustible gas or vapor.[34] For example, purified gas loaded with titanic chloride vapors is ignited in air. Highly dispersed titanium dioxide results, and owing to its high specific volume it is carried along with the gas currents and separated in a chamber provided with several filtering materials or in precipitators of the Cottrell type.

Titanium dioxide pigments of the anatase and rutile form are manufactured by the combustion in full displacement flow of a stream of carbon tetrachloride and oxygen with carbon monoxide and additional oxygen present to supply supplemental heat.[35] The mixture is supplied at a temperature of 650° C through an aluminum tube to a refractory combustion tube. En route, it passes in full displacement flow, streamlined or turbulent, into a stationary, self-sustaining flame, formed by the combustion of fuel gas within the tube. In this tube the mixture is burned, thereby oxidizing the titanium tetrachloride to the dioxide. The combustion mixture has an average flow rate of from 0.67 to 1.33 of its axial flow rate in the center of the conduit. As shown in electron photomicrographs the particles of the pigment are well-formed, compact, approximately spherical in form, and very uniform in size. Krchma[36] heated the tetrachloride vapor by direct contact with the combustion of carbon monoxide with oxygen. If one mole of carbon monoxide at 18° C is burned in an insulated chamber with 0.5 mole of oxygen preheated to 900° C, sufficient heat is evolved that the mole of carbon dioxide produced is heated to 900° C with 61,600 calories left over. This amount of heat is sufficient to raise the temperature of 3.6 moles of titanium tetrachloride vapors from 200 to 900° C and will produce a gas mixture at 900° C of 78.3 per cent titanium tetrachloride and 21.7 per cent of carbon dioxide. Powdered coke, carbon black, or petroleum coke may be used in place of the carbon monoxide. In an example, liquid titanium tetrachloride was quickly volatilized and superheated to 250° C in a conventionel electrically heated tube or coil preheater constructed of silica. The tetrachloride vapor was conducted to a mixing chamber also constructed of silica. Carbon monoxide was burned with air in a separate ceramic chamber. Air was introduced and combustion was initiated by a spark. The hot carbon dioxide product gases were immediately conducted to the silica mixing chamber where the tetrachloride vapor was introduced and rapidly preheated to 950° C. This resulting hot mixture of titanium tetrachloride and carbon dioxide was at once passed to an oxidation chamber where it was reacted with air preheated to 975° C to produce titanium dioxide pigment.

The exothermic heat of the reaction of titanium tetrachloride with oxygen is not adequate to reach the required temperature of 950 to 1100° C or even to maintain this temperature so that auxiliary heat must be supplied in the reaction chamber.[37] The preheated oxygen-titanium tetrachloride mixture is introduced through a rotating center jet. Auxiliary fuel such as carbon monoxide and hydrogen with the oxygen required for combustion are injected through ring-shaped outlets surrounding the center jet. A rotating motion in a direction opposite to that of the reaction mixture is imparted to the auxiliary combustion zone by vanes in the ring-shaped outlets. The particle size of the titanium dioxide is controlled by variations in the gas flow rate, temperature, and component reactions. In a typical operation, a mixture of 1 volume titanium tetrachloride vapor and 1.5 volumes of oxygen at 250° C was injected through the center jet at a velocity of 20 meters per second. At the same time 0.8 volume of carbon monoxide was introduced through the inner nozzle and 0.4 volume of oxygen through the outer ring nozzle. The mixture ignited at a distance of 5 to 10 cm from the nozzle. A titanium dioxide pigment with a particle size of 0.5 micron was obtained at a yield of 99 per cent.

A suitable unit for the oxidation of titanium tetrachloride with oxygen in an auxiliary flame comprises three concentric tubes connected to a reaction chamber.[38] The titanium tetrachloride vapor passes through the innermost tube. Oxygen with controlled water vapor content (0.005 volume per cent for titanium tetrachloride) passes through the intermediate tube. Gases for the auxiliary flame such as hydrogen, carbon monoxide, illuminating gas, gasoline vapor, oil vapor, or mixtures pass through the outside tube. The temperature of the reaction chamber may vary from 500 to 1200° C, and the amount of material produced ranges from 0.5 to 50 kg per hour with a particle size less than 0.1 micron.

Mixed vapors of titanium tetrachloride and oxygen are sent through a burner tube having one or two outer concentric tubes carrying carbon monoxide and oxygen.[39] To mix the flammable gases thoroughly with the charge vapors, the ends of the tubes are constricted. The auxiliary flame from these gases aids the reaction and titanium dioxide of 0.2 to 0.4 micron particle size is obtained. By the reaction of a dispersion of titanium tetrachloride in oxygen or air at 1200° C in a vertical reactor into which carbon monoxide and acetylene or carbon monoxide and hydrogen are also admitted the finely divided oxide is obtained.[40] The pigment powders obtained in the process have an average particle size of 0.5 to 2.0 microns. The properly proportioned mixture is ignited in the reaction furnace.[41]

Titanium dioxide of exceptional fineness and high purity is produced by introducing the vaporized tetrachloride in a stream of preheated air into an insulated, tubular reactor so that the vapor is surrounded by an annular blanket of flame.[42] The flame is maintained by burning a gaseous mixture of hydrocarbon with oxygen. Sufficient hydrocarbon is present to ensure on combustion complete hydrolysis of the tetrachloride. Sufficient oxygen is present for complete combustion. One per cent carbon dioxide vapor is added to the hydrogen auxiliary fuel in the oxidation of titanium tetrachloride to produce a pigment containing more than 50 per cent rutile.[43] Titanium dioxide pigments containing at least 60 per cent rutile are made by passing a mixture of titanium tetrachloride vapor and an oxygen-containing gas into a reactor in the form of a layer 1 to 10 mm thick.[44] The reaction is kindled by a flame surrounding the reactant gas mixture which is preheated to 135° C.

Pigment is produced by thermally decomposing a gaseous mixture of titanium tetrachloride containing free oxygen under flame conditions in the presence of 0.01 to 10.0 per cent of a volatile silicon compound.[45] To avoid premature oxidation a maximum preheating temperature of 500° C is employed. The process is carried out by utilizing an auxiliary flame and maintaining a temperature of 750 to 1200° C of the atmosphere immediately ambient to the auxiliary flame. By this process titanium dioxide containing more than 90 per cent anatase is produced. To eliminate the stoppages caused by titanium dioxide deposits on exit orifices and reactor walls, the hot gases from the oxidation of the tetrachloride are evacuated from the chamber center by cooled aluminum-magnesium alloy tubes without deforming the flame pattern.[46] Recycled cold gas is brought to the center by the suction created by the flow of hot gases in the tubes.

By the reaction of titanium tetrachloride, carbon monoxide, and oxygen in a cylindrical chamber preheated to 1000° C, spherical titanium dioxide particles are obtained.[47] The gas ignites spontaneously, producing 80 per cent anatase having a particle size of 0.2 micron, and chlorine. This titanium dioxide product is separated from the chlorine in a cyclone separator and in a Cottrell precipitator. The hot gases from the combustion of low hydrogen content carbonaceous material are passed into a flowing stream of titanium tetrachloride to produce titanium dioxide and chlorine.[48] Fuels include carbon black, coke, charcoal, and coal. A combustion temperature between 1400 and 2200° F is produced. Between 1 and 5 mole per cent aluminum or a compound may be added with the carbonaceous material to deposit alumina on the pigment particles.

A mixture of titanium tetrachloride and air preheated to 200 to 600° C is heated rapidly in a closed refractory lined reactor by a flame of sulfur vapor to 800 to 1200° C to produce titanium dioxide.[49] The reaction products are cooled, titanium tetrachloride is removed, and the gases are led through a catalyst chamber with added oxygen to form sulfuric acid. This sulfur vapor can be mixed with the titanium tetrachloride and oxygen, separately injected through nozzles or through a fluidized bed of sand or rutile. Titanium dioxide of pigment grade is produced by introducing titanium tetrachloride vapor into a free oxygen containing hot reaction zone and at the same time introducing into the reactor tangent to the flow of the tetrachloride separate streams of a burning fuel and an auxiliary gas which is inert to the pigment particles.[50] Suitable gases are chlorine, nitrogen and carbon dioxide.

During the vapor phase oxidation of titanium tetrachloride in a fluidized bed of titanium dioxide, carbon monoxide is burned in the bed to maintain the operating temperature at 950 to 1250° C.[51] Blocking of the bed by partial fusion near the carbon monoxide inlet is prevented by dilution with nitrogen, chlorine, or carbon dioxide to lower the temperature. Additional chlorine is added to the vapor of purified titanium tetrachloride before heating with oxygen to produce the dioxide of the proper particle size and reduce the loss of chlorine.[52] The excess oxygen in the final gas mixture from the oxidation step is removed by burning a carbonaceous material. This also serves to heat the chlorine recovered in the process for reaction with more ore.

## FLUIDIZED BED OXIDATION

Titanium tetrachloride and air preheated to 1000° C are passed into a fluidized bed of an inert material, as rutile or zircon, where they react to form pigment-grade titanium dioxide, which is carried away by the product gases.[53] The material comprising the fluidized bed has a particle size of 40 to 1000 microns. A velocity is maintained at five times the minimum fluidizing velocity. Materials suitable for the bed are rutile, alumina, silica, and zircon which have been treated to remove any objectionable impurities. The reaction chamber consists of a shaft furnace having as a base a perforated plate with porous diaphragms above to allow the gas to pass upward but to prevent passage of solid particles through the plate. Temperatures of 800 to 1100° C are maintained in the furnace. As an alternative the titanium tetrachloride and air may be premixed at temperatures below 500° C before they are passed into the fluidizing bed. Crystals of titanium dioxide formed are deposited on the bed solids rather than on the walls of the reactor.

In an example, a silica tube 5 inches in diameter and 36 inches long mounted vertically was heated in an electrically wound furnace. A porous silica disc was cemented in the base of the tube to support the fluidized bed. The lower part of the tube was sealed with two inlet tubes. One supplied the air into the chamber below the disc; the other supplied the titanium tetrachloride and passed through the disc. Silica sand having a grain size of 250 to 350 microns to a static height of 7 inches constituted the fluidized bed. The bed was heated to 920° C while maintained in a fluidized state by the admission of air at the rate of 12 liters a minute. Titanium tetrachloride was admitted at a rate of 0.18 mole per minute and the air was adjusted so that the ratio of the tetrachloride to oxygen was 1 to 3. The reaction was practically instantaneous and the time of contact in the fluidized bed zone was approximately 2 seconds. At least 99.9 per cent of the titanium tetrachloride was converted.

Gaseous titanium tetrachloride is passed with oxygen through a hot bed of granular silica or rutile to effect reaction.[54] The very fine titanium dioxide particles are carried out by the gas stream. The reaction may be carried out in the presence of 0.5 to 7 weight per cent, calculated as the oxides, of zirconium chloride as a modifying agent.[55] By preheating the reactants, the rate of the reaction in a fluidized bed is increased.[56] Deposition of titanium dioxide on the fluidizing particles is inhibited by the preheating. The yield of pigment recovered from the effluent gases from the fluidized bed oxidation of the tetrachloride is improved by introducing metallic aluminum into the bed.[57] Less titanium dioxide is left in the bed. This effect is caused by alumina deposition replacing titanium dioxide on preferred sites on the bed particles. A similar effect is obtained by introducing chlorine and carbon monoxide or carbon into the bed at a location shielded from the oxygen.[58]

The vapor phase oxidation of titanium tetrachloride is carried out at 750 to 1500° C in a bed of fluidized refractory particles under conditions that at least part of the titanium dioxide formed remains within the bed.[59] By gradual withdrawal of particles from the bed, accompanied by gradual addition of fluidizable particles of alumina, silica, zirconium oxide, or titanium dioxide, the size distribution of the particles is controlled. Particles withdrawn from the bed have a larger mean size than that of the particles added. The finely divided titanium dioxide which escapes from the reaction bed with the fluidizing gases may be used as a pigment or may be returned to the bed to control the particle size of a later product. Titanium dioxide of larger particle size withdrawn from the bed is especially suitable for the manufacture of vitreous enamels.

Pigment-grade titanium dioxide is produced by the reaction of the tetrachloride vapor with an oxygen-containing gas and a chloride vapor conditioning agent in a fluidized bed composed of 0.001 to 0.1 inch particles of a heat resistant material such as titanium dioxide, alumina, or silica at 750 to 1250° C.[60] Small amounts of water in the gas mixture accelerate the reaction. A gas velocity is maintained high enough to keep the bed fluid and to remove most of the titanium dioxide formed. This is recovered in cyclone separators. Effective conditioning agents are titanium trichloride, aluminum chloride, and silicon tetrachloride. Titanous chloride and aluminum chloride promote rutile formation. The bed is heated by introducing carbon monoxide or other fuel gas.

As an illustration, in a fluidized bed of 40 to 60 mesh titanium dioxide 16 cm deep and 5.1 cm diameter in an externally heated silica tube, a temperature of 1080° C was maintained with an upward flow of a gas mixture providing purified titanium tetrachloride at a rate of 12.4 ml per minute, aluminum chloride 0.21 g per minute, oxygen containing 4.53 mg of water per liter 15 liters per minute, and nitrogen 6.1 liters per minute. Twenty per cent of the titanium dioxide formed remained in the bed. The remainder separated from the effluent gas in a cyclone was 100 per cent rutile pigment having a Reynolds tinting strength of 1625.

The larger particle sizes of titanium dioxide are withdrawn from the fluidized bed and chlorinated.[61] A gas stream carries off the fine pigment particles and the larger particles fall to the bottom. The reactor is a refractory lined vertical cylinder arranged for gravity charging of titanium dioxide and with four angled pipes for feed of titanium tetrachloride, oxygen, chlorine, methane, carbon monoxide, and nitrogen.

## NUCLEI

Pigment-quality titanium dioxide of the rutile crystal form is produced by reacting at 900 to 1200° C in the vapor phase over a period of 0.1 to 1 second purified titanium tetrachloride with air in the presence of 0.1 to 5 per cent water vapor based on the total volume of gases to continuously form within the reacting gases small amounts of titanium dioxide nucleating agent which promotes and insures the production of a high quality product.[62] Titanium tetrachloride vapor and air enriched with 0.1 to 3 per cent water vapor are separately and continuously introduced into an oxidation zone where they are rapidly and thoroughly mixed. These reactants are preheated before introduction into the reaction chamber and the reaction is carried out at a constant temperature of from 900 to 1200° C. The product is cooled by mixing with cold gases from a previous reaction to below 600° C. This should be effected in 1 second and does not exceed 10 seconds. The rutile titanium dioxide is separated from the gaseous mixture by a cyclone, electrostatic separator, filtration through a porous medium, or any conventional means.

As an illustration, titanium tetrachloride vapor preheated to 800° C was admitted to the reaction chamber at a continuous rate of 100 parts by weight per hour into the upper portion of a reaction chamber maintained at 1050° C. Simultaneously, air preheated to 800° C and containing 0.95 per cent water vapor by volume was continuously admitted through a separate inlet adjacent to the titanium tetrachloride inlet. The air flow was 20 parts by weight of oxygen per hour. Inside the reactor the gas streams converged immediately to become rapidly mixed. Flow rates were adjusted to get a retention time of 0.5 second. The reaction products at 1000° C were cooled quickly by introducing sufficient cold chlorine gas to drop the temperature to 300° C in less than 2 seconds. The titanium dioxide particles were separated in filter containers. In the production of the seeding agent titanium tetrachloride reacted with water to form very finely divided titanium dioxide and hydrochloric acid.

To prepare finely divided titanium dioxide, 1 volume of the tetrachloride vapor and at least 3 volumes of air or oxygen containing a small proportion of water vapor are separately preheated in concentric tubes so that on mixing a temperature of at least 600° C is attained.[62a] The gases are mixed in a chamber large enough to permit complete reaction to take place before the mixture strikes the chamber walls. Dilution of the chloride with nitrogen or carbon dioxide and high gas velocities during the mixing favor small particle size.

Titanium dioxide pigment having the rutile crystal form is prepared by feeding a stream of the tetrachloride into the center of a slowly moving stream of oxygen in a concentric burner.[63] Both gases are preheated and both move in streamline flow. Addition of up to 5 per cent water vapor in the oxygen and of up to 0.25 per cent silicon tetrachloride in the titanium tetrachloride yields a better pigment. The mean particle size of the pigment is between 0.1 and 0.35 micron. Oxygen and titanium tetrachloride preheated to 850° C are fed to the reactor in a mole ratio of 4 to 1. With the addition of 2.3 per cent water vapor to the oxygen and 0.07 per cent silicon tetrachloride to the titanium tetrachloride, the pigment has a rutile content of 91.5 per cent and a tinting strength of 1800.

Further improvement in pigment properties is obtained if the water for forming the nucleating agent is produced in situ by the oxidation of hydrogen present during the reaction.[64] Also, more efficient control over the reaction results. Pure anhydrous titanium tetrachloride vapor, air, and from 0.0062 to 0.186 mole of hydrogen per mole of the tetrachloride to form from 0.1 to 3 per cent water vapor by volume are separately and continuously introduced into a reaction vessel at a constant temperature of 900 to 1200° C. The reactants are preheated before introducing into the furnace. Retention time in the furnace is from 0.1 to 1 second to allow time for complete reaction but not enough time to result in undesired particle growth of the titanium dioxide. In an example, 5330 g of vaporized titanium tetrachloride preheated to 1050° C was introduced per hour into the reaction vessel. Through separate openings air preheated to 1050° C was introduced at a rate of 3400 liters per hour. Through a third opening methane preheated to 1050° C was admitted at a rate of 52.6 liters per hour, equivalent to 9.4 liters of hydrogen per hour. Within the reaction chamber the temperature was 1080° C. The collected titanium dioxide after removal of absorbed chlorine was of the rutile crystal form, had a particle size of 0.17 micron, and possessed excellent pigment properties. To produce rutile rather than anatase, the conditions are: a minimum water content, minimum preheating temperature, minimum retention time, and quick cooling.

By the oxidation of the tetrachloride with atomic oxygen in the gas phase or in a fluidized bed titanium dioxide in a finely divided state is prepared.[65] Water vapor is present during the reaction.

In the vapor phase oxidation process preheated titanium tetrachloride vapor is rapidly mixed with preheated air containing 0.5 to 20 per cent of an oxide of nitrogen.[66] The oxides of nitrogen react rapidly with titanium tetrachloride to form nuclei which act as growth centers for the pigment particles. Effective oxides are nitrous oxide, nitric oxide, and nitrogen dioxide. The oxidation reaction is carried out at a temperature of 900 to 1350° C which requires preheating of the reactants to temperatures above 600° C. The oxide of nitrogen is added to the air before or after preheating. For example, titanium tetrachloride vapor preheated to 750° C was admitted through one inlet to a reaction chamber maintained at 1000° C. An amount of air 5.1 times the volume of the tetrachloride vapor preheated to 770° C was admitted through a second inlet so arranged that mixing took place rapidly. Nitrogen dioxide was added to the air just prior to introduction into the mixing zone in an amount sufficient to give 0.2 per cent of the oxide of nitrogen. After a retention time in the furnace of 0.25 second the reaction product was removed from the reaction zone and cooled quickly to prevent particle growth. From the cooled mixture the titanium dioxide was recovered as pigment grade.

The particle size of titanium dioxide pigments produced by the vapor phase oxidation of titanium tetrachloride is controlled through the use of nucleating agents.[67] For this high temperature oxidation fragile nonmetallic equipment such as fused quartz is required to withstand the high corrosive gases. Ordinarily water vapor is added to the air used to develop the desired particle size and aluminum chloride is added to the titanium tetrachloride to prevent discoloration of white paints and enamels caused by the titanium dioxide. By adding a suitable nucleating agent the preheating temperature of the titanium tetrachloride may be lowered thereby simplifying the process and at the same time pigment of excellent properties is obtained. A portion of the titanium tetrachloride vapor is preheated with a reducing agent to form a small proportion of lower valent chlorides such as titanium trichloride and dichloride in the vapor or as finely divided solids prior to or at the point of the introduction of the vapor into the oxidation reactor. The resulting vaporized mixture of titanium tetrachloride and lower valent chlorides is continuously oxidized to titanium dioxide of pigment grade by reaction with air at 800 to 1350° C. The titanium tetrachloride is reduced to form the lower valent chloride by heating with a granular metal as titanium, with hydrogen, or with a hydride at 136 to 400° C. After discharge from the reactor the reaction products are cooled quickly and the titanium dioxide is separated from the gaseous medium, calcined at 500 to 700° C, and milled in a fluid energy mill. Metals for reducing the titanium tetrachloride are sodium, magnesium, aluminum, titanium, and zinc. Hydrides of titanium, sodium, aluminum, and zinc may be used.

In an example, 100 parts by weight of titanium tetrachloride vapor were preheated to 765° C in silica equipment. A portion of this stream was continuously admitted at a rate equivalent to 10 parts by weight per hour to a lower chlorides generator where 10 per cent of the diverted titanium tetrachloride was passed over titanium sponge granules to form titanium trichloride in the vapor as a nucleator. This mixture was united with the main tetrachloride stream, and the nucleated vapor was charged immediately and continuously into the reaction chamber of a tubular form of vertical reactor of corrosion resistant construction. Simultaneously, 17 parts by weight of oxygen as air preheated to 750° C were continuously charged from a separate inlet into the reactor and mixed with the titanium tetrachloride. Reaction took place at a temperature maintained at 1000° C. The low rate provided an average retention time of the reactant and product gases within the reactor of 0.15 second. Titanium trichloride made up 2.5 parts by weight of the mixed vapor. After cooling the titanium dioxide product rapidly to below 600° C, the titanium dioxide was separated, calcined at 650° C for 2 hours, and ground in a fluid energy mill.

Titanium dioxide of small, uniform particle size and improved tinting strength is made by the vapor phase oxidation of the tetrachloride with air or oxygen at 700 to 1500° C in the presence of 0.01 to 20 mole per cent of an aromatic organic compound.[68] The mole ratio of titanium tetrachloride to oxygen is from 1 to 1 to a value of 1 to 10. Nucleating centers that control the titanium dioxide particle size are formed by the decomposition of the organic compound in the reaction zone. Benzene, phenyl chloride, trichlorobenzene, biphenyl, and naphthalene are preferred organic compounds. An orifice-annulus burner is used in the process.

To decrease the abrasive effect titanium dioxide of spherical form is produced by the vapor phase oxidation of titanium tetrachloride at a temperature of 900 to 1200° C in the presence of a small amount, 2 to 10 per cent by weight, of a sulfur chloride or phosphorus chloride.[69] In paint production, iron particles are ground off the mills and discolor the paint film. In the pigment process the reacting vapors and oxygen or air remain in the heated chamber for 0.1 to 5 seconds. During the reaction the titanium tetrachloride and sulfur or phosphorus chlorides are converted to their respective oxides with chlorine being formed as a by-product. The products withdrawn from the reaction chamber are cooled rapidly to prevent further growth of the titanium dioxide particles. From the gas stream titanium dioxide can be separated and recovered by cyclone separators, settling chambers, glass cloth filters, or other suitable means. As an illustration, a mixture of 2 parts by weight of sulfur monochloride and 98 parts of titanium tetrachloride was vaporized, preheated to 1000° C, and then introduced through a central orifice of a concentric type nozzle to the reaction chamber maintained at 1000° C. Dry oxygen similarly preheated to 1000° C was concurrently admitted through the outer annulus of the concentric nozzle into the reaction chamber at a rate such that 38 parts by weight of oxygen were admitted for each 100 parts of the chloride mixture. The gases were allowed to remain in the furnace for 3 seconds, after which they were rapidly cooled. Titanium dioxide recovered from the product by passage through a glass cloth filter consisted predominantly of spherical particles.

Anatase of particle size 0.1 to 0.5 micron is obtained by combustion at 900° C of a mixture containing titanium tetrachloride 25, silicon tetrachloride 0.25, oxygen 44.75, and nitrogen 30 per cent.[70] A particle size of 0.1 micron results from the combustion of a mixture of titanium tetrachloride 1, oxygen 1.5, nitrogen 1.6, and silicon tetrachloride 0.05 parts. The products with 2 to 10 per cent silica have limited use as pigments.

Specified vapor mixtures are formed by transferring liquid mixtures of the components to a surface heated to a temperature above the boiling point of the highest boiling constituent.[71] Thus 99 parts of liquid titanium tetrachloride mixed with 1 part of liquid silicon tetrachloride was agitated rapidly and dropped at a rate of 2 parts per minute onto a surface heated to 250° C to effect vaporization. The vaporized mixture is oxidized to give titanium dioxide pigments. Likewise small amounts of chlorides of aluminum, phosphorus, tin, and antimony can be added to the titanium tetrachloride.

## ALUMINA

Pigment-grade titanium dioxide is produced by heating vaporized anhydrous titanium tetrachloride containing 1 to 2 per cent anhydrous aluminum chloride and up to 5 per cent anhydrous chlorine at 136 to 1000° C with oxygen.[72] The vapor mixture is heated to 350° C in contact with an electrically heated tubular resistor composed of amorphous carbon. From the heater the vapors are passed through a second resistor furnace consisting of elecrically heated solid graphite and then into a reaction chamber for interaction with oxygen. Titanium dioxide having a particle size of 0.75 micron and a rutile content of 95 per cent is obtained by the reaction at 1200° C of the vapors of titanium tetrachloride and aluminum chloride with an oxygen containing gas.[73] The mixed vapors are produced by passing a mixture of oxygen and nitrogen through an evaporator containing the two components at 95° C. A mixture obtained by treating 4.5 volumes of bromine and 10 volumes of nitrogen with pure aluminum at 600° C together with titanium tetrachloride 100, silicon tetrachloride 0.5, oxygen 180, and nitrogen 140 volumes yields on oxidation particles of 0.5 micron diameter comprising 95 per cent rutile. Rutile titanium dioxide pigment is prepared by the thermal decomposition of a gaseous reaction mixture of titanium tetrachloride and oxygen in the presence of 0.01 to 10 per cent of a volatile aluminum compound which is converted to the oxide in the process.[74] Between 0.1 and 1 per cent ferric chloride based on the weight of the titanium tetrachloride is included to avoid discoloration.

Finely divided titanium dioxide of pigment grade is produced by the decomposition of vaporized titanium tetrachloride with gases containing free oxygen at a temperature above 600° C in the presence of aluminum chloride and silicon tetrachloride.[75] For instance, a reaction gas mixture consisting of titanium tetrachloride, aluminum chloride, silicon tetrachloride, oxygen, and nitrogen is ignited at 1200° C to produce a pigment containing 0.9 per cent aluminum oxide and 0.35 per cent silica. Pigment-grade titanium dioxide admixed with alumina and nonpigment-grade titanium dioxide is prepared by oxidizing titanium tetrachloride in a fluidized bed at 900° C.[76] Carbon monoxide is introduced to supply heat. Aluminum chloride vapor is added through ports above the fluidized bed. Fine particles of pigment-grade titanium dioxide are carried from the bed with the gas stream. Air or oxygen preheated to 500 to 1200° C is mixed with similarly preheated aluminum chloride within 0.1 to 3 seconds before reaction with titanium tetrachloride vapor at 500 to 1200° C to give a pigment of uniformly fine particle size.[77] The aluminum chloride, 0.1 to 10 mole per cent of the titanium tetrachloride, forms alumina nuclei between 0.05 and 0.5 micron suspended in the air stream which serve as centers of growth for the titanium dioxide.

Pigmentary-grade titanium dioxide is produced by the vapor phase oxidation of the tetrachloride in the presence of salts or oxides of calcium, strontium, barium, or cerium at 900 to 1400° C by means of air containing water vapor.[78] The reaction products leaving the furnace at 1150° C are quickly quenched to 600° C to prevent increase in particle size.

## METHODS OF PREVENTING THE FORMATION OF TITANIUM DIOXIDE SCALE ON THE WALLS OF THE REACTORS AND CONDUITS

The high-temperature vapor phase oxidation of purified titanium tetrachloride with air is carried out under conditions which suppress the deposition of titanium dioxide particles on the walls of the reacting chamber and the conduits.[79] Titanium tetrachloride vapor reacts with oxygen at 900 to 1600° C to produce titanium dioxide and the resulting suspension is cooled quickly to 600° C at which temperature the particles lose their tendency to grow in size. This cooling may be effected by injecting cold reaction product gases, cold titanium tetrachloride or liquid chlorine. The suspension may be cooled below 600° C by indirect means and passed through a cyclone and Cottrell precipitator to remove the titanium dioxide. A principal difficulty arises from the tendency of the titanium dioxide particles to adhere to the walls of the equipment. Coarse particles are formed on the hot surfaces of the reaction chamber.

If the titanium tetrachloride reacts with oxygen in a combustion chamber having walls of a porous refractory material and carbon monoxide passes transversely through the porous walls into the chamber, formation of the coarse material and of the deposit is prevented. The porous walls of the apparatus are surrounded by a gas-tight steel jacket with space between through which the gas is admitted at a slight pressure. Alundum is a suitable material for the porous refractory walls. In addition to repelling the preformed particles of titanium dioxide the carbon monoxide reacts chemically with any free oxygen adjacent to the surface of the combustion zone to prevent its reaction with titanium tetrachloride. By supplying the reducing agent, the carbon monoxide favors the reaction of chlorine with any titanium dioxide that may have plated on the walls of the reactor.

Objectionable titanium dioxide scale deposition and buildup on the internal surface of the reactor used in the vapor phase oxidation of titanium tetrachloride with air is prevented by periodically removing the scale during the reaction by passing a rotating, reciprocally movable, internally cooled scraping device over the surface.[80] By internally cooling the scraping unit it is maintained at a temperature low enough to avoid corrosion by the chlorine gas. Consequently the pigment is not discolored.

As an application of the process, a vertical cylindrical reactor constructed of nickel and having a 9 inch internal diameter was utilized in the cooxidation of titanium tetrachloride and aluminum chloride. The water cooled scraper was provided with an open ring type of cutting head, the internal diameter of which was slightly less than the internal diameter of the reactor so that it could be moved up and down in the reactor. An electric motor attached to the shaft outside the reactor rotated the scraper. Purified titanium tetrachloride at the rate of 1250 pounds per hour was separately preheated to 950° C and aluminum chloride vapor equivalent to 5 per cent per hour of alumina was mixed with it. The combined vapors were charged continuously into the reactor. Dry air at a rate of 1050 pounds per hour was preheated to 1030° C and water amounting to 6.2 pounds per hour was added by combustion of an equivalent quantity of acetylene in the preheated air and charged into the reactor. The two streams, the chloride vapor and the moist air, became quickly mixed in the reaction zone where they reacted to form titanium dioxide, alumina, and chlorine.

The reactor for the production of pigmentary titanium dioxide is designed so that the titanium tetrachloride and oxygen injection tips are cooled to maintain their surface temperature above the vaporization temperature of the tetrachloride but below the oxidation temperature.[81] A surface temperature of 700 to 800° C is suitable and prevents buildup of objectionable scale deposits on the injection tubes.

In the continuous vapor phase oxidation of titanium tetrachloride to produce pigments, to prevent scale formation on the internal surfaces of the reactor, a wall of rigid, porous, refractory material is used.[82] The wall is kept below 300° C. A shielding gas is maintained over the surface by charging an inert normally gaseous fluid in the liquid state through the pores of the reactor walls for vaporization on the surfaces within the oxidation zone. To avoid incrustations with titanium dioxide the gaseous mixture of titanium tetrachloride and oxygen at 500° C is directed upwards into the center of the reactor.[83] Here it encounters the cascade-like descending, coarse titanium dioxide which has been obtained by separation from the finer particles produced in the oxidation reaction. It is cooled to 250° C before re-entry.

Vaporized titanium tetrachloride is reacted with humidified air at 1000 to 1350° C in a porous carbon tube with chlorine gas diffusing through the walls of the porous tube.[84] This gas shields the surface from contact with the reactants so that undesirable titanium dioxide scale or film does not collect on the surface of the reactor. In the former processes a portion of the titanium dioxide formed deposits on and tenaciously adheres to the internal surfaces of the reactor. A portion of the product is lost since it cannot be recovered in useful form. Even if the titanium dioxide forms in a desirable state, the conditions at the surface of the reactor cause an undesirable particle size growth which renders the product unfit for pigment use. Even where loss of material can be tolerated, segments of the scale frequently dislodge from the surface to contaminate the product. A wide variety of proposals for removing deposits of this type have been suggested, including mechanical scraping, chemical removal, and inert gas shielding.

In an example, employing a reactor provided with a porous carbon reaction zone tube having an internal diameter of 12 inches, a length of 24 inches, and an average pore radius of 48 microns, titanium tetrachloride vapor heated to 850° C was supplied at a rate of 41,000 pounds per hour where it was oxidized upon admixture with a stream of oxygen heated to 1150° C. Twenty-five per cent excess oxygen or 8300 pounds per hour was admitted. The oxygen contained 7.5 pounds of water per 100 pounds of oxygen. Chlorine at room temperature was supplied to the exterior walls of the porous carbon reactor tube at a rate of 3060 pounds, the equivalent of 3.38 pounds per square foot of wall area, per hour. The oxidation reaction proceeded over an extended period of time without serious deterioration of the porous carbon reactor wall and without buildup of titanium dioxide scale.

In decomposing volatile halides of titanium with humidified air at high temperature, some of the oxide remains in the apparatus and interferes with the full flow of gases. These oxides can be removed by reaction at 800 to 1300° C with a mixture of equal parts by volume of chlorine and carbon monoxide.[85] In this process the vapor phase oxidation of titanium tetrachloride with air at 900 to 1200° C is continued until the accumulation of scale on the walls begins to seriously impair the operation of the furnace. At this stage the introduction of reactants is interrupted and a mixture of equal parts by volume of carbon monoxide and chlorine is introduced to convert the titanium dioxide scale to the gaseous tetrachloride which leaves the furnace with the carbon dioxide formed.

An inert gas is injected as a separator between the preheated titanium tetrachloride vapor and oxygen-containing gas forced into the reaction chamber to prevent clogging of the concentric nozzle openings by the titanium dioxide.[86] A spinning or spiral motion is given to the gas components to assure complete mixing. The concentric injection tubes for the gas components are provided with spiral separating strips in such a manner that a clockwise tangential motion is given to the gases from the central and the outer tubes, while the gas from the middle tube has a counter-clockwise motion. Pigmentary titanium dioxide is obtained by the vapor phase oxidation of the tetrachloride in a porous, tubular type reaction furnace.[87] To prevent contact of the reactants with the internal surface of the reactor, an inert gas is diffused through the porous walls from an external source maintained under pressure. In the vapor phase oxidation of titanium tetrachloride, incrustation of the burner nozzle is prevented by using hydrogen instead of air as the scavenging gas.[88]

## COOLING AND SEPARATING TITANIUM DIOXIDE FROM CHLORINE AND OTHER PRODUCT GASES

The products formed in the vapor phase oxidation of titanium tetrachloride are cooled quickly from 1000° C to 600° C or lower by adding cold gases from a previous reaction, and the titanium dioxide is recovered from the cooled mixture.[89] Final cooling is effected against the aluminum walls of the apparatus which have a high thermal conductivity. The metallic walls may be cooled with water. After cooling the chlorine has little action on the aluminum of the container, primarily a result of the aluminum oxide coating which forms on the surface. Although dry dust-free air, inert gases, or liquefied inert gases may be employed, cooled gas from a previous reaction is more suitable.

In an example, 20 kg per hour of titanium tetrachloride are decomposed with 9 kg per hour of air, 3 kg per hour of carbon dioxide, and 1.7 kg per hour of oxygen of 98 per cent purity in a reaction furnace. The reaction products with titanium dioxide suspended leave the furnace at the bottom at 1000° C. The gases are composed of 41 per cent chlorine, 20.5 per cent carbon dioxide, 29.5 per cent nitrogen, and 9 per cent oxygen, and carry in suspension 750 g per cubic meter (S.T.P.) of titanium dioxide. Immediately after leaving the reaction furnace double the quantity of the reaction gas formed in the process of oxidation but free from titanium dioxide and having a temperature of 20° C is admitted at the exit of the furnace with the hot reaction gases through a nozzle. This decreases the temperature of the mixture to 400° C and decreases the concentration of the titanium dioxide to 250 g per cubic meter. The gas composition is cooled to 150° C in water cooled aluminum tubes. The titanium dioxide is recovered from the gas by a water cooled cyclone and a water cooled electric filter made of aluminum. After removal of the titanium dioxide, the gas current is divided in two parts; two-thirds is cooled to 20° C and reused, while the other one-third is used to chlorinate more ore.

The hot product gases containing titanium dioxide are fed with regulated amounts of cold particles of silica sand or agglomerated titanium dioxide particles into the water jacketed cooling conduit to prevent accumulation of titanium dioxide on the inside of the cooling conduits.[90]

In this process from 5 to 15 per cent of coarse particles are added. The suspension of produced and added titanium dioxide particles is maintained at a high enough velocity through the heat exchangers that the particles remain in suspension. The coarse particles abrasively scour the internal walls of the conduit and prevent deposition of titanium dioxide particles. First the larger abrading particles are removed in the settling chamber, cyclone, or other means and recycled. Thereafter the finer pigment particles are separated in cyclones, precipitators, or filters.

For example, titanium tetrachloride and aluminum chloride were mixed in proportion to yield 99 parts of titanium dioxide to 1 part of alumina and heated to 895° C at a rate equivalent to 2500 pounds per hour of titanium dioxide. Air containing 1 per cent water vapor by volume was heated to 1060° C and quickly mixed with the chloride vapors under 9 psi gauge in an oxidation reactor provided with a scraper for removing oxide scale deposit from the inner walls. The reaction products were continuously fed into a 1500 foot length of water cooled aluminum pipe, the first one-third of which was 12 inches in diameter and the remaining portion was 10 inches in diameter, and then into the collection apparatus. Simultaneously 1100 pounds per hour of cool, recycled gas from a previous reaction was introduced into the system before passing through the water cooled coil. At 2.5 minute intervals 10 pounds of titanium dioxide pigment in the form of aggregates below 0.25 inch in diameter were added to the reaction product at the entrance to the cooling coil.

The hot titanium dioxide suspended in gases produced in the vapor phase oxidation of the tetrachloride is cooled by passing the suspension through a conduit surrounded by water.[91] A pressure is maintained on the water less than the pressure of the gases so that in case of leaks the gas escapes into the water. The water in each compartment is continuously checked for pH to locate any leaks in the system. Water jackets of the first 14 legs are constructed to give a 5 foot head of water; the last five legs are equipped with 3 foot jackets because of the lower gas pressure toward the discharge end. Water exerts a pressure of 0.45 psi per foot of head. The cooling water is supplied at 35° C. The reaction product consisting of 3500 pounds of titanium dioxide, 6230 pounds of chlorine, 1050 pounds of nitrogen, 317 pounds of oxygen, and 170 pounds of hydrochloric acid per hour were at a temperature of 1300° C. Cooled recycled product gases were continuously fed at 3500 pounds per hour to cool the products to 1000 to 1200° C before entering the first leg of the heat exchanger. In operation 3900 pounds per hour of titanium dioxide were separated from the product gases and sent to the calciner.

A hot suspension of titanium dioxide in the gaseous products at 900 to 1350° C obtained from the vapor phase oxidation of the tetrachloride is cooled quickly to 100° C by direct contact with liquid chlorine or with liquid titanium tetrachloride within 10 seconds to insure titanium dioxide particles of 0.1 to 0.25 micron.[92] In a specific operation, a gaseous suspension at 1000° C comprising the reaction products from the vapor phase oxidation of titanium tetrachloride with air was continuously fed at a rate of 82 cubic feet per minute directly into a conventional spray cooler for passage upward. By volume the suspension was composed of 30 per cent chlorine, 68.5 per cent nitrogen, 1.4 per cent oxygen, 0.09 per cent hydrochloric acid, and 0.29 pound of solid titanium dioxide suspension per pound of gases. The spray cooler was constructed of refractory bricks and the spraying equipment was made of nickel. Concurrently liquid titanium tetrachloride at a feed rate of 2.6 tons a day or 215 pounds per hour was continuously discharged from a spray nozzle into the upper part of the cooler for passage downward for direct contact with the hot gaseous suspension. As a result, the suspension was quenched to 300° C within 10 seconds. After cooling, the suspension was passed through a conventional cyclone separator and thence into a Cottrell precipitator to separate the pigment particles. To condense out and recover the titanium tetrachloride for reuse, the gases were passed through a condenser which cooled them to 25° C. In another example, liquid chlorine was used instead of the titanium tetrachloride.

The separation of titanium dioxide, obtained by the decomposition of the tetrachloride with an oxygen containing gas, from the gaseous products is facilitated by diluting the reaction mixture with a cold inert gas to decrease the temperature rapidly to 400° C.[93] This is followed by further cooling in a water cooled aluminum tube to 150° C. The titanium dioxide is then separated in a cyclone separator and an aluminum electrofilter, both of which are water cooled.

Walmsley[94] cooled the reaction product by passing the suspension through a fluidized bed of coarse titanium dioxide granules which are cooled by contact with tubes around which a cold liquid circulates. The coarse granules in motion prevent the fine suspended particles from adhering to the walls of the bed. Granules of the cooling bed have a diameter at least 100 times that of the suspended titanium dioxide particles, and the gas velocity is at least 1.5 times the critical velocity for fluidizing. Cooling is better effected in stages in several beds. In an example in which 75 pounds of titanium tetrachloride vapor was oxidized per hour at 1040° C in a fluidized bed of 14 to 36 mesh titanium dioxide, 40 per cent of the pigment passed out in the effluent gases which contained per hour 56 pounds of chlorine and 12.8 pounds of titanium dioxide of 0.86 micron average grain size. This gaseous suspension at a velocity of 1.7 feet per second passed upward through a water cooled 8 inch diameter aluminum tube containing a bed of 39 pounds of 44 to 60 mesh titanium dioxide which was fluidized to a 10.5 inch height. The bed weight increased to 39.2 pounds in 2 hours by deposition of fine titanium dioxide, but the remainder of the fines passed out with the gases at 226° C.

A hot suspension of titanium dioxide in the product gases is cooled to 300 to 400° C by blowing into it a cold suspension from a previous reaction having the same particle size.[95] Similarly the hot gaseous suspension is evacuated continuously from the combustion chamber through a cooled tube which opens into the central part of the chamber.[96] After removal of the titanium dioxide the gas is recycled into the reactor by another cooled tube opposite the first. Booge[97] cooled the pigment suspension in a cyclic process outside the reaction vessel. The hot suspension is brought in contact with cool chlorine-containing gas from a previous reaction from which the solid titanium dioxide has been removed. After cooling, the titanium dioxide is separated from the mixture and a part of the gas is recycled for cooling further quantities of the hot reaction product.

Titanium dioxide obtained by the vapor phase oxidation of the tetrachloride is recovered from the hot gases by cooling first by dilution with a chilled gas and then passing the suspension over cooled aluminum surfaces coated with aluminum oxide.[98] From the cooled suspension the titanium dioxide is removed by cyclones or by electric precipitators. The dust-free gas is cooled and recycled to the process as a cooling agent. In an example, titanium tetrachloride is oxidized at 1000° C with an oxygen-nitrogen-carbon monoxide mixture to give a suspension of titanium dioxide in a gas containing chlorine, carbon dioxide, nitrogen, and oxygen. The mixture is cooled to 400° C by adding recycle gas at 20° C and further cooled in water cooled aluminum tubes. Solid titanium dioxide is recovered first in a water-cooled cyclone and then in a water-cooled electric precipitator. Two-thirds of the exit gas is cooled and recycled; the remainder is used to chlorinate more titanium ore.

Rapid cooling below 600° C to reduce corrosion is effected before separating the titanium dioxide pigment.[99] To eliminate clogging by the particles the hot gases are evacuated from the chamber center by cooled aluminum-magnesium alloy tubes without deforming the flame pattern. By another process the titanium dioxide is separated from the vehicular gas in a vertical cylindrical chamber having a high velocity, longitudinal paddle stirrer.[100] Separation is also effected in a fluidized bed of titanium dioxide of the same origin.[101]

In the process of cooling a suspension of titanium dioxide particles in the gaseous products of reaction, primarily chlorine and nitrogen, the inner surfaces of the heat exchangers are kept free of titanium dioxide particles by introducing particulate solid carbon dioxide which sublimes in the cooling zone.[102] The reaction product at 1150° C containing finely divided pigmentary titanium dioxide, chlorine, nitrogen, and a small amount of hydrogen chloride is continuously discharged into an elongated, water cooled, tubular, corrosion-resistant metal (aluminum, nickel, or stainless steel) heat exchanger consisting of several vertical sections connected in series by 180-degree bends. The product is cooled to below 600° C in not more than 5 to 10 seconds and discharged to a cyclone, filter, or Cottrell precipitator. At the point of introduction of the reaction products to the chamber a stream of solid carbon dioxide particles 0.3 to 0.4 inch in diameter in an amount of 1 to 10 per cent by weight of the titanium dioxide is introduced. These solid particles of carbon dioxide completely evaporate during passage with the suspension through the cooler within 20 seconds. As a result of the concurrent abrading and cooling functions of the solid carbon dioxide, the heat transfer surfaces of the cooler are kept free of compacted titanium dioxide particles.

## REMOVAL OF CHLORINE AND HYDROCHLORIC ACID FROM TITANIUM DIOXIDE

According to Kitamura [103] chlorine is completely removed from titanium dioxide produced by the vapor phase oxidation of the tetrachloride by treating the product with steam at 300° C. To remove adsorbed chlorine, Werner [104] mixed 100 parts of titanium dioxide prepared by the vapor phase oxidation of the tetrachloride with 1.8 parts by weight of boric acid and heated it for 1 hour at 600° C. The pigment obtained was very fine grained and soft. It had a pH of 7.6, an oil absorption of 21.9 and high resistance to discoloration in baked enamels. Similar results were obtained by passing steam containing 0.1 per cent boric acid through the stirred, dry pigment powder at 150° C.[105] The residual acid content of titanium dioxide prepared by the vapor phase oxidation of titanium tetrachloride is neutralized by treating the reaction products with boron chloride at 300 to 900° C.[106] After treatment the solid titanium dioxide is separated from the gas stream and heated at 100 to 200° C in air or in air-containing steam. The boron trichloride calculated as boric oxide is 0.87 per cent of the titanium dioxide.

To avoid the residual acidity of titanium dioxide pigments, from 0.1 to 2.0 per cent by weight of the tetrachloride of molten aluminum is sprayed into the furnace and oxidized to aluminum oxide at the same time.[107] From 0.1 to 5 per cent by volume of water vapor is added to the air admitted. The air enriched with water vapor is preheated to 1350° C, and the oxidation reaction takes place at 900 to 1300° C. A slot jet type of mixing apparatus is employed. For best results the retention time of the reactants in the furnace is from 0.05 to 1 second. The product gases with the suspended solids are cooled to below 600° C by introducing chilled gas from a previous run or by introducing chilled sand. A cyclone, a porous filter medium or an electrostatic separator is used to remove the titanium dioxide particles.

For example, titanium tetrachloride vapor at 300° C was continuously admitted at a rate equivalent to 100 parts by weight per hour to the upper portion of a vertical reaction chamber constructed of nickel and cooled by means of a circulating molten salt mixture. Simultaneously air containing 1 per cent moisture by volume, preheated to 1050° C, was admitted continuously to the furnace at a rate equivalent to 19 parts by volume of oxygen per hour through a separate inlet. At the same time molten aluminum at 775° C was sprayed into the converging streams of air and titanium tetrachloride in the reaction zone through a ceramic lined nozzle by means of an annular stream of air. Aluminum was introduced at a rate of 0.95 parts by weight per hour. The three reactant streams were rapidly mixed in the upper part of the reactor where the heat of reaction maintained a temperature of 1260° C in the reaction zone. The flow rates provided a retention time of reactants and products within the reaction chamber of 0.15 second. The gaseous suspension of titanium dioxide and alumina was discharged at the lower end of the furnace at 1180° C and cooled quickly by quenching at the reactor outlet by mixing with it cooled product gases in a ratio of 30 parts by weight per hour. After initial cooling the gaseous suspension was passed through a water-jacketed cooling conduit. The pigment of titanium dioxide and alumina was separated from the gaseous products by filtration, calcined to 650° C for 2 hours, and dry ground in a fluid energy mill.

Since the titanium tetrachloride was preheated only to 400° C corrosion problems did not exist. Conversion of titanium tetrachloride was 100 per cent. The finished pigment containing 95.7 per cent titanium dioxide and 4.3 per cent alumina had an excellent resistance to yellowing in enamel finishes.

Titanium dioxide obtained by the vapor phase oxidation of the tetrachloride is neutralized by mixing in the dry state with an oxide or hydroxide of an alkaline earth metal followed by moistening with a limited amount of water and finally drying.[108] Titanium dioxide recovered from the product gases contains 0.1 to 0.3 per cent by weight of hydrochloric acid along with chlorine. The chlorine may be removed by aeration but the hydrochloric acid adheres firmly to the many small particles. Although this closely held hydrochloric acid may be removed by calcination at 600° C, an agglomeration of the particles results, so that the product must be milled. About 0.1 per cent of the alkaline earth compound is mixed with the pigment dry at 100° C to avoid lump formation. From 1 to 3 per cent of water is then added as a spray or as steam to the agitated mixture. After neutralization the pigment is dried without further treatment.

In an example, titanium tetrachloride is decomposed with an excess of oxygen in a flame of carbon monoxide and oxygen, and the titanium dioxide formed is separated by an electrostatic filter at a temperature of 200° C. The titanium dioxide is passed into a rotating mixing drum fitted with a jacket for heating and for cooling. A vacuum is applied to free the material from chlorine held in the voids. To the resulting product, which contains 0.1 per cent hydrochloric acid, 0.15 per cent screened calcium hydroxide is added and mixed in the rotating drum for 30 minutes. During the mixing operation the temperature drops from 100 to 80° C. By blowing in steam or a spray of water with the drum rotating the mixture is adjusted to a moisture content of 1 per cent. The drum is rotated for 30 minutes more, after which it is evacuated and filled with carbon dioxide. After 30 minutes the drum is heated to 100° C and a suction is applied to remove the moisture and unreacted carbon dioxide.

# 22

# Special Processes of Pigment Manufacture

**FLUORIDE PROCESS**

Fluoride compounds of titanium, unlike the sulfates, chlorides, and nitrates, do not hydrolyze in solution on dilution or on boiling. Consequently the titanium is precipitated by adding ammonium hydroxide or other base to the carefully purified solution. Complete removal of the heavy metal compounds, particularly those of iron, vanadium, and chromium is required since these are precipitated along with the titanium to injure the color and other properties of the pigment. Certainly on a laboratory scale fluoride solutions yield pigments possessing excellent properties but a commercial plant based on this process proved unsuccessful.

For removing silica from oxide compounds of titanium and vanadium and from clays the crude materials were treated with nonaqueous ammonium fluoride in a proportion less than that corresponding to the fluosilicate to produce a volatile compound ($SiF_4 \cdot 2NH_3$) which was distilled off.[1] By a related process, materials containing alumina, silica, titanium oxide, and vanadium oxide were transformed into complex mixtures of fluosilicates and fluorides by the action of fluorine or ammonium bifluoride.[2] The fluosilicates of ammonium, vanadium, and titanium were driven off by heating, leaving a residue of aluminum and potassium fluorides which were converted into chlorides by treatment with gaseous hydrochloric acid and ammonia. Titaniferous material containing siliceous impurities was treated with calcium fluoride and sulfuric acid at temperatures below 140° C to form fluorides of all base constituents except the titanium.[3] These compounds were then removed by washing with acidified water. Silica and titania were removed from bauxites or ores of chromium and manganese by heating with hydrofluoric acid or a mixture of fluorite and sulfuric acid to form the corresponding fluorides, followed by distillation,[4] leaving the other constituents in the residue.

Doremus[5] treated ilmenite with 20 per cent hydrofluoric acid and heated the mix with live steam to initiate and maintain the reaction. The product was dissolved in water and after separation of the insoluble residue an oxidizing agent such as hydrogen peroxide was added to the solution to convert all the iron to the ferric state, and potassium hydroxide was then added to precipitate potassium fluotitanate, fluozirconate, and fluosilicate. Of these, only the fluotitanate was soluble, and it was separated from the other constituents by leaching the mass with hot water. Titanium hydroxide, precipitated from the solution by adding ammonium hydroxide, sodium carbonate, or potassium hydroxide, was filtered, washed, and calcined. Hydrofluoric acid was recovered from the residue. The ferric fluoride solution from the first operation was treated with calcium chloride to obtain insoluble calcium fluoride and ferric chloride, or with potassium hydroxide to form ferric oxide and potassium fluoride. In the latter procedure the potassium fluoride solution was treated with lime to form calcium fluoride and potassium hydroxide.

According to a method developed by Svendsen,[6] the titanium component was separated from ores by volatilization as fluorides, dissolved in water, and precipitated as hydrous oxide of pigment grade by adding an aqueous alkali. On heating titanic oxide compounds with hydrofluoric acid or ammonium fluoride, a titanium-fluorine compound was formed, and this product at higher temperatures evolved titanium tetrafluoride or diaminotetrafluoride, depending upon the agent used in the treatment. However, a nonvolatile titanium oxyfluoride compound formed to some extent. If sodium-titanium sulfate was heated with a metallic fluoride, double decomposition took place, yielding titanium tetrafluoride and metallic sulfate. The normal titanium sulfate is unstable if it exists at all, and for this reason the stable double sulfate with an alkali metal or ammonium was used. Employing the ammonium double sulfate, a sufficient amount of the fluoride compound was added to prevent the formation of ammonium bisulfate, which would lead to the formation of oxyfluoride compounds. Metallic fluorides gave best results, since hydrofluoric acid and ammonium fluoride caused the formation of sulfuric acid or ammonium sulfate. The double sulfates were produced by heating titanium-oxygen compounds with a mixture of ammonium sulfate and alkali metal sulfate, either normal or acid. Reaction took place above 280° C, but to prevent reduction by ammonia the temperature was held below 350° C, although toward the end of the operation it was raised to 400° C. The ammonium double sulfate was similarly produced by employing ammonium sulfate alone. Since ferrous iron produced troublesome ammonium sulfite, the ore (ilmenite) was roasted before the sulfate treatment.

Vaporization of titanium tetrafluoride took place about the same temperature at which the double sulfate was formed, but a higher temperature could be used, provided that it was held below the dissociation point of ferric sulfate. The tetrafluoride vapors evolved condensed, and at room temperature solidified to a colorless mass. Unlike silicon tetrafluoride it was soluble in water without hydrolysis, but the hydroxide was thrown down by aqueous ammonia. The precipitate obtained from dilute solutions in this manner had colloidal properties, although dense, readily filtered products were obtained by increasing the density of the original liquors either by increasing the concentration of titanium tetrafluoride or by adding a neutral salt other than titanium, such as ammonium fluoride. Similar results were obtained by dissolving ammonium fluoride in the ammoniacal liquor used in the precipitation.

The titanium tetrafluoride vapors were contacted directly with the ammonium hydroxide solution to precipitate titania, or they were first liquefied or solidified and then treated. By mixing the ammonia gas from the sulfate reaction with the titanium tetrafluoride vapors, titanium diaminotetrafluoride was formed.

As an example, calcined and pulverized titanium dioxide was mixed with 2 moles of ammonium sulfate and 1 mole of sodium sulfate for each mole of titanium dioxide, and heated in a rotating or muffle type of furnace at 300° to 350° C to form sodium titanium sulfate. The escaping ammonia and water vapors were collected and used later in the operation. After reaction had ceased, sodium fluoride was added in the proportion of 4 molecular parts to 1 part of titanic oxide; the temperature was raised to 350° to 400° C and held within this range until the evolution of fluoride vapors ceased. During the process the temperature did not exceed 400° C.

The furnace residue, consisting primarily of sodium sulfate, was used to regenerate the sodium fluoride and the sodium and ammonium sulfates used in these reactions. The titanium tetrafluoride vapors, together with the ammoniacal vapors first evolved, were contacted with an aqueous ammoniacal solution of ammonium fluoride in a scrubber tower constructed of acid-proof stone or non-corrosive alloy (18 per cent chromium and 8 per cent nickel stainless steel). A temperature below 34° C was maintained to prevent dissociation of normal ammonium fluoride into the bifluoride and ammonia, and by carrying out the reaction at even lower temperature denser precipitates were obtained. A considerable excess of circulating liquor was used. The product obtained from the scrubber, consisting of a suspension of hydrous titanium oxide in ammonium fluoride solution, was filtered, and the cake was washed, dried, and calcined to produce pigment.

A part of the filtrate, containing approximately the same amount of fluorine as that used in the reaction to vaporize the titanium, was used to regenerate sodium fluoride and ammonium sulfate, and the remainder was recirculated through the scrubber. The ammonium fluoride solution withdrawn was mixed with the sodium sulfate of the furnace residue to precipitate sodium fluoride, while sodium and ammonium sulfates remained in solution and were recovered by evaporation. After once establishing the cyclic process it was necessary to supply new reagents only to replace small mechanical losses.

If silica was present in the ore in a form that could be attacked by the fluoride, it was volatilized along with the titanium, but the water liberated by the silicon fluoride caused the formation of oxyfluoride compounds of titanium. For this reason the silicon was removed before distillation of the titanium tetrafluoride, and this was accomplished during the sulfate formation by addition of calcium sulfate. Important reactions involved in the process are the following:

$$TiO_2 + Na_2SO_4 + 2(NH_4)_2SO_4 \rightarrow Na_2Ti(SO_4)_3 + 4NH_3 + 2H_2O$$
$$Na_2Ti(SO_4)_3 + 4NaF \rightarrow TiF_4 + 3Na_2SO_4$$
$$TiF_4 + 4NH_4OH \rightarrow Ti(OH)_4 + 4NH_4F$$
$$4NH_4F + 3Na_2SO_4 \rightarrow 4NaF + 2(NH_4)_2SO_4 + Na_2SO_4$$

Similarly, the titanium content of ores may be separated from other metallic constituents by distillation, after conversion into a complex volatile fluoride.[7] Silica, if present, is also converted to a volatile fluoride, but this compound may be readily separated since it is a gas at ordinary temperatures, while titanium tetrafluoride boils at about 290° C. Furthermore, the vaporization temperatures of the diaminotetrafluorides of silicon and of titanium are about 100° C apart. Also, silicon tetrafluoride and its amino compound are decomposed by water, and the silica is completely precipitated by the addition of a stoichiometrically equivalent amount of ammonia, while the corresponding titanium compounds dissolve in water without hydrolysis and the titanium is precipitated only by addition of ammonia.

In operation, the ore is subjected to the action of a reactive fluoride compound (ammonium fluoride) under conditions so regulated as to produce a volatile titanium-fluorine complex. Such a compound, formed by gradually adding an excess of ammonium fluoride to the titanium oxide compound at a temperature between 110° and 225° C, contains an appreciably higher proportion of fluorine than that corresponding to the tetrafluoride, and is probably a complex titanium tetrafluoride-amino-ammonium fluoride. To insure complete conversion of the titanium component into the volatile form, ammonium fluoride should be added in considerable excess, 50 per cent or more above the chemically equivalent proportion. The complex compound thus formed below 250° C begins to volatilize at 290° C and dissociates into ammonia, ammonium fluoride, and titanium diaminotetrafluoride. Instead of adding ammonium fluoride as such to the charge, a reactive mixture of ammonium sulfate and fluorspar capable of producing it may be employed. Either the normal or acid sulfate may be used. Ammonium fluoride has the advantage that it can be regenerated in the process and that it yields a primary complex titanium tetrafluoride compound which is not attacked by water vapor generated in the reaction.

The primary reaction is carried out at temperatures above 110° C so that any water formed is removed from the mass, and to aid in maintaining a dry state the reactive fluoride may be added gradually. This avoids any possible accumulation or sintering of the charge. The temperature is maintained below 250° C during the formation of the primary complex titanium tetrafluoride compound to avoid losses due to dissociation. Any silicon diaminotetrafluoride begins to volatilize at about 230° C, and practically all distills over below 290° C. The reaction product is then heated to 290° to 350° C, at which temperature the primary complex titanium tetrafluoride compound vaporizes and dissociates to form the titanium diaminotetrafluoride, along with other products, which may be condensed and collected. This compound in the vapor phase is contacted with water and ammonia, in excess, to precipitate hydrous titanium oxide. A denser and more readily filtrable form of titania may be obtained by employing an ammoniacal solution of concentrated ammonium fluoride instead of aqueous ammonia, and by carrying out the reaction at reduced temperatures, the density of the precipitate may be further increased.

The solid residue from the primary reaction, consisting primarily of iron fluoride, is heated with ammonium sulfate at 200° to 250° C to form ammonium fluoride, which passes off as a vapor, leaving ferrous sulfate. This recovered ammonium fluoride may be dissolved in the solution used in decomposing the titanium diaminotetrafluoride. Losses of ammonium fluoride in the process may be replaced by adding ammonium sulfate and fluorspar to the reaction residue just before its decomposition by ammonium sulfate. Either the normal ammonium fluoride or the bifluoride may be employed with equal success.

As an example of the process, ilmenite ground to pass a 150-mesh screen was heated at 150° to 200° C in an externally fired enclosed furnace equipped with stirring devices, a feeding arrangement, and outlets for discharge of the vapors and residue, and 2 to 2.5 parts by weight of ammonium fluoride crystals were gradually added over a period of several hours. The ammonia and water vapor liberated were recovered, and after the reaction had gone to completion, evidenced by a change in color of the powdery mass from black to grayish white, the temperature was raised to 350° C to vaporize and distill off the titanium-fluorine compounds. Ammonium sulfate equal to the weight of the original ore was then added to the residue, and the mixture was heated at 300° to 350° C to recover the fluoride content as ammonium fluoride, which was distilled off, leaving ferrous sulfate and a small proportion of other metallic sulfates. During this operation a mixture of fluorspar and ammonium sulfate was added to replace mechanical losses of ammonium fluoride in the previous cycle.

The combined furnace vapors, containing titanium tetrafluoride, ammonium fluoride, and ammonia, were led into a spray of ammoniacal ammonium fluoride in a cooled scrubbing tower to effect precipitation of the titanium. The reaction was carried out below 34° C to obtain better yields, and a denser, more readily filtrable product was obtained at temperatures below 10° C. In general, the density of the precipitate was proportional to the fluoride content of the solution and inversely proportional to the temperature of the scrubber liquor. The titania was separated from the ammoniacal liquor by filtration, washed, dried, and calcined to produce pigment. A part of the filtrate was evaporated to recover crystalline ammonium fluoride, and the ammonia given off was used for regeneration of ammonium sulfate. Wash water from the filtration step was added to the remainder of the filtrate and the liquor was recirculated.

According to an improved process devised by the same workers, the ore was heated with ammonium fluoride to produce tetrafluoride compounds of titanium and ammonium which were dissolved in water or nonalkaline ammonium fluoride solution.[8] To secure a high yield of these primary compounds, the ammonium fluoride was added to the titaniferous material in excess of the quantity theoretically required for conversion into simple fluorides. Thorough mixing of the ore and reagent was found necessary to insure complete reaction, and this was accomplished by grinding the components in a ball mill or edge-runner mill, or by adding the ammonium fluoride in the form of a concentrated aqueous solution. The water was then driven off and the mass was heated at 150° C until the reaction proceeded to completion and ammonia ceased to come off. Rutile required a higher temperature than ilmenite and either normal ammonium fluoride or the bifluoride gave satisfactory results. At this stage the silicon content was volatilized as a diaminotetrafluoride by heating the residue below 290° C, for at this temperature titanium compounds began to distill over. Because of the wide difference in the boiling points of the corresponding fluoride compounds (titanium and silicon) complete separation was effected. Although the reaction mass was kept below 290° C for a considerable period of time to volatilize as much ammonium fluoride as possible for separate recovery, in general the temperature after reaction at 150° C was raised as rapidly as possible to 300° C to cut the operating time and increase the output of the unit.

The titanium was then largely or completely separated from the reaction mass by leaching with water or a nonalkaline solution of ammonium fluoride. In some instances there was excess ammonium fluoride in the reaction product, sufficient to provide part of or all the requirements of the leaching solution, and water or dilute liquor was used. After filtration, to separate the suspended solids, the solution was neutralized with ammonia to the point of incipient precipitation, and treated with a soluble sulfide (ammonium, sodium, potassium, or hydrogen sulfide) at 70° C to precipitate the iron and other heavy metals present as impurities. If hydrogen sulfide was employed, additional alkali was required to maintain the necessary hydrogen ion concentration. During the precipitation, ammonium sulfide reduced any ferric iron to the ferrous state, and care was taken to prevent reoxidation of this component, since the ferric compound redissolved in the fluoride liquor. This purified solution was mixed with aqueous ammonia, and the resulting liquor, containing the hydrous titanic oxide in suspension, was cooled to below 34° C to effect complete precipitation. Denser and more readily filtered titania was obtained by further reducing the temperature to below 10° C and increasing the concentration of fluorides in the liquor from which it was precipitated. The hydrous oxide was filtered, washed, dried, and calcined, and the filtrate containing ammonium fluoride and ammonium hydroxide was reused in the process.

After leaching the titanium from the primary reaction product, 0.8 parts ammonium sulfate was added to the residue for each part of the original ore, and the mixture was heated to 300° to 350° C to drive off the ammonium fluoride, which was collected for reuse.

In an actual operation, 100 kg of finely ground ilmenite was mixed with 400 kg of ammonium fluoride in the form of a strong aqueous solution and introduced into a closed reaction vessel fitted with a stirrer. The mass was heated to 110° to 180° C and held at the latter temperature, with agitation, until the reaction had gone to completion. All ammonia, ammonium fluoride, and water vapor liberated were drawn off and collected for recycling. Hot water was then added to the reaction product, and after digestion the insoluble ferrous fluoride was filtered hot from the leach liquor. Ammonia was added to the hot filtrate, containing the titanium tetrafluoride compound just short of the amount required to effect precipitation, and then ammonium sulfide was added to convert the iron and other heavy metals present to insoluble sulfides. These were filtered off and the iron-free solution was diluted with water and introduced into aqueous ammonia to precipitate the titanium as hydrous oxide. If on the other hand, the ammonium hydroxide was poured into the solution, some of the crystalline complex titanium ammonium fluoride compound formed. The hydrous oxide was filtered, washed, and calcined at 900° C. Ammonia and ammonium fluoride were recovered from the filtrate in a later cycle.

The iron double-fluoride residue was heated in the presence of steam in a closed furnace at 400° to 550° C, and the volatilized ammonium fluoride was condensed, leaving residual iron oxide suitable for pigment or abrasive uses. This ordinarily gave a black product, but if a red oxide was desired, the double fluoride was first decomposed at 400° to 450° C, and after volatilization of the ammonium fluoride the residue was heated with moist air at 500° to 550° C. Ammonia and ammonium fluoride produced in the various reactions were collected and used again in the process, so that in continuous operation it was only necessary to make up for mechanical losses of these materials from time to time. In certain reactions ammonium sulfate was employed, and the process with reference to this reagent was also cyclic.

As already pointed out, the titanium solution obtained by leaching the reaction product of ilmenite with ammonium fluoride usually contains a small proportion of salts of iron and other heavy metals, and unless removed, these impurities would be precipitated with the titanium and have a detrimental effect upon the color and other properties of the resulting pigment. However, by regulating the hydrogen ion concentration of the fluoride solution within narrow limits, the iron may be precipitated quantitatively by adding a soluble sulfide, with no significant loss of titanium.[9] The tendency of the added sulfide to form titanium oxide or other insoluble compound may be depressed further by the presence of an additional soluble fluoride salt such as ammonium fluoride.

To illustrate, finely ground ilmenite was heated with an excess of ammonium fluoride as a concentrated solution, 50 per cent or more above the stoichiometric proportion, until the water had been driven off, and finally at 150° to 180° C to complete the reaction. This large excess of reagent served the double purpose of insuring complete conversion of the ilmenite into fluorides and of supplying the desired ammonium fluoride to the leach liquor. After the reaction had ceased, the product was lixiviated with water, and the solution of titanium and ammonium fluorides, containing also a small amount of iron fluoride as impurity, was filtered to remove the practically insoluble double fluoride of iron and ammonium. Solutions prepared in this manner were acidic, and before treatment were neutralized with ammonia to the point just short of a permanent precipitate of titanium oxide. This condition corresponded to a pH of approximately 6.8, and it was closely maintained until the insoluble sulfides were removed from the solution. Ammonium sulfide was then added to precipitate the iron and heavy metal compounds. Sodium, potassium, or hydrogen sulfide could be used, however, but if the latter agent was employed additional ammonia was required to maintain the necessary neutrality, otherwise the iron sulfide redissolved. Best results were obtained at 50° to 60° C, and also at this temperature the insoluble sulfides formed were readily removed by filtration. The iron was precipitated as ferrous sulfide, since any ferric ions would be reduced to the ferrous state by the ammonium sulfide, but precautions were taken to prevent reoxidation before or during filtration, for ferric sulfide was soluble in the fluoride solution. In general the ratio of ammonium fluoride to titanium fluorides in the solution was greater than 1 to 1 but less than 5 to 1, and the specific gravity was 1.110 to 1.125 at 60° C.

Solid particles suspended in an aqueous solution of a titanium fluoride salt, such as is obtained by the reaction of ilmenite with ammonium fluoride, were coagulated so as to facilitate settling and filtration, by agitating the liquor with a small amount of a flocculent titanium compound.[10] Such an agent was prepared by adding sodium phosphate, sodium carbonate, or potassium hydroxide to the fluoride solution, under such conditions that a stable, absorbent, flocculent precipitate of hydrous titanium phosphate or oxide formed. The precipitate could be formed in place.

Ammonium fluotitanate solution, as obtained in digesting ilmenite with ammonium fluoride, was added to an excess aqueous ammonia to precipitate the titanium as hydrous oxide,[11] or the two solutions were fed simultaneously into the mixing tank.[12]

In a two-stage process,[13] finely pulverized ilmenite was heated with a solution of ammonium fluoride of more than 20 per cent strength under pressure of 2 to 20 atmospheres at the corresponding temperature. More ammonium fluoride was added to replace losses and the reaction product in the form of a slurry was filtered. Titanium hydroxide was precipitated from the solution, filtered, calcined, and milled in the usual manner to yield pigment. The filter cake was heated with 20 per cent ammonium fluoride under pressure, as before. More ammonium fluoride was added to replace losses and the resulting slurry was filtered. The filtrate was used for the initial treatment of a fresh batch of ore.

Titanium dioxide, hydrolytically precipitated from an aqueous solution of titanium fluoride to which had been added a small proportion of sodium citrate, arsenate, or carbonate, gave on calcination a pigment having rutile crystal structure.[14] Improved pigment properties [15] were developed by adding a small proportion of zinc, beryllium, or magnesium sulfate to the filtered and washed pulp before calcination.

According to Svendsen,[16] a pigment consisting of an intimate mixture of titanium dioxide and silica, prepared by coprecipitation from fluoride solutions, had the same characteristics as the pure titanium dioxide product. In decomposing ilmenite or other ores with an excess of acid ammonium fluoride in both the wet and dry procedures, the silica and silicates usually present as impurities were also converted to soluble form. If the reaction was carried out in the dry state, the ammonium tetrafluorides of titanium and silicon formed were volatilized by heating the reaction mass at temperatures above 270° C, and collected in water to form pure solutions which were treated with ammonia to precipitate the mixed hydroxides. However, if the original reaction product was kept below the temperature of volatilization and the product was leached with water, the titanium and part of the silicon dissolved as fluoride compounds, along with a small amount of iron and other impurities. Heavy metals were precipitated with ammonium sulfide after the liquor had been made almost neutral by adding ammonium hydroxide and the purified solution was added to aqueous ammonia to coprecipitate the hydroxides of titanium and silicon. This procedure was followed to insure the presence of ammonium hydroxide throughout the precipitation step. The mixed hydroxides were filtered, washed, calcined, and pulverized for pigment use.

Individual particles of coprecipitated titania and silica usually had an average diameter of less than 1 micron and ranged down to below 0.25 micron. With silica up to 10 per cent, each separate calcined particle appeared under the microscope as crystallized titanium dioxide, and separate silica particles could not be detected under the most powerful petrographic microscope. The index of refraction of the composition was practically that of pure titanium dioxide, although the specific gravity was substantially lower.

Calcination temperature required to effect crystallization of the titania and develop the required pigmentary properties was found to increase with the proportion of silica, but varied from 850° to 1100° C. For example, with 10 per cent silica the product was heated at 1025° C for 1 hour to effect anatase crystallization, but if the temperature was raised well above this value, for instance, 1050° C, the product became hard and yellow. With 3 per cent silica, calcination at about 985° C was sufficient, although with 20 per cent of this component 1045° C was required. Pure titanium oxide became hard and assumed a gray-brown color even at these temperatures, which seemed to be associated with the rutile crystal modification. The silica in intimate association apparently raised the temperature at which crystallization of the titania component took place. Pigments having from 5 to 15 per cent silica possessed these improved qualities to the highest degree, although this component was in some tests raised to 75 per cent.

A similar pigment containing a smaller proportion of silica was prepared by calcining the composite precipitate obtained by treating an aqueous solution of an alkali fluosilicate of titanium and silicon with ammonium hydroxide.[17] By another approach, pigments of similar properties were made by blending at least 5 per cent silica with titanium dioxide.[18] Silicon dioxide of the proper texture was produced by calcining the precipitate obtained by treating the solution of an alkali metal fluosilicate with ammonium hydroxide.

By carrying out the hydrolytic precipitation in the presence of small proportions (1 per cent) of such coagulating agents as acids or salts that, under the conditions employed, yielded polyvalent negative radicals or ions, titania of a nonpeptized, flocculated form was obtained, with the result that it was easily filtered and washed.[19] The more effective agents were found to be oxalic, citric, sulfuric, phosphoric, and tartaric acids, and their alkali metal and ammonium salts; these could be added either to the titanium fluoride solution or to the aqueous ammonia used as precipitant. In an actual operation, 4.0 g of ammonium sulfate was added to 10 kg of titanium ammonium fluoride solution containing 800 g titanium dioxide, and this mixture was run into 50 kg of 10 per cent ammonium hydroxide solution over a period of 40 minutes. The highly flocculated precipitate was washed and calcined at 1000° C for 2 hours to produce a titanium dioxide pigment of the rutile crystal form possessing superior tinting strength and color.

Ilmenite or rutile, ground to 100 mesh, was heated with $K_3FeF_6$ or the corresponding compound of nickel, cobalt, or zinc at 700° C for 1 hour. The sintered product was crushed and leached with water, and the resulting compound, $K_2TiF_6$, was purified by crystallization and converted to the oxide, chloride, or sulfate of titanium.[20]

## BROMIDE PROCESS

Titanium dioxide pigment of the rutile crystal form is produced by the vapor phase oxidation of titanium tetrabromide at elevated temperature.[21] For example, briquets composed of 70 per cent titaniferous ore, 20 per cent coke, and 10 per cent resin binder are subjected to a stream of bromine at 1100° C. The titanium tetrabromide formed is cooled to 50° C, decanted from sediment, and heated under reflux in an oxygen-free reducing atmosphere to remove the free bromine. Final purification is effected by fractional distillation. The 220 to 230° C fraction is mixed at 180° C with 800 liters per minute of an equimolecular mixture of oxygen and nitrogen heated to 250° C in a burner to which is added 800 liters per minute of carbon monoxide and 400 liters per minute of oxygen to raise the combustion temperature to 1000° C. From the cooled products the finely divided titanium dioxide is separated by electrostatic filters. After removal of the solid titanium dioxide, bromine is recovered from the gases and recycled. The bromination and distillation process eliminates chromium and vanadium bromides.

## NITRATE PROCESS

McKinney [22] reported an interesting process for producing rutile-type pigments, employing nitrate solutions which involved heating anatase titanium dioxide with barium hydroxide to form the titanate, followed by dissolution in nitric acid and reprecipitation by thermal hydrolysis of the solution, after separation of most of the barium nitrate by crystallization. Keats [23] developed a cyclic operation based on this process, according to which the nitric acid and barium hydroxide were recovered and reused. Anatase pigments are produced commercially by processes involving the thermal hydrolysis of sulfuric acid solutions because of the smoothness of operation and relative simplicity of handling such solutions. Since the hiding power of a white pigment is proportional to the refractive index of the particles, it follows that rutile, with an index of refraction of 2.71, will have a higher hiding power than anatase with a refractive index of 2.55, other factors being equal. From this standpoint the production of rutile pigment is desirable, but methods consuming halide solutions (chlorides and fluorides) are relatively expensive. Furthermore, the dilute hydrochloric acid recovered in the hydrolysis step is, as a general rule, salable in large amounts at prices competitive with sulfuric acid, if at all, and regeneration of chlorine from the liquor would not be economical, since the recovery cost would equal or exceed the original cost of manufacture of chlorine.

This cyclic process avoids the use of halide solutions and the corrosion problem encountered in handling the corresponding acids. Nitric acid is not nearly so corrosive, and plant design is therefore simpler and more economical. This method is largely supplemental to the conventional process based on sulfate solutions, and most of the equipment would remain in use.

Hydrous titanic oxide of the anatase crystal modification, as obtained commercially by the thermal hydrolysis of sulfate solutions of ilmenite, was washed free of iron, neutralized with an alkaline agent, and again washed to remove the salts formed in the reaction. This sulfate-free pulp was suspended in a chemically equivalent amount of aqueous barium hydroxide and heated near the boiling point of the solution to form barium titanate. For each mole of this compound, 5 to 10 moles of nitric acid of 30 to 50 per cent strength was added to form barium nitrate and titanium nitrate. Barium nitrate was only slightly soluble in the strong liquor so that a greater part of it crystallized and was separated readily by decantation, filtration, or centrifuging. The filtrate, consisting largely of titanium nitrate, was adjusted to a concentration of 80 g to 120 g titanium dioxide equivalent per liter, and after proper seeding was hydrolyzed by boiling. The precipitate was filtered, washed, and calcined to produce titanium dioxide of pigment grade having the rutile crystal structure. In carrying out the hydrolysis reaction, solutions of concentration above 40 g titanium dioxide equivalent per liter were used, for otherwise the anatase modification resulted.

The nitric acid filtrate from the hydrolysis product was collected for reuse, and the barium nitrate crystals were decomposed by heat to give oxides of nitrogen and barium oxide, all of which were recovered and used again. The residue of barium oxide was added to water to regenerate the hydroxide and used for attacking another batch of anatase pulp. Chemical reactions involved in the process are as follows:

$$Ba(OH)_2 + TiO_2 \text{ (anatase)} \rightarrow BaTiO_3 + H_2O$$
$$BaTiO_3 + 6HNO_3 \rightarrow Ba(NO_3)_2 + Ti(NO_3)_4 + 3H_2O$$
$$Ba(NO_3)_2 \rightarrow BaO + N_2O_5$$
$$N_2O_5 + H_2O \rightarrow 2HNO_3$$
$$Ti(NO_3)_4 + 2H_2O \rightarrow TiO_2 \text{ (rutile)} + 4HNO_3$$

According to a specific procedure, iron-free anatase pulp obtained by the conventional hydrolysis of sulfate solutions of ilmenite was slurried in water, neutralized with ammonium hydroxide, and washed free from sulfates and other salts formed in the reaction. The sulfate-free cake was mixed with a chemically equivalent amount of barium hydroxide in aqueous solution; the slurry was heated to the boiling point and maintained for 1 hour to bring about formation of barium titanate. This reaction product was dewatered and the cake was treated with cold 50 per cent nitric acid, with agitation. After several hours the suspension lost its opalescence and then consisted of crystals of barium nitrate suspended in the solution of titanium nitrate. Upon filtration, 80 per cent of the nitrate solution was recovered before washing, and the remainder by washing with cold acidified, saturated barium nitrate solution. The initial filtrate was combined to give a solution having more than 80 g titanium dioxide per liter, which was subjected to thermal hydrolysis after addition of a seeding material prepared after the method of Kubelka and Srbek.[24] The precipitate was filtered, washed, and calcined according to accepted methods to produce rutile titanium dioxide of pigment grade.

Rutile titanium dioxide is obtained by the reaction of titanium tetrachloride with nitric acid in aqueous solution.[25] The gaseous products of the reaction are passed into an aqueous slurry of manganese chloride to convert the nitrosyl chloride formed to manganous nitrate and manganous chloride. Chlorine recovered is used for the manufacture of titanium tetrachloride. Manganous chloride reacts with nitric acid to form additional chlorine and manganous nitrate. The manganous nitrate is in turn decomposed with air and water to form manganese dioxide and nitric acid which are recycled in the process.

## SULFIDE PROCESS

A unique process of producing titanium dioxide, which involved a sulfide stage, was developed by Pipereaut and Helbronner.[26] Ilmenite was heated with coal and sulfuric acid or pyrite in a closed vessel, and the iron was dissolved from the reaction product with dilute sulfuric acid, leaving a titanium dioxide residue which was mixed with coal and subjected to the action of chlorine at a dull red heat. The titanium tetrachloride still in the vapor form was passed with dry hydrogen sulfide through tubes externally heated at 400° C to form the sulfide. Alternatively, the tetrachloride was condensed and cooled to 100° C, and the hydrogen sulfide was admitted. In either modification the titanium sulfide produced was roasted alone to produce the dioxide or with zinc sulfate to produce a composite pigment consisting of oxides of titanium and zinc. By carrying out the roasting operation in closed vessels in the absence of oxygen, the zinc component was obtained in the form of sulfide rather than oxide. Fillers such as barium sulfate were added before the last roasting operation to obtain extended pigments.

Similarly, titanium sulfide and hydrochloric acid were produced by the reaction at a dull red heat of a gaseous mixture of equal volumes of carbon dioxide and hydrogen sulfide with titanium tetrachloride. The sulfide was then calcined to remove the sulfur and roasted to yield titanium dioxide.[27]

Ilmenite concentrate or high titania slag is smelted with sufficient sodium sulfide to give a readily fusible product which is slurried in water to give metallic sulfides and a basic hydrated titanium compound suspended in the sodium hydroxide solution.[28] The solids are filtered, washed, and treated with ammonium hydrogen sulfate at a pH of 2 to dissolve the sulfides, leaving the titanium compound as a residue. Hydrogen sulfide evolved is passed into the sodium hydroxide filtrate to regenerate sodium sulfide for smelting. To this residue is added an excess of ammonium hydrogen sulfate of a concentration to dissolve 85 per cent of the titanium. The remainder is recycled to the smelter. After partial neutralization the solution is heated to hydrolytically precipitate 95 per cent of the titanium as a hydrated basic oxide. This precipitate is calcined to a product containing 75 per cent titanium dioxide suitable as a raw material for pigment manufacture. To the impure iron sulfate solution obtained from the first ammonium hydrogen sulfate treatment is added ammonium carbonate to precipitate the iron as a useful carbonate.

The filtrate is dehydrated and fused at 350° C to produce ammonia and ammonium hydrogen sulfate for recycling. The recovered ammonia is mixed with carbon dioxide and water in a coke-packed scrubber to produce ammonium carbonate for precipitating the iron. Smelting is essential for utilizing many otherwise wasted titanium-containing materials.

Titanium dioxide is obtained by treating heated ilmenite with hydrogen sulfide in a rotating cylinder and leaching the iron sulfide formed with dilute acid.[29] The cylinder is rotated to facilitate complete reaction between the mineral and the hydrogen sulfide.

In a cyclic process the titanium ore is ground with coal and calcium sulfate or sodium sulfate and heated in a rotary kiln.[30] The product, after leaching with warm water to dissolve the calcium sulfide formed, is washed with hydrochloric acid and then with dilute sulfuric acid. From this step ferrous sulfate and hydrogen sulfide are recovered as by-products. The titanium dioxide is dissolved with oleum. Sulfuric acid is added to the calcium sulfate leach solution to precipitate calcium sulfate for recycling. The ferrous sulfate is fired to produce sulfur dioxide for the sulfuric acid plant. More sulfur dioxide is produced by burning the hydrogen sulfide gas.

Titanium dioxide with the crystal form of rutile is prepared from ilmenite by a process which includes heating the ore with sulfur.[31] The ilmenite is heated at 500 to 1000° C in the presence of sulfur vapor produced in situ from pyrite and carbon in the absence of oxygen. After cooling to 100 to 200° C out of contact with oxygen, water is added to the product and oxygen is introduced at 10 to 100 psi giving rutile aggregates, globular sulfur, and ferrous hydroxide sludge. Compounds of iron, chromium, manganese, aluminum, and silicon remain as discrete particles fraction and can be separated by normal ore-dressing methods.

A product obtained by heating titania slag at 1400° C with soda-lime equal in amount to the titanium dioxide content is plunged into water to give spongy granules, 70 per cent of which dissolve. After aeration with sulfur dioxide gas, the solution is filtered.[32] The filtrate, after removal of the gelatinous silica which forms at 70° C, is hydrolyzed at 90° C to precipitate titanium hydroxide, which is filtered and washed. After dissolving the alumina with 1 normal sodium hydroxide the product is calcined to obtain titanium dioxide of 98.6 per cent purity at 92 per cent recovery.

## CHLOROACETATE PROCESS

Young[33] prepared the chloroacetate of titanium and heated the compound to 800° C or above for 2 hours to form the dioxide of the rutile crystal modification. This product had a petal-like structure with a surface area many times that of the initial material, and the coalesced particles were readily broken up by light rubbing.

## FIBROUS TITANIUM DIOXIDE

Titanium dioxide in the form of fibers 0.5 to 3 mm long and 0.7 to 0.8 micron in diameter is produced by the oxidation of titanium subhalides in a fused bath of one or more halides of an alkali or alkaline earth metal at 600 to 800° C.[34] The titanium in the fused salt mixture should have a valence of two or three and titanium subchlorides can constitute 8 to 10 per cent of the bath by weight. Fluorides in the bath lead to the production of longer fibers, but other halides should also be present in greater amount. The oxidation can be produced by dry air or oxygen either at the surface or bubbled through the bath. After oxidation of the fused salt mixture for a period of time from several hours to several days, the titanium dioxide which is partly fibrous is separated from the salt by decantation or filtering, or by dissolving the salts in water. The fibers are then separated from irregular particles or aggregates of titanium dioxide by agitation in water followed by settling. The fibers remain longer in suspension. These fibers are useful in reinforcing ceramics, plastics, and paper. They are readily matted for refractory fibers and thermal insulators. For example, in a quartz reaction tube, an anhydrous mixture of potassium chloride 167, sodium chloride 78, titanium metal 10, and a 47 to 53 mixture of sodium chloride and lower titanium chlorides, $TiCl_{2.5}$, 53 g was fused at 700° C. A dry mixture of 30 to 60 ml per minute of oxygen and 5 to 15 ml per minute of nitrogen was introduced for 88 hours at a pressure of 1.26 atmospheres. After cooling and dissolving the salts in water, 29.5 g of titanium dioxide was recovered, 13 per cent of which was in the form of fibers 0.2 to 0.3 mm long with diameters below 5 microns.

Fibrous titanium dioxide of the rutile crystal form is produced by the reaction of titanium tetrachloride with a melt consisting of boric oxide, alkali metal borates, or mixtures of these.[35] A temperature sufficient to maintain the boric oxide or borate in a liquid condition, from 650 to 850° C, is employed. After the melt has cooled, the fibers are readily isolated by agitation in boiling water. Any nonfibrous titanium dioxide is removed from the suspension after flocculating the fibrous material. Fibers produced by this method have a cross section of less than 5 microns and a length of 0.2 to 5 mm or more. In an example, nitrogen gas dried over phosphorus pentoxide was bubbled at a rate of 17 ml per minute through liquid titanium tetrachloride at 27° C. The resulting vapor mixture was passed through a silica tube containing 9.4 g of molten boric acid in a platinum boat, maintained at a temperature of 600° C by external heating. After passage over the molten boric oxide, the vapor mixture was passed through a trap cooled to −78° C with solid carbon dioxide. During 24 hours of operation, 3.1 g of titanium tetrachloride was vaporized, and unchanged titanium tetrachloride and boron trichloride collected in the trap. After cooling, the surface of the boric acid was found to contain fan and rosette shaped clusters of fibrous crystals. These were recovered by dissolving the boric oxide in boiling water and separating the crystals by filtration. At a temperature of 800° C fibers 0.5 to 0.7 mm long and less than 5 microns in diameter were produced. A similar product was obtained employing a melt of sodium chloride and potassium chloride.[36]

Pure needle-like fibrous crystals of titanium dioxide are separated from fused solutions in boric oxide and sodium oxide by slow stepwise cooling.[37] Crystalline fibers of pure titanium dioxide are prepared by cooling a melt containing boric oxide, sodium tetraborate, and titanium dioxide.[38] Fibers up to 4 inches long are obtained by cooling the melt from 2500 to 70° F.

## SMALL PARTICLE SIZE

Peterson[39] developed a process for the production of titanium dioxide of uniform particle size, ranging from 0.02 to 0.05 micron, readily dispersible in organic media. Hydrous titanium dioxide prepared by the hydrolysis of titanium sulfate solution and containing 8 to 10 per cent sulfuric acid on the titanium dioxide basis is slurried in water and sodium hydroxide is added to a pH of 7. After removing the sodium sulfate by washing with water, the product is reslurried to a titanium dioxide concentration of 275 gpl. To 11.9 liters of the slurry is added 250 ml of concentrated hydrochloric acid. Then sufficient dilute ammonium hydroxide is added to the colloidal suspension to give a pH of 7.0 to 7.2. The neutral product is dried at 110° C and calcined at 600° C for 30 minutes to give a pigment having individual particles of about 0.25 micron in diameter suitable for use as a reinforcing agent for natural rubber compositions.

A high surface area titanium dioxide practically free from impurities is prepared by the reaction of tetraisopropyl titanate with acetic acid followed by hydrolyzing the ester and calcining the dried gel.[40]

Rutile titanium dioxide in laminar shapes 0.01 to 0.50 micron thick by 1 to 20 microns diameter is formed by spraying a solution containing 1 per cent to saturation of titanium sulfate together with 5 to 20 per cent free sulfuric acid in a flame at 890 to 1800° C for 0.001 to 0.5 second.[41] Thus an aqueous solution containing titanium sulfate equivalent to 10 per cent titanium dioxide with 12.7 per cent free sulfuric acid was sprayed in a stream of natural gas into a burner at the rate of 4.6 pounds per hour. Oxygen was introduced at the rate of 2 pounds per hour and air was introduced peripherally at 155 pounds per hour. The laminar product was collected on a wetted wall in a vertical tower through a venturi collector. This material is used with heavy lubricating oil to prepare a grease formulation which is retained on bearing surfaces operating at high temperature.

Titanium dioxide having a surface area of 1000 square meters per gram and particles ranging in size from 25 to 50 Ångstrom units is prepared by hydrolyzing a titanium sulfate, chloride, or nitrate compound in a water solution intimately mixed with carbon black at an elevated temperature.[42] The particles are separated and dried, after which the carbon is burned out in air, hydrogenated, or removed in other convenient manner. The oxide is then broken into a fine powder.

### COARSE PARTICLE SIZE

Free flowing titanium dioxide for enamel frits is obtained by carrying out the hydrolysis so as to get coarse particles.[43] In an example, 1000 gallons of a sulfate solution containing titanium dioxide 180, iron 60, and sulfuric acid 460 g per liter at 65° C was poured at a uniform rate in 30 minutes into 1250 gallons of water at 90° C. The mixture was boiled for 2 hours to effect hydrolysis. Calcining the washed hydrate yielded titanium dioxide particles of 10 to 20 microns in diameter.

### DUAL PROCESS

Ilmenite is digested in sulfuric acid and the resulting solution is hydrolyzed to obtain a coarse, rapid filtering hydrous titanium dioxide.[44] This washed hydrate is redissolved in sulfuric acid and diluted with water to a titanium dioxide concentration of 200 to 300 gpl. The solution is hydrolyzed by boiling to produce a hydrate which on calcination yields titanium dioxide of pigment grade.

### FUSION AND SOLUTION

Sodium titanate is washed with water at a temperature below 40° C to obtain titanium hydroxide, which is dissolved in 5.3 parts of 50 per cent sulfuric acid per part of titanium dioxide in the presence of 0.5 to 5.0 per cent sodium fluoride and an alkali metal bromide or iodide as a promoter.[45] After reducing the ferric ions with scrap iron, the titanium sulfate solution is hydrolyzed to give titanium dioxide of pigment grade. For example, a mixture of 100 parts of titanium dioxide and 366 parts of 50 per cent sulfuric acid containing 2.8 parts of sodium fluoride was heated for 25 minutes just below the boiling point. The solution was cooled, filtered, or centrifuged and reduced by heating with scrap iron. At least 90 per cent of the titanium was reduced to the trivalent form as a bluish black viscous solution. This solution held at 80° C was poured into boiling water to give a blue precipitate of $Ti_3O_5$. By agitating and boiling the mixture the greater part of the precipitate was transformed systematically to white titanium dioxide having a uniform particle size of 0.8 micron. No inoculation was needed and no iron was adsorbed by the precipitate. In 2.5 hours 95 per cent of the titanium dioxide was precipitated. The pulp obtained was dried and calcined with complete oxidation of possible residues of lower oxides.

Ito [46] heated at 1300° C a mixture of ilmenite with coke and soda ash to obtain a product which was powdered and again heated with soda ash. This material was stirred first with 10 per cent sodium hydroxide solution and then with 30 per cent hydrogen peroxide. After separation from the residue the extract was heated at 90° C; the precipitate was heated at 300° C, washed to remove the alkali, and then heated to 900° C for 2 hours to produce titanium dioxide.

A high titania slag is heated at 500° C with sodium hydroxide and cooled rapidly.[47] Hydrated titanium dioxide is extracted from the mass at 7 to 8° C with hydrogen peroxide and sodium hydroxide. Boiling the solution of sodium peroxypertitanate gives a yellow precipitate of hydrated titanium peroxide. The precipitate is washed with dilute acid to remove the sodium ion and calcined for 1 to 2 hours at 900 to 950° C to give a rutile type pigment of good quality.

A solution containing 8 per cent titanium dioxide and 5.5 per cent iron obtained by the reaction of ilmenite with sulfuric acid is treated with iron powder to reduce the ferric ions to ferrous, and hydrolyzed by the addition of 1 to 2 parts of Glauber's salt and 0.1 to 0.2 part sodium pyrophosphate per part of titanium dioxide.[48] The precipitate washed with water gave titanium dioxide of 99.6 per cent purity at a yield of 90 per cent or more.

A mixed precipitate of ammonium and titanium chloride and oxalate obtained by adding oxalic acid and ammonium chloride to titanium tetrachloride solution is calcined at 600 to 700° C to obtain finely divided titanium dioxide of the anatase crystal structure.[49]

## COMPOSITE PIGMENTS

Washburn and Aagaard [50] prepared a composite pigment of greatly improved properties by chemically precipitating the calcium sulfate separately and converting it to the anhydrite form before mixing it with the ilmenite solution.

Dehydrated calcium sulfate was prepared by adding a slurry of hydrated lime or calcium carbonate to concentrated sulfuric acid under controlled conditions of concentration, temperature, and time of heating to produce an extremely finely divided product having the anhydrite structure. As a result of this extreme state of subdivision, each unit of weight of the extender supplied more particles and much more surface to act as adsorption nuclei for the precipitated compounds of titanium. This not only resulted in an accelerated rate of hydrolysis, but also yielded pigments of superior color and covering power, even with higher proportions of titanium dioxide. Considerable heat was evolved by the reaction between the lime and sulfuric acid, and this was conserved by adding the hot suspension to the ilmenite solution. The mixture was further heated to effect hydrolysis, and the titanium compounds, on precipitation, largely coalesced with the calcium sulfate and the two components could be present in any proportions, whereas former methods were only applicable to composites containing relatively low proportions of titanium dioxide. The coalesced precipitate was washed and calcined at 900° to 1000° C.

In an actual operation, ilmenite ore was brought into solution after digestion with strong sulfuric acid; all the ferric iron and a small part of the titanium were reduced to their next lower valences by introducing metallic iron, and the residue was allowed to settle after coagulation with glue. A slurry of 95 pounds hydrated lime in 110 gallons of water at 70° C was added slowly, with agitation, to 372 pounds of 78 per cent sulfuric acid at 20° C. Considerable heat was evolved by the reaction, and the resulting suspension of calcium sulfate in dilute sulfuric acid (2 to 5 per cent) was further heated near the boiling point to effect practically complete dehydration. To this thick pulp of anhydrite, in the tank in which it was formed, was added 900 pounds of the clear supernatant ilmenite liquor containing 7.02 per cent titanium dioxide, 6.85 per cent ferrous oxide as sulfates, and 2.26 per cent uncombined (free) sulfuric acid. The mixture was then heated to boiling and maintained until 95 per cent of the titanium was hydrolyzed, after which the composite precipitate was washed, dried, and calcined at 900° C to yield a pigment containing 29.6 per cent titanium dioxide and 70.3 per cent calcium sulfate, with only traces of other substances. The effective specific gravity was 3.2.

Calcium sulfate and barium sulfate extended pigments in which the titanium dioxide component was in the rutile crystal modification were obtained by treating the regular coalesced product with a small proportion of zinc oxide or other zinc compound before calcination.[51] Hydrous titanic oxide was coprecipitated from a sulfuric acid solution of ilmenite on a calcium sulfate or barium sulfate extender, so that the final pigment contained 30 per cent titanium dioxide. After thorough washing, 1 per cent zinc oxide was added and the mixture was calcined at 775° to 1000° C. Such pigments were characterized by improved resistance to chalking and high tinting strength characteristic of rutile. High resistance to chalking was developed by continuing the normal calcination period or by raising the temperature, but at the expense of tinting strength. The desired balance of properties was obtained by regulating the degree of calcination.

A composite precipitate of calcium sulfate and titanium dioxide is prepared which can be calcined without a promoter to the rutile form or can be calcined to a pigment containing the anatase modification of titanium dioxide in a form which can be easily converted to rutile.[52] A titaniferous ore is leached with hydrochloric acid to remove most of the iron. As little as 4 per cent of the leached ore residue mixed with ilmenite or slag is reacted with sulfuric acid and the product is dissolved in water to form a solution. After clarification the solution is mixed with anhydrite slurry and boiled to precipitate titanium dioxide on the extender in a form which can be calcined in part or all to rutile. For example, 1000 parts of leached ore having a particle size such that 0.9 per cent were retained on a 200 mesh screen and 1700 parts of 96 per cent sulfuric acid were heated to 110 to 120° C. Cold water was added to reduce the sulfuric acid concentration to 88 per cent and initiate the reaction. Heating was continued for 0.5 hour after which the resulting digestion cake was dissolved in water. After filtration, the solution was added to a slurry of anhydrite containing 30 per cent solids in the ratio of 1 part of titanium dioxide to 2.3 parts of calcium sulfate. Sulfuric acid was added to adjust the acidity to 18 per cent after which the mixture was boiled for 3.5 hours to precipitate the titanium dioxide on the anhydrite extender. The composite precipitate was filtered and washed. Calcination for 1 hour at 925° C followed by 1 hour at 950° C converted the titanium dioxide component entirely to the rutile form. A solution of titanium sulfate mixed with anhydrite having a particle size of 0.3 to 0.7 micron is hydrolyzed by boiling to obtain a composite product.[53] This mixed product is filtered and washed, after which 0.3 to 5 per cent of a colloidal solution of titanium dioxide and 0.5 to 2 per cent of zinc oxide are added. Calcination is continued until over 90 per cent of the titanium dioxide possesses the rutile structure.

Anhydrite suitable for the manufacture of composite calcium sulfate-titanium dioxide pigment is produced from limestone.[54] Thus, 30,000 pounds of limestone ground to 200 mesh and containing only 0.008 per cent of color degrading impurities was mixed with 45,000 pounds of water. This slurry was added continuously at a rate of 225 pounds per minute to a tank into which was added simultaneously 2250 pounds per minute of waste 6 per cent sulfuric acid at 50° C containing 6 per cent ferrous sulfate to form gypsum. The temperature of the reacted mixture in the tank was 60° C. This gypsum slurry containing 15 gpl sulfuric acid was passed over a rotary vacuum type filter and washed with water to produce 103,200 pounds of cake containing 50 per cent solids. To adjust the slurry to 45 per cent solids and a pH of 6.0, 11,800 pounds of water was added. The average particle size of the gypsum in the slurry was 2.5 microns. An anhydrite seed was formed by adding with rapid agitation 23,000 pounds of the washed gypsum slurry at 35° C to 7500 pounds of 78 per cent sulfuric acid at 40° C at a rate of 1000 pounds per minute. During the first 5 minutes the temperature rose to 55° C and remained at this value until the end of the strike period. This anhydrite seed slurry was transferred to a large tank equipped with a sweep agitator and the remainder of the gypsum slurry, 92,000 pounds, was added during 92 minutes. The mixture was heated to 100° C with steam during this addition to convert all of the gypsum to anhydrite. Remaining in the liquor of the anhydrite slurry was 7 per cent sulfuric acid. The anhydrite of the slurry had an average particle size of 0.58 micron. This slurry was used for the production of coalesced calcium sulfate-titanium dioxide pigment possessing excellent whiteness and high hiding power.

## BLENDED COMPOSITE PIGMENTS

Barton [55] blended mineral fillers having relatively low indices of refraction (barium sulfate, silica, gypsum, clay, and asbestine) with pigments consisting of titanium dioxide alone or coalesced with alkaline earth metal sulfates by grinding the constituents together in the dry state.

A hydrophillic calcium sulfate extended rutile pigment contains water insensitive porous anhydrite with a high ratio of specific surface area to particle size.[56] Gypsum is precipitated by the reaction of lime slurry with dilute sulfuric acid, filtered, washed and calcined at 600 to 825° C to obtain the porous anhydrite. This special anhydrite is blended with previously calcined titanium dioxide in the ratio of 70 parts by weight of anhydrite to 30 parts by weight of titanium dioxide. The mixture is then ground first by pressure milling and finally in a fluid energy mill.

Booge [57] made blended pigments in which the individual particles of titanium dioxide and calcium sulfate existed independently side by side by mechanically mixing selected components of specific particle size. These blends, although physically different, were equal and within certain limits of composition superior in hiding power, tinting strength, brightness, and reflectance to coalesced pigments having the same components in the same proportions.

Any desired method of blending the constituents can be followed so long as a most thorough distribution of one component throughout the mass of the other is obtained. Although this can be accomplished in the dry state, in general wet blending is more convenient to carry out. Aqueous pastes of the pigment-grade titanium dioxide and of the alkaline earth metal sulfate were mixed and thoroughly agitated for several hours until the individual particles were completely dispersed and thoroughly blended. The mixture was then filtered, dried, and in some cases recalcined, and disintegrated. If desired, the product was subjected to a controlled dry-milling process to reduce the oil absorption. Such grinding did not reduce the particle size of the components, however, and the lowering of oil absorption value was caused by some change in surface or adsorption characteristics of the particles.

Calcium sulfate of the anhydrite form having pigmentary properties within the specific limits was produced easily by chemical precipitation. For example, a slurry containing 200 g slaked lime per liter was gradually added to 60° Bé sulfuric acid until 85 per cent of the acid was neutralized. During the initial stages of the precipitation the temperature rose rapidly to approximately 113° C, then fell off slowly. The reaction mass was held close to the boiling point, with thorough agitation, until a very few acicular crystals (gypsum) remained, as determined by observation under a microscope at 400 magnifications. The slurry of anhydrite was then filtered and washed until the filtrate showed a pH of at least 4. The cake was repulped in fresh water, made alkaline by adding a small amount of sodium hydroxide, and again filtered. This product was blended with titanium dioxide without further treatment or after calcination. Since the presence of alkali metal compounds during calcination lowered the oil absorption of the product, traces of sodium hydroxide remaining with the calcium sulfate had such an effect. A product of excellent color, and a frequency particle size average of 0.5 micron with more than 90 per cent of the particles below 1.5 microns, was obtained.

Calcium sulfate produced in this manner may be calcined separately and then mechanically mixed with titanium dioxide of specified properties, or, for the sake of simplicity and ease of large-scale operation, the two components may be blended prior to calcination of the calcium sulfate.

## COLORED PIGMENTS

Titanium dioxide pigments of distinctive colors and tints may be produced by calcining the hydrous oxide, as obtained by thermal hydrolysis of ilmenite solutions, with small proportions of compounds of the heavy metals having atomic numbers between 50 and 64 inclusive, such as vanadium, chromium, iron, cobalt, nickel, manganese, and copper.[58] Chromium and vanadium are particularly effective for the main reason that their power of imparting color to the resulting pigment is exceedingly strong; only very small amounts of these agents are required to produce distinct colorations. Such products consist of particles of titanium dioxide containing a small amount of a heavy metal compound (usually oxide) physically or chemically associated therewith. Colors obtainable range from the extreme of red to the other extreme of green, through such shades or tints as grays, buffs, yellows, light browns, and tans.

These pigments have the valuable properties of straight titanium dioxide and differ in chemical properties and constitution from titanates. The small amount of colored compounds does not adversely affect the tinting strength, chemical resistance, or particle size of the straight titanium dioxide which forms the basis of the pigment. On the other hand, the phenomenon of fading in paint films is greatly decreased or eliminated altogether. Resistance to chalking is greatly improved, and if the products chalk on excessive exposure, the film is colored and not conspicuous. This property is so pronounced that the tinted products protect white pigments or extenders with which they are mixed in controlled amounts. Tinted pigments of this type are composed of particles which under the microscope appear homogeneous and show the same color throughout. For this reason they are readily distinguished from mechanical mixtures of the components. In general the desired tints are obtained within the range of 0.3 to 7 mole per cent of the heavy metal compound.

# 23
# Chalking and Discoloration of Titanium Dioxide Pigments

Paint, enamel, and lacquer films, pigmented with regular (unmodified) titanium dioxide alone, in outdoor service usually exhibit a type of rapid failure known as "chalking," [1] which manifests itself as a tendency of the surface layer to disintegrate to a powdery chalk. As this material is removed by the eroding action of wind and rain the underlying section is exposed to further attack so that the wearing away is progressive. The powdery surface assumes an unpleasing dull appearance and tends to collect dirt and mildew. In mixtures with coloring agents to form tinted coating compositions, titanium dioxide prepared by ordinary methods produces a whitish chalk on the surface of the tinted film upon exposure to atmospheric conditions, thereby producing an unsightly and faded appearance.

Furthermore, such pigments incorporated in compositions with organic dyestuffs frequently accelerate fading of the added color, particularly on exposure to sunlight. However, by blending the titanium dioxide with pigments of the chemically active type such as zinc oxide and white lead, the degree of chalking can be minimized and controlled. At the same time the types of failure characteristic of the reactive pigments, such as checking and cracking, are greatly reduced or eliminated altogether.

On the other hand, coating compositions pigmented with titanium dioxide do not exhibit this objectionable property of chalking in indoor service, but tend to discolor, characterized by a progressive yellowing of the film.

Many methods have been proposed for producing titanium dioxide pigments free from these limitations. The most common practice is to incorporate with the hydrous oxide before calcination, small proportions of conditioning agents such as antimony trioxide, zinc oxide, and rare earth compounds, or to coat the individual particles of titanium dioxide after calcination with small amounts of such agents as hydrous alumina, chromic oxide, silica, and zirconium compounds, or combinations of these. Such treatments increase the resistance to chalking, discoloration, and fading to an appreciable extent, but in some cases at some sacrifice in color, brightness, and tinting strength. The alumina treatment is particularly effective in overcoming the tendency toward discoloration.

Various explanations have been offered to account for this poor resistance to chalking. This phenomenon occurs only in the presence of solar radiation, and moisture is also an important factor.[2] Some authorities hold that it is the result of lack of molecular or other contact or attrition between the so-called nonpolar pigment particles and the vehicle. This would appear to be only a secondary factor, however, since films of this type last indefinitely in indoor service. Thus the primary cause for destruction of the loose contact appears to be solar radiation not necessarily in the visible range.

Resistance to actinic light appears in many cases to be a function of the atomic weight and of the size of the molecules of the various pigment materials. Metallic lead, for instance, is impervious to X rays, while aluminum with a very much lower atomic weight is transparent. Barium sulfate containing barium, an element of high atomic weight, is also relatively impervious, in contrast to sodium sulfate or even calcium sulfate. Pigments containing lead compounds, as for example the carbonate, are more resistant than those consisting of compounds of the lighter metals having lower atomic weights, and similarly polymerized linseed oil with a large molecule gives more resistant films than does the untreated material. As an analogy, these factors may be interpreted as meaning that the destructive rays have to be stopped by an armor plate of high atomic weight or by a "sandbag" containing a multitude of smaller atoms.

According to this hypothesis, titanium with a relatively low atomic weight of 48 should not be resistant to radiation, but it is exceptional in having a very high index of refraction. Radiant energy falling on a moderately opaque film is reflected gradually from the particles throughout the whole depth of the film, but the same radiation hitting a highly opaque film pigmented with titanium dioxide is reflected more from the surface layer, and the shock of impact must therefore be borne by a relatively small number of particles containing a metal of low atomic weight. Thus it appears reasonable that titanium dioxide alone may never be so resistant as lead pigments in a linseed oil vehicle.

Two methods of overcoming this difficulty suggest themselves. The first is to employ titanium dioxide, along with other pigments having a high atomic weight, either in chemical combination or as a mechanical mixture. Lead titanate, for example, is little affected by solar radiation. In a study of mixtures of titanium dioxide with carbonates of barium, strontium, and calcium, the resistance to chalking of the composite pigments was found to be of the same order, showing that the compound containing the metal of high atomic weight was superior. The second method, employing vehicles of high molecular weight such as bodied linseed oil, also gave good results in practice.

Wagner [3] concluded that the initial destruction of a paint film is due to the action of ultraviolet light on the oil and that chalking appears later. Water, if present, penetrates the disrupted surface, causing swelling, and thus accelerates the complete destruction. Both these defects are related to the oil absorption of the pigment, and fine grinding should be avoided. Maximum durability is obtained from compositions in which neither the pigment nor any soap formed from it by reaction with the vehicle undergoes hydrolysis. Further improvement may be obtained by sealing with dispersed fillers the pore spaces of pigments having low swelling power.

Chalking has also been ascribed to the inactive nature [4] of the titanium dioxide. In preparing coating compositions, surface active pigments such as zinc oxide, basic lead carbonate, and iron oxides become an inherent part of the oil gel and give dense impervious films while the inactive pigments as titanium dioxide produce films which are more porous and more easily penetrated by water. Films of this type naturally tend to fail more readily by chalking, and consequently show poor tint retention.

Zhukova and Sovalova [5] concluded that such disintegration is caused by instability of the system titanium dioxide-linseed oil to atmospheric action, which in turn is the result of inertness and poor wettability of solid particles in the vehicle. The system can be so stabilized as to prevent the chalking by the addition of soaps such as barium oleate and sodium salts of linseed oil fatty acids, which by forming pigment-enveloping aggregates facilitate dispersion in the oil. Fillers such as nephelite, talc, and barium carbonate are effective in retarding chalking of the composite films since these form dipolar compounds with the vehicle and thus favor the formation of adsorption aggregates.

According to another hypothesis, the chalking of films pigmented with titanium dioxide is ascribed to the action of atmospheric factors on the oil of the vehicle, catalyzed by the titanic oxide.[6] The process is inhibited by covering the surface with a second layer containing another and more stable pigment. Alternatively the same favorable results may be obtained by coating the particles of the commercial pigment with an adsorbed layer of aluminum hydrate, ferric hydroxide, zirconium oxide, and other nonpigment material before suspending them in the oil vehicle. Lyophile substances such as aluminum soaps are not effective.

Some authorities have suggested that pigments which absorb strongly in the ultraviolet should show good resistance to chalking failure, since their absorption would shield the organic constituent of the film from this harmful radiation. Such a theory is lent credence by the relatively low ultraviolet absorption of regular titanium dioxide. However, on the other hand, some of the special grades, for instance those treated with small proportions of antimony oxide, alumina, and zirconium compounds, and possessing normal tinting strength, have practically the same absorption characteristics but widely different chalking resistance. According to an explanation by Rubin,[7] the action of ultraviolet light on drying oils produces peroxides and these oxidize the titanium dioxide to pertitanic acid, which in turn oxidizes the oil of the film to produce chalking. Rutter[8] carried out experiments with titanium dioxide pigmented paints on glass and concluded that the chief cause of breakdown was the oxidation of the medium with the pigment acting as catalyst. Rutile proved more resistant to weathering than the anatase modification.

Titanium dioxide pigments prepared commercially by the usual sulfate precipitation process have several defects as far as their paint-making properties are concerned; however, by suitable after-treatments these defects can be greatly reduced.[9] Oversize particles are removed by grinding in series with classification. Surface treatment to give better dispersibility consists of precipitating on the surface of the pigment particles one or more of the white inorganic hydroxides, silicates, or similar compounds of barium, calcium, aluminum, magnesium, lead, and zinc. After this treatment the pigment is washed to remove the soluble salts. There is an improvement in fineness of grind, viscosity, and flow characteristics of pastes and paints made from these pigments. Blister- and water-resistance is improved. Resistance to chalking of the rutile type pigment is increased greatly.

The necessity for modifying titanium dioxide pigment is brought about by its specific properties and the processes taking place during the reaction of titanium dioxide with various binding substances.[10] Because of the denser packing of atoms in the crystal lattice, titanium dioxide of the rutile type has a greater weathering resistance than anatase. However, a disadvantage of the rutile type is the greater hardness and abrasion of its particles. Weather-resistant titanium pigments easily miscible with binding agents are obtained by heating the particles with various inorganic and organic substances to modify the surface depending on the specific purpose and also by fine grinding, mainly in fluid energy mills utilizing superheated steam. Hydrophobic properties may be imparted to titanium dioxide by coating the particles with methyltrichlorosilane, dimethyltrichlorosilane, polysiloxane oil, monoisocyanate, and other agents. Surfactants of various types are more recent developments. The greatest use is made of cation-active agents which are strongly adsorbed on the hard pigment surfaces forming a more or less continuous monomolecular film which facilitates wetting of the pigment with binding agents. Cation active substances are used not only to increase wettability and dispersion of the pigments, but also to decrease the settling of pigments in prepared paints, to improve the durability of water emulsion, to check the corrosive properties, and also to increase the resistance of paints to chalking and discoloration. Surfactants are introduced during the pigment manufacture, or during the process of dispersing the pigment in the vehicle in the paint manufacture.

The development of rutile type pigments has made possible a wider variety of tinted finishes with good glass retention, chalk resistance, and durability in outdoor use.[11] In general such finishes were not possible with the older anatase type pigments because of their poor resistance to chalking.

According to Krasnovskii and Gurevich [12] the formation of peroxides in paint films containing titanium dioxide leads to accelerated destruction of the binder around the particles of the pigment of the upper layer of the film as a result of which chalking takes place. The peroxides formed during accelerated weathering were measured colorimetrically. Todisco [13] showed the formation of benzene in exposed slides of titanium dioxide ground in glycerol with mandelic acid added. This demonstrates that chalking results from the reduction of titanium dioxide to the trioxide with the simultaneous oxidation of the vehicle catalyzed by light. Pigment grinds in shallow Plexiglass dishes covered with 0.2 mm cover glasses and sealed with a solution of Plexiglass in acetone were exposed at close range under a tubular electroviolet source of 300 to 600 millimicrons (maximum emission at 400 millimicrons). The change was followed by means of changes in reflectance (green filter) of the samples. Reflectance was plotted against the log of exposure time. The method was not very sensitive but allowed the ready recognition of highly chalking and nonchalking pigments.

Vondjidis and Clark [14] explained chalking and photosensitivity of rutile in terms of the bonding of atomic oxygen, formed by the action of ultraviolet rays, to the surface of the particles by an electron sharing process. The rutile surface contains a significant concentration of titanium III ions. Electrical neutrality of the surface is maintained by quasi-free electrons localized in surface states at these ions. Pigmentary rutile has a high specific surface area so that the influence of surface electron states on its photo-oxidizing behavior should be great. Atomic oxygen liberated by absorption of ultraviolet rays is bonded to the surface by an electron sharing process between an oxygen atomic orbital and the localized surface states. The very low diffusion rate of oxygen at room temperature indicates that it originates either in the immediate sub-surface layers or is obtained from pigment environment as part of a catalytic process. For photooxidation the attraction of the pigment environment for the bonded oxygen has to be strong enough to overcome this electron-sharing force. The bonding may be modified by impurity-induced changes in the electron surface states. Aluminum ions calcined into the pigment or absorbed on the surface furnish accepted levels at the surface. Valence electrons raised to these levels behave as oxygen-trapping centers with a consequent improvement in chalking resistance. Pentavalent ions at the surface do not form bonding states but introduce donor impurity states near the conduction bond. In the bulk lattice they inhibit oxygen diffusion, but this process is likely to be insignificant as a result of the low diffusion rates at room temperature.

An explanation for the mechanism of chalking proposed by Van Hoek [15] is that selective adsorption by the titanium dioxide of the vehicle decomposition products during weathering causes noncompatibility between the pigment particles and the binder which results in the formation of a powdered surface. Another view is that titanium dioxide acts as a photosensitizer for accelerating normal oxidation of the vehicle thereby releasing pigment particles.[16] The formation of pertitanic acid which in turn oxidizes the vehicle has been suggested as the cause of chalking.

Renz [17] observed that titanium dioxide, if subjected to actinic rays in the presence of glycerol or other oxidizable medium, is photoreduced to a lower, colored oxide with resultant discoloration of the paste, and that this discoloration disappears with removal of the rays or on exposure to air. This shows that titanium dioxide is a photochemically reactive material.

Based on this basic information Jacobsen [18] developed a photochemical reaction test which gives results that correlate with outdoor chalking behavior of both anatase and rutile grades of titanium dioxide pigment. This laboratory test depends essentially on the photochemical reduction of titanium dioxide to a lower oxide with simultaneous oxidation of the organic medium. Typical media arranged in decreasing order of activity are aqueous solutions of stannous chloride, glycerol, aqueous solutions of mandelic acid, octyl alcohol, mineral oil, alkyl paint vehicle protected from air by glycerol, and alkyl paint vehicle exposed to air. Although glycerol produces the fastest reaction, mandelic acid is employed because of operational advantages, such as workability of the paste. Dyes are unsatisfactory because of their fugitive nature. Photochemically reduced rutile titanium dioxide in the test panels has been identified by its electron diffraction pattern as alpha titanium III oxide. Since the results of these tests correlate with outdoor chalking characteristics of the pigments, it appears that chalking may also be attributed largely to photochemical reactions involving the pigment and vehicle.

In the paint film, the titanium dioxide subjected to actinic light is reduced to the unstable alpha titanous oxide and at the same time organic material is oxidized. The lower oxide reverts to the dioxide by reaction with air but the organic reaction cannot be reversed. By this process the continuity of the film is gradually destroyed so that water can enter and leach out the soluble constituents. This gives a layer of unprotected pigment chalk at the surface of the film. The rate of chalking depends on the reactivity of the titanium dioxide, the nature of the vehicle employed, and the severity of the exposure conditions (brightness of the sun).

The fading of paints tinted with an inert pigment is due to the masking of the tinting material by the chalk. With oxidizable tinting material, fading may be attributed to a photoreduction-oxidation cycle involving titanium dioxide and the tinting material.

Although the General Electric S-1 type sun lamp was used because of its efficiency and convenience of operation, other sources of actinic rays including the carbon arc, the high intensity ultraviolet mercury arc lamp, and direct sunlight effect photoreaction. The samples are located on the outer edge of a rotating table 22 inches in diameter placed 29 inches below the sun lamp. Exposure time is varied in accordance with the resistance of the samples to discoloration. The apparent reflectance readings are made with a Hunter multipurpose reflectometer. To study the rate of photochemical reduction of titanium dioxide mixed with the reducing agent, the percentage change in initial apparent reflectance, that is, discoloration of the paste, is plotted against time of exposure under the standard actinic light source. It is customary to expose a pigment of known photoreaction behavior as a control to permit slight corrections to compensate for any variations in the intensity of the light with age of the lamp or fluctuations in the current supply. Along the edges of the test slides the paste does not discolor. Here the titanium dioxide is in contact with air so that it is reoxidized as rapidly as it is reduced by the mandelic acid.

To carry out the test, 10 g of pigment are made into a soft paste with a 0.5 molar solution of mandelic acid in water by working the mixture with a spatula on a smooth glass plate. After thorough mixing the paste is placed on a glass plate 5 by 5 by 0.04 inches, covered by another plate of the same size and pressed out to a surface area 4 inches in diameter. The edges of the plate are bound with ⅜ inch cellulose adhesive tape to prevent drying of the paste. An initial apparent reflectance is made and recorded. For exposure, the slide is placed on the outer edge of the rotating table below the sun lamp. Additional readings are made at suitable intervals of time. The percentage decrease in reflectance is plotted against time of exposure on semilog paper. In general the curves are not straight lines. For some pigments there is an induction period during which photoreduction does not take place. Consequently, for uniformity the point on the curve corresponding to 6 per cent decrease in the initial reflectance on the ordinate is selected for correlation with outdoor exposure. The paints for outdoor exposure contain 10 per cent titanium dioxide by volume in fatty acid modified alkyd resin vehicle. These paints are tinted with carbon black to give an apparent reflectance of 35 per cent. Steel panels coated with these compositions are baked at 125° C for 2 hours and exposed for outdoor chalking. Other paints may be used. Chalking is indicated by the white powder appearing on the gray surface.

The time usually expressed in weeks of outdoor exposure required to develop "considerable" chalking plotted logarithmically against the time in minutes for the slides to decrease 6 per cent in apparent reflectance may be represented by a straight line. The line is really a median. From this curve the weeks of outdoor exposure required to develop chalking can be read directly from the time in minutes required for the slide to develop a 6 per cent decrease in apparent reflectance.

Some agreement is found between results obtained with laboratory methods such as Weatherometer tests, mandelic acid tests, and Lithol red fading tests, and chalking results from outdoor exposure.[19] To speed up chalking the pigment volume concentration is increased to 33 per cent. The resulting paint is made gray by the addition of carbon black for easier evaluation of the degree of chalking. Chalking is attributed to photoreduction of the pigment, and the anatase crystal unit is more labile than the rutile crystal. With finger rubbing it is difficult to distinguish between heavy chalking paints. The reproducibility of the mandelic acid test is not too good so that several tests have to be run to obtain an average value. With the Lithol red fading test better results are obtained with "free chalking" grades than with the more resistant pigments.

Tests of 500 hours in the Marr Weatherometer corresponded to 2200 hours or 3 months of outdoor exposure, but only for the breakdown of the film by chalking and not for cracking or scaling. With linseed oil paints, chalking is heavier and starts earlier because the oil has less resistance to photochemical breakdown. Alkyd paints are more suitable for testing purposes. Two coats of the gray alkyd paint ground on a three roller mill to 25 microns are brushed on an iron panel 280 by 190 by 2 millimeters and exposed on a roof at a 45-degree angle facing south from May to October. The photochemical reaction under the influence of sunlight leads to the formation of carbon dioxide gas films between the pigment grains and the binder. After losing its adhesion the titanium dioxide particles come off the paint like chalk, masking the surface, and giving increased light reflectance. The chalking rate can thus be followed colorimetrically.

To develop a rapid test for chalking, experiments were carried out with aniline blue FF which is rapidly destroyed in the presence of titanium dioxide by ultraviolet rays.[20] The products of oxidation are weakly colored so that the photosensitizing action of the pigment may be judged from the extent of fading.

## IMPROVED CHALKING RESISTANCE BY CALCINATION WITH ADDED AGENTS

In normal calcination practice the pigment is subjected to a temperature of 900° to 1000° C, depending upon the exact nature of the original material and the properties desired in the final product. The intensity of the calcination is a function of both the temperature employed and the time of exposure, and although some properties improve progressively with the severity of this treatment, others increase to a maximum value and then fall off. For instance, the tinting strength of the resulting pigment increases with the intensity of the calcination to a maximum value and then decreases, while on the other hand there is a progressive improvement in the ability of the pigment to give stable and durable coatings in the usual paint vehicles. If calcination is intensified to increase the tinting strength to the fullest extent, there is a tendency for the pigment to acquire a grayish, yellowish, or reddish tone and often a lack of fastness to light. A still further increase in the degree of calcination to obtain a product that will show great resistance to chalking in paint films normally results in a drop in tinting strength and the material becomes excessively hard so as to render it unsuited for normal use in coating compositions.

Pigments of particularly good durability, brightness, stability, and fastness to light may be obtained by incorporating from 0.1 to 20 per cent antimony trioxide with the hydrolysis product before calcination,[21] although 1 per cent of this agent gave optimum results. In general, the antimony can be more conveniently added as a suspension of the trioxide, although water-soluble compounds (potassium antimony tartrate) may be employed. The principal advantage gained by using the relatively large proportion of antimony is that the intensity of the calcination may be increased so as to enhance the chalking resistance of the resulting pigment without excessive hardness, but on the other hand the degree of brightness resulting from the use of small proportions of this conditioning agent is not always reached, although the tone is usually attractive and varies from pure yellow through neutral white to bluish. The neutral and bluish tones are obtained under particularly intense conditions of calcination. Although the titanium dioxide with the higher proportions of antimony compounds can be calcined at higher temperatures than were formerly possible without causing excessive hardening, the tinting strength of the product will not always be the maximum value and may be lower than could be obtained at a lower temperature. For many purposes, however, it is more important to produce a pigment that yields films of high durability than one having the maximum covering power.

Improvements in the softness of the pigments and the durability of films are noticeable with an amount of antimony trioxide as low as 0.1 per cent of the titanium dioxide, and there is a progressive improvement in these properties as the proportion is increased to about 20 per cent. With proportions above this value the product again hardens and tends to assume a grayish to greenish tone. Under some conditions, antimony trioxide, even in very small amounts, has a tendency to impart a bluish cast to the pigment. By employing an equivalent amount of the pentoxide a less pronounced tone is obtained without adversely affecting the other properties.

A careful study of the reactions involved indicated that the antimony combines with the titanium dioxide to form an antimony-titanium-oxygen compound which is practically insoluble in water, dilute acids, and alkalies. This conclusion is substantiated by the fact that the full amount of antimony added appears in the finished pigment, although the calcination temperature was much higher than that at which antimony oxide would normally be volatilized. Further improvement in the softness of the calcined pigment may be effected by adding the antimony compound along with other conditioning agents, such as potassium carbonate, but phosphoric acid and phosphates in general tend to counteract the beneficial effect of the antimony and should be used with care if at all. This process is also applicable to the production of other pigments in which titanium dioxide is an important constituent as composites and titanates.

Although the antimony compound is normally mixed with the washed pulp, it may be added at any earlier stage of the process; for instance, the trioxide or tartar emetic may be added directly to the titanium salt solution prior to the hydrolysis step. Such compounds may be mixed with the titanium ore before it is reacted with sulfuric acid and the antimony will remain in the solution and precipitate along with the titanium. According to another modification, the antimony compound is mixed with an already calcined titanium dioxide pigment and the mixture is subjected to a second calcination. Regardless of the method, the antimony compound must be thoroughly and uniformly incorporated with the titanium oxide.

In an example, thoroughly washed hydrous titanium oxide pulp obtained by hydrolytic precipitation was intimately mixed with 1 per cent antimony trioxide and 0.4 per cent potassium carbonate, based on the titanium dioxide, and calcined at 980° C. The product was white with a neutral tone and had a high fastness to light. It was very soft, and in ordinary vehicles gave films of great durability in outdoor service. All the antimony originally added was retained in the calcined product, indicating that combination had taken place.

Similarly, titanium dioxide pigment, calcined at a temperature above 900° C with 0.1 to 10 per cent of an oxide of lead or a compound that yields the oxide under the conditions employed, showed improved stability and weather resistance.[22] To improve resistance to chalking, Pall [23] calcined titanic oxide, hydrolytically precipitated from sulfate solutions, first in a reducing atmosphere and then in an oxidizing atmosphere. Additional improvements were obtained by treating the pulp with potassium carbonate as well as with 0.01 to 0.03 per cent ferric oxide and 0.0004 to 0.0025 per cent copper before calcination. By calcining gamma titanic acid under different conditions with varying proportions of zinc oxide, the product consisted of zinc titanate with titanium dioxide in the anatase or rutile form or in solid solution.[24] Such pigments had excellent hiding power and resistance to chalking. Pigments of rutile type possessing superior color, brightness, and resistance to light and weathering were produced by carrying out the calcination in the presence of minor proportions of an alkali metal compound and a water-soluble compound of a metal having a valence greater than 1, such as aluminum, zinc, magnesium, and beryllium.[25]

Pigments of markedly improved resistance to chalking and fading were produced on a commercial scale by heating a mixture of titanium dioxide with approximately 1 per cent zinc oxide or an equivalent amount of other compounds of zinc at 850° to 1050° C until the crystal structure, as determined by X-ray examination, showed a large proportion of rutile.[26] Employing anatase titanium dioxide as the starting material, better results were obtained if the hydrous product was preliminarily calcined in the usual manner to develop first such desired pigmentary properties as tinting strength, color, and oil absorption, and then recalcined with zinc oxide to convert the crystal structure to the rutile form and impart the durability characteristics. The zinc component appeared in the final product as the titanate. Finished pigments were characterized by uniform relatively coarse particle size. This method was also applicable to normal pigments of the rutile type such as those obtained from chloride solutions, and also to calcium sulfate, barium sulfate, and other composites. In an actual operation, 100 kg of previously calcined anatase titanium dioxide was suspended in water to a volume of 400 liters, and a solution of zinc sulfate containing the equivalent of 1 kg of the oxide was added, with vigorous agitation. The slurry was evaporated to dryness, calcined for 30 minutes at 950° C in a rotary kiln, wet-ground, filtered, dried, and pulverized. Examination showed that 90 per cent of the titanium dioxide had been converted to rutile; its average particle-size diameter was 0.45 micron; and color and tinting strength approached those properties of the standard pigment. Peterson [27] found that after proper treatment the same results were obtained with a single calcination. To an aqueous slurry of hydrolytically precipitated anatase titanium dioxide was added 0.5 to 0.75 per cent of a mixture of potassium sulfate and sodium sulfate and 0.25 per cent zinc sulfate. The product was heated at 975° C until conversion to rutile was practically complete as determined by X-ray examination. At the same time the desired pigment properties were developed. Aluminum sulfate, magnesium sulfate, and barium sulfate could be substituted for the zinc sulfate.

Ancrum [28] produced a pigment of the rutile type, having improved brightness and whiteness, by incorporating with the hydrous oxide before calcination potassium silicate corresponding to 0.25 per cent potassium oxide and 0.18 per cent silica, based on the dry weight. Similar improvements were obtained by adding an insoluble white silicate.[29] The finely divided rutile form of titanium dioxide was suspended in water, and the treating agent was precipitated in place by the reaction between sodium silicate and the water-soluble metallic compound. From 2 to 5 per cent of the insoluble silicate gave optimum improvement.

Titanium dioxide pigments of the rutile crystal form, having very high durability and resistance toward chalking and fading in paint films, were produced by a two-stage calcination involving conversion from the anatase modification at elevated temperatures.[30] Hydrous titanium oxide, as obtained by the hydrolysis of sulfate solutions, was first calcined in the usual manner to develop pigmentary properties in the presence of such negative catalysts as alkali metal compounds which inhibit or retard rutile crystallization. The product obtained was recalcined after removing the added agent to effect conversion to the rutile crystal form and at the same time develop the property of resistance to chalking. From 0.3 to 1.0 per cent of the alkali metal compound (potassium sulfate), based on the weight of the pigment, was required in the first calcination, but in the second the proportion of this agent was reduced to below 0.1 per cent. The compound was removed during the usual washing and filtering operations normally employed in pigment manufacture. This process was also applicable to composite pigments.

In an example, hydrous titanium oxide was calcined in the presence of 0.5 per cent, based on the weight of the titanium dioxide, of potassium sulfate at a temperature of 985° C to form anatase titanium dioxide having good pigment properties such as color, tinting strength, and oil absorption. This product was dispersed in water with 0.18 per cent sodium hydroxide and ground in a pebble mill to break up aggregates. The aqueous slurry was coagulated with magnesium sulfate, approximately 0.25 per cent, neutralized with sulfuric acid to a pH of 7, and filtered, dried, and pulverized. This finely divided pigment was then subjected to a second calcination at 1000° C for several hours to develop the rutile crystal modification. At the same time the size of the crystals increased. The product of this second calcination was in a finely divided form, and after being ground by wet or dry methods was suitable for direct use in coating compositions. Average particle-size diameter ranged from 0.35 to 1.8 microns, but the major portion of the pigment was within the range 0.40 to 0.80 micron.

Without sacrifice of the resistance to chalking of zinc-conditioned rutile pigments, the zinc is selectively removed by leaching with a 10 to 25 per cent solution of a mineral acid as sulfuric.[31] A temperature of 70° C is employed for the leaching operation to accelerate the rate of the reaction. The leaching time is from 1 to 2 hours depending on the amount of zinc originally present in the pigment, the particular leaching agent used, the concentration of the acid and the temperature employed. A 100 per cent excess of the acid is used in leaching. A final zinc content of 0.25 per cent is preferred, although values as low as 0.05 per cent can be obtained. After the leaching operation, the titanium dioxide is finished in the usual manner for use as pigment. Coating compositions containing pigments prepared in this manner show better flow and leveling characteristics. The pigments have less tendency to flocculate in organic media. Paints containing these pigments display longer shelf life and have little or no tendency to liver.

Chalking-resistant titanium dioxide pigment having the anatase crystal structure and possessing good color and brightness is prepared by calcining hydrous titanium dioxide at 900 to 1100° C in the presence of compounds of potassium, phosphorus, and barium.[32] The barium and phosphorus are added as barium carbonate and orthophosphoric acid in amounts of 0.2 to 0.5 per cent each calculated as the oxide based on the titanium dioxide. Potassium sulfate in the amount of 0.6 to 1.0 per cent by weight of the titanium dioxide is added to supply the potassium. To complex the quadrivalent titanium which is responsible for the chalking of titanium dioxide pigmented paints, Petit and Poisson [33] incorporated with the pigment before or after calcination from 0.1 to 2.0 per cent by weight of the calcined oxide of an alkali metal fluoride.

## IMPROVING CHALKING RESISTANCE BY COATING

Gardner [34] found that the chalking resistance could be improved by coating the particles with inorganic cementation substances such as hydraulic cement, hydraulic lime, or plaster of Paris. After dispersing the calcined titanium dioxide in a slurry of the coating material, the system hardened by hydration and the solidified mass was crushed and powdered to obtain a particle size suitable for use in film-forming coating compositions.

Pigments suitable for external use were prepared by suspending calcined and ground titanium dioxide in dilute aqueous solutions of salts (sulfates) of the rare earths, of tetravalent compounds of titanium, zirconium, hafnium, thorium, and cerium, and precipitating the hydroxide or hydrated oxide on the suspended particles in such a manner as to fill and seal the interstices or pores of the original material.[35] Precipitation was accomplished by hydrolysis on boiling the solution or by adding sodium carbonate, ammonium hydroxide, potassium hydroxide, or a similar agent, with vigorous agitation. Alternatively the coating or filling material was added to a suspension of the pigment in water as a colloidal oxide or hydroxide and deposited on the individual particles of titanium dioxide by adding a coagulating agent or by evaporating the water. After neutralization of any residual acid, the mixed product was dried and calcined at a red heat. For example, 10 parts by weight of finely ground calcined titanium dioxide was suspended in 100 parts of a solution of titanyl sulfate containing the equivalent of 200 g per liter of the dioxide. The mixture was then heated to effect hydrolytic precipitation of the titanium as hydrous oxide on the suspended particles, and the pore-filled product was separated from the solution, washed, and calcined at 800° to 1000° C.

Patterson[36] prepared a pigment characterized by high resistance to chalking and discoloration by coating the individual particles with 1 to 10 mole per cent of cadmium oxide, hydroxide, or carbonate. According to a typical operation, 4000 parts by weight of regular pigment-grade titanium dioxide was suspended in 16,138 parts of an aqueous solution of 183 parts cadmium chloride. The mixture was agitated until a smooth slurry was obtained, and 80 parts of sodium hydroxide, dissolved in 720 parts of water, was added to precipitate all the cadmium as hydroxide; this compound remained attached to the pigment particles. After thorough mixing the suspension was filtered, washed, dried, and disintegrated to obtain a product containing 2 mole per cent of cadmium hydroxide. The treated pigment was a fine white powder, and paint films pigmented with it exhibited excellent resistance to chalking in outdoor service and to progressive yellowing in indoor use.

Addition of titanium phthalate[37] was found to increase the chalking resistance and also to retard the rate of fading of organic dyes adsorbed on the pigment on exposure to sunlight. This agent was made by adding a titanium salt solution, adjusted to a slight acidity, to aqueous sodium phthalate prepared by neutralizing phthalic acid or anhydride with sodium hydroxide. One mole of phthalic acid reacted with 2 moles of alkali, and the resulting salt reacted with ½ mole of titanic sulfate. A dense white precipitate formed, which usually contained more or less titanium oxide, depending upon the degree of hydrolysis. On drying at 105° C the phthalate became gelatinous and showed a tendency to form semivitreous or horny aggregates resembling silica gel or a hard resin. Because of this property, better results were obtained by precipitating it directly upon suspended particles of titanium dioxide, and the dried material consisted of individual pigment particles coated with a semivitreous shell of titanium phthalate. Linseed oil paints, enamels based on phenolic-tung-oil varnishes, or alkyd resin vehicles and nitrocellulose lacquers pigmented with this composite product showed greatly improved resistance to chalking and color fading as compared with equivalent untreated pigments. From 5 to 10 per cent of the titanium phthalate gave the maximum improvement in these properties. The phthalate did not have any great degree of opacity, however, and thus contributed relatively little to the hiding power of the coating composition.

The most important factor in the destruction of films is ultraviolet radiations; the fact that titanium phthalate transmits none, or a very small proportion, of ultraviolet light may be one of the reasons for its unusual weathering property.

Titanium phthalate, on the other hand, contributed little to the durability of clear varnishes.[38]

In preparing a treated pigment of this type, 2000 g regular calcined titanium dioxide was suspended in 10,000 g of water, and 1000 g of slightly acid sulfate solution containing the equivalent of 42.5 g titanium dioxide was added. Separately, 160 g of phthalic anhydride (a slight excess) was dissolved in 3000 g of water containing 87 g sodium hydroxide and poured gradually into the suspension with the result that titanium phthalate precipitated immediately on the surface of the pigment particles. Thus, approximately 200 g titanium phthalate was precipitated on 2000 g of regular titanium dioxide and after thorough agitation the product was washed, dried, and pulverized.

Similar improvement in the chalking resistance of titanium dioxide pigment was obtained by coating the individual particles with from 1 to 10 per cent of a basic titanium phthalate having approximately the composition indicated by the formula $C_8H_4O_4 \cdot (TiO(OH))_2 \cdot X H_2O$.[39] In carrying out the process, a solution of titanyl sulfate was first prepared. Commercial titanium sulfate cake containing the equivalent of 20 per cent titanium dioxide and 50 per cent sulfuric acid was dissolved in an equal weight of water at about 60° C. The cooled solution was treated with lime slurry to neutralize one third of the total sulfuric acid, and the calcium sulfate formed was filtered off. The filtrate was then analyzed and sodium hydroxide solution was added to adjust the titanium dioxide-sulfuric acid ratio to that corresponding to titanyl sulfate. Titanium dioxide pigment was suspended in the clear solution and a 0.5 mole proportion of aqueous sodium phthalate was added to precipitate the basic compound on the individual particles. The proportions were so adjusted that the final washed and dried pigment contained 3.5 per cent of the coating compound. Although the basic titanium phthalates contained 40 to 50 per cent titanium dioxide, they possessed very little hiding power and were of a brownish color. From 0.5 to 2.5 per cent of a polysubstituted derivative of phthalic acid, such as dialkyl or diphenyl phthalate, also proved effective.[40]

A pigment composition consisting of titanium dioxide intimately associated with from 0.5 to 10.0 per cent of an insoluble salt of a metal from groups III or IV of the Periodic Table with a saturated, dibasic, aliphatic acid containing at least three carbon atoms to the molecule gave paint films having high durability and improved resistance to chalking and fading.[41] The more effective acids of the group were found to be malonic, succinic, glutaric, adipic, and pimelic. In a typical operation, 1000 parts by weight of calcined and ground titanium dioxide pigment was suspended in 6000 parts of water, and 89 parts of titanic sulfate in 300 parts of slightly acid solution was added. To the agitated mixture was then introduced a separately prepared solution of 66 parts succinic acid dissolved in water containing 45 parts sodium hydroxide. The titanium succinate, approximately 100 parts, precipitated immediately on the surface of the pigment particles, and after thorough agitation the composite product was filtered, washed, dried, and pulverized for use in film-forming coating compositions.

These desirable chalk- and fade-resistant properties were also obtained by mixing regular titanium dioxide pigment with a small quantity of a dibasic acid or anhydride at any stage of the manufacturing process following calcination.[42] Phthalic acid and anhydride were found to be the most efficient agents of the group.

New[43] observed that the addition of 3 per cent chromium oleate to paints pigmented with titanium dioxide greatly increased the stability of the resulting films. In some tests the period of exposure to weathering required to cause chalking to reach a standard value was doubled. The effectiveness of the oleate or resinate of tin was less conclusive, for although 3 per cent of these agents showed a similar improvement with one paint, no improvement was detected with another of similar composition. Beneficial results were also obtained by treating the pigment particles with the chromium compound before incorporation into the vehicle.

The resistance to weathering of paint films containing titanium dioxide was improved by incorporating a small proportion of an oxidizing agent such as a peroxide or per salt of an alkali or alkaline earth metal (barium peroxide) which was capable of liberating hydrogen peroxide by reaction with the acid of the vehicle.[44] Neutralizing agents, such as oxides, carbonates, or hydroxides of zinc, magnesium, and aluminum were also added for this purpose. The resistance of titanium dioxide pigments to the action of light and atmospheric agencies was improved by heating them in alkaline solutions as an added step in the manufacturing process.[45]

Goodeve[46] observed a sharp drop in the reflecting power of titanium dioxide and zinc oxide at 400 and at 385 millimicrons, respectively. These oxides were reported to be a source of photoactivation for the oxidation of oil vehicles or for the bleaching of adsorbed dyes, and the action of each was specific. For example, titanium dioxide promoted oxidation of linseed oil while zinc oxide did not. From these data Goodeve ascribed the reduction of chalking by zinc oxide to inner filter action.

Acicular zinc oxide mixed with titanium dioxide pigments was particularly effective in arresting chalking without cracking.[47] Jaeger[48] added 5 per cent zinc oxide to a titanium dioxide pigment containing 2.5 per cent combined phosphoric acid to prevent disintegration of the resultant paint film on outside exposure. Similar results were obtained by adding 5 to 25 per cent of this agent to relatively pure titanium dioxide or alkaline earth sulfate base composites.[49] A nonchalking pigment consisting of 5 to 25 per cent titanium dioxide and 75 to 95 per cent zinc carbonate was obtained by adding sodium carbonate to a suspension of finely divided calcined titanium dioxide in a zinc sulfate solution, with continuous agitation.[50] The composite precipitate was washed, dried, and pulverized.

White pigments consisting of titanium dioxide of the rutile crystal form together with zinc orthotitanate proved very resistant to chalking.[51] Fillers and extenders could be included in the composition. The mixture was heated at 840° C to effect combination with the zinc oxide to form the titanate and to convert the unreacted titanium dioxide to the rutile modification as determined by X-ray examination. Part of the zinc could be replaced by magnesium. Large quantities of anatase were transformed to rutile with a small proportion of zinc, since the titanium dioxide seemed to pass through the titanate stage and then break up to yield rutile and zinc oxide to form more titanate. As an illustration of the process, 1 part of zinc oxide was intimately mixed with 3 parts of titanium dioxide of the anatase type and heated for 12 hours at 900° C. The product was a white pigment consisting of zinc orthotitanate with 60 mole per cent rutile, and it possessed good covering power and resistance to disintegration on weathering. In another example, a mixture of hydrous titanium oxide and barium sulfate, containing 25 per cent titanic oxide on the dry basis, was heated with 7 per cent zinc oxide for 10 hours at 900° C to produce a chalk-resistant pigment consisting of zinc orthotitanate, rutile, and barium sulfate.

Titanium oxide pigment treated with 0.5 to 10 per cent of a metal salt of a water-soluble, relatively volatile organic acid was found to impart to paint films color stability and improved resistance to chalking and weathering.[52] Salts of aluminum, zinc, cadmium, beryllium, magnesium, calcium, barium, and strontium with acetic, formic, propionic, butyric, citric, and oxalic acids were particularly effective. Aluminum acetate, for example, is quite unstable in solution and can be readily hydrolyzed by heating, thereby depositing relatively insoluble basic aluminum acetate on the surface of suspended pigment particles.

In actual operation of the process, 4000 g of ground pigment was mixed with 3500 ml of solution containing 183 g zinc acetate. The paste was thoroughly mixed to break up aggregates and to procure a uniform distribution of the components, after which it was dried, and the product was pulverized in a hammer mill.

Titanium dioxide pigments, in association with chromium naphthenate, possessed markedly reduced chalking and fading characteristics.[53] The chromium naphthenate was applied as a surface coating to the pigment particles or by mixing chromium naphthenate into the film-forming composition pigmented with titanium dioxide. In the former the pigment was suspended in an aqueous solution of an alkali metal naphthenate, and water-soluble chromium chloride or sulfate was added. Thus by double decomposition there was precipitated on the suspended particles a coating of chromium naphthenate. Proportions of 0.2 to 5 per cent were used, although the lower limit usually gave the desired results.

The resistance to fading, chalking, and loss of gloss-retention of titanium dioxide in paint films is appreciably improved by coating the individual particles of the calcined pigment with very small amounts of hydrous oxides of both aluminum and chromium.[54] Furthermore, dyed artificial fibers delustered with such a pigment showed good light resistance and color retention. Between 0.05 and 0.15 per cent chromium oxide and between 0.5 and 1.5 per cent aluminum oxide, both in the hydrated form and based upon the weight of the pigment, gave optimum results, and these agents were added directly to the suspension or precipitated in place. The pH of the suspension was then adjusted to a value of 6 or 8 by adding a neutralizing agent, and the mixture was heated to 60° to 90° C and held at this temperature, with stirring, to insure an intimate association of the components. Subsequently the slurry was dewatered and the pigment was dried at a temperature not over 250° C and dry milled.

In a typical operation, a 20 per cent aqueous slurry of the calcined wet-milled titanium dioxide was heated to 75° C and solutions of aluminum sulfate and chromium sulfate equivalent to 1.0 per cent and 0.1 per cent, respectively, of the corresponding oxides on the pigment basis were added. The mixture was maintained at a temperature of 80° to 90° C and stirred continuously for 2 hours. At this stage heating was discontinued, and ammonium hydroxide was added with agitation until a pH of 7 was attained, thereby precipitating the aluminum and chromium as hydrous oxides on the surface of the pigment particles. The suspension was agitated for an additional hour, after which it was filtered and the treated product was dried at 160° C and disintegrated by passage through a rotary-hammer mill.

The beneficial effect of some of these agents was found to be additive. A combination of 0.05 per cent chromium, as oxide, 2.0 per cent zirconium silicate, and 1 per cent alumina, separately precipitated upon the particles of calcined titanium dioxide, was very effective in increasing the resistance to chalking and to discoloration.[55] The colored chromium oxide did not impair the brightness of the finished pigment.

Incorporation with the titanium dioxide, after calcination, of 0.02 to 1.5 per cent precipitated beryllium oxide or hydroxide inhibited discoloration in baked enamels.[56] Similar improvements were effected by incorporating 2 to 3 per cent elemental molybdenum[57] of 150-mesh particle size, and by coating the pigment particles with 0.002 to 2.0 per cent chemically precipitated cobalt oxide.[58] Calcined pigment particles in suspension were treated with zirconium sulfate solution in a predetermined proportion, and phosphoric acid was added to precipitate the zirconium as phosphate. The mixture was simply washed to remove any residual acid, dried at 110° C, and pulverized in the usual manner.[59] With appropriate modifications the treatment was found to be applicable to hydrous titanium oxide.

Increased durability in outdoor service was developed by coating the individual particles with fluorides that were soluble only with difficulty.[60] This treatment was effected by precipitating the fluoride compound directly onto the pigment, or by suspending the titanium dioxide in a solution of a complex fluoride, e.g., a salt of fluosilicic acid and heating the mixture to convert the complex into the corresponding fluoride. For instance, finely divided pigment was stirred into aqueous ammonium fluoride containing magnesium chloride or a 40 per cent fluosilicic acid solution, and aqueous ammonium or barium hydroxide was added to effect precipitation, whereby the individual particles were coated with basic magnesium fluoride or with barium fluorosilicate, according to the method followed. The product was filtered, washed, and pulverized.

Improved drying rate, tint retention, and chalk resistances are imparted to rutile titanium dioxide pigments by precipitation of aluminum orthophosphate on the particles.[61] To an aqueous suspension of titanium dioxide is added the desired amount of an alkali metal orthophosphate. After thorough mixing, a solution of an aluminum salt is added to precipitate and introduce in the pigment 1.0 to 3.5 per cent by weight on an anhydrous basis of an aluminum orthophosphate in which the molar ratio is at least 1 to 1. For example, to obtain a titanium dioxide pigment containing 2.226 per cent of an aluminum orthophosphate product having a molar ratio of aluminum oxide to phosphate of 3 to 1, 1 liter of a slurry containing 300 g of rutile titanium dioxide was well stirred, and 11.5 g of crystalline trisodium phosphate dissolved in 150 ml of water was added slowly. After stirring the mixture for 15 minutes, a solution of 30.25 g of crystalline aluminum sulfate in 200 ml of water was added slowly. Stirring was continued for 5 minutes, after which the pH was raised to 7.2 by the slow addition of 79 to 80 ml of a 10 per cent sodium hydroxide solution. The slurry was then heated to 60° C, maintained at this temperature for 1 hour with continuous agitation, and filtered. The filter cake was washed with 1800 ml of water, repulped in 1000 ml of water, allowed to stand overnight, and again filtered. Finally the product was dried at 105° C and pulverized. An anatase titanium dioxide pigment treated in the same manner showed the same degree of improvement in drying time and stability.

For comparison, a control with no treatment required 5.5 hours drying time, chalked moderately, and had a low tint retention after 30 days outdoor exposure. A sample treated with aluminum silicate (1.33 per cent alumina and 1.19 per cent silica) required 4.5 hours to dry, chalked slightly, and had a tint retention of 8.3 after 30 days exposure. Another sample treated with aluminum orthophosphate (1.33 per cent alumina and 0.93 per cent phosphorus pentoxide) required only 4 hours drying time, showed no chalking, and had a tint retention of 1.1 after 30 days outside exposure. Less than 1 per cent aluminum orthophosphate gave little improvement and more than 3.5 per cent caused a loss of hiding power.

## ALUMINA COATINGS

The discoloration or progressive yellowing of films pigmented with unmodified anatase grade of titanium dioxide in indoor service has been ascribed to the formation of peroxide linkages at the unsaturated double bonds in the drying oil of the vehicle.[62]

Pigments coated with aluminum oxide produce a minimum of discoloration in resin vehicles under conditions of both air drying and baking. In fact, such films actually bleach on exposure to diffused light.

Farup [63] developed a process of coating the individual particles of titanium dioxide with insoluble compounds, for example oxides, hydroxides, and sulfates of aluminum, zinc, and lead, as a protection against destructive external influences, without at the same time materially impairing their pigmentary values. This treatment rendered the pigment more resistant to physical and chemical changes and prevented discoloration of paint films on exposure to light and heat. The coating was effected by precipitating the desired compound directly on the titanium dioxide particles from an aqueous solution in which the pigment was suspended. Suitable precipitants were sodium carbonate, sodium hydroxide, ammonium hydroxide, and a soluble white sulfate. Since only a small proportion of the protecting material was necessary to effect stabilization, larger amounts were used only to modify the inherent characteristics of the pigment.

In a typical operation of the process, calcined titanium dioxide was finely ground and suspended in water. A concentrated solution of aluminum sulfate was added in amount to yield alumina equal to 2.0 per cent of the suspended pigment, and after mixing thoroughly, aqueous sodium carbonate was added to precipitate the aluminum as hydrous oxide. Agitation was continued until the reaction proceeded to completion. The solid product, consisting of particles of titanium dioxide covered with hydrated aluminum oxide, was washed thoroughly, dried, and pulverized to yield a pigment stable against the action of weathering, heat, and light.

In a modified method of applying the coating, a suspension of the pigment in a solution of aluminum sulfate containing the equivalent of 1 per cent of the oxide based on the titanium dioxide was heated to 80° C and agitated for 16 hours, after which the temperature was again brought to 80° C.[64] Ammonium hydroxide solution was then added to raise the pH to 7.4, and agitation was continued for an additional 2 hours. The coated pigment particles were filtered, dried at 130° C, and pulverized. By another procedure, a suspension of the pigment, a solution of aluminum sulfate, and aqueous sodium carbonate were introduced simultaneously into a reaction vessel.[65] Seidel[66] treated an aqueous suspension of calcined titanium pigment with 2 per cent (on the solid basis) of a soluble compound of titanium and of aluminum or chromium, and adjusted the pH to 7 to effect precipitation. The treated product was washed, dried, and pulverized.

Patterson[67] produced a pigment having high stability and improved resistance to weathering, discoloration, chalking, and lack of tint retention by blending separately prepared and washed hydrous aluminum oxide with a slurry of calcined and ground titanium dioxide. This process had the advantage over methods by which the alumina was precipitated in the presence of suspended particles of pigment, in that it avoided the introduction of soluble salts into the product during the treatment. Similar improvements in properties were obtained by exposing finely divided titanium dioxide to vapors of aluminum chloride at a temperature above 300° C.[68] Enamels pigmented with this product did not darken on exposure to ultraviolet light. Because of the adsorptive nature of such pigments, the soluble salts cannot be completely removed by washing, and unless these are reduced to extremely low proportions, 0.25 per cent, they exert an injurious effect upon the durability of exterior paints and coating compositions in which the pigment is employed.

Alumina of the desired quality was prepared by adding an alkaline agent to a solution of aluminum sulfate, with agitation in the cold, until the mixture was neutral as determined by pH measurements. The precipitate was washed free of soluble salts by decantation or filtration and blended as an aqueous slurry with a suspension of the titanium pigment. To obtain uniform incorporation of the components, the suspension was agitated thoroughly, after which the treated pigment was filtered, dried, and pulverized. Maximum benefits were in general obtained by applying 1 per cent of the agent to the titanium dioxide, although amounts from 0.1 to 2 per cent were employed to suit individual requirements. There are numerous uses for which treated pigments are not required, and it is apparent that several different products can be prepared from the same batch of raw pigment.

In an example, 15 parts by weight of crystalline aluminum sulfate was dissolved in 250 parts water, and dilute ammonium hydroxide was added slowly, with agitation, until the pH of the slurry was brought to 7.0 to 7.2. The precipitated hydrated aluminum oxide was then washed by decantation and filtration to remove soluble salts, after which it was slurried in water to 250 parts by weight. This suspension was added to a water slurry of 230 parts calcined and ground titanium dioxide, and after thorough mixing the treated pigment was filtered, dried, and pulverized in the conventional manner.

Precipitation and washing of the alumina were carried out in the cold, and the product was not allowed to dry before use, otherwise its effectiveness was greatly diminished. Better results were obtained by blending the agent immediately after precipitation with pigment slurry.

Hanahan and McKinney [69] precipitated hydrous alumina upon titanium dioxide pigments by reacting aluminum sulfate with an alkaline earth metal hydroxide, the sulfate of which was insoluble in water. Pigment produced in this manner was practically free from soluble salts and showed improved resistance to chalking, discoloration, and fading of tint. A solution of aluminum sulfate was added to a water suspension of titanium pigment, and the pH of the slurry was brought to 7 by adding, with vigorous stirring, an equivalent amount of barium hydroxide solution. This resulted in precipitation of the aluminum on the surface of the pigment particles in the form of hydrous oxide and at the same time the sulfate was converted to insoluble barium sulfate. The slurry was then filtered, and the pigment was dried and pulverized for use in coating compositions. Alumina content of the final product ranged from 0.02 to 2 per cent, but in general 1 per cent gave optimum results.

The resistance to chalking and discoloration was also increased by suspending the calcined and finely divided pigment particles in an aqueous solution of an alkali metal aluminate and allowing it to remain until a small amount of an aluminum compound was adsorbed by or on the individual pigment particles.[70] According to a more direct method, a neutralizing acid (sulfuric) was added to the solution to decompose the aluminate and precipitate the aluminum hydroxide on the particles, after which the suspension was heated at 70° to 90° C for 3 to 5 hours. The coated product was filtered, washed, dried, and pulverized. In an example, 1 kg calcined titanium dioxide was ball-milled for 2 hours at room temperature with 3 kg of a sodium aluminate solution containing the equivalent of 10 g aluminum oxide, 1 per cent, based on the titanium dioxide. The aluminate was prepared by adding sodium hydroxide to a solution of aluminum sulfate to precipitate the hydroxide, and an additional amount to redissolve the precipitate. After milling was completed, the slurry was discharged into a suitable vessel, diluted to 15 per cent solids, and heated for ½ hour at 80° C. The titanium dioxide thus coated with alumina was separated from the liquor, washed to remove the sodium sulfate, dried, and pulverized.

Better weathering characteristics are obtained by adding sulfuric acid to the slurry of titanium dioxide before adding the sodium aluminate than by adding the reagents in the usual reverse order.[71] After coating the particles in this manner the pigment is filtered, washed, dried, heated at 800° C for 10 minutes, and jet milled.

Similar results may be obtained by stirring the calcined and ground titanium dioxide in a solution of a basic aluminum salt such as a highly basic chloride, formate, or acetate.[72] Pigment particles treated in this manner acquired a coating of aluminum oxide and showed improved resistance to weathering and discoloration. Basic aluminum chloride was prepared by dissolving freshly precipitated aluminum hydroxide in aluminum chloride solution or by adding sodium carbonate to an aluminum chloride solution. Composition may be expressed according to the degree of substitution, for instance, $Al_2Cl_5OH$, $Al_2Cl_4(OH)_2$, and $Al_2Cl_2(OH)_4$. According to a specific operation, 10 parts by weight of titanium dioxide pigment was stirred for 3 hours at room temperature with 80 parts by weight of a clear 0.10 per cent solution of highly basic aluminum chloride containing 1 mole of chloride for each 2 moles of aluminum. The suspension was then filtered, washed, and dried.

Krasnovskii and Kiselev [73] added oxides of aluminum, zinc, magnesium, calcium, and barium to increase the resistance to chalking and discoloration. The improvement increased with ionic radius of the oxide.

Hanahan and McKinney [74] suspended calcined titanium dioxide in a solution of barium aluminate and added aluminum sulfate to precipitate small proportions of aluminum hydroxide and barium sulfate on the individual pigment particles. Similarly, a pigment suspension was mixed with a solution of aluminum acetate and calcium hydroxide was added to precipitate a basic aluminum acetate compound and acetic acid was added to neutralize the solution exactly.[75] Surface coatings of hydroxides of aluminum, antimony, tin, or bismuth were deposited effectively upon the surface of the pigment particles by hydrolysis of the corresponding chloride, formate, arsenate, propionate, or butyrate solution.[76]

A pigment consisting of a mixture of titanium dioxide and hydrated or anhydrous alumina gave paint films of superior whiteness and improved resistance to chalking.[77]

To improve the chalking resistance of titanium dioxide pigment, the individual particles are coated with 0.5 to 5.0 per cent of hydrous alumina, calculated as aluminum oxide on the weight of the titanium dioxide followed by calcination at 500 to 800° C for 10 minutes to at least partially dehydrate the alumina.[78] An aqueous solution of an ionizable aluminum salt such as the sulfate is added to the titanium dioxide slurry with stirring. Ammonium hydroxide is then added to precipitate the hydrous aluminum oxide on the pigment particles, after which the coated product is filtered, washed, and calcined.

Rutile type titanium dioxide pigment is produced by the vapor phase oxidation of titanium tetrachloride in the presence of 0.01 to 10.0 per cent of a volatile aluminum compound which is converted to alumina under the conditions of the process.[79] Between 0.1 and 1.0 per cent of ferric chloride based on the weight of the aluminum chloride is included to avoid discoloration. Pigment produced by the vapor phase oxidation of titanium tetrachloride with oxygen or air is improved in durability by the deposition of 0.1 to 10.0 per cent of alumina on the particles by the oxidation of aluminum chloride at a lower temperature.[80] The effluent gas from the primary reaction at 1000° C is cooled to 250 to 750° C and mixed in the gas stream with the aluminum chloride vapor. This reaction produces aluminum oxide as a solid and titanium tetrachloride as a vapor. The alumina thus made is associated with and collects on the particles of titanium dioxide from the primary reaction. Product gases with entrained titanium dioxide-alumina from the secondary reaction are then treated at 200 to 500° C with sufficient steam to convert the titanium tetrachloride to the dioxide. In addition to decreasing the absorption of acids by the finished pigment mixture, the alumina makes the pigment more stable to the action of light.

Vaporized titanium tetrachloride is oxidized with air in the presence of aluminum chloride equivalent to from 0.3 to 3.0 per cent of alumina based on the weight of the titanium dioxide to produce a pigment which on normal calcination has a pH of 6.5 to 7.5 so that conventional neutralization is not required.[81] In the process the aluminum chloride is oxidized to white aluminum oxide which joins the pigment. The usual amount of water vapor, 0.1 to 5.0 per cent by volume based on the total volume of gases present, is added to the air. In an example, aluminum chloride vapor at a temperature of 860° C was added to a stream of titanium tetrachloride at 920° C to form a mixture consisting of 97.7 per cent by volume of titanium tetrachloride and 2.3 per cent by volume of aluminum chloride. This vapor mixture was continuously fed to a slot jet type of mixing and reaction vessel constructed of silica in the form of a sheeted stream at a rate of 30.4 moles per hour. Simultaneously air preheated to 930° C, to which 0.5 per cent by volume of water vapor had previously been added, was separately and continuously fed into the device through a separate inlet of the tubular element remote from the slot inlet to flow downward past the peripheral slot and into the reaction zone. The reaction zone was maintained at a temperature of 1050° C, and the reactants remained within the zone for only 0.21 second. The resulting gaseous suspension of composite titanium dioxide pigment was discharged at 1030° C and quickly quenched to 250° C in less than 2 seconds by introducing cold chlorine gas. After separation in filter containers the pigment was calcined at 600° C for 1.5 hours and ground in a Raymond mill.

The titanium tetrachloride and aluminum chloride for oxidation are proportioned by a process including distillation.[82] For example, 0.1 to 5.0 per cent aluminum chloride is dissolved in titanium tetrachloride and the mixture is distilled. The distillate is then continuously flash volatilized. The ratio of titanium tetrachloride to aluminum chloride in the gas stream is the same as that in the distillate.

Titanium dioxide pigments produced by the vapor phase oxidation of titanium tetrachloride are improved in color retention by coating with alumina during the fluid energy milling operation.[83] A finely divided titanium dioxide pigment as is obtained from the vapor phase oxidation of titanium tetrachloride at 950 to 1100° C with air containing 0.1 to 10.0 per cent water vapor is first intimately mixed with 1 to 4 per cent by weight of an aluminum alcoholate such as aluminum isopropoxide or aluminum phenate. Admixture is conveniently effected by adding a dilute solution of the aluminum compound in an inert solvent as carbon tetrachloride by spraying onto the pigment spread out on a tray, on a moving belt, or in a tumbling barrel. A more uniform distribution of the aluminum isopropoxide is obtained by subjecting the treated pigment to mechanical blending treatment in a hammer mill, high speed mixer, or other type of disintegrating mill. The resulting aluminum isopropoxide treated pigment is dry ground in a fluid energy mill. In this mill the pigment particles are conveyed in streams by superheated steam into the outer portion of an inwardly spiraling vortex at high velocity and in a manner which maintains the vortex at a high rotative speed and a relatively low inward speed to cause the pigment particles to rub or strike against each other. The superheated steam thus serves simultaneously to hydrolyze the aluminum isopropoxide on the surfaces of the finely divided pigment particles as they are formed and to furnish the energy required for the grinding.

In an example, 3600 parts of titanium dioxide produced by the vapor phase oxidation of titanium tetrachloride and calcined at 700° C to remove absorbed chlorine was spread out on a tray and sprayed with a solution of 72 parts of aluminum isopropoxide in 160 parts of carbon tetrachloride. The treated pigment was then passed for further mixing through a hammer mill pulverizer with the screen removed. With the screen removed the unit functioned as a mixer rather than as a pulverizer. After thorough mixing in this manner the treated pigment was passed through an 8 inch jet pulverizer at a rate of 300 parts a minute, using steam at 100 psi and 550° F for the feed jet and ring. Alumina from the hydrolysis of the aluminum isopropoxide precipitated on the surfaces of the particles of the titanium dioxide pigment. Baked enamel films pigmented with this alumina-coated pigment did not yellow as did the untreated control pigment.

## SILICA COATINGS

Robertson [84] coated the pigment particles with 2 to 5 per cent of a silicate of zirconium, aluminum, magnesium, cerium, zinc, or barium to improve the resistance to weathering and to discoloration. In an example, calcined titanium dioxide was slurried in a solution of basic zirconium chloride and aqueous sodium silicate was added as the precipitating agent. Similar improvements were obtained by coating the titanium dioxide with a small proportion of an alkali metal silicate.[85] A slurry of the pigment was mixed with sodium silicate, and barium chloride was added to precipitate barium silicate.

Kinzie [86] coated or impregnated the individual particles of titanium dioxide with small proportions of the oxide or silicon complexes of zirconium to increase the resistance to chalking. The zirconium compound was incorporated with the hydrous titanic oxide or with calcined material, but in either case the mixture was calcined to develop the desired properties. Proportions of the added agent were employed, expressed as the oxide, from 0.5 to 5 per cent of the base pigment, but values of 1 per cent in general gave optimum improvement.

Calcined titanium dioxide pigment was treated with a relatively small proportion (below 5 per cent) of a silicon complex such as sodium zirconium silicon citrate, zirconium silicon chloride, or sodium zirconium silicon chloride, and the mixture was lightly calcined to 700° C. On heating the mixture, zirconium oxide and silicon oxide were formed in the same proportions in which they occurred in the complex, $ZrSiO_4$. The sodium chloride remained as such and was removed by washing.

Rutile pigments of improved brightness and tone result from the addition of an alkali metal silicate or silicon dioxide to the hydrolyzate before calcination.[87] The proportion of the agent added is such that the final pigment contains 0.02 to 0.5 per cent silicon dioxide. Calcination is carried out at 800 to 1000° C. The method is more applicable to hydrolyzates which yield rutile directly on calcination. Potassium silicate is a very suitable agent. Titanium dioxide pigment containing silica is formed by the vapor phase oxidation of titanium tetrachloride with free oxygen under flame conditions in the presence of 0.01 to 10.0 per cent of a volatile silica compound.[88] To avoid premature oxidation of the tetrachloride a maximum preheating temperature of 500° C is employed. The reaction is carried out by utilizing an auxiliary flame and at a temperature of 750 to 1200° C in the atmosphere immediately ambient to the flame. Pigments produced by this process contain more than 90 per cent anatase.

Titanium dioxide is ground in a fluid energy mill in the presence of 0.05 to 5.0 per cent of an alumina aerogel or of a silica aerogel to improve the whiteness of the finished pigment and to prevent excessive wear on the steel parts of the grinding equipment.[89] Hydrolyzate calcined at 800 to 1000° C to develop pigment properties, as well as the anhydrous product obtained directly by the vapor phase oxidation of the tetrachloride at approximately 1000° C, contains some hard gritty particles which must be disintegrated by grinding. Although wet milling methods are effective, dry grinding in fluid energy mills is employed widely because of greater effectiveness and economy. By this process a more uniformly textured pigment is produced from which a more light-stable and heat-stable paint film is obtained. The aerosol can be intimately mixed with the pigment before dry milling by mechanical blending in standard equipment or by passing the comixture through high speed mixers or disintegrators. If desired, the pigment and aerogel can be separately fed simultaneously in the desired proportions to the dry grinding stage of the pigment producing operation. The term aerogel refers to the particularly voluminous dried gel prepared from a large number of colloidal systems by removing the liquid from the gel under conditions that prevent shrinkage. Typical silica aerogels have bulk densities from 5 to 10 pounds per cubic foot.

For example, 100 pounds of titanium dioxide of pigment grade produced by the oxidation of the tetrachloride was thoroughly mixed with 1 pound of commercial silica aerogel. The resulting mixture was then fed into an 8 inch diameter stainless steel fluid energy mill at a rate of 0.625 pound per minute. In addition to improved brightness, the pigment treated with the alumina aerogel showed a marked resistance to discoloration in enamels baked at high temperatures.

## PHOTOSENSITIVITY

Renz [90] found that titanium dioxide yielded dark-colored reaction products in the sunlight and in the presence of organic acceptors. Similarly, Williamson [91] noted that certain samples of titanium dioxide darkened reversibly on exposure to sunlight. Ceramic glazes prepared from these samples had similar properties. In some cases photosensitive glazes were prepared from materials which before combination did not show this phenomenon. This photosensitivity appeared to be a function of the source of the raw material from which the oxide was prepared. Ilmenite from Norway, for instance, tended to produce sensitive oxide. Extremely small proportions of certain metallic compounds as impurities, such as ferric oxide, aluminum oxide, and rubidium oxide, intensified this tendency toward photosensitivity.

According to Williamson,[92] the photoluminiscence of impure rutile is probably connected with the entry of impurities, particularly iron compounds, into the crystals during calcination. For example, a pure sample of nonphotosensitive titanium dioxide mixed with enough of an iron salt to add 1 part in 500 of ferric oxide, ground, and calcined for 3 hours at a final temperature of 1000° C, gave a product which exhibited reversible darkening in daylight.[93]

The photoactivity of titanium dioxide and other oxides in the presence of glycerin was studied by Renz, who showed that the titanium dioxide was reduced to the sesquioxide, while the glycerin was oxidized to aldehyde and acids.[94] A paint prepared from linseed-oil varnish and titanium dioxide pigment tinted with Prussian blue showed considerable bleaching after exposure to sunlight for 35 days. The effect could not be explained by the presence of residual sulfuric acid, for lithopone, which may develop acid on weathering, did not bleach Prussian blue.[95] Wagner [96] concluded, however, that such fading was caused by residual sulfuric acid. In some cases the bleaching was attributed to the formation of titanium sulfate, and in others to the trioxide. Addition of zinc oxide in proper proportion produced a stable pigment which did not bleach the dyes.

A series of paints in vehicles of 15 per cent dextrin solution alone, and mixed with 5 per cent glycerol, casein solution, and oil, and tinted with azo dyes which contained no sulfo group and also with some basic dyestuffs (madder lake K), was pigmented with titanium dioxide, zinc oxide, lithopone, and white lead. Films containing titanium dioxide faded more rapidly than the others on exposure to the sun or to ultraviolet light.[97] The bleaching by light of a powder made by dyeing titanium dioxide with "Chlorazal Sky Blue" was studied by Goodeve and Kitchener.[98] After an initial slow period, the quantum efficiency depended upon the concentration of vulnerable dye molecules.

Titanic oxide, sulfate, and chloride in alcohol or oxalic acid solution turned blue or brown slowly on exposure as a result of formation of simple or complex salts.[99]

A pigment which prevents the discoloration of melamine and urea resins and laminates by ultraviolet light is prepared by coating calcined rutile with hydrated silica and alumina followed by calcination at 650 to 800° C for 1 to 3 hours to effect partial dehydration.[100] The coating materials are precipitated on the suspended particles from water solutions of sodium silicate and aluminum sulfate. Paper laminates are prepared by impregnation with a suspension of the coated rutile and the resin dissolved in alcohol. Samples of anatase pigment coated with aluminum hydroxide, aluminum phosphate, and silica showed decreased photochemical activity and enhanced hydrophobic properties.[101] All modified samples exhibited higher resistance to chalking.

Rutile pigments having improved photochemical stability are produced by coating the titanium dioxide particles with 0.05 to 1.0 per cent cerium oxide, 0.5 to 3.0 per cent alumina, and 0.5 to 3.0 per cent silica.[102] Resinous materials such as melamine resins and paper laminates, containing conventional rutile pigments show a strong gray discoloration on exposure to light. Such pigments treated with silica and alumina show improved resistance toward chalking but are not suitable for incorporation in plastics. Calcination of the hydrous titanium dioxide with cerium salts also improves the resistance of the pigment to chalking. By employing these three agents in combination a considerable increase in the photochemical stability is achieved. To apply the coatings, rutile pigment is slurried in water, a white water soluble silicate, an aluminum, and a cerium salt are added as aqueous solutions. An alkali is then added to precipitate the silicon, aluminum, and cerium as hydrous oxides on the pigment particles. The coated product is filtered, washed, dried, and milled.

As an example, a rutile pigment was slurried in water with the aid of a dispersing agent. By wet milling in a ball mill and classification in a centrifuge, the coarser particles were removed and subjected to additional grinding. One liter of this dispersion containing 300 g of titanium dioxide was heated to 60° C with continuous stirring. Then 28.5 ml of a sodium silicate solution containing 190 g of silicon per liter, corresponding to 1.8 per cent silica by weight of the pigment, was added. The mixture was stirred for 10 minutes. A solution of 37.0 g of crystalline aluminum sulfate in 100 ml of water, corresponding to 2.1 per cent alumina based on the weight of the pigment, was added. Stirring was continued for an additional period of 10 minutes. At this stage a solution of 2.12 g of ceric sulfate hexahydrate, corresponding to 0.3 per cent ceric oxide by weight of the pigment was added. The mixture was stirred for 20 minutes. With the temperature of the slurry still at 60° C, dilute aqueous ammonia was added to adjust the pH to 8.1. The mixture was then stirred for 30 minutes with the temperature maintained at 60° C. After separation from the suspension by filtration, the pigment was washed, dried for 15 to 20 hours at 120° C, and dry milled. The pigment aftertreated in this way showed a considerable improvement in photochemical stability as well as greater resistance to chalking. It was especially suitable for incorporation in resinous materials such as melamine resin and paper laminates, and for lightening colored pigments.

The photochemical activity of anatase or rutile pigments is determined by the rate of decomposition of methylene blue.[103] If irradiated with ultraviolet light in the presence of photochemically active pigments, methylene blue is converted to the colorless leuco base. The photodegradation of melamine-formaldehyde films pigmented with titanium dioxide was followed by changes in their infrared absorption.[104]

According to Vondjidis and Clark [105] the titanium dioxide surface contains a significant concentration of titanium III ions and the electrical neutrality is maintained by quasifree electrons localized in surface states at these ions. Because of the high surface area of titanium dioxide, the influence of surface electron states should be considerable. Atomic oxygen liberated by ultraviolet absorption is bonded to the surface between oxygen atom orbitals and the localized surface states. The attraction of the pigment environment for the bonded oxygen has to overcome the electron sharing force. Aluminum ions in the pigment create acceptor levels at the surface. Valency electrons raised to these levels act as oxygen trapping centers and thus improve resistance to chalking.

## IMPROVED DISPERSION IN PAINT VEHICLES

The addition of 0.1 to 0.3 per cent of a water-soluble salt of a tertiary amine with an organic acid of low solubility to titanium dioxide pigment particles improves the dispersibility in paint vehicles.[106] Effective salts

## CHALKING AND DISCOLORATION

include tribenzylamine, 2-ethylhexanoate, triethanolamine benzoate. To improve the dispersibility the pigments are suspended in water containing 0.25 to 0.75 per cent of sodium silicate, sodium phosphate, sodium aluminate or sodium tripoly phosphate and flocculated by the addition of aluminum sulfate.[107] Then sodium hydroxide or sodium aluminate is added to precipitate the alumina on the pigment particles. Organophillic pigments are prepared by suspending titanium dioxide particles in an aqueous ammonium stearate solution at 60° C and adjusting to a pH of 5 by adding dilute hydrochloric acid to precipitate stearic acid on the individual particles.[108] The product is washed and dried at 80 to 100° C.

Titanium dioxide is made easily dispersible in latex paints by adding from 7 to 65 per cent by weight of the pigment of alkaryl polyether alcohols or thio esters together with a small amount of an anionic surfactant as alkylarenesulfonates to the wet pigment filter cake.[109] Pigments of improved chalking resistance, color stability, and dispersion are obtained by precipitating on the calcined titanium dioxide particles, hydrous oxides of titanium and alumina.[110] The purified solution of the mixed sulfates is added to a suspension of titanium dioxide and the hydrous oxides are precipitated by adding sodium hydroxide or ammonium hydroxide to raise the pH to 7. Pure titanyl sulfate is prepared by heating hydrous titanium dioxide obtained as an intermediate product in the sulfate pigment process with from 93 per cent sulfuric acid to 20 per cent oleum at a minimum temperature of 120° C.

Pigment developed hydrous titanium dioxide is mixed with 0.01 to 0.15 per cent by weight of commercial triethanolamine and from 0.1 to 2.0 per cent by weight of a silica aerogel or alumina aerogel before dry grinding in a fluid energy type mill using high pressure steam as the source of energy to improve the film fineness, gloss, and whiteness or color characteristics of the products.[111] At the same time the metallic corrosion of the equipment is reduced or eliminated. Intimate association can be effected by pumping the triethanolamine onto the pigment flowing along a conveyor, depending on the conveyor action and associated tumbling for mixing. Since such a small amount is used, the triethanolamine is added as a water solution for better distribution. Up to 2.0 per cent of a silica aerogel is added just prior to the triethanolamine addition. In addition to enhancing the desirable results obtained from the triethanolamine, the aerogel reduces erosion of the metal of the grinding equipment. The treated pigment is ground in a fluid energy mill. Best results are obtained with aerogels having 500 or more square meters of surface area per gram. To prepare the aerogel a gel is placed under a pressure equal to or greater than the critical pressure of the liquid phase, and the temperature is raised to or above the critical value. The vapor is drawn off and replaced by air leaving the solid phase in the form of an aerogel which in the case of silica occupies 80 per cent of the volume of the original gel.

Triethanolamine has been found superior to all other organic agents tested, and it is of particular value in combination with the aerogel which assists in the maintenance of the color of the pigment in its unground condition. The superior results are attributed to the effectiveness of the triethanolamine as a wetting and dispersing agent, and to its resistance to decomposition or carbonization under the conditions encountered in fluid energy mills operated with steam as the fluid. The action of the triethanolamine in connection with an aerosol is to improve the dispersion of the pigment in the grinding mill, thereby facilitating removal of the fines and permitting more of the available energy to be expended in the grinding of the coarser particles. Although aerogels of silica and alumina are of special interest, aerogels of a number of white inorganic oxides such as those of titanium and zirconium yield improved pigments.

As an example of the process, 2000 parts of titanium dioxide produced by the oxidation of titanium tetrachloride were thoroughly mixed with 1 part by weight of commercial triethanolamine and 10 parts of commercial silica aerogel. Undiluted triethanolamine was added and distributed throughout the pigment so that no large pasty masses resulted. This mixture was fed to a 36 inch fluid energy mill at the rate of 30 pounds per minute. For purposes of comparison, pigment treated with the aerosol only and untreated pigment were ground in the same machine at the same rate. The fully treated pigment had a tinting strength of 196, a fineness of 9, and a gloss of 8, while the pigment treated with the aerogel only had a tinting strength of 196, a film fineness of 8, and a gloss of 7.25. For the untreated pigment a tinting strength of 178, a film fineness of 7.5, and a gloss of 7.25 were obtained.

## GLOSS CHARACTERISTICS OF PIGMENTED FILMS

Kingsbury [112] treated titanium dioxide pigment with perchloric acid or a perchlorate to improve the gloss of enamels. For example, a solution of 0.3 g of magnesium perchlorate in 10 ml of water was mixed with 300 g of titanium dioxide containing 5 per cent water and dried. A washed titanium dioxide hydrolyzate is partially dried to 52 per cent solids to form voids and then densified until the voids are filled before final calcination.[113] The calcined and finished pigment gives baked enamels of improved color and gloss. Surface to surface contact is avoided by coating the titanium dioxide particles with 0.25 to 5.0 per cent by weight of a nondrying, water insoluble alkyd resin having an acid number of 10 to 30.[114] The coating is applied from an aqueous resin emulsion containing a volatile organic base. The emulsion is mixed with the hydroclassified titanium dioxide dispersion by simple agitation prior to the filtering step. Particles of titanium dioxide adhere to the surface of globules of emulsified coating material so that filtration and washing are facilitated. In the drying process the coating material spreads over the surfaces to avoid cementation of the pigment particles to form aggregates. Great savings in manufacturing time are accomplished. Products produced with the coated pigment mix faster, have a lower consistency and have superior texture, lightness, gloss, and hiding power as compared with untreated pigments.

The resistance of titanium dioxide pigments to efflorescence is increased by precipitating a small amount of insoluble hydrated titanium phosphate particles of approximately 0.01 micron diameter on the surface of the pigment particles which have a diameter of 0.2 to 0.4 micron.[115] Grinding does not separate the titanium phosphate from the primary pigment particles. Other agents usually added to avoid efflorescence are also introduced in the process. For example, 4 kg of a suspension containing 20 weight per cent of titanium dioxide was heated at 30° C and stirred with 64 ml of a sulfate solution containing 125 gpl titanium dioxide and 400 gpl sulfuric acid. To this mixture was added progressively 56 g of an aqueous solution containing 100 g of phosphoric acid per liter. After thorough mixing, the pH of the suspension was brought to 8 by the addition of sodium hydroxide solution. The treated pigment was filtered, washed, and dried at 110° C.

Industrial enamels are subject to a wider variety of problems leading to flocculation than any other system because of the wide variety of combinations of resins and solvents used.[116] Good dispersion is necessary to develop the maximum hiding power, color, and gloss from a given formulation. Hiding power is a function of the titanium dioxide used in a white enamel. Poor hiding may be caused by incomplete dispersion, poor wetting, poor or careless manufacturing procedure, critical vehicle, improper solvent or multiple solvent balance, and improper formulation. Maximum hiding power is obtained from pure titanium dioxide at approximately 30 per cent pigment volume concentration. The presence of anything in the paint system causing incompatibility between different vehicle components may cause the pigment to flocculate.

A series of tests showed that best gloss was obtained with highest alkyd modification. An increase in melamine or urea resulted in decreased gloss. Melamine produced better gloss than urea in either type alkyd. The nonoxidizing alkyd gave better gloss than did the oxidizing alkyd in comparable percentages of modifications. High solvency resins such as the amines can tolerate only a limited amount of oil acids present in the alkyd portion of the vehicle. The ultimate color which can be developed by an industrial enamel containing titanium dioxide is a compromise between color and hiding power. Full realization of the maximum color depends on proper incorporation of the pigment into the vehicle. Maximum color development depends on the initial color of the vehicle, its color retention, and the film thickness of the paint. The most outstanding resin for color and color retention are the acrylics.

Contraction of baked alkyd resin films on cooling produces surface irregularities above flocculated particles with a resulting loss of gloss.[117] The gloss is improved by the formulation of surface active salts from the acids of the alkyd. It is proposed that the cation of this salt is adsorbed within the water layer present as an adsorbed film on the pigment surface. This cation confers a positive electrical charge on the pigment while the high molecular weight anion serves as counter ion. This accounts for the gloss increase obtained by treatment with alumina hydrate and other basic agents. Metal naphthenates were found to be most effective in dispersing titanium dioxide pigments.[118]

The hiding power of titanium dioxide pigments in alkyd enamels varies with the pigment volume concentration at which it is used.[119] Higher hiding power is obtained at the lower pigment volume concentration. With pure titanium dioxide the highest hiding power is obtained at 30 per cent pigment volume concentration. With extended titanium dioxide, the highest hiding power is obtained at the highest practical pigment volume concentration. The most economical pigment volume concentration for pure rutile titanium dioxide is 22 per cent. Minimum loss of hiding power from the wet to the dry state is achieved at the lowest possible pigment volume concentration and the highest possible solids. Thus, the amount of thinner in the vehicle is of importance in the hiding power properties of the enamel.

The maximum of tinting strength and hiding power is found in pigments of 0.2 to 0.3 micron.[120] Below this particle size, undesirable effects such as poor hiding power, colloidal solubility, tendency to agglomerate, high reactivity, poor flow, and poor gloss may result. Colors also vary with the particle size of a given pigment. Acicular pigments may exhibit pleochromism, may show brush marks, and give films of poor gloss. For the study of hiding power with particle size, a meter is used which measures the light that penetrates a film of the pigment in linseed oil.

The optimum particle size for rutile pigments to give the highest hiding power and a pure gray color with carbon black in paint films is 0.21 micron.[121] If the particles are smaller than this, a blue gray tone results. Particles larger than 0.21 micron give a brown gray tone. The hiding power of the pigment decreases very rapidly as the particle size is decreased below the optimum 0.21 micron, but more gradually as the particles become larger.

Although acicular titanium dioxide pigments are not available commercially, acicular zinc oxide 0.2 micron in diameter and several microns long not only gives higher hiding power but better durability in outside paints than pigments having spherical particles of the same diameter. These needles being of optimum diameter behave as if they had been cut off into pieces of the exact optimum length.

That the color or tone of pigments vary with the particle size may be illustrated with iron oxides. Red iron oxide pigments may vary in color from orange through pure red to maroon depending on the size of the particles.

# 24

# The Titanium Pigment Industry

In 1869, Ryland, in England, pulverized a sample of ilmenite and ground the dried material with a linseed-oil vehicle to produce a black paint. The following year, Overton, in the United States, patented an anti-fouling coating composition for ships' bottoms which was prepared by mixing finely ground rutile in a bituminous material such as tar or pitch. These appear to be the first recorded attempts to utilize titanium compounds as paint pigments, but these pioneer investigators seemed to have entirely missed the advantages of separating the oxide in the pure condition, and it was more than 30 years later before this problem received serious consideration.

About 1870, A. J. Rossi, a young French metallurgist and chemist, came to the United States and successfully smelted titaniferous iron ores at Boonton, New Jersey. As an outgrowth of this work he became very much interested in utilizing the titanium content as such, and he developed methods of producing titanium alloys in the electric furnace. The Titanium Alloy Manufacturing Company was incorporated in 1906, and a plant was built at Niagara Falls, New York, for the production of ferro-carbontitanium according to these processes, to supply the demands of the iron and steel industry. Encouraged by this success, Rossi's interests spread to other fields, and in 1908 he prepared a relatively pure titanium dioxide and proved its high opacity as a pigment by mixing the material with salad oil and brushing out the resulting composition. This was probably the first instance of the use of chemically prepared titanium dioxide as a white pigment.

In 1912 L. E. Barton joined Rossi in a systematic program of research on the pigment possibilities of titanium compounds, and shortly worked out a method of separating the dioxide from ilmenite, titaniferous iron ores, and rutile, and demonstrated the adaptability and outstanding properties of this material. The composite barium sulfate and calcium sulfate pigments, containing 25 per cent titanium dioxide, were conceived by these workers, and development work progressed with such success that the Titanium Pigment Company was organized and a factory was begun at Niagara Falls, New York, in 1916, for large-scale production according to their method. Wartime conditions delayed operation, however, and it was not until after the armistice in 1918 that commercial production began. The National Lead Company purchased a substantial interest in the company in 1920, and in 1932 acquired the outstanding shares of common stock. In 1937 the operating units were designated as Titanium Division, National Lead Company, and the name Titanium Pigment Corporation was applied to the sales agent. A second and larger plant was built in St. Louis, Missouri, in 1923 for making the composite pigment. Production of relatively pure (96 to 99 per cent) titanium dioxide was begun at the Niagara Falls plant in 1925, and demand for this pigment soon exceeded the production capacity so that a new factory was built in Sayreville Township, New Jersey, in 1934. The first unit was placed in operation in April, 1935, and the following year the Niagara Falls plant was discontinued. Several additions have since been made to the Sayreville plant and St. Louis plant.

Very extensive and readily worked deposits of titaniferous magnetites and ilmenite occur near the southwest coast of Norway, and in 1908 a government commission was appointed to investigate the utilization of these ores. Farup and Jebsen first conducted a joint study on the smelting of these ores for iron, and they came to realize the difficulties involved in such an operation. A search for other methods of utilization focused attention on the titanium content, and this investigation led to the development of titanium dioxide pigments. The first product, made from ilmenite and known as titanium ocher, had great opacity but was of a reddish yellow color far short of the desired goal. However, methods for the recovery of pure white titanium dioxide of pigment quality were developed by 1912, and production of the composite pigments on a commercial scale was begun at Fredrikstad almost simultaneously with the initial output in the United States. According to this process, the finely ground ilmenite ore was heated with strong sulfuric acid to produce soluble titanium and iron sulfates, and this was a decided improvement over the fusion methods of Rossi and Barton.

To facilitate the commercial development of the pigments, the American and Norwegian interests, which had worked independently up to this time, came together in 1920 and agreed to a plan for cross licensing of patents and mutual exchange of technical information and operating experience. In 1927 the National Lead Company bought a controlling interest in the original Norwegian company, Titan Co., A.-S., including its ore holdings and French affiliate, Société Industrielle du Titane, a sales company.

Later in the same year an agreement was made with the I. G. Farbenindustrie, A. G., for the formation of a jointly owned German company, Titangesellschaft m.b.H., to supply the European and Asiatic markets. A new manufacturing plant was built at Leverkusen, Germany, to supersede the original Fredrikstad unit, which was abandoned. It is now owned fully by the National Lead Co.

In 1933 the British Titan Products Company, Ltd. was organized under the joint ownership of the Titan Company, Inc., a holding company of Delaware, representing the interests of the National Lead Company and Dr. G. Jebsen in the Norwegian, German and French companies, and several British interests, the Imperial Chemical Industries, Ltd., Imperial Smelting Corporation, Ltd., and Goodlass Wall and Lead Industries, Ltd., for the manufacture and sale of titanium pigments throughout the British Empire except Canada. Production began in 1934.

In connection with certain of its European associates, the Titan Company, Inc., along with Japanese interests, organized the Titan Kogyo Kabushki Kaisha in 1936 and built a plant at Kobe to supply the demand for titanium compounds in the Japanese Empire.

In Europe, Blumenfeld early turned his attention to the production of relatively pure, 96 to 99 per cent, titanium dioxide pigments, and in 1920 he devised an improved method of hydrolysis which now bears his name. Patents were granted in England in 1923 and in the United States in 1924. By 1926 three plants employing this hydrolysis procedure had been built and placed in operation in Europe by the Fabriques de Produits Chimiques de Thann et de Mulhouse, in France, by the Montecatini interests in Italy, and by the Verein für Chemische und Metallurgische Produktion in Czechoslovakia.

At about this time the Commercial Pigments Company in the United States concluded a license agreement with the Société de Produits Chimiques des Terres Rares, in France, the European licensee of the Blumenfeld process, and a plant was built at Baltimore, Maryland, in 1927. Production began in 1928, and by 1931 an output of 20,000 tons yearly of pure titanium dioxide pigment was reached. In 1932 this company was taken over by the Krebs Pigment and Color Corporation, and in 1943 it became the pigment department of E. I. du Pont de Nemours and Company.

Another plant employing the Blumenfeld process of hydrolysis was built at Edge Moor, Delaware, in 1935. Initially the plant produced blended composite pigments, although the relatively pure titanium dioxide has become the principal product.

In mid-1939 the E. I. du Pont de Nemours and Company began production of titanium dioxide pigments at its new 45,000 ton per year plant at New Johnsonville, Tennessee, by the chloride process. This first plant to produce titanium dioxide pigments by the vapor phase oxidation of titanium tetrachloride was a radical departure from the conventional sulfate process. Perhaps 10,000 tons a year of titanium pigment are produced in the Edge Moor plant by the chloride process.

A smaller plant based on the chloride process was built and placed in operation at Antioch, California, early in 1964 to supply the demands of titanium dioxide pigments in the Pacific Coast area. This was the first unit west of the Mississippi River.

Since their initial construction, these plants have been enlarged periodically to keep abreast of the steadily increasing demand for titanium dioxide pigments.

In 1933 the Chemical and Pigment Company (Glidden Company) and the Metal and Thermit Corporation, both of which had been conducting independent research relative to the production of titanium dioxide pigments for a number of years, joined in forming the American Zirconium Corporation for the purpose of manufacturing such pigments. A plant was constructed in Baltimore, Maryland, and placed in operation in approximately one year.

Another plant was built at Hawkins Point also in Baltimore, Maryland, in 1955. Glidden acquired all of the assets of the company and it became simply, The Glidden Company. In 1958 production of titanium dioxide pigments in the original plant was discontinued and the capacity of the Hawkins Point plant was expanded.

In 1933 the Southern Mineral Products Corporation was organized for the purpose of making titanium dioxide pigments from the ilmenite obtained along with rutile and apatite in mining the nelsonite ore of Virginia. A plant was built at Piney River, Virginia, and placed in operation in 1936. After overcoming a variety of difficulties, a product of high quality was obtained and the output was increased in 1938. This property was taken over by the Virginia Chemical Corporation, a subsidiary of the Interchemical Corporation, and in 1945 the pigment plant together with the ore property was sold to the American Cyanamid Company. A larger plant was built by the American Cyanamid Company at Savannah, Georgia, in 1955.

Early in 1938 the Sherwin-Williams Company bought a site at Gloucester, New Jersey, and built a plant for the production of titanium dioxide pigments by a fluoride based process. The fluoride process proved unsuccessful although it had been developed through the pilot plant stage. In 1945 the pigment plant and developments at Gloucester were sold to the American Cyanamid Company. The New Jersey Zinc Company bought the property in 1956 and converted the plant to the sulfate process.

In 1963 the Cabot Corporation bought a titanium tetrachloride produc-

## TABLE 24-1
### Titanium Pigment Plants of the World

| | | |
|---|---|---|
| **North America** | | |
| **United States** | | |
| E. I. du Pont de Nemours and Co. | Edge Moor, Delaware | 126,000 |
| | Baltimore, Maryland | 40,000 |
| | New Johnsonville, Tennessee | 77,000 |
| | Antioch, California | 27,000 |
| National Lead Co. | Sayreville, New Jersey | 183,000 |
| | St. Louis, Missouri | 136,000 |
| The Glidden Co. | Baltimore, Maryland | 56,000 |
| American Cyanamid Co. | Piney River, Virginia | 18,000 |
| | Savannah, Georgia | 92,000 |
| New Jersey Zinc Co. | Gloucester City, New Jersey | 48,000 |
| Cabot Corp. | Ashtabula, Ohio | 20,000 |
| American Potash and Chemical Corp. | Hamilton, Mississippi | 25,000 |
| **Canada** | | |
| Canadium Titanium Pigments, Ltd. | Varennes, Quebec | 25,000 |
| British Titan Products (Canada) Ltd. | Sorel, Quebec | 22,000 |
| **Mexico** | | |
| Pigmentos y Productos Quimicos S. A. de C. V. | Tampico, Tamaulipos | 8,000 |
| **South America** | | |
| **Brazil** | | |
| Cia Quimica Industrial S. A. | São Paulo, São Paulo | 4,600 |
| **Europe** | | |
| **Belgium** | | |
| Soc. Chimiques des Derives du Titane | Langerbrugge | 16,000 |
| **Finland** | | |
| Vuorikemnia Oy | Otammaki | 25,000 |
| **France** | | |
| Fabriques de Produits Chimiques de Thann et de Mulhouse | Strasbourg | 36,000 |
| Le Produits du Titane, S. A. | Le Havre | 36,000 |
| **Germany** | | |
| Titangesellschaft m. b. H. | Leverkusen | 88,000 |
| Farbenfabriken Bayer A. G. | Uerdingen | 56,000 |
| Pigment Chemie G. m. b. H. | Homberg | 40,000 |
| **Italy** | | |
| Soc. Montecatini | Bovisa (Milan) | 10,000 |
| | Spinetta Morengo | 40,000 |
| **Netherlands** | | |
| N. V. Titandioxydefabriek | Botiek (Rotterdam) | 20,000 |

## TABLE 24-1 (Continued)

| | | |
|---|---|---|
| Norway | | |
|   Titan Company A./S. | Fredrikstad | 15,000 |
| Portugal | | |
|   La Pigmentos de Titanium, Lda. | Sines, Estremadura | 16,000 |
| Spain | | |
|   Union Quimica del Norte de España, S. A. | Axpe, Erandio, Biscay | 13,000 |
|   Chromogenia y Quimica, S. A. | Barcelona | 2,800 |
| United Kingdom | | |
|   British Titan Products Co. | Grimsby | 130,000 |
| | Billingham | 29,000 |
|   Laporte Titanium, Ltd. | Stallingborough | 56,000 |
| Asia | | |
| India | | |
|   Travancore Titanium Products, Ltd. | Trivandrum, Kerala | 25,000 |
| Japan | | |
|   Fuji Titanium Industry Co., Ltd. | Hiratsuka, Kanagawa | 4,800 |
| | Kobe, Hyogo | 4,600 |
|   Furukawa Mining Co., Ltd. | Osaka, Osaka | 15,000 |
|   Ishihara Sangyo Kaisha, Ltd. | Yokkaichi, Mie | 45,000 |
|   Sakai Chemical Industry Co. | Salai, Osaka | 10,000 |
|   Teikoku Kako Company, Ltd. | Saidaiji, Okayama | 15,000 |
|   Titanium Industry Co., Ltd. | Ube, Yamaguchi | 8,000 |
| Africa | | |
| South Africa | | |
|   Republic of South African Titan Products (Pty.) Ltd. | Umbogintiwini | 11,000 |
| Oceania | | |
| Australia | | |
|   Australian Titan Products Co., Ltd. | Burnie, Tasmania | 27,000 |
|   Laporte Titanium, Ltd. | Bunbury, Western Australia | 10,000 |

ing plant at Ashtabula, Ohio, from the National Distillers Corporation and added the facilities to produce 25,000 tons of titanium dioxide pigment a year by the vapor phase oxidation of the tetrachloride.

The American Potash and Chemical Corporation built a plant in 1965 at Hamilton, Mississippi, for the manufacture of titanium dioxide pigments by the vapor phase oxidation of the tetrachloride. Production began in 1966.

The companies throughout the world engaged in the manufacture of titanium dioxide pigments together with the location and capacity of the plants in tons per year are given in Table 24–1.

New titanium dioxide pigment plants based on the vapor phase oxidation of the tetrachloride under construction, with the yearly capacity in tons and the expected date of completion, are listed in Table 24–2.

### TABLE 24–2

Titanium Dioxide Pigment Plants in Process

| Company | Location | Capacity | Date of Completion |
|---|---|---|---|
| American Cyanamid Co. | Savannah, Ga. | 20,000 | 1966 |
| E. I. du Pont de Nemours and Co. | Edge Moor, Del. | 36,000 | 1966 |
| National Lead Co. | Sayreville, N.J. | 36,000 | 1966 |
| Pittsburgh Plate Glass Co. | Natrium, W. Va. | 25,000 | 1967 |

Outside the United States, titanium pigment plants based on the chloride process are in operation or are planned in France, Japan, and the United Kingdom.

The first Polish titanium dioxide pigment plant is planned for 1969 and it will have a capacity of 30,000 to 40,000 tons a year.[1] Ilmenite from the Union of Soviet Socialist Republics will be digested with sulfuric acid.

It should be remembered that the production, on a commercial scale, of pure white titanium dioxide having good pigment properties, from ilmenite which contains from 35 to 55 per cent oxides of iron, is one of the most precise of chemical operations, and staffs of high order have failed through long periods of time to master the technique of the interrelated steps. It seems paradoxical that one of the blackest minerals should yield one of the whitest pigments.

Although the original titanium dioxide pigments were of the anatase crystal form, rutile type pigments first placed on the American market on a commercial scale in 1941 now account for 70 per cent of the total production. These pigments possess higher hiding power and greater resistance to chalking as well as the other desirable properties of titanium dioxide pigments in general. However, the older anatase pigments are still preferred for some specific uses particularly in paper and in self-cleaning outside white house paints.

In 1959 the E. I. du Pont de Nemours and Company placed on the market rutile type titanium dioxide pigments produced by the vapor phase oxidation of titanium tetrachloride. These pigments possessed appreciably higher tinting strength, better color, greater ease of dispersion, and smaller and more uniform particle size than the pigments produced at the time by the conventional sulfate process. A significant advance in titanium dioxide technology had been made. However, the producers by the sulfate process began an intensive research program based on the belief that the improved pigment properties were not inherent in the chloride process. After two or three years this work culminated in the production of a series of titanium dioxide pigments made by the sulfate process which matched for all commercial purposes the pigments produced by the newer chloride process. The final outcome was a general improvement in the quality of all titanium dioxide pigments whether made by the chloride or by the sulfate process.

The most reliable estimates indicate that the tetrachloride based pigment plants cost only 60 to 70 per cent of the 600 to 1000 dollars per annual ton investment for sulfate process units.[2] More recent investment for sulfate plants is probably closer to the 1000 dollar mark. But the tetrachloride process requires rutile ore (95 per cent titanium dioxide) priced at 105 dollars a ton, while the sulfate method uses ilmenite ore (40 to 60 per cent titanium dioxide) costing 25 dollars a ton. Recent entrants in the field say the tetrachloride route does not hold any particular economic advantage over the sulfate route despite the lower cost of plants using the tetrachloride process. Regardless of the process the larger plants have a lower unit cost.

However, it is generally accepted that the cost of producing titanium dioxide pigment by the chloride process is 5.5 to 6 cents a pound less than by the sulfate process.

The processing cost of titanium dioxide by the sulfate process comes out at about 380 dollars a ton. Raw materials make up a substantial part of the charge. The individual costs of the items per ton of titanium dioxide pigment are given in Table 24–3.

**TABLE 24–3**

Production Costs of Titanium Dioxide in Dollars per Ton

| Item | Dollars per Ton |
|---|---|
| Ilmenite | 25 |
| Sulfuric acid | 67 |
| Utilities | 42 |
| Labor, supervision, maintenance, and overhead | 98 |
| Depreciation and interest | 98 |
| Research and technical services | 17 |
| Disposal of waste | 3 |
| Total | 350 |

As shown in Table 24–4, the production of titanium dioxide pigment in the United States increased from 4000 tons in 1925 to 555,211 tons in 1964. Production of 523,201 tons in 1962 was up 4 per cent from 1961. At this rate the industry was operating at almost 80 per cent of the on-paper capacity of 657,000 tons. However, since many of the plants are old and the equipment requires a high degree of maintenance this may be near the actual effective capacity of the plants.

**TABLE 24–4**

Production of Titanium Dioxide Pigment in the United States

| Year | Tons |
|---|---|
| 1925 | 4,000 |
| 1930 | 11,000 |
| 1935 | 58,000 |
| 1940 | 100,000 |
| 1945 | 220,000 |
| 1950 | 330,000 |
| 1955 | 445,000 |
| 1960 | 455,000 |
| 1961 | 502,879 |
| 1962 | 523,201 |
| 1963 | 519,458 |
| 1964 | 555,211 |

World production of titanium dioxide pigments in 1962 was well over 1,000,000 tons with consumption around 850,000 tons. Demand is estimated to rise during the 1960's at about 6 per cent a year.

**TABLE 24–5**

Consumption of Titanium Dioxide Pigments in the United States in 1963 by Uses and Projected Consumption in 1968, in Tons

| Uses | 1963 | 1968 |
|---|---|---|
| Paints, varnishes, and lacquers | 320,000 | 380,000 |
| Paper | 95,000 | 185,000 |
| Floor coverings (linoleum, felt base) | 29,000 | 31,000 |
| Rubber | 26,000 | 27,000 |
| Coated fabrics and textiles (oilcloth, shade cloth, artificial leather) | 19,000 | 22,000 |
| Printing inks | 11,000 | 16,000 |
| Plastics | 13,000 | 24,000 |
| Export | 18,000 | 18,000 |
| Others (roofing granules, welding rods, synthetic fabrics, ceramics, cosmetics) | 33,000 | 42,000 |
| Total | 564,000 | 745,000 |

## TABLE 24-6

Value of the Titanium Dioxide Pigments Consumed in the United States by Industries in 1963

| Industries | Dollars |
| --- | --- |
| Paints, varnishes and lacquers | 188,630,000 |
| Paper | 37,250,000 |
| Floor coverings (linoleum, felt base) | 12,810,000 |
| Rubber | 11,920,000 |
| Plastics | 8,640,000 |
| Roofing granules | 6,260,000 |
| Coated fabrics and textiles (oilcloth, shade cloth, artificial leather) | 5,960,000 |
| Printing inks | 4,770,000 |
| Ceramics | 3,580,000 |
| Others (cosmetics, welding rods, dielectrics) | 5,360,000 |
| Export | 12,820,000 |
| Total | 298,000,000 |

Consumption of titanium dioxide pigments, based on the titanium dioxide content, in the United States by uses in 1964 is given in Table 24-7 in percentage of the total consumption.[3]

## TABLE 24-7

Consumption of Titanium Dioxide Pigments in the United States in 1964 by Uses in Percentages

| Uses | Percentage |
| --- | --- |
| Paints, varnishes, and lacquers | 56.8 |
| Paper | 15.2 |
| Floor coverings | 4.7 |
| Rubber | 3.7 |
| Coated fabrics and textiles (oilcloth, shade cloth, artificial leather) | 1.4 |
| Printing inks | 2.1 |
| Roofing granules | 1.9 |
| Ceramics | 1.9 |
| Plastics (except floor coverings and vinyl coated fabrics and textiles) | 5.4 |
| Others (cosmetics, dielectrics, textiles) | 6.9 |
| Total | 100 |

# Notes

## CHAPTER 1

[1] *Annals of Philosophy*, 1818, **11**, 112.
[2] *Berzelius' Letters,* Almquist and Wiksell, Uppsala, 1912–1914. (Letter to Thomson.)
[3] *Annals of Philosophy*, 1818, **11**, 112; Crell, *Chemische Annalen*, 1791, **1**, 40, 103.
[4] Thenard, *Traité de Chimie Elémentaire Théoretique et Pratique*, C. J. de Mat, Brussels, 1829, **1**, 119.
[5] M. H. Klaproth, *Analytical Essays Toward Promoting the Chemical Knowledge of Mineral Substances* (English translation), Cadell and Davies, London, 1801, 200.

## CHAPTER 2

[1] F. W. Clarke, *U.S. Geol. Survey Bull.* 1924, **770**, 36.
[2] C. Krugel and A. Retter, *Superphosphate*, 1932, **5**, 95; *Chem. Abs.*, 1932, **26**, 4123.
[3] W. L. Hill, H. L. Marshall, and K. D. Jacob, *Ind. Eng. Chem.*, 1932, **24**, 1306; *Chem. Abs.*, 1933, **27**, 363.
[4] G. Aminoff, *Chem. Zentr.*, 1943, **II**, 995; *Chem. Abs.*, 1944, **38**, 6244.
[5] L. Tung and C. L. Jao, *Ti Chih Hsueh Pao*, 1963, **43**, 271; *Chem. Abs.*, 1964, **61**, 4090.
[6] E. P. Dunnington, *Am. J. Sci.*, 1892, **42**, 491; *J. Chem. Soc.*, 1892, **A**, 791.
[7] W. Geilmann, *Biedermanns Zentr.*, 1921, **50**, 122; *Chem. Abs.*, 1922, **16**, 3999.
[8] S. E. Kaminskaya, *Compt. rend. acad. sci. U.R.S.S.*, 1941, **33**, 50; *Chem. Abs.*, 1943, **37**, 2501. A. P. Vinogradov, *Pedology U.S.S.R.*, 1945, 348; *Chem. Abs.*, 1946, **40**, 1892.
[9] H. O. Askew, *New Zealand J. Sci. Technol.*, 1930, **12**, 173; *Chem. Abs.*, 1931, **25**, 1932.
[10] W. H. Hamilton and L. I. Grange, *New Zealand Dep. Sci. Ind. Research Bull.* **61**, 1938, 593; *Chem. Abs.*, 1938, **32**, 7178.
[11] E. P. Dunnington, *Chem. News*, 1897, **76**, 221; *J. Chem. Soc.*, 1898, **A**, 122.
[12] V. Charrin, *Argile*, 1937, **174**, 1; *Chem. Abs.*, 1937, **31**, 8840.
[13] E. Blanck and W. Geilmann, *J. Landw.*, 1922, **70**, 253; *Chem. Abs.*, 1923, **17**, 1683.
[14] J. H. Heathman, *Geol. Survey Wyoming Bull.* **28**, 1939; *Chem. Abs.*, 1939, **33**, 7702.
[15] B. M. Burchfield and H. Mulryan, *Am. Ceram. Soc. Bull.* **19**, 1940, 161; *Chem. Abs.*, 1940, **34**, 4871.
[16] R. J. Havell and T. N. McVay, *Am. Ceram. Soc. Bull.* **18**, 1939, 429; *Chem. Abs.*, 1940, **34**, 597.
[17] A Salminer, *Suomen Kemistilehti*, 1936, **9A**, 1; *Chem. Abs.*, 1936, **30**, 7504.
[18] A. Fioletov, *Keram. Rundschau*, 1929, **37**, 659; *Chem. Abs.*, 1930, **24**, 5955.
[19] M. M. Odinstov, *Soviet Geol.*, 1938, **8**, 83; *Chem. Abs.*, 1939, **33**, 4916.

[20] G. Banchi and A. Maffei, *Ann. chim. applicata,* 1932, **22**, 138; *Chem. Abs.,* 1932, **26**, 5185.
[21] E. Riley, *J. Chem. Soc.,* 1862, **15**, 311.
[22] C. K. Wentworth, R. I. Wells, and V. J. Allen, *Am. Mineral.,* 1940, **25**, 1; *Chem. Abs.,* 1940, **34**, 2294.
[23] W. E. Petrascheck, *Neues Jahrb. Mineral. Geol.,* 1939, **II**, 913; *Chem. Abs.,* 1940, **34**, 3211.
[24] M. Isomatsu, *J. Japan. Ceram. Assoc.,* 1937, **45**, 21; *Chem. Abs.,* 1938, **32**, 1881.
[25] A. Orlov, *Neues Jahrb. Mineral. Geol., Beilage Bb.,* **A**, 1939, **74**, 251; *Chem. Abs.,* 1939, **33**, 7242; *Rozpravy II, Tr. Csl. Akademie,* 1937, **47**, 1; *Chem. Abs.,* 1939, **33**, 946.
[26] M. da S. Pinto, *Boll. do S. F. P. M.* (Rio de Janeiro), No. **22**, 1938; *Chem. Abs.,* 1940, **34**, 1945.
[27] I. A. Lyubimov, *Tsvetnye Metal,* 1946, **19**, No. 4, 1; *Chem. Abs.,* 1947, **41**, 1959.
[28] M. A. A. Alwar, *Indian Minerals,* 1962, **16**, 30; *Chem. Abs.,* 1963, **59**, 1253.
[29] P. S. Dean and J. W. Whittemore, *J. Am. Ceram. Soc.,* 1940, **23**, 77; *Chem. Abs.,* 1940, **34**, 3457.
[30] F. Hart, *Zement,* 1924, **13**, 377; *Chem. Abs.,* 1924, **18**, 3462.
[31] F. W. Clarke, *U.S. Geol. Survey Bull.* **770**, 1924.
[32] H. J. Rose, *Trans. Ann. Anthracite Conf. Lehigh Univ.,* 1938, **25**; *Chem. Abs.,* 1938, **32**, 6834.
[33] J. H. Jones and J. M. Miller, *Chemistry & Industry,* 1939, **58**, 237; *Chem. Abs.,* 1939, **33**, 6023.
[34] F. Heide, *Naturwissenschaften,* 1938, **26**, 693; *Chem. Abs.,* 1939, **33**, 2317.
[35] W. H. Durum and Haffty, *Geochim. Cosmochim. Acta.,* 1963, **27**, 1, *Chem. Abs.,* 1963, **58**, 7718.
[36] S. E. Kaminskaya, *Trav. lab. biogéochim. acad. sci. U.R.S.S.,* 1944, **7**, 130; *Chem. Abs.,* 1946, **40**, 5781.
[37] A. B. Griffiths, *Chem. News,* 1903, **88**, 231.
[38] F. Zambonini and L. Coniglio, *Atti accad. nazl. Lincei,* 1926, **3**, 521; *Chem. Abs.,* 1926, **20**, 2969.
[39] A. R. Alderman, *Records Australian Museum,* 1936, **5**, 537; *Chem. Abs.,* 1938, **32**, 6185. J. L. Lonsdale, *Am. Mineral.,* 1937, **22**, 877; *Chem. Abs.,* 1938, **32**, 4912. M. Ishibashi, *Centr. Mineral. Geol.,* 1930, **A**, 457; *Chem. Abs.,* 1932, **26**, 4561.
[40] A. L. Coulson, *Mem. Geol. Survey India,* 1936, **71**, Pt. 2, 123; *Chem. Abs.,* 1938, **32**, 85.
[41] L. S. Selivanov, *Compt. rend. acad. sci. U.R.S.S.,* 1940, **26**, 389; *Chem. Abs.,* 1940, **34**, 5381.
[42] J. A. Dunn, *Records Geol. Survey India,* 1939, **74**, 260; *Chem. Abs.,* 1940, **34**, 5793.
[43] A. Hadding, *Geol. Fören. i Stockholm Förh.,* 1940, **62**, 148; *Chem. Abs.,* 1940, **34**, 5381.
[44] J. N. Lockyer, *Chemistry of the Sun,* The Macmillan Co., New York, 1887, 89.
[45] M. N. Saha, *Phil. Mag.,* 1920, **40**, 72; *Chem. Abs.,* 1921, **15**, 468. S. A. Mitchell and E. T. R. Williamson, *Astrophys. J.,* 1933, **77**, 1; *Chem. Abs.,* 1933, **27**, 1569. H. D. Babcock, *Astrophys. J.,* 1945, **102**, 145; *Chem. Abs.,* 1946, **40**, 275. P. Bruggencate, *Z. Astrophys.,* 1944, **23**, 119; *Chem. Abs.,* 1946, **40**, 6986. R. Wildt, *Astrophys. J.,* 1947, **105**, 36; *Chem. Abs.,* 1947, **41**, 2641.
[46] W. Petrie, *J. Roy. Astron. Soc. Can.,* 1944, **38**, 137; *Chem. Abs.,* 1944, **38**, 6193.
[47] W. O. Robinson, L. A. Steinkoenig, and C. F. Miller, *U.S. Dep. Agr. Bull.* **600**, 1917; *Chem. Abs.,* 1918, **12**, 197.
[48] W. Geilmann, *J. Landw.,* 1920, **68**, 107; *Chem. Abs.,* 1920, **14**, 3739. *Biedermanns Zentr.,* 1921, **50**, 122; *Chem. Abs.,* 1922, **16**, 3999.
[49] W. P. Headden, *Science,* 1925, **61**, 590; *Chem. Abs.,* 1925, **19**, 2361.
[50] W. P. Headden, *Colorado Agr. Exp. Sta. Bull.* **267**, 1921; *Chem. Abs.,* 1922, **16**, 1447.

# NOTES

[51] K. Koniski and T. Tsuge, *Mem. Coll. Agr., Kyoto Imp. Univ.*, No. 37, 1936; *Chem. Abs.*, 1936, **30**, 4541.
[52] G. Bertrand and C. Voronca-Spirt, *Compt. rend.*, 1929, **188**, 1199; *Chem. Abs.*, 1929, **23**, 3952.
[53] C. Baskerville, *J. Am. Chem. Soc.*, 1899, **21**, 1099.
[54] L. C. Maillard and J. Ettori, *Bull. acad. med.*, 1936, **115**, 631; *Chem. Abs.*, 1937, **31**, 3545.
[55] F. Dutoit and C. Zbinden, *Compt. rend.*, 1930, **190**, 172; *Chem. Abs.*, 1930, **24**, 3552.
[56] G. Bertrand and Mme. Voronca-Spirt, *Compt. rend.*, 1929, **189**, 221; *Chem. Abs.*, 1929, **23**, 5225.
[57] V. T. Chuiko and A. O. Voinar, *Biochem. J. (Ukraine)*, 1939, **14**, 191; *Chem. Abs.*, 1940, **34**, 5510.
[58] J. M. Newell and E. V. McCollum, *U.S. Bur. Fisheries*, Investigational Report **5**, 1931; *Chem. Abs.*, 1932, **26**, 1356.
[59] S. E. Kaminskaya, *Trav. lab. biogéochim. acad. sci. U.R.S.S.*, 1944, **7**, 130; *Chem. Abs.*, 1946, **40**, 5781.
[60] F. Lowater and M. M. Murray, *Biochem. J.*, 1937, **31**, 837; *Chem. Abs.*, 1937, **31**, 6311.
[61] R. Berg, *Biochem. Z.*, 1925, **165**, 461; *Chem. Abs.*, 1926, **20**, 2508.
[62] S. A. Borovik and S. K. Kalinin, *Trav. lab. biogéochim. acad. sci. U.R.S.S.*, 1944, **7**, 114; *Chem. Abs.*, 1946, **40**, 5128.
[63] V. R. Soroka and E. V. Sabadash, *Vopr. Pitaniya*, 1963, **22**, No. 1, 87; *Chem. Abs.*, 1963, **58**, 12939.
[64] E. C. Tabor and W. V. Warren, *A.M.A. Arch. Ind. Health*, 1958, **17**, 145; *Chem. Abs.*, 1958, **52**, 7585.

## CHAPTER 3

[1] J. P. Iddings, *Igneous Rocks*, John Wiley and Sons, Inc., New York, 1909, Vol. I, 60, 145.
[2] N. Sundius, *Geol. Fören. i Stockholm Förh.*, 1945, **67**, No. 2, 271; *Chem. Abs.*, 1946, **40**, 7090.
[3] P. H. Lundegardh, *Nature*, 1946, **157**, 625; *Chem. Abs.*, 1946, **40**, 5368.
[4] D. Wait and I. E. Weber, *J. Oil & Colour Chemists' Assoc.*, 1934, **17**, No. 17, 257.
[5] T. L. Watson and S. Taber, *Virginia Geol. Survey Bull.* **III-A**, 1913.
[6] H. Ries, *Economic Geology*, John Wiley and Sons, New York, 1937, 698.
[7] F. W. Clarke, *U.S. Geol. Survey Bull.* **770**, 1925, 351.
[8] J. P. Iddings, *op. cit.*, 498.
[9] L. T. Walker, *Univ. Toronto Studies, Geol. Ser.*, 1930, No. **29**, 17; *Chem. Abs.*, 1931, **25**, 2667.
[10] K. Rankama, *Bull. comm. géol. Finlande*, 1944, No. 133, 1; *Chem. Abs.*, 1946, **40**, 1122.
[11] F. W. Clarke, *op. cit.*, 351. T. L. Watson and S. Taber, *loc. cit.*
[12] J. T. Singewald, Jr., *U.S. Bur. Mines Bull.* **64**, 1913, 34.
[13] L. v. Hamos and V. Shcherbina, *Science Abstracts*, 1934, **36A**, 1121; *Chem. Abs.*, 1934, **28**, 4009.
[14] W. Lindgren, *Econ. Geol.*, 1907, **2**, 125. H. W. Emmons, *Econ. Geol.*, 1908, **3**, 621.
[15] F. L. Watson and S. Taber, *loc. cit.*
[16] F. W. Clarke, *op. cit.*, 21, 351.
[17] W. Lindgren, *Ore Deposits*, McGraw-Hill Book Co., Inc., New York, 1933, 786.
[18] F. L. Hess and J. L. Gillson, *loc. cit.*
[19] F. W. Clarke, *op. cit.*, 352.
[20] F. L. Hess, and J. L. Gillson, *op. cit.*, 893.

## NOTES

[21] A. F. Buddington and D. H. Lindsley, *J. Petrol.*, 1964, **5**, 310; *Chem. Abs.*, 1965, **61**, 10455.

[22] M. G. Dyadchenko and A. Y. Khatuntseva, *Dokl. Akad. Nauk S.S.S.R.*, 1960, **132**, 435; *Chem. Abs.*, 1962, **57**, 16171.

[23] *Trans., Am. Inst. Min. Eng.*, 1954, 821.

[24] K. P. Yanulov and I. V. Chulkova, *Doklady Akad. Nauk S.S.S.R.*, 1961, **140**, 215; *Chem. Abs.*, 1962, **56**, 4175.

[25] J. W. Gruner, *Econ. Geol.* 1959, **54**, 1315; *Chem. Abs.*, 1961, **55**, 5245.

[26] P. P. Pilipenko, *Mineral Syr'e*, 1930, **5**, 981; *Chem. Abs.*, 1932, **26**, 4560.

[27] R. S. Mitchell, *Am. Mineralogist*, 1964, **49**, 1136; *Chem. Abs.*, 1964, **61**, 13047.

[28] H. Meixner, *Aufschluss*, 1962, **13**, 300; *Chem. Abs.*, 1963, **58**, 8478.

[29] L. L. Leonova and N. S. Klassova, *Geokhimiya*, 1964, 384; *Chem. Abs.*, 1964, **61**, 483.

## CHAPTER 4

[1] F. L. Hess and J. L. Gillson, *Am. Inst. Mining Met. Engrs., Industrial Minerals and Rocks*, New York, 1937, 893. C. S. Ross, *U.S. Geol. Survey Profess. Paper* **198**, 1941; *Chem. Abs.*, 1942, **36**, 2815. C. S. Ross, Section in *Ore Deposits as Related to Structural Features*, ed. W. H. Newhouse, Princeton University Press, 1942. T. L. Watson and S. Taber, *Virginia Geol. Survey Bull.* **III-A**, 1913.

[2] C. S. Ross, *Am. Mineral.*, 1936, **21**, 143; *Chem. Abs.*, 1936, **30**, 3370: *Trans. Am. Geophys. Union*, 1934, 245; *Chem. Abs.*, 1934, **28**, 7213.

[3] D. M. Davidson, F. F. Grout, and G. M. Schwartz, *Econ. Geol.*, 1946, **41**, 738; *Chem. Abs.*, 1948, **42**, 8724.

[4] F. L. Hess and J. L. Gillson, *op. cit.*, 893.

[5] H. E. Dunn, U.S. 2,109,917, Mar. 1, 1938, to Southern Mineral Products Corp.; *Chem. Abs.*, 1938, **32**, 3644.

[6] S. D. Broadhurst, Asst. State Geologist of North Carolina, private communication. J. T. Singlewald, Jr., *U.S. Bur. Mines Bull.* **64**, 1913, 84.

[7] A. Stanley, *Trans. Am. Inst. Min. Met. Engrs.*, Tech. Pub. No. 3410-H; *Chem. Abs.*, 1952, **46**, 11061.

[8] I. D. Hagar, *Paint Ind. Mag.*, 1941, **56**, 410; *Am. Paint J.*, 1941, **26**, 54; *Chem. Abs.*, 1942, **36**, 1197.

[9] O. Herres, *Mining and Met.*, 1943, **24**, 509; *Chem. Abs.*, 1944, **38**, 317. J. R. Balsley, Jr., *U.S. Geol. Surv. Bull.* **940D**, 1943; *Chem. Abs.*, 1944, **38**, 3223. R. C. Stephenson, *Am. Inst. Mining Met. Eng. Contrib. Tech. Pub.* No. **1789**, 1945; *Chem. Abs.*, 1945, **39**, 1120. I. D. Hagar, *Mining and Met.*, 1942, **23**, 594; *Eng. Mining J.*, 1942, **143**, No. 12, 47. R. C. Stephenson, *N.Y. State Museum Bull.* No. **340**, 1945; *Chem. Abs.*, 1946, **40**, 6026.

[10] F. J. Oliver, *Iron Age*, 1942, **149**, No. 10, 53; *Chem. Abs.*, 1942, **36**, 3125.

[11] E. P. Youngman, *U.S. Bur. Mines Inform. Circ.* **6386**, 1930.

[12] E. Frey, *U.S. Bur. of Mines Rept. Invest.* No. **3968**, 1946; *Chem. Abs.*, 1947, **41**, 4072.

[13] N. L. Wimmler, *U.S. Bur. Mines Rept. Invest.* **3981**, 1946; *Chem. Abs.*, 1947, **41**, 4072.

[14] W. T. Benson, A. L. Engel, and H. J. Heinen, *U.S. Bur. Mines*, Rept. Invest. No. 5962, 1962; *Chem. Abs.*, 1962, **56**, 15184.

[15] C. A. Merritt, *Proc. Oklahoma Acad. Sci.*, 1938, **18**, 51; *Chem. Abs.*, 1939, **33**, 2852; *Econ. Geol.*, 1939, **34**, 268; *Chem. Abs.*, 1939, **33**, 4914.

[16] C. C. Coulter, *Mining J. (Phoenix, Ariz.)*, 1939, **23**, No. 9, 7; *Chem. Abs.*, 1940, **34**, 1281.

[17] J. L. Gillson, *Econ. Geol.*, 1932, **27**, 554; *Chem. Abs.*, 1932, **26**, 5517.

[18] M. E. Hurst, *Ann. Rept. Ontario Dept. Mines*, 1932, **40**, IV, 105; *Chem. Abs.*, 1932, **26**, 3459.

[19] *Minerals Yearbook*, U.S. Bureau of Mines, 1951, 1281.

[20] G. B. Mellon, *Geol. Div. Bull.* No. 9. 1961; *Chem. Abs.*, 1963, **58**, 2282.

[21] V. A. Kulibin, *Sovet. Met.*, 1934, **6**, 123; *Chem. Abs.*, 1935, **29**, 87.
[22] V. A. Kulibin, *Gorno-Obogatitel' Zhur.*, 1936, **8**, 10; *Chem. Abs.*, 1937, **31**, 3838.
[23] N. Pervushin, *Redkie Metal*, 1935, **2**, 27; *Chem. Abs.*, 1938, **32**, 6190.
[23a] J. A. Miller, *U.S. Bur. Mines, Inf. Circular* 7791, 1957, 51.
[24] D. S. Shteinberg and V. G. Fominykh, *M.M.M.U., A.N.S.S.S.R., U.F.G.I., T.P.U.P.S.*, 1961, 513; *Chem. Abs.*, 1964, **61**, 15850.
[25] M. Weibel and H. V. Bambauer, *Schweiz. Mineral. petrog. Mitt.*, 1958, **38**, 475; *Chem. Abs.*, 1959, **53**, 8953.
[26] P. Evrand, *Ann. soc. géol. Belg. Bull.* **67B**, 1943, 110; *Chem. Abs.*, 1946, **40**, 5363. F. L. Hess and J. L. Gillson, *op. cit.*, 893. E. P. Youngman, *U.S. Bur. Mines Inform. Circ.* **6386**, 1930, 2, 36.
[27] G. Gruder, I. Nicu, I. Giurcanu, and C. Raducanu, *Rev. chim.*, 1958, **9**, 382; *Chem. Abs.*, 1961, **55**, 12790.
[28] F. Morgensen, *Geol. Fören i Stockholm Förh.*, **68**, 579; *Chem. Abs.*, 1947, **41**, 1959.
[29] F. G. Geiler, *Metallwirtschaft*, 1940, **19**, 343; *Chem. Abs.*, 1941, **35**, 56.
[30] J. D. Pallett, *Bull. Imp. Inst.*, 1937, **35**, 333; *Chem. Abs.*, 1940, **34**, 1943.
[31] M. Legraye, *Ann. soc. géol. Belg. Bull.* 63, 1939–1940, 167; *Chem. Abs.*, 1940, **34**, 3210.
[32] G. M. Stockley, *Mining Mag.*, 1945, **73**, 265; *Chem. Abs.*, 1946, **40**, 813.
[33] V. S. Swanimathan, *Proc. Indian Sci. Congr.*, 15th Congr., 1928, 287; *Chem. Abs.*, 1931, **25**, 2946.
[34] A. Stella, *Atti accad. nazl. Lincei*, 1932, **15**, 336; *Chem. Abs.*, 1932, **26**, 5517.
[35] H. G. Higgins and D. Carroll, *Geol. Mag.*, 1940, **77**, 145; *Chem. Abs.*, 1940, **34**, 3630.
[36] *Titanox Handbook*, Titanium Pigment Co., New York, 1938, Part 1, 3.
[37] J. E. Thoenen and J. D. Warne, *U.S. Bur. Mines, Rept. Invest.* 4515, 1949.
[38] J. R. Thoenen and J. D. Warne, *U.S. Bur. Mines, Rept. Invest. No. 4515*, 1949; *Chem. Abs.*, 1950, **44**, 82.
[39] J. D. Bates, *Econ. Geol.*, 1963, **58**, 1237; *Chem. Abs.*, 1964, **60**, 9033.
[40] W. F. Tanner, A. Mullins, and J. D. Bates, *Econ. Geol.*, 1961, **56**, 1079; *Chem. Abs.*, 1962, **56**, 6940.
[41] J. V. Beall, *Mining Eng.*, 1962, **14**, No. 1, 38.
[42] R. D. Foxworth, R. R. Priddy, W. B. Johnson, and W. S. Moore, *Miss. State Geol. Survey*, Bull. 93, 1962; *Chem. Abs.*, 1962, **56**, 15198.
[43] D. C. Holt, *U.S. Bur. Mines, Rept. Invest.* No. 6365, 1964; *Chem. Abs.*, 1964, **60**, 15614.
[44] R. S. Houston, *Geol. Surv. Wyoming, Bull.* No. 49, 1962; *Chem. Abs.*, 1964, **60**, 11766.
[45] R. H. Storch and D. C. Holt, *U.S. Bur. Mines, Rept. Invest.* No. 6319, 1963; *Chem. Abs.*, 1964, **60**, 3888.
[46] R. V. Berryhill, *U.S. Bur. Mines, Rept. Invest.* No. 6214, 1963; *Chem. Abs.*, 1963, **59**, 1391.
[47] D. F. Holbrook, *Arkansas Resources and Development Commission, Div. Geol. Bull.* No. 12, 16, 1948; *Chem. Abs.*, 1948, **42**, 8724.
[48] E. I. Evdokimov, *G.i G., A.N.S.S.S.R., S.O.*, 1964, **4**, 131; *Chem. Abs.*, 1964, **61**, 14386.
[49] A. I. Imshenetskii, *Sov. Geol.*, 1964, **7**, 140; *Chem. Abs.*, 1964, **61**, 10470.
[50] F. L. Hess, *Mining and Met.*, 1938, **19**, 5; *Chem. Abs.*, 1938, **32**, 1621.
[51] D. N. Wadia, *Rec. Dept. Mineral. Ceylon, Profess. Paper* No. 1, 1943; *Chem. Abs.*, 1945, **39**, 1372.
[52] C. O. Hutton, *New Zealand J. Sci. Technol.*, 1945, **26B**, 291; *Chem. Abs.*, 1945, **39**, 5216.
[53] C. O. Hutton, *New Zealand J. Sci. Technol.*, 1940, **21B**, 190; *Chem. Abs.*, 1947, **41**, 1959.
[54] N. H. Fisher, *Australian Mining Res. Survey Sum. Rept.* No. 1, 1945; *Chem. Abs.*, 1947, **41**, 3019.

[55] T. Nakano, *Kwagaku Kōgyō Siryō*, 1941, **14**, 30; *Chem. Abs.*, 1941, **35**, 7662.
[56] R. Ando and I. Nitto, *J. Chem. Soc. Japan*, 1941, **62**, 978; *Chem. Abs.*, 1947, **41**, 2662.
[57] T. Fujiwara, *Rept. Geol. Surv. Hokkaido*, No. 27, 1962; *Chem. Abs.*, 1963, **59**, 1391.
[58] T. Fujiwara and K. Futa, *Rept. Geol. Surv. Hokkaido*, No. 25, 1961. *Chem. Abs.*, 1962, **56**, 6940.
[59] F. B. Esguerra, *Philippines Bur. Mines, Pert. Investigation* No. 32, 1961; *Chem. Abs.*, 1962, **56**, 6940.
[60] P. Legoux and R. Faucheux, *Congr. intern. mines, mét. et géol. appl., 7e session*, Paris, Oct., 1935, Geol. 1, 187; *Chem. Abs.*, 1936, **30**, 8093.
[61] J. T. Pacheco, *E.C.H.J.C. da C.*, Lisbon, 1962, 109; *Chem. Abs.*, 1963, **58**, 11113.
[62] L. A. J. Williams, *Colony, Protect. Kenya, Geol. Surv. Rept.* No. 52, 1962; *Chem. Abs.*, 1964, **61**, 463.
[63] *S. African Mining Eng. J.*, 1943, **54**, 137; *Chem. Abs.*, 1944, **38**, 3388.
[64] G. Gutzert and P. Kovaliv, *Arch. sci. phys. et nat.*, 1939, **21**, 260; *Chem. Abs.*, 1940, **34**, 3139.
[65] W. O. Johnston, Jr., *Rev. quím. ind. (Rio de Janeiro)*, 1944, **13**, 19; *Chem. Abs.*, 1945, **39**, 3096.
[66] S. F. Abreu, *Ministerio agr. Estação exptl. combustiveis e minerios (Rio de Janeiro)*, separate, 1933; *Chem. Abs.*, 1936, **30**, 5909.
[67] F. L. Hess and J. L. Gillson, *loc. cit.*
[68] V. C. Fryklund, Jr. and D. F. Holbrook, *Ark. Resources and Dev. Com., Div. Geol. Bull.* 16, 1950; *Chem. Abs.*, 1952, **46**, 68.
[69] F. A. Vogel, Jr., *U.S. Bur. Mines Inform. Circ.* **7293**, 1944; *Chem. Abs.*, 1944, **38**, 6249.
[70] T. L. Watson and S. Taber, *op. cit.*, 31.
[71] *U.S. Bur. Mines Minerals Yearbook*, 1937, 782.
[72] S. F. Abreu, *Rev. quím. ind. (Rio de Janeiro)*, 1936, **5**, 16; *Chem. Abs.*, 1936, **30**, 6692.
[73] L. L. Shilin, *Trudy Inst. Geol. Nauk S.S.S.R.*, No. 10, 1940; *Chem. Abs.*, 1941, **35**, 1358.
[74] A. v. Moos, *Medd. Grønland.*, 1938, No. 4, 103; *Chem. Abs.*, 1940, **34**, 1946.
[75] *U.S. Bur. Mines Foreign Minerals Quarterly*, 1939, **2**, No. 3, 10.
[76] F. Schumacher, *Metallwirtschaft*, 1938, **17**, 399; *Chem. Abs.*, 1940, **34**, 3211.
[77] P. Bellair, *Compt. rend. soc. géol. France*, 1940, 75; *Chem. Abs.*, 1940, **34**, 4705.
[78] *Eng. and Mining J.*, Aug. 1964, 74; *Chemical Week*, Nov. 21, 1964, 74.
[79] E. P. Youngman, *op. cit.*, 23.
[80] A. W. Fahrenwald, *Mining Congr. J.*, 1940, **26**, 22.
[81] D. S. Phelps, U.S. 2,257,808, Oct. 7, 1941, to Edgar Plastic Kaolin Co.; *Chem. Abs.*, 1942, **36**, 385. L. Perruche, *Nature, La*, **1941**, 411; *Chem. Abs.*, 1943, **37**, 2965.
[82] R. A. Pickens, U.S. 2,387,856, Oct. 30, 1945, to American Cyanamid Co.; *Chem. Abs.*, 1946, **40**, 47.
[83] J. C. Detwiller, *Mining Eng.*, 1952, **4**, 560; *Chem. Abs.*, 1952, **46**, 6563.
[84] V. F. Swanson and J. E. Shelton, *U.S. Bur. Mines, Rept. Invest.* No. 5953, 1962; *Chem. Abs.*, 1962, **56**, 13869.
[85] G. E. Fish Jr. and V. F. Swanson, *U.S. Bur. Mines, Rept. Invest.* 6429, 1964; *Chem. Abs.*, 1964, **61**, 2846.
[86] V. F. Swanson, *U.S. Bur. Mines, Rept. Invest.* No. 6172, 1963; *Chem. Abs.*, 1963, **58**, 9899.
[87] E. J. Michal, U.S. 2,904,177, Sept. 15, 1959, to National Lead Co.; *Chem. Abs.*, 1960, **54**, 842.
[88] K. Yamazaki, Y. O. Nakamura, and M. Nakamura, Japanese 13,803, 1960, to Mitsubishi Metal Mining Co., Ltd.; *Chem. Abs.*, 1962, **56**, 8369.
[89] R. E. Baarson, U.S. 2,792,940, May 21, 1957, to Armour and Co.; *Chem. Abs.*, 1957, **51**, 14221.
[90] W. B. Lenhart, *Rock Products*, 1949, **52**, No. 2, 102; *Chem. Abs.*, 1949, **43**, 3248.

[91] I. V. Chipanin, M. T. Ivanova, and A. N. Kozhukhovskaya, *S.N.T.I., Nauchn.-Issled. Inst. Met.*, 1959, No. 8, 162; *Chem. Abs.*, 1962, **56**, 15213.

[92] L. G. Podkosov, K. S. Akopova, and N. E. Romanovskaya, *Tr. V. N.-I., Inst. Mineraln. Syrga*, 1961, No. 6, 158; *Chem. Abs.*, 1962, **57**, 3078.

[93] E. A. Savari, A. A. Frolova, and L. I. Bandenok, *Sb. M. po G. D., O. i Met. T. N.-I. G.-R. Inst.*, 1961, No. 6, 70; *Chem. Abs.*, 1962, **57**, 16217.

[94] S. A. Sysolyatin, *Nauch. I. i P. Inst.*, 1958, No. 4, 136; *Chem. Abs.*, 1960, **54**, 20112.

[95] S. A. Sysolyatin, *Titan i ego Splavy, Akad. Nauk. S.S.S.R., Inst. Met.*, 1963, 24; *Chem. Abs.*, 1963, **59**, 14927.

[96] C. Macarovici, E. Motiu, and E. Perte, *Acad. rep. polulare Romanie, Filiala Cluj, Studii chim.*, 1960, **11**, 91; *Chem. Abs.*, 1961, **55**, 10820.

[97] U.S.S.R. 98,944, Nov. 22, 1958; *Chem. Abs.*, 1959, **53**, 14438.

[98] Y. M. Chikin and B. V. Levinskii, U.S.S.R. 160,477, Jan. 31, 1964; *Chem Abs.*, 1964, **60**, 14166.

[99] Y. M. Chikin, *N. T., I. G. N.-I. I. R. M.*, 1963, **11**, 155; *Chem. Abs.*, 1964, **60**, 12925.

[100] L. G. Podkovsov, *M. S. V. N. I. I. M. S.*, 1963, No. 9, 3; *Chem. Abs.*, 1964, **61**, 9202.

[101] R. K. Neuman, *Australian Inst. Mining Met. Proc.*, 1963, **205**, 105; *Chem. Abs.*, 1963, **59**, 14927.

[102] J. Biernat, Polish 44,549, May 29, 1961, to Akademia Gorniczo-Hutnicza; *Chem. Abs.*, 1964, **60**, 6532.

[103] M. K. Hussein and S. Z. El-Tawil, *Z. Erzbergbau Metallhuettenw.*, 1964, **17**, 192; *Chem. Abs.*, 1964, **61**, 3973.

[104] O. Mokelbust, G. Olavesen, and H. Bjoranesset, U.S. 2,990,250, June 27, 1961, to National Lead Co.; *Chem. Abs.*, 1961, **55**, 25182.

[105] A. S. Ivoilov and T. B. Tarasova, *Nauch. Tr. I. N.-I. Inst. Redkikh Metal.*, 1961, No. 10, 262; *Chem. Abs.*, 1963, **58**, 8685.

[106] W. K. Finn, *Indian Mining J., Spec. Issue*, 1957, **5**, 95; *Chem. Abs.*, 1960, **54**, 1199.

## CHAPTER 5

[1] *Minerals Yearbook*, U.S. Bur. Mines, yearly volumes.

## CHAPTER 6

[1] M. A. Hunter, *J. Am. Chem. Soc.*, 1910, **32**, 330; *Chem. Abs.*, 1910, **4**, 1275.

[2] M. H. Klaproth, *Analytical Essays Toward Promoting the Chemical Knowledge of Mineral Substances* (English translation) Cadell and Davies, London, 1801, 200.

[3] J. J. Berzelius, *Pogg. Ann.*, 1825, **4**, 3.

[4] W. H. Wollaston, *Phil. Trans.*, 1828, **17**, 113.

[5] F. Wohler, *Compt. rend.*, 1849, **29**, 505.

[6] H. St. C. Deville, *Comp. rend.*, 1859, **45**, 480.

[7] L. F. Nilson and O. Petersson, *Z. physik. Chem.*, 1887, **1**, 25.

[8] H. Moissan, *Compt. rend.*, 1895, **120**, 290.

[9] H. Le Chatelier, *Rev. mét.*, 1911, **8**, 367; *Chem. Abs.*, 1911, **5**, 2804.

[10] Huppertz, *Electrochem. and Metallurgical Ind.*, 1905, **3**, 35.

[11] M. A. Hunter, *J. Am. Chem. Soc.*, 1910, **32**, 330; *Chem. Abs.*, 1910, **4**, 1275.

[12] D. Lely and L. Hamburger, *Z. anorg. Chem.*, 1914, **87**, 209; *Chem. Abs.*, 1914, **8**, 2855.

[13] G. F. Comstock and V. V. Efinoff, U.S. 2,273,834, Feb. 24, 1942, to Titanium Alloy Manufacturing Co.; *Chem. Abs.*, 1942, **36**, 3774.

[14] R. Potvin and G. S. Farnham, *Trans. Can. Inst. Mining Met.*, 1946, **49**, 516; *Chem. Abs.*, 1947, **41**, 66.

[15] L. Weiss and Kaiser, Z. anorg. Chem., 1910, 65, 345; Chem. Abs., 1910, 4, 1276.

[16] A. C. Vaurnasos, Z. anorg. Chem., 1913, 81, 364; Chem. Abs., 1913, 7, 2913.

[17] H. Freudenberg, U.S. 2,148,345, Feb. 21, 1939, to Deutsche Gold-und-Silber-Schneideanstalt Roessler; Chem. Abs., 1939, 33, 3981: British 479,014, Jan. 28, 1938, to Deutsche Gold-und-Silber-Schneideanstalt Roessler; Chem. Abs., 1938, 32, 4933.

[18] W. Kroll, Trans. Electrochem. Soc., 1940, 78; Chem. Abs., 1940, 34, 5759.

[19] O. Ruff and H. Brintzinger, Z. anorg. u. allgem. Chem., 1923, 129, 267; Chem. Abs., 1923, 17, 3143.

[20] W. Kroll, Z. anorg. u. allgem. Chem., 1937, 234, 42, Chem. Abs., 1937, 31, 8416.

[21] H. Lohmann, Elektrochem. Z., 1919, 26, 29; Chem. Abs., 1920, 14, 1081.

[22] J. M. Merle, U.S. 2,395,286, Feb. 19, 1946; Chem. Abs., 1946, 40, 2432.

[23] K. Iwase and N. Nasu, Kinzoku-no-Kenkyu, 1937, 14, 87; Chem. Abs., 1939, 33, 7229.

[24] A. Stahler and F. Bachman, Ber., 1912, 44, 2906; Chem. Abs., 1912, 6, 461.

[25] N. Parravano and C. Mazzetti, Rec. trav. chim., 1923, 42, 821; Chem. Abs., 1923, 17, 3651.

[26] A. E. van Arkel and J. H. de Boer, Z. anorg. u. allgem. Chem., 1925, 148, 345; Chem. Abs., 1926, 20, 881: Metallwirtschaft, 1934, 13, 405; Chem. Abs., 1934, 28, 5011.

[27] J. D. Fast, Z. anorg. u. allgem. Chem., 1939, 241, 42; Chem. Abs., 1939, 33, 4540.

[28] F. H. Dotterweich, Refiner Natural Gasoline Mfr., 1942, 21, 135; Chem. Abs., 1942, 36, 3757.

[29] W. J. Kroll, Mining and Met., 1946, 27, 262; Chem. Abs., 1946, 40, 3375. W. J. Kroll and A. W. Schlechter, Metal Ind., 1946, 69, 319; Chem. Abs., 1947, 41, 355.

[30] R. S. Dean, J. R. Long, F. S. Wartman, and E. L. Anderson, Am. Inst. Mining Met. Engrs., Tech. Pub. 1961, 1946. R. S. Dean and B. Silkes, U.S. Bur. Mines, Inform. Circ. 7381, 1946. W. H. Waggman and E. A. Gee, Chem. Eng. News, 1948, 26, 377; Chem. Abs., 1948, 42, 2213.

[31] K. R. Hanna and H. W. Wormer, J. Council Sci. Ind. Research, 1946, 19, 449; Chem. Abs., 1947, 41, 4746.

[32] Metal Ind., 1947, 70, 363; Chem. Abs., 1947, 41, 4421.

[33] P. P. Alexander, U.S. 2,427,338, Sept. 16, 1947, to Metal Hydrides Inc.; Chem. Abs., 1948, 42, 332: U.S. 2,284,551, May 26, 1942; Chem. Abs., 1942, 36, 6708.

[34] D. B. Keyes and S. Swann, Univ. of Illinois Expt. Sta., Bull. 206, 1930; Chem. Abs., 1930, 24, 2954.

[35] E. Porkony, German 605,551, Nov. 13, 1934, to I. G. Farbenind. A-G.; Chem. Abs., 1935, 29, 1725.

[36] M. Haissinsky and H. E. Zauizziano, Compt. rend., 1936, 203, 161; Chem. Abs., 1936, 30, 6287: J. Chim. Phys., 1937, 34, 641; Chem. Abs., 1938, 32, 3701: Compt. rend., 1937, 204, 759; Chem. Abs., 1937, 31, 3392.

[37] R. Groves and A. S. Russell, J. Chem. Soc., 1931, 2805; Chem. Abs., 1932, 26, 681. A. S. Russell, Nature, 1931, 127, 273; Chem. Abs., 1931, 25, 2352.

[38] N. N. Gratsianskii and A. P. Vovkogon, Zapiski Inst. Khim. Akad. Nauk U.R.S.S., 1940, 7, 159; Chem. Abs., 1941, 35, 3530.

[39] A. Travers, Chimie & industrie, 1932, Special number, 347; Chem. Abs., 1932, 26, 3470.

[40] B. Diethelm and F. Forester, Z. physik. Chem., 1908, 62, 129; Chem. Abs., 1908, 2, 1655. F. F. Dresden, Chem. Ztg., 1907, 31, 922; Chem. Abs., 1908, 2, 226.

[41] S. Zeltner, Collection Czechoslov. Chem., Commun., 1932, 4, 319; Chem. Abs., 1932, 26, 5853.

[42] O. Esin, Acta Physicochim. U.R.S.S., 1940, 13, 429; Chem. Abs., 1941, 35, 2056.

[43] E. D. Botts and F. C. Krauskoff, J. Phys. Chem., 1927, 31, 1404; Chem. Abs., 1927, 21, 3807.

[44] W. G. Burgers, A. Claasen, and I. Zernike, Z. Physik, 1932, 74, 593; Chem. Abs., 1932, 26, 3190.

[45] Jelks Barksdale, General Chemistry for Colleges, Longmans, Green & Co., New York, 1949.

[46] W. Kunitz, *Neues Jahrb. Mineral. Geol., Beilage Bd.*, 1936, **70A**, 385; *Chem. Abs.*, 1936, **30**, 8101.

[47] P. M. Tyler, *Metals & Alloys*, 1935, **6**, 93.

[48] R. S. Dean and B. Silkes, *U.S. Bur. Mines, Inform. Circ.* 7381, 1946, J. W. Mellor, *A Comprehensive Treatise on Inorganic and Theoretical Chemistry*, Longmans, Green & Co., New York, 1927, 1. G. Fownes and R. Bridges, *A Manual of Elementary Chemistry*, Henry C. Lee, Philadelphia, 1870, 393.

[49] F. W. Aston, *Nature*, 1934, **133**, 684; *Chem. Abs.*, 1934, **28**, 4306: *Proc. Roy. Soc.*, 1935, **A149**, 396; *Chem. Abs.*, 1935, **29**, 3910.

[50] A. O. Nier, *Phys. Rev.*, 1938, **53**, 282; *Chem. Abs.*, 1938, **32**, 3258. T. Okuda and K. Ogata, *Phys. Rev.*, 1941, **60**, 690; *Chem. Abs.*, 1942, **36**, 333.

[51] H. Walke, *Phys. Rev.*, 1937, **52**, No. 8, 777; *Chem. Abs.*, 1938, **32**, 41.

[52] J. J. O'Conner, M. L. Pool, and J. B. Kurbatou, *Phys. Rev.*, 1942, **62**, 413; *Chem. Abs.*, 1943, **37**, 4007.

[53] S. Caillere and J. Noetzlin, *Compt. rend.*, 1942, **215**, 22; *Chem. Abs.*, 1944, **38**, 5455.

[54] O. Huber, O. Lienhard, P. Scherrer, and H. Waffler, *Helv. Phys. Acta*, 1943, **16**, 33; *Chem. Abs.*, 1946, **40**, 5636.

[55] H. Mark and R. Wierl, *Z. Elektrochem.*, 1930, **36**, 675; *Chem. Abs.*, 1931, **25**, 9.

[56] E. Wedekind and P. Hausknecht, *Ber.*, 1914, **46**, 3763; *Chem. Abs.*, 1914, **8**, 1384.

[57] M. A. Hunter, *J. Am. Chem. Soc.*, 1910, **32**, 330; *Chem. Abs.*, 1910, **4**, 1275.

[58] J. D. Fast, *Metallwirtschaft*, 1938, **17**, 459; *Chem. Abs.*, 1938, **32**, 5748.

[59] H. W. Gillett, *Foote Prints*, 1940, **13**, 1; *Chem. Abs.*, 1941, **35**, 3582.

[60] R. S. Dean, J. R. Long, F. S. Wartman, and E. T. Hayes, *Am. Inst. Mining Met. Engrs., Tech. Pub.* **1965**, 1946; *Chem. Abs.*, 1946, **40**, 2425. R. S. Dean, J. R. Long, E. T. Hayes, and D. C. Root, *Am. Inst. Mining Met. Engrs., Tech. Pub.* **2102**, 1946. R. S. Dean and B. Silkes, *U.S. Bur. Mines, Inform. Circ.* 7381, 1946; *Chem. Abs.*, 1947, **41**, 4080. *Nat. Bur. Standards, Circ.* **C447**. *Titanium Metal Sponge and Ingot*, E. I. du Pont de Nemours and Co., 1948. R. S. Dean, J. R. Long, F. S. Wartman, and E. L. Anderson, *Am. Inst. Mining Met. Engrs., Tech. Pub.* **1961**, 1946. *Report of Symposium on Titanium*, sponsored by Office of Naval Research, Washington, March 1949.

[61] F. M. Jaeger, E. Rosenbohm, and R. Fonteyne, *Rec. trav. chim.*, 1936, **55**, 615; *Chem. Abs.*, 1936, **30**, 7021: *Proc. Acad. Sci. Amsterdam*, 1936, **39**, 442; *Chem. Abs.*, 1936, **30**, 5111.

[62] H. H. Potter, *Proc. Phys. Soc. (London)*, 1941, **53**, 695; *Chem. Abs.*, 1942, **36**, 948.

[63] J. H. de Boer, W. G. Burgers, and J. D. Fast, *Proc. Acad. Sci. Amsterdam*, 1936, **39**, 519; *Chem. Abs.*, 1936, **30**, 5107.

[64] J. D. Fast, *Rec. trav. chim.*, 1939, **58**, 973; *Chem. Abs.*, 1940, **34**, 932.

[65] A. W. Hull, *Phys. Rev.*, 1921, **18**, 88; *Chem. Abs.*, 1922, **16**, 3563.

[66] R. A. Patterson, *Phys. Rev.*, 1925, **26**, 56; *Chem. Abs.*, 1925, **19**, 3180.

[67] W. G. Burgers and F. M. Jacobs, *Z. Krist.*, 1936, **94**, 299; *Chem. Abs.*, 1936, **30**, 7947.

[68] A. W. Hull, *Science*, 1920, **52**, 227; *Chem. Abs.*, 1920, **14**, 3363.

[69] J. Fitzwilliam, A. P. Kaufman, and C. F. Squire, *J. Chem. Phys.*, 1941, **9**, 678; *Chem. Abs.*, 1941, **35**, 7248.

[70] P. W. Bridgman, *Phys. Rev.*, 1935, **48**, 825; *Chem. Abs.*, 1936, **30**, 916.

[71] A. Schulze, *Z. Metallkunde*, 1931, **23**, 261; *Chem. Abs.*, 1932, **26**, 638.

[72] P. W. Bridgman, *Proc. Natl. Acad. Sci.*, 1920, **6**, 505; *Chem. Abs.*, 1921, **15**, 1447: *Proc. Am. Acad. Arts Sci.*, 1929, **64**, 19; *Chem. Abs.*, 1930, **24**, 1556.

[73] P. Clausing and G. Moubis, *Physics*, 1927, **7**, 245; *Chem. Abs.*, 1929, **23**, 1320.

[74] J. C. McLennan, L. E. Howlett, and J. O. Wilhelm, *Trans. Roy. Soc. Can.*, 1930, **23**, 287; *Chem. Abs.*, 1930, **24**, 2352.

[75] P. Kapitza, *Proc. Roy. Soc.*, 1929, **A123**, 292; *Chem. Abs.*, 1929, **23**, 2860.

[76] W. Meissner, *Z. Physik*, 1930, **60**, 181; *Chem. Abs.*, 1930, **24**, 2350. R. T. Webber, and J. M. Reynolds, *Phys. Rev.*, 1948, **73**, 640; *Chem. Abs.*, 1948, **42**, 4015.

[77] W. G. Mixter, *Am. J. Sci.*, 1909, **27**, 393; *Chem. Abs.*, 1909, **3**, 1712: *Am. J. Sci.*, 1912, **33**, 45; *Chem. Abs.*, 1912, **6**, 704.

[78] K. K. Kelley, *Ind. Eng. Chem.*, 1944, **36**, 865; *Chem. Abs.*, 1944, **38**, 5724.

[79] P. Hindnert, *J. Research Natl. Bur. Standards,* 1943, **30**, 101; *Chem. Abs.*, 1943, **37**, 2317.

[80] W. Kroll, *Z. Metallkunde,* 1937, **29**, 189; *Chem. Abs.*, 1938, **32**, 1225: *Z. anorg. u. allgem. Chem.*, 1937, **234**, 42; *Chem. Abs.*, 1937, **31**, 8416.

[81] A. C. Fieldner and W. E. Rice, *U.S. Bur. of Mines, Inform. Circ.* **7241**, 1943; *Chem. Abs.*, 1943, **37**, 6844. I. Hartmann and H. P. Greenwald, *Mining and Met.*, 1945, **26**, 331; *Chem. Abs.*, 1946, **40**, 2629.

[82] P. H. Brace, *J. Electrochem. Soc.*, 1948, **94**, 170; *Chem. Abs.*, 1948, **42**, 8763.

[83] H. E. Duckworth and R. S. Preston, *Phys. Rev.*, 1951, **82**, 468; *Chem. Abs.*, 1951, **45**, 6063.

[84] J. W. Oliver and J. W. Ross, Jr., *J. Am. Chem. Soc.*, 1963, **85**, 2565; *Chem. Abs.*, 1963, **59**, 8359.

[85] B. F. Markov and B. P. Podafa, *Ukr. Khim. Zh.*, 1963, **29**, 600; *Chem. Abs.*, 1963, **59**, 9591.

[86] I. A. Menzies, G. J. Hills, L. Young and J. O'M. Bockris, *Trans. Faraday Soc.*, 1959, **55**, 1580; *Chem. Abs.*, 1960, **54**, 9561.

[87] M. A. Maurakh, *Trans. Indian Inst. Metals,* 1964, **14**, 209; *Chem. Abs.*, 1964, **61**, 10399.

[88] P. Costa and G. Cizeron, *Comp. rend.*, 1958, **246**, 2261; *Chem. Abs.*, 1958, **52**, 19806.

[89] M. J. Bibly and J. D. Parr, Jr., *J. Inst. Metals,* 1964, **92**, 341; *Chem. Abs.*, 1964, **61**, 8020.

[90] V. I. Lakomskii and N. N. Kalinyrek, *Avtomat. Svarka,* 1963, 31; *Chem. Abs.*, 1964, **60**, 1406.

[91] V. F. Ulyanov and M. I. Vinogradov, U.S.S.R. 154,963, July 20, 1963; *Chem. Abs.*, 1964, **60**, 2561.

[92] J. D. Jackson, P. D. Miller, W. K. Boyd, and F. W. Funk, *U.S. Office of Tech. Serv.*, 1962, **55**, 19675; *Chem. Abs.*, 1963, **59**, 1350.

[93] F. E. Littman and F. M. Church, *U.S. Atomic Energy Comm.*, SRIA-29, 1960; *Chem. Abs.*, 1961, **55**, 8230.

[94] S. A. Nikolaeva and Zashikhina, *Tsvetn. Metal.*, 1964, **37**, 54; *Chem. Abs.*, 1964, **61**, 400.

[95] D. I. Lainer, A. S. Bai, and M. I. Tsypin, *Tr. G. N. I. i P. I. po O. T. M.*, 1963, 62; *Chem. Abs.*, 1964, **61**, 6741.

[96] D. I. Lainer and A. S. Bai, *Fiz. Metal. i Metalloved,* 1962, **14**, 283; *Chem. Abs.*, 1963, **58**, 316.

[97] V. V. Andreeva and E. A. Yakovleva, *J. Physik. Chem.*, 1964, **226**, 232; *Chem. Abs.*, 1964, **61**, 12946.

[98] J. B. Rittenhouse, *Trans. Am. Soc. Metals,* 1957, Preprint No. 68; *Chem. Abs.*, 1958, **62**, 1015.

[99] L. B. Golden, J. R. Lane, Jr., and W. L. Ackerman, *Ind. Eng. Chem.*, 1952, **44**, 1930; *Chem. Abs.*, 1952, **46**, 11081.

[100] K. Mori, *Kogyo Kagaku Zasshi,* 1963, **66**, 1750; *Chem. Abs.*, 1964, **61**, 1577.

[101] A. M. Shroff and P. A. Moutou, *Vide.*, 1963, **18**, 417; *Chem. Abs.*, 1964, **60**, 1343.

[102] R. W. Roberts and R. S. Owens, *Nature,* 1963, **200**, 357; *Chem. Abs.*, 1964, **60**, 2590.

[103] K. Jordan and R. W. Fischer, *Tech. Mitt. Krupp,* 1955, **13**, No. 2, 44; *Chem. Abs.*, 1955, **49**, 10825.

[104] B. J. Connolly, *Light Metals,* 1960, **23**, 286; *Chem. Abs.*, 1961, **55**, 3392. *Chemical and Process Eng.*, 1958, **39**, 247; *Chem. Abs.*, 1958, **52**, 16810. E. I. Rybakov and N. A. Fomicheva, *Lakokrasochnye Materialy i ikh Primenenie,* 1962, No. 2, 60; *Chem. Abs.*, 1962, **57**, 3173.

[105] W. Simm and W. Kwasnick, German 1,113,444, Oct. 13, 1958, to Farbenfabriken Bayer A.-G.; *Chem. Abs.*, 1962, **56**, 7051.

## CHAPTER 7

[1] J. Bohm, Z. anorg. u. allgem. Chem., 1925, 149, 217; Chem. Abs., 1926, 20, 527.
[2] L. Wohler, Kolloid-Z., 1926, 38, 97; Chem. Abs., 1926, 20, 3268.
[3] K. Borneman and H. Schirmeister, Metallurgie, 1910, 7, 646; J. Chem. Soc., 1910, A, 1073.
[4] C. Reichard, Chem. Ztg., 1903, 27, 1; J. Chem. Soc., 1903, A, 217.
[5] P. Baumgarten and W. Bruns, Ber., 1941, 74B, 1232; Chem. Abs., 1942, 36, 5105.
[6] G. P. Luchinskie, J. Gen. Chem. (U.S.S.R.), 1940, 10, 769; Chem. Abs., 1941, 35, 2432.
[7] F. W. Clarke, U.S. Geol. Survey, Bull. 770, 1924, 354.
[8] F. Barblan, E. Brandenberger, and P. Niggli, Helv. Chim. Acta, 1944, 27, No. 1, 88; Chem. Abs., 1944, 38, 3179.
[9] A. V. Pamfilov and E. G. Ivancheva, J. Gen. Chem. (U.S.S.R.), 1940, 10, 154; Chem. Abs., 1940, 34, 7199.
[10] E. N. Bunting, Bur. Standards J. Research, 1933, 11, 719; Chem. Abs., 1934, 28, 1595.
[11] W. Knop, J. Chem. Soc., 1871, 24, 200.
[12] D. Baraleff, J. prakt. Chem., 1921, 102, 283; Chem. Abs., 1921, 15, 3235.
[13] W. A. Roth and G. Becker, Z. physik. Chem., 1931, 55; Chem. Abs., 1931, 25, 5343. W. A. Roth and U. Wolf, Rec. trav. chim., 1940, 59, 511; Chem. Abs., 1941, 35, 4667.
[14] W. A. Roth and G. Becker, Z. physik. Chem., 1932, A159, 1; Chem. Abs., 1932, 26, 3152.
[15] C. Kroger and H. Kunz, Z. anorg. u. allgem. Chem., 1934, 218, 379; Chem. Abs., 1934, 28, 7135.
[16] H. J. McDonald and H. Seltz, J. Am. Chem. Soc., 1939, 61, 2405; Chem. Abs., 1939, 33, 9111.
[17] H. Buttner and J. Engle, Z. tech. Physik., 1937, 18, 113; Chem. Abs., 1937, 31, 4863.
[18] C. Schusterius, Z. tech. Physik., 1935, 16, 640; Chem. Abs., 1936, 30, 2439.
[19] L. J. Berberick and M. E. Bell, J. Applied Phys., 1940, 11, 681; Chem. Abs., 1940, 34, 7676.
[20] D. M. Gans, U.S. Brooks, and E. E. Boyd, Ind. Eng. Chem., Anal. Ed., 1942, 14, 396; Chem. Abs., 1942, 36, 4002. G. Jura and W. D. Harkins, J. Am. Chem. Soc., 1944, 66, 1356; Chem. Abs., 1944, 38, 5125.
[21] M. Foex, Bull. soc. chim., 1944, II, 6; Chem. Abs., 1945, 39, 8.
[22] M. D. Earle, Phys. Rev., 1942, 61, 56; Chem. Abs., 1942, 36, 1841.
[23] C. H. Shomate, J. Am. Chem. Soc., 1947, 69, 220; Chem. Abs., 1947, 41, 2633.
[24] F. R. Coheur, Bull. soc. roy. sci. Liége, 1943, 12, 98; Chem. Abs., 1944, 38, 6193.
[25] E. L. Nichols, Phys. Rev., 1923, 22, 420; Chem. Abs., 1924, 18, 1949.
[26] S. Izawa, J. Soc. Chem. Ind., Japan, 1933, 36, 43; Chem. Abs., 1933, 27, 2871.
[27] J. P. Blewett and E. P. Jones, Phys. Rev., 1936, 50, 464; Chem. Abs., 1936, 30, 8001.
[28] P. D. St. Pierre, J. Am. Ceramic Soc., 1952, 35, 188; Chem. Abs., 1952, 46, 9274.
[29] G. Brauer and W. Littke, J. Inorg. and Nuclear Chem., 1960, 16, 76; Chem. Abs., 1961, 55, 5099.
[30] F. Hund, German 1,080,245, April 21, 1960, to Farbenfabriken Bayer A.-G.; Chem. Abs., 1961, 55, 18136.
[31] C. N. R. Rao, S. R. Yoganarasimhan, and P. D. Faeth, Trans. Faraday Soc., 1961, 57, 504; Chem. Abs., 1961, 55, 23193.
[32] M. E. Straumanis, T. Ejima, and W. J. James, Acta Cryst., 1961, 14, 493; Chem. Abs., 1961, 55, 17152.
[33] C. Legrand and J. Delville, Compt. rend., 1953, 236, 944; Chem. Abs., 1953, 47, 5208.
[33a] W. H. Baur, Acta Cryst., 1956, 9, 515; Chem. Abs., 1956, 50, 12588.

[34] D. T. Cromer and K. Herrington, *J. Am. Chem. Soc.*, 1955, **77**, 4708; *Chem. Abs.*, 1956, **50**, 3164.
[35] P. Coufova and H. Arend, *Czech. J. Phys.*, 1961, **11**, 845; *Chem. Abs.*, 1962, **57**, 4131.
[36] T. Hurlen, *Acta Chem. Scand.*, 1959, **13**, 365; *Chem. Abs.*, 1960, **54**, 20781.
[37] G. V. Jere and C. C. Patel, *J. Sci. Ind. Research* (India), 1961, **20B**, 292; *Chem. Abs.*, 1962, **56**, 4347.
[38] S. Wilska, *Acta Chem. Scand.*, 1954, **8**, 1796; *Chem. Abs.*, 1955, **49**, 13725.
[39] L. Merker and H. Espenschied, U.S. 2,521,392, Sept. 5, 1950, to National Lead Co.; *Chem. Abs.*, 1956, **50**, 11046.
[40] E. M. Gladrow and H. G. Ellert, *J. Chem. Eng. Data*, 1961, **6**, 318; *Chem. Abs.*, 1961, **55**, 25180.
[41] V. A. Dorin, *Zhur. Tekh. Fiz.*, 1955, **25**, 577; *Chem. Abs.*, 1956, **50**, 642.
[42] V. M. Kharitonov and L. A. Morgun, *Khim. Volokna*, 1962, No. 2, 20; *Chem. Abs.*, 1962, **57**, 9252.
[43] C. Kurylenko, *Cahiers Phys.*, 1958, **92**, 163; *Chem. Abs.*, 1962, **57**, 163.
[44] R. Haul and G. Duembgen, *Z. Electrochem.*, 1962, **66**, 636; *Chem. Abs.*, 1963, **58**, 943.
[45] A. S. Barker, Jr., *J. Chem. Phys.*, 1963, **38**, 2257; *Chem. Abs.*, 1963, **58**, 13316.
[46] V. A. Reznichenko and F. B. Khalimov, *Tiu. E.S., A-N.S.S.S.R., I.M. i A-A.B.*, 1959, No. 2, 11; *Chem. Abs.*, 1960, **54**, 15859.
[47] J. Nakazumi, Japanese 233, Jan. 23, 1962, to Fuji Titanium Industry C., Ltd.; *Chem. Abs.*, 1963, **59**, 3630.
[48] E. Belyakova, A. Komar, and V. Mikhailov, *Metallurg*, 1939, **14**, 23; *Chem. Abs.*, 1940, **34**, 2685. E. Friederick and L. Sittig, *Z. anorg. u. allgem. Chem.*, 1925, **145**, 127; *Chem. Abs.*, 1925, **19**, 2580.
[49] G. Lunde, *Z. anorg. u. allgem. Chem.*, 1927, **164**, 341; *Chem. Abs.*, 1927, **21**, 3845.
[50] N. Nasu, *J. Chem. Soc. Japan*, 1935, **56**, 542; *Chem. Abs.*, 1935, **29**, 5726.
[51] V. M. Goldschmidt, U.S. 1,389,191, Aug. 20, 1922; *Chem. Abs.*, 1922, **16**, 167.
[52] J. Irwin and R. H. Monk, British 267,788, Aug. 6, 1926, to E. C. R. Marks; *Chem. Abs.*, 1928, **22**, 1244.
[53] G. Meder, U.S. 2,138,384, Nov. 29, 1938, to I. G. Farbenind., A-G.; *Chem. Abs.*, 1939, **33**, 1973.
[54] M. Billy, *Comp. rend.*, 1913, **155**, 777; *Chem. Abs.*, 1913, **7**, 737.
[55] E. Junker, *Z. anorg. u. allgem. Chem.*, 1936, **228**, 97; *Chem. Abs.*, 1936, **30**, 7983.
[56] O. Ruff, *Z. anorg. u. allgem. Chem.*, 1913, **82**, 373; *Chem. Abs.*, 1913, **7**, 3935.
[57] M. Billy, *Ann. chim.*, 1921, **16**, 5; *Chem. Abs.*, 1922, **16**, 1051.
[58] S. S. Sklyarenko, Y. M. Lipkes, F. D. Iozefovich, and V. V. Mukhantseva, *J. Applied Chem.* (*U.S.S.R.*), 1940, **13**, 51; *Chem. Abs.*, 1940, **34**, 7756.
[59] E. P. Belyakova, A. Komar, and V. Mikhailov, *Metallurg*, 1940, No. 4, 5; *Chem. Abs.*, 1943, **37**, 1358.
[60] C. H. Shomate, *J. Am. Chem. Soc.*, 1946, **68**, 310; *Chem. Abs.*, 1946, **40**, 2729.
[61] B. F. Naylor, *J. Am. Chem. Soc.*, 1946, **68**, 1077; *Chem. Abs.*, 1946, **40**, 5326.
[62] A. K. Breger, *Acta Physiochim. U.R.S.S.*, 1940, **13**, 723; *Chem. Abs.*, 1941, **35**, 2390.
[63] N. Nasu, *J. Chem. Soc. Japan*, 1935, **56**, 659; *Chem. Abs.*, 1935, **29**, 7170.
[64] T. Watanabe, Japanese 4851, April 26, 1963; *Chem. Abs.*, 1964, **60**, 15472.
[65] H. Watanabe and S. Narita, Japanese 3101, April 12, 1963; *Chem. Abs.*, 1964, **60**, 11650.
[66] E. Belyakova, A. Komar, and V. Mikhailov, *Metallurg*, 1939, **14**, 23; *Chem. Abs.*, 1940, **34**, 2685.
[67] M. Billy, *Ann. chim.*, 1921, **16**, 5; *Chem. Abs.*, 1922, **16**, 1051.
[68] W. Dawihl and K. Schroter, *Z. anorg. u. allgem. Chem.*, 1937, **233**, 178; *Chem. Abs.*, 1937, **31**, 7781.
[69] H. Brakken, *Z. Krist.*, 1928, **67**, 547; *Chem. Abs.*, 1928, **22**, 4020.

[70] A. Chrétien and R. Wyss, *Compt. rend.*, 1947, **224**, 1642; *Chem. Abs.*, 1947, **41**, 6831.

[71] R. Wyss, *Ann. chim.*, 1948, **3**, 215; *Chem. Abs.*, 1948, **42**, 7187.

[72] E. V. Snopova and N. I. Rotkov, *Chem. Zentr.*, 1937, **11**, 2972; *Chem. Abs.*, 1939, **33**, 5606. M. Billy, *Ann. chim.*, 1921, **16**, 5; *Chem. Abs.*, 1922, **16**, 1051.

[73] M. A. Steinberg and E. Wainer, U.S. 2,750,259, June 12, 1956, to Horizons Titanium Corp.; *Chem. Abs.*, 1956, **50**, 14192: German 966,322, July 25, 1957, to Horizons Titanium Corp.; *Chem. Abs.*, 1959, **53**, 14438.

[74] W. A. Barber, U.S. 3,078,149, Feb. 19, 1963, to American Cyanamid Co.; *Chem. Abs.*, 1963, **58**, 12206.

[75] V. Biciste and J. Vrana, Czechoslovakia 105,873, Dec. 15, 1962; *Chem. Abs.*, 1964, **60**, 3757.

[76] A. Lecerf, *Ann. Chim.*, 1962, **7**, 513; *Chem. Abs.*, 1963, **58**, 212.

[77] V. A. Reznichenko, F. B. Khalimov, and T. P. Ukolova, *Titan i ego Splavy*, Akad. Nauk S.S.S.R., Inst. Met., 1963, 42; *Chem. Abs.*, 1963, **59**, 14864.

[78] G. L. Humphrey, *J. Am. Chem. Soc.*, 1951, **73**, 1587; *Chem. Abs.*, 1951, **45**, 6918.

[79] A. Mazzucchelli, *Atti acad. nazl. Lincei*, 1908, **16**, 349; *Chem. Abs.*, 1908, **2**, 2912.

[80] G. H. Ayres and E. M. Vienneau, *Ind. Eng. Chem., Anal. Ed.*, 1939, **12**, 96; *Chem. Abs.*, 1940, **34**, 1935.

[81] D. M. Nicholson and M. A. Reiter, *J. Am. Chem. Soc.*, 1937, **59**, 151; *Chem. Abs.*, 1937, **31**, 4221.

[82] M. E. Kumpf, *Ann. Chim.*, 1937, **8**, 456; *Chem. Abs.*, 1938, **32**, 1596.

[83] R. Schwarz and W. Sexauer, *Ber.*, 1927, **60B**, 500; *Chem. Abs.*, 1927, **21**, 1418.

[84] M. Billy, *Compt. rend.*, 1921, **172**, 1411; *Chem. Abs.*, 1921, **15**, 3596. M. Billy and I. San-Galli, *Compt. rend.*, 1932, **194**, 1126; *Chem. Abs.*, 1932, **26**, 3198.

[85] M. Billy, *Compt. rend.*, 1928, **186**, 760; *Chem. Abs.*, 1928, **22**, 2334.

[86] R. M. McKinney and W. H. Madson, *J. Chem. Education*, 1936, **13**, 155.

[87] S. Hakomori, *J. Chem. Soc. Japan*, 1929, **50**, 231; *Chem. Abs.*, 1932, **26**, 40.

[88] S. Katzoff and R. Roseman, *J. Am. Chem. Soc.*, 1935, **57**, 1384; *Chem. Abs.*, 1935, **29**, 5763.

[89] L. Weiss and M. Landecker, *Z. anorg. u. allgem. Chem.*, 1909, **64**, 71; German 221,429, Apr. 25, 1909; *Chem. Abs.*, 1910, **4**, 2718.

[90] A. Krause and G. Schmidt, *Ber.*, 1936, **69B**, 656; *Chem. Abs.*, 1936, **30**, 4420.

[91] F. Rivenq, *Bull. soc. chim.*, 1945, **12**, 283; *Chem. Abs.*, 1946, **40**, 7038.

[92] G. R. Levi, *Giorn. chim. ind. applicata*, 1925, **7**, 410; *Chem. Abs.*, 1926, **20**, 2947.

[93] O. Glemser, *Z. Elektrochem.*, 1939, **45**, 820; *Chem. Abs.*, 1940, **34**, 1898.

[94] A. M. Brusilovskii, B. E. Boguslovskaya, and E. G. Nosovick, *Trudy Inst. Lakov i Krasok*, 1939, **2**, 23; *Chem. Abs.*, 1940, **34**, 2142.

[95] A. M. Morley and J. K. Wood, *J. Soc. Dyers and Colourists*, 1923, **39**, 100; *Chem. Abs.*, 1923, **17**, 2647.

[96] A. M. Morley and J. K. Wood, *J. Chem. Soc.*, 1924, **125**, 1626; *Chem. Abs.*, 1924, **18**, 3330.

[97] H. B. Weiser and W. O. Milligan, *Chem. Rev.*, 1939, **25**, 1; *Chem. Abs.*, 1939, **33**, 9088.

[98] H. B. Weiser and W. O. Milligan, *J. Phys. Chem.*, 1940, **44**, 1081; *Chem. Abs.*, 1941, **35**, 951: *Science*, 1942, **I**, 227; *Chem. Abs.*, 1942, **36**, 3414.

[99] H. B. Weiser, W. O. Milligan, and W. C. Simpson, *J. Phys. Chem.*, 1942, **46**, 1051; *Chem. Abs.*, 1943, **37**, 1911.

[100] R. M. McKinney and W. H. Madson, *J. Chem. Education*, 1936, **13**, 155. M. Fladmark, U.S. 1,288,863, Dec. 24, 1918, to Titan Co., A-S.; *Chem. Abs.*, 1919, **13**, 641.

[101] J. Blumenfeld, U.S. 1,504,673, Aug. 12, 1924; *Chem. Abs.*, 1924, **18**, 3282: Reissue 16,956, May 8, 1928; *Chem. Abs.*, 1928, **22**, 2282. C. Weizmann and J. Blumenfeld, British 247,296, July 12, 1922; *Chem. Abs.*, 1927, **21**, 659.

[102] J. Blumenfeld, U.S. 1,851,487, Mar. 29, 1932, to Krebs Pigment and Color Corp.; *Chem. Abs.*, 1932, **26**, 3080.
[103] H. Kuzel, German 186,980, April 28, 1906, to Baden bei Wien; *Chem. Abs.*, 1908, **2**, 599: British 6,109, Mar. 13, 1907; *Chem. Abs.*, 1908, **2**, 2124.
[104] H. Kuzel, French 371,799, Nov. 26, 1906, to Baden bei Wien; *Chem. Abs.*, 1908, **2**, 2606: German 197,379, Dec. 13, 1905, to Baden bei Wien; *Chem. Abs.*, 1908, **2**, 2320.
[105] A. Guthier and H. Weithase, *Z. anorg. u. allgem. Chem.*, 1928, **169**, 264; *Chem. Abs.*, 1928, **22**, 2502.
[106] J. Muller, British 115,236, April 22, 1918; *Chem. Abs.*, 1918, **12**, 1820.
[107] N. Parravano and V. Caglioti, *Gazz. chim. ital.*, 1934, **64**, 450; *Chem. Abs.*, 1935, **29**, 23.
[108] S. K. Majumdar, *J. Indian Chem. Soc.*, 1929, **6**, 357; *Chem. Abs.*, 1929, **23**, 4867.
[109] A. V. Dumanskii and A. G. Kniga, *Kolloid-Z.*, 1928, **44**, 273; *Chem. Abs.*, 1928, **22**, 4308.
[110] S. C. Lyons, British 505,709, May 16, 1939; *Chem. Abs.*, 1939, **33**, 9158.
[111] H. Brintzinger and W. Brintzinger, *Z. anorg. u. allgem. Chem.*, 1931, **196**, 44; *Chem. Abs.*, 1931, **25**, 2347.
[112] O. Joseph and S. M. Mahta, *J. Univ. Bombay*, 1935, 4, pt. 2, 123; *Chem. Abs.*, 1936, **30**, 7418: *J. Indian Chem. Soc.*, 1933, **10**, 177; *Chem. Abs.*, 1933, **27**, 4986.
[113] A. Gutbier, R. Ottenstein, E. Leutheusser, K. Lassen, and F. Allan, *Z. anorg. u. allgem. Chem.*, 1927, **162**, 87; *Chem. Abs.*, 1927, **21**, 3511.
[114] L. S. Bhatia and S. Ghosh, *J. Indian Chem. Soc.*, 1930, **7**, 687; *Chem. Abs.*, 1931, **25**, 1721.
[115] A. W. Thomas and W. G. Stewart, *Kolloid-Z.*, 1939, **86**, 279; *Chem. Abs.*, 1939, **33**, 4849.
[116] P. S. Vasilev and N. M. Deshalit, *J. Phys. Chem. (U.S.S.R.)*, 1936, **7**, 707; *Chem. Abs.*, 1936, **30**, 7957.
[117] T. V. Gatovskaya and P. S. Vasilev, *J. Phys. Chem. (U.S.S.R.)*, 1936, **7**, 697; *Chem. Abs.*, 1936, **30**, 7959.
[118] V. Kargin, *Acta Physiochim. U.R.S.S.*, 1934, **1**, 64; *Chem. Abs.*, 1935, **29**, 660.
[119] A. V. Dumanskii and B. S. Puchkovskii, *J. Russ. Phys. Chem. Soc.*, 1930, **62**, 469; *Chem. Abs.*, 1930, **24**, 5201.
[120] W. A. Patrick, U.S. 1,682,242, Aug. 28, 1928, to Silica Gel Corp.; *Chem. Abs.*, 1928, **22**, 3965.
[121] G. C. Connolly and E. B. Miller, British 303,138, Dec. 30, 1927; *Chem. Abs.*, 1929, **23**, 4541: French 652,269, Apr. 5, 1926, to Silica Gel Corp.; *Chem. Abs.*, 1929, **23**, 3547.
[122] E. B. Miller and G. C. Connolly, British 314,398, Mar. 13, 1927, to Silica Gel Corp.; *Chem. Abs.*, 1930, **24**, 1477. E. B. Miller, British 287,066, Mar. 12, 1927, to Silica Gel Corp.; *Chem. Abs.*, 1929, **23**, 495: French 650,695, Mar. 9, 1928, to Silica Gel Corp.; *Chem. Abs.*, 1929, **23**, 3315.
[123] E. V. Alekseevskii and G. M. Belotzerkovskii, *J. Gen. Chem. (U.S.S.R.)*, 1936, **6**, 370; *Chem. Abs.*, 1936, **30**, 5481.
[124] F. Stowener, German 566,081, Aug. 20, 1926, to I. G. Farbenind., A-G.; *Chem. Abs.*, 1933, **27**, 1112.
[125] British 313,242, Mar. 30, 1928, to Silica Gel Corp.; *Chem. Abs.*, 1930, **24**, 926.
[126] E. H. Barclay, U.S. 1,822,848, Sept. 8, 1931; *Chem. Abs.*, 1931, **25**, 5959.
[127] S. Klosky and C. Marzano, *J. Phys. Chem.*, 1925, **29**, 1125; *Chem. Abs.*, 1925, **19**, 3401.
[128] O. Krause, *Kolloid-Z.*, 1935, **72**, 18; *Chem. Abs.*, 1935, **29**, 7753.
[129] L. Passerini, *Gazz. chim. ital.*, 1935, **65**, 518; *Chem. Abs.*, 1936, **30**, 353.
[130] P. N. Laschenko and D. I. Kompanskii, *J. Applied Chem. (U.S.S.R.)*, 1935, **8**, 628; *Chem. Abs.*, 1936, **30**, 3308.
[131] V. Auger, *Compt. rend.*, 1923, **177**, 1302; *Chem. Abs.*, 1924, **18**, 638.

[132] I. Higuchi, *Bull. Inst. Phys. Chem. Research (Tokyo)*, 1935, **14**, 853; *Chem. Abs.*, 1936, **30**, 2074.

[133] N. I. Nikitin and V. I. Yuryev, *Z. anorg. u. allgem. Chem.*, 1928, **171**, 281; *Chem. Abs.*, 1928, **22**, 3813: *J. Russ. Phys. Chem. Soc.*, 1929, **61**, 1029; *Chem. Abs.*, 1930, **24**, 539.

[134] I. Higuchi, *Bull. Inst. Phys. Chem. Research (Tokyo)*, 1936, **15**, 266; *Chem. Abs.*, 1937, **31**, 4870.

[135] I. Higuti, *Bull. Inst. Phys. Chem. Research (Tokyo)*, 1939, **18**, 657; *Chem. Abs.*, 1940, **34**, 4959. R. S. Rao, *Current Sci. (India)*, 1939, **8**, 468; *Chem. Abs.*, 1940, **34**, 4960.

[136] S. Klosky and A. J. Burggraff, *J. Am. Chem. Soc.*, 1928, **50**, 1045; *Chem. Abs.*, 1928, **22**, 2093.

[137] I. Higuti, *Bull. Inst. Phys. Chem. Research (Tokyo)*, 1940, **19**, 951; *Chem. Abs.*, 1940, **34**, 7159.

[138] K. S. Rao, *J. Phys. Chem.*, 1941, **45**, 500; *Chem. Abs.*, 1941, **35**, 3138. T. Krishnappa, K. S. Rao, and B. S. Rao, *Proc. Indian Acad. Sci.*, 1947, **25A**, 162; *Chem. Abs.*, 1947, **41**, 5358.

[139] C. B. Hurd, W. J. Jacober, and D. W. Godfrey, *J. Am. Chem. Soc.*, 1941, **63**, 723; *Chem. Abs.*, 1941, **35**, 2772.

[140] S. Klosky, *J. Phys. Chem.*, 1930, **34**, 2621; *Chem. Abs.*, 1931, **25**, 450.

[141] L. C. Drake and L. P. Evans, U.S. 2,378,290, June 12, 1945, to Socony-Vacuum Oil Co.; *Chem. Abs.*, 1945, **39**, 3889.

[142] S. Suzaki, *J. Japan Ceramic Assoc.*, 1941, **49**, 530; *Chem. Abs.*, 1950, **44**, 10282.


CHAPTER 8

[1] F. von Bichowsky, U.S. 1,915,393, June 27, 1933; *Chem. Abs.*, 1933, **27**, 4359.

[2] C. Weizmann and J. Blumenfeld, British 209,480, July 12, 1922; *Chem. Abs.*, 1924, **18**, 1884.

[3] British 471,397, Sept. 3, 1937, to Titan Co., Inc.; *Chem. Abs.*, 1938, **32**, 1415: French 813,785, June 8, 1937, to Titan Co., Inc.; *Chem. Abs.*, 1938, **32**, 1060.

[4] A. Kirkham and H. Spence, British 263,886, July 1, 1925, to P. Spence and Sons, Ltd.; *Chem. Abs.*, 1928, **22**, 143.

[5] B. C. Boguslavskaya and O. M. Ottamanovskaya, *J. Gen. Chem. (U.S.S.R.)*, 1940, **10**, 673; *Chem. Abs.*, 1941, **35**, 2432.

[6] F. Raspe and P. Weise, U.S. 1,850,154, Mar. 22, 1932, to I. G. Farbenind., A-G.; *Chem. Abs.*, 1932, **26**, 2831: French 671,639, Mar. 18, 1929, to I. G. Farbenind., A-G.; *Chem. Abs.*, 1930, **24**, 2253: German 479,491, Apr. 4, 1928, to I. G. Farbenind., A-G.; *Chem. Abs.*, 1929, **23**, 4783: British 309,047, Apr. 3, 1928, to I. G. Farbenind., A-G.; *Chem. Abs.*, 1930, **24**, 474.

[7] F. Raspe and P. Weise, U.S. 1,849,153, Mar. 15, 1932, to I. G. Farbenind., A-G.; *Chem. Abs.*, 1932, **26**, 2831: British 309,090, Apr. 4, 1928, to I. G. Farbenind., A-G.; *Chem. Abs.*, 1930, **24**, 474: French 672,142, Mar. 27, 1929, to I. G. Farbenind., A-G.; *Chem. Abs.*, 1930, **24**, 2253.

[8] H. Plant, Canadian 285,355, Dec. 4, 1928, to I. G. Farbenind., A-G.; *Chem. Abs.*, 1929, **23**, 673: *French* 653,893, May 5, 1928, to I. G. Farbenind., A-G.; *Chem. Abs.*, 1929, **23**, 3655: German 508,110, May 7, 1927, to I. G. Farbenind., A-G.; *Chem. Abs.*, 1931, **25**, 675: British 290,174, May 7, 1928, to I. G. Farbenind., A-G.; *Chem. Abs.*, 1929, **23**, 1224.

[9] L. Wohler and K. Flick, *Ber.*, 1934, **67B**, 1679; *Chem. Abs.*, 1935, **29**, 65.

[10] W. B. Llewellyn and H. Spence, U.S. 1,980,812, Nov. 13, 1935, to P. Spence and Sons, Ltd.; *Chem. Abs.*, 1935, **29**, 563: British 364,613, July 5, 1930, to P. Spence and Sons, Ltd.; *Chem. Abs.*, 1933, **27**, 1999: German 599,502, July 3, 1934, to P. Spence and Sons, Ltd.; *Chem. Abs.*, 1935, **29**, 3790.

[11] N. H. McCoy, U.S. 1,559,113, Oct. 27, 1925, to Lindsay Light Co.; *Chem. Abs.*, 1926, **20**, 97.

[12] T. Sagawa, *J. Soc. Chem. Ind. Japan*, 1939, **41**, 27; *Chem. Abs.*, 1938, **32**, 2820.
[13] W. Reinders and H. L. Kies, *Rec. trav. chim.*, 1940, **59**, 785; *Chem. Abs.*, 1941, **35**, 4665.
[14] V. S. Miyamoto, *J. Chem. Soc. Japan*, 1933, **54**, 85; *Chem. Abs.*, 1933, **27**, 2387.
[15] J. Ettori, *Compt. rend.*, 1936, **202**, 852; *Chem. Abs.*, 1936, **30**, 7604.
[16] V. I. Belokoskov, *Zh. Neorgan. Khim.*, 1962, **7**, 279; *Chem. Abs.*, 1962, **56**, 13603.
[17] K. Shinriki, T. Kubo, and M. Kato, *J. Chem. Soc. Japan, Ind. Chem. Section*, 1953, **56**, 832; *Chem. Abs.*, 1954, **48**, 14137.
[18] V. I. Belokoskov, *Zhur. Neorg. Khim.*, 1961, **6**, 1443; *Chem. Abs.*, 1962, **56**, 2032.
[19] K. Shinriki, T. Kubo, and M. Kato, *J. Chem. Soc. Japan, Ind. Chem. Section*, 1954, **57**, 259; *Chem. Abs.*, 1955, **49**, 4950.
[20] A. V. Pamfilov and T. A. Khudyakova, *J. Gen. Chem. U.S.S.R.*, 1949, **19**, 1443; *Chem. Abs.*, 1950, **44**, 1354. *Zhur. Obshchei Khim.*, 1949, **19**, 1443; *Chem. Abs.*, 1951, **45**, 1891.
[21] N. G. Klimenko and V. S. Syrokomskii, *Zavodskaya Lab.*, 1947, **13**, 1029; *Chem. Abs.*, 1949, **43**, 4083.
[22] A. I. Ulyanov, *Izvest. Akad. Nauk, S.S.S.R., Otdel. Khim. Nauk*, 1960, 580; *Chem. Abs.*, 1962, **56**, 10989.
[23] E. Takakura and H. Hikuchi, *Kogyo Kagaku Zasshi*, 1961, **64**, 1179; *Chem. Abs.*, 1962, **57**, 4272.
[24] V. L. Varshavskii, U.S.S.R. 122,743, Oct. 10, 1959; *Chem. Abs.*, 1960, **54**, 9229.
[25] V. M. Polyakova and E. I. Chernyavskaya, *Tr. I.K K., A. N. S.S.S.R., U. F.*, 1963, 33; *Chem. Abs.*, 1964, **61**, 5187.
[26] B. L. Nabivanets, *Zh. Neorgan Khim.*, 1962, **7**, 417; *Chem. Abs.*, 1962, **57**, 2880.
[27] E. S. Hanrahan, *J. Inorg. Nucl. Chem.*, 1964, **26**, 1757; *Chem. Abs.*, 1964, **61**, 14173.
[28] R. M. McKinney, U.S. 1,922,816, Aug. 15, 1933, to Krebs Pigment and Color Corp.; *Chem. Abs.*, 1933, **27**, 5156.
[29] H. Spence and T. J. I. Craig, British 13,260, June 2, 1911; *Chem. Abs.*, 1912, **6**, 3316.
[30] R. Roseman and W. M. Thornton, *J. Am. Chem. Soc.*, 1935, **57**, 328; *Chem. Abs.*, 1935, **29**, 2466.
[31] C. R. Hager, U.S. 2,125,340, Aug. 2, 1938, to E. I. du Pont de Nemours and Co.; *Chem. Abs.*, 1938, **32**, 7687.
[32] E. Kneckt, *Mem. Pro. Manchester Lit. & Phil. Soc.*, 1904, **48**, 1; *J. Chem. Soc.*, 1904, **A**, 448. R. Jonnard, *Chimie & industrie*, 1947, **57**, 551; *Chem. Abs.*, 1947, **41**, 6374.
[33] A. Piccini, *Gazzetta*, 1895, 542; *J. Chem. Soc.*, 1896, **A**, 365.
[34] A. Piccini, *Z. anorg. u. allgem. Chem.*, 1898, **17**, 355; *J. Chem. Soc.*, 1898, **A**, 521.
[35] P. Ehrlich, *Z. anorg. u. allgem. Chem.*, 1941, **247**, 53; *Chem. Abs.*, 1942, **36**, 6072.
[36] E. V. Vigoroux and G. A. Arrivant, *Compt. rend.*, 1907, **144**, 485; *Chem. Abs.*, 1907, **1**, 1362.
[37] O. v. d. Pfordten, *Annalen*, 1887, **237**, 201; *J. Chem. Soc.*, 1887, **A**, 337.
[38] A. V. Pamfilov and E. G. Shtandel, *J. Gen. Chem. (U.S.S.R.)*, 1937, **7**, 258; *Chem. Abs.*, 1937, **31**, 4609.
[39] F. W. Meister, U.S. 2,416,191, Feb. 18, 1947, to National Lead Co.; *Chem. Abs.*, 1947, **41**, 2543.
[40] F. Gage, U.S. 2,178,685, Nov. 7, 1939, to Pittsburgh Plate Glass Co.; *Chem. Abs.*, 1940, **34**, 1448.
[41] L. G. Jenness and R. L. Annis, U.S. 2,230,538, Feb. 4, 1941, to Vanadium Corp. of America; *Chem. Abs.*, 1941, **35**, 3400.
[42] W. F. Meister, U.S. 2,344,319, Mar. 4, 1944, to National Lead Co.; *Chem. Abs.*, 1944, **38**, 3428.

[43] A. Pechukas, U.S. 2,289,327 and 2,287,328, July 7, 1942, to Pittsburgh Plate Glass Co.; *Chem. Abs.*, 1943, **37**, 237.

[44] B. de Witt, U.S. 2,370,525, Feb. 27, 1945, to Pittsburgh Plate Glass Co.; *Chem. Abs.*, 1945, **39**, 3132.

[45] A. Pechukas, U.S. 2,207,597, July 9, 1940, to Pittsburgh Plate Glass Co.; *Chem. Abs.*, 1940, **34**, 8190.

[46] L. Neunier, P. Sisley, and F. Genin, *Chimie & industrie*, 1932, **27**, 1017; *Chem. Abs.*, 1932, **26**, 4521.

[47] T. E. Thorpe, *J. Chem. Soc.*, 1880, **I**, 329.

[48] T. Sagawa, *Complete Abs. Japanese Chem. Literature*, 1932, **6**, 341; *Chem. Abs.*, 1933, **27**, 8.

[49] G. P. Luchinskii, *J. Phys. Chem. (U.S.S.R.)*, 1935, **6**, 607; *Chem. Abs.*, 1935, **29**, 7739.

[50] N. Nasu, *Bull. Chem. Soc. Japan*, 1934, **9**, 198; *Chem. Abs.*, 1934, **28**, 4656.

[51] K. Arii, *Bull. Inst. Phys. Chem. Research (Tokyo)*, 1929, **8**, 719; *Chem. Abs.*, 1930, **24**, 277: *Science Reports Tôhoku Imp. Univ.*, 1933, **22**, 182; *Chem. Abs.*, 1933, **27**, 3865.

[52] H. M. Spencer, *J. Am. Chem. Soc.*, 1945, **67**, 1859; *Chem. Abs.*, 1946, **40**, 783.

[53] S. Bhagavantam, *Indian J. Phys.*, 1932, **7**, 79; *Chem. Abs.*, 1932, **26**, 4538.

[54] R. Wierl, *Ann. Physik.*, 1931, **8**, 521; *Chem. Abs.*, 1931, **25**, 2886.

[55] M. W. Lister and L. E. Sutton, *Trans. Faraday Soc.*, 1941, **37**, 393; *Chem. Abs.*, 1942, **36**, 50.

[56] E. Bergman and L. Engel, *Z. physik. Chem.*, 1931, **B13**, 232 and 247; *Chem. Abs.*, 131, **25**, 5803 and 5804.

[57] V. I. Vaidyanothan, *Nature*, 1931, **128**, 189; *Chem. Abs.*, 1931, **25**, 5805. V. I. Vaidyanothan and B. Singh, *Indian J. Phys.*, 1932, **7**, 19; *Chem. Abs.*, 1932, **26**, 4738.

[58] L. H. Siertsema, *Verslag Gewone Vergader. Afdeel. Natuurk. Nederland, Akad. Wetenshap.*, 1915, **23**, 1259; *Chem. Abs.*, 1915, **9**, 3165.

[59] F. B. Garner and S. Sugden, *J. Chem. Soc.*, **1929**, 1298; *Chem. Abs.*, 1929, **23**, 5362.

[60] M. L. Delwaulle and F. François, *Compt. rend.*, 1945, **220**, 173; *Chem. Abs.*, 1946, **40**, 2072.

[61] O. v. d. Pfordten, *Annalen*, 1887, **237**, 201; *J. Chem. Soc.*, 1887, **A**, 337.

[62] E. Wertyporock and B. Altmann, *Z. physik. Chem.*, 1934, **A168**, 1; *Chem. Abs.*, 1934, **28**, 2975.

[63] E. Urbain and C. Schol, *Compt. rend.*, 1919, **168**, 887; *Chem. Abs.*, 1919, **13**, 3057.

[64] S. S. Kirstler and K. Kearby, *Acta Physiochim. U.R.S.S.*, 1934, **1**, 354; *Chem. Abs.*, 1935, **29**, 6487.

[65] W. A. Roth and G. Becker, *Z. physik. Chem.*, 1932, **A159**, 1; *Chem. Abs.*, 1932, **26**, 3152.

[66] D. M. Yost and C. Blair, *J. Am. Chem. Soc.*, 1933, **55**, 2610; *Chem. Abs.*, 1933, **27**, 3389.

[67] A. Chrétien and G. Varga, *Compt. rend.*, 1935, **201**, 558; *Chem. Abs.*, 1935, **29**, 7847.

[68] M. G. Reader, *Kgl. Norske Videnskab. Selskabs Skrifter*, 1929, **3**, 1; *Chem. Abs.*, 1931, **25**, 17.

[69] D. G. Nicholson, *Trans. Illinois State Acad. Sci.*, 1936, **29**, 97; *Chem. Abs.*, 1937, **31**, 3307.

[70] G. P. Luchinskii and A. I. Likhacheva, *Z. anorg. u. allgem. Chem.*, 1936, **226**, 333; *Chem. Abs.*, 1936, **30**, 4422.

[71] F. Clausnizer, *Ber.*, 1879, **11**, 2011; *J. Chem. Soc.*, 1879, **A**, 201.

[72] G. P. Luchinskii, *J. Gen. Chem. (U.S.S.R.)*, 1937, **7**, 207; *Chem. Abs.*, 1937, **31**, 4613.

[73] H. Sorum, *Kgl. Norske Videnskab. Selskabs Förh.*, 1944, **17**, 17 and 24; *Chem. Abs.*, 1946, **40**, 5974.

[74] J. Gnezda, *Festschrift Feier zweijährigen Bestandes Unabhängigen Staates Kroatien*, Zagreb, 1943; *Chem. Abs.*, 1946, **40**, 6318.

[75] L. Hock and W. Knauff, *Z. anorg. u. allgem. Chem.*, 1936, **228**, 193; *Chem. Abs.*, 1936, **30**, 8061.

[76] H. Rheinboldt and R. Wasserfuhr, *Ber.*, 1927, **60B**, 732; *Chem. Abs.*, 1927, **21**, 3573.

[77] A. W. Ralston and J. A. Wilkinson, *J. Am. Chem. Soc.*, 1928, **50**, 258; *Chem. Abs.*, 1928, **22**, 1519.

[78] D. N. Tarasenkov and A. V. Komandin, *J. Gen. Chem.* (*U.S.S.R.*), 1940, **10**, 1319; *Chem. Abs.*, 1941, **35**, 2381.

[79] T. Sagawa, *Science Reports Tôhoku Imp. Univ.*, 1933, **22**, 959; *Chem. Abs.*, 1934, **28**, 1904.

[80] N. Nasu, *Bull. Chem. Soc. Japan*, 1933, **8**, 195; *Chem. Abs.*, 1933, **27**, 5236.

[81] N. N. Ruban and V. D. Ponomarev, *Tr. Inst. Met. i O., Akad. Nauk. Kaz S.S.R.*, 1962, **4**, 19; *Chem. Abs.*, 1963, **58**, 10745.

[82] P. Brand and H. Sackmann, *Z. anorg. u. allgem. Chem.*, 1963, **321**, 262; *Chem. Abs.*, 1963, **59**, 90.

[83] K. Dehnicke, *Angew. Chem.*, 1963, **75**, 419; *Chem. Abs.*, 1963, **59**, 1263.

[84] C. L. Hildreth, U.S. 3,063,797, Nov. 13, 1962, to Allied Chemical Corp.; *Chem. Abs.*, 1963, **58**, 7639.

[85] P. Ehrlich and W. Engel, *Naturwissenschaften*, 1961, **48**, 716; *Chem. Abs.*, 1962, **56**, 11171: *Z. anorg. u. allgem. Chem.*, 1962, **317**, 21; *Chem. Abs.*, 1962, **57**, 14681.

[86] H. Roehl, German 1,141,626, Dec. 27, 1962, to Chemische Werke Huels A.-G.; *Chem. Abs.*, 1963, **58**, 6480.

[87] K. F. Guenther, *Inorg. Chem.*, 1964, **3**, 923; *Chem. Abs.*, 1964, **61**, 3895.

[88] M. Billy and P. Brasseur, *Compt. rend.*, 1935, **200**, 1765; *Chem. Abs.*, 1935, **29**, 4689.

[89] O. Ruff and F. Neumann, *Z. anorg. u. allgem. Chem.*, 1923, **128**, 81; *Chem. Abs.*, 1923, **17**, 2842.

[90] C. Friedel, *Chem. Zentr.*, **1874**, 315.

[91] W. C. Schumb and R. F. Sundstrom, *J. Am. Chem. Soc.*, 1933, **55**, 596; *Chem. Abs.*, 1933, **27**, 2106.

[92] A. Stahler and F. Bachman, *Ber.*, 1912, **44**, 2906; *Chem. Abs.*, 1912, **6**, 461.

[93] H. Georges and A. Stahler, *Ber.*, 1910, **42**, 3200; *Chem. Abs.*, 1910, **4**, 154.

[94] F. Bock and L. Moser, *Monatsh.*, 1914, **34**, 1825; *Chem. Abs.*, 1914, **8**, 1394.

[95] F. Bock and L. Moser, *Monatsh.*, 1913, **33**, 1407; *Chem. Abs.*, 1913, **7**, 1333.

[96] German 154,542, to P. Spence and Sons, Ltd.; *J. Chem. Soc.*, 1904, **A**, 823.

[97] G. Patscheke and W. Schaller, *Z. anorg. u. allgem. Chem.*, 1938, **235**, 257; *Chem. Abs.*, 1938, **32**, 4097.

[98] A. J. Kolk, Jr., and D. E. Davis, U.S. 2,894,887, July 14, 1959, to Horizons Titanium Corp.; *Chem. Abs.*, 1959, **53**, 22790.

[99] E. Wainer and A. J. Kolk, Jr., German 1,035,623, Aug. 7, 1958, to Horizons Titanium Corp.; *Chem. Abs.*, 1960, **54**, 25628.

[100] R. A. Ruehrwein and G. B. Skinner, U.S. 2,940,825, June 14, 1960, to Monsanto Chemical Co.; *Chem. Abs.*, 1960, **54**, 18912.

[101] S. Sommer and S. Wagener, German 1,054,974, April 4, 1959, to Farbwerke Hoechst Akt.-Ges.; *Chem. Abs.*, 1961, **55**, 7778.

[102] H. Tadenuma and S. Kadota, Japanese 3067, May 9, 1955, to Sumimoto Chemical Industries Co.; *Chem. Abs.*, 1957, **51**, 14221.

[103] T. Ishino, H. Tamura, and O. Nakagawa, *Kagyo Kagaku Zasshi*, 1961, **64**, 1344; *Chem. Abs.*, 1962, **57**, 4288.

[104] V. S. Etlis, A. I. Kirillov, and O. P. Baranova, U.S.S.R. 115,998, Nov. 22, 1958; *Chem. Abs.*, 1959, **53**, 14438.

[105] W. E. Shaw, R. M. Spenceley, and F. M. Teetzel, U.S. 2,898,187, Aug. 4, 1959, to U.S. Atomic Energy Commission; *Chem. Abs.*, 1959, **53**, 22791.

[106] E. Fornasieri and A. Forni, German 1,072,975, Jan. 14, 1960, to Montecatini Società Generale per l'industria mineraria e chimica; *Chem. Abs.*, 1961, **55**, 22734.

[107] D. E. Brown, J. P. Keller, and A. T. Watson, U.S. 2,904,486, Sept. 15, 1959, to Esso Research and Engineering Co.; *Chem. Abs.*, 1960, **54**, 1134.

[108] Belgian 625,131, March 14, 1963, to Titangesellschaft m. b. H.; *Chem. Abs.*, 1963, **59**, 1312.

[109] R. B. Beyer and D. M. Mason, *Ind. Eng. Chem.*, 1963, **2**, 78; *Chem. Abs.*, 1963, **58**, 2880.

[110] V. G. Gopienko and A. I. Ivanov, U.S.S.R. 149,571, Aug. 28, 1962; *Chem. Abs.*, 1963, **58**, 5272.

[111] Belgian 625,131, March 14, 1963, to Titangesellschaft m. b. H.; *Chem. Abs.*, 1963, **59**, 1312.

[112] W. H. Johnson, A. A. Gilliland, and E. J. Prosen, *J. Research Natl. Bur. Standards*, 1960, **A64**, 515; *Chem. Abs.*, 1961, **55**, 8026.

[113] C. Eden and H. Feilchenfeld, *J. Soc. Chem.*, 1962, 2066; *Chem. Abs.*, 1962, **57**, 4284.

[114] D. E. Brown and H. C. Williams, U.S. 2,993,009, July 18, 1961, to Esso Research and Engineering Co.; *Chem. Abs.*, 1961, **55**, 27810.

[115] British 725,572, March 9, 1955, to Titan Co., Inc.; *Chem. Abs.*, 1955, **49**, 11297.

[116] H. A. E. Mackenzie and F. C. Tompkins, *Trans. Faraday Soc.*, 1942, **38**, 465; *Chem. Abs.*, 1943, **37**, 1320.

[117] H. von Euler and H. Hellstrom, *Svensk Kem. Tid.*, 1929, **11**, 41; *Chem. Abs.*, 1929, **23**, 4227.

[118] P. Karrer, Y. Yen, and I. Reichstein, *Helv. Chim. Acta*, 1930, **13**, 1308; *Chem. Abs.*, 1931, **25**, 952.

[119] W. C. Schumb and R. F. Sundstrom, *J. Am. Chem. Soc.*, 1933, **55**, 596; *Chem. Abs.*, 1933, **27**, 2106.

[120] T. Dreisch and O. Kallscheuer, *Z. physik. Chem.*, 1939, **B45**, 19; *Chem. Abs.*, 1940, **34**, 940.

[121] A. K. Bose, *Indian J. Phys.*, 1944, **18**, 199; *Chem. Abs.*, 1945, **39**, 3985.

[122] F. Meyer and H. Kerstein, U.S. 1,173,012, Feb. 22, 1916; *Chem. Abs.*, 1916, **10**, 1013: German 281,094, Apr. 30, 1913; *Chem. Abs.*, 1915, **9**, 1457.

[123] F. Meyer, A. Bauer, and R. Schmidt, *Ber.*, 1923, **56B**, 1908; *Chem. Abs.*, 1923, **17**, 3650.

[124] R. Schmidt, *Ber.*, 1925, **58B**, 400; *Chem. Abs.*, 1925, **19**, 1386.

[125] G. Patscheke and W. Schaller, *Z. anorg. u. allgem. Chem.*, 1938, **235**, 257; *Chem. Abs.*, 1938, **32**, 4097.

[126] W. C. Schumb and R. F. Sundstrom, *J. Am. Chem. Soc.*, 1933, **55**, 596; *Chem. Abs.*, 1933, **27**, 2106.

[127] H. Tadenuma, Japanese 2168, March 26, 1956, to Sumimoto Chemical Industries Co.; *Chem. Abs.*, 1957, **51**, 7667.

[128] R. D. Blue and M. P. Neipert, U.S. 2,943,033, June 28, 1960, to Dow Chemical Co.; *Chem. Abs.*, 1960, **54**, 21679.

[129] M. J. Cooper and J. T. Richmond, British 925,848, May 8, 1963, to Laporte Titanium Ltd.; *Chem. Abs.*, 1963, **59**, 3579.

[130] S. Okudairo, Japanese 7509, June 14, 1961, to Toho Titanium Co., Ltd.; *Chem. Abs.*, 1963, **59**, 6052.

[131] H. Roehl and E. Langer, German 1,175,657, Aug. 13, 1964, to Chemische Werke Huels A.-G.; *Chem. Abs.*, 1964, **61**, 14230.

[132] R. H. Singleton and P. J. Clough, U.S. 2,783,142, Feb. 26, 1957, to National Research Corp.; *Chem. Abs.*, 1957, **51**, 7996.

[133] R. C. Wade, U.S. 2,830,888, April 15, 1958, to National Distillers and Chemical Corp.; *Chem. Abs.*, 1958, **52**, 13208: German 1,047,182, Dec. 24, 1958, to National Distillers and Chemical Corp.; *Chem. Abs.*, 1960, **54**, 25638.

[134] A. G. Stromberg and A. I. Kartushinskaya, *F. K. A. A. N. S.S.S.R., S. O. I. N. K., T. Y. K. N.*, 1960, 315; *Chem. Abs.*, 1964, **61**, 15667.

[135] W. Klemm and L. Grimm, *Z. anorg. u. allgem. Chem.*, 1941, **249**, 198 and 1942, **249**, 209; *Chem. Abs.*, 1942, **36**, 5073 and 1943, **37**, 3984.

[136] S. S. Svendsen, U.S. 1,995,334, Mar. 25, 1935, to C. F. Burgess Laboratories, Inc.; *Chem. Abs.*, 1935, **29**, 3122.

[137] N. G. Natta, *Gazz. chim. ital.*, 1930, **60**, 911; *Chem. Abs.*, 1931, **25**, 2342.

[138] S. Hartmann, *Z. anorg. u. allgem. Chem.*, 1926, **155**, 355; *Chem. Abs.*, 1926, **20**, 3658.

[139] A. Piccini, *Compt. rend.*, 1884, **97**, 1064; *J. Chem. Soc.*, 1884, **A**, 264.

[140] A. Piccini, *Chem. Zentr.*, 1890, **II**, 544; *J. Chem. Soc.*, 1891, **A**, 271.

[141] J. A. Schaeffer, *J. Am. Chem. Soc.*, 1909, **30**, 1862; *Chem. Abs.*, 1909, **3**, 519.

[142] J. C. Schaefer, U.S. 2,722,510, Nov. 1, 1955, to Republic Steel Corp.; *Chem. Abs.*, 1956, **50**, 5250.

[143] E. Wainer, U.S. 2,694,616, Nov. 16, 1954, to Horizons Titanium Corp.; *Chem. Abs.*, 1955, **49**, 3490.

[144] H. C. Kawecki, U.S. 2,475,287, July 5, 1949, to The Beryllium Corp.; *Chem. Abs.*, 1949, **43**, 7652.

[145] K. Aotani, *Kogyo Kagaku Zasshi*, 1959, **62**, 1368; *Chem. Abs.*, 1962, **57**, 8184.

[146] Q. M. McKenna, U.S. 2,772,946, Dec. 4, 1956, to Horizons Titanium Corp.; *Chem. Abs.*, 1957, **51**, 7667.

[147] R. C. Young, in *Inorganic Syntheses*, II, McGraw-Hill Book Co., New York, 1946, 294; *Chem. Abs.*, 1946, **40**, 6355.

[148] E. Lange and H. Roehl, German 1,123,303, Feb. 8, 1962, to Chemische Werke Huels A.-G.; *Chem. Abs.*, 1962, **56**, 13803.

[149] K. Fumaki, K. Uchimura, and Y. Kuniya, *Kogyo Kagaku Zasshi*, 1961, **64**, 1914; *Chem. Abs.*, 1962, **57**, 2923.

[150] P. Fritsch, *Compt. rend.*, 1943, **217**, 447; *Chem. Abs.*, 1945, **39**, 2679.

[151] S. Seki, *J. Chem. Soc. Japan*, 1941, **62**, 789; *Chem. Abs.*, 1947, **41**, 5353.

[152] R. C. Young, *J. Chem. Education*, 1943, **20**, 378; *Chem. Abs.*, 1944, **38**, 2859.

[153] J. D. Fast, *Rec. trav. chim.*, 1939, **58**, 174; *Chem. Abs.*, 1939, **33**, 5311.

[154] G. Vaughan and E. Catterall, British 945,766, Jan. 8, 1964, to Dunlop Rubber Co., Ltd.; *Chem. Abs.*, 1964, **60**, 8932.

[155] M. L. Nielsen, U.S. 2,904,397, Sept. 15, 1959, to Monsanto Chemical Co.; *Chem. Abs.*, 1960, **54**, 6062.

[156] British 801,415, Sept. 19, 1958, to Monsanto Chemical Co.; *Chem. Abs.*, 1959, **53**, 5611.

[157] K. Funaki, K. Uchimura, and H. Matsunaga, *Kogyo Kagaku Zasshi*, 1961, **64**, 129; *Chem. Abs.*, 1962, **57**, 4279.

[158] M. B. Reifman, A. I. Gribov, V. N. Dumitriev, and M. A. Losikova, *Tsvetnye Metally.*, **34**, No. 5, 49; *Chem. Abs.*, 1961, **55**, 26922.

[159] British 662,577, Dec. 5, 1951, to Chilean Nitrate Sales Corp.; *Chem. Abs.*, 1952, **46**, 5778.

[160] A. C. Loonan, U.S. 2,519,385, Aug. 22, 1950, to Chilean Nitrate Sales Corp.; *Chem. Abs.*, 1951, **45**, 316; British 662,578, Dec. 5, 1951, to Chilean Nitrate Sales Corp.; *Chem. Abs.*, 1952, **46**, 4186.

[161] J. M. Blocher and I. E. Campbell, *J. Am. Chem. Soc.*, 1947, **69**, 2100; *Chem. Abs.*, 1947, **41**, 7184.

[162] L. Malaprade, *Bull. soc. chim.*, 1939, **6**, 223; *Chem. Abs.*, 1939, **33**, 8518.

[163] G. Jander and H. Wendt, *Z. anorg. Chem.*, 1949, **258**, 1; *Chem. Abs.*, 1949, **43**, 6495.

[164] E. Friederich, *Z. physik.*, 1925, **31**, 813; *Chem. Abs.*, 1925, **19**, 1644.

[165] O. Ruff and F. Eisen, *Ber.*, 1908, **41**, 2250; *Chem. Abs.*, 1908, **2**, 2764.

[166] C. Bosch, U.S. 957,842, May 10, 1910, to Badische Anilin und Soda Fabrik.; *Chem. Abs.*, 1910, **4**, 2358: British 1,842, Jan. 27, 1908, to Badische Anilin und Soda Fabrik.; *Chem. Abs.*, 1909, **3**, 2207: German 203,750, Aug. 31, 1907, to Badische Anilin und Soda Fabrik.; *Chem. Abs.*, 1909, **3**, 585.

[167] L. Hock and W. Knauff, *Z. anorg. u. allgem. Chem.*, 1936, **228**, 193; *Chem. Abs.*, 1936, **30**, 8061.

[168] F. von Bichowsky, U.S. 1,391,147 and 8, Sept. 30, 1921; *Chem. Abs.*, 1922, **16**, 319.

[169] V. P. Remin, *Vestnik Metalloprom.*, **1938**, No. 7, 54; *Chem. Abs.*, 1939, **33**, 2665.

[170] C. Bosch and A. Mittasch, U.S. 1,102,715, July 7, 1914; *Chem. Abs.*, 1914, **8**, 3353.

[171] A. J. Rossi, U.S. 1,032,432, July 16, 1912; *Chem. Abs.*, 1912, **6**, 2666.

[172] F. von Bichowsky and J. Hartman, U.S. 1,408,661, Mar. 7, 1922; *Chem. Abs.*, 1922, **16**, 1641: British 188,558, Dec. 7, 1921; *Chem. Abs.*, 1923, **17**, 1695.

[173] S. Janssom, *Jernkontorets Ann.*, 1940, **124**, 274; *Chem. Abs.*, 1941, **35**, 7670.

[174] L. Aagaard and H. Espenschied, U.S. 2,822,246, Feb. 4, 1958, to National Lead Co.; *Chem. Abs.*, 1958, **52**, 10521.

[175] H. Espenschied, U.S. 2,898,193, Aug. 4, 1959, to National Lead Co.; *Chem. Abs.*, 1960, **54**, 18912.

[176] M. Kitamura and T. Shyo, Japanese 9706, June 20, 1963, to Osaka Titanium Co., Ltd.; *Chem. Abs.*, 1964, **60**, 11653.

[177] G. Heymer and H. Harnisch, German 1,160,831, Jan. 9, 1964, to Knapsack-Griesheim A.-G.; *Chem. Abs.*, 1964, **60**, 8932.

[178] W. Opperman, *M. D. A. W.*, Berlin, 1964, **6**, 92: *Chem. Abs.*, 1964, **61**, 10292.

[179] W. H. Jeitschko, H. Nowotny, and F. Benesovsky, *Monatsh.*, 1963, **94**, 1198; *Chem. Abs.*, 1964, **60**, 8883.

[180] S. Peacock, U.S. 1,088,359, Feb. 24, 1914; *Chem. Abs.*, 1914, **8**, 1490. H. W. Lamb, U.S. 1,123,763, Jan. 5, 1915; *Chem. Abs.*, 1915, **9**, 696.

[181] F. Wohler, *J. Chem. Soc.*, 1849, **2**, 352.

[182] E. A. Rudge and F. Arnall, *J. Soc. Chem. Ind.* (London), 1928, **47**, 376; *Chem. Abs.*, 1929, **23**, 1585.

[183] F. von Bichowsky, *Ind. Eng. Chem.*, 1929, **21**, 1061; *Chem. Abs.*, 1930, **24**, 470.

[184] F. von Bichowsky, U.S. 1,742,674, Jan. 7, 1930; *Chem. Abs.*, 1930, **24**, 1187.

[185] F. von Bichowsky, U.S. 1,828,710, Oct. 20, 1931, to Titania Corp.; *Chem. Abs.*, 1932, **26**, 812.

[186] F. von Bichowsky, Canadian 306,145, Nov. 25, 1930, to Titania Corp., *Chem. Abs.*, 1931, **25**, 782.

[187] British 2,414 and 2,525, Feb. 3, 1908, to Badische Anilin und Soda Fabrik.; *Chem. Abs.*, 1909, **3**, 2207.

[188] P. Guignard, U.S. 1,411,087, Mar. 28, 1922; *Chem. Abs.*, 1922, **16**, 1840: Canadian 217,443, April 4, 1922; *Chem. Abs.*, 1922, **16**, 1840.

[189] F. von Bichowsky, U.S. 1,835,829, Dec. 8, 1931, to Titania Corp.; *Chem. Abs.*, 1932, **26**, 1074.

[190] E. Riley, *J. Chem. Soc.*, 1862, **15**, 324.

[191] P. Farup, U.S. 1,539,996, June 2, 1925; *Chem. Abs.*, 1925, **19**, 2263: British 199,713, June 1, 1923, to Titan Co., A-S.; *Chem. Abs.*, 1924, **18**, 309: Norwegian 40,986, Mar. 2, 1925, to Titan Co., A-S.; *Chem. Abs.*, 1925, **19**, 2322.

[192] P. Andreu and R. Paquet, U.S. 1,487,521, Mar. 18, 1924; *J. Soc. Chem. Ind.* (London), 1924, **B**, 380: British 175,989, Feb. 23, 1922; *Chem. Abs.*, 1922, **16**, 2201.

[193] P. Farup, U.S. 1,343,441, June 15, 1920; *Chem. Abs.*, 1920, **14**, 2424: British 115,020, Mar. 1, 1918, to Titan Co., A-S.: *Chem. Abs.*, 1918, **12**, 1839.

[194] T. Da-Tchang and L. Houng, *Compt. rend.*, 1935, **200**, 2173; *Chem. Abs.*, 1935, **29**, 5369.

[195] W. B. Llewellyn and S. F. W. Crundall, U.S. 1,876,065, Sept. 6, 1933, to P. Spence and Sons, Ltd.; *Chem. Abs.*, 1933, **27**, 170: Canadian 280,393, May 22, 1928, to P. Spence and Sons, Ltd.; *Chem. Abs.*, 1928, **22**, 2818: British 261,051, May 8, 1925, to P. Spence and Sons, Ltd.; *Chem. Abs.*, 1927, **21**, 3473: German 569,697, Feb. 6, 1933, to P. Spence and Sons, Ltd.; *Chem. Abs.*, 1933, **27**, 3835.

[196] B. D. Saklatwalla, H. E. Dunn, and A. E. Marshall, U.S. 1,953,777, Apr. 3, 1934, to Southern Mineral Products Corp.; *Chem. Abs.*, 1934, **28**, 3920.

[197] H. Spence and S. F. W. Crundall, U.S. 2,068,877, Jan. 26, 1937, to P. Spence and Sons, Ltd.; *Chem. Abs.*, 1937, **31**, 2022: British 419,522, Nov. 7, 1934, to P. Spence and Sons, Ltd.; *Chem. Abs.*, 1935, **29**, 2671: French 768,168, Aug. 1, 1934,

to P. Spence and Sons, Ltd.; *Chem. Abs.*, 1935, **29**, 561: German 631,460, June 20, 1936, to P. Spence and Sons, Ltd.; *Chem. Abs.*, 1936, **30**, 6145.

[198] J. Barnes, British 6,329, Mar. 16, 1906; *Chem. Abs.*, 1907, **1**, 929.

[199] G. R. Levi and G. Peyronel, *Z. Krist.*, 1935, **92**, 190; *Chem. Abs.*, 1936, **30**, 2450.

[200] G. V. Samsonov and L. L. Vereikina, U.S.S.R. 127,028, March 10, 1960; *Chem. Abs.*, 1960, **54**, 18912.

[201] F. W. Garrett and S. McCann, U.S. 3,079,229, Feb. 26, 1963, to Union Carbide Corp.; *Chem. Abs.*, 1963, **58**, 12210.

[202] W. L. Alderson and J. T. Maynard, U.S. 2,728,637, Dec. 27, 1955, to E. I. du Pont de Nemours and Co.; *Chem. Abs.*, 1956, **50**, 6759.

[203] M. Picon, *Compt. rend.*, 1933, **197**, 1415; *Chem. Abs.*, 1934, **28**, 2292: *Compt. rend.*, 1934, **198**, 1415; *Chem. Abs.*, 1935, **29**, 5034.

[204] X. Siebers and E. J. Kohlmeyer, *Arch. Erzbergbau Erzaufbereit., Metallhüttenw.*, 1931, **1**, 97; *Chem. Abs.*, 1934, **28**, 582.

[205] A. Clearfield, U.S. 3,148,998, Sept. 15, 1964, to National Lead Co.; *Chem. Abs.*, 1964, **61**, 12979.

[206] A. Clearfield, *Acta. Cryst.* 1963, **16**, 135; *Chem. Abs.*, 1963, **58**, 9695.

[207] M. Cheveton and M. S. Brunie, *Bull. Soc. Franc. Mineral. Crist.*, 1964, **87**, 277; *Chem. Abs.*, 1964, **61**, 15441.

[208] G. V. Samsonov and V. S. Sinelnikova, *Tsvetn. Metal.*, 1962, **35**, No. 11, 92; *Chem. Abs.*, 1963, **58**, 11045.

[209] H. Moissan, *Compt. rend.*, 1897, **125**, 839; *J. Chem. Soc.*, 1898, **A**, 161.

[210] R. R. Ridgway, U.S. 2,155,682, April 25, 1939, to Norton Co.; *Chem. Abs.*, 1939, **33**, 5755: British 483,248, April 14, 1938, to Norton Grinding Wheel Co., Ltd.; *Chem. Abs.*, 1938, **32**, 7354.

[211] B. Fetkenheuer, German 571, 292, Sept. 28, 1929, to Siemens und Halske, A-G.; *Chem. Abs.*, 1933, **27**, 2769.

[212] A. J. Parker, U.S. 2,102,214, Dec. 14, 1937; *Chem. Abs.*, 1938, **32**, 1233.

[213] K. Kieffer, German 661,842, June 28, 1938, to Deutsche Edelstahlwerke, A-G.; *Chem. Abs.*, 1938, **32**, 8354: British 471,862, Sept. 9, 1937, to Vereinigte Edelstahl, A-G.; *Chem. Abs.*, 1938, **32**, 1235.

[214] British 473,510, Oct. 14, 1937, to I. G. Farbenind., A-G.; *Chem. Abs.*, 1938, **32**, 2892.

[215] C. J. Kinzie and D. S. Hake, U.S. 2,149,939, Mar. 7, 1939, to Titanium Alloy Mfg. Co.; *Chem. Abs.*, 1939, **33**, 4138.

[216] G. A. Meerson, *Redkie Metal.*, 1935, **4**, No. 4, 6; *Chem. Abs.*, 1936, **30**, 3173.

[217] C. Ballhausen, German 745,269, Dec. 2, 1943, to Deutsche Edelstahlwerke, A-G.; *Chem. Abs.*, 1946, **40**, 1775.

[218] W. G. Burgers and J. C. M. Basart, *Z. anorg. u. allgem. Chem.*, 1934, **216**, 209; *Chem. Abs.*, 1934, **28**, 3021.

[219] W. R. Mott, *Trans. Am. Electrochem. Soc.*, 1919, **34**, *Chem. Abs.*, 1919, **13**, 1052.

[220] A. F. Scott, *J. Phys. Chem.*, 1925, **29**, 304; *Chem. Abs.*, 1925, **19**, 1358.

[221] L. P. Molkov and I. V. Vikker, *Chem. Zentr.*, 1936, **11**, 1679; *Chem. Abs.*, 1938, **32**, 4061.

[222] G. A. Meerson, *Redkie Metal.*, 1935, **4**, No. 4, 6; *Chem. Abs.*, 1936, **30**, 3173.

[223] E. Friederich and L. Sittig, *Z. anorg. u. allgem. Chem.*, 1925, **144**, 169; *Chem. Abs.*, 1925, **19**, 2313.

[224] G. A. Meerson and Y. M. Lipkes, *J. Applied Chem. (U.S.S.R.)*, 1945, **18**, 251; *Chem. Abs.*, 1946, **40**, 3575.

[225] B. F. Naylor, *J. Am. Chem. Soc.*, 1946, **68**, 370; *Chem. Abs.*, 1946, **40**, 2729.

[226] S. Peacock, British, 11,393, May 13, 1912; *Chem. Abs.*, 1913, **7**, 3644.

[227] C. J. Kinzie and D. S. Hake, U.S. 2,040,854, May 19, 1936, to Titanium Alloy Mfg. Co.; *Chem. Abs.*, 1936, **30**, 4411.

[228] P. M. McKenna, *Ind. Eng. Chem., News Ed.*, 1939, **17**, 476.

[229] A. E. Kovalskii and Y. S. Umanskii, *J. Phys. Chem. (U.S.S.R.)*, 1946, **20**, 929; *Chem. Abs.*, 1947, **41**, 2312.

[230] T. Ploetz, H. Caspar, and W. Hinrichs, German 1,013,634, Aug. 14, 1957, to Feldmühle Papier und Zellstoffwerke A.-G.; *Chem. Abs.*, 1960, **54**, 6062.

[231] T. Kubo, K. Shinriki, and T. Hanagawa, *Kogyo Kagaku Zasshi*, 1961, **64**, 619; *Chem. Abs.*, 1962, **57**, 4291.

[232] Y. Horiguchi and Y. Nomura, *Bull. Chem. Soc. Japan*, 1963, **36**, 486; *Chem. Abs.*, 1963, **59**, 210.

[233] R. A. Mercuri, J. M. Finn, Jr., and E. M. Nelson, U.S. 2,998,302, Aug. 29, 1961, to Union Carbide Corp.; *Chem. Abs.*, 1962, **57**, 4318.

[234] V. F. Funke, S. I. Yudkovskii, and G. V. Samsonov, *Zhur. Pricklad. Khim.*, 1960, **33**, 831; *Chem. Abs.*, 1960, **54**, 15859.

[235] W. Rutkowski and T. Giboa, *Szklo Ceram.*, 1962, **13**, 341; *Chem. Abs.*, 1964, **61**, 10410.

[236] H. B. Weiser, W. O. Milligan, and J. B. Bates, *J. Phys. Chem.*, 1942, **46**, 99; *Chem. Abs.*, 1942, **36**, 2460.

[237] F. R. Archibald and P. P. Alexander, Canadian 435,003, May 28, 1946, to Metal Hydrides, Inc.; *Chem. Abs.*, 1946, **40**, 5213: P. P. Alexander, U.S. 2,427,338, Sept. 16, 1947, to Metal Hydrides, Inc.; *Chem. Abs.*, 1948, **42**, 332.

[238] A. Klauber, *Z. anorg. u. allgem. Chem.*, 1921, **117**, 243; *Chem. Abs.*, 1921, **15**, 3796.

[239] H. Huber, L. Kirschfield, and A. Sieverts, *Ber.*, 1926, **59B**, 2891; *Chem. Abs.*, 1927, **21**, 682.

[240] K. Iwase and M. Fukushina, *Nippon Kinzoku Gakukai-Shi.*, 1937, **I**, 203; *Chem. Abs.*, 1938, **32**, 5750. G. Borelius, *Metallwirtschaft*, 1929, **8**, 105; *Chem. Abs.*, 1929, **23**, 1850.

[241] L. Kirschfield and A. Sievertz, *Z. physik. Chem.*, 1929, **A145**, 227; *Chem. Abs.*, 1930, **24**, 4207.

[242] Y. V. Baimakov and O. A. Lebedev, *Tr. Leningr. Politekhn. Inst.* No. 223, 1963; *Chem. Abs.*, 1964, **61**, 1495.

[243] P. Breisacher and B. Siegel, *J. Am. Chem. Soc.*, 1963, **85**, 1705; *Chem. Abs.*, 1963, **59**, 2378.

[244] P. P. Alexander, U.S. 2,168,185, Aug. 1, 1939, to Metal Hydrides, Inc.; *Chem. Abs.*, 1939, **33**, 9278.

[245] E. W. Washburn and E. M. Bunting, *Bur. Standards J. Research*, 1934, **12**, 239; *Chem. Abs.*, 1934, **28**, 2983.

[246] L. W. Ryan, U.S. 1,697,929, Jan. 8, 1929; *Chem. Abs.*, 1929, **23**, 1222.

[247] J. Rockstroh, Canadian 363,010, Dec. 29, 1936, to Titanium Pigment Co., Inc.; *Chem. Abs.*, 1937, **31**, 1560.

[248] French 743,851, Apr. 7, 1933, to Titanium Pigment Co., Inc.; *Chem. Abs.*, 1933, **27**, 4036: British 408,215, Apr. 3, 1934, to Titanium Pigment Co., Inc.; *Chem. Abs.*, 1934, **28**, 5938.

[249] G. W. Coggeshall, U.S. 987,544, Mar. 21, 1921.

[250] M. F. Freise, British 403,025, Dec. 14, 1933, to Metallges., A-G.; *Chem. Abs.*, 1934, **28**, 3194.

[251] D. Giusca and I. Popescu, *Bull. soc. roumaine phys.*, 1940, **40**, No. 73, 13; *Chem. Abs.*, 1940, **34**, 7695.

[252] F. A. A. Gregoire, French 1,108,062, Jan. 9, 1956; *Chem. Abs.*, 1959, **53**, 6555.

[253] S. A. Kutolin and A. I. Vulikh, U.S.S.R. 157,967, Oct. 18, 1963; *Chem. Abs.*, 1964, **60**, 11650.

[254] S. Taki and K. Tanaka, *Kogyo Kagaku Zasshi*, 1963, **66**, 417; *Chem. Abs.*, 1964, **60**, 1309.

[255] F. Muto and M. Kumitomi, *Kogyo Kagaku Zasshi*, 1962, **65**, 1775; *Chem. Abs.*, 1963, **58**, 8621.

[256] F. Muto, *Y.D.K.K.H.*, 1961, **12**, 187; *Chem. Abs.*, 1964, **61**, 12923.

[257] T. E. Gier and P. L. Salzberg, U.S. 2,833,620, May 6, 1958, to E. I. du Pont de Nemours and Co.; *Chem. Abs.*, 1958, **52**, 15853.

[258] H. C. Gulledge, *Ind. Eng. Chem.*, 1960, **52**, 117; *Chem. Abs.*, 1960, **54**, 12426.

[259] R. M. McKinney, U.S. 2,193,563, Mar. 13, 1940, to E. I. du Pont de Nemours and Co.; *Chem. Abs.*, 1940, **34**, 4930.

[260] H. D. Megaw, *Nature*, 1946, **157**, 20: *Experientia*, 1946, **2**, 183; *Chem. Abs.*, 1946, **40**, 5975.

[261] H. Blattner and W. Merz, *Helv. Phys. Acta*, 1948, **21**, 210; *Chem. Abs.*, 1948, **42**, 8603.

[262] G. G. Blowers and F. E. Welsby, British 933,059, July 31, 1963, to Erie Resistor Ltd.; *Chem. Abs.*, 1963, **59**, 12490.

[263] C. Bousquet, M. Lambert, A. M. Quittet, and A. Guinier, *U.S. Office Tech. Serv. A. D.* 278,109, 1962; *Chem. Abs.*, 1964, **60**, 1193.

[264] M. Schenk, *Phys. Status Solidi*, 1964, **4**, 25; *Chem. Abs.*, 1964, **61**, 1331.

[265] K. E. Nelson and H. Thurnauer, U.S. 2,908,579, Oct. 13, 1959, to American Lava Corp.; *Chem. Abs.*, 1960, **54**, 12532.

[266] A. Bergstein, *Collection Czechoslovak Chem. Communs.*, 1955, **20**, 1040; *Chem. Abs.*, 1956, **50**, 6856: *Chem. Listy.*, 1955, **49**, 1117; *Chem. Abs.*, 1955, **49**, 14401.

[267] H. T. Arend, Czechoslovakia 91,151, July 15, 1959; *Chem. Abs.*, 1960, **54**, 9228.

[268] W. B. Blumenthal, U.S. 2,827,360, March 18, 1955, to National Lead Co.; *Chem. Abs.*, 1958, **52**, 15854: German 1,026,287, March 20, 1958, to National Lead Co.; *Chem. Abs.*, 1960, **54**, 9228.

[269] H. J. Argenta and J. L. Johnston, U.S. 2,838,430, June 10, 1958, to Ford Motor Co.; *Chem. Abs.*, 1958, **52**, 15405.

[270] G. W. Cleek and E. H. Hamilton, Jr., *J. Research, Natl. Bur. Standards*, 1956, **57**, 317; *Chem. Abs.*, 1957, **51**, 6109.

[271] J. A. Hedvall, *Z. anorg. u. allgem. Chem.*, 1915, **93**, 313; *Chem. Abs.*, 1916, **10**, 727.

[272] H. J. Reusch and H. von Wartenberg, *Heraeus Vacuumschmelze Tenth Anniv.*, 1933, 347; *Chem. Abs.*, 1934, **28**, 3974.

[273] Y. Tanaka, *J. Chem. Soc. Japan*, 1940, **61**, 345; *Chem. Abs.*, 1940, **34**, 5332.

[274] I. Parga-Pondal and K. Bergt, *Anales soc. españ. fís y quím.*, 1933, **31**, 623; *Chem. Abs.*, 1934, **28**, 61.

[275] L. E. Lynd and L. Merker, German 1,014,972, Sept. 5, 1957, to Titangesellschaft m. b. H.; *Chem. Abs.*, 1960, **54**, 13576.

[276] A. I. Borisenko and P. V. Shirokova, *Zhur. Neorg. Khim.*, 1956, **1**, 615; *Chem. Abs.*, 1956, **50**, 15313.

[277] L. Merker, U.S. 2,751,279, June 19, 1956, to National Lead Co.; *Chem. Abs.*, 1956, **50**, 16057.

[278] L. Merker, U.S. 2,985,520, May 23, 1961, to National Lead Co.; *Chem. Abs.*, 1961, **55**, 27810.

[279] M. Mizuno, T. Noguchi, and S. Naka, *N.K.G.S.H.*, 1963, **12**, 385; *Chem. Abs.*, 1964, **60**, 2530.

[280] L. H. Brixner, *Inorg. Chem.*, 1964, **3**, 1065; *Chem. Abs.*, 1964, **61**, 6611.

[281] E. Pouillard and A. Michael, *Compt. rend.*, 1949, **228**, 1232; *Chem. Abs.*, 1949, **43**, 8296.

[282] I. N. Belyaev, N. P. Smolyaninov, and N. R. Kalmitskii, *Zh. Neorgan. Khim.*, 1963, **8**, 384; *Chem. Abs.*, 1963, **58**, 12003.

[283] F. Queyroux, *Compt. rend.*, 1964, **259**, 1527; *Chem. Abs.*, 1964, **61**, 15646.

[284] S. J. Buckman and J. D. Pera, German 1,145,732, March 21, 1963, to Buckman Laboratories, Inc.; *Chem. Abs.*, 1963, **58**, 14319.

[285] French 820,035, Oct. 30, 1937, to Titangesellschaft m. b. H.; *Chem. Abs.*, 1938, **32**, 3176.

[286] Y. Tanaka, *Bull. Chem. Soc. Japan*, 1941, **16**, 428; *Chem. Abs.*, 1947, **41**, 4393.

[287] S. S. Cole and W. K. Nelson, *J. Phys. Chem.*, 1938, **42**, 245; *Chem. Abs.*, 1938, **32**, 3249.

[288] S. S. Cole and H. Espenschied, *J. Phys. Chem.*, 1937, **41**, 445; *Chem. Abs.*, 1937, **31**, 4557.

[289] D. G. Nicholson, *Ind. Eng. Chem.*, 1937, **29**, 716; *Chem. Abs.*, 1937, **31**, 6033.

[290] A. V. Pamfilov and K. S. Fridman, *J. Gen. Chem. (U.S.S.R.)*, 1940, **10**, 210; *Chem. Abs.*, 1940, **34**, 7166.
[291] T. Ernst, *Z. angew. Mineral.*, 1943, **4**, 394; *Chem. Abs.*, 1946, **40**, 6327.
[292] F. O. Rummery, U.S. 2,607,659, Aug. 19, 1952, to The Glidden Co.; *Chem. Abs.*, 1952, **46**, 11710.
[293] V. V. Klimor, U.S.S.R. 159,809, Jan. 14, 1964; *Chem. Abs.*, 1964, **60**, 11651.


CHAPTER 9

[1] H. Spence and H. Wrigley, British 23,089, Nov. 26, 1914, to P. Spence and Sons, Ltd.; *Chem. Abs.*, 1916, **10**, 1411.
[2] E. Wainer, U.S. 2,316,141, Apr. 6, 1943, to Titanium Alloy Mfg. Co.; *Chem. Abs.*, 1943, **37**, 5561.
[3] S. F. W. Crundall, U.S. 2,027,812, Jan. 14, 1936, to P. Spence and Sons, Ltd.; *Chem. Abs.*, 1936, **30**, 1393: French 735,213, April 14, 1932, to P. Spence and Sons, Ltd.; *Chem. Abs.*, 1933, **27**, 1103: British 378,906, Aug. 15, 1932, to P. Spence and Sons, Ltd.; *Chem. Abs.*, 1933, **27**, 4107.
[4] German 497,626, Nov. 17, 1926, to Titan Co., A-S.; *Chem. Abs.*, 1930, **24**, 4127: British 289,111, Nov. 17, 1926, to Titan Co., A-S.; *Chem. Abs.*, 1929, **23**, 673.
[5] E. Pechard, *Compt. rend.*, 1893, **116**, 1513; *J. Chem. Soc.*, 1893, **A**, 625.
[6] A. Stahler, German 248,251, July 7, 1911, to Kunkein and Co.; *Chem. Abs.*, 1912, **6**, 2823.
[7] H. J. H. Fenton, *J. Chem. Soc.*, 1908, **I**, 1064.
[8] W. W. Plechner, U.S. 2,132,997, Oct. 11, 1938, to National Lead Co.; *Chem. Abs.*, 1939, **33**, 886.
[9] V. Caglioti and G. Sartori, *Gazz. chim. ital.*, 1936, **66**, 741; *Chem. Abs.*, 1937, **31**, 2956.
[10] D. W. MacCorquodale and H. Adkins, *J. Am. Chem. Soc.*, 1928, **50**, 1938; *Chem. Abs.*, 1928, **22**, 3131.
[11] H. A. Gardner, *Ind. Eng. Chem.*, 1936, **28**, 1020: *Ind. Eng. Chem.*, 1937, **29**, 640: U.S. 2,038,836, April 28, 1936; *Chem. Abs.*, 1936, **30**, 4339: U.S. 2,172,505, Sept. 12, 1939; *Chem. Abs.*, 1940, **34**, 649.
[12] W. K. Nelson and A. O. Ploetz, U.S. 2,234,681, Mar. 11, 1941, to National Lead Co.; *Chem. Abs.*, 1941, **35**, 4230: U.S. 2,244,258, June 3, 1941, to National Lead Co.; *Chem. Abs.*, 1941, **35**, 5731.
[13] C. E. Every, British 569,054, May 2, 1945, to Titanium Alloy Mfg. Co.; *Chem. Abs.*, 1947, **41**, 4900.
[14] A. V. Endredy and F. Brugger, *Z. anorg. u. allgem. Chem.*, 1942, **249**, 263; *Chem. Abs.*, 1943, **37**, 4982.
[15] J. Nelles, U.S. 2,187,821, Jan. 23, 1940, to I. G. Farbenind., A-G.; *Chem. Abs.*, 1940, **34**, 3764: British 479,470, Feb. 7, 1938, to I. G. Farbenind., A-G.; *Chem. Abs.*, 1938, **32**, 5003: French 818,570, Sept. 29, 1937, to I. G. Farbenind., A-G.; Chem. Abs., 1938, **32**, 2545: German 729,759, Nov. 26, 1942, to I. G. Farbenind., A-G.; *Chem. Abs.*, 1944, **38**, 378: German 720,080, March 26, 1942, to I. G. Farbenind., A-G.; *Chem. Abs.*, 1943, **37**, 2014.
[16] J. S. Jennings, W. Wardlow, and W. J. R. Way, *J. Chem. Soc.*, **1936**, 637; *Chem. Abs.*, 1938, **30**, 5180.
[17] R. L. McCleary, U.S. 2,410,119, Oct. 29, 1946, to E. I. du Pont de Nemours and Co.; *Chem. Abs.*, 1947, **41**, 573.
[18] F. Funk and E. Rogler, *Z. anorg. u. allgem. Chem.*, 1944, **252**, 323; *Chem. Abs.*, 1946, **40**, 4311.
[19] W. K. Nelson, U.S. 2,227,508, Jan. 7, 1941, to National Lead Co.; *Chem. Abs.*, 1941, **35**, 2533: French 840,907, May 8, 1939, to Titan Co., Inc.; *Chem. Abs.*, 1940, **34**, 1828.
[20] H. Meerwein, B. von Boch, B. Kirschnick, W. Lenz, and A. Migge, *J. prakt. Chem.*, 1936, **147**, 211; *Chem. Abs.*, 1937, **31**, 656.

[21] A. Benrath and A. Oblander, Z. wiss. Phot., 1922, **22**, 65; Chem. Abs., 1923, **17**, 1382.

[22] British 512,452, Sept. 15, 1939, to I. G. Farbenind., A-G.; Chem. Abs., 1941, **35**, 463.

[23] G. P. Luchinskii and E. S. Altman, Z. anorg. u. allgem. Chem., 1935, **225**, 321; Chem. Abs., 1936, **30**, 2512.

[24] G. P. Luchinskii, J. Gen. Chem. (U.S.S.R.), 1937, **7**, 2044; Chem. Abs., 1938, **32**, 519.

[25] F. Evard, Compt. rend., 1933, **196**, 2007; Chem. Abs., 1933, **27**, 4186.

[26] A. Rosenbeim, B. Raibmann, and G. Schendel, Z. anorg. u. allgem. Chem., 1931, **196**, 160; Chem. Abs., 1931, **25**, 3926.

[27] E. Demarcay, Compt. rend., 1873, **26**, 1414; J. Chem. Soc., 1873, 1015.

[28] D. J. Loden and K. E. Walker, U.S. 2,052,889, Sept. 1, 1936, to E. I. du Pont de Nemours and Co.; Chem. Abs., 1936, **30**, 7255.

[29] A. Bertrand, Bull. soc. chim., 1881, **34**, 631; J. Chem. Soc., 1881, **A**, 273.

[30] G. Scagliarini and G. Tartarini, Atti accad. naz. Lincei, 1926, **4**, 318; Chem. Abs., 1927, **21**, 739.

[31] A. G. Demetrios and E. Ladikos, Praktika Akad. Athenon, 1930, **5**, 449; Chem. Abs., 1933, **27**, 3160.

[32] F. Arloing, A. Morel, and A. Josserand, French 823,732, Jan. 25, 1938; Chem. Abs., 1938, **32**, 6010.

[33] C. W. J. Wende, U.S. 2,184,538, Dec. 26, 1939, to E. I. du Pont de Nemours and Co.; Chem. Abs., 1940, **34**, 2603.

[34] H. Gilmer, R. R. Burtner, N. O. Calloway, and J. A. V. Turck, Jr., J. Am. Chem. Soc., 1935, **57**, 907; Chem. Abs., 1935, **29**, 4355.

[35] O. C. Dermer and W. C. Fernelius, Z. anorg. u. allgem. Chem., 1934, **221**, 83; Chem. Abs., 1935, **29**, 1350. M. Dubien, Rev. gén. sci., 1926, **37**, 366; Chem. Abs., 1926, **20**, 3156.

[36] A. K. Macbeth and J. R. Price, J. Chem. Soc., **1935**, 151; Chem. Abs., 1935, **29**, 2931.

[37] N. A. Pushkin, Ann., 1942, **551**, 259 and 1942, **553**, 278; Chem. Abs., 1943, **37**, 5037.

[38] British 547,148, Aug. 17, 1942, to Titanium Alloy Mfg. Co.; Chem. Abs., 1943, **37**, 5417.

[39] H. Gilman and R. G. Jones, J. Org. Chem., 1945, **10**, 505; Chem. Abs., 1946, **40**, 1780.

[40] M. Polyani, Nature, 1946, **157**, 520; Chem. Abs., 1946, **40**, 4282.

[41] E. Pascu and F. B. Cramer, J. Am. Chem. Soc., 1937, **59**, 1059.

[42] T. M. Reynolds, J. Proc. Roy. Soc. N. S. Wales, 1932, **66**, 167; Chem. Abs., 1933, **27**, 959.

[43] H. S. Rothrock, U.S. 2,258,718, Oct. 14, 1941, to E. I. du Pont de Nemours and Co.; Chem. Abs., 1942, **36**, 595.

[44] I. Kraitzer, K. M. McTaggart, and G. Winter, Australian Department of Munitions, 1947, **2**, 348; Chem. Abs., 1948, **42**, 2114.

[45] A. Thomas, Peintures, pigments, vernis, 1952, **28**, 457; Chem. Abs., 1953, **47**, 4102.

[46] M. Kronstein, Paint and Varnish Production, 1950, **30**, No. 8, 10; Chem. Abs., 1950, **44**, 10344.

[47] V. L. Varshavskii, L. S. Krupitskaya, M. A. Kosykh, M. P. Zhukhareva, F. P. Kochanin, and V. A. Rozhlestvenskaya, U.S.S.R. 132,204, Oct. 5, 1960; Chem. Abs., 1961, **55**, 8787.

[48] D. F. Herman and W. K. Nelson, J. Am. Ceramic Soc., 1952, **74**, 2693; 1953, **75**, 3877; 1953, **75**, 3882; Chem. Abs., 1953, **47**, 6810; 1954, **48**, 8725.

[49] M. Deribère, Ann. hyg., 1941, **18**, 133; Chem. Abs., 1946, **40**, 7531: Chimie & industrie, 1942, **47**, 201; Chem. Abs., 1946, **40**, 7531.

[50] N. Heaton, Chem. Trade J., 1929, **85**, 439; Chem. Abs., 1930, **24**, 513. E. P. Youngman, U.S. Bur. Mines, Inform. Circ. 6365, 1930, 32.

[51] H. Udluft and J. H. Hellmers, *Arbeitsschutz, Unfallverhüt. Gewerbehyg. Reichsarbeitsblattes*, **1937**, 214; *Chem. Abs.*, 1938, **32**, 7601. G. Gerstel, *Arch. Gewerbepath. Gewerbehyg.*, 1936, **6**, 304; *Chem. Abs.*, 1937, **31**, 752.
[52] K. B. Lehman and L. Herget, *Chem. Ztg.*, 1927, **51**, 793; *Chem. Abs.*, 1928, **22**, 3053.
[53] H. O. Askew, *New Zealand J. Sci. Technol.*, 1931, **13**, 76; *Chem. Abs.*, 1932, **26**, 4861.
[54] Schoofs, *Bull. acad. roy. méd. Belg.*, 1922, **2**, 473; *Chem. Abs.*, 1923, **17**, 419.
[55] E. Marui, *Folia Pharmacol. Japon.*, 1928, **8**, 20; *Chem. Abs.*, 1930, **24**, 659.
[56] G. Carteret, *Bull. soc. chim.*, 1935, **2**, 159; *Chem. Abs.*, 1935, **29**, 3313.
[57] B. G. Gould, *Proc. Soc. Exptl. Biol. Med.*, 1936, **34**, 381; *Chem. Abs.*, 1938, **30**, 8251.
[58] F. Arloing, A. Morel, and A. Josserand, *Compt. rend.*, 1937, **204**, 824; *Chem. Abs.*, 1937, **31**, 4370.
[59] J. Pick, *Med. Klin. (Munich)*, **1911**, 4033; *Chem. Abs.*, 1912, **6**, 514.
[60] D. Abragam, *Compt. rend.*, 1935, **200**, 990; *Chem. Abs.*, 1935, **29**, 3728.
[61] H. H. Beard and V. C. Myers, *J. Biol. Chem.*, 1931, **94**, 89; *Chem. Abs.*, 1932, **26**, 759.
[62] B. L. Meredith and W. G. Christiansen, *J. Am. Pharm. Assoc.*, 1929, **18**, 607; *Chem. Abs.*, 1929, **23**, 4745.
[63] Moynier, *Paris Med.*, 1935, **1**, 258; *Chem. Abs.*, 1937, **31**, 7537.
[64] C. Kahone, *J. pharm. chim.*, 1932, **16**, 202; *Chem. Abs.*, 1933, **27**, 1991.
[65] F. Arloing, A. Morel, A. Josserand, L. Thevonot, and R. Caille, *Compt. rend. soc. biol.*, 1937, **126**, 5; *Chem. Abs.*, 1938, **32**, 207.
[66] J. Pick, *Med. Klin. (Munich)*, **1911**, 4033; *Chem. Abs.*, 1912, **6**, 514.
[67] P. J. Hanzlick and J. Tarr, *J. Pharmacology*, 1919, **14**, 221; *Chem. Abs.*, 1920, **14**, 1161.
[68] B. Malac, *Věstník Československ. Akad. Zemědělské*, 1931, **7**, 665; *Chem. Abs.*, 1932, **26**, 4120.
[69] Y. Yasue, *Science (Japan)*, 1950, **20**, 184; *Chem. Abs.*, 1957, **51**, 10467.
[70] I. V. Santoskii and M. D. Babina, *Toksikol. Nouykh. Khim.*, 1962, No. 4, 128; *Chem. Abs.*, 1963, **58**, 9547.
[71] *Toksikol. Nougykh Prom. Khim. Veshchestv*, 1961, No. 2, 69; *Chem. Abs.*, 1963, **58**, 869.
[72] G. Buenemann, W. Klosteckoette, and W. Ritzerfeld, *Arch. Hyg. Bakteriol.*, 1963, **147**, 58; *Chem. Abs.*, 1963, **59**, 5534.

## CHAPTER 10

[1] W. M. Thornton, *Titanium*, Chemical Catalog Co., New York, 1927, 58, 90. R. H. McKinney and W. H. Madson, *J. Chem. Education*, 1936, **13**, 155.
[2] J. A. Rahm, *Analytical Chemistry*, 1952, **24**, No. 11, 1832; *Chem. Abs.*, 1953, **47**, 1002.
[3] A. Guerreiro, *Brazil ministerio agr., Dept. nacl. producao mineral, Lab. producao mineral*, Bol. No. 24, 1946; *Chem. Abs.*, 1949, **43**, 7864.
[4] J. D. Stetkewicz, Dissertation, Columbia University, 1939.
[5] J. Bancelin and Y. Crimail, *Chimie & industrie*, 1939, **42**, 20; *Chem. Abs.*, 1939, **33**, 9680.
[6] A. D. Whitehead, *J. Oil & Colour Chemists' Assoc.*, 1939, **22**, 139; *Chem. Abs.*, 1939, **33**, 9009.
[7] J. V. A. Novak, *Chem. Anal.* (Warsaw), 1956, **1**, 123; *Chem. Abs.*, 1957, **51**, 4766.
[8] E. I. Chuikin, *Lakokrasochnye Materialy i ikh Primenenie*, 1962, No. 5, 64; *Chem. Abs.*, 1963, **58**, 4196.
[9] J. R. Wells and L. Carpenter, *U.S. Bur. Mines, Rept. Invest.* No. 6105, 1962; *Chem. Abs.*, 1963, **58**, 2830.

[10] W. Lamprecht, *Farbe u. Lack*, 1957, **63**, 342; *Chem. Abs.*, 1957, **51**, 15965.
[11] R. A. Spurr and H. Meyers, *Anal. Chem.*, 1957, **29**, 760; *Chem. Abs.*, 1957, **51**, 11171.
[12] G. Kortuem and G. Herzog, *Z. Anal. Chem.*, 1962, **190**, 239; *Chem. Abs.*, 1963, **58**, 5040.
[13] R. S. Dantuma, *Verfkroniek*, 1954, **27**, 64; *Chem. Abs.*, 1954, **48**, 11243.
[14] A. J. van Soest, *Verfkroniek*, 1955; **28**, 195; *Chem. Abs.*, 1955, **49**, 14573.
[15] R. S. Dantuma, *Verfkroniek*, 1955, **28**, 39; *Chem. Abs.*, 1955, **49**, 7445.
[16] V. A. Malevannyi, *Lakokrasochnye Materialy i ikh Primenenie*, 1962, No. 2. 54; *Chem. Abs.*, 1963, **58**, 9331.

## CHAPTER 11

[1] M. G. Mastin, *Ordnance*, 1956, **41**, 432.
[2] C. T. Baroch, T. B. Kaczmarek, W. D. Barnes, L. W. Galloway, W. W. Mack, and G. A. Lee, *U.S. Bur. Mines, Rept. Invest.* 5141, 1955; *Chem. Abs.*, 1956, **50**, 131.
[3] W. W. Mark, S. Yih, C. L. Lo, and D. H. Baker, Jr., *U.S. Bur. Mines, Rept. Invest.* No. 5665, 1960; *Chem. Abs.*, 1961, **55**, 3370.
[4] F. L. Kingsbury, U.S. 2,903,000, July 18, 1961, to National Lead Co.; *Chem. Abs.*, 1961, **55**, 24507; U.S. 2,835,568, May 20, 1958, to National Lead Co.; *Chem. Abs.*, 1958, **52**, 14509.
[5] W. J. Kroll, British 658,213, Oct. 3, 1951; *Chem. Abs.*, 1952, **46**, 404.
[6] R. Holst and R. Proft, *Neue Huette*, 1959, **4**, 106; *Chem. Abs.*, 1963, **59**, 14931.
[7] D. S. Chisholm, U.S. 3,102,807, Sept. 3, 1963, to Dow Chemical Co.; *Chem. Abs.*, 1963, **59**, 12410.
[8] D. S. Chisholm, U.S. 3,067,025, April 5, 1957, to Dow Chemical Co.; *Chem. Abs.*, 1963, **58**, 3177.
[9] J. Smolinski, British 814,181, June 3, 1959, to Minister of Supply; *Chem. Abs.*, 1960, **54**, 234.
[10] S. Takeuchi, M. Tezuka, T. Kurosawa, and S. Eda, *Met. Soc. Conf.*, 1961, **8**, 745; *Chem. Abs.*, 1962, **56**, 5704.
[11] S. Takeuchi, *Trans. Japan. Inst. Metals*, 1961, **2**, 53 and 57; *Chem. Abs.*, 1962, **56**, 3172.
[12] G. V. Forsblom and R. A. Sandler, *Tsvetnye Metally*, 1960, **33**, No. 10, 62; *Chem. Abs.*, 1961, **55**, 15269.
[13] C. H. Winter, Jr., U.S. 2,997,385, Aug. 22, 1961, to E. I. du Pont de Nemours and Co.; *Chem. Abs.*, 1961, **55**, 26984.
[14] A. R. Conklin, U.S. 2,787,539, April 2, 1957, to E. I. du Pont de Nemours and Co.; *Chem. Abs.*, 1957, **51**, 7996: British 827,398, Feb. 3, 1960, to E. I. du Pont de Nemours and Co.; *Chem. Abs.*, 1960, **54**, 10800.
[15] H. K. Najarian, U.S. 2,840,466, June 24, 1956, to St. Joseph Lead Co.; *Chem. Abs.*, 1958, **52**, 14503: British 776,124, June 5, 1957, to St. Joseph Lead Co.; *Chem. Abs.*, 1957, **51**, 14529.
[16] R. S. Dean, U.S. 2,904,428, Sept. 15, 1959, to Chicago Development Corp.; *Chem. Abs.*, 1960, **54**, 1161: U.S. 2,783,192, Feb. 26, 1957, to Chicago Development Corp.; *Chem. Abs.*, 1957, **51**, 7912.
[17] S. Takeuchi, British 837,905, June 15, 1960, to Metal Resources Research Institute; *Chem. Abs.*, 1960, **54**, 20809.
[18] M. M. Wright, Canadian 596,092, April 12, 1960, to Consolidated Mining and Smelting Company of Canada, Ltd.; *Chem. Abs.*, 1960, **54**, 19432.
[19] C. L. Schmidt and C. K. Stoddard, U.S. 2,882,143, April 14, 1959, to National Lead Co.; *Chem. Abs.*, 1959, **53**, 13969.
[20] C. D. Atkinson, U.S. 2,860,966, Nov. 18, 1958, to E. I. du Pont de Nemours and Co.; *Chem. Abs.*, 1959, **53**, 4097.
[21] H. Ishizuka, Japanese 9251, Oct. 18, 1958; *Chem. Abs.*, 1959, **53**, 4096.

[22] P. Gregoire, French 1,094,987, May 25, 1955, to Société de brevets d'études et de recherches metallurgiques "Soberti"; *Chem. Abs.*, 1959, **53**, 1079.

[23] S. Nakao, M. Ito, and S. Akimoto, Japanese 3953, June 20, 1957, to Nippon Soda Co.; *Chem. Abs.*, 1958, **52**, 13600.

[24] W. H. Deitz, U.S. 2,812,250, Nov. 5, 1957, to E. I. du Pont de Nemours and Co.; *Chem. Abs.*, 1958, **52**, 3652.

[25] S. Tekeuchi, T. Kurosawa, and M. Tezuka, Japanese 7301, Sept. 9, 1957, to Metal Resources Research Institute; *Chem. Abs.*, 1958, **52**, 19872.

[26] D. S. Chisholm and D. F. Hall, U.S. 2,840,465, June 24, 1958, to Dow Chemical Co.; *Chem. Abs.*, 1958, **52**, 15403.

[27] C. M. Olson, U.S. 2,839,385, June 17, 1958, to E. I. du Pont de Nemours and Co.; *Chem. Abs.*, 1958, **52**, 14509. O. B. Wilcox, U.S. 2,835,567, May 20, 1958, to E. I. du Pont de Nemours and Co.; *Chem. Abs.*, 1958, **52**, 14508.

[28] N. Morash, U.S. 2,826,492, March 11, 1958, to National Lead Co.; *Chem. Abs.*, 1958, **52**, 8924.

[29] T. Takubo, Japanese 7252, Aug. 24, 1956; *Chem. Abs.*, 1958, **52**, 8923.

[30] K. Nishida, E. Fugita, and A. Kimura, Japanese 6705 and 6706, Aug. 10, 1956, to Agency of Industrial Science and Technol.; *Chem. Abs.*, 1958, **52**, 8923.

[31] R. J. Fletcher, U.S. 2,820,722, Jan. 21, 1958; *Chem. Abs.*, 1958, **52**, 7101.

[32] W. Schaller, P. Tillman, and F. Raspe, German 832,205, Feb. 21, 1952, to Titangesellschaft m. b. H.; *Chem. Abs.*, 1958, **52**, 7101.

[33] K. Ishizuka, Japanese 8252, Nov. 16, 1955; *Chem. Abs.*, 1957, **51**, 17533.

[34] British 776,124, June 5, 1957, to St. Joseph Lead Co.; *Chem. Abs.*, 1957, **51**, 14529.

[35] A. R. Conklin, U.S. 2,787,539, April 2, 1957, to E. I. du Pont de Nemours and Co.; *Chem. Abs.*, 1957, **51**, 7996.

[36] C. H. Winter, Jr., U.S. 2,763,542, Sept. 18, 1956, to E. I. du Pont de Nemours and Co.; *Chem. Abs.*, 1957, **51**, 4254.

[37] C. H. Winter, Jr., U.S. 2,744,006, May 1, 1956, to E. I. du Pont de Nemours and Co.; *Chem. Abs.*, 1956, **50**, 12796.

[38] C. E. Ricks, U.S. 2,753,254, July 3, 1956, to E. I. du Pont de Nemours and Co.; *Chem. Abs.*, 1956, **50**, 12796.

[39] British 734,166, July 27, 1955, to Dow Chemical Co.; *Chem. Abs.*, 1955, **49**, 15715.

[40] E. M. Smith, U.S. 2,708,158, May 10, 1955; *Chem. Abs.*, 1955, **49**, 11536.

[41] J. P. Levy, D. H. Pickard, and L. Pickard, British 722,184, Jan. 19, 1955; *Chem. Abs.*, 1955, **49**, 8082.

[42] T. Ishino, H. Tamura, and Y. Takimoto, Japanese 406, Jan. 29, 1954, to Sakai Chemical Industries Co.; *Chem. Abs.*, 1954, **48**, 13609.

[43] S. Takeuchi, U.S. 3,039,866, June 19, 1962, to Research Institute for Iron, Steel and Other Metals; *Chem. Abs.*, 1962, **57**, 6944.

[44] K. Kamibayashi, Japanese 6257, Dec. 7, 1953; *Chem. Abs.*, 1954, **48**, 12012.

[45] C. K. Stoddard and J. L. Wyatt, U.S. 2,663,634, Dec. 22, 1953, to National Lead Co.; *Chem. Abs.*, 1954, **48**, 6367: British 702,771, Jan. 20, 1954, to Titan Co., Inc.; *Chem. Abs.*, 1954, **48**, 8721.

[46] J. F. Jordan, U.S. 2,667,413, Jan. 26, 1954; *Chem. Abs.*, 1954, **48**, 4414.

[47] C. H. Winter, Jr., U.S. 2,607,674, Aug. 19, 1952, to E. I. du Pont de Nemours and Co.; *Chem. Abs.*, 1952, **46**, 10085.

[48] H. J. Friedrich, German 812,117, Aug. 27, 1951, to Herman C. Starck A.-G.; *Chem. Abs.*, 1953, **47**, 5866.

[49] H. Ishizuka, Japanese 7704, Nov. 22, 1954; *Chem. Abs.*, 1956, **50**, 7044.

[50] British 748,615, May 9, 1956, to Titangesellschaft m. b. H.; *Chem. Abs.*, 1956, **50**, 12706.

[51] S. V. Ogurtsov, U.S.S.R. 156,296, Aug. 21, 1963; *Chem. Abs.*, 1964, **60**, 6534.

[52] R. D. Blue, U.S. 2,567,838, Sept. 11, 1951, to Dow Chemical Co.; *Chem. Abs.*, 1951, **45**, 10183.

[53] C. T. Baroch, T. B. Kaczmarek, and J. F. Lenc, *U.S. Bur. Mines, Rept. Invest.* 5253, 1956; *Chem. Abs.*, 1956, **50**, 13686.

[54] L. Gillemot, German 1,124,247, Feb. 22, 1962, to Fermipari Kutato Suterget; *Chem. Abs.*, 1962, **57**, 1916.

[55] C. M. Olson, U.S. 2,753,256, July 3, 1956, to E. I. du Pont de Nemours and Co.; *Chem. Abs.*, 1956, **50**, 12796.

[56] W. H. Keller and I. S. Zonis, U.S. 2,848,319, Aug. 19, 1958, to National Research Corp.; *Chem. Abs.*, 1959, **53**, 180.

[57] L. J. Derham, British 846,490, Aug. 31, 1960, to U.K. Atomic Energy Authority; *Chem. Abs.*, 1961, **55**, 10286.

[58] S. V. Ogurtsov, V. A. Reznichenko, and A. I. Dedkov, *Titan i Ego Splavy, Akad. Nauk. S.S.S.R. Inst. Met.*, 1962, 145; *Chem. Abs.*, 1963, **59**, 9624.

[59] M. S. Shneerson and E. M. Braverman, *S.M.P.A.P.P. i D.*, 1958, No. 3, 22; *Chem. Abs.*, 1961, **55**, 6306.

[60] A. G. Arkadev, M. M. Zaretskii, and V. A. Pyankov, *TVNIA Magn. Inst.*, 1957, No. 40, 413; *Chem. Abs.*, 1961, **55**, 24147.

[61] V. V. Sergeev, U.S.S.R. 151,469, Oct. 31, 1962; *Chem. Abs.*, 1963, **58**, 12265.

[62] G. M. Vorobev and B. N. Svetlichnyi, U.S.S.R. 109,795, Feb. 25, 1958; *Chem. Abs.*, 1958, **52**, 14508.

[63] M. Pirner, B. Pivny, and J. Rakosuik, Czechoslovakia 92,294, Oct. 15, 1959; *Chem. Abs.*, 1962, **57**, 8320.

[64] C. H. Winter, Jr., and J. N. Tully, U.S. 2,778,726, Jan. 22, 1957, to E. I. du Pont de Nemours and Co.; *Chem. Abs.*, 1957, **51**, 7286.

[65] H. Okeda, Japanese 8109, Sept. 18, 1956, to Nippon Soda Co.; *Chem. Abs.*, 1958, **52**, 13600.

[66] K. C. Wasberg, Norwegian 91,109, March 3, 1958, to Norsk Hydro Elektrisk Kvaelstofaktieselskab; *Chem. Abs.*, 1959, **53**, 13969.

[67] H. v. Zeppelin, German 1,029,166, Apr. 30, 1958; *Chem. Abs.*, 1960, **54**, 14084.

[68] R. M. Fowler and A. R. Gahler, British 838,916, June 22, 1960, to Union Carbide Corp.; *Chem. Abs.*, 1960, **54**, 20809.

[69] A. Boozenny, M. H. Kleineman, and A. R. Tarsey, U.S. 2,992,098, July 11, 1961, to Titanium Metals Corp. of America; *Chem. Abs.*, 1961, **55**, 24502.

[70] J. T. Kelley, U.S. 3,085,874, Apr. 16, 1963, to Union Carbide Corp.; *Chem. Abs.*, 1963, **58**, 13568.

[71] British 726,367, March 16, 1955, to National Lead Co.; *Chem. Abs.*, 1955, **49**, 11536.

[72] J. P. Levy, D. H. Pickard, and L. Pickard, British 757,894, Sept. 26, 1956; *Chem. Abs.*, 1957, **51**, 7666.

[73] V. V. Sergeev, *Tsvetnye Metal*, 1956, **29**, No. 11, 63; *Chem. Abs.*, 1957, **51**, 17674.

[74] T. P. Whaley, U.S. 2,727,817, Dec. 20, 1955, to Ethyl Corp.; *Chem. Abs.*, 1956, **50**, 4761.

[75] S. Sato, M. Awata, K. Ishikawa, T. Iwata, and A. Sakuma, *Rept. Sci. Research Inst.* (Japan), 1955, **31**, 410; *Chem. Abs.*, 1956, **50**, 11153.

[76] P. Himmelstein, German 1,090,863, Oct. 13, 1960, to Deutsche Gold und Silber Scheideanstalt vorm. Roessler; *Chem. Abs.*, 1961, **55**, 26985.

[77] D. H. Baker, Jr., and V. E. Homme, U.S. 3,069,255, Dec. 18, 1962, to U.S. Dept. of the Interior; *Chem. Abs.*, 1963, **58**, 4265.

[78] P. J. Lynskey, U.S. 2,895,823, July 21, 1959, to Peter Spence and Sons; *Chem. Abs.*, 1961, **55**, 24507.

[79] G. D. Bagley, British 805,730, Dec. 10, 1958, to Union Carbide Corp.; *Chem. Abs.*, 1959, **53**, 4097.

[80] M. G. Mastin, *Eng. Mining J.*, 1956, **157**, No. 6, 78; *Chem. Abs.*, 1956, **50**, 13686.

[81] T. A. Henrice and D. H. Baker, Jr., *Met. Soc. Conf.*, 1961, **8**, 721; *Chem. Abs.*, 1962, **56**, 8366.

[82] A. Boettcher and P. Himmelstein, German 1,042,901, Nov. 6, 1958, to Deutsche Gold und Silber Scheideanstalt vorm. Roessler; *Chem. Abs.*, 1961, **55**, 337.

[83] A. Boettcher and P. Himmelstein, German 1,051,007, Feb. 19, 1959, to Deutsche Gold und Silber Scheideanstalt vorm. Roessler; Chem. Abs., 1961, **55**, 20884.

[84] K. F. Griffiths, U.S. 3,085,871, April 16, 1963; *Chem. Abs.*, 1963, **58**, 13569.

[85] S. M. Shelton, German 1,067,222, Oct. 15, 1959, to Oregon Metallurgical Corp.; *Chem. Abs.*, 1961, **55**, 15319.

[86] British 816,017, July 8, 1959, to National Distillers and Chemical Corp.; *Chem. Abs.*, 1961, **55**, 3406.

[87] J. L. Vaughan, U.S. 2,936,232, May 10, 1960, to National Research Corp.; *Chem. Abs.* 1960, **54**, 16362.

[88] G. D. Bagley and D. E. Barbour, British 829,108, Feb. 24, 1960, to Union Carbide Corp.; *Chem. Abs.*, 1960, **54**, 15211.

[89] R. M. Sarla and D. H. Barbour, British 820,119, Sept. 16, 1959, to Union Carbide Corp.; *Chem. Abs.*, 1960, **54**, 2137.

[90] J. J. Gray, U.S. 2,903,350, Sept. 8, 1959, to Imperial Chemical Industries, Ltd.; *Chem. Abs.*, 1960, **54**, 1234.

[91] S. V. Ogurtsov, A. V. Revyakin, and V. A. Reznichenko, *Akad. Nauk. S.S.S.R.*, 1961, No. 6, 41, 50; *Chem. Abs.*, 1962, **56**, 9782.

[92] J. P. Quinn, U.S. 2,944,888, July 12, 1960, to Imperial Chemical Industries Ltd.; *Chem. Abs.*, 1960, **54**, 20809.

[93] W. W. Gullett, U.S. 2,901,411, Aug. 25, 1959, to Chicago Development Corp.; *Chem. Abs.*, 1959, **53**, 21293.

[94] J. J. Gray, British 797,155, June 25, 1958, to Imperial Chemical Industries Ltd.; *Chem. Abs.*, 1959, **53**, 179.

[95] D. C. Fleck, M. M. Wong, and D. H. Baker, Jr., *U.S. Bur. Mines Rept. Invest.* No. 5596, 1960; *Chem. Abs.*, 1960, **54**, 15166.

[96] J. Smolinski, J. C. Hannam, and A. L. Leach, *J. Appl. Chem.* (London), 1958, **8**, 375; *Chem. Abs.*, 1958, **52**, 19804.

[97] W. O. Di Petro, U.S. 2,847,297, Aug. 12, 1958, to National Research Corp.; *Chem. Abs.*, 1958, **52**, 19873.

[98] J. E. Booge, U.S. 2,843,477, July 15, 1958, to E. I. du Pont de Nemours and Co.; *Chem. Abs.*, 1958, **52**, 17070.

[99] J. C. Smart, British 792,300, March 26, 1958, to Imperial Chemical Industries Ltd.; *Chem. Abs.*, 1958, **52**, 17070.

[100] J. P. Quinn, British 791,783, March 12, 1958, to Imperial Chemical Industries Ltd.; *Chem. Abs.*, 1958, **52**, 17070.

[101] G. R. Findley, U.S. 2,826,491, March 11, 1958, to National Research Corp.; *Chem. Abs.*, 1958, **52**, 9931.

[102] J. P. Quinn, British 936,811, Sept. 11, 1963, to Imperial Chemical Industries Ltd.; *Chem. Abs.*, 1963, **59**, 14937.

[103] A. J. Kallfelz, U.S. 3,022,158 and 3,022,159, Sept. 24, 1959, to Allied Chemical Corp.; *Chem. Abs.*, 1962, **56**, 12600.

[104] S. Schott and V. A. Hansley, U.S. 2,950,963, Aug. 30, 1960, to National Distillers and Chemical Corp.; *Chem. Abs.*, 1960, **54**, 24313.

[105] F. W. Garrett and R. A. Skimin, U.S. 2,826,493, March 11, 1958, to Union Carbide Corp.; *Chem. Abs.*, 1958, **52**, 9931.

[106] F. J. Langmyhr, Norwegian 90,840, Jan. 27, 1958, to Sentralinstitutt fur Industriell Forskning; *Chem. Abs.*, 1959, **53**, 13969.

[107] B. B. Raney, U.S. 2,922,712, Jan. 26, 1960, to Chicago Development Corp.; *Chem. Abs.*, 1960, **54**, 10601.

[108] W. O. Di Petro, U.S. 2,951,021, Aug. 30, 1960, to National Research Corp.; *Chem. Abs.*, 1961, **55**, 1245.

[109] British 857,346, Dec. 29, 1960, to Deutsche Gold and Silber Scheideanstalt vorm. Roessler; *Chem. Abs.*, 1961, **55**, 14273.

[110] W. C. Muller, U.S. 3,012,878, Sept. 16, 1958, to National Distillers and Chemical Corp.; *Chem. Abs.*, 1962, **56**, 6970.

[111] W. C. Muller, U.S. 2,865,738, Dec. 23, 1958, to National Distillers and Chemical Corp.; *Chem. Abs.*, 1959, **53**, 4097.

[112] L. J. Derham, British 762,519, Nov. 28, 1956, to National Smelting Co., Ltd.; *Chem. Abs.*, 1957, **51**, 6493.

[113] R. S. Hood, U.S. 2,782,118, Feb. 19, 1957, to Monsanto Chemical Co.; *Chem. Abs.*, 1957, **51**, 7287.

[114] V. E. Homme, M. M. Wong, and D. H. Baker, Jr., *U.S. Bur. Mines, Rept. Invest.* 5398, 1958; *Chem. Abs.*, 1958, **52**, 12715.

[115] S. Yamaguchi, *Angew. Chem.*, 1957, **67**, 748; *Chem. Abs.*, 1958, **52**, 16110.

[116] P. Himmelstein, German 1,090,863, Oct. 13, 1960, to Deutsche Gold und Silber Scheideanstalt vorm. Roessler; *Chem. Abs.*, 1961, **55**, 26985.

[117] J. Kamlet, U.S. 2,823,991, Feb. 18, 1958, to National Distillers and Chemical Corp.; *Chem. Abs.*, 1958, **52**, 7102.

[118] K. L. Strelts, V. S. Khomyakov, and V. N. Zaitsev, U.S.S.R. 109,880, Feb. 25, 1958; *Chem. Abs.*, 1959, **53**, 13029.

[119] Y. Uebayashi, Japanese 4012, June 14, 1955; *Chem. Abs.*, 1957, **51**, 12808.

[120] R. G. Hellier, N. Beecher, and P. L. Raymond, U.S. 2,915,382, Dec. 1, 1959, to National Research Corp.; *Chem. Abs.*, 1960, **54**, 4337.

[121] R. D. Blue and G. B. Cobel, U.S. 2,983,600, May 9, 1961, to Dow Chemical Co.; *Chem. Abs.*, 1961, **55**, 19728.

[122] E. Davies, U.S. 2,839,386, June 17, 1958, to Imperial Chemical Industries Ltd.; *Chem. Abs.*, 1958, **52**, 16170.

[123] J. Ferguson, J. J. Gray, and W. N. Howell, British 750,355, June 13, 1956, to Imperial Chemical Industries Ltd.; *Chem. Abs.*, 1956, **50**, 12796.

[124] J. P. Quinn, U.S. 2,816,020, and 2,816,021, Dec. 10, 1957, to Imperial Chemical Industries Ltd.; *Chem. Abs.*, 1958, **52**, 7101. British 788,308, and 788,309, Dec. 23, 1957, to Imperial Chemical Industries Ltd.; *Chem. Abs.*, 1958, **52**, 8923.

[125] R. B. Mooney, British 697,917, Sept. 30, 1953, to Imperial Chemical Industries Ltd.; *Chem. Abs.*, 1954, **48**, 6367.

[126] J. J. Gray, U.S. 2,944,887, July 12, 1960, to Imperial Chemical Industries Ltd.; *Chem. Abs.*, 1960, **54**, 24314. British 788,174, Dec. 23, 1957, to Imperial Chemical Industries Ltd.; *Chem. Abs.*, 1958, **52**, 8923.

[127] J. P. Quinn, U.S. 2,827,371, March 18, 1958, to Imperial Chemical Industries Ltd.; *Chem. Abs.*, 1958, **52**, 9932. British 717,930, Mar. 3, 1954, to Imperial Chemical Industries Ltd.; *Chem. Abs.*, 1955, **49**, 2288.

[128] R. L. Yamartino, U.S. 2,915,383, Dec. 1, 1959, to National Research Corp.; *Chem. Abs.*, 1960, **54**, 5416.

[129] J. J. Gray, U.S. 2,864,691, Dec. 16, 1958, to Imperial Chemical Industries Ltd.; *Chem. Abs.*, 1959, **53**, 4097. British 776,739, June 12, 1957, to Imperial Chemical Industries Ltd.; *Chem. Abs.*, 1957, **51**, 13727.

[130] V. D. Savin and V. A. Reznichenko, *Titan i ego S. A. N. S.S.S.R. Inst. Met.*, 1963, 172; *Chem. Abs.*, 1964, **60**, 190.

[131] T. Oshiba, Japanese 2765, Mar. 17, 1964, to Showa Denko K.K.; *Chem. Abs.*, 1964, **61**, 10356.

[132] P. D'Aragon, Canadian 500,594, March 16, 1954; *Chem. Abs.*, 1956, **50**, 5432.

[133] R. S. Hood, U.S. 2,773,759, Dec. 11, 1956, to Monsanto Chemical Co.; *Chem. Abs.*, 1957, **51**, 2516.

[134] T. Tomonari, Japanese 7808, Sept. 8, 1956, to Tohoku Electrical Chemical Industries Co.; *Chem. Abs.*, 1958, **52**, 9931.

[135] R. S. Dean, U.S. 2,838,393, June 10, 1958, to Chicago Development Corp.; *Chem. Abs.*, 1958, **52**, 19873.

[136] R. M. McKinney, U.S. 2,839,383, June 17, 1958, to E. I. du Pont de Nemours and Co.; *Chem. Abs.*, 1958, **52**, 19873.

[137] R. Holst, F. Thum, and R. Hollivitz, German 1,086,898, Aug. 11, 1960, to VEB Elektrochemisches Kombinat Bitterfeld; *Chem. Abs.*, 1961, **55**, 16381.

[138] M. P. Hnilicka, U.S. 2,847,205, Aug. 12, 1958, to National Research Corp.; *Chem. Abs.*, 1959, **53**, 180.

[139] K. Schwabe and L. Krause, East German 27,247, Feb. 25, 1964; *Chem. Abs.*, 1964, **61**, 12992.

[140] R. S. Dean, U.S. 2,830,893, April 15, 1958; *Chem. Abs.*, 1958, **52**, 15403.

[141] R. Maillet, German 1,070,387, Dec. 3, 1959, to Soberti; *Chem. Abs.*, 1961, **55**, 15314.

[142] British 762,541, Nov. 28, 1956, to National Research Corp.; *Chem. Abs.*, 1957, **51**, 6497.

[143] British 694,921, July 29, 1953, to Titan Co., Inc.; *Chem. Abs.*, 1954, **48**, 8721.

[144] C. H. Winter, Jr., U.S. 2,817,585, Dec. 24, 1957, to E. I. du Pont de Nemours and Co.; *Chem. Abs.*, 1958, **52**, 3653.

[145] C. M. Olson, U.S. 2,982,645, May 2, 1961, to E. I. du Pont de Nemours and Co.; *Chem. Abs.*, 1961, **55**, 20885.

[146] H. S. Dombrowski and C. H. Winter, Jr., U.S. 2,922,710, Jan. 26, 1960, to E. I. du Pont de Nemours and Co.; *Chem. Abs.*, 1960, **54**, 10797.

[147] C. A. Hampel and J. Glasser, U.S. 2,618,550, Nov. 18, 1952, to Armour Research Foundation; *Chem. Abs.*, 1953, **47**, 2116.

[148] J. M. Avery, U.S. 2,956,872, Oct. 18, 1960, to Ethyl Corp.; *Chem. Abs.*, 1961, **55**, 4332.

[149] H. Hohn, E. Pitzer, and K. Komarek, Austrian 196,624, March 25, 1958; *Chem. Abs.*, 1958, **52**, 7109.

[150] J. P. Quinn, British 720,543 and 720,517, Dec. 22, 1954, to Imperial Chemical Industries, Ltd.; *Chem. Abs.*, 1955, **49**, 6810.

[151] J. Glasser and C. A. Hampel, U.S. 2,703,752, March 8, 1955, to Kennecott Copper Corp.; *Chem. Abs.*, 1955, **49**, 6809.

[152] B. W. Whitehurst, U.S. 3,058,820, Oct. 10, 1962; *Chem. Abs.*, 1963, **58**, 2249.

[153] W. Schmidt, U.S. 2,813,787, Nov. 19, 1957, to Reynolds Metal Co.; *Chem. Abs.*, 1958, **52**, 2719.

[154] E. Fitzer, H. Hofbrauer, and H. Hohn, Austrian 194,616, Jan. 10, 1958; *Chem. Abs.*, 1958, **52**, 3652. F. Kohler, Austrian 181,430, March 25, 1955; *Chem. Abs.*, 1955, **49**, 6754.

[155] G. G. Hatch, U.S. 2,676,882, April 27, 1954, to Kennecott Copper Corp.; *Chem. Abs.*, 1954, **48**, 8162.

[156] T. Oshiba, Japanese 6601, Aug. 24, 1957, to Showa Electric Industry Co.; *Chem. Abs.*, 1958, **52**, 19872.

[157] T. Oshiba, Japanese 5509, July 26, 1958, to Showa Electric Industry Co.; *Chem. Abs.*, 1959, **53**, 1079.

[158] Norwegian 100,610, July 21, 1962, to Titan Co. A./S.; *Chem. Abs.*, 1964, **60**, 5079.

[159] M. B. Alpert, U.S. 3,114,685, Dec. 17, 1963, to National Lead Co.; *Chem. Abs.*, 1964, **60**, 6491.

[160] B. W. Whitehurst, U.S. 3,058,820, Oct. 16, 1962; *Chem. Abs.*, 1963, **58**, 2249.

[161] L. J. Reinert, U.S. 3,082,159, March 19, 1963, to New Jersey Zinc Co.; *Chem. Abs.*, 1963, **58**, 10981.

[162] G. Truempler, U.S. 3,067,112, Dec. 4, 1962, to Lonza E. u. C. F. A. G.; *Chem. Abs.*, 1963, **58**, 6458.

[163] H. L. Slatin, U.S. 2,994,650, Aug. 1, 1961, to U.S. Atomic Energy Commission; *Chem. Abs.*, 1961, **55**, 26798.

[164] R. B. Head, *J. Electrochem. Soc.*, 1961, **108**, 806; *Chem. Abs.*, 1961, **55**, 24325.

[165] P. Wolski, German 1,005,278, March 28, 1957, to Farbenfabriken Bayer Akt.-Ges.; *Chem. Abs.*, 1961, **55**, 183.

[166] R. S. Dean and W. W. Gullett, U.S. 2,948,663, Aug. 9, 1960, to Chicago Development Corp.; *Chem. Abs.*, 1961, **55**, 182.

[167] W. R. Opie and O. W. Moles, *Trans. A. I. M. E.*, 1960, **218**, 646; *Chem. Abs.*, 1960, **54**, 20584.

[168] H. L. Slatin, German 1,139,985, Nov. 22, 1962, to Timax Associates; *Chem. Abs.*, 1963, **58**, 5271.

[169] A. Brenner and J. M. Sherfey, U.S. 3,002,905, Jan. 30, 1962, to U.S. Dept. of the Army; *Chem. Abs.*, 1962, **57**, 13546.

[170] R. L. Bickerdike and G. T. Brown, British 834,792, May 11, 1960, to Peter Spence and Sons, Ltd.; *Chem. Abs.*, 1960, **54**, 24034.

[171] L. W. Gendvil, O. W. Moles, and H. R. Palmer, U.S. 2,936,267, May 10, 1960, to National Lead Co.; *Chem. Abs.*, 1960, **54**, 17122.

[172] T. Tomonari, Japanese 53, Jan. 17, 1959, to Osaka Titanium Co., Ltd.; *Chem. Abs.*, 1959, **53**, 19642.

[173] S. Takumoto, Japanese 8254, Sept. 17, 1958; *Chem. Abs.*, 1959, **53**, 18699.

[174] G. E. Snow, U.S. 2,898,276, Aug. 4, 1959, to New Jersey Zinc Co.; *Chem. Abs.*, 1959, **53**, 18699.

[175] E. W. Andrews, U.S. 2,898,275, Aug. 4, 1959, to New Jersey Zinc Co.; *Chem. Abs.*, 1959, **53**, 18699.

[176] F. A. Howard, U.S. 2,880,150, March 31, 1959; *Chem. Abs.*, 1959, **53**, 11069.

[177] H. Ishizuka, Japanese 5053, July 17, 1957; *Chem. Abs.*, 1958, **52**, 16095.

[178] E. Wainer, U.S. 2,833,706, May 6, 1958, to Horizons Titanium Corp.; *Chem. Abs.*, 1958, **52**, 19623.

[179] British 756,943, Sept. 12, 1956, to Reynolds Metals Co.; *Chem. Abs.*, 1957, **51**, 6403.

[180] G. E. Snow and A. T. McCord, U.S. 2,780,593, Feb. 5, 1957, to New Jersey Zinc Co.; *Chem. Abs.*, 1957, **51**, 6403.

[181] W. M. Normore and A. G. Scobie, U.S. 2,755,240, July 17, 1956, to Shawinigan Water and Power Co., Ltd.; *Chem. Abs.*, 1956, **50**, 16487.

[182] J. C. Smart, British 698,151, Oct. 7, 1953; *Chem. Abs.*, 1954, **48**, 6296.

[183] H. Ishizuka, Japanese 3859, Sept. 26, 1952; *Chem. Abs.*, 1953, **47**, 8561.

[184] British 678,807, Sept. 10, 1953, to Shawinigan Water and Power Co.; *Chem. Abs.*, 1953, **47**, 5282.

[185] M. B. Alpert, German 1,026,969, March 27, 1958, to Titangesellschaft m.b.H.; *Chem. Abs.*, 1960, **54**, 22111.

[186] S. Mellgren, U.S. 2,748,073, May 29, 1956, to National Lead Co.; *Chem. Abs.*, 1956, **50**, 11864.

[187] R. S. Dean, U.S. 2,951,795, Sept. 6, 1960, to Chicago Development Corp.; *Chem. Abs.*, 1961, **55**, 1246.

[188] G. Ervin, Jr., and H. F. G. Ueltz, U.S. 2,837,478, June 3, 1958, to Norton Co.; *Chem. Abs.*, 1959, **53**, 928.

[189] W. E. Reid, Jr., *J. Electrochem. Soc.*, 1961, **108**, 393; Chem. Abs., 1961, **55**, 12113.

[190] H. L. Slatin, U.S. 2,864,749, Dec. 16, 1958, to Timax Corp.; *Chem. Abs.*, 1959, **53**, 5926.

[191] J. Smolinski, British 836,888, June 9, 1960, to Minister of Supply; *Chem. Abs.*, 1960, **54**, 19233.

[192] G. D. P. Cordner and H. W. Worner, *Australian J. Applied Sci.*, 1951, **2**, 358; *Chem. Abs.*, 1952, **46**, 4927.

[193] British 682,919, Nov. 19, 1952, to Titan Co., Inc.; *Chem. Abs.*, 1953, **47**, 5826.

[194] British 712,742, July 28, 1954, to Titan Co., Inc.; *Chem. Abs.*, 1955, **49**, 2910.

[195] R. D. Blue and M. P. Neipert, U.S. 2,943,033, June 28, 1960, to Dow Chemical Co.; *Chem. Abs.*, 1964, **60**, 21679.

[196] C. J. Carignan, U.S. 2,848,395, Aug. 19, 1958, to E. I. du Pont de Nemours and Co.; *Chem. Abs.*, 1958, **52**, 19623.

[197] T. Tomonari, Japanese 503, Jan. 28, 1957, to Tohoku Electro Chemical Industries Co.; *Chem. Abs.*, 1958, **52**, 5172.

[198] M. Takahashi, T. Okada, and M. Kawane, Japanese 7553, Oct. 19, 1955, to Teikoku Kako Co.; *Chem. Abs.*, 1958, **52**, 8803.

[199] M. B. Alpert, F. J. Schultz, and W. F. Sullivan, *J. Electrochem. Soc.*, 1957, **104**, 555; *Chem. Abs.*, 1957, **51**, 17525.

[200] H. Tadenuma, H. Ikeda, and K. Fujita, *Kogyo Kagaku Zasshi*, 1956, **59**, 356; *Chem. Abs.*, 1957, **51**, 10273.

[201] H. N. Sinha and H. K. Worner, *Trans. Indian Inst. Metals*, 1956, **9**, 123; *Chem. Abs.*, 1957, **51**, 9375.

[202] H. L. Slatin, U.S. 3,137,641, June 16, 1964, to Timax Associates; *Chem. Abs.*, 1964, **61**, 5208.

[203] V. E. Homme and M. W. Wong, *U.S. Bur. Mines, Rept. Invest.* No. 6360, 1964; *Chem. Abs.*, 1964, **60**, 14151.

[204] V. E. Homme, U.S. 3,113,017, Dec. 3, 1963, to U.S. Dept. of the Interior; *Chem. Abs.*, 1964, **60**, 5108.

[205] T. A. Henrie, E. K. Kleespies, and D. H. Baker, Jr., *U. S. Bur. Mines, Rept. Invest.* No. 6162, 1963; *Chem. Abs.*, 1963, **58**, 8439.

[206] K. Nishida, E. Fugita, and T. Kimura, Japanese 8154, Sept. 25, 1957, to Bureau of Industrial Technics; *Chem. Abs.*, 1959, **53**, 180.

[207] E. Ustan, U.S. 2,938,783, May 31, 1960; *Chem. Abs.*, 1960, **54**, 19230.

[208] M. B. Alpert and R. L. Powell, U.S. 2,741,588, April 10, 1956, to National Lead Co.; *Chem. Abs.*, 1956, **50**, 8351: British 712,742, July 28, 1954, to Titan Co., Inc.; *Chem. Abs.*, 1955, **49**, 2910.

[209] C. E. Barnett, U.S. 2,908,619, Oct. 13, 1959, to New Jersey Zinc Co.; *Chem. Abs.*, 1960, **54**, 2052: L. J. Reinert and E. A. Fatzinger, U.S. 2,848,397, Aug. 19, 1958, to New Jersey Zinc Co.; *Chem. Abs.*, 1958, **52**, 19623.

[210] E. Wainer, U.S. 2,731,404, Jan. 17, 1956, to Horizons Titanium Corp.; *Chem. Abs.*, 1956, **50**, 6229.

[211] P. Drossbach, *Z. Elektrochem*, 1953, **57**, 548; *Chem. Abs.*, 1954, **48**, 5680.

[212] J. G. Wurn, L. Gravel, and R. J. A. Potvin, *J. Electrochem. Soc.*, 1957, **104**, 301; *Chem. Abs.*, 1957, **51**, 9375.

[213] H. Ishizuka, Japanese 7554, Oct. 19, 1955; *Chem. Abs.*, 1957, **51**, 17532.

[214] T. Tomonari, Japanese 5702, July 14, 1956, to Tokoku Electrochemical Industries Co.; *Chem. Abs.*, 1958, **52**, 5172.

[215] M. A. Steinberg and A. A. Topinka, U.S. 2,813,068, Nov. 12, 1957, to Horizons Titanium Corp.; *Chem. Abs.*, 1958, **52**, 6985.

[216] British 812,817, April 29, 1959, to Solar Aircraft Co.; *Chem. Abs.*, 1959, **53**, 21293.

[217] A. Brenner and J. M. Sherfey, U.S. 3,002,905, May 27, 1955, to U.S. Dept. of the Army; *Chem. Abs.*, 1962, **56**, 1204.

[218] A. A. Topinka, Q. H. McKenna, and S. S. Carlton, U.S. 2,731,402, Jan. 17, 1956, to Horizons Titanium Corp.; *Chem. Abs.*, 1956, **50**, 6228.

[219] E. Wainer and M. E. Sibert, U.S. 2,714,575, Aug. 2, 1955, to Horizons Titanium Corp.; *Chem. Abs.*, 1955, **49**, 14543. British 736,567, Sept. 7, 1955, to Horizons Titanium Corp.; *Chem. Abs.*, 1956, **50**, 3987.

[220] British 727,088, March 30, 1955, to Horizons Titanium Corp.; *Chem. Abs.*, 1955, **49**, 11535.

[221] E. Wainer, U.S. 2,707,170, April 26, 1955, to Horizons Titanium Corp.; *Chem. Abs.*, 1955, **49**, 12163: British 740,849, Nov. 23, 1955, to Horizons Titanium Corp.; *Chem. Abs.*, 1956, **50**, 8351.

[222] B. C. Raynes, U.S. 2,786,809, March 26, 1957, to Horizons Titanium Corp.; *Chem. Abs.*, 1957, **51**, 7910.

[223] A. Clausaru, V. Craiu, and P. Anghel, *Rev. Chim.*, 1960, **11**, 90; *Chem. Abs.*, 1963, **58**, 4166.

[224] R. L. Bickerdike and G. T. Brown, British 834,792, May 11, 1960, to Peter Spence and Sons, Ltd.; *Chem. Abs.*, 1960, **54**, 24034.

[225] M. M. Wong, R. C. Campbell, D. C. Fleck, and D. H. Baker, Jr., *U.S. Bur. Mines, Rept. Invest.* No. 6161, 1963; *Chem. Abs.*, 1963, **58**, 8692.

[226] S. Okada, M. Kawane, and T. Hashino, *Kogyo Kagaku Zasshi*, 1960, **63**, 48; *Chem. Abs.*, 1962, **56**, 1290.

[227] S. Takeuchi, Japanese 7809, Sept. 8, 1956, to Institute for Study of Metallic Materials; *Chem. Abs.*, 1958, **52**, 9821.

[228] R. A. Keiffer and F. Benesovsky, U.S. 2,848,315, Aug. 19, 1958, to Schwarzkopf Development Corp.; *Chem. Abs.*, 1959, **53**, 180.

[229] S. Takeuchi, Japanese 1504, March 7, 1957, to Metal Resources Research Institute; *Chem. Abs.*, 1958, **52**, 6985.

[230] T. Tomonari, Japanese 1406, Feb. 27, 1957, to Tohoku Electro Chemical Industries Co.; *Chem. Abs.*, 1958, **52**, 8804.

[231] S. Takeuchi, Japanese 2357, Aug. 18, 1957, to Metal Resources Research Institute; *Chem. Abs.*, 1958, **52**, 8804.

[232] S. Hill, British 713,446, Aug. 11, 1954, to Peter Spence and Sons, Ltd.; *Chem. Abs.*, 1955, **49**, 2910.

[233] H. L. Slatin, U.S. 3,003,934, Jan. 8, 1959, to Timax Associates; *Chem. Abs.*, 1962, **56**, 4439.

[234] M. A. Steinberg, M. E. Sibert, and A. A. Topinka, U.S. 2,707,169, April 26, 1955, to Horizons Titanium Corp.; *Chem. Abs.*, 1955, **49**, 8716: British 701,289, Dec. 23, 1953, to Horizons Titanium Corp.; *Chem. Abs.*, 1954, **48**, 8094: E. Wainer and M. E. Sibert, U.S. 2,707,168, April 26, 1955, to Horizons Titanium Corp.; *Chem. Abs.*, 1955, **49**, 8716: British 701,288, Dec. 23, 1953, to Horizons Titanium Corp.; *Chem. Abs.*, 1954, **48**, 8093.

[235] I. Igami, Japanese 5503, 5506, and 5507, July 26, 1958; *Chem. Abs.*, 1959, **53**, 15829.

[236] Y. K. Delimarskii and G. G. Buderaskaya, *Ukr. Khim. Zh.*, 1962, **28**, 565; *Chem. Abs.*, 1963, **58**, 3960.

[237] British 786,460, Nov. 20, 1957, to Norton Grinding Wheel Co.; *Chem. Abs.*, 1958, **52**, 6985.

[238] British 792,716, April 2, 1958, to Norton Grinding Wheel Co., Ltd.; *Chem. Abs.*, 1958, **52**, 16951.

[239] K. Schwabe, East German 14,571, April 3, 1958; *Chem. Abs.*, 1959, **53**, 6975.

[240] G. Ervin, Jr., and H. F. G. Ueltz, German 1,108,923, to Deutsche Norton G. m. b. H.; *Chem. Abs.*, 1962, **56**, 4438.

[241] H. F. G. Ueltz, U.S. 2,910,021, Jan. 5, 1960, to Norton Co.; *Chem. Abs.*, 1960, **54**, 6369.

[242] E. Wainer, U.S. 2,722,509, Nov. 1, 1955, to Horizons Titanium Corp.; *Chem. Abs.*, 1956, **50**, 3950: British 743,695, Jan. 18, 1956, to Horizons Titanium Corp.; *Chem. Abs.*, 1956, **50**, 8351.

[243] G. Ervin, Jr., and H. F. G. Ueltz, U.S. 3,098,805, July 23, 1963, to Norton Co.; *Chem. Abs.*, 1963, **59**, 8365.

[244] G. Ervin, Jr., and H. F. G. Ueltz, German 1,072,393, Dec. 31, 1959, to Deutsche Norton G. m. b. H.; *Chem. Abs.*, 1961, **55**, 15191.

[245] E. Wainer, U.S. 2,904,426, Sept. 19, 1959, to Horizons Titanium Corp.; *Chem. Abs.*, 1960, **54**, 1134: U.S. 2,868,703, Jan. 13, 1959, to Horizons Titanium Corp.; *Chem. Abs.*, 1959, **53**, 8894.

[246] British 795,416, May 21, 1958, to Horizons Titanium Corp.; *Chem. Abs.*, 1959, **53**, 2893.

[247] M. E. Sibert, Q. H. McKenna, M. A. Steinberg, and E. Wainer, *J. Electrochem. Soc.*, 1955, **102**, 252; *Chem. Abs.*, 1956, **50**, 87.

[248] S. Takeuchi, Japanese 6213, Aug. 9, 1958, to Metal Resources Research Institute; *Chem. Abs.*, 1959, **53**, 15829.

[249] E. Pruvot, C. Boulanger, and P. L. A. Belon, French 1,146,248, Nov. 7, 1957, to Soberti; *Chem. Abs.*, 1960, **54**, 116.

[250] C. M. Olson, U.S. 2,917,440, Dec. 15, 1959, to E. I. du Pont de Nemours and Co.; *Chem. Abs.*, 1960, **54**, 7383.

[251] P. Herasymenko, U.S. 3,036,961, May 29, 1962; *Chem. Abs.*, 1962, **57**, 7023.

[252] W. Schaller, A. Ehringfeld, and P. Tillman, German 1,061,081, July 9, 1959, to Titangesellschaft m. b. H.; *Chem. Abs.*, 1961, **55**, 9122.

[253] J. J. Casey and J. W. Berham, U.S. 3,123,464, March 3, 1964, to Amalgamated Growth Industries, Inc.; *Chem. Abs.*, 1964, **60**, 11653.

[254] K. Ishizuka, Japanese 7408, Oct. 15, 1955; *Chem. Abs.*, 1958, **52**, 15312.

[255] T. Waku, Japanese 6854, May 23, 1963, to Tokai Electrode Manuf. Co.; *Chem. Abs.*, 1963, **59**, 14995.

[256] T. A. Ferraro, Jr., U.S. 2,874,040, Feb. 17, 1959, to U.S. Government; *Chem. Abs.*, 1959, **53**, 9018.

[257] R. B. Eaton, U.S. 2,891,857, Jan. 23, 1959, to E. I. du Pont de Nemours and Co.; *Chem. Abs.*, 1960, **54**, 231.

[258] O. Kubaschewski, H. Villa, and W. A. Dench, *Trans Faraday Soc.*, 1956, **52**, 214; *Chem. Abs.*, 1956, **50**, 14423.

[259] A. Oka, T. Kubo, and N. Oshima, *Nippon Kagaku Zasshi*, 1964, **85**, 394; *Chem. Abs.*, 1964, **61**, 7777.

[260] P. R. C. Dure and J. J. A. Wilkins, German 1,078,843, March 31, 1960, to Société belge de l'azote; *Chem. Abs.*, 1961, **55**, 19726.

[261] H. Harnisch, A. Mehne, and F. Rodis, German 1,142,159, Jan. 10, 1963, to Knapsack-Griesheim A.-G.; *Chem. Abs.*, 1963, **58**, 8681.

[262] H. Roehl, German 1,164,998, March 12, 1964, to Chemische Werke Huels A.-G. *Chem. Abs.*, 1964, **60**, 14157.

[263] W. Freundlich and M. Bichara, *Compt-rend.*, 1954, **238**, 1324; *Chem. Abs.*, 1954, **48**, 8104.

[264] British 807,889, Jan. 21, 1959, to Farbenfabriken Bayer A.-G.; *Chem. Abs.*, 1959, **53**, 9018.

[265] D. Goerrig, E. Walaschewski, and V. Lowowski, German 974,210, Oct. 20, 1960, to Farbenfabriken Bayer A.-G.; *Chem. Abs.*, 1961, **55**, 26984.

[266] G. A. Meerson and O. P. Kolchin, *Referat. Zhur. Met.*, 1956, No. 1012; *Chem. Abs.*, 1956, **50**, 15369.

[267] C. Sheer and S. Korman, U.S. 2,979,449, April 11, 1961; *Chem. Abs.*, 1961, **55**, 17458.

[268] S. L. Rushmore, U.S. 2,934,481, April 26, 1960, to Union Carbide Corp.; *Chem. Abs.*, 1960, **54**, 16696.

[269] J. R. Spraul and D. Batzer, U.S. 3,047,477, July 31, 1962, to General American Transportation Corp.; *Chem. Abs.*, 1962, **57**, 14777.

[270] British 823,999, Nov. 18, 1959, to Barium Steel Corp.; *Chem. Abs.*, 1960, **54**, 6494.

[271] F. Gregoire and R. Ricard, French 1,167,261, Feb. 27, 1957, to Société des blancs de zinc de la Mediterranée; *Chem. Abs.*, 1962, **56**, 1205.

[272] British 797,616, July 2, 1958, to Eltro G. m. b. H.; *Chem. Abs.*, 1958, **52**, 19872.

[273] W. Dawhill and L. Wesch, German 917,034, Sept. 12, 1955; *Chem. Abs.*, 1958, **52**, 8021.

[274] L. W. Coffer, U.S. 2,789,896, April 23, 1957, to Climax Molybdenum Co.; *Chem. Abs.*, 1957, **51**, 9463.

[275] T. A. Henrie, H. Dolezal, and E. K. Kleespies, U.S. 3,140,170, July 7, 1964, to U.S. Dept. of the Interior; *Chem. Abs.*, 1964, **61**, 6656.

[276] D. W. Rostron, U.S. 2,834,667, May 13, 1958, to Dominion Magnesium Ltd.; *Chem. Abs.*, 1958, **52**, 14508.

[277] O. Kubaschewski and W. A. Dench, *Bull. Inst. Mining Met.*, 1956, **599**, 1; *Chem. Abs.*, 1957, **51**, 974.

[278] W. W. Gleave and J. P. Quinn, U.S. 2,757,135, July 31, 1956, to Imperial Chemical Industries Ltd.; *Chem. Abs.*, 1956, **50**, 15296: British 724,198, Feb. 16, 1955, to Imperial Chemical Industries Ltd.; *Chem. Abs.*, 1955, **49**, 11473.

[279] N. I. Anufrieva and A. I. Ivanov, *Izvest. Akad. Nauk S.S.S.R.*, 1960, No. 4, 9; *Chem. Abs.*, 1961, **55**, 7106.

[280] W. Juda, U.S. 2,902,360, Sept. 1, 1959, to Ionics Inc.; *Chem. Abs.*, 1960, **54**, 15207.

[281] V. A. Reznichenko, V. I. Lukashin, and V. I. Solovev, *Akad. Nauk S.S.S.R.*, 1961, No. 6, 104; *Chem. Abs.*, 1962, **56**, 9782.

[282] K. Kamijo, Japanese 501, March 8, 1962, to Tekkosha Co., Ltd.; *Chem. Abs.*, 1963, **58**, 13525.

[283] W. C. Lilliendahl and E. D. Gregory, U.S. 2,707,679, May 3, 1955, to Westinghouse Electric Corp.; *Chem. Abs.*, 1955, **49**, 11536.

[284] K. M. Bowling, *Australian J. Chem.*, 1963, **16**, 66; *Chem. Abs.*, 1963, **58**, 11987.

[285] A. W. Petersen and L. A. Bromley, *J. Metals*, 1956, **8**, 284; *Chem. Abs.*, 1956, **50**, 4744.

[286] S. Suva and A. Tietz, Czechoslovakia 88,791 and 88,175, Jan. 15, 1959; *Chem. Abs.*, 1961, **55**, 19562.

[287] H. M. Weir, U.S. 2,978,316, April 4, 1961; *Chem. Abs.*, 1961, **55**, 18537.

[288] R. F. Rolston, *Z. anorg. u. allgem Chem.*, 1960, **305**, 25; *Chem. Abs.*, 1961, **55**, 195.

[289] E. J. Dunn, U.S. 2,855,331, Oct. 7, 1958, to U.S. Government; *Chem. Abs.*, 1959, **53**, 4098.

[290] I. Iidaka and R. Otsuka, Japanese 8454, Nov. 21, 1958, to Institute for Physics and Chemistry; *Chem. Abs.*, 1958, **52**, 3652.

[291] T. Shikauchi, Japanese 1802, March 20, 1957; *Chem. Abs.*, 1958, **52**, 19872.

[292] E. J. Dunn, U.S. 2,792,438, May 14, 1957; *Chem. Abs.*, 1957, **51**, 11225.

[293] A. E. Loonan, U.S. 2,694,652, 2,694,653, and 2,694,654, Nov. 16, 1955, to Chilean Nitrate Sales Corp.; *Chem. Abs.*, 1955, **49**, 2288: British 698,232, 698,233, 698,234, and 698,235, Oct. 14, 1953, to Chilean Nitrate Sales Corp.; *Chem. Abs.*, 1954, **48**, 5779. U.S. 2,714,564, Aug. 2, 1955, to Chilean Nitrate Sales Corp.; *Chem. Abs.*, 1955, **49**, 14629.

[294] F. J. Langmyhr, Norwegian 94,293, July 20, 1959, to Christiania Spigerverk; *Chem. Abs.*, 1960, **54**, 6369.

[295] K. J. Korpi and R. C. Johnson, U.S. 3,015,555, Oct. 16, 1958, to Lummus Co.; *Chem. Abs.*, 1962, **56**, 8371.

[296] T. Goldenberg, U.S. 2,986,502, May 30, 1961; *Chem. Abs.*, 1961, **55**, 18398.

[297] J. O. Bockris, I. A. Menzies, and L. Young, British 825,951, Dec. 23, 1959, to Minister of Supply; *Chem. Abs.*, 1960, **54**, 11773.

[298] P. Ehrlich, W. Gutsche, and H. Kuehnl, *Z. Anorg. Allgem. Chem.*, 1961, **312**, 70; *Chem. Abs.*, 1962, **56**, 5760.

[299] H. Kuhnl, P. Ehrlich, and R. D. Uihlein, *Z. anorg. u. allgem. Chem.*, 1960, **306**, 243; *Chem. Abs.*, 1961, **55**, 13130.

[300] S. S. Carlton and B. C. Raynes, U.S. 2,880,149, May 31, 1959, to Horizons Titanium Corp.; *Chem. Abs.*, 1959, **53**, 11069.

[301] E. Wainer, U.S. 2,821,506, Jan. 28, 1958, to Horizons Titanium Corp.; *Chem. Abs.*, 1958, **52**, 6985.

[302] H. Ishizuka, Japanese 606, Jan. 30, 1957; *Chem. Abs.*, 1958, **52**, 5172.

[303] T. Tomonari, Japanese 5405, July 24, 1957, to Osaka Titanium Co.; *Chem. Abs.*, 1958, **52**, 19622.

[304] J. R. Nettle, T. E. Hill, Jr., and H. D. Baker, Jr., *U.S. Bur. Mines Rept. Invest.* No. 5410, 1958; *Chem. Abs.*, 1958, **52**, 18018.

[305] S. Takeuchi, Japanese 2655, April 11, 1956, to Metal Resources Research Institute; *Chem. Abs.*, 1957, **51**, 8556.

[306] G. R. Couch and W. E. Mooz, U.S. 2,779,727, Jan. 29, 1957, to National Lead Co.; *Chem. Abs.*, 1957, **51**, 4851.

[307] B. C. Raynes and M. E. Sibert, U.S. 2,773,023, Dec. 4, 1956, to Horizons Titanium Corp.; *Chem. Abs.*, 1957, **51**, 4254.

[308] R. S. Dean, U.S. 2,817,630, Dec. 24, 1957, to Chicago Development Co.; *Chem. Abs.*, 1958, **52**, 3570.

[309] R. S. Dean and F. X. McCawley, U.S. 2,951,794, Sept. 6, 1960, to Chicago Development Corp.; *Chem. Abs.*, 1961, **55**, 1243. British 856,588, Dec. 21, 1960, to International Metallurgical Corp.; *Chem. Abs.*, 1961, **55**, 12116.

[310] R. S. Dean and W. W. Gullett, U.S. 2,901,410, Aug. 25, 1959, to Chicago Development Co., *Chem. Abs.*, 1959, **53**, 21293.

[311] W. W. Gullett, U.S. 2,817,631, Dec. 24, 1957, to Chicago Development Co.; *Chem. Abs.*, 1958, **52**, 3570.

[312] K. Kuehnl and G. Flischhauer, *Chem. Ingr. Tech.*, 1964, **36**, 729; *Chem. Abs.*, 1964, **61**, 10318.

[313] P. Gross and D. L. Levi, U.S. 2,785,973, March 19, 1957, to Fulmer Research Institute, Ltd.; *Chem. Abs.*, 1957, **51**, 7996: British 722,091, Feb. 2, 1955, to Fulmer Research Institute, Ltd.; *Chem. Abs.*, 1955, **49**, 8777.

[314] C. E. Rick, U.S. 2,773,787, Dec. 11, 1956, to E. I. du Pont de Nemours and Co.; *Chem. Abs.*, 1957, **51**, 6493.

[315] R. B. Mooney and J. P. Quin, British 698,753, Oct. 21, 1953, to Imperial Chemical Industries, Ltd.; *Chem. Abs.*, 1954, **48**, 6367.

[316] J. P. Quin, British 697,487, Sept. 23, 1953, to Imperial Chemical Industries, Ltd.; *Chem. Abs.*, 1954, **48**, 6367.

[317] D. W. Rostrom, British 798,750, July 23, 1958, to Dominion Magnesium Ltd.; *Chem. Abs.*, 1959, **53**, 179.

[318] E. Wainer, U.S. 2,844,499, July 22, 1958, to Horizons Titanium Corp.; *Chem. Abs.*, 1958, **52**, 17070.

[319] F. E. Edlin, U.S. 2,904,427, Sept. 15, 1959, to E. I. du Pont de Nemours and Co.; *Chem. Abs.*, 1960, **54**, 234.

[320] H. S. Dombrowski and F. E. Edlin, U.S. 3,114,626, Dec. 17, 1963, to E. I. du Pont de Nemours and Co.; *Chem. Abs.*, 1964, **60**, 6579.

[321] N. T. Kudryavtsev and R. G. Golovchanskaya, *Dokl. Akad. Nauk. S.S.S.R.*, 1963, **148**, 1339; *Chem. Abs.*, 1963, **59**, 1276.

[322] V. Stein and W. Zschaage, German 973,578, March 31, 1960; *Chem. Abs.*, 1961, **55**, 19561.

[323] P. Csokan and F. Simon, *Magyar Kem Lapja*, 1960, **15**, 442; *Chem. Abs.*, 1961, **55**, 4198.

[324] N. T. Kudryavstev and R. G. Golovchanskaya, U.S.S.R. 127,121, March 10, 1960; *Chem. Abs.*, 1960, **54**, 18139.

[325] S. Morioka, T. Shibata, and A. Umezono, *Nippon Kinzoku Gakkaishi*, 1957, **21**, 32; *Chem. Abs.*, 1960, **54**, 17109.

[326] S. Sato, Japanese 15,202, Oct. 13, 1960; *Chem. Abs.*, 1962, **56**, 11363.

[327] M. Kawakami, Japanese 10,502, Dec. 14, 1956; *Chem. Abs.*, 1958, **52**, 16095.

[328] K. Takada, J. Iida, and Y. Shimura, Japanese 7303, Sept. 9, 1957; *Chem. Abs.*, 1958, **52**, 19623.

[329] T. Osihiba, Japanese 7304, Sept. 9, 1957, to Showa Electric Industry Co.; *Chem. Abs.*, 1958, **52**, 19622.

[330] R. Hamada, Japanese 4665, June 18, 1956; *Chem. Abs.*, 1957, **51**, 17531.

[331] T. Hamada, Japanese 5159, July 27, 1955; *Chem. Abs.*, 1957, **51**, 16154.

[332] T. Hamada, Japanese 2357, April 11, 1955; *Chem. Abs.*, 1957, **51**, 11132.

[333] S. Sato, Japanese 1707, March 16, 1955, to Scientific Research Institute, Ltd.; *Chem. Abs.*, 1957, **51**, 112.

[334] S. Sato and K. Yamane, *Repts. Sci. Research Inst.* (Japan), 1955, **31**, 345; *Chem. Abs.*, 1956, **50**, 10570.

[335] F. D. Waldron-Trowman and J. K. Wilson, British 824,253, Nov. 29, 1959, to D. Napier and Sons, Ltd.; *Chem. Abs.*, 1961, **55**, 181.

[336] M. Koyama, *Rikagaku Kenkyusho Hokoku*, 1962, **38**, 546; *Chem. Abs.*, 1964, **61**, 3912.

[337] V. Takeuchi, Japanese 9976, Nov. 21, 1956; *Chem. Abs.*, 1958, **52**, 16096.

[338] V. V. Usova and V. I. Lainer, *I.V.U.Z.T.M.*, 1963, **66**, 132; *Chem. Abs.*, 1964, **60**, 2595.

[339] T. Oshiba, Japanese 6404, July 31, 1956, to Showa Electric Industry Co.; *Chem. Abs.*, 1958, **52**, 8802.

[340] T. Oshiba, *Kogyo Kagaku Zasshi*, 1959, **62**, 998; *Chem. Abs.*, 1962, **57**, 13533.

[341] F. von Bichowsky, U.S. 2,820,745, Jan. 21, 1958; *Chem. Abs.*, 1958, **52**, 8803.

[342] British 788,295, Dec. 23, 1957, to Horizons Titanium Corp.; *Chem. Abs.*, 1958, **52**, 7913.

[343] A. W. Schlechten, M. E. Straumanis, and C. B. Gill, *J. Electrochem. Soc.*, 1955, **102**, 81; *Chem. Abs.*, 1956, **50**, 705.

[344] M. E. Sibert and J. T. Burwell, Jr.; U.S. 2,828,251, March 25, 1958, to Horizons Titanium Corp.; *Chem. Abs.*, 1958, **52**, 12626: British 788,295, Dec. 23, 1957, to

Horizons Titanium Corp.; *Chem. Abs.*, 1958, **52**, 7913; German 1,034,446, July 17, 1958, to Horizons Titanium Corp.; *Chem. Abs.*, 1960, **54**, 17122: British 778,218, July 3, 1957, to Horizons Titanium Corp.; *Chem. Abs.*, 1957, **51**, 13623.

[345] A. R. Stetson, *Mater Desigh Eng.*, 1963, **57**, No. 3, 81; *Chem. Abs.*, 1963, **58**, 12166.

[346] L. W. McGraw and J. L. Stockdale, U.S. 3,075,896, Jan. 29, 1963, to Shuron Optical Co.; *Chem. Abs.*, 1963, **58**, 7615.

[347] F. W. Drosten, S. V. Weglars, and R. C. Robinson, U.S. 3,058,841, Oct. 16, 1962, to Republic Steel Corp.; *Chem. Abs.*, 1963, **58**, 299.

[348] British 868,011, May 17, 1961, to Union Carbide Corp.; *Chem. Abs.*, 1961, **55**, 22075.

[349] F. Pearlstein, R. Wick, and A. Galaccio, *J. Electrochem. Soc.*, 1963, **110**, 843; *Chem. Abs.*, 1963, **59**, 8357.

[350] F. C. Wagner, U.S. 2,908,966, Oct. 20, 1959, to Horizons Titanium Corp.; *Chem. Abs.*, 1960, **54**, 2144.

[351] British 945,452, Jan. 2, 1964, to E. I. du Pont de Nemours and Co.; *Chem. Abs.*, 1964, **61**, 2799.

[352] R. H. Singleton, U.S. 3,122,413, Feb. 25, 1964, to National Research Corp.; *Chem. Abs.*, 1964, **60**, 11653.

[353] H. W. Jacobson, W. A. Jenkins, C. M. Olson, and O. B. Wilcox, German 1,162,572, Feb. 6, 1964, to E. I. du Pont de Nemours and Co.; *Chem. Abs.*, 1964, **60**, 14218.

[354] K. J. Korpi and R. C. Johnson, U.S. 3,001,867, June 23, 1958, to Lummus Co.; *Chem. Abs.*, 1962, **56**, 1205.

[355] M. Miksits and W. Schaller, German 1,168,088, April 16, 1964, to Titangesellschaft m. b. H.; *Chem. Abs.*, 1964, **61**, 4032.

[356] P. Galvin, H. Cartoux and L. Septier, French 1,321,508, March 22, 1963, to P. C. de P. C. et E.; *Chem. Abs.*, 1963, **59**, 1352.

[357] R. Schwarz and A. Koster, *Z. anorg. u. allgem. Chem.*, 1956, **285**, 1; *Chem. Abs.*, 1956, **50**, 16607.

[358] H. von Zeppelin, German 1,057,786, May 21, 1959; *Chem. Abs.*, 1961, **55**, 7257.

[359] M. P. Neipert and R. D. Blue, U.S. 2,913,332, Nov. 17, 1959, to Dow Chemical Co.; *Chem. Abs.*, 1960, **54**, 3149.

[360] S. Takeuchi, Japanese 6212, Aug. 9, 1964, to Metal Resources Research Institute; *Chem. Abs.*, 1959, **53**, 15828.

[361] S. T. Jazwinski and J. A. Sisto, U.S. 2,975,049, March 14, 1961, to Phoenix Steel Corp.; *Chem. Abs.*, 1961, **55**, 17460.

[362] P. Thome, French 1,138,554, June 17, 1957, to R.B.V.R.I.; *Chem. Abs.*, 1960, **54**, 1234.

[363] M. A. Brooks, P. W. Hyde, and J. R. Lee, British 939,317, Oct. 9, 1963, Westland Aircraft Ltd.; *Chem. Abs.*, 1963, **59**, 14995.

[364] N. C. Welsh, British 860,563, Feb. 8, 1961, to Associated Electrical Industries Ltd.; *Chem. Abs.*, 1951, **55**, 14272.

[365] A. J. Griest, W. W. Parris, and P. D. Frost, U.S. 2,892,743, June 30, 1959, to U.S. Dept. of the Army; *Chem. Abs.*, 1959, **53**, 16908.

[366] C. M. Brown and R. L. Folkman, U.S. 3,024,102, March 6, 1962, to Union Carbide Corp.; *Chem. Abs.*, 1962, **56**, 12642.

[367] W. E. Kuhn, *J. Electrochem. Soc.*, 1952, **99**, 89; *Chem. Abs.*, 1953, **47**, 10379.

[368] S. L. Asmus, F. W. Wood, and R. A. Beall, *U.S. Bur. Mines. Rept. Invest.* No. 5686, 1960; *Chem. Abs.*, 1961, **55**, 6211.

[369] S. Kawakatsu and Y. Nishiura, Japanese 2360, March 7, 1964, to Osaka Titanium Co., Ltd.; *Chem. Abs.*, 1964, **61**, 10408.

[370] H. R. Spendelow, R. L. Folkman, and C. R. Allenbach, U.S. 3,054,166, Sept. 18, 1964, to Union Carbide Corp.; *Chem. Abs.*, 1962, **57**, 13542.

[371] M. Becker, German 1,027,406, April 3, 1958, to Deutsche Gold und Silber Scheideanstalt vorm. Roessler; *Chem. Abs.*, 1960, **54**, 17122.

[372] R. A. Beall, F. W. Wood, and A. H. Roberson, *J. Metals*, 1955, **7**, 801; *Chem. Abs.*, 1955, **49**, 10816.
[373] S. F. Radthe, R. M. Scriver, and J. A. Snyder, *J. Metals*, 1951, **3**, 620; *Chem. Abs.*, 1951, **45**, 7895.
[374] G. A. Pagonis, U.S. 3,116,998, Jan. 7, 1964, to Light Metals Research Laboratory, Inc.; *Chem. Abs.*, 1964, **60**, 6579.
[375] A. R. C. Westwood, *Mod. Castings*, 1960, **37**, 36; *Chem. Abs.*, 1964, **60**, 8965.
[376] British 861,846, March 1, 1961, to New Jersey Zinc Co.; *Chem. Abs.*, 1961, **55**, 16381.
[377] C. F. Wilford and R. F. Tylecote, *Brit. Welding*, 1960, **7**, 708; *Chem. Abs.*, 1961, **55**, 6345.
[378] D. F. Bowman and S. A. Hays, U.S. 3,108,919, Oct. 29, 1963, to North American Aviation, Inc.; *Chem. Abs.*, 1964, **60**, 278.
[379] French 1,334,498, Aug. 9, 1963, to Société Nationale de Constructions Aéronautique; *Chem. Abs.*, 1964, **60**, 6488.
[380] Jelks Barksdale, section on Titanium in Economic Geography of Industrial Materials, Reinhold Publishing Corp., New York, 1956, 209.
[381] W. A. Wooster and G. L. Macdonald, *Nature*, 1947, **160**, 260; *Chem. Abs.*, 1948, **42**, 4108.
[382] J. Herenguel, *Usine*, 1942, **51**, 35; *Chem. Abs.*, 1943, **37**, 6615.

## CHAPTER 12

[1] P. Farup, U.S. 966,815, Aug. 9, 1910; *Chem. Abs.*, 1910, **4**, 3015: British 3,649, Feb. 14, 1910; *Chem. Abs.*, 1911, **5**, 2980.
[2] G. Jebsen, Norwegian 21,693, Feb. 25, 1910; *Chem. Abs.*, 1912, **6**, 2153.
[3] L. E. Barton, U.S. 1,201,541, Oct. 17, 1916; *Chem. Abs.*, 1916, **10**, 3141.
[4] L. E. Barton, U.S. 1,206,796, Dec. 5, 1917; *Chem. Abs.*, 1917, **11**, 279: French 483,780, Aug. 8, 1917, to Titanium Alloy Mfg. Co.; *Chem. Abs.*, 1918, **12**, 1001: British 106,585, Nov. 22, 1916, to Titanium Alloy Mfg. Co.; *Chem. Abs.*, 1917, **11**, 2575: Norwegian 29,194, Nov. 4, 1918, to Titanium Alloy Mfg. Co.; *Chem. Abs.*, 1920, **14**, 1418.
[5] French 483,781, Aug. 8, 1917, to Titanium Alloy Mfg. Co.; *Chem. Abs.*, 1918, **12**, 1002: British 106,584, Nov. 22, 1916, to Titanium Alloy Mfg. Co.; *Chem. Abs.*, 1917, **11**, 2574.
[6] L. E. Barton, U.S. 1,206,798, Dec. 5, 1916, to Titanium Alloy Mfg. Co.; *Chem. Abs.*, 1917, **11**, 279: British 106,428, Nov. 22, 1916, to Titanium Alloy Mfg. Co.; *Chem. Abs.*, 1917, **11**, 2574: French 483,782, Aug. 8, 1917, to Titanium Alloy Mfg. Co.; *Chem. Abs.*, 1918, **12**, 1002: Norwegian 28,673, Mar. 25, 1918, to Titanium Alloy Mfg. Co.; *Chem. Abs.*, 1919, **13**, 367.
[7] A. J. Rossi and L. E. Barton, U.S. 1,166,547, Jan. 14, 1916; *Chem. Abs.*, 1916, **10**, 668: Norwegian 29,193, Nov. 4, 1918, to Norske Aktieselskab för Elektrokemisk Industrie; *Chem. Abs.*, 1920, **14**, 1419.
[8] Belgian 447,709, Nov. 30, 1942, to Titangesellschaft m. b. H.; *Chem. Abs.*, 1945, **39**, 1026. Belgian 499,758, Apr. 1943, to Titangesellschaft m. b. H.; *Chem. Abs.*, 1947, **41**, 7064.
[9] H. F. Johnstone and W. E. Winsche, *Ind. Eng. Chem.*, 1944, **36**, 435; *Chem. Abs.*, 1944, **38**, 3187.
[10] W. L. Moss and A. W. Dye, Australian 111,758, Oct. 18, 1940; *Chem. Abs.*, 1942, **36**, 2833.
[11] C. Rees, Canadian 391,704, Oct. 1, 1940; *Chem. Abs.*, 1941, **35**, 1370.
[12] A. M. Brusilovskii, M. A. Shtem, and I. I. Pankova, U.S.S.R. 57,165, May 31, 1940; *Chem. Abs.*, 1944, **38**, 5094.
[13] F. von Bichowsky, U.S. 1,902,203, Mar. 21, 1933; *Chem. Abs.*, 1933, **27**, 3349.
[14] G. Jebsen, Norwegian 21,693, Feb. 25, 1910; *Chem. Abs.*, 1912, **6**, 2153.

[15] P. Farup, U.S. 1,852,510, Apr. 5, 1932, to Titan Co., A-G.; *Chem. Abs.*, 1932, **26**, 3080: Norwegian 52,774, July 10, 1933, to Titan Co., A-G.; *Chem. Abs.*, 1934, **28**, 2137.

[16] J. E. Booge, U.S. 2,183,365, Dec. 12, 1939, to E. I. du Pont de Nemours and Co.; *Chem. Abs.*, 1940, **34**, 2314.

[17] P. Farup, U.S. 1,341,307, May 25, 1920, to Titan Co., A-G.; *Chem. Abs.*, 1920, **14**, 2269: Canadian 201,703, July 6, 1920; *Chem. Abs.*, 1920, **14**, 2721: Norwegian 27,293, Oct. 9, 1916, to Norske Aktieselskab för Elektrokemisk Industrie; *Chem. Abs.*, 1917, **11**, 279.

[18] R. J. O'Dea, *U.S. Bur. Mines, Rept. Invest.* **3886**, 1946, 19; *Chem. Abs.*, 1946, **40**, 5677.

[19] A. J. Ravenstad and O. Moklebust, U.S. 2,339,808, Jan. 25, 1944, to Titan Co., Inc.; *Chem. Abs.*, 1944, **38**, 3949: British 547,898, Sept. 16, 1942, to Titan Co., Inc.; *Chem. Abs.*, 1943, **37**, 6417: Belgian 440,807, Apr. 30, 1941, to Titan Co., A-G.; *Chem. Abs.*, 1942, **36**, 5762: Norwegian 64,744, Dec. 6, 1943, to Aktieselskapet Titania; *Chem. Abs.*, 1946, **40**, 1440.

[20] H. A. Brassert, U.S. 2,313,044, Mar. 9, 1943, to Minerals and Metals Corp.; *Chem. Abs.*, 1943, **37**, 5009.

[21] R. Asak, Norwegian 62,713, July 22, 1940; *Chem. Abs.*, 1946, **40**, 541.

[22] H. E. Dunn, U.S. 2,109,917, Mar. 1, 1938, to Southern Mineral Products Corp.; *Chem. Abs.*, 1938, **32**, 3644.

[23] L. E. Barton, U.S. 1,348,843, Aug. 10, 1920; *Chem. Abs.*, 1920, **14**, 3133.

[24] E. Stahl, German 478,740, Jan. 17, 1928, to Metallgesellschaft, A-G.; *Chem. Abs.*, 1929, **23**, 4781.

[25] F. L. Clark, British, 339,608, Sept. 7, 1928, to Imperial Chemical Ind., Ltd.; *Chem. Abs.*, 1931, **25**, 2649.

[26] M. E. Zborovskii and E. V. Germogenova, *Trans. All-Union Sci. Research Inst. Econ. Mineral. (U.S.S.R.)*, 1935, **68**, 13; *Chem. Abs.*, 1936, **30**, 2711.

[27] B. H. Moore, *School of Mines, Western Australia, Bull.* **3**; *Chem. Abs.*, 1930, **24**, 5680.

[28] R. J. Traill, W. R. McClelland, and E. A. Thompson, *Can. Dept. Mines Resources, Bur. Mines*, 1926, No. **670**, 72: R. J. Traill, 1928, No. **688**, 95; *Chem. Abs.*, 1927, **21**, 724.

[29] A. Folliet and N. Sainderichin, French 698,516, May 28, 1930; *Chem. Abs.*, 1931, **25**, 3183.

[30] G. J. Bancroft, U.S. 1,745,732, Feb. 4, 1930; *Chem. Abs.*, 1930, **24**, 1610.

[30a] M. H. Tikkanen, Finnish 32,894, April 30, 1963; *Chem. Abs.*, 1964, **61**, 10356.

[31] H. H. Hoekje and R. A. Kearley, German 1,058,463, June 4, 1959, to Columbia-Southern Chemical Corp.; *Chem. Abs.*, 1961, **55**, 10281. British 846,468, Aug. 31, 1960, to Columbia-Southern Chemical Corp.; *Chem. Abs.*, 1961, **55**, 5892.

[32] A. J. Gaskin and A. E. Ringwood, U.S. 2,954,278, Sept. 27, 1960; *Chem. Abs.*, 1961, **55**, 4906.

[33] T. Ishino, T. Tanaka, Y. Tanaka, and Y. Takimoto, Japanese 8771, Oct. 13, 1956, to Sakai Chemical Industry Co.; *Chem. Abs.*, 1958, **52**, 16709.

[34] A. A. Fortiev and V. M. Andreev, *Izvest. Sibir, Otdel. Akad. Nauk S.S.S.R.*, 1959, No. 7, 71; *Chem. Abs.*, 1960, **54**, 7991.

[35] Y. Takimoto, M. Tamaka, and H. Hattori, Japanese 1176, March 6, 1959, to Sakai Chemical Industry Co., Ltd.; *Chem. Abs.*, 1959, **53**, 22790: *J. Chem. Soc. Japan*, 1955, **58**, 654; *Chem. Abs.*, 1956, **50**, 8978.

[36] R. J. Wigginton, British 881,808, Nov. 8, 1961, to Laporte Titanium Ltd.; *Chem. Abs.*, 1962, **57**, 418.

[37] A. A. Gregoire, French 1,125,534, Oct. 31, 1956, to Société des blancs de zinc de la Mediterranée; *Chem. Abs.*, 1959, **53**, 20724.

[38] F. Horiuchi, M. Fukuda, and S. Kitabayashi, Japanese 6869, Sept. 27, 1955; *Chem. Abs.*, 1957, **51**, 18514.

[39] T. Ishino, T. Tanaka, Y. Tanaka, and Y. Takimoto, Japanese 7316, Sept. 9, 1957, to Sakai Chemical Industry Co.; *Chem. Abs.*, 1958, **52**, 20947.

[40] E. Wainer, U.S. 2,941,863, June 21, 1960, to Horizons Inc.; *Chem. Abs.*, 1960, **54**, 21686.

[41] M. C. S. Chang, U.S. 2,912,320, Nov. 10, 1959, to Crucible Steel Company of America; *Chem. Abs.*, 1960, **54**, 3147.

[42] R. H. Walsh, H. W. Hocking, H. R. Brandt, P. L. Deitz, and P. R. Girardot, *Trans. A.I.M.E.*, 1960, **218**, 994; *Chem. Abs.*, 1961, **55**, 5268.

[43] M. Mukaiyama, Japanese 10,031, Nov. 30, 1957; *Chem. Abs.*, 1959, **53**, 9594.

[44] E. H. Goda, U.S. 2,496,993, Feb. 7, 1950, to Ferro Enamel Co.; *Chem. Abs.*, 1950, **44**, 4212.

[45] A. K. Sharova and A. A. Fotiev, *Isvest. Sibir. Otdel. Akad. Nauk S.S.S.R.*, 1959, **4**, 52; *Chem. Abs.*, 1959, **53**, 20717.

[46] S. Prasad and J. B. Tripathi, *Indian J. Appl. Chem.*, 1958, **21**, 162; *Chem. Abs.*, 1959, **53**, 16486.

[47] French 1,142,679, Sept. 20, 1957, to Centre national de la recherche scientifique; *Chem. Abs.*, 1960, **54**, 3888.

[48] W. W. Anderson and L. W. Rowe, U.S. 2,770,529, Nov. 13, 1956, to National Lead Co.; *Chem. Abs.*, 1957, **51**, 4670: German 1,025,846, March 13, 1958, to Titangesellschaft m. b. H.; *Chem. Abs.*, 1960, **54**, 21686.

[49] E. P. Belyakova and A. A. Dvernyakova, *Ukr. Khim. Zh.*, 1964, **30**, 880; *Chem. Abs.*, 1964, **61**, 14238.

[50] W. W. Anderson and L. W. Roe, U.S. 2,731,327, Jan. 17, 1956, to National Lead Co.; *Chem. Abs.*, 1956, **50**, 9759.

[51] M. N. Pedorva, *Titan i ego Splavy, Akad. Nauk S.S.S.R., Inst. Met.*, 1963, 36; *Chem. Abs.*, 1963, **59**, 14929.

[52] British 105,853, Aug. 14, 1917, to Titanium Alloy Mfg. Co.; *Chem. Abs.*, 1917, **11**, 2392.

[53] L. W. Ryan and J. R. Knoff, U.S. 1,929,521, Oct. 10, 1934, to Titanium Pigment Co., Inc.; *Chem. Abs.*, 1934, **28**, 87: British 408,215, Oct. 1, 1932, to Titanium Pigment Co., Inc.; *Brit. Chem. Abstracts*, 1934, **B**, 498.

[54] A. J. Rossi, U.S. 1,184,131, May 23, 1916; *Chem. Abs.*, 1916, **10**, 1779.

[55] R. H. Monk, U.S. 1,695,341, Dec. 18, 1929; *Chem. Abs.*, 1929, **23**, 939. R. H. Monk and J. Irwin, British 266,211, Aug. 6, 1926; *Chem. Abs.*, 1928, **22**, 505.

[56] L. E. Barton and C. J. Kinzie, U.S. 1,695,270, Dec. 18, 1928, to Titanium Pigment Co., Inc.; *Chem. Abs.*, 1929, **23**, 939.

[57] French 35,377, May 25, 1928, to I. G. Farbenind., A-G.; *Chem. Abs.*, 1930, **24**, 3865.

[58] H. W. Richter, U.S. 1,932,087, Oct. 24, 1934; *Chem. Abs.*, 1934, **28**, 588.

[59] German 624,446, Jan. 21, 1936; *Chem. Abs.*, 1936, **30**, 2542.

[60] S. J. Lubowsky, U.S. 1,640,952, Aug. 30, 1927; *Chem. Abs.*, 1927, **21**, 3429: U.S. 1,793,501, Feb. 24, 1931; *Chem. Abs.*, 1931, **25**, 2252: British 351,841, Sept. 25, 1930, to Metal and Thermit Corp.; *Brit. Chem. Abstracts*, 1931, **B**, 804: French 702,642, Sept. 25, 1930, to Metal and Thermit Corp.; *Chem. Abs.*, 1931, **25**, 4368.

[61] B. D. Saklatwalla and H. E. Dunn, U.S. 1,911,396, May 30, 1933, to Southern Mineral Products Corp.; *Brit. Chem. Abstracts*, 1934, **B**, 152.

[62] K. H. S. Lofquist, Swedish 88,995, Apr. 13, 1937; *Chem. Abs.*, 1937, **31**, 6424.

[63] L. Pellereau, M. Jacmart, and G. Le Bris, French 690,348, Feb. 21, 1930; *Chem. Abs.*, 1931, **25**, 1108.

[64] L. Pellereau, M. Jacmart, and G. Le Bris, French 663,068, Oct. 23, 1928; *Chem. Abs.*, 1930, **24**, 696.

[65] M. R. Raffin, French 655,085, Oct. 14, 1927, to Soc. minière "La Barytine"; *Chem. Abs.*, 1929, **23**, 4088.

[66] M. R. Raffin, French 677,370, Oct. 23, 1928, to Soc. minière "La Barytine"; *Chem. Abs.*, 1930, **24**, 3122.

[67] C. A. Klein and R. S. Brown, British 243,081, Aug. 25, 1924; *Brit. Chem. Abstracts*, 1926, **B**, 99.

[68] A. W. Gregory, British 256,836, Nov. 30, 1925; *Chem. Abs.*, 1927, **21**, 3108. I. Shikhutzkii, *J. Chem. Ind. (U.S.S.R.)*, 1930, **7**, 543; *Chem. Abs.*, 1932, **26**, 55.

[69] H. H. Buckman, U.S. 1,396,924, Nov. 15, 1921; *Chem. Abs.*, 1922, **16**, 653.

[70] A. J. Rossi and L. E. Barton, U.S. 1,106,406, Aug. 11, 1914; *Chem. Abs.*, 1914, **8**, 3285.

[71] M. Tatarskii, *J. Applied Chem. (U.S.S.R.)*, 1934, **7**, 1375; *Chem. Abs.*, 1935, **29**, 5998.

[72] H. Yamamoto, Japanese 99,016, Jan. 13, 1933; *Chem. Abs.*, 1934, **28**, 2475.

[73] F. E. Bachman, U.S. 1,489,417, Apr. 4, 1924; *J. Soc. Chem. Ind. (London)*, 1924, **B**, 566.

[74] H. Ginsberg, German 525,908, Dec. 15, 1927, to Vereinigte Aluminiumwerke, A-G.; *Chem. Abs.*, 1931, **25**, 4670.

[75] C. R. Whittemore, Canadian 258,871, Mar. 9, 1926; *Chem. Abs.*, 1926, **20**, 2395.

[76] British 111,668, Nov. 19, 1917, to Soc. P. Raffin et Fils; *Chem. Abs.*, 1918, **12**, 605.

[77] P. Farup, U.S. 1,831,852, Nov. 17, 1932, to Titanium Pigment Co., Inc.; *Chem. Abs.*, 1932, **26**, 683.

[78] F. A. Fitzgerald and P. M. Bennie, U.S. 921,686, May 18, 1909, to General Electric Co.; *Chem. Abs.*, 1909, **3**, 2040.

[79] F. von Bichowsky, U.S. 1,902,203, Mar. 21, 1933; *Chem. Abs.*, 1933, **27**, 3349.

[80] A. J. Rossi and L. E. Barton, U.S. 1,196,029, Aug. 29, 1916; *Chem. Abs.*, 1916, **10**, 2624.

[81] C. J. Kinzie and E. Wainer, U.S. 2,129,161, Sept. 16, 1938, to Titanium Alloy Mfg. Co.; *Chem. Abs.*, 1938, **32**, 8714.

[82] V. S. Suirokomskii, E. V. Snopova, and N. I. Rotkov, *Mineral. Syr'e*, 1931, **6**, 522; *Chem. Abs.*, 1932, **26**, 3462.

[83] J. Brode and G. Kab, U.S. 1,891,911, Dec. 27, 1933, to I. G. Farbenind., A-G.; *Chem. Abs.*, 1933, **27**, 1855: German 490,600, Mar. 12, 1926, to I. G. Farbenind., A-G.; *Chem. Abs.*, 1930, **24**, 2250: British 267,547, Mar. 11, 1926, to I. G. Farbenind., A-G.; *Chem. Abs.*, 1928, **22**, 1128.

[84] P. Farup, U.S. 1,325,561, Dec. 23, 1920; *Chem. Abs.*, 1920, **14**, 455.

[85] W. B. Llewellyn, British 409,847, Jan. 18, 1933, to P. Spence and Sons, Ltd.; *Brit. Chem. Abstracts*, 1934, **B**, 757.

[86] E. E. Creitz and H. G. Iverson, U.S. 2,750,255, June 12, 1956; *Chem. Abs.*, 1956, **50**, 14494.

[87] F. A. A. Gregoire, British 789,104, Jan. 15, 1958, to Société des blancs de zinc de la Mediterranée; *Chem. Abs.*, 1958, **52**, 16709.

[88] T. Ishino and Y. Takimoto, Japanese 1718, March 15, 1957, to Tadayoshi Tanaka; *Chem. Abs.*, 1958, **52**, 8919.

[89] A. W. Evans and N. G. Gray, British 696,501, Sept. 2, 1953, to British Titan Products Co., Ltd.; *Chem. Abs.*, 1954, **48**, 4857.

[90] British 711,833, July 14, 1954, to National Titanium Pigments Ltd.; *Chem. Abs.*, 1955, **49**, 2040.

[91] British 616,426, Oct. 15, 1962, to British Titan Products Co. Ltd.; *Chem. Abs.*, 1963, **58**, 8694.

[92] C. A. Tanner, Jr. and W. J. Cauwenberg, U.S. 2,842,428, July 8, 1958, to American Cyanamid Co.; *Chem. Abs.*, 1958, **52**, 17753.

[93] T. P. Campbell, U.S. 2,417,101, Mar. 11, 1947; *Chem. Abs.*, 1947, **41**, 3035.

[94] K. Iwase, U.S. 2,135,466, Nov. 1, 1938, to Kinzokuzairyo Kenkyusyo; *Chem. Abs.*, 1939, **33**, 957.

[95] N. P. Chizhevskii, A. P. Vlasov, and V. I. Shmeley, *Sbornik Trudov Moskov. Inst. Stal. Novoe. Tekhnol. Protzessakh Metallurg. Proizvodstva*, **1935**, 92; *Chem. Abs.*, 1936, **30**, 2889.

[96] L. D. Fetterolf, U.S. 2,919,982, Jan. 5, 1960, to Quebec Iron and Titanium Corp.; *Chem. Abs.*, 1960, **54**, 7509.

[97] N. V. Gueko, *Sb. Nauchn*, 1961, No. 5, 299; *Chem. Abs.*, 1963, **58**, 284.

[98] R. Garlach, German 164,440, March 1, 1964, to B. und P. G. m. b. H.; *Chem. Abs.*, 1964, **61**, 348.

[99] S. Mitsui and Y. Yamada, Japanese 7551, Oct. 19, 1955, to Osaka Titanium Co.; *Chem. Abs.*, 1958, **52**, 8923.

[100] Z. Orman and Z. Kozielska, *Prace Inst. Hutniczych*, 1960, **12**, No. 2, 79; *Chem. Abs.*, 1960, **54**, 23221.

[101] D. L. Armant and H. S. Sigurdson, U.S. 2,751,307, June 19, 1956, to National Lead Co.; *Chem. Abs.*, 1956, **50**, 13708: British 710,065, to Titan Co., Inc., *Chem. Abs.*, 1954, **48**, 11282.

[102] P. Tardieu, U.S. 2,851,350, Sept. 9, 1958, to Pechiney compagnie de produits et chimiques electrometallurgique; *Chem. Abs.*, 1958, **52**, 19869.

[103] F. L. Turbett, U.S. 2,453,050, Nov. 2, 1948, to Eagle-Picher Lead Co.; *Chem. Abs.*, 1949, **43**, 1306.

[104] M. Mukaiyama, Japanese 631, Jan. 31, 1957; *Chem. Abs.*, 1958, **52**, 5268.

[105] R. T. McMillan, R. A. Heindl, and J. E. Conley, *U.S. Bur. Mines, Rept. Invest.* No. 4912, 1952; *Chem. Abs.*, 1953, **47**, 457.

[106] M. J. Udy and M. C. Udy, German 1,091,761, Oct. 27, 1960, to Strategic Udy Metallurgical and Chemical Processes, Ltd.; *Chem. Abs.*, 1961, **55**, 23288.

[107] P. H. Royster, U.S. 2,471,242, May 24, 1949, to Pickands Mather Co.; *Chem. Abs.*, 1949, **43**, 6142.

[108] S. S. Cole, U.S. 2,631,941, March 17, 1953, to National Lead Co.; *Chem. Abs.*, 1953, **47**, 4827.

[109] D. L. Armant and H. S. Sigurdson, U.S. 2,815,272, Dec. 3, 1957, to National Lead Co.; *Chem. Abs.*, 1958, **52**, 5768: British 752,713, July 11, 1956, to National Lead Co.; *Chem. Abs.*, 1957, **51**, 1807.

[110] F. A. A. Gregoire, U.S. 2,814,557, Nov. 26, 1957, to Société des blancs de zinc de la Mediterranée; *Chem. Abs.*, 1958, **52**, 6736: British 802,336, Oct. 1, 1958, to Société des blancs de zinc de la Mediterranée; *Chem. Abs.*, 1959, **53**, 3625: German 1,023,890, Feb. 6, 1958, to Société des blancs de zinc de la Mediterranée; *Chem. Abs.*, 1960, **54**, 5030.

[111] R. T. McMillan, J. I. Dinnin, and J. E. Conley, *U.S. Bur. Mines, Rept. Invest.* No. 4638, 1950; *Chem. Abs.*, 1950, **44**, 1862.

[112] R. S. I. McLaren, U.S. 2,537,229, Jan. 9, 1951, to Shawinigan Water and Power Co.; *Chem. Abs.*, 1951, **45**, 3318.

[113] C. K. Stoddard, S. S. Cole, L. J. Eck, and C. W. Davis, *U.S. Bur. Mines, Rept. Invest.* No. 4750, 1950; *Chem. Abs.*, 1951, **45**, 1926.

[114] M. J. Udy and M. C. Udy, U.S. 2,878,114, March 17, 1959, to Strategic Udy Metallurgical and Chemical Processes, Ltd.; *Chem. Abs.*, 1959, **53**, 9018.

[115] V. M. Andreev, V. Y. Kramnik, V. A. Mishenev, Y. G. Nemeryuk, Y. A. Tsabolov, and V. P. Shvets, U.S.S.R. 119,678, May 10, 1959; *Chem. Abs.*, 1960, **54**, 3149.

[116] A. E. Bach, C. J. Chindgren, and R. G. Peterson, *U.S. Bur. Mines, Rept. Invest.* No. 4902, 1952; *Chem. Abs.*, 1952, **46**, 10063.

[117] M. Fukuda, H. Taga, and T. Asada, Japanese 9953, Nov. 19, 1958; *Chem. Abs.*, 1959, **53**, 5616.

[118] M. H. A. Tikkanen, British 805,438, Dec. 3, 1958; *Chem. Abs.*, 1959, **53**, 9018.

[119] Italian 436,715, June 14, 1948, to Titan Co., Inc.; *Chem. Abs.*, 1950, **44**, 5554.

[120] W. Freundlich, *Bull. soc. chim. France*, 1952, 490; *Chem. Abs.*, 1951, **46**, 11061.

[121] D. L. Armant and H. S. Sigurdson, U.S. 2,804,384, Aug. 27, 1957, to National Lead Co.; *Chem. Abs.*, 1958, **52**, 676.

[122] O. Mokelbust, G. Olavesen, and H. Bjoranesset, U.S. 2,990,250, June 27, 1961, to National Lead Co.; *Chem. Abs.*, 1961, **55**, 25182.

[123] W. Freundlich, *Bull. soc. chim. France*, 1952, No. 5, 655; *Chem. Abs.*, 1952, **46**, 11062.

[124] A. V. Rudneva, M. S. Model, and T. Y. Malysheva, *Titan i Ego Splavy*, Akad. Nauk S.S.S.R., Inst. Met. im A. A. Baikova, 1959, No. 2, 50; *Chem. Abs.*, 1960, **54**, 14028.

[125] I. P. Bardin, I. A. Karyazin, and V. A. Reznichenko, *I. A. N. SSSR., O. T. N., M. i T.*, 1959, No. 5, 35; *Chem. Abs.*, 1961, **55**, 14222.

[126] T. Fukasawa, *Kogyo Kagaku Zasshi*, 1959, **62**, 1505; *Chem. Abs.*, 1962, **57**, 16222.

[127] O. Mokelbust and H. Bjoranesset, German 1,041,021, Oct. 16, 1958, to Titan Co. A./S.; *Chem. Abs.*, 1960, **54**, 25637.

[128] H. Takei, *J. Electrochem. Soc. Japan*, 1955, **23**, 433; *Chem. Abs.*, 1956, **50**, 10629.

[129] E. I. Khazanov and V. S. Maltsev, *T. i E. S., A. N. S.S.S.R., I. M. i A. A. Baikova*, 1959, No. 2, 6; *Chem. Abs.*, 1960, **54**, 14028.

[130] O. Kippe, U.S. 2,842,434, July 8, 1958, to Paul O. Tobeler; *Chem. Abs.*, 1958, **52**, 15402.

[131] A. W. Knoerr, *Eng. Mining J.*, 1952, **153**, No. 3, 72; *Chem. Abs.*, 1952, **46**, 4445.

[132] P. J. Ensio, U.S. 2,853,375, Sept. 23, 1958, to Quebec Iron and Titanium Corp.; *Chem. Abs.*, 1959, **53**, 1079.

[133] G. E. Viens, R. A. Campbell, and R. R. Rogers, *Canadian Mining and Met., Bull.* No. 543, 1957; *Chem. Abs.*, 1957, **51**, 13686.

[134] T. Ichikawa, *Suiyokwai Shi*, 1958, **13**, 451; *Chem. Abs.*, 1959, **53**, 2982.

[135] Y. Takimoto and M. Tanaka, *Kogyo Kagaku Zasshi*, 1956, **59**, 339; *Chem. Abs.*, 1957, **51**, 9441.

[136] H. Ishizuka, Japanese 2215, May 20, 1953; *Chem. Abs.*, 1954, **48**, 5451.

[137] A. H. Roberson and L. H. Banning, *J. Metals*, 7, *A.I.M.E. Trans.*, 1955, **203**, 1335; *Chem. Abs.*, 1956, **50**, 744.

[138] V. Y. Kramnik, *Tsvetnye Metally*, 1960, **33**, No. 5, 40; *Chem. Abs.*, 1961, **55**, 4295.

[139] J. H. Brennan, British 868,717, May 25, 1961, to Union Carbide Corp.; *Chem. Abs.*, 1961, **55**, 23130.

[140] L. D. Fetterolf, U.S. 2,919,982, Jan. 5, 1960, to Quebec Iron and Titanium Co.; *Chem. Abs.*, 1960, **54**, 7509.

[141] S. Mitsui, Y. Yamada, and M. Setoguchi, Japanese 3553, June 11, 1957, to Osaka Titanium Co.; *Chem. Abs.*, 1958, **52**, 16098.

[142] R. S. Miller and G. G. Hatch, U.S. 2,808,324, Oct. 1, 1957, to Quebec Iron and Titanium Corp.; *Chem. Abs.*, 1958, **52**, 1034.

## CHAPTER 13

[1] T. P. Forbath, *Chem. Eng.*, 1958, **65**, No. 2, 98; *Chem. Abs.*, 1958, **52**, 6808.

[2] W. H. Coates, *Soc. Chem. Ind. (London), Chem. Eng. Group*, February, 1950; *Chem. Abs.*, 1950, **44**, 3266.

[3] *Titanox Handbook* (New York: National Lead Co., 1964), p. 10.

[4] *U.S. Bur. Mines, Inform. Circ.* 7791, 1957, 82.

[5] *Simplified Flow Sheet*, Adrian Joyce Works, The Glidden Co.

[6] G. Weintraub, U.S. 1,014,793, Jan. 16, 1912, to General Electric Co.; *Chem. Abs.*, 1912, **6**, 671.

[7] P. Farup, U.S. 1,087,575, Feb. 17, 1914; *Chem. Abs.*, 1914, **8**, 1354.

[8] P. Farup, U.S. 1,156,220, Oct. 12, 1915; *Chem. Abs.*, 1915, **9**, 3215: German 276,025, July 5, 1913; *Chem. Abs.*, 1915, **9**, 362: Canadian 164,510 Aug. 24, 1915; *Chem. Abs.*, 1915, **9**, 3121.

[9] A. J. Rossi and L. E. Barton, U.S. 1,196,030 and 1,196,031, Aug. 29, 1916; *Chem. Abs.*, 1916, **10**, 2624.

[10] J. Blumenfeld, U.S. 1,504,669, Aug. 12, 1924; *Chem. Abs.*, 1924, **18**, 3257.

[11] J. Blumenfeld, U.S. 1,504,671, Aug. 12, 1924; *Chem. Abs.*, 1924, **18**, 3257.

[12] C. Weizmann and J. Blumenfeld, British 203,352, Mar. 1, 1922; *Chem. Abs.*, 1924, **18**, 573.

[13] C. Weizmann and J. Blumenfeld, British 209,441, July 12, 1924; *Chem. Abs.*, 1924, **18**, 1884.

[14] C. Weizmann and J. Blumenfeld, British 210,033, July 12, 1918; *Chem. Abs.*, 1924, **18**, 1884. J. Blumenfeld, U.S. 1,504,672, Aug. 12, 1924; *Chem. Abs.*, 1924, **18**, 3257.

[15] G. Jebsen, U.S. 1,333,819, Mar. 16, 1920; *Chem. Abs.*, 1920, **14**, 1416.

[16] C. L. Schmidt, U.S. 2,130,565, Sept. 30, 1938, to National Lead Co.; *Chem. Abs.*, 1938, **32**, 9414: British 470,154, Aug. 10, 1937, to British Titan Products Co., Ltd.; *Chem. Abs.*, 1938, **32**, 1060: French 814,417, June 23, 1937, to Titan Co., Inc.; *Chem. Abs.*, 1938, **32**, 1057: German 671,279, Feb. 4, 1939, to Titan Co., Inc.; *Chem. Abs.*, 1939, **33**, 3542.

[17] W. B. Llewellyn and H. Spence, British 200,848, Jan. 17, 1922, to P. Spence and Sons, Ltd.; *Chem. Abs.*, 1924, **18**, 310: F. Doerinckel, German 497,931, Dec. 18, 1926, to I. G. Farbenind., A-G.; *Chem. Abs.*, 1930, **24**, 3981.

[18] British 102,059, Oct. 30, 1916, to Norske Aktieselskab för Elektrokemisk Industrie; *Chem. Abs.*, 1917, **11**, 527: French 483,311, June 26, 1917, to Det. Norske Aktieselskab för Elektrokemisk Industrie; *Chem. Abs.*, 1918, **12**, 980: Holland 3,076, May 1, 1919, to Norske Aktieselskab för Elektrokemisk Industrie; *Chem. Abs.*, 1920, **14**, 1414.

[19] J. d'Ans and F. Sommer, U.S. 1,655,940, Jan. 10, 1928; *Chem. Abs.*, 1928, **22**, 1019. A. T. Chornii, *Ukraïn Khem. Zhur.*, 1937, **12**, 137; *Chem. Abs.*, 1937, **31**, 5956.

[20] R. H. Monk and A. S. Ross, U.S. 2,092,132, Sept. 7, 1937, to American Zinc, Lead and Smelting Co.; *Chem. Abs.*, 1937, **31**, 7612: British 466,384, May 27, 1937, to American Zinc, Lead and Smelting Co.; *Chem. Abs.*, 1937, **31**, 8132.

[21] W. Wrigley and H. Spence, U.S. 1,338,473, Apr. 27, 1920, to P. Spence and Sons; *Chem. Abs.*, 1920, **14**, 1875: British 133,336, Oct. 16, 1917, to P. Spence and Sons; *Chem. Abs.*, 1920, **14**, 805.

[22] F. L. Kingsbury and W. Grave, U.S. 2,154,130, Apr. 11, 1939, to National Lead Co.; *Chem. Abs.*, 1939, **33**, 5348.

[23] E. N. Kramer, U.S. 2,180,961, Nov. 21, 1939, to E. I. du Pont de Nemours and Co.; *Chem. Abs.*, 1940, **34**, 1826.

[24] F. Raspe and P. Weise, German 504,843, Sept. 2, 1928, to I. G. Farbenind., A-G.; *Chem. Abs.*, 1931, **25**, 480.

[25] W. G. Moran, U.S. 2,278,709, Apr. 7, 1942, to National Lead Co.; *Chem. Abs.*, 1942, **36**, 4794.

[26] R. Dahlstrom, U.S. 2,344,288, Mar. 14, 1944, to National Lead Co.; *Chem. Abs.*, 1944, **38**, 3494.

[27] L. G. Bousquet and M. J. Brooks, U.S. 2,327,166, Aug. 16, 1943, to General Chemical Co.; *Chem. Abs.*, 1944, **38**, 622.

[28] E. G. Shtandel, *J. Gen. Chem. (U.S.S.R.)*, 1935, **5**, 1629; *Chem. Abs.*, 1936, **30**, 2870.

[29] E. N. Bogoyavlenskii and B. E. Boguslavskaya, *Lakokrasochnuyu Ind. Za*, 1934, No. 3, 20; *Chem. Abs.*, 1935, **29**, 6712.

[30] E. V. Germogenova and S. I. Shur, *Trans. All-Union Sci. Research Inst. Econ. Mineral. (U.S.S.R.)*, 1935, **68**, 63; *Chem. Abs.*, 1936, **30**, 2711.

[31] T. Matsubara, *The Osaka Industrial Lab., Bull.* **IV**, No. 13, 1923: *Chem. Abs.*, 1925, **19**, 1475.

[32] P. A. Mackay, U.S. 1,613,234, Jan. 4, 1927; *Chem. Abs.*, 1927, **21**, 630; Canadian 269,983, Apr. 19, 1927; *Chem. Abs.*, 1927, **21**, 3341: British 256,734, May 28, 1925; *Chem. Abs.*, 1927, **21**, 3108.

[33] K. Y. Grachev, *J. Applied Chem. (U.S.S.R.)*, 1940, **13**, 1126; *Chem. Abs.*, 1941, **35**, 2286.

[34] W. J. O'Brien. *Chem. Eng. Progress*, 1948, **44**, 809; *Chem. Abs.*, 1949, **43**, 870.

[35] S. Wilska, *Suomen Kemistilehti*, 1954, **27B**, No. 4, 25; *Chem. Abs.*, 1955, **49**, 16369.

[36] F. Majdik and B. Balla, *Magyar Kem. Lapja*, 1959, **14**, 197; *Chem. Abs.*, 1960, **54**, 3878.

[37] T. P. Tan and Y. C. Chang, *Union. Ind. Research Inst. Rept.*, 1957, No. 26; *Chem. Abs.*, 1960, **54**, 20112.

[38] R. J. Wigginton, British 873,611, Aug. 30, 1956, to Laporte Titanium Ltd., *Chem. Abs.*, 1962, **56**, 170.

[39] T. P. Tan and Y. H. Su, *Union. Ind. Research Inst. Rept.*, 1957, No. 27: *Chem. Abs.*, 1960, **54**, 20112.

[40] A. G. Pusko, V. D. Ponomarev, and G. I. Titova, *Zh. Prikl. Khim.*, 1963, **36**, 1665; *Chem. Abs.*, 1964, **60**, 2563.

[41] V. Jara, Czechoslovakia 106,153, Jan. 15, 1963; *Chem. Abs.*, 1964, **60**, 3774.

[42] S. Sato, Japanese 4708, April 16, 1964; *Chem. Abs.*, 1964, **61**, 6677.

[43] British 874,339, Aug. 2, 1961, to National Lead Co.; *Chem. Abs.*, 1962, **57**, 14777: British 874,336, Aug. 2, 1961, to National Lead Co.; *Chem. Abs.*, 1962, **57**, 14777.

[44] L. B. Knudsen, U.S. 3,071,435, Jan. 1, 1963, to National Lead Co.; *Chem. Abs.*, 1963, **58**, 7639.

[45] British 789,497, Jan. 22, 1958, to Farbenfabriken Bayer A.-G.; *Chem. Abs.*, 1958, **52**, 10855.

[46] H. Espenschied, U.S. 2,617,724, Nov. 11, 1952, to National Lead Co.; *Chem. Abs.*, 1953, **47**, 1027.

[47] I. L. Bagbanly, K. L. H. Zeinalova, and T. R. Mirzoeva, *Tr. Inst. Khim., Akad. Nauk Azerb. S. S. R.*, 1960, **18**, 55; *Chem. Abs.*, 1962, **57**, 1851.

[48] Y. Kozaki, Japanese 173,790, Oct. 4, 1946; *Chem. Abs.*, 1952, **46**, 2472.

[49] A. G. Oppegaard, U.S. 2,774,650, Dec. 18, 1956, to National Lead Co.; *Chem. Abs.*, 1957, **51**, 4922.

[50] W. H. Daiger, U.S. 2,875,107, Feb. 24, 1959, to E. I. du Pont de Nemours and Co.; *Chem. Abs.*, 1959, **53**, 9989.

[51] T. S. Griffin and W. Rodgers, U.S. 2,767,053, Oct. 16, 1956, to National Lead Co.; *Chem. Abs.*, 1957, **51**, 1807.

[52] T. S. Griffin and W. Rodgers, U.S. 2,822,241, Feb. 4, 1958, to National Lead Co.; *Chem. Abs.*, 1958, **52**, 7101.

[53] A. Stanley, U.S. 2,724,637, Nov. 22, 1955, to National Lead Co.; *Chem. Abs.*, 1956, **50**, 3189: German 930,502, July 18, 1955, to National Lead Co.; *Chem. Abs.*, 1958, **52**, 1034.

[54] S. J. Lubowsky, British 351,841, Sept. 28, 1929, to Metal and Thermit Corp.; *Chem. Abs.*, 1932, **26**, 3340: French 702,642, Sept. 25, 1930; *Chem. Abs.*, 1931, **25**, 4368.

[55] L. W. Ryan and J. R. Knoff, U.S. 1,929,521, Oct. 10, 1934, to Titanium Pigment Co., Inc.; *Chem. Abs.*, 1934, **28**, 87: British 408,215, Oct. 1, 1932, to Titanium Pigment Co., Inc.; *Brit. Chem. Abstracts*, 1934, **B**, 498.

[56] G. N. Giraud, French 1,030,734, June 16, 1952; *Chem. Abs.*, 1958, **52**, 14192.

[57] H. V. Alessandroni, U.S. 2,167,626, Aug. 1, 1939, to National Lead Co.; *Chem. Abs.*, 1939, **33**, 9021.

[58] I. V. Riskin and R. G. Zotova, *U.S.S.R.* 92,951, July 22, 1964; *Chem. Abs.*, 1964, **61**, 14226.

[59] Y. G. Goroshchenko, V. I. Belokosov, Y. A. Fomin, and M. I. Andreeva, *S. T. K. T. M. S. K. P.*, 1959, **1**, 5; *Chem. Abs.*, 1961, **55**, 20347.

[60] Y. G. Goroshchenko, D. L. Motov, G. V. Trofimov, and V. I. Belokoskov, *I. K. i K. F., Akad. Nauk S.S.S.R.*, 1959, No. 4, 135; *Chem. Abs.*, 1960, **54**, 12513.

[61] M. A. Shtern, *J. Applied Chem. (U.S.S.R.)*, 1938, **11**, 1155; *Chem. Abs.*, 1939, **33**, 4386.

[62] V. E. Tishchenko, Y. M. Pesin, A. L. Kostroh, and S. S. Freidlin, *J. Applied Chem. (U.S.S.R.)*, 1932, **5**, 685; *Chem. Abs.*, 1933, **27**, 167.

[63] H. V. Alessandroni, U.S. 2,167,627, Aug. 1, 1939, to National Lead Co.; *Chem. Abs.*, 1939, **33**, 9021.

[64] F. J. Anderson and D. R. Williams, Canadian 430,168, Sept. 18, 1945, to Monolith Portland Cement Co.; *Chem. Abs.*, 1946, **40**, 1294.

[65] K. Kikukawa, M. Masutoski, and F. Chiba, Japanese 111,269, June 21, 1935; *Chem. Abs.*, 1936, **30**, 2333.

[66] S. C. Chakravarty, *Bull. Indian Ind. Research Bur., Govt. India,* No. 15, 1939; *Chem. Abs.,* 1941, **35,** 3042.

[67] D. Swarup and A. S. Sharma, *Trans. Indian Ceram. Soc.,* 1945, **4,** 75; *Chem. Abs.,* 1946, **40,** 4857.

[68] S. S. Bhatmagar, S. Parthasarathy, G. C. Singh, and A. L. Rao, *J. Sci. Ind. Research (India),* 1945, **4,** 378; *Chem. Abs.,* 1946, **40,** 3238.

[69] R. D. Desai and F. R. Peermahomed, *J. Indian Chem. Soc., Ind. News Ed.,* 1945, **8,** 9; *Chem. Abs.,* 1946, **40,** 3575.

[70] M. Orliac and L. Capdecomme, French 1,152,128, Feb. 12, 1958, to Centre National de la recherche scientifique; *Chem. Abs.,* 1960, **54,** 11512.

[71] H. Imai and S. Kadota, Japanese 1825, April 7, 1954, to Sumitomo Chem. Industries Co.; *Chem. Abs.,* 1955, **49,** 1292.

[72] S. M. Mehta and S. R. Patel, *J. Am. Chem. Soc.,* 1951, **73,** 226; *Chem. Abs.,* 1951, **45,** 5375.

[73] S. C. Niyogy, *J. Proced. Inst. Chemists (India),* 1956, **28,** 226; *Chem. Abs.,* 1957, **51,** 7033.

[74] V. Damodaran and J. Gupta, *J. Sci. Ind. Research (India),* 1955, **14B,** 292; *Chem. Abs.,* 1956, **50,** 3775.

[75] C. G. Marcarovici, L. Czeledi, H. Roth, and I. Soos, *Acad. rep. populare Romine,* 1955, **6,** No. 3, 87; *Chem. Abs.,* 1958, **52,** 1564.

[76] T. P. Forbath, *Chem. Eng.,* 1958, **65,** No. 2, 98; *Chem. Abs.,* 1958, **52,** 6808.

[77] W. M. Peirce, R. K. Waring, and L. D. Fetterolf, U.S. 2,476,453, July 19, 1949, to Quebec Iron and Titanium Co.; *Chem. Abs.,* 1949, **43,** 9482.

[78] F. Y. Irkov and V. A. Reznichenko, *Titan i ego Splavy, Akad. Nauk S.S.S.R., Inst. Met.,* 1961, No. 5, 279; *Chem. Abs.,* 1962, **57,** 12169.

[79] S. P. Todd, M. L. Meyers, and W. J. Cauwenberg, U.S. 2,531,926, Nov. 28, 1950, to American Cyanamid Co.; *Chem. Abs.,* 1951, **45,** 2232.

[80] M. L. Borodina, S. B. Shaikevich, N. K. Piktorinskaya, and N. A. Gubareva, *Lakokrasochny Materialy i ikh Primenenie,* 1961, No. 1, 33; *Chem. Abs.,* 1961, **55,** 25180.

[81] S. Sato, Japanese 13,315, Aug. 14, 1961; *Chem. Abs.,* 1962, **57,** 16136.

[82] I. Slama, *Chem. prumysl,* 1960, **10,** 337; *Chem. Abs.,* 1961, **55,** 284.

[83] M. L. Borodina, T. A. Velikoslavinskaya, and B. L. Davydovskaya, *Titan i Ego Splavy, Akad. Nauk, S.S.S.R., Inst. Met. A.A. Baikova,* 1959, No. 2, 73; *Chem. Abs.,* 1960, **54,** 15866.

[84] T. Sato, H. Kameko, and H. Suedo, *Bull. Research Inst. Mineral Dressing and Metallurgy, Tohoku Univ.,* 1954, **10,** 29; *Chem. Abs.,* 1955, **49,** 15188.

[85] R. M. McKinney and G. S. Reeder, Jr., U.S. 2,953,435, Sept. 20, 1960, to E. I. du Pont de Nemours and Co.; *Chem. Abs.,* 1961, **55,** 3940.

[86] O. Mokelbust, G. Olavesen, and H. Bjoranesset, U.S. 2,990,250, June 27, 1961, to National Lead Co.; *Chem. Abs.,* 1961, **55,** 25182: British 809,753, March 4, 1959, to Titan Co., A./S.; *Chem. Abs.,* 1959, **53,** 12611.

[87] J. H. Weikel, U.S. 2,589,909, March 18, 1952, to New Jersey Zinc Co.; *Chem. Abs.,* 1952, **46,** 5798: British 671,728, May 7, 1952, to New Jersey Zinc Co.; *Chem. Abs.,* 1952, **46,** 8339.

[88] B. W. Allan, F. O. Rummery, and J. M. Gilbert, U.S. 2,794,702, June 4, 1957, to Glidden Co.; *Chem. Abs.,* 1957, **51,** 14221: German 1,051,261, Feb. 26, 1959, to Glidden Co.; *Chem. Abs.,* 1960, **54,** 21679.

[89] E. L. Schneider, U.S. 2,589,910, March 18, 1952, to New Jersey Zinc Co.; *Chem. Abs.,* 1952, **46,** 5798: British 671,729, May 7, 1952, to New Jersey Zinc Co.; *Chem. Abs.,* 1952, **46,** 8339.

[90] R. J. Magri, Jr., and G. C. Marcot, U.S. 2,715,501, Aug. 16, 1955, to American Cyanamid Co.; *Chem. Abs.,* 1955, **49,** 16376.

[91] B. W. Allan, F. O. Rummery, and F. L. Appel, German 1,052,378, March 12, 1959, to Glidden Co.; *Chem. Abs.,* 1961, **55,** 2039.

[92] R. M. McKinney, U.S. 2,631,924, March 17, 1953, to E. I. du Pont de Nemours and Co.; *Chem. Abs.,* 1953, **47,** 5649.

[93] W. F. Washburn, U.S. 1,889,027, Nov. 29, 1933, to Titanium Pigment Co., Inc.; *Chem. Abs.*, 1933, **27**, 1457: British 288,569, Apr. 12, 1927, to Titanium Pigment Co., Inc.; *Chem. Abs.*, 1929, **23**, 590: French 652,357, Apr. 7, 1928; *Chem. Abs.*, 1929, **23**, 3432: Canadian 299,992, May 6, 1930, to Titanium Pigment Co., Inc.; *Chem. Abs.*, 1930, **24**, 3091.

[94] W. J. Cauwenberg, U.S. 2,066,093, Dec. 29, 1936, to United Color and Pigment Co.; *Chem. Abs.*, 1937, **31**, 992: British 457,719, Jan. 27, 1936, to United Color and Pigment Co.; *Chem. Abs.*, 1937, **31**, 2991: German 710,495, Aug. 7, 1941, to Interchemical Corp.; *Chem. Abs.*, 1943, **37**, 3889.

[95] A. T. Coffelt, U.S. 2,138,090, Nov. 29, 1938, to E. I. du Pont de Nemours and Co.; *Chem. Abs.*, 1939, **33**, 1649.

[96] J. Blumenfeld, French 1,129,654, Jan. 24, 1957, to Fabriques de produits chimiques de Thann et de Mulhouse; *Chem. Abs.*, 1959, **53**, 19406.

[97] H. H. Buckman, British 206,284, Aug. 29, 1922; *Chem. Abs.*, 1924, **18**, 1209.

[98] British 267,547, Mar. 11, 1927, to I. G. Farbenind., A-G.; *Brit. Chem. Abstracts*, 1928, **B**, 198.

[99] D. H. Dawson, I. J. Krchma, and R. M. McKinney, Canadian 377,396, Nov. 1, 1938, to Canadian Industries, Ltd.; *Chem. Abs.*, 1939, **33**, 2292.

[100] Japanese 156,845, June 2, 1943, to Japanese High Frequency Wave Heavy Industries Co.; *Chem. Abs.*, 1949, **43**, 6377.

[101] J. E. Booge, I. J. Krchma, and R. M. McKinney, U.S. 2,098,025, Nov. 2, 1937, to E. I. du Pont de Nemours and Co.; *Chem. Abs.*, 1938, **32**, 100: Canadian 390,014, July 16, 1940, to E. I. du Pont de Nemours and Co.; *Chem. Abs.*, 1940, **34**, 6419.

[102] J. E. Booge, I. J. Krchma, and R. M. McKinney, U.S. 2,098,026, Nov. 2, 1937, to E. I. du Pont de Nemours and Co.; *Chem. Abs.*, 1938, **32**, 100: Canadian 390,015, July 16, 1940, to E. I. du Pont de Nemours and Co.; *Chem. Abs.*, 1940, **34**, 6419.

[103] F. H. McBerty, U.S. 2,098,054, Nov. 2, 1937, to E. I. du Pont de Nemours and Co.; *Chem. Abs.*, 1938, **32**, 101: Canadian 390,016, July 16, 1940, to E. I. du Pont de Nemours and Co.; *Chem. Abs.*, 1940, **34**, 6419.

[104] F. H. McBerty, U.S. 2,098,055, Nov. 2, 1937, to E. I. du Pont de Nemours and Co.; *Chem. Abs.*, 1938, **32**, 101: Canadian 390,017, July 16, 1940, to E. I. du Pont de Nemours and Co.; *Chem. Abs.*, 1940, **34**, 6419.

[105] B. D. Saklatwalla, H. E. Dunn, and A. E. Marshall, U.S. 1,977,208, Oct. 16, 1934, to Southern Mineral Products Corp.; *Brit. Chem. Abstracts*, 1936, **B**, 239.

[106] W. G. Moran and J. E. Nelson, U.S. 2,329,641, Sept. 14, 1943, to National Lead Co.; *Chem. Abs.*, 1944, **38**, 1082.

[107] E. W. Andrews, U.S. 2,557,528, June 19, 1951, to New Jersey Zinc Co.; *Chem. Abs.*, 1951, **45**, 7313: British 677,172, Aug. 13, 1952, to New Jersey Zinc Co.; *Chem. Abs.*, 1952, **46**, 11085.

[108] W. Moschel, H. Zirngible, and A. Meyer, German 952,711, Nov. 22, 1956, to Farbenfabriken Bayer A.-G.; *Chem. Abs.*, 1959, **53**, 4097.

[109] T. S. Griffin, U.S. 2,982,613, May 2, 1961, to National Lead Co.; *Chem. Abs.*, 1961, **55**, 16378.

[110] H. Zirngibl and G. Heinze, U.S. 2,849,289, Aug. 26, 1958, to Farbenfabriken A.-G.; *Chem. Abs.*, 1958, **52**, 18156.

[111] A. G. Pusko, L. G. Getskin, E. Y. Benyash, and V. G. Feldman, U.S.S.R. 142,027, Nov. 15, 1961; *Chem. Abs.*, 1962, **56**, 13803.

[112] P. Farup, U.S. 1,919,425, July 25, 1933, to Titan Co., A-S.; *Chem. Abs.*, 1933, **27**, 4761: German 534,968, May 25, 1927, to Titan Co., A-G.; *Chem. Abs.*, 1932, **26**, 1399.

[113] W. F. Washburn, U.S. 1,889,027, Nov. 29, 1933, to Titanium Pigment Co., Inc.; *Chem. Abs.*, 1933, **27**, 1457.

[114] I. J. Krchma, U.S. 2,049,504, Aug. 4, 1936, to E. I. du Pont de Nemours and Co.; *Chem. Abs.*, 1936, **30**, 6316.

[115] H. Spence, U.S. 758,710, May 3, 1904: L. E. Barton, U.S. 1,235,638, Aug. 7, 1917; *Chem. Abs.*, 1917, **11**, 2642.

[116] H. Olsen and T. R. Forland, U.S. 1,333,849, Mar. 16, 1920; *Chem. Abs.*, 1920, **14**, 1450: Canadian 201,718, July 6, 1920; *Chem. Abs.*, 1920, **14**, 2722.

[117] W. H. Evans, *Mem. Proc. Manchester Lit. & Phil. Soc.*, 1905, **49**, 1; *J. Chem. Soc.*, 1905, **A**, 169. L. E. Barton, U.S. 1,233,357 and 1,233,358, Apr. 24, 1917, to Titanium Alloy Mfg. Co.; *Chem. Abs.*, 1917, **11**, 1912.

[118] G. Weintraub, U.S. 1,014,793, Jan. 16, 1912, to General Electric Co.; *Chem. Abs.*, 1912, **6**, 671.

[119] British 173,774, Jan. 5, 1922, to Chemische Werke vorm. A-G.; *Chem. Abs.*, 1922, **16**, 2392.

[120] P. Farup, German 492,685, May 21, 1927, to Titan Co., A-S.; *Chem. Abs.*, 1930, **24**, 2418: British 279,786, July 2, 1927, to Titan Co., A-S.; *Brit. Chem. Abstracts*, 1928, **B**, 261.

[121] I. J. Krchma, U.S. 2,047,208, July 14, 1936, to E. I. du Pont de Nemours and Co.; *Chem. Abs.*, 1936, **30**, 6145.

[122] C. Rau and F. E. Swartz, Jr., U.S. 2,416,216, Feb. 18, 1947, to National Lead Co.; *Chem. Abs.*, 1947, **41**, 2543.

[123] L. W. Ryan and W. J. Cauwenberg, U.S. 2,309,988, Feb. 2, 1943, to Interchemical Corp.; *Chem. Abs.*, 1943, **37**, 4211.

[124] I. J. Krchma, U.S. 2,049,504, Aug. 4, 1936, to E. I. du Pont de Nemours and Co.; *Chem. Abs.*, 1936, **30**, 6316.

[125] British 470,266, Aug. 11, 1937, to British Titan Products Co., Ltd.; *Chem. Abs.*, 1939, **32**, 1059.

[126] M. J. Brooks, U.S. 2,231,181, Feb. 11, 1941, to General Chemical Co.; *Chem. Abs.*, 1941, **35**, 3468.

[127] British 108,850, May 18, 1917, to Norske Aktieselskab för Elektrokemisk Industrie; *Chem. Abs.*, 1918, **12**, 204.

[128] V. M. Goldschmidt, U.S. 1,276,644, Aug. 20, 1918; *Chem. Abs.*, 1918, **12**, 2114: Canadian 188,808, Feb. 25, 1919; *Chem. Abs.*, 1919, **13**, 774.

[129] C. Weizmann and J. Blumenfeld, British 228,814, Aug. 4, 1923; *Chem. Abs.*, 1925, **19**, 3004.

[130] A. A. Milinskii, *J. Chem. Ind.* (*U.S.S.R.*), 1936, **13**, 1098; *Chem. Abs.*, 1937, **31**, 221.

[131] British 473,054, Oct. 5, 1937, to British Titan Products Co., Ltd.; *Chem. Abs.*, 1938, **32**, 1811.

[132] J. B. Leibert, British 249,647, Feb. 19, 1925, to Spencer, Chapman and Messel, Ltd.; *Chem. Abs.*, 1927, **21**, 995.

[133] E. N. Bogoyavlenskii and A. Y. Drinberg, U.S.S.R. 41,612, Feb. 28, 1935; *Chem. Abs.*, 1936, **30**, 8538.

[134] I. J. Krchma, U.S. 2,280,590, Apr. 21, 1942, to E. I. du Pont de Nemours and Co.; *Chem. Abs.*, 1942, **36**, 5457.

[135] C. Weizmann and J. Blumenfeld, British 227,143, Aug. 3, 1923; *Chem. Abs.*, 1925, **19**, 2730.

[136] A. Zhukova, *J. Chem. Ind.* (*U.S.S.R.*), 1932, No. **8**, 17; *Chem. Abs.*, 1933, **27**, 432.

[137] Van de Lande, French 818,905, Oct. 6, 1937, to N. V. Industrielle Maatschappij, voorheen Noury; *Chem. Abs.*, 1938, **32**, 2695.

[138] W. J. Cauwenberg, Canadian 381,422, May 16, 1939, to Interchemical Corp.; *Chem. Abs.*, 1939, **33**, 5612: British 487,100, June 15, 1938, to United Color and Pigment Co., Inc.; *Chem. Abs.*, 1938, **32**, 9529: French 822,994, Jan. 11, 1938, to United Color and Pigment Co., Inc.; *Chem. Abs.*, 1938, **32**, 4291.

[139] W. J. Cauwenberg, Canadian 381,423, May 16, 1939, to Interchemical Corp.; *Chem. Abs.*, 1939, **33**, 5612: British 479,082, Jan. 31, 1938, to United Color and Pigment Co., Inc.; *Chem. Abs.*, 1938, **32**, 5165: French 822,959 and 822,994, Jan. 11, 1938, to United Color and Pigment Co., Inc.; *Chem. Abs.*, 1938, **32**, 4291.

[140] F. E. Bachman, U.S. 1,354,940, Oct. 5, 1921; *Chem. Abs.*, 1921, **15**, 151.

[141] R. M. McAdam, U.S. 2,413,640, Dec. 31, 1946, to E. I. du Pont de Nemours and Co.; *Chem. Abs.*, 1947, **41**, 1467.

[142] R. M. McAdam, U.S. 2,413,641, Dec. 31, 1946, to E. I. du Pont de Nemours and Co.; *Chem. Abs.*, 1947, **41**, 1467.

[143] L. G. Bousquet and M. J. Brooks, U.S. 2,292,506, Aug. 11, 1942, to General Chemical Co.; *Chem. Abs.*, 1943, **37**, 734.

[144] L. G. Bousquet and D. W. Young, U.S. 2,313,615, Mar. 9, 1943, to General Chemical Co.; *Chem. Abs.*, 1943, **37**, 5204.

[145] L. G. Bousquet, D. W. Young, and A. W. Low, U.S. 2,349,936, May 30, 1944, to General Chemical Co.; *Chem. Abs.*, 1945, **39**, 1517.

[146] L. G. Bousquet and M. J. Brooks, U.S. 2,298,032, Oct. 6, 1942, to General Chemical Co.; *Chem. Abs.*, 1943, **37**, 1568.

[147] K. Y. Grachev, *J. Applied Chem. (U.S.S.R.)*, 1942, **15**, 336; *Chem. Abs.*, 1943, **37**, 4017.

[148] C. R. Wicker, U.S. 2,326,592, Aug. 10, 1943, to E. I. du Pont de Nemours and Co.; *Chem. Abs.*, 1944, **38**, 623.

[149] H. Schlecht and L. Schlecht, German 709,091, July 3, 1941, to I. G. Farbenind., A-G.; *Chem. Abs.*, 1943, **37**, 3235.

[150] J. J. Libera, E. J. Puetz, and A. V. Schopp, Jr., U.S. 2,839,364, June 17, 1958, to National Lead Co.; *Chem. Abs.*, 1958, **52**, 17063.

[151] P. Weise and I. Raspe, German 513,291, July 23, 1929, to I. G. Farbenind., A-G.; *Chem. Abs.*, 1931, **25**, 1344: British 309,834, Apr. 13, 1928, to I. G. Farbenind., A-G.; *Chem. Abs.*, 1930, **24**, 696: French 673,024, Apr. 12, 1929, to I. G. Farbenind., A-G.; *Chem. Abs.*, 1930, **24**, 2253.

[152] W. G. Luttger, *Rayon Textile Monthly*, 1941, **22**, 410; *Chem. Abs.*, 1941, **35**, 6156.

[153] E. Ermilov, *Khim. Prom.*, 1944, No. **6**, 20; *Chem. Abs.*, 1945, **39**, 653.

[154] F. H. McBerty, C. H. Evans, and R. O. Waugh, U.S. 2,265,386, Dec. 9, 1941, to E. I. du Pont de Nemours and Co.; *Chem. Abs.*, 1942, **36**, 1816.

[155] L. G. Bousquet, H. W. Richter, and B. W. Allan, U.S. 2,089,180, Aug. 10, 1937, to American Zirconium Corp.; *Chem. Abs.*, 1937, **31**, 6835: British 467,046, June 10, 1937, to Titan Co., Inc.; *Chem. Abs.*, 1937, **31**, 8846.

[156] G. L. Lewis, U.S. 2,414,049, Jan. 7, 1947, to E. I. du Pont de Nemours and Co.; *Chem. Abs.*, 1947, **41**, 2915.

[157] Norwegian 27,619, Feb. 5, 1917, to Norske Aktieselskab för Elektrokemisk Industrie; *Chem. Abs.*, 1917, **11**, 1889. H. Olsen, Canadian 201,711, July 6, 1920; *Chem. Abs.*, 1920, **14**, 2722. J. Blumenfeld, U.S. 1,707,248, Apr. 2, 1929, to Commercial Pigments Corp.; *Chem. Abs.*, 1929, **23**, 2538: British 253,550, June 11, 1925; *Chem. Abs.*, 1927, **21**, 2538.

[158] R. S. Long, R. S. Olson, and J. P. Surls, Jr., U.S. 3,067,010, Dec. 4, 1962, to Dow Chemical Co.; *Chem. Abs.*, 1963, **58**, 5290. Belgian 613,489, Aug. 6, 1962, to Dow Chemical Co.; *Chem. Abs.*, 1963, **58**, 2178.

[159] N. M. Sobinyakova and M. A. Soinova, *R. B. p. S. R. Metal*, 1962, 134; *Chem. Abs.*, 1963, **58**, 1152.

[160] I. K. Tsitovich, *Z. V. K. O. im D. J. M.*, 1961, **6**, 233; *Chem. Abs.*, 1961, **55**, 23017.

[161] B. L. Nabivanets, *Zh. Neorgan. Khim.*, 1962, **7**, 417; *Chem. Abs.*, 1962, **57**, 2880.

[162] L. A. Kenworthy, U.S. 3,001,854, May 18, 1959, to Glidden Co.; *Chem. Abs.*, 1962, **56**, 3123.

[163] D. A. Ellis, U.S. 3,104,950, Sept. 24, 1963, to Dow Chemical Co.; *Chem. Abs.*, 1963, **59**, 14936.

[164] L. G. Bousquet, H. W. Richter, and B. W. Allan, U.S. 2,089,180, Aug. 10, 1937, to American Zirconium Corp.; *Chem. Abs.*, 1937, **31**, 6835.

## CHAPTER 14

[1] Jelks Barksdale, Section on titanium pigments, *Protective and Decorative Coatings*, Vol. II, John Wiley and Sons, Inc., New York, 1942, 397. A. Liberti, E. Chiantella, and F. Corigliano, *J. Inorg. and Nuclear Chem.*, 1963, 25, No. 4, 415.

[2] N. N. Mironov, *Tr. po Khim. i. Khim. Tekhnol.*, 1963, 85; *Chem. Abs.*, 1964, 60, 8870.

[3] A. Peter, *Chem. Tech. (Berlin)*, 1962, 14, 103; *Chem. Abs.*, 1962, 56, 13747: L. W. Robinson, *Paint Manuf.*, 1955, 25, 25; *Chem. Abs.*, 1955, 49, 4999: S. A. Ray, *Verfkroniek*, 1953, 26, 46; *Chem. Abs.*, 1953, 47, 4388: W. Barkhouse, *Australia Dept. Munitions, Paint Notes*, 1948, 3, No. 6. 193; *Chem. Abs.*, 1949, 43, 4023: S. J. Johnstone, *Ind. Chemist*, 1948, 24, 750; *Chem. Abs.*, 1951, 45, 3963.

[4] A. V. Pamfilov and E. G. Ivancheva, *J. Gen. Chem. (U.S.S.R.)*, 1939, 9, 1739; *Chem. Abs.*, 1940, 34, 3607.

[5] N. Parravano and V. Caglioti, *Congr. intern. quím. pura y aplicada. 9th Congr.*, Madrid, 1934, 3, 304; *Chem. Abs.*, 1936, 30, 4698.

[6] H. B. Weiser and W. O. Milligan, *J. Phys. Chem.*, 1934, 38, 513; *Chem. Abs.*, 1934, 28, 4328.

[7] A. V. Pamfilov and E. G. Ivancheva, *J. Gen. Chem. (U.S.S.R.)*, 1940, 10, 736; *Chem. Abs.*, 1941, 35, 2431: *J. Gen. Chem. (U.S.S.R.)*, 1937, 7, 2774; *Chem. Abs.*, 1938, 32, 3715.

[8] *Ibid.*, *J. Applied Chem. (U.S.S.R.)*, 1940, 13, 1310; *Chem. Abs.*, 1941, 35, 2432.

[9] N. Parravano and V. Caglioti, *Congr. intern. quím. pura y aplicada. 9th congr.*, Madrid, 1934, 3, 304; *Chem. Abs.*, 1936, 30, 4698.

[10] S. S. Cole, U.S. 2,316,840, Apr. 20, 1943, to National Lead Co.; *Chem. Abs.*, 1943, 37, 5879: U.S. 2,290,539, July 21, 1942, to National Lead Co.; *Chem. Abs.*, 1943, 37, 505: British 517,742, Aug. 6, 1938, to British Titan Products Co.: French 841,794, Aug. 6, 1938, to Titan Co., Inc.; *Chem. Abs.*, 1940, 34, 4596. J. E. Booge, U.S. 2,253,551, Aug. 26, 1941, to E. I. du Pont de Nemours and Co.; *Chem. Abs.*, 1941, 35, 8327. A. N. C. Bennett, British 553,135, May 10, 1943, to National Titanium Pigments, Ltd.; *Chem. Abs.*, 1944, 38, 5094.

[11] R. W. Ancrum, U.S. 2,365,135, Dec. 12, 1944, to Titan Co., Inc.; *Chem. Abs.*, 1945, 39, 3680: French 865,801, Apr. 2, 1940, to Titan Co., Inc.

[12] A. V. Pamfilov and S. V. Peltikhin, *Zhur. Priklad. Khim.*, 1949, 22, 245; *Chem. Abs.*, 1949, 43, 5970.

[13] G. Weintraub, U.S. 1,014,793, Jan. 16, 1912, to General Electric Co.; *Chem. Abs.*, 1912, 6, 671.

[14] A. J. Rossi and L. E. Barton, U.S. 1,196,030 and 1,196,031, Aug. 29, 1916; *Chem. Abs.*, 1916, 10, 2624: French 482,699, Apr. 12, 1917, to Titanium Alloy Mfg. Co.; *Chem. Abs.*, 1918, 12, 205.

[15] L. E. Barton, U.S. 1,223,358, Apr. 24, 1917; *Chem. Abs.*, 1917, 11, 1912: British 108,693, June 15, 1916, to Titanium Alloy Mfg. Co.; *Chem. Abs.*, 1918, 12, 84: French 483,197, June 6, 1917, to Titanium Alloy Mfg. Co.; *Chem. Abs.*, 1918, 12, 1002.

[16] British 348,724, Feb. 24, 1930, to I. G. Farbenind., A-G.; *Chem. Abs.*, 1932, 26, 2022.

[17] French 818,906, Oct. 6, 1937, to N. V. Industrielle Maatschappij voorheen Noury; *Chem. Abs.*, 1938, 32, 2767: British 491,541, Sept. 5, 1938, to N. V. Industrielle Maatschappij voorheen Noury; *Chem. Abs.*, 1939, 33, 1525.

[18] C. Weizmann and J. Blumenfeld, British 209,441, July 12, 1924; *Chem. Abs.*, 1924, 18, 1884: British 203,352, Mar. 1, 1922; *Chem. Abs.*, 1924, 18, 573.

[19] M. Fladmark, U.S. 1,288,863, Dec. 14, 1918, to Titan Co., A-S.; *Chem. Abs.*, 1919, 13, 641: Canadian 201,716, July 6, 1920; *Chem. Abs.*, 1920, 14, 2722: British 110,153, May 16, 1917, to Norske Aktieselskab för Elektrokemisk Industrie; *Chem. Abs.*, 1918, 12, 408: Norwegian 27,917, May 14, 1917, to Norske Aktieselskab för Elektrokemisk Industrie; *Chem. Abs.*, 1917, 11, 2720.

[20] N. Specht, U.S. 1,649,496, Nov. 15, 1927; *Chem. Abs.*, 1928, **22**, 505: British 234,518, May 24, 1924, to Deutsche Gasglühlicht, G.m.b.H.; *Chem. Abs.*, 1926, **20**, 803.

[21] German 542,334, Nov. 23, 1926, to Deutsche Gasglühlicht, G.m.b.H.; *Chem. Abs.*, 1932, **26**, 2282.

[22] H. H. Buckman, U.S. 1,402,256, Jan. 3, 1922; *Chem. Abs.*, 1922, **16**, 1018: U.S. 1,410,056, Mar. 21, 1922; *Chem. Abs.*, 1922, **16**, 1874: British 195,181, Jan. 4, 1922; *Chem. Abs.*, 1923, **17**, 3615.

[23] F. E. Bachman, U.S. 1,489,417 and 1,489,418, Apr. 4, 1924; *Chem. Abs.*, 1924, **18**, 1916; Reissue 15,973, Dec. 23, 1924; *Chem. Abs.*, 1925, **19**, 739.

[24] H. W. Richter, U.S. 1,947,226, Feb. 13, 1934; *Chem. Abs.*, 1934, **28**, 2553: British 406,580, Mar. 1, 1934, to Metal and Thermit Corp.; *Chem. Abs.*, 1934, **28**, 4923: French 760,035, Feb. 15, 1934, to Metal and Thermit Corp.; *Chem. Abs.*, 1934, **28**, 3604.

[25] J. d'Ans and F. Sommer, U.S. 1,655,940, Jan. 10, 1928; *Chem. Abs.*, 1928, **22**, 1019: British 214,483, May 30, 1923, to Chemische Werke vorm. A-G.; *Chem. Abs.*, 1924, **18**, 2584.

[26] A. G. Oppegaard and C. J. Stopford, U.S. 2,200,373, May 14, 1940, to Titan Co., Inc.; *Chem. Abs.*, 1940, **34**, 6110.

[27] British 257,259, Aug. 20, 1925, to Deutsche Gasglühlicht, A-G.; *Chem. Abs.*, 1927, **21**, 3107: German 549,407, Aug. 21, 1925, to Deutsche Gasglühlicht, A-G.; *Chem. Abs.*, 1932, **26**, 3883.

[28] J. L. Keats, U.S. 2,345,980, Apr. 4, 1944, to E. I. du Pont de Nemours and Co.; *Chem. Abs.*, 1944, **38**, 4817.

[29] E. C. Loth and C. S. Inscho, U.S. 2,344,553, Mar. 21, 1944; *Chem. Abs.*, 1944, **38**, 3494.

[30] M. E. Zborovskii and E. V. Germogenova, *Trans. All-Union Sci. Research Inst. Econ. Mineral (U.S.S.R.)*, 1935, **68**, 29; *Chem. Abs.*, 1936, **30**, 2711.

[31] W. Mecklenburg, U.S. 1,758,528, May 13, 1930; *Chem. Abs.*, 1930, **24**, 3330: Reissue 18,790, Apr. 4, 1933, to Krebs Color and Pigment Corp.; *Chem. Abs.*, 1933, **27**, 3299.

[32] J. Blumenfeld, British 310,949, May 5, 1928, to Verein für Chemische und Metallurgische Produktion; *Chem. Abs.*, 1930, **24**, 696: German 540,863, May 8, 1928, to Verein für Chemische und Metallurgische Produktion; *Chem. Abs.*, 1932, **26**, 2285.

[33] J. Blumenfeld, U.S. 1,851,487, Mar. 29, 1932, to Krebs Pigment and Color Corp.; *Chem. Abs.*, 1932, **26**, 3080.

[34] R. Dahlstrom and L. W. Ryan, U.S. 2,098,278, Nov. 9, 1937, to National Lead Co.; *Chem. Abs.*, 1938, **32**, 374: British 447,059, May 7, 1936, to British Titan Products Co., Ltd.; *Chem. Abs.*, 1936, **30**, 6969: French 780,944, May 6, 1936, to Titan Co., Inc.; *Chem. Abs.*, 1935, **29**, 6082.

[35] British 453,188, July 18, 1935, to British Titan Products Co., Ltd.; *Brit. Chem. Abstracts*, 1936, **B**, 1109.

[36] P. Weise and F. Raspe, German 591,007, Jan. 15, 1934, to I. G. Farbenind., A-G.; *Chem. Abs.*, 1934, **28**, 2478: French 738,620, June 14, 1932, to I. G. Farbenind., A-G.; *Chem. Abs.*, 1933, **27**, 2050: British 388,978, Mar. 9, 1933, to I. G. Farbenind., A-G.; *Chem. Abs.*, 1933, **27**, 4636.

[37] C. L. Schmidt, U.S. 2,034,923, Mar. 24, 1936, to Titanium Pigment Co., Inc.; *Chem. Abs.*, 1936, **30**, 3261: French 792,827, Jan. 11, 1936, to Titan Co., Inc.; *Chem. Abs.*, 1936, **30**, 4702: British 453,188, Sept. 7, 1936, to British Titan Products Co., Ltd.; *Chem. Abs.*, 1937, **31**, 1168.

[38] B. W. Allan and L. G. Bousquet, U.S. 2,182,420, Dec. 5, 1939, to American Zirconium Corp.; *Chem. Abs.*, 1940, **34**, 2193.

[39] A. J. Ravenstad, U.S. 2,055,221, Sept. 22, 1936, to Titan Co., Inc.; *Chem. Abs.*, 1936, **30**, 7884: British 392,194, May 9, 1933, to Titan Co., A-S.; *Chem. Abs.*, 1933, **27**, 5559: Norwegian 51,236, Aug. 1, 1932, to Titan Co., A-S.; *Chem. Abs.*, 1933,

**27**, 1457: French 726,177, Feb. 23, 1932, to Titan Co., Inc.; *Chem. Abs.,* 1932, **26**, 4969: German 605,859, Nov. 19, 1934; *Chem. Abs.,* 1935, **29**, 2005.

[40] B. W. Allan, U.S. 2,143,851, Jan. 17, 1939, to American Zirconium Corp.; *Chem. Abs.,* 1939, **33**, 3084. R. Neubner, German 711,427, Aug. 28, 1941, to Bayerische Stickstoff-Werke, A-G.; *Chem. Abs.,* 1943, **37**, 4211.

[41] K. Leuchs, U.S. 1,853,626, Apr. 12, 1932; *Chem. Abs.,* 1932, **26**, 3340: Reissue 19,594, May 28, 1935, to Krebs Pigment and Color Corp.; *Chem. Abs.,* 1935, **29**, 4958: German 533,326, Nov. 28, 1925; *Chem. Abs.,* 1932, **26**, 566.

[42] British 388,978, June 15, 1932, to I. G. Farbenind., A-G.; *Brit. Chem. Abstracts,* 1933, **B**, 386.

[43] I. E. Webber and A. N. C. Bennett, British 546,283, July 6, 1942, to National Titanium Pigments Ltd.; *Chem. Abs.,* 1943, **37**, 3285.

[44] J. L. Keats and H. M. Stark, U.S. 2,321,490, June 8, 1943, to E. I. du Pont de Nemours and Co.; *Chem. Abs.,* 1943, **37**, 6826.

[45] French 795,484, March 14, 1936, to Titangesellschaft m. b. H.; *Chem. Abs.,* 1936, **30**, 5820: British 449,327, June 24, 1936, to Titangesellschaft m. b. H.; *Chem. Abs.,* 1936, **30**, 8663: German 718,510, Feb. 19, 1942, to Titangesellschaft m. b. H.; *Chem. Abs.,* 1943, **37**, 1839.

[46] J. Brode and K. Klein, German 493,815, Jan. 13, 1926, to I. G. Farbenind., A-G.; *Chem. Abs.,* 1930, **24**, 2903: French 691,458, Mar. 7, 1930, to I. G. Farbenind., A-G.

[47] A. Carpmael, British 449,327, Dec. 24, 1934, to Titangesellschaft m. b. H.; *Brit. Chem. Abstracts,* 1937, **B**, 263.

[48] French 738,620, June 14, 1932, to I. G. Farbenind., A-G.; *Chem. Abs.,* 1933, **27**, 2050.

[49] British 473,470, Oct. 13, 1937, to British Titan Products Co., Ltd.; *Chem. Abs.,* 1938, **32**, 3176.

[50] B. W. Allan, U.S. 2,040,823, May 19, 1936; *Brit. Chem. Abstracts,* 1937, **B**, 909: French 794,037, Feb. 6, 1936, to Titan Co., Inc.; *Chem. Abs.,* 1936, **30**, 4632: British 451,890, Aug. 13, 1936, to Titan Co., Inc.; *Chem. Abs.,* 1937, **31**, 561.

[51] B. W. Allan, U.S. 2,143,850, Jan. 17, 1939, to American Zirconium Corp.; *Chem. Abs.,* 1939, **33**, 3084.

[52] B. W. Allan, U.S. 2,143,530, Jan. 10, 1939, to American Zirconium Corp.; *Chem. Abs.,* 1939, **33**, 3084: French 819,086, Oct. 9, 1937, to Titan Co., Inc.; *Chem. Abs.,* 1938, **32**, 2768: German 710,866, Aug. 14, 1941, to Titangesellschaft m. b. H.: *Chem. Abs.,* 1943, **37**, 3889.

[53] J. E. Pollack, British 463,966, Apr. 8, 1937, to N. V. Industrielle Maatschappij voorheen Noury; *Chem. Abs.,* 1937, **31**, 6907.

[54] H. W. Richter, U.S. 1,947,226, Feb. 13, 1934; *Brit. Chem. Abstracts,* 1935, **B**, 277.

[55] H. W. Richter, U.S. 2,078,279, Apr. 27, 1931, to American Zirconium Corp.; *Chem. Abs.,* 1937, **31**, 4515.

[56] W. T. Little, U.S. 2,029,881, Feb. 4, 1936, to American Zirconium Corp.; *Brit. Chem. Abstracts,* 1937, **B**, 668. W. J. Tennant, British 462,998, Oct. 30, 1935, to Titan Co., Inc.; *Brit. Chem. Abstracts,* 1937, **B**, 909: British 462,998, Mar. 17, 1937, to Titan Co., Inc.; *Chem. Abs.,* 1937, **31**, 6038: French 797,922, Mar. 6, 1936, to Titan Co., Inc.; *Chem. Abs.,* 1936, **30**, 7291: German 666,063, Oct. 12, 1938, to Titan Co., Inc.; *Chem. Abs.,* 1939, **33**, 1974.

[57] S. F. W. Crundall, British 447,744, May 21, 1936, to P. Spence and Co., Ltd.; *Chem. Abs.,* 1936, **30**, 7290: French 793,139, Jan. 16, 1936, to P. Spence and Sons, Ltd.; *Chem. Abs.,* 1936, **30**, 4632.

[58] W. J. Cauwenberg, U.S. 1,957,528, May 8, 1934, to Titanium Pigment Co., Inc.; *Chem. Abs.,* 1934, **28**, 4186: British 427,339, Apr. 23, 1935, to Titan Co., A-S.; *Chem. Abs.,* 1935, **29**, 6375: French 762,013, Apr. 23, 1934, to Titan Co., Inc.; *Chem. Abs.,* 1934, **28**, 5189: Canadian 345,236, Oct. 6, 1934, to Titanium Pigment Co., Inc.; *Chem. Abs.,* 1935, **29**, 2375.

[59] P. Weise, U.S. 2,091,799, Aug. 31, 1937, to National Lead Co.; *Chem. Abs.,* 1937, **31**, 7677: German 661,522, June 20, 1938, to Titangesellschaft m. b. H.; *Chem. Abs.,* 1938, **32**, 8808: French 768,585, Aug. 8, 1934, to Titangesellschaft m. b. H.; *Chem. Abs.,* 1935, **29**, 626: British 433,960, Aug. 30, 1935, to Titangesellschaft m. b. H.; *Chem. Abs.,* 1936, **30**, 885.

[60] Norwegian 27,665, Feb. 26, 1917, to Norske Aktieselskab för Elektrokemisk Industrie; *Chem. Abs.,* 1917, **11**, 1889.

[61] M. J. Brooks, U.S. 2,292,507, Aug. 11, 1942, to General Chemical Co.; *Chem. Abs.,* 1943, **37**, 734.

[62] B. W. Allan, U.S. 2,133,941, Oct. 25, 1938, to American Zirconium Corp.; *Chem. Abs.,* 1939, **33**, 819: German 668,242, Nov. 20, 1938, to Titan Co., Inc.; *Chem. Abs.,* 1939, **33**, 2292.

[63] C. A. Tanner, U.S. 2,503,692, April 11, 1950, to American Cyanamid Co.; *Chem. Abs.,* 1950, **44**, 5612.

[64] L. Aagaard, U.S. 2,886,415, May 12, 1959, to National Lead Co.; *Chem. Abs.,* 1959, **53**, 20837.

[65] F. L. Kingsbury, U.S. 2,564,365, Aug. 14, 1951, to National Lead Co.; *Chem. Abs.,* 1951, **45**, 10518: British 677,216, Aug. 13, 1952, to Titan Co., Inc.; *Chem. Abs.,* 1952, **46**, 11600.

[66] J. Blumenfeld, U.S. 1,504,669, 1,504,671 and 1,504,672, Aug. 12, 1924; *Chem. Abs.,* 1924, **18**, 3257: Reissue 17,429 and 17,430, Sept. 10, 1929, to Commercial Pigments Corp.; *Chem. Abs.,* 1929, **23**, 5282. J. Blumenfeld and C. Weizmann, British 225,593, July 31, 1925; *Chem. Abs.,* 1925, **19**, 1784.

[67] J. Blumenfeld, U.S. 1,795,467, Mar. 10, 1931, to Commercial Pigments Corp.; *Chem. Abs.,* 1931, **25**, 2530: Reissue 18,854, May 30, 1933, to Krebs Pigment and Color Corp.; *Chem. Abs.,* 1933, **27**, 4038: British 275,672, Aug. 9, 1926; *Chem. Abs.,* 1928, **22**, 2474: French 640,181, Aug. 9, 1927; *Chem. Abs.,* 1929, **23**, 999.

[68] C. M. Olson, U.S. 2,331,496, Oct. 13, 1946, to E. I. du Pont de Nemours and Co.; *Chem. Abs.,* 1944, **38**, 1614: British 558,285, Dec. 30, 1943, to E. I. du Pont de Nemours and Co.; *Chem. Abs.,* 1945, **39**, 3636.

[69] L. W. Ryan, U.S. 1,820,987, Sept. 1, 1931, to Titanium Pigment Co., Inc.; *Chem. Abs.,* 1931, **25**, 5958: French 672,002, Mar. 25, 1929, to Titanium Pigment Co., Inc.; *Chem. Abs.,* 1930, **24**, 2250.

[70] L. W. Ryan, U.S. 1,820,988, Sept. 1, 1931, to Titanium Pigment Co., Inc.; *Chem. Abs.,* 1931, **25**, 5958: French 672,003, Mar. 25, 1929, to Titanium Pigment Co., Inc.; *Chem. Abs.,* 1930, **24**, 2250: British 308,725, Mar. 27, 1928, to Titanium Pigment Co., Inc.; *Chem. Abs.,* 1930, **24**, 473: German 542,007, Mar. 16, 1932, to Titanium Pigment Co., Inc.; *Chem. Abs.,* 1932, **26**, 2283: U.S. 1,916,236, July 4, 1933, to Titanium Pigment Co., Inc.; *Brit. Chem. Abstracts,* 1934, **B**, 275.

[71] F. G. C. Stephens, L. J. Anderson, and W. A. Cash, U.S. 1,748,429, Feb. 25, 1930, to National Metal and Chemical Bank Ltd.; *Chem. Abs.,* 1930, **24**, 1995: British 309,051, Oct. 1, 1927; *Chem. Abs.,* 1930, **24**, 474: French 672,175, Mar. 24, 1929, to National Metal and Chemical Bank Ltd.; *Chem. Abs.,* 1930, **24**, 2316.

[72] R. H. Monk and A. S. Ross, U.S. 2,108,723, Feb. 15, 1938, to American Zinc, Lead and Smelting Co.; *Chem. Abs.,* 1938, **32**, 3105: British 466,970, June 9, 1937, to American Zinc, Lead and Smelting Co.; *Chem. Abs.,* 1937, **31**, 8230. R. H. Monk, U.S. 2,353,918, July 18, 1944, to American Zinc, Lead and Smelting Co.; *Chem. Abs.,* 1944, **38**, 6058: German 690,302, Apr. 22, 1940, to American Zinc, Lead and Smelting Co.; *Chem. Abs.,* 1941, **35**, 4164.

[73] R. H. Monk and A. S. Ross, U.S. 2,028,292, Jan. 21, 1936, to American Zinc, Lead and Smelting Co.; *Chem. Abs.,* 1936, **30**, 1528: British 456,544, Nov. 11, 1936, to American Zinc, Lead and Smelting Co.; *Chem. Abs.,* 1937, **31**, 2370: German 690,302, Mar. 28, 1940, to American Zinc, Lead and Smelting Co.; *Chem. Abs.,* 1941, **35**, 4164: French 792,825, Jan. 21, 1936, to American Zinc, Lead and Smelting Co.; *Chem. Abs.,* 1936, **30**, 4278.

[74] R. W. Ancrum and A. G. Oppegaard, British 529,596, May 24, 1939, to British Titan Products Co., Ltd.; *Chem. Abs.,* 1941, **35**, 7739.

[75] W. W. Plechner and J. M. Jarmus, U.S. 2,344,265, Mar. 14, 1944, to National Lead Co.; *Chem. Abs.*, 1944, **38**, 3494.

[76] H. J. Wood, U.S. 2,253,595, Aug. 26, 1941, to E. I. du Pont de Nemours and Co.; *Chem. Abs.*, 1941, **35**, 8328.

[77] G. Orsenigo, Italian 504,993, Dec. 14, 1954, to Montecatini Società generale per l'industria mineraria e Chimica; *Chem. Abs.*, 1957, **51**, 10020.

[78] Ko Hsuek, *Chin Chan*, 1958, No. 2, 11; *Chem. Abs.*, 1962, **56**, 5631.

[79] C. de Rohden, *Chimie & industrie*, 1956, **75**, 287; *Chem. Abs.*, 1956, **50**, 10423.

[80] Z. Y. Berestneva and V. A. Kargin, *Uspekhi Khim*, 1955, **24**, 249; *Chem. Abs.*, 1955, **49**, 12084.

[81] N. Sakai, K. Yoshikawa, M. Suzuki, and S. Kobashi, *Kogyo Kagaku Zasshi*, 1961, **64**, 613; *Chem. Abs.*, 1962, **57**, 3077.

[82] J. S. Han and J. S. Kim, *Bull. Sci. Research Inst. (Korea)*, 1959, **4**, No. 1, 5; *Chem. Abs.*, 1960, **54**, 837.

[83] P. Farup, U.S. 1,773,727, Aug. 26, 1930, to Titanium Pigment Co., Inc.; *Chem. Abs.*, 1930, **24**, 5115: British 271,085, May 12, 1926, to Titan Co., A-S.; *Chem. Abs.*, 1928, **22**, 1658: Canadian 280,396, May 22, 1928; *Chem. Abs.*, 1928, **22**, 2818: German 496,257, Mar. 6, 1927, to Titan Co., A-S.; *Chem. Abs.*, 1930, **24**, 3329.

[84] B. D. Saklatwalla, H. E. Dunn, and A. E. Marshall, U.S. 1,978,228, Oct. 23, 1935, to Southern Mineral Products Corp.; *Chem. Abs.*, 1935, **29**, 304.

[85] M. Schetka, U.S. 1,758,472, May 12, 1930, to Titanium Pigment Co., Inc.; *Chem. Abs.*, 1930, **24**, 3330: British 296,730, Sept. 6, 1927, to I. G. Farbenind., A-G.; *Chem. Abs.*, 1929, **23**, 2538: French 655,701, June 13, 1928, to I. G. Farbenind., A-G.; *Chem. Abs.*, 1929, **23**, 4027.

[86] French 676,281, June 6, 1929, to Soc. minière "La Barytine"; *Chem. Abs.*, 1930, **24**, 2844.

[87] D. Doerinckel and L. Mehler, U.S. 1,738,765, Dec. 10, 1930, to Titanium Pigment Co., Inc.; *Chem. Abs.*, 1930, **24**, 928: British 303,468, Oct. 4, 1927, to I. G. Farbenind., A-G.; *Chem. Abs.*, 1929, **23**, 4539: French 653,985, May 8, 1928, to I. G. Farbenind., A-G.; *Chem. Abs.*, 1929, **23**, 3779.

[88] S. J. Lubowsky, U.S. 1,888,993, Nov. 29, 1932, to Metal and Thermit Corp.; *Brit. Chem. Abstracts*, 1933, **B**, 800.

[89] French 653,985, May 8, 1928, to I. G. Farbenind., A-G.; *Chem. Abs.*, 1929, **23**, 3779.

[90] P. Weise and F. Raspe, U.S. 2,121,215, June 21, 1938, to Titangesellschaft m. b. H.; *Chem. Abs.*, 1938, **32**, 6407: German 675,408, May 8, 1939, to Titangesellschaft m. b. H.; *Chem. Abs.*, 1939, **33**, 5611: British 471,830, Sept. 13, 1937, to Titan Co., Inc.; *Chem. Abs.*, 1938, **32**, 1413: French 810,541, Mar. 23, 1937, to Titangesellschaft m. b. H.; *Chem. Abs.*, 1937, **31**, 8843.

[91] E. C. Smith, U.S. 2,147,370, Mar. 7, 1939, to E. I. du Pont de Nemours and Co.; *Chem. Abs.*, 1939, **33**, 4388.

[92] B. W. Saklatwalla and H. E. Dunn, U.S. 1,959,765, May 22, 1934, to Southern Mineral Products Corp.; *Chem. Abs.*, 1934, **28**, 4548: French 770,883, Sept. 22, 1934, to Southern Mineral Products Corp.; *Chem. Abs.*, 1935, **29**, 894. A. H. Stevens, British 418,798, Mar. 22, 1934, to Southern Mineral Products Corp.; *Brit. Chem. Abstracts*, 1935, **B**, 148.

[93] *Chemical Week*, July 22, 1961, 70: N. Soloducha, Canadian 610,334, Dec. 13, 1960.

[94] A. V. Pamfilov, E. G. Ivancheva, and I. M. Soboleva, *J. Applied Chem. (U.S.S.R.)*, 1939, **12**, 226; *Chem. Abs.*, 1939, **33**, 6742.

[95] L. T. Work, S. B. Tuwiner, and A. J. Gloster, *Ind. Eng. Chem.*, 1934, **26**, 1263; *Chem. Abs.*, 1935, **29**, 945. L. T. Work and S. B. Tuwiner, *Ind. Eng. Chem.*, 1934, **26**, 1266; *Chem. Abs.*, 1935, **29**, 945.

[96] C. Ligorio and L. T. Work, *Ind. Eng. Chem.*, 1937, **29**, 213; *Chem. Abs.*, 1937, **31**, 2026.

[97] A. W. Hixon and W. W. Plechner, *Ind. Eng. Chem.*, 1933, **25**, 262; *Chem. Abs.*, 1933, **27**, 1995. W. W. Plechner, Thesis, Columbia University, 1932.

[98] H. J. Braun, *Metallbörse*, 1931, **21**, 507; *Chem. Abs.*, 1931, **25**, 3180.

[99] A. V. Pamfilov, S. V. Peltikhim, and I. M. Soboleva, *J. Applied Chem.* (*U.S.S.R.*), 1947, **20**, 63; *Chem. Abs.*, 1947, **41**, 5365.

[100] A. W. Hixon and R. F. C. Fredrickson, *Ind. Eng. Chem.*, 1945, **37**, 678; *Chem. Abs.*, 1945, **39**, 3720.

[101] L. Kayser, *Z. anorg. u. allgem. Chem.*, 1924, **138**, 43; *Chem. Abs.*, 1925, **19**, 450.

[102] N. V. Parravano and V. Caglioti, *Gazz. chim. ital.*, 1934, **64**, 429; *Chem. Abs.*, 1935, **29**, 22: *Gazz. chim. ital.*, 1934, **64**, 703; *Chem. Abs.*, 1935, **29**, 2002: *Congr. intern. quím. pura y aplicada. 9th Congr.*, Madrid, 1934, **3**, 304; *Chem. Abs.*, 1936, **30**, 4698.

[103] B. A. Tsarev and T. A. Velikoslavinskaya, *Byull. Obmena Opyt. Lakokrasoch. Prom.* 1939, No. 6, 7, 26; *Chem. Abs.*, 1940, **34**, 6415.

[104] T. Sagawa, *Kinzoku-no-Kenkyu*, 1935, **12**, 543; *Chem. Abs.*, 1936, **30**, 2514: *Science Repts., Tôhoku Imp. Univ.*, **1936**, 480; *Chem. Abs.*, 1937, **31**, 1687.

[105] H. B. Weiser and W. O. Milligan, *J. Phys. Chem.*, 1934, **38**, 513; *Chem. Abs.*, 1934, **28**, 4328.

[106] S. Wilska, *Suomen Kemistilehti*, 1954, **27B**, No. 7, 57; *Chem. Abs.*, 1955, **49**, 4303.

[107] W. F. Washburn and L. Aagaard, U.S. 1,906,729, May 2, 1933, to Titanium Pigment Co., Inc.; *Chem. Abs.*, 1933, **27**, 3626: British 364,562, Dec. 23, 1931, to Titanium Pigment Co., Inc.; *Chem. Abs.*, 1933, **27**, 2050: German 588,446, Nov. 23, 1933, to Titanium Pigment Co., Inc.; *Chem. Abs.*, 1934, **28**, 1555: French 704,044, Sept. 22, 1930, to Titanium Pigment Co., Inc.; *Chem. Abs.*, 1931, **25**, 4723.

[108] W. F. Washburn and F. L. Kingsbury, U.S. 2,055,222, Sept. 22, 1936, to Titanium Pigment Co., Inc.; *Chem. Abs.*, 1936, **30**, 7884: British 387,720, Feb. 13, 1933, to Titanium Pigment Co., Inc.; *Chem. Abs.*, 1933, **27**, 4698: British 462,206, Mar. 4, 1937, to Titanium Pigment Co., Inc.; *Chem. Abs.*, 1937, **31**, 5605: French 721,188, Aug. 10, 1931, to Titanium Pigment Co., Inc.; *Chem. Abs.*, 1932, **26**, 4189: French 47,345, Mar. 16, 1937, addition to 721,188, to Titanium Pigment Co., Inc.; *Chem. Abs.*, 1937, **31**, 8230.

[109] L. G. Bousquet, H. W. Richter, and B. W. Allan, U.S. 2,089,180, Aug. 10, 1937, to American Zirconium Corp.; *Chem. Abs.*, 1937, **31**, 6835: British 470,266, Aug. 11, 1937, to British Titan Products Co., Ltd.; *Chem. Abs.*, 1938, **32**, 1059: L. G. Bousquet and B. W. Allan, French 805,353 and 805,354, Nov. 18, 1936, to Titan Co., Inc.; *Chem. Abs.*, 1937, **31**, 3645.

[110] J. E. Booge, U.S. 2,269,139, Jan. 6, 1942, to E. I. du Pont de Nemours and Co.; *Chem. Abs.*, 1942, **36**, 3056.

[111] O. Lederer and H. Kassler, British 267,491, Mar. 9, 1926; *Chem. Abs.*, 1928, **22**, 1218.

[112] M. S. Platonov, B. A. Tsarev, and T. A. Velikoslavinskaya, *Khim. Referat. Zhur.*, 1939, **2**, No. 5, 113; *Chem. Abs.*, 1940, **34**, 3512: 1939, **2**, No. 5, 107; *Chem. Abs.*, 1940, **34**, 860.

[113] H. L. Rhodes, U.S. 1,922,328, Aug. 15, 1933; *Brit. Chem. Abstracts*, 1934, **B**, 402.

[114] P. Tillman and F. Raspe, German 700,918, Dec. 5, 1940, to Titangesellschaft m. b. H.; *Chem. Abs.*, 1941, **35**, 7739: German 718,169, Feb. 12, 1942, to Titangesellschaft m. b. H.; *Chem. Abs.*, 1944, **38**, 2458: British 513,867, Oct. 24, 1939, to Titangesellschaft m. b. H.; *Chem. Abs.*, 1941, **35**, 2287: French 837,238, Feb. 6, 1939, to Titangesellschaft m. b. H.; *Chem. Abs.*, 1939, **33**, 7133.

[115] Jelks Barksdale, U.S. 2,285,485, June 9, 1942, to National Lead Co.; *Chem. Abs.*, 1942, **36**, 6761: British 536,141, May 5, 1941, to Titan Co., Inc.; *Chem. Abs.*, 1942, **36**, 1446: French 861,320, to Titan Co., Inc.

[116] Jelks Barksdale and W. W. Plechner, U.S. 2,285,486, June 9, 1942, to National Lead Co.; *Chem. Abs.*, 1942, **36**, 6761: Norwegian 65,233, Sept. 21, 1942, to Titan Co., A-S.; *Chem. Abs.*, 1946, **40**, 686.

[117] A. T. McCord and H. F. Saunders, U.S. 2,333,660, Nov. 9, 1943, to Sherwin-Williams Co.; *Chem. Abs.*, 1944, **38**, 2514.

[118] P. Tillman, F. Raspe, and J. Heinen, U.S. 2,303,306, Nov. 24, 1942, to Titan Co., Inc.; *Chem. Abs.*, 1943, **37**, 2525: British 542,579, Jan. 16, 1942, to Titan Co., Inc.; *Chem. Abs.*, 1942, **36**, 3918: British 533,227, Feb. 10, 1941, to National Titanium Pigments Ltd.; *Chem. Abs.*, 1942, **36**, 1510.

[119] French 863,766, April 9, 1941, to Fabriques de produits chimiques de Thann et de Mulhouse; *Chem. Abs.*, 1948, **42**, 9202.

[120] A. N. C. Bennett, British 566,499, Jan. 2, 1945, to National Titanium Pigments, Ltd.; *Chem. Abs.*, 1947, **41**, 873.

[121] British 562,617, July 10, 1944, to British Titan Products Co., Ltd.; *Chem. Abs.*, 1946, **40**, 434.

[122] I. E. Weber and A. N. C. Bennett, British 405,669, Feb. 12, 1934, to B. Laporte, Ltd.; *Chem. Abs.*, 1934, **28**, 4844.

[123] R. M. McKinney and H. M. Stark, U.S. 2,345,985, Apr. 4, 1944, to E. I. du Pont de Nemours and Co.; *Chem. Abs.*, 1944, **38**, 4818: British 568,232, Mar. 26, 1945, to E. I. du Pont de Nemours and Co.; *Chem. Abs.*, 1947, **41**, 3638.

[124] W. J. Cauwenberg and L. Aagaard, U.S. 2,433,597, Dec. 30, 1947, to American Cyanamid Co.; *Chem. Abs.*, 1948, **42**, 2064.

[125] M. J. Mayer, U.S. 2,444,939 and 2,444,940, July 13, 1948; *Chem. Abs.*, 1948, **42**, 6550: British 596,656, Jan. 8, 1948; *Chem. Abs.*, 1948, **42**, 3919.

[126] H. M. Stark, U.S. 2,369,262, Feb. 13, 1945, to E. I. du Pont de Nemours and Co.; *Chem. Abs.*, 1945, **39**, 3680.

[127] H. M. Stark and J. L. Keats, U.S. 2,346,091, Apr. 4, 1944, to E. I. du Pont de Nemours and Co.; *Chem. Abs.*, 1944, **38**, 5093. R. M. McKinney, U.S. 2,237,764, Apr. 8, 1941, to E. I. du Pont de Nemours and Co.; *Chem. Abs.*, 1941, **35**, 4619.

[128] C. M. Olson, U.S. 2,342,483, Feb. 22, 1944, to E. I. du Pont de Nemours and Co.; *Chem. Abs.*, 1944, **38**, 4818.

[129] J. L. Keats and H. M. Stark, U.S. 2,301,412, Nov. 10, 1942, to E. I. du Pont de Nemours and Co.; *Chem. Abs.*, 1943, **37**, 2146.

[130] C. A. Tanner, Jr., U.S. 2,516,604, July 25, 1950, to American Cyanamid Co.; *Chem. Abs.*, 1950, **44**, 11122.

[131] R. M. McKinney, U.S. 2,507,729, May 16, 1950, to E. I. du Pont de Nemours and Co.; *Chem. Abs.*, 1952, **46**, 9321.

[132] C. M. Olson, U.S. 2,479,637, Aug. 23, 1949, to E. I. du Pont de Nemours and Co.; *Chem. Abs.*, 1950, **44**, 3722.

[133] C. M. Olson and J. E. Booge, U.S. 2,511,218, June 13, 1950, to E. I. du Pont de Nemours and Co.; *Chem. Abs.*, 1950, **44**, 9164.

[134] C. M. Olson, U.S. 2,452,390, Oct. 26, 1948, to E. I. du Pont de Nemours and Co.; *Chem. Abs.*, 1949, **43**, 1579.

[135] British 854,933, Nov. 23, 1960, to Glidden Co.; *Chem. Abs.*, 1961, **55**, 10919.

[136] J. T. Richmond and R. J. Wigginton, U.S. 2,589,964, March 18, 1952, to National Titanium Pigments, Ltd.; *Chem. Abs.*, 1952, **46**, 5339.

[137] L. Aagaard, U.S. 2,488,755, Nov. 22, 1949, to National Lead Co.; *Chem. Abs.*, 1950, **44**, 1662.

[138] L. A. Kenworthy, U.S. 3,068,068, Dec. 11, 1962, to Glidden Co.; *Chem. Abs.*, 1963, **58**, 5290.

[139] M. Solomka, U.S. 3,071,439, Jan. 1, 1963, to Dow Unquinesa, S. A.; *Chem. Abs.*, 1963, **58**, 8146.

[140] German 903,619, Feb. 8, 1954, to Titangesellschaft m. b. H.; *Chem. Abs.*, 1958, **52**, 13210: British 697,673, Sept. 30, 1953, to Titangesellschaft m. b. H.; *Chem. Abs.*, 1954, **48**, 5524.

[141] Y. Arai, S. Yoshii, and Y. Kawane, Japanese 7066, Oct. 30, 1955, to Sakai Chemical Industry Co.; *Chem. Abs.*, 1958, **52**, 15922.

[142] M. J. Mayer, U.S. 2,480,869, Sept. 6, 1949; *Chem. Abs.*, 1949, **43**, 8624.

[143] N. S. Rassudova, G. A. Ermakova, and V. N. Istomina, *Lakokrasochnyl Materialy i ikh Primenenie*, 1961, No. 1, 30; *Chem. Abs.*, 1961, **55**, 20350.

[144] M. J. Mayer, U.S. 2,519,389, Aug. 22, 1950; *Chem. Abs.*, 1950, **44**, 10274: French 968,476, Nov. 28, 1950; *Chem. Abs.*, 1952, **46**, 6296.

[145] M. J. Mayer, U.S. 2,571,150, Oct. 16, 1951; *Chem. Abs.*, 1952, **46**, 2826.

[146] B. Sebestova, Czechoslovakia 94,679, March 15, 1960; *Chem. Abs.*, 1961, **55**, 14936.
[147] B. W. Allan and F. O. Rummery, U.S. 2,533,208, Dec. 12, 1950; *Chem. Abs.*, 1951, **45**, 3265.
[148] R. M. McKinney, U.S. 2,193,563, Mar. 13, 1940, to E. I. du Pont de Nemours and Co.; *Chem. Abs.*, 1940, **34**, 4930.
[149] J. L. Keats, U.S. 2,193,559, Mar. 12, 1940, to E. I. du Pont de Nemours and Co.; *Chem. Abs.*, 1940, **34**, 4930.
[150] R. M. McKinney and H. M. Stark, U.S. 2,304,110, Dec. 8, 1942, to E. I. du Pont de Nemours and Co.; *Chem. Abs.*, 1943, **37**, 2948.

## CHAPTER 15

[1] P. Weise, German 552,776, Mar. 3, 1929, to I. G. Farbenind., A-G.; *Chem. Abs.*, 1932, **26**, 5776: British 346,116, Feb. 28, 1930, to I. G. Farbenind., A-G.; *Brit. Chem. Abstracts*, 1931, **B**, 642.
[2] French 690,764, Feb. 27, 1930, to I. G. Farbenind., A-G.; *Chem. Abs.*, 1931, **25**, 1399.
[3] W. F. Washburn, U.S. 2,148,283, Feb. 2, 1939, to National Lead Co.; *Chem. Abs.*, 1939, **33**, 4065.
[4] R. K. Whitten, U.S. 2,305,368, Dec. 15, 1942, to E. I. du Pont de Nemours and Co.; *Chem. Abs.*, 1943, **37**, 3285: British 519,785, Apr. 5, 1940, to E. I. du Pont de Nemours and Co.; *Chem. Abs.*, 1942, **36**, 670.
[5] C. de Rohden, U.S. 1,885,187, Nov. 1, 1933, to Krebs Pigment and Color Corp.; *Chem. Abs.*, 1933, **27**, 1216.
[6] C. de Rohden, U.S. 1,846,188, Feb. 23, 1932, to Krebs Pigment and Color Corp.; *Chem. Abs.*, 1932, **26**, 2608.
[7] French 641,712, Oct. 1, 1927, to I. G. Farbenind., A-G.; *Chem. Abs.*, 1929, **23**, 1222.
[8] German 411,723, Oct. 30, 1921, to Deutsche Gasglühlicht, G.m.b.H.; *J. Soc. Chem. Ind. (London)*, 1925, **B**, 631.
[9] W. F. Washburn, U.S. 2,148,283, Feb. 21, 1939, to National Lead Co.; *Chem. Abs.*, 1939, **33**, 4065.
[10] R. H. Monk and A. S. Ross, French 804,225, Oct. 19, 1936, to American Zinc, Lead and Smelting Co.; *Chem. Abs.*, 1937, **31**, 2760.
[11] British 519,785, Sept. 30, 1938, to E. I. du Pont de Nemours and Co.; *Chem. Abs.*, 1942, **36**, 670.
[12] J. Ourisson, *Bull. soc. ind. Mulhouse*, 1934, **100**, 565; *Chem. Abs.*, 1935, **29**, 1267.
[13] N. Specht, U.S. 1,845,633, Feb. 16, 1932, to Krebs Pigment and Color Corp.; *Chem. Abs.*, 1932, **26**, 2285: British 309,598, Apr. 14, 1928, to Deutsche Gasglühlicht, G.m.b.H.; *Chem. Abs.*, 1930, **24**, 696: German 523,015, Apr. 15, 1928, to Deutsche Gasglühlicht, G.m.b.H.; *Chem. Abs.*, 1931, **25**, 3447: French 673,074, Apr. 13, 1929, to Deutsche Gasglühlicht, G.m.b.H.; *Chem. Abs.*, 1930, **24**, 2556.
[14] B. W. Allan, U.S. 2,188,259, Jan. 23, 1940, to American Zirconium Corp.; *Chem. Abs.*, 1940, **34**, 3936: German 714,230, Oct. 30, 1941, to Titangesellschaft m.b.H.; *Chem. Abs.*, 1944, **38**, 1854.
[15] B. K. Brown, *Chem. & Met. Eng.*, 1928, **35**, 427; *Chem. Abs.*, 1928, **22**, 3539.
[16] B. M. Hokhberg, *Khim. Referat. Zhur.*, 1939, **2**, 113; *Chem. Abs.*, 1940, **34**, 3510.
[17] L. E. Olmstead and E. J. Puetz, U.S. 2,999,011, Sept. 5, 1961, to National Lead Co.; *Chem. Abs.*, 1961, **55**, 27915.
[18] J. T. Richmond, R. J. Wigginton, and G. G. Durrant, British 846,085, Aug. 24, 1960, to Laporte Titanium Ltd.; *Chem. Abs.*, 1961, **55**, 4987.
[19] K. Ohyagi, Japanese 1518, May 2, 1952; *Chem. Abs.*, 1953, **47**, 6620.
[20] F. von Bichowsky, U.S. 2,807,578, Sept. 24, 1957; *Chem. Abs.*, 1958, **52**, 934.

## NOTES

[21] L. A. Kenworthy, U.S. 3,063,807, Nov. 13, 1962, to Glidden Co.; *Chem. Abs.*, 1963, **58**, 6582.

[22] G. Jebsen, U.S. 1,361,867, Dec. 14, 1921; *Chem. Abs.*, 1921, **15**, 579: British 104,885, Mar. 15, 1917, to Norske Aktieselskab för Elektrokemisk Industrie; *Chem. Abs.*, 1917, **11**, 2050.

[23] British 590,743, Feb. 15, 1924, to Fabr. de produits chimiques de Thann et de Mulhouse; *Brit. Chem. Abstracts*, 1926, **B**, 156.

[24] R. H. Monk and A. S. Ross, U.S. 2,069,554, Feb. 2, 1937, to American Zinc, Lead and Smelting Co.; *Chem. Abs.*, 1937, **31**, 1967: British 450,797, July 24, 1936, to American Zinc, Lead and Smelting Co.; *Chem. Abs.*, 1937, **31**, 227: German 665,974, Oct. 8, 1938, to American Zinc, Lead and Smelting Co.; *Chem. Abs.*, 1939, **33**, 1110.

[25] British 447,246, May 14, 1936, to Klockner-Werke, A-G.; *Chem. Abs.*, 1936, **30**, 7291.

[26] N. Specht, U.S. 1,750,287, Mar. 11, 1930, to Deutsche Gasglühlicht, G.m.b.H.; *Chem. Abs.*, 1936, **24**, 2316.

[27] French 748,714, July 8, 1933, to Titanium Pigment Co., Inc.; *Chem. Abs.*, 1933, **27**, 5559: British 392,194, May 9, 1933, to Titan Co., A-S.; *Chem. Abs.*, 1932, **26**, 4969. G. Jebsen, U.S. 1,361,866, Dec. 14, 1921; *Chem. Abs.*, 1921, **15**, 579: Canadian 201,710, July 6, 1920; *Chem. Abs.*, 1920, **14**, 2722: French 484,712, Oct. 31, 1917, to Norske Aktieselskab för Elektrokemisk Industrie; *Chem. Abs.*, 1918, **12**, 1128: Norwegian 30,311, Jan. 19, 1920, to Titan Co., A-S.; *Chem. Abs.*, 1921, **15**, 1630: Holland 3,319, June 2, 1919, to Titan Co., A-S.; *Chem. Abs.*, 1920, **14**, 1903.

## CHAPTER 16

[1] H. Rose, *Ann. chim. et phys.*, 1844 (3) **12**, 176.

[2] N. Heaton, *J. Roy. Soc. Arts,* 1922, **70**, 552; *Chem. Abs.*, 1922, **16**, 3003.

[3] V. M. Goldschmidt, U.S. 1,348,129, July 27, 1920; *Chem. Abs.*, 1920, **14**, 2865: Canadian 201,715, July 6, 1920; *Chem. Abs.*, 1920, **14**, 2721: Danish 23,199, July 15, 1918, to Titan Co., A-S.; *Chem. Abs.*, 1919, **13**, 383: French 485,392, Jan. 4, 1918, to Titan Co., A-S.; *Chem. Abs.*, 1919, **13**, 1938: British 110,535, May 18, 1917, to Norske Aktieselskab för Elektrokemisk Industrie; *Chem. Abs.*, 1918, **12**, 536: Norwegian 28,216, Sept. 24, 1917, to Norske Aktieselskab för Elektrokemisk Industrie; *Chem. Abs.*, 1918, **12**, 869.

[4] L. E. Barton and C. E. Reynolds, British 374,420, May 31, 1932, to Titanium Pigment Co., Inc.; *Chem. Abs.*, 1933, **27**, 4107: German 607,395, Dec. 27, 1934, to Titanium Pigment Co., Inc.; *Chem. Abs.*, 1935, **29**, 2005: French 721,188, Aug. 10, 1931, to Titanium Pigment Co., Inc.; *Chem. Abs.*, 1932, **26**, 4189: French 722,035, Jan. 10, 1931, to Titanium Pigment Co., Inc.; *Chem. Abs.*, 1932, **26**, 4189.

[5] J. Blumenfeld and M. Mayer, U.S. 1,892,693, Jan. 3, 1933, to Krebs Pigment and Color Corp.; *Chem. Abs.*, 1933, **27**, 2318: J. Blumenfeld, French 720,810, July 29, 1931; *Chem. Abs.*, 1932, **26**, 4189: German 575,841, May 4, 1933, to Verein für Chemische und Metallurgische Produktion; *Chem. Abs.*, 1933, **27**, 4941: British 360,436, Aug. 2, 1930; *Chem. Abs.*, 1933, **27**, 615.

[6] J. H. Paterson, U.S. 2,266,260, Dec. 16, 1941, to E. I. du Pont de Nemours and Co.; *Chem. Abs.*, 1942, **36**, 2436.

[7] R. H. Monk, U.S. 2,304,947, Dec. 15, 1942, to American Zinc, Lead and Smelting Co.; *Chem. Abs.*, 1943, **37**, 2948.

[8] H. L. Rhodes, U.S. 1,931,683, Oct. 24, 1933, to The Glidden Co.; *Brit. Chem. Abstracts*, 1934, **B**, 576: U.S. 2,078,278, Apr. 27, 1937, to The Glidden Co.; *Chem. Abs.*, 1937, **31**, 4515.

[9] W. F. Washburn, R. Dahlstrom, and A. T. McCord, U.S. 2,166,082, July 11, 1939, to National Lead Co.; *Chem. Abs.*, 1939, **33**, 8428: French 825,897, Mar. 16, 1938, to Titan Co., Inc.; *Chem. Abs.*, 1938, **32**, 6892: British 499,153, Jan. 19, 1939, to

Titan Co., Inc.; *Chem. Abs.*, 1939, **33**, 5209: German 715,279 and 718,526, Nov. 27, 1941, to Titangesellschaft m. b. H.; *Chem. Abs.*, 1944, **38**, 2226.

[10] R. Dahlstrom, U.S. 2,170,940, Aug. 29, 1939, to National Lead Co.; *Chem. Abs.*, 1940, **34**, 276: French 831,178, Aug. 25, 1938, to Titan Co., Inc.; *Chem. Abs.*, 1939, **33**, 1973: British 507,506, June 16, 1939, to Titan Co., Inc.; *Chem. Abs.*, 1940, **34**, 649.

[11] N. Heaton, *J. Roy. Soc. Arts*, 1922, **70**, 552; *Chem. Abs.*, 1922, **16**, 3003.

[12] P. Farup, U.S. 1,360,737, Nov. 30, 1921; *Chem. Abs.*, 1921, **15**, 605: Canadian 201,704, July 6, 1920; *Chem. Abs.*, 1920, **14**, 2722.

[13] W. F. Washburn, U.S. 1,412,027, Apr. 4, 1922; *Chem. Abs.*, 1922, **16**, 2035: Canadian 225,981, Nov. 14, 1922; *Chem. Abs.*, 1923, **17**, 888: British 149,316, Apr. 30, 1920, to Titan Co., A-S.; *Chem. Abs.*, 1921, **15**, 442.

[14] P. Farup, Canadian 221,526, July 25, 1922; *Chem. Abs.*, 1922, **16**, 3405.

[15] British 439,312, Dec. 4, 1935, to Titan Co., Inc.; *Chem. Abs.*, 1936, **30**, 3261.

[16] Norwegian 27,517, Jan. 2, 1917, to Norske Aktieselskab för Elektrokemisk Industrie; *Chem. Abs.*, 1917, **11**, 1757. F. G. C. Stephens, L. J. Anderson, and W. A. Cash, U.S. 1,714,408, May 21, 1929, to National Metal and Chemical Bank, Ltd.; *Chem. Abs.*, 1929, **23**, 3587: British 273,017, Mar. 26, 1926, to National Metal and Chemical Bank, Ltd.; *Chem. Abs.*, 1928, **22**, 1862.

[17] B. W. Allan, U.S. 2,091,955, Sept. 7, 1937, to American Zirconium Corp.; *Chem. Abs.*, 1937, **31**, 7677: French 805,152, Nov. 13, 1936; *Chem. Abs.*, 1937, **31**, 3718.

[18] I. E. Weaver and A. N. C. Bennett, U.S. 2,304,719, Dec. 8, 1942; *Chem. Abs.*, 1943, **37**, 2948: British 528,955, Nov. 11, 1940, to National Titanium Pigments, Ltd.; *Chem. Abs.*, 1941, **35**, 8327.

[19] German 708,381, June 12, 1941, to I. G. Farbenind., A-G.; *Chem. Abs.*, 1943, **37**, 3285.

[20] British 418,269, Oct. 22, 1934, to Titan Co., Inc.; *Chem. Abs.*, 1935, **29**, 1666.

[21] J. E. Booge and L. Stewart, U.S. 2,361,987, Nov. 7, 1944, to E. I. du Pont de Nemours and Co.; *Chem. Abs.*, 1945, **39**, 2892.

[22] J. Drucker, U.S. 2,118,916, May 31, 1938, to I. G. Farbenind., A-G.; *Chem. Abs.*, 1938, **32**, 5647: German 675,686, May 13, 1939, to I. G. Farbenind., A-G.; *Chem. Abs.*, 1939, **33**, 7133: British 454,324, Sept. 28, 1936, to I. G. Farbenind., A-G.; *Chem. Abs.*, 1937, **31**, 1643.

[23] R. W. Ancrum and A. G. Oppegaard, British 473,312, Oct. 11, 1937, to British Titan Products Co., Ltd.; *Chem. Abs.*, 1938, **32**, 1954: French 820,351, Nov. 9, 1937, to British Titan Products Co., Ltd.; *Chem. Abs.*, 1938, **32**, 3176: Canadian 386,623, Jan. 30, 1940, to National Lead Co.; *Chem. Abs.*, 1940, **34**, 2194.

[24] F. L. Kingsbury and C. L. Schmidt, U.S. 2,211,828, Aug. 20, 1940, to National Lead Co.; *Chem. Abs.*, 1941, **35**, 640: French 844,525, July 26, 1939, to Titan Co., Inc.; *Chem. Abs.*, 1940, **34**, 7634.

[25] A. G. Oppegaard and C. J. Stopford, U.S. 2,200,373, May 14, 1940, to Titan Co., Inc.; *Chem. Abs.*, 1940, **34**, 6110: British 516,369, June 23, 1938, to British Titan Products Co., Ltd.; *Chem. Abs.*, 1941, **35**, 6133: German 703,182, Jan. 30, 1941, to Titangesellschaft m. b. H.; *Chem. Abs.*, 1942, **36**, 284.

[26] L. W. Ryan and W. J. Cauwenberg, French 762,983, Apr. 21, 1934, to Titan Co., Inc.; *Chem. Abs.*, 1934, **28**, 5262: British 415,602, Aug. 20, 1934, to Titan Co., Inc., *Chem. Abs.*, 1935, **29**, 1269.

[27] R. H. Monk, U.S. 1,605,851, Nov. 2, 1927; *Chem. Abs.*, 1927, **21**, 184: J. Irwin and R. H. Monk, British 267,788, Aug. 6, 1926, to E. C. R. Marks; *Chem. Abs.*, 1928, **22**, 1244.

[28] French 686,440, Dec. 11, 1929, to Soc. minière "La Barytine"; *Chem. Abs.*, 1931, **25**, 609.

[29] R. H. Monk and L. Firing, British 345,668, Nov. 18, 1929; *Brit. Chem. Abstracts*, 1931, **B**, 597: French 685,893, Dec. 2, 1929, to Titanium Ltd.; *Chem. Abs.*, 1930, **24**, 6040.

[30] J. Rockstroh, F. Raspe, and H. Kircher, German 633,973, Aug. 13, 1936, to I. G. Farbenind., A-G.; *Chem. Abs.*, 1937, **31**, 512.

[31] H. H. Buckman, U.S. 1,411,839, Apr. 4, 1922; *Chem. Abs.*, 1922, **16**, 2035.

[32] R. M. McKinney, U.S. 2,192,501, Mar. 5, 1940, to E. I. du Pont de Nemours and Co.; *Chem. Abs.*, 1940, **34**, 4595: British 500,012, Feb. 1, 1939, to E. I. du Pont de Nemours and Co.; *Chem. Abs.*, 1939, **33**, 5685.

[33] N. Parravano and V. Caglioti, *Gazz. chim. ital.*, 1934, **64**, 429; *Chem. Abs.*, 1935, **29**, 22: *Gazz. chim. ital.*, 1934, **64**, 703; *Chem. Abs.*, 1935, **29**, 2002.

[34] A. Zhukova and A. Sovalova, *Lakokrasochnuyu Ind., Za*, 1934, **4**, 26; *Chem. Abs.*, 1935, **29**, 6442.

[35] I. D. Hagar, *American Society for Testing Materials, Symposium on Paints and Paint Materials*, 1935.

[36] W. F. Sullivan and S. S. Cole, *J. Am. Ceram. Soc.*, 1959, **42**, 127; *Chem. Abs.*, 1959, **53**, 8907.

[37] J. E. Latty, *J. Appl. Chem.* 1958, **8**, 96; *Chem. Abs.*, 1958, **52**, 14189.

[38] P. Conjeaud, *Compt. Rend.*, 1954, **238**, 2075; *Chem. Abs.*, 1954, **48**, 13320.

[39] T. Kubo, K. Shinriki, K. Kutaka, and M. Kitahara, *Kogyo Kagaku Zasshi*, 1958, **61**, 831; *Chem. Abs.*, 1961, **55**, 20349.

[40] W. D. Ross, U.S. 2,865,622, Dec. 23, 1958, to E. I. du Pont de Nemours and Co.; *Chem. Abs.*, 1959, **53**, 6648.

[41] J. T. Richmond and R. J. Wigginton, British 853,959, Nov. 16, 1960, to Laporte Titanium Ltd.; *Chem. Abs.*, 1961, **55**, 11879.

[42] W. J. Cauwenberg and C. A. Tanner, Jr., U.S. 2,516,548, July 25, 1950, to American Cyanamid Co.; *Chem. Abs.*, 1950, **44**, 11122.

[43] J. A. Luethge, U.S. 3,105,744, Oct. 1, 1963, to National Lead Co.; *Chem. Abs.*, 1964, **60**, 223.

[44] J. E. Booge, U.S. 2,253,551, Aug. 26, 1941, to E. I. du Pont de Nemours and Co.; *Chem. Abs.*, 1941, **35**, 8327. S. S. Cole, U.S. 2,316,841, Apr. 20, 1943, to National Lead Co.; *Chem. Abs.*, 1943, **37**, 5879. A. N. C. Bennett, British 553,135, May 10, 1943, to National Titanium Pigments, Ltd.; *Chem. Abs.*, 1944, **38**, 5094.

[45] S. S. Cole, U.S. 2,316,840, Apr. 20, 1943, to National Lead Co.; *Chem. Abs.*, 1943, **37**, 5879. J. L. Keats, U.S. 2,307,048, Jan. 5, 1943, to E. I. du Pont de Nemours and Co.; *Chem. Abs.*, 1943, **37**, 3622.

[46] S. S. Cole, U.S. 2,290,539, July 21, 1942, to National Lead Co.; *Chem. Abs.*, 1943, **37**, 505: French 841,794, Aug. 6, 1938, to Titan Co., Inc.; *Chem. Abs.*, 1940, **34**, 4596: British 517,742, Aug. 6, 1938, to British Titan Products Co., Ltd.: A. N. C. Bennett, British 553,135, May 10, 1943, to National Titanium Pigments, Ltd.; *Chem. Abs.*, 1944, **38**, 5094.

[47] R. W. Ancrum, U.S. 2,365,135, Dec. 12, 1944, to Titan Co., Inc.; *Chem. Abs.*, 1945, **39**, 3680: French 865,801, Apr. 2, 1940, to Titan Co., Inc.

[48] E. Lederle and R. Brill, U.S. 2,275,856, Mar. 10, 1942; *Chem. Abs.*, 1942, **36**, 4359.

[49] R. W. Ancrum, British 561,142, May 8, 1944, to British Titan Products Co., Ltd.; *Chem. Abs.*, 1945, **39**, 5092.

[50] J. L. Keats, U.S. 2,406,465, Aug. 27, 1946, to E. I. du Pont de Nemours and Co.; *Chem. Abs.*, 1946, **40**, 6846: British 566,920, Jan. 19, 1945, to E. I. du Pont de Nemours and Co.; *Chem. Abs.*, 1947, **41**, 1467: British 565,349, Nov. 7, 1944, to E. I. du Pont de Nemours and Co.; *Chem. Abs.*, 1946, **40**, 4533.

[51] J. L. Keats and J. E. Booge, U.S. 2,358,167, Sept. 12, 1944, to E. I. du Pont de Nemours and Co.; *Chem. Abs.*, 1945, **39**, 1552.

[52] F. von Bichowsky and R. M. McKinney, U.S. 2,286,882, June 16, 1942, to E. I. du Pont de Nemours and Co.; *Chem. Abs.*, 1942, **36**, 7334.

[53] J. E. Booge, U.S. 2,253,551, Aug. 26, 1941, to E. I. du Pont de Nemours and Co.; *Chem. Abs.*, 1941, **35**, 8327.

[54] J. H. Peterson, U.S. 2,369,246, Feb. 13, 1945, to E. I. du Pont de Nemours and Co.; *Chem. Abs.*, 1945, **39**, 5092.

[55] J. E. Booge, U.S. 2,273,431, Feb. 17, 1942, to E. I. du Pont de Nemours and Co.; *Chem. Abs.*, 1942, **36**, 3976: British 530,877, Dec. 24, 1940, to E. I. du Pont de Nemours and Co.; *Chem. Abs.*, 1942, **36**, 1200.

[56] H. B. Weiser, W. O. Milligan, and E. L. Cook, *J. Phys. Chem.*, 1941, **45**, 1227; *Chem. Abs.*, 1942, **36**, 955.

[57] F. Schlossberger, *Z. Krist.*, 1942, **104**, 358; *Chem. Abs.*, 1943, **37**, 5559.
[58] C. A. Tanner and W. J. Cauwenberg, U.S. 2,427,165, Sept. 9, 1947; *Chem. Abs.*, 1948, **42**, 387.
[59] R. Neilsen and V. M. Goldschmidt, U.S. 1,343,469, June 15, 1920; *Chem. Abs.*, 1920, **14**, 2424: Canadian 201,717, July 6, 1920; *Chem. Abs.*, 1920, **14**, 2722: Canadian 201,713, July 6, 1920; *Chem. Abs.*, 1920, **14**, 2722: Canadian 201,719, July 6, 1920; *Chem. Abs.*, 1920, **14**, 2721.
[60] British 116,266, Apr. 10, 1918, to Titan Co., A-G.; *Chem. Abs.*, 1919, **13**, 64.
[61] G. Carteret and M. Devaux, British 184,132, June 27, 1921; *Chem. Abs.*, 1923, **17**, 188.
[62] British 104,885, Mar. 15, 1917, to Norske Aktieselskab för Elektrokemisk Industrie; *Chem. Abs.*, 1917, **11**, 2050.
[63] French 676,280, June 6, 1929, to Soc. minière "La Barytine"; *Chem. Abs.*, 1930, **24**, 2903.
[64] H. V. Alessandroni, *Official Digest, Federation Paint and Varnish Production Clubs*, 1949, No. 298, 504; *Chem. Abs.*, 1949, **43**, 8696.
[65] A. W. Czanderna, C. N. R. Rao, and J. M. Honig, *Trans. Faraday Soc.*, 1958, **54**, 1069; *Chem. Abs.*, 1959, **53**, 8952.
[66] H. Bach, *Naturwissenschaften*, 1964, **51**, 10; *Chem. Abs.*, 1964, **60**, 8714.
[67] O. Glemser and E. Schwarzmann, *Angew. Chem.*, 1956, **68**, 791; *Chem. Abs.*, 1957, **51**, 12720.
[68] R. D. Shannon, *J. Appl. Phys.*, 1964, **35**, 3414; *Chem. Abs.*, 1964, **61**, 15433.
[69] H. Knoll and U. Kuhnhold, *Naturwissenschaften*, 1957, **44**, 394; *Chem. Abs.*, 1958, **52**, 5214.
[70] C. A. Tanner, Jr., U.S. 2,771,345, Nov. 20, 1956, to American Cyanamid Co.; *Chem. Abs.*, 1957, **51**, 4023.
[71] L. C. Copeland and C. W. Farbex, U.S. 2,486,465, Nov. 1, 1949, to New Jersey Zinc Co.; *Chem. Abs.*, 1950, **44**, 2256.
[72] S. R. Yoganarasimham and C. N. R. Rao, *Trans. Faraday Soc.*, 1962, **58**, 1579; *Chem. Abs.*, 1963, **58**, 7444.
[73] T. Kubo and K. Shinriki, *J. Chem. Soc. Japan. Ind. Chem. Sect.*, 1953, **56**, 149; *Chem. Abs.*, 1954, **48**, 9635.
[74] K. Ohashi and S. Takeuchi, *Kagaku*, 1954, **24**, 473; *Chem. Abs.*, 1954, **48**, 13178.
[75] J. T. Richmond and G. G. Durrant, British 865,327, April 12, 1961, to Laporte Titanium Ltd.; *Chem. Abs.*, 1961, **55**, 26392.
[76] British 877,734, Nov. 14, 1958, to Farbenfabriken Bayer A.-G.; *Chem. Abs.*, 1962, **56**, 6905.
[77] Y. Iida and S. Ozaki, *J. Am. Chem. Soc.*, 1961, **44**, 120; *Chem. Abs.*, 1961, **55**, 11791.
[78] T. Kubo, K. Shinriki, K. Kulaka, and M. Kitahara, *Kogyo Kagaku Zasshi*, 1958, **61**, 831; *Chem. Abs.*, 1961, **55**, 23242.
[79] W. J. Cauwenberg and H. L. Sanders, U.S. 2,477,559, Aug. 2, 1949, to American Cyanamid Co.; *Chem. Abs.*, 1950, **44**, 349.
[80] E. Wainer, U.S. 2,489,246, Nov. 22, 1949, to National Lead Co.; *Chem. Abs.*, 1950, **44**, 1662.
[81] L. E. Ross and C. A. Tanner, Jr., U.S. 2,494,492, Jan. 10, 1950, to American Cyanamid Co.; *Chem. Abs.*, 1956, **50**, 4266.
[82] W. J. Cauwenberg and C. A. Tanner, Jr., U.S. 2,389,026, Nov. 13, 1945, to American Cyanamid Co.; *Chem. Abs.*, 1946, **40**, 999.
[83] L. Aagaard, U.S. 2,488,755, Nov. 22, 1949, to American Cyanamid Co.; *Chem. Abs.*, 1950, **44**, 1662.
[84] L. Aagaard, U.S. 2,723,186, Nov. 8, 1955, to National Lead Co.; *Chem. Abs.*, 1956, **50**, 2989.
[85] C. A. Tanner, U.S. 2,505,344, April 25, 1950, to American Cyanamid Co.; *Chem. Abs.*, 1950, **44**, 7498.
[86] F. Majdik and B. Balla, *Magyar Kem. Lapja*, 1960, **15**, No. 4, 168; *Chem. Abs.*, 1960, **54**, 23221.

[87] M. L. Hanahan, U.S. 2,044,941, June 23, 1936, to E. I. du Pont de Nemours and Co.; *Chem. Abs.*, 1936, **30**, 5821: U.S. 1,937,037, Nov. 28, 1934, to E. I. du Pont de Nemours and Co.; *Chem. Abs.*, 1934, **28**, 1206: British 525,472, Aug. 29, 1940, to E. I. du Pont de Nemours and Co.; *Chem. Abs.*, 1941, **35**, 6819.

[88] J. Blumenfeld, British 385,315, Dec. 19, 1932; *Chem. Abs.*, 1933, **27**, 4358: German 560,979, Apr. 18, 1931, to Verein für Chemische und Metallurgische Produktion; *Chem. Abs.*, 1933, **27**, 1108: French 735,338, Apr. 18, 1932, to Soc. de produits chimiques des terres rares; *Chem. Abs.*, 1933, **27**, 1216.

[89] R. M. McKinney, U.S. 2,378,148, June 12, 1945, to E. I. du Pont de Nemours and Co.; *Chem. Abs.*, 1945, **39**, 3946: British 525,472, Feb. 23, 1938, to E. I. du Pont de Nemours and Co.; *Chem. Abs.*, 1941, **35**, 6819.

[90] B. W. Allan, U.S. 2,299,120, Oct. 20, 1942, to American Zirconium Co.; *Chem. Abs.*, 1943, **37**, 1883: French 862,077, Feb. 26, 1941, to Titan Co., Inc.

[91] W. K. Nelson, U.S. 2,084,917, June 22, 1937, to National Lead Co.; *Chem. Abs.*, 1937, **31**, 6038: British 440,611, Jan. 2, 1936, to Titan Co., Inc.; *Chem. Abs.*, 1936, **30**, 4025: French 790,190, Nov. 15, 1935, to Titan Co., Inc.; *Chem. Abs.*, 1936, **30**, 3261: German 666,409, Oct. 19, 1938, to Titan Co., Inc.; *Chem. Abs.*, 1939, **33**, 2354.

[92] R. W. Ancrum and A. G. Oppegaard, U.S. 2,246,030, June 17, 1941, to Titan Co., Inc.; *Chem. Abs.*, 1941, **35**, 6133.

[93] L. C. Eckels, U.S. 2,216,879, Oct. 8, 1940, to E. I. du Pont de Nemours and Co.; *Chem. Abs.*, 1941, **35**, 1253.

[94] A. G. Oppegaard and R. W. Ancrum, British 499,228, Jan. 20, 1939, to British Titan Products Co., Ltd.; *Chem. Abs.*, 1939, **33**, 5209.

[95] S. C. Lyons, U.S. 2,131,841, Oct. 4, 1938, to Bird Machine Co.; *Chem. Abs.*, 1938, **32**, 9529: British 507,700, June 2, 1939, to Bird Machine Co.; *Chem. Abs.*, 1940, **34**, 649.

[96] R. M. McKinney, U.S. 2,150,236, Mar. 14, 1938, to E. I. du Pont de Nemours and Co.; *Chem. Abs.*, 1939, **33**, 4805: French 844,855, Aug. 2, 1939, to E. I. du Pont de Nemours and Co.; *Chem. Abs.*, 1940, **34**, 8312: British 508,530, July 3, 1939, to E. I. du Pont de Nemours and Co.; *Chem. Abs.*, 1940, **34**, 2622: German 717,077, Jan. 15, 1942, to E. I. du Pont de Nemours and Co.; *Chem. Abs.*, 1944, **38**, 2515.

[97] C. de Rohden, U.S. 1,797,760, Mar. 24, 1931, to Commercial Pigments Co.; *Chem. Abs.*, 1931, **25**, 2821.

[98] D. L. Gamble and L. D. Grady, Jr., British 472,001, Sept. 15, 1937, to Imperial Smelting Corp., Ltd.; *Chem. Abs.*, 1938, **32**, 1498.

[99] A. N. Dempster and W. K. Nelson, U.S. 2,933,408, April 19, 1960, to National Lead Co.; *Chem. Abs.*, 1960, **54**, 14721.

[100] F. L. Kingsbury, U.S. 2,744,029, May 1, 1956, to National Lead Co.; *Chem. Abs.*, 1956, **50**, 12502.

[101] W. R. Whatley, U.S. 2,464,192, March 8, 1949, to American Cyanamid Co.; *Chem. Abs.*, 1949, **43**, 4028.

[102] W. R. Whatley, U.S. 2,480,092, Aug. 23, 1949, to American Cyanamid Co.; *Chem. Abs.*, 1949, **43**, 9482.

[103] J. T. Richmond and G. G. Durrant, German 1,023,165, Jan. 23, 1958, to Laporte Titanium Ltd.; *Chem. Abs.*, 1960, **54**, 17911.

[104] C. A. Tanner, Jr. and D. C. Hall, U.S. 2,628,919, Feb. 17, 1953, to American Cyanamid Co.; *Chem. Abs.*, 1953, **47**, 4630.

[105] W. R. Whatley, U.S. 2,626,707, Jan. 27, 1953, to American Cyanamid Co.; *Chem. Abs.*, 1953, **47**, 3581.

[106] I. D. Hagar, *op. cit.*

[107] M. A. Lissman, *Chem. & Met. Eng.*, 1938, **45**, No. 5.

[108] French 849,675, Nov. 29, 1939, to Emile Ch. de Stubner; *Chem. Abs.*, 1941, **35**, 6847.

[109] R. W. Sullivan, U.S. 2,254,630, Sept. 2, 1941, to E. I. du Pont de Nemours and Co.; *Chem. Abs.*, 1941, **35**, 8327.

[110] French 848,406, Oct. 30, 1939, to E. I. du Pont de Nemours and Co.; *Chem. Abs.*, 1941, **35**, 5730.

[111] C. E. Berry, U.S. 2,274,521, Feb. 24, 1943, to E. I. du Pont de Nemours and Co.; *Chem. Abs.*, 1942, **36**, 4359.

[112] C. M. Downs and H. F. Saunders, U.S. 2,386,885, Oct. 16, 1945, to Sherwin-Williams Co.; *Chem. Abs.*, 1946, **40**, 477.

[113] F. L. Kingsbury and F. Schultz, U.S. 2,365,559 and 2,365,560, Dec. 19, 1944, to National Lead Co.; *Chem. Abs.*, 1945, **39**, 4499.

[114] L. E. Barton and C. E. Reynolds, British 374,420, May 31, 1932, to Titanium Pigment Co., Inc.; *Chem. Abs.*, 1933, **27**, 4107: German 607,395, Dec. 27, 1934, to Titanium Pigment Co., Inc.; *Chem. Abs.*, 1935, **29**, 2005: French 722,035, Jan. 10, 1931, to Titanium Pigment Co., Inc.; *Chem. Abs.*, 1932, **26**, 4189: French 721,188, Aug. 10, 1931, to Titanium Pigment Co., Inc.; *Chem. Abs.*, 1932, **26**, 4189.

[115] H. L. Rhodes, U.S. 1,931,682, Oct. 24, 1934, to The Glidden Co.; *Chem. Abs.*, 1934, **28**, 662.

[116] H. L. Rhodes, U.S. 2,078,278, Apr. 27, 1937, to The Glidden Co.; *Chem. Abs.*, 1937, **31**, 4515.

[117] R. M. McKinney, U.S. 2,150,235, Mar. 14, 1938, to E. I. du Pont de Nemours and Co.; *Chem. Abs.*, 1939, **33**, 4805: French 838,919, Mar. 20, 1939, to E. I. du Pont de Nemours and Co.; *Chem. Abs.*, 1939, **33**, 7133.

[118] S. Kubin, U.S. 1,885,921, Nov. 1, 1932, to Krebs Pigment and Color Corp.; *Chem. Abs.*, 1933, **27**, 1216: British 354,799, Feb. 16, 1929, to Soc. de produits chimiques des terres rares; *Chem. Abs.*, 1932, **26**, 3942.

[119] L. T. Work and S. B. Tuwiner, *Ind. Eng. Chem.*, 1934, **26**, 1266; *Chem. Abs.*, 1935, **29**, 945.

[120] J. E. Booge, U.S. 2,046,054, June 30, 1936, to E. I. du Pont de Nemours and Co.; *Chem. Abs.*, 1936, **30**, 5821.

[121] L. E. Barton, French 768,435, Aug. 6, 1934, to Titan Co., Inc.; *Chem. Abs.*, 1935, **29**, 625.

[122] E. Sauer and W. Sussmann, *Kolloid-Z.*, 1938, **82**, 253; *Chem. Abs.*, 1938, **32**, 4360.

[123] M. L. Hanahan and J. D. Prince, U.S. 2,214,815, Sept. 17, 1940, to E. I. du Pont de Nemours and Co.; *Chem. Abs.*, 1941, **35**, 919: French 850,882, Dec. 28, 1939, to E. I. du Pont de Nemours and Co.; *Chem. Abs.*, 1942, **36**, 2169.

[124] R. H. Sawyer, U.S. 2,346,085, Apr. 4, 1944, to E. I. du Pont de Nemours and Co.; *Chem. Abs.*, 1944, **38**, 5093.

[125] N. Parravano and D. Caglioti, *Chimica e industria (Milan)*, 1935, **17**, 141; *Chem. Abs.*, 1935, **29**, 5288.

[126] H. K. Carter, U.S. 2,232,164, Feb. 18, 1941, to E. I. du Pont de Nemours and Co.; *Chem. Abs.*, 1941, **35**, 3468.

[127] J. B. Sutton, U.S. 2,226,147, Dec. 24, 1940, to E. I. du Pont de Nemours and Co.; *Chem. Abs.*, 1941, **35**, 2348.

[128] J. A. Geddes, U.S. 2,219,129, Oct. 22, 1940, to E. I. du Pont de Nemours and Co.; *Chem. Abs.*, 1941, **35**, 1253.

[129] M. L. Hanahan, U.S. 2,231,467, Feb. 11, 1941, to E. I. du Pont de Nemours and Co.; *Chem. Abs.*, 1941, **35**, 3467.

[130] W. H. Madson and W. H. Daiger, U.S. 2,216,536, Oct. 1, 1940, to E. I. du Pont de Nemours and Co.; *Chem. Abs.*, 1941, **35**, 919.

[131] A. E. Williams, *Paint Manuf.*, 1960, **30**, 153; *Chem. Abs.*, 1961, **55**, 2133.

## CHAPTER 17

[1] F. H. McBerty, U.S. 2,098,056, Nov. 2, 1937, to E. I. du Pont de Nemours and Co.; *Chem. Abs.*, 1938, **32**, 101.

[2] S. F. Spangler, *Chem. & Met. Eng.*, 1935, **42**, 139. See also R. R. Smith and J. Belding, U.S. 2,185,095, Dec. 26, 1939, to Chemical Construction Co.; *Chem. Abs.*, 1940, **34**, 2778.

[3] J. M. O'Shaughnessy, U.S. 2,394,470, Feb. 5, 1946, to National Lead Co.; *Chem. Abs.*, 1946, **40**, 2595.

[4] D. R. Clarkson, U.S. 2,255,445, Sept. 9, 1941, to Ella van Gorder; *Chem. Abs.*, 1942, **36**, 228.

[5] P. de Lattre, British 523,241, July 14, 1940; *Chem. Abs.*, 1941, **35**, 6072.

[6] F. Heinrich, *Stahl u. Eisen*, 1938, **58**, 617; *Chem. Abs.*, 1938, **32**, 6995.

[7] J. Ourisson, *Bull. soc. ind. Mulhouse*, 1934, **100**, 565; *Chem. Abs.*, 1935, **29**, 1267.

[8] N. Specht, U.S. 1,827,691, Oct. 13, 1932, to Krebs Pigment and Color Corp.; *Chem. Abs.*, 1932, **26**, 414: French 673,076, Apr. 13, 1930, to Deutsche Gasglühlicht, G.m.b.H.; *Chem. Abs.*, 1930, **24**, 2418.

[9] L. G. Bousquet and M. J. Brooks, U.S. 2,280,508, Apr. 21, 1942, to General Chemical Co.; *Chem. Abs.*, 1942, **36**, 5621.

[10] W. Hardiek, U.S. 2,038,078, Apr. 21, 1936, to Davison Chemical Corp.; *Chem. Abs.*, 1936, **30**, 3951.

[11] C. A. Klein and R. S. Brown, British 277,769, July 8, 1926; *Chem. Abs.*, 1928, **22**, 2628.

[12] R. L. McClarey, U.S. 2,192,687, Mar. 5, 1940, to E. I. du Pont de Nemours and Co.; *Chem. Abs.*, 1940, **34**, 4529: French 848,405, Oct. 27, 1939, to E. I. du Pont de Nemours and Co.; *Chem. Abs.*, 1941, **35**, 5655.

[13] P. Parrish, British 500,193, Jan. 30, 1939; *Chem. Abs.*, 1939, **33**, 5610.

[14] T. C. Oliver, B. D. Long, and L. H. Crosson, U.S. 2,333,672, Nov. 2, 1943, to Charlotte Chemical Laboratory, Inc.; *Chem. Abs.*, 1944, **38**, 2306.

[15] M. S. Platonov, V. A. Zakharova, and S. M. Efros, *J. Applied Chem.* (*U.S.S.R.*), 1940, **13**, 1119; *Chem. Abs.*, 1941, **35**, 2286.

[16] Belgian 567,504, May 31, 1958, to Société Belge du Titane S. A.; *Chem. Abs.*, 1959, **53**, 14435.

[17] V. Stein and R. Michaels, German 1,129,140, May 10, 1962, to *Metallgesellschaft A.-G.*; *Chem. Abs.*, 1962, **57**, 4314.

[18] Z. Cizinsky, *Chem. prumysl*, 1960, **10**, 231; *Chem. Abs.*, 1960, **54**, 23213.

[19] Belgian 617,513, Aug. 31, 1962, to Titangesellschaft m. b. H.; *Chem. Abs.*, 1963, **58**, 11003.

[20] British 959,333, May 27, 1964, to Farbenfabriken Bayer A.-G.; *Chem. Abs.*, 1964, **61**, 3962.

[21] Y. Ozaki, *J. Soc. Chem. Ind. Japan*, 1944, **47**, 853; *Chem. Abs.*, 1949, **43**, 1924: *J. Soc. Chem. Ind. Japan*, 1944, **47**, 855; *Chem. Abs.*, 1949, **43**, 1924.

[22] T. Kubo, M. Taniguchi, K. Yamaguchi, and T. Hagashi, *Kogyo Kagaku Zasshi*, 1959, **62**, 340; *Chem. Abs.*, 1962, **57**, 14700.

[23] T. Kubo and M. Taniguchi, *Kogyo Kagaku Zasshi*, 1962, **65**, 312; *Chem. Abs.*, 1962, **57**, 10771.

[24] T. Kubo and M. Taniguchi, *Kogyo Kagaku Zasshi*, 1962, **65**, 315; *Chem. Abs.*, 1962, **57**, 10771.

[25] M. C. Chang, U.S. 2,867,524, Jan. 6, 1959, to Crucible Steel Company of America; *Chem. Abs.*, 1959, **53**, 6555.

[26] K. Kubo, Japanese 4469, May 30, 1959, to Chiyoda Chemical Eng. and Construction Co., Ltd.; *Chem. Abs.*, 1960, **54**, 8008.

[27] K. Kubo and M. Taniguchi, Japanese 15,921, Sept. 11, 1961, to Chiyoda Chemical Eng. and Construction Co., Ltd.; *Chem. Abs.*, 1963, **59**, 8382.

[28] T. Kubo, *Kogyo Kagaku Zasshi*, 1964, **67**, 863; *Chem. Abs.*, 1964, **61**, 11643.

[29] E. Junger, *Chem. Tech.*, 1963, **15**, 280; *Chem. Abs.*, 1963, **59**, 13617.

[30] H. E. Broughton, U.S. 2,452,517, Oct. 26, 1948, to Chemical Construction Corp.; *Chem. Abs.*, 1949, **43**, 1927. F. J. Bartholomew, *Chem. Eng.*, 1950, **57**, No. 8, 118; *Chem. Abs.*, 1950, **44**, 8846.

[31] *Chemical Age (London)*, 1960, **84**, No. 2164, 1075.

[32] British 650,722, Feb. 28, 1951, to Spolek pro chemichou a hutni vyrobu, narodni podnik; *Chem. Abs.*, 1951, **45**, 8727.

[33] T. H. Milliken, Jr., U.S. 2,926,070, Feb. 23, 1960, to Houdry Process Corp.; *Chem. Abs.*, 1960, **54**, 13575.

[34] F. L. Kingsbury and F. J. Schultz, U.S. 2,576,462, Nov. 27, 1951, to National Lead Co.; *Chem. Abs.*, 1952, **46**, 2314.

[35] A. L. Nugey, U.S. 2,642,334, June 16, 1953; *Chem. Abs.*, 1953, **47**, 8298.

[36] J. Schulmann, *Chem. Prumysl*, 1953, 3(28), 193; *Chem. Abs.*, 1954, **48**, 7826.

[37] A. C. Redfield and L. A. Walford, *Natl. Research Council, Natl. Acad. Sci. Publ.* No. 201, 1951; *Chem. Abs.*, 1954, **48**, 8993.

## CHAPTER 18

[1] W. B. Llewellyn, U.S. 2,088,913, Aug. 3, 1937, to P. Spence and Sons, Ltd.; *Chem. Abs.*, 1937, **31**, 6834.

[2] J. Brode and C. Wurster, German 507,151, Dec. 17, 1925; *Chem. Abs.*, 1931, **25**, 479.

[3] E. Riley, *J. Chem. Soc.*, 1862, **15**, 339.

[4] E. E. Dutt, British 189,700, Feb. 10, 1922; *Chem. Abs.*, 1923, **17**, 2481.

[5] H. V. Alessandroni, U.S. 2,167,626, Aug. 1, 1939, to National Lead Co.; *Chem. Abs.*, 1939, **33**, 9021: B. A. Rubin, U.S.S.R. 52,528, Jan. 31, 1938; *Chem. Abs.*, 1940, **34**, 3936.

[6] D. H. Dawson, I. J. Krchma, and R. M. McKinney, Canadian 377,396, Nov. 1, 1938, to Canadian Industries, Ltd.; *Chem. Abs.*, 1939, **33**, 2292.

[7] E. Vigoroux and G. Arrivant, *Compt. rend.*, 1907, **144**, 485; *Chem. Abs.*, 1907, **1**, 1362.

[8] L. E. Barton, U.S. 1,189,229, July 4, 1916; *Chem. Abs.*, 1916, **10**, 2129.

[9] H. Zirngible and H. J. Kappey, German 1,170,385, May 21, 1964, to Farbenfabriken Bayer A.-G.; *Chem. Abs.*, 1964, **61**, 3982.

[10] H. Zirngible and H. J. Kappey, Belgian 620,418, Nov. 14, 1963, to Farbenfabriken Bayer A.-G.; *Chem. Abs.*, 1963, **59**, 243.

[11] H. Ruter, F. Weingarten, and E. Cherdron, German 1,083,244, June 15, 1960, to Gebs. Giulini G. m. b. H.; *Chem. Abs.*, 1961, **55**, 15310.

[12] F. Strain, British 801,424, Sept. 10, 1958, to Columbia-Southern Chemical Co.: *Chem. Abs.*, 1959, **53**, 5611. German 1,045,382, Dec. 4, 1958, to Columbia-Southern Chemical Co.; *Chem. Abs.*, 1960, **54**, 25638.

[13] F. J. Schultz, U.S. 2,576,483, Nov. 27, 1951, to National Lead Co.; *Chem. Abs.*, 1952, **46**, 874.

[14] L. M. Schechter and J. J. Leddy, U.S. 3,069,235, March 5, 1963, to Dow Chemical Co.; *Chem. Abs.*, 1963, **58**, 11012.

[15] A. W. Evans, A. N. C. Bennett, and P. J. Russell, British 714,857, Sept. 1, 1954, to British Titan Products Co., Ltd.; *Chem. Abs.*, 1955, **49**, 3553.

[16] F. deCarli, Italian 481,533, June 3, 1953; *Chem. Abs.*, 1955, **49**, 3769.

[17] R. M. Comstack, F. V. Schossberger, and F. A. Ficulka, U.S. 3,032,393, May 1, 1963, to Armour Research Foundation of Illinois Inst. of Tech.; *Chem. Abs.*, 1962, **57**, 3082.

[18] H. Schaper and W. Fuhi, *Z. anorg. Allgem. Chem.*, 1962, **319**, 52; *Chem. Abs.*, 1963, **58**, 10750.

[19] C. M. Olson and I. J. Krchma, U.S. 2,486,572, Nov. 1, 1949, to E. I. du Pont de Nemours and Co., *Chem. Abs.*, 1950, **44**, 3223.

[20] P. Kubelka, U.S. 1,899,574, Feb. 28, 1933, to Krebs Pigment and Color Corp.; *Chem. Abs.*, 1933, **27**, 3043.

[21] P. Kubelka, U.S. 1,899,573, Feb. 28, 1933, to Krebs Pigment and Color Corp.; *Chem. Abs.*, 1933, **27**, 3043.

[22] P. Kubelka, U.S. 1,899,572, Feb. 28, 1933, to Krebs Pigment and Color Corp.; *Chem. Abs.*, 1933, **27**, 3043: German 578,791, June 17, 1933, to Verein für Chemische und Metallurgische Produktion; *Chem. Abs.*, 1934, **28**, 866.

[23] W. Y. Agnew, U.S. 2,393,582, Jan. 29, 1946, to National Lead Co.; *Chem. Abs.*, 1946, **40**, 2395.

[24] G. M. Toptygina and I. S. Morozov, *Zhur. Neorg. Khim.*, 1961, **6**, 1685; *Chem. Abs.*, 1962, **56**, 3107.

[24a] G. Yuravskii, *Compt. rend.*, 1936, **203**, 373; *Chem. Abs.*, 1936, **30**, 7501.

[25] P. Kubelka and J. Srbek, U.S. 2,062,133 and 2,062,134, Nov. 24, 1936, to E. I. du Pont de Nemours and Co.; *Chem. Abs.*, 1937, **31**, 893: J. Blumenfeld, British 384,875, Dec. 15, 1932, to Verein für Chemische und Metallurgische Produktion; *Chem. Abs.*, 1933, **27**, 4359: German 568,255, Mar. 21, 1931; *Chem. Abs.*, 1933, **27**, 2769: French 733,960, Mar. 21, 1932, to Verein für Chemische und Metallurgische Produktion; *Chem. Abs.*, 1933, **27**, 1216.

[26] P. Tillmann and F. Raspe, U.S. 2,303,305 and 2,303,307, Nov. 24, 1942, to Titan Co., Inc.; *Chem. Abs.*, 1943, **37**, 2525: German 700,918, Dec. 5, 1940, to Titangesellschaft m. b. H.; *Chem. Abs.*, 1941, **35**, 7739: German 718,169, Feb. 12, 1942, to Titangesellschaft m. b. H.; *Chem. Abs.*, 1944, **38**, 2458: German 732,235, Jan. 28, 1943, to Titangesellschaft m. b. H.; *Chem. Abs.*, 1944, **38**, 841: German 707,020, May 8, 1941, to Titangesellschaft m. b. H.; *Chem. Abs.*, 1942, **36**, 2097: British 513,867, Oct. 24, 1939, to Titangesellschaft m. b. H.; *Chem. Abs.*, 1941, **35**, 2287: French 837,238, Feb. 6, 1939, to Titangesellschaft m. b. H.

[27] Jelks Barksdale, U.S. 2,285,485, June 9, 1942, to National Lead Co.; *Chem. Abs.*, 1942, **36**, 6761: British 536,141, May 5, 1941, to Titan Co., Inc.; *Chem. Abs.*, 1942, **36**, 1446: French 861,320, Feb. 6, 1941, to Titan Co., Inc.; British 549,326, Nov. 16, 1943, to British Titan Products Co.; *Chem. Abs.*, 1944, **38**, 623.

[28] Jelks Barksdale and W. W. Plechner, U.S. 2,285,486, June 9, 1942, to National Lead Co.; *Chem. Abs.*, 1942, **36**, 6762: Norwegian 65,233, Sept. 21, 1942, to Titan Co., A-S.; *Chem. Abs.*, 1946, **40**, 686.

[29] F. von Bichowsky, U.S. 2,337,215, Dec. 21, 1943, to E. I. du Pont de Nemours and Co.; *Chem. Abs.*, 1944, **38**, 3427.

[30] F. von Bichowsky, U.S. 2,286,881, June 16, 1942, to E. I. du Pont de Nemours and Co.; *Chem. Abs.*, 1942, **36**, 7246.

[31] Jelks Barksdale, British 576,588, Apr. 11, 1946, to British Titan Products Co., Ltd.; *Chem. Abs.*, 1948, **42**, 2064: British 550,995, Feb. 3, 1943, to British Titan Products Co., Ltd.; *Chem. Abs.*, 1944, **38**, 1892.

[32] D. B. Pall, U.S. 2,426,788, Sept. 2, 1947, to American Cyanamid Co.; *Chem. Abs.*, 1947, **41**, 7774.

[33] W. W. Plechner and A. W. Hixson, U.S. 2,113,945 and 2,113,946, Apr. 12, 1938, to National Lead Co.; *Chem. Abs.*, 1938, **32**, 4291: Reissue 21,693, Jan. 14, 1941, to National Lead Co.; *Chem. Abs.*, 1941, **35**, 3044: British 481,892, Mar. 8, 1938, to Titan Co., Inc.; *Chem. Abs.*, 1938, **32**, 6891: British 497,694, Dec. 23, 1938, to Titan Co., Inc.; *Chem. Abs.*, 1939, **33**, 3980: French 813,757, June 8, 1937, to Titan Co., Inc.; *Chem. Abs.*, 1938, **32**, 1059: French 828,741, May 27, 1938, to Titan Co., Inc.; *Chem. Abs.*, 1939, **33**, 329.

[34] W. W. Plechner and J. M. Jarmus, *Ind. Eng. Chem., Anal. Ed.*, 1934, **6**, 447.

[35] A. V. Pamfilov, V. E. Kiseleva, and G. V. Millinskaya, *J. Applied Chem. (U.S.S.R.)*, 1938, **11**, 621; *Chem. Abs.*, 1938, **32**, 6569.

[36] E. R. Darling, U.S. 1,315,011, Sept. 2, 1918; *Chem. Abs.*, 1919, **13**, 2746.

[37] L. G. Jenness, U.S. 2,044,753, June 16, 1936, to Intermetal Corp.; *Chem. Abs.*, 1936, **30**, 5738: British 420,105, May 26, 1933, to Intermetal Corp.; *Chem. Abs.*, 1935, **29**, 3538: German 660,744, June 2, 1938, to Intermetal Corp.; *Chem. Abs.*, 1938, **32**, 6891: French 756,605, Dec. 13, 1933, to Intermetal Corp.; *Chem. Abs.*, 1934, **28**, 2477.

[38] H. V. Alessandroni, U.S. 2,167,628, Aug. 1, 1939, to National Lead Co.; *Chem. Abs.*, 1939, **33**, 9021: French 844,360, July 24, 1940, to Titan Co., Inc.; *Chem. Abs.*, 1940, **34**, 7552.

[39] J. L. Turner and W. W. Plechner, U.S. 2,441,856, May 18, 1948, to National Lead Co.; *Chem. Abs.*, 1948, **42**, 5626.

[40] E. E. Dutt, British 189,700, Feb. 10, 1922; *Chem. Abs.*, 1923, **17**, 2481.

[41] L. E. Barton, U.S. 1,189,229, July 4, 1916; *Chem. Abs.*, 1916, **10**, 2129.

[42] L. E. Barton, U.S. 1,223,356, Apr. 24, 1917, to Titanium Alloy Mfg. Co.; *Chem. Abs.*, 1917, 11, 1912.

[43] German 568,309, May 4, 1927, to Soc. française du titane; *Chem. Abs.*, 1933, 27, 2049.

[44] G. Carteret and M. Devaux, British 185,374, June 27, 1921; *Chem. Abs.*, 1923, 17, 188.

[45] G. Carteret, *L'age de fer*, 1920, 36, 670; *Chem. Abs.*, 1921, 15, 928.

[46] A. Gutbier, B. Ottenstein, E. Leutheusser, and F. Allam, *Z. anorg. u. allgem. Chem.*, 1927, 162, 87; *Chem. Abs.*, 1927, 21, 3511.

[47] French 638,051, July 20, 1927, to I. G. Farbenind., A-G.; *Chem. Abs.*, 1929, 23, 244.

[48] J. A. Atanasiu, *Bul. Chim. Soc. Română Stiinte*, 1924, 27, 81; *Chem. Abs.*, 1927, 21, 1742.

[49] N. Parravano and V. Caglioti, *Gazz. chim. ital.*, 1934, 64, 429; *Chem. Abs.*, 1935, 29, 22.

[50] W. Grave and J. P. Urbanek, Canadian 506,093, Sept. 28, 1954, to Canadian Titanium Pigments, Ltd.; *Chem. Abs.*, 1956, 50, 9037.

[51] A. Hloch and G. Wehner, *Chem. Tech.* (*Berlin*), 1956, 8, 677; *Chem. Abs.*, 1957, 51, 11941.

[52] H. Sueta, Japan 269, Jan. 20, 1954, to Mitsubishi Chemical Industry Co.; *Chem. Abs.*, 1954, 48, 14141.

[53] T. A. Ermolaeva and N. S. Anufrieva, *Lakokrasochnye Material in ikh Primenenie*, 1960, No. 1, 38; *Chem. Abs.*, 1960, 54, 23221.

[54] K. Fumaki and Y. Saeki, *Kogyo Kagaku Zasshi*, 1956, 59, 1291; *Chem. Abs.*, 1958, 52, 16105.

[55] L. Aagaard and F. J. Schultz, U.S. 2,622,964, Dec. 23, 1952, to National Lead Co.; *Chem. Abs.*, 1953, 47, 2947.

[56] A. K. Babko, G. I. Gridchina, and B. I. Nabivanets, *Zhur. Neorg. Khim.*, 1962, 7, No. 1, 132; *Chem. Abs.*, 1962, 56, 13619.

[57] K. Funaki and Y. Saeki, *Kogyo Kagaku Zasshi*, 1956, 59, 1295; *Chem. Abs.*, 1958, 52, 16105.

[58] J. L. Turner and W. W. Plechner, U.S. 2,441,856, May 18, 1948, to National Lead Co.; *Chem. Abs.*, 1948, 42, 5626. Canadian 506,094, Sept. 28, 1954, to Canadian Titanium Pigments Corp.; *Chem. Abs.*, 1956, 50, 9037.

[59] British 572,866, Oct. 26, 1945, to British Titan Products Co.; *Chem. Abs.*, 1949, 43, 7358.

[60] J. Katsurai and A. Kitahara, *Kolloid Z.*, 1956, 148, 155; *Chem. Abs.*, 1957, 51, 9390.

[61] H. Haber and P. Kubelka, U.S. 1,913,380, Oct. 17, 1933, to Krebs Pigment and Color Corp.; *Chem. Abs.*, 1934, 28, 268. J. Blumenfeld, British 358,492, July 7, 1930; *Chem. Abs.*, 1932, 26, 4969: German 551,448, July 8, 1930, to Verein für Chemische und Metallurgische Produktion; *Chem. Abs.*, 1932, 26, 4424.

[62] R. H. McInery, E. P. Williams, and H. L. Glaze, U.S. 1,842,620, Jan. 24, 1932, to Minerals Increment Co.; *Chem. Abs.*, 1932, 26, 1727.

[63] M. Mayer, U.S. 1,885,934, Dec. 1, 1932, to Krebs Pigment and Color Corp.; *Brit. Chem. Abstracts*, 1933, B, 785. J. Blumenfeld, French 671,106, Mar. 8, 1929; *Chem. Abs.*, 1930, 24, 1942: British 307,881, Mar. 15, 1928, to Verein für Chemische und Metallurgische Produktion; *Chem. Abs.*, 1930, 24, 210: German 533,836, Mar. 16, 1928, to Verein für Chemische und Metallurgische Produktion; *Chem. Abs.*, 1932, 26, 812.

[64] B. Simek, L. Jaeger, and O. Hajek, Czechoslovakia 106,001, Dec. 15, 1962; *Chem. Abs.*, 1964, 60, 6517.

[65] R. F. Hudson, *Proc. Intern. Congr. Pure and Applied Chem.*, 1947, 11, 297; *Chem. Abs.*, 1950, 44, 7697.

[66] K. Fumaki and Y. Saeki, *Kogyo Kagaku Zasshi*, 1956, 59, 1297; *Chem. Abs.*, 1958, 52, 16105.

[67] P. Hagenmuller, A. Lecref, and M. Tourmoux, *Compt. Rend.*, 1959, **248**, 2009; *Chem. Abs.*, 1962, **56**, 11198.
[68] British 574,818, Jan. 22, 1946, to British Titan Products Co. Ltd.; *Chem. Abs.*, 1949, **43**, 1994.
[69] Belgian 616,426, Oct. 15, 1962, to British Titan Products Co. Ltd.; *Chem. Abs.*, 1963, **58**, 8694.
[70] S. Wilska, *Suomen Kemistilehti*, 1956, **29A**, 247; *Chem. Abs.*, 1957, **51**, 3946.
[71] R. E. Robinson, G. S. James, C. N. van Zyl, and H. J. Bovey, *S. African Ind. Chemist*, 1960, **14**, No. 5, 84; *Chem. Abs.*, 1961, **55**, 15852.
[72] H. S. Cooper, U.S. 2,752,300, June 26, 1956, to Walter M. Weil; *Chem. Abs.*, 1956, **50**, 12796.
[73] N. N. Murach and L. G. Povedskaya, U.S.S.R. 116,155, Nov. 22, 1958; *Chem. Abs.*, 1959, **53**, 14438.
[74] British 715,257, Sept. 8, 1954, to Titan Co. Inc.; *Chem. Abs.*, 1955, **49**, 3769.
[75] J. Kamlet, U.S. 2,804,375, Aug. 27, 1957, to National Distillers and Chemical Corp.; *Chem. Abs.*, 1958, **52**, 7101.
[76] I. E. P. Belykova and A. A. Dnernyakova, *Ukr. Khim. Zh.*, 1963, **29**, 220; *Chem. Abs.*, 1963, **59**, 3575.
[77] J. Kamlet, U.S. 2,804,375, Aug. 27, 1957, to National Distillers and Chemical Co.; *Chem. Abs.*, 1958, **52**, 7101.
[78] British 421,308, Mar. 27, 1934, to Titan Co., Inc.; *Brit. Chem. Abstracts*, 1935, **B**, 226: French 770,915, Sept. 24, 1934, to Titan Co., Inc.; *Chem. Abs.*, 1935, **29**, 894: German 704,506, Feb. 27, 1941, to Titan Co., Inc.; *Chem. Abs.*, 1942, **36**, 1446.
[79] A. V. Pamfilov and E. G. Shtandel, *J. Gen. Chem. (U.S.S.R.)*, 1936, **6**, 300; *Chem. Abs.*, 1937, **31**, 1317.
[80] D. H. Dawson, I. J. Krchma, and R. M. McKinney, U.S. 2,127,247, Aug. 16, 1938, to E. I. du Pont de Nemours and Co.; *Chem. Abs.*, 1938, **32**, 7883.
[81] J. H. Bachman, German 1,014,534, Aug. 29, 1957, to Columbia-Southern Chemical Corp.; *Chem. Abs.*, 1960, **54**, 12517.
[82] S. Wilska, *Suomen Kemistilehti*, 1958, **B31**, 156; *Chem. Abs.*, 1958, **52**, 13571.
[83] Y. Takimoto, *J. Chem. Soc. Japan, Ind. Chem. Section*, 1955, **58**, 732; *Chem. Abs.*, 1956, **50**, 9255.
[84] R. P. Green and H. Judd, U.S. 3,043,655, July 10, 1963, to Champion Papers, Inc.; *Chem. Abs.*, 1962, **57**, 12105.
[85] E. P. Belyakova and A. A. Dvernyakova, *Ukr. Khim. Zh.*, 1963, **29**, 633; *Chem. Abs.*, 1963, **59**, 12426.
[86] R. Sturgeon and E. Seddon, British 652,576, April 25, 1951, to United Glass Bottle Manufacturers Ltd.; *Chem. Abs.*, 1951, **45**, 10526.
[87] J. Kamlet, U.S. 2,804,375, Aug. 27, 1957, to National Distillers and Chemical Co.; *Chem. Abs.*, 1958, **52**, 7101.
[88] J. J. Leddy and D. L. Schechter, U.S. 3,060,002, Oct. 23, 1962, to Dow Chemical Co.; *Chem. Abs.*, 1963, **58**, 2178.
[89] E. Wainer, U.S. 2,960,387, Nov. 15, 1960, to Horizons, Inc.; *Chem. Abs.*, 1961, **55**, 5887.
[90] C. Pascaud, *Compt. rend.*, 1950, **231**, 1232; *Chem. Abs.*, 1951, **45**, 2639.
[91] G. Nakazaiva and K. Terunuma, Japan 1117, Feb. 21, 1958, to Bureau of Industrial Technics; *Chem. Abs.*, 1959, **53**, 9594.
[92] E. P. Belyakova and A. A. Dvernyakova, *Titan i Ego Splavy*, Akad. Nauk S.S.S.R., Inst. Met., 1962, 124; *Chem. Abs.*, 1963, **59**, 8390.
[93] British 944,905, Dec. 18, 1963, to Australia Ministry of Mines; *Chem. Abs.*, 1964, **60**, 8946.
[94] H. Judd, U.S. 3,112,178, Nov. 23, 1963, to Champion Papers Inc.; *Chem. Abs.*, 1964, **60**, 6516.
[95] S. Nakao and T. Sasaki, Japanese 2972, April 30, 1955, to Nippon Soda Co.; *Chem. Abs.*, 1957, **51**, 14220.
[96] H. S. Cooper, U.S. 2,802,721, Aug. 13, 1957, to Walter M. Weil; *Chem. Abs.*, 1957, **51**, 18509.

[97] O. Moklebust, U.S. 2,811,434, Oct. 29, 1957, to National Lead Co.; *Chem. Abs.*, 1958, **52**, 225: U.S. 2,778,724, Jan. 22, 1957, to National Lead Co.; *Chem. Abs.*, 1957, **51**, 6491: British 754,453, Aug. 8, 1956, to Titan Co., A. S.; *Chem. Abs.*, 1957, **51**, 3423; British 791,366, Feb. 20, 1958, to Titan Co., A. S.; *Chem. Abs.*, 1958, **52**, 14504.

[98] D. L. Armant and H. S. Sigurdson, U.S. 2,680,681, June 8, 1954, to National Lead Co.; *Chem. Abs.*, 1954, **48**, 10522.

## CHAPTER 19

[1] A. V. Pamfilov and E. G. Shtandel, *J. Gen. Chem. (U.S.S.R.)*, 1937, **7**, 258; *Chem. Abs.*, 1937, **31**, 4609.

[2] Y. Saeki and K. Fumaki, *Kogyo Kagaku Zasshi*, 1957, **60**, 403; *Chem. Abs.*, 1959, **53**, 6740.

[3] L. E. Barton, U.S. 1,179,394, Apr. 18, 1916; *Chem. Abs.*, 1916, **10**, 1584.

[4] R. J. McInery, E. F. Williams, and H. L. Glaze, U.S. 1,888,996, Dec. 29, 1932, to Minerals Increment Co.; *Brit. Chem. Abstracts*, 1934, **B**, 865.

[5] T. E. Thorpe, *J. Chem. Soc.*, 1885, **T**, 119.

[6] German 531,400, Aug. 14, 1926, to I. G. Farbenind., A-G.; *Chem. Abs.*, 1931, **25**, 5521.

[7] G. G. Urazov, I. S. Morozov, and M. P. Shmantsan, *J. Applied Chem. (U.S.S.R.)*, 1937, **10**, 6; *Chem. Abs.*, 1937, **31**, 4460.

[8] W. Kangro and R. Jahn, *Z. anorg. u. allgem. Chem.*, 1933, **210**, 325; *Chem. Abs.*, 1933, **27**, 1838.

[9] A. V. Pamfilov and M. G. Shikher, *J. Gen. Chem. (U.S.S.R.)*, 1937, **7**, 2760; *Chem. Abs.*, 1938, **32**, 3715.

[10] I. N. Godnev and A. V. Pamfilov, *J. Gen. Chem. (U.S.S.R.)*, 1937, **7**, 1264; *Chem. Abs.*, 1937, **31**, 6088.

[11] F. K. McTaggart, *J. Council Sci. Ind. Research*, 1945, **18**, 5; *Chem. Abs.*, 1945, **39**, 3127.

[12] A. V. Pamfilov, A. S. Khudyakov, and E. G. Shtandel, *J. Prakt. Chem. (U.S.S.R.)*, 1935, **142**, 232; *Chem. Abs.*, 1935, **29**, 5762: *J. Applied Chem. (U.S.S.R.)*, 1936, **9**, 1781; *Brit. Chem. Abstracts*, 1937, **B**, 34: *J. Applied Chem. (U.S.S.R.)*, 1936, **9**, 1770; *Chem. Abs.*, 1937, **31**, 2538.

[13] D. Helbig, Italian 274,540.

[14] W. Y. Agnew and S. S. Cole, U.S. 2,378,675, June 13, 1945, to National Lead Co.; *Chem. Abs.*, 1945, **39**, 3887.

[15] P. S. Brallier, U.S. 2,401,543, June 4, 1946, to Stauffer Chemical Co.; *Chem. Abs.*, 1946, **40**, 5214.

[16] S. S. Cole, U.S. 2,479,904, Aug. 23, 1949, to National Lead Co.; *Chem. Abs.*, 1949, **43**, 9398.

[17] J. A. Rogers, U.S. 2,635,036, Apr. 14, 1953, to Dow Chemical Co.; *Chem. Abs.*, 1953, **47**, 6331.

[18] L. R. Lyons, U.S. 2,753,243, July 3, 1956, to Titanium Metals Corp. of America; *Chem. Abs.*, 1956, **50**, 16058.

[19] J. Rakosnik, *Hutnicke Listy*, 1954, **9**, 268; *Chem. Abs.*, 1954, **48**, 12002.

[20] A. Stefanesw, L. Ornstrat, and C. Mina, *Rev. Chim. (Bucharest)*, 1958, **9**, 387; *Chem. Abs.*, 1961, **55**, 11774.

[21] H. M. Cyr, F. S. Griffith, and C. M. McFarland, U.S. 2,723,903, Nov. 15, 1955, to New Jersey Zinc Co.; *Chem. Abs.*, 1956, **50**, 3718. British 768,867, Feb. 20, 1957, to New Jersey Zinc Co.; *Chem. Abs.*, 1957, **51**, 9108. German 1,029,354, May 8, 1958, to New Jersey Zinc Co.; *Chem. Abs.*, 1960, **54**, 25637.

[22] K. Anazarva and E. Tsuruta, Japanese 2269, April 5, 1958, to Furukarva Mining Co.; *Chem. Abs.*, 1959, **53**, 12607.

[23] British 929,207, June 19, 1963 to Montecatini; *Chem. Abs.*, 1963, **59**, 7181.

[24] A. S. Berengard, V. A. Kazhemyakin, and N. A. Filatova, *Titan i ego Splavy,* Akad. Nauk, S.S.S.R. Inst. Met., 1961, No. 5, 181; *Chem. Abs.,* 1962, **57**, 12100.

[25] G. B. Cobel and J. A. Rogers, U.S. 2,963,360, Dec. 6, 1960, to Dow Chemical Co.; *Chem. Abs.,* 1961, **55**, 6802.

[26] C. E. Lesher, U.S. 2,794,728, June 4, 1957, to Lesher and Associates; *Chem. Abs.,* 1957, **51**, 12804.

[27] British 744,415, Feb. 8, 1956, to National Lead Co.; *Chem. Abs.,* 1956, **50**, 14192.

[28] Y. Takimoto, *J. Chem. Soc. Japan, Ind. Chem. Sect.,* 1955, **58**, 494; *Chem. Abs.,* 1955, **49**, 16368.

[29] A. M. Thomasen, U.S. 2,858,193, Oct. 28, 1958, to Champion Papers and Fiber Co.; *Chem. Abs.,* 1959, **53**, 2556.

[30] Italian 631,849, Jan. 16, 1962, to Montecatini; *Chem. Abs.,* 1963, **58**, 3128.

[31] T. Takahashi, *J. Chem. Soc. Japan, Ind. Chem. Sect.,* 1954, **57**, 339; *Chem. Abs.,* 1955, **49**, 4386.

[32] R. Cortez, German 1,126,143, May 22, 1962, to Pittsburgh Plate Glass Co.; *Chem. Abs.,* 1963, **58**, 272.

[33] W. H. Coates and J. Hayden, U.S. 2,957,757, Oct. 25, 1960, to British Titan Products Co., Ltd.; *Chem. Abs.,* 1961, **55**, 2042.

[34] British 772,695, April 17, 1957, to Horizons Titanium Corp.; *Chem. Abs.,* 1957, **51**, 11673.

[35] L. W. Rowe and S. S. Cole, U.S. 2,555,374, June 5, 1951, to National Lead Co.; *Chem. Abs.,* 1951, **45**, 10524.

[36] C. H. Gorski, *J. Metals,* 191, *Trans.,* 1951, 131; *Chem. Abs.,* 1951, **45**, 2637.

[37] C. C. Patel and D. P. Kharkan, *J. Sci. Ind. Research (India),* 1961, **20D**, 60; *Chem Abs.,* 1961, **55**, 20346.

[38] D. P. Kharkan and C. C. Patel, *J. Mines, Metals and Fuels,* 1960, **8**, No. 7, 129; *Chem. Abs.,* 1961, **55**, 14223.

[39] R. M. McKinney, British 776,295, June 5, 1957, to E. I. du Pont de Nemours and Co.; *Chem. Abs.,* 1957, **51**, 15084.

[40] A. Belchetz, U.S. 2,486,912, Nov. 1, 1949, to Stauffer Chemical Co.; *Chem. Abs.,* 1950, **44**, 2714.

[41] A. W. Evans and J. D. Groves, British 724,193, Feb. 16, 1955, to British Titan Products Co. Ltd.; *Chem. Abs.,* 1955, **49**, 12790.

[42] K. Anazarva, Japanese 6867, Oct. 23, 1954, to Furukarva Mining Co.; *Chem. Abs.,* 1956, **50**, 3718.

[43] W. Kangro, German 1,011,151, June 27, 1957; *Chem. Abs.,* 1960, **54**, 7510.

[44] British 893,067, April 4, 1962, to Titanium Metals Corp. of America; *Chem. Abs.,* 1962, **56**, 13875.

[45] R. M. McKinney, U.S. 2,701,179, Feb. 1, 1955, to E. I. du Pont de Nemours and Co.; *Chem. Abs.,* 1955, **49**, 8572.

[46] A. W. Evans and J. D. Groves, U.S. 2,855,273, Oct. 7, 1958, to British Titan Products Co. Ltd.; *Chem. Abs.,* 1959, **53**, 5611. British 724,193, to British Titan Products Co., Ltd.; *Chem. Abs.,* 1955, **49**, 12790.

[47] A. N. C. Bennett and J. D. Groves, U.S. 2,868,622, Jan. 13, 1959, to British Titan Products Co., Ltd.; *Chem. Abs.,* 1959, **53**, 7531: British 792,151, March 10, 1958, to British Titan Products Co., Ltd.; *Chem. Abs.,* 1958, **52**, 19043: German 1,032,727, June 26, 1958, to British Titan Products Co., Ltd.; *Chem. Abs.,* 1960, **54**, 23223.

[48] A. W. Evans and J. D. Groves, British 861,991, March 1, 1961, to British Titan Products Co., Ltd.; *Chem. Abs.,* 1961, **55**, 15857.

[49] W. E. Dunn, U.S. 2,856,264, Oct. 14, 1958, to E. I. du Pont de Nemours and Co.; *Chem. Abs.,* 1959, **53**, 785.

[50] British 872,198, Aug. 27, 1957, to Stora Koppasbergs Bergslags Aktiebolag; *Chem. Abs.,* 1962, **56**, 5634.

[51] K. Anzuwa and E. Tsuruta, U.S. 2,777,756, Jan. 15, 1957, to Furukawa Electro-Industries Co.; *Chem. Abs.,* 1957, **51**, 5602.

[52] W. S. Coates and J. Hayden, German 1,035,901, Aug. 7, 1958, to British Titan Products Co., Ltd.; *Chem. Abs.*, 1960, **54**, 25637.

[53] W. H. Coates and J. Hayden, British 890,944, March 7, 1962, to British Titan Products Co.; *Chem. Abs.*, 1962, **56**, 15323.

[54] M. C. Irani, German 1,081,426, April 24, 1958, to Metal Chlorides Corp.; *Chem. Abs.*, 1962, **56**, 1150.

[55] G. A. Fredenmarch and L. J. R. Lindestrom, Swedish 151,418, Sept. 13, 1955, to Aktiebolaget Nynas-Petroleum; *Chem. Abs.*, 1956, **50**, 2932.

[56] J. D. Groves, German 1,083,243, June 15, 1960, to British Titan Products Co., Ltd.; *Chem. Abs.*, 1961, **55**, 22734.

[57] British 871,433, June 28, 1961, to Metal Chlorides Corp.; *Chem. Abs.*, 1961, **55**, 25707.

[58] E. Wainer, U.S. 3,006,728, Sept. 10, 1959, to Horizons Inc.; *Chem. Abs.*, 1962, **56**, 2170.

[59] L. R. Barrett, F. H. Clews, and A. T. Green, *Gas Research Board*, 1940, No. 2, 11; *Chem. Abs.*, 1941, **35**, 4167.

[60] I. E. Muskat and R. H. Taylor, U.S. 2,184,887, Dec. 26, 1939, to Pittsburgh Plate Glass Co.; *Chem. Abs.*, 1940, **34**, 2776.

[61] French 817,502, Sept. 4, 1937, to Deutsche Gold und Silber Scheideanstalt vorm. Roessler; *Chem. Abs.*, 1938, **32**, 2037.

[62] K. H. Donaldson, U.S. 2,120,602, June 14, 1938; *Chem. Abs.*, 1938, **32**, 6015.

[63] C. Rohden, U.S. 1,707,257, Apr. 4, 1929, to Commercial Pigments Corp; *Chem. Abs.*, 1929, **23**, 2538.

[64] K. F. Beloglagov, *Contributions to Study of Mineral Resources of U.S.S.R.*, 1926, No. **56**, 20; *Chem. Abs.*, 1927, **21**, 2053.

[65] A. V. Pamfilov and E. G. Shtandel, *J. Applied Chem. (U.S.S.R.)*, 1936, **9**, 1770; *Chem. Abs.*, 1937, **31**, 2538.

[66] B. D. Saklatwalla, U.S. 1,845,342, Feb. 16, 1932, to Vanadium Corp.; *Brit. Chem. Abstracts*, 1932, **B**, 1029.

[67] I. E. Muskat and R. H. Taylor, U.S. 2,184,884, Dec. 26, 1939, to Pittsburgh Plate Glass Co.; *Chem. Abs.*, 1940, **34**, 2776: British 533,379, Feb. 12, 1941, to Pittsburgh Plate Glass Co.; *Chem. Abs.*, 1942, **36**, 1581.

[68] I. E. Muskat and R. H. Taylor, U.S. 2,184,885, Dec. 26, 1939, to Pittsburgh Plate Glass Co.; *Chem. Abs.*, 1940, **34**, 2776: British 533,378, Feb. 12, 1941, to Pittsburgh Plate Glass Co.; *Chem. Abs.*, 1942, **36**, 999.

[69] F. J. Cleveland, British 553,056, May 6, 1943, to Pittsburgh Plate Glass Co.; *Chem. Abs.*, 1944, **38**, 4392: British 535,179; *Chem. Abs.*, 1942, **36**, 1446.

[70] F. J. Cleveland, British 548,352, Oct. 7, 1942, to Pittsburgh Plate Glass Co.; *Chem. Abs.*, 1944, **38**, 460: British 548,145, Sept. 28, 1942, to Pittsburgh Plate Glass Co.; *Chem. Abs.*, 1943, **37**, 6417.

[71] I. E. Muskat and R. H. Taylor, U.S. 2,245,076, June 10, 1941, to Pittsburgh Plate Glass Co.; *Chem. Abs.*, 1941, **35**, 5845.

[72] W. Kangro, German 666,351, Oct. 17, 1938; *Chem. Abs.*, 1939, **33**, 2096.

[73] D. P. Kharkar and C. C. Patel, *Current Sci. (India)*, 1952, **21**, 98; *Chem. Abs.*, 1952, **46**, 8817.

[74] P. P. Phatnagar and T. Benerjee, Indian 58,244, Feb. 19, 1958, to Council of Scientific and Industrial Research; *Chem. Abs.*, 1958, **52**, 12343.

[75] S. Nakao, Japanese 3905, June 9, 1955, to Nippon Soda Co.; *Chem. Abs.*, 1957, **51**, 12803.

[76] C. E. Lesher, *Trans. Am. Inst. Mining Met. Engrs.*, 202, Tech. Pub. 4058–CF, 1955; *Chem. Abs.*, 1956, **50**, 133.

[77] E. K. Plant, U.S. 2,805,120, Sept. 3, 1957, to Columbia-Southern Chemical Corp.; *Chem. Abs.*, 1958, **52**, 2352: British 783,006, Sept. 18, 1957, to Columbia-Southern Chemical Corp.; *Chem. Abs.*, 1958, **52**, 9538.

[78] A. Bergholm, *Jernkontorets Ann.*, 1961, **145**, 205; *Chem. Abs.*, 1961, **55**, 23236.

[79] Indian 43,073, Oct. 6, 1951, to Indian Institute of Science; *Chem. Abs.*, 1952, **46**, 2986.

[80] A. B. Bezukladnikov, *Zhur. Priklad. Khim.* 1960, **33,** 1240; *Chem. Abs.,* 1960, **54,** 20112.

[81] A. W. Evans, A. N. C. Bennett, and J. D. Groves, British 762,583, Nov. 28, 1956, to British Titan Products Co., Ltd.; *Chem. Abs.,* 1957, **51,** 7667.

[82] C. E. Lesher, U.S. 2,794,728, June 4, 1957, to Lesher and Associates, Inc.; *Chem. Abs.,* 1957, **51,** 12804.

[83] E. W. Nelson and W. E. Teer, U.S. 3,067,005, Dec. 4, 1962, to American Cyanamid Co.; *Chem. Abs.,* 1963, **58,** 7637.

[84] E. C. Perkins and R. S. Lang, *U.S. Bur. Mines, Rept. Invest.,* No. 5763; *Chem. Abs.,* 1961, **55,** 16921.

[85] J. D. Richards, F. O. Rommery, and R. Roseman, U.S. 2,118,732, Jan. 21, 1964; *Chem. Abs.,* 1964, **60,** 10257.

[86] Y. I. Ivashentsev and N. A. Akhmazova, *Tr. T. G. U., S. K.,* 1962, **154,** 184; *Chem. Abs.,* 1964, **60,** 5063.

[87] A. Wataribe, J. Fukada, and A. Goko, Japanese 10,921, Dec. 28, 1956, to Nippon Soda Co.; *Chem. Abs.,* 1958, **52,** 14,999.

[88] G. Carteret and M. Devaux, U.S. 1,528,319, Mar. 2 1925; *Chem. Abs.,* 1925, **19,** 1502: British 184,948, June 27, 1921; *Chem. Abs.,* 1923, **17,** 1113: Canadian 221,537, Aug. 1, 1922; *Chem. Abs.,* 1922, **16,** 3368: *L'age de fer,* 1920, **36,** 670; *Chem. Abs.,* 1921, **15,** 928.

[89] C. C. Patel and G. V. Jere, Trans. *A.I.M.E.,* 1960, **218,** 219; *Chem. Abs.,* 1960, **54,** 13569.

[90] L. W. Rowe and R. P. Smith, U.S. 2,657,976, Nov. 3, 1953, to National Lead Co.; *Chem. Abs.,* 1954, **48,** 2555: British 678,998, Sept. 10, 1952, to Titan Co., Inc.; *Chem. Abs.,* 1954, **48,** 100.

[91] J. M. Daubenspeck and C. L. Schmidt, U.S. 2,852,362, Sept. 16, 1958, to National Lead Co.; *Chem. Abs.,* 1959, **53,** 1079.

[92] W. Frey, Swiss 325,452, Dec. 31, 1957 to Saurefabrik Schweizerhall; *Chem. Abs.,* 1958, **52,** 15861.

[93] G. V. Jere and C. C. Patel, *J. Mines, Metals and Fuels,* 1960, **8,** No. 7, 133; *Chem. Abs.,* 1961, **55,** 14223.

[94] D. P. Kharkar and C. C. Patel, *J. Sci. Ind. Research,* 1958, **17B,** 367; *Chem. Abs.,* 1959, **53,** 7892.

[95] F. E. Love, L. R. Lyons, and J. C. Priscu, U.S. 2,933,373, April 19, 1960, to Titanium Metals Corp. of America; *Chem. Abs.,* 1960, **54,** 15207.

[96] S. Wilska, *Suomen Kemistilehti,* 1956, **29A,** 220; *Chem. Abs.,* 1957, **51,** 4801.

[97] K. B. Lee and J. H. Shim, *Bull. Sci. Research Inst. (Korea),* 1959, **4,** No. 1, 12; *Chem. Abs.,* 1960, **54,** 837.

[98] B. Wilcox, U.S. 2,589,466, March 18, 1952; *Chem. Abs.,* 1952, **46,** 5798.

[99] A. P. Engelmann, U.S. 3,144,303, Aug. 11, 1964, to E. I. du Pont de Nemours and Co.; *Chem. Abs.,* 1964, **61,** 12979.

[100] A. V. Pamfilov and E. G. Shtandel, *J. Applied Chem. (U.S.S.R.),* 1936, **9,** 1781; *Chem. Abs.,* 1937, **31,** 2539.

[101] T. Ishino and Y. Takimoto, *Technol. Repts. Osaka Univ.,* 1955, **5,** 185; *Chem. Abs.,* 1956, **50,** 8147.

[102] H. Ishizuka, Japanese 1869, March 22, 1955; *Chem. Abs.,* 1957, **51,** 2242.

[103] L. D. Petterolf and C. M. McFarland, *Chem. Eng. Progr.,* 1960, **56,** No. 5, 68; *Chem. Abs.,* 1960, **54,** 19363.

[104] M. M. Barr, H. L. Gilbert, and D. D. Harper, *U.S. Bur. of Mines. Rept. Invest.* 5431, 1958; *Chem. Abs.,* 1959, **53,** 5604.

[105] J. W. Anderson, U.S. 2,936,217, May 10, 1960, to Monsanto Chemical Co.; *Chem. Abs.,* 1960, **54,** 18912.

[106] I. N. Tselik and N. S. Ukshe, *Tsvetnye Metally,* 1959, **32,** No. 9, 49; *Chem. Abs.,* 1960, **54,** 9226.

[107] Y. Takimoto and H. Hattori, *J. Chem. Soc. Japan, Ind. Chem. Sect.,* 1955, **58,** 250; *Chem. Abs.,* 1955, **49,** 14283.

[108] Y. Takimoto and H. Hattori, *J. Chem. Soc. Japan, Ind. Chem. Sect.*, 1955, **58**, 253; *Chem. Abs.*, 1955, **49**, 14283.

[109] H. Ishizuka, U.S. 2,805,919, Sept. 10, 1957; *Chem. Abs.*, 1958, **52**, 226.

[110] *Chemical Engineering Progress*, 1960, **56**, No. 5, 68.

[111] I. M. Cheprasov, U.S.S.R. 159,990, Jan. 14, 1964; *Chem. Abs.*, 1964, **60**, 11653.

[112] E. C. Perkins, H. Dolezal, D. M. Taylor, and R. S. Lang, *U.S. Bur. Mines, Rept. Invest.* No. 6317, 1963; *Chem. Abs.*, 1964, **60**, 3764.

[113] E. C. Perkins, H. Dolezal, H. Leitch, and R. S. Lang, *U.S. Bur. Mines, Rept. Invest.* No. 5983, 1962; *Chem. Abs.*, 1962, **57**, 3115.

[114] A. W. Evans, A. N. C. Bennett, and J. D. Groves, British 921,531, March 20, 1963, to British Titan Products Co., Ltd.; *Chem. Abs.*, 1963, **58**, 12210.

[115] E. I. Machkasov, E. N. Suleimenov, and V. D. Ponomarev, *Tr. T. M. i O., A. N. K. S. S. R.*, 1963, **8**, 32; *Chem. Abs.*, 1964, **60**, 14164.

[116] I. J. Krchma, U.S. 2,701,180, Feb. 1, 1955, to E. I. du Pont de Nemours and Co.; *Chem. Abs.*, 1955, **49**, 7820.

[117] S. Kiuchi and Y. Okahara, *Nippon Kogyo Kaishi*, 1961, **77**, 115; *Chem. Abs.*, 1962, **56**, 11288.

[118] R. Cortes, U.S. 3,074,777, Jan. 22, 1963, to Pittsburgh Plate Glass Co.; *Chem. Abs.*, 1963, **58**, 8679.

[119] Z. L. Hair, U.S. 2,784,058, March 5, 1957, to E. I. du Pont de Nemours and Co.; *Chem. Abs.*, 1957, **51**, 9108.

[120] E. A. Mason and C. M. Cobb, U.S. 2,928,724, March 15, 1960, to Ionics, Inc.; *Chem. Abs.*, 1960, **54**, 14605.

[121] E. I. Machkasov, V. D. Ponomarev, and Y. M. Spivak, *Isvest. Akad. Nauk. Kazakh S. S. R.*, 1961, No. 1, 41; *Chem. Abs.*, 1962, **56**, 8365.

[122] J. N. Anderson, U.S. 2,842,425, July 8, 1958, to Stauffer Chemical Co.; *Chem. Abs.*, 1958, **52**, 15404.

[123] R. Onoda, J. Sakai, and S. Kinoshita, Japanese 7873, Oct. 28, 1955, to Toa Synthetic Chemical Industry Co.; *Chem. Abs.*, 1958, **52**, 10521.

[124] J. T. Bashour, H. L. Bikofsky, and J. N. Haimsohn, U.S. 2,974,009, Sept. 9, 1957, to Stauffer Chemical Co.; *Chem. Abs.*, 1962, **56**, 4369.

[125] J. W. Anderson, U.S. 2,936,217, May 10, 1960, to Monsanto Chemical Co.; *Chem. Abs.*, 1960, **54**, 18912.

[126] A. G. Oppegaard, H. Aas, and B. Hauge, Norwegian 92,999, Dec. 8, 1958, to Titan Co., A./S.; *Chem. Abs.*, 1959, **53**, 15505: British 836,079, June 1, 1960, to Titan Co., A./S.; *Chem. Abs.*, 1960, **54**, 25638.

[127] T. N. Chang and M. H. Wang, *Chem. World*, 1958, **13**, 361; *Chem. Abs.*, 1960, **54**, 23221.

[128] Z. Kozielska and Z. Orman, *Prace Inst. Hutniczych*, 1960, **12**, No. 2, 85: *Chem. Abs.*, 1960, **54**, 23216.

[129] R. Cortes, British 894,250, April 18, 1962, to Columbia-Southern Chemical Co.; *Chem. Abs.*, 1962, **57**, 12105.

[130] J. W. Andersen, German 1,064,931, Sept. 10, 1959, to Monsanto Chemical Co.; *Chem. Abs.*, 1961, **55**, 10825.

[131] J. M. Daubenspeck and R. S. McNeil, U.S. 2,747,987, May 29, 1956, to National Lead Co.; *Chem. Abs.*, 1956, **50**, 11922.

[132] L. Aagaard and S. S. Cole, U.S. 2,622,005, Dec. 16, 1952, to National Lead Co.; *Chem. Abs.*, 1953, **47**, 2947; British 702,109, Jan. 6, 1954, to Titan Co., Inc.; *Chem. Abs.*, 1954, **48**, 6664: German 1,042,554, Nov. 6, 1958, to Titangesellschaft m. b. H.; *Chem. Abs.*, 1960, **54**, 18912.

[133] L. Aagaard and L. W. Rowe, U.S. 2,622,006, Dec. 16, 1952, to National Lead Co.; *Chem. Abs.*, 1953, **47**, 2947: British 702,108, Jan. 6, 1954, to Titan Co., Inc.; *Chem. Abs.*, 1954, **48**, 6664.

[134] H. Higashihara and Omoda, Japanese 1467, March 5, 1958, to Tao Synthetic Chemical Industries Co.; *Chem. Abs.*, 1959, **53**, 10682.

[135] K. Sugimura, Japanese 2784, April 17, 1958, to Mitsubishi Chemical Industries Co.; *Chem. Abs.*, 1959, **53**, 13525.

[136] K. Sugimura, Japanese 2785, April 17, 1958, to Mitsubishi Chemical Industries Co.; *Chem. Abs.*, 1959, **53**, 13525.

[137] R. Cortes, U.S. 3,074,777, Jan. 22, 1963, to Pittsburgh Plate Glass Co.; *Chem. Abs.*, 1963, **58**, 8679.

[138] E. C. Perkins, H. Dolezal, H. Leitch, and R. S. Lang, *U.S. Bur. Mines, Rept. Invest.* No. 5983, 1962; *Chem. Abs.*, 1962, **57**, 3115.

[139] V. A. Ilichev and A. M. Vladimirova, *Titan i ego Splavy, Akad. Nauk S.S.S.R., Inst. Met.*, 1961, No. 5, 245; *Chem. Abs.*, 1962, **57**, 9443.

[140] H. Nishimura, I. Kushima, J. Moriyama, A. Yamaguchi, and S. Nishimura, *Suiyokaishi*, 1955, **13**, 31; *Chem. Abs.*, 1957, **51**, 14455.

[141] S. Yagi, D. Kunii, T. Yoshida, and H. Umeyama, *Kagaku Kogaku*, 1959, **23**, 566; *Chem. Abs.*, 1960, **54**, 193.

[142] H. Nishimura, I. Kushima, J. Moriyama, and K. Ishi, *Bull. Eng. Research Inst. Kyoto University*, 1954, **6**, 9; *Chem. Abs.*, 1955, **49**, 4950.

[143] L. Terebesi and J. Kornyei, *Kohaszati Lapok*, 1957, **90**, 460; *Chem. Abs.*, 1958, **52**, 19035.

[144] A. Aradi, *Femipari Kutato Intezet Kozlemenyei*, 1958, No. 2, 48; *Chem. Abs.*, 1959, **53**, 15819.

[145] A. V. Serebryakova, *Trudy Inst. Met., Akad. Nauk S.S.S.R., Ural Filial*, 1958, No. 2, 55; *Chem. Abs.*, 1960, **54**, 15859.

[146] A. P. Giraitis, U.S. 2,868,621, Jan. 13, 1959, to Ethyl Corporation; *Chem. Abs.*, 1959, **53**, 10678.

[147] S. Sawada and K. Fuji, Japanese 264, Jan. 20, 1956, to Nippon Soda Co.; *Chem. Abs.*, 1957, **51**, 4667: British 781,220, Aug. 14, 1957, to Nippon Soda Co.; *Chem. Abs.*, 1957, **51**, 17119: German 1,026,286, March 20, 1958, to Nippon Soda Co., Ltd.; *Chem. Abs.*, 1960, **54**, 9230.

[148] A. B. Bezukladnikov and Y. E. Vilnyanskii, *Zhur. Priklad. Khim.* 1961, **34**, 49; *Chem. Abs.*, 1961, **55**, 10250.

[149] Y. A. Polyakov and N. V. Baryshnikov, *Titan i ego Splavy, Akad. Nauk S.S.S.R., Inst. Metals*, 1961, No. 5, 143; *Chem. Abs.*, 1962, **57**, 10775.

[150] A. V. Serebryakova and V. V. Efremkin, *Titan i Ego Splavy, Akad. Nauk S.S.S.R. Inst. Met. im. A. A. Baikova*, 1959, No. 2, 78; *Chem. Abs.*, 1960, **54**, 14027.

[151] C. T. Hill, U.S. 2,970,887, Feb. 7, 1961, to Texas Gulf Sulfur Co.; *Chem. Abs.*, 1961, **55**, 11782.

[152] A. B. Bezukladnikov, *Zh. Prikl. Khim.*, 1963, **36**, 451; *Chem. Abs.*, 1963, **58**, 13422.

[153] I. M. Cheprasov, I. N. Korobov, and M. K. Baibekov, U.S.S.R. 161,492, March 19, 1964; *Chem. Abs.*, 1964, **61**, 5233.

[154] S. I. Oreshkin, *Contributions to Study of Mineral Resources of U.S.S.R.*, 1926, No. **56**, 14; *Chem. Abs.*, 1927, **21**, 2053.

[155] A. Stahler, *Ber.*, 1905, **38**, 2619; *J. Chem. Soc.*, 1905, **A**, 595.

[156] S. E. Verkhlovskii, *J. Applied Chem. (U.S.S.R.)*, 1938, **11**, 12; *Chem. Abs.*, 1938, **32**, 4804: *Novosti Tekhniki*, 1939, No. **18**, 38; *Chem. Abs.*, 1939, **33**, 8931.

[157] P. S. Brallier, *Trans. Am. Electrochem. Soc.*, 1926, **49**; *Chem. Abs.*, 1926, **20**, 1693.

[158] W. Juda and M. C. Cretella, U.S. 2,870,071, Jan. 20, 1959, to Ionics, Inc.; *Chem. Abs.*, 1959, **53**, 5926.

[159] Y. Uebayashi, Japanese 4522, June 30, 1955; *Chem. Abs.*, 1957, **51**, 14221.

[160] J. N. Haimsohn, U.S. 2,962,352, Nov. 29, 1960, to Stauffer Chemical Co.; *Chem. Abs.*, 1961, **55**, 7778.

[161] G. A. Favre, French 800,688, July 16, 1936; *Chem. Abs.*, 1936, **30**, 8541: British 458,892, Dec. 29, 1936, to Société des produits chim. de Saint Bueil; *Chem. Abs.*, 1937, **31**, 4065.

[162] E. Vigoroux and G. Arrivant, *Bull. soc. chim.*, 1907, **1**, 19; *Chem. Abs.*, 1907, **1**, 824.

[163] F. de Carli, *Atti congr. naz. chim. pura applicata*, **1923**, 399; *Chem. Abs.*, 1924, **18**, 2849.

[164] H. R. Ellis, *Chem. News*, 1907, **95**, 122; *Chem. Abs.*, 1907, **1**, 1226.

[165] A. Koster, German 1,045,985, Dec. 11, 1958, to Henkel & Cie G. m. b. H.; *Chem. Abs.*, 1960, **54**, 25627.

[166] V. G. Gopienko, U.S.S.R. 139,658, March 2, 1961; *Chem. Abs.*, 1963, **56**, 8299.

[167] A. R. Globus, U.S. 2,858,189, Oct. 28, 1958, to United International Research, Inc.; *Chem. Abs.*, 1959, **53**, 2556.

[168] J. F. Gall, G. Barth-Wehrenalp, and A. Kowalski, U.S. 2,797,980, July 2, 1957, Pennsalt Chemical Corp.; *Chem. Abs.*, 1957, **51**, 15080: German 1,040,516, Oct. 9, 1958, to Pennsalt Chemical Corp.; *Chem. Abs.*, 1960, **54**, 25627.

[169] G. Barth-Wehrenalp, U.S. 2,819,148, Jan. 7, 1958, to Pennsalt Chemical Corp.; *Chem. Abs.*, 1958, **52**, 7634: German 1,025,404, March 6, 1958, to Pennsalt Chemical Corp.; *Chem. Abs.*, 1960, **54**, 11413.

[170] W. E. Dunn, Jr., *Trans. A. I. M. E.*, 1960, **218**, 6; *Chem. Abs.*, 1960, **54**, 6459.

[171] J. F. Wythe, British 827,603, Feb. 10, 1960, to Laporte Titanium Ltd.; *Chem. Abs.*, 1960, **54**, 20116.

[172] J. Hille and W. Durrwachter, German 1,056,098, April 30, 1959, to Badische Anilin & Soda Fabrik Akt. Ges.; *Chem. Abs.*, 1961, **55**, 7778.

[173] G. Nakazawa and Y. Okahara, *Nippon Kogyo Kaishi*, 1958, **74**, 183; *Chem. Abs.*, 1959, **53**, 2909.

[174] W. Frey, U.S. 2,790,703, April 30, 1957, to Fabrique de produits chimiques de Thann et de Mulhouse; *Chem. Abs.*, 1957, **51**, 12452.

[175] A. Bergholm, *Trans. A. I. M. E.*, 1961, **221**, 1121; *Chem. Abs.*, 1962, **56**, 6900.

[176] K. Okasaki, J. Moriyama, and I. Krchma, *Kyoto Univ.*, 1957, **19**, 291; *Chem. Abs.*, 1958, **52**, 10792.

[177] C. de Rohden and R. Mas, French 1,125,152, Oct. 25, 1956, to Fabriques de produits chimiques de Thann et de Mulhouse; *Chem. Abs.*, 1960, **54**, 842.

[178] L. K. Doraiswamy, H. C. Bijawat, and M. V. Kunte, *Chem. Eng. Progress*, 1959, **55**, No. 10, 80; *Chem. Abs.*, 1960, **54**, 7999.

[179] S. P. Ohuja, S. P. Garg, R. T. Thampy, and N. R. Kuloor, Indian 64,238, Oct. 10, 1959, to Shri Ram Institute for Industrial Research; *Chem. Abs.*, 1960, **54**, 9228.

[180] C. Renz, *Ber.*, 1906, **39**, 249; *J. Chem. Soc.*, 1906, A, 173.

[181] E. Chauvenet, *Compt. rend.*, 1911, **152**, 87; *Chem. Abs.*, 1911, **5**, 1036.

[182] L. G. Jenness, U.S. 1,923,094, Aug. 22, 1932, to Intermetal Corp.; *Chem. Abs.*, 1933, **27**, 5046.

[183] A. Stahler, *Ber.*, 1905, **38**, 2619; *J. Chem. Soc.*, 1905, A, 595.

[184] I. Nakray, Hungarian 115,530, Dec. 1, 1936; *Chem. Abs.*, 1937, **31**, 2370.

[185] H. Ramsetter and F. Kogler, German 628,953, Apr. 20, 1936, to Consolidirte Alkaliwerke; *Chem. Abs.*, 1936, **30**, 6144.

[186] British 716,681, Oct. 13, 1954, to Saurefabrik Schweizerhall; *Chem. Abs.*, 1955, **49**, 3489.

[187] Q. Belchetz, U.S. 2,486,912, Nov. 1, 1949, to Stauffer Chemical Co.; *Chem. Abs.*, 1950, **44**, 2714.

[188] M. Gaines and E. Arnold, *Proc. West Virginia Acad. Sci.*, 1951, **23**, 216; *Chem. Abs.*, 1952, **46**, 10030.

[189] Y. I. Ivashentsev, *Tr. T. G. U. S. K.*, 1962, **154**, 56; *Chem. Abs.*, 1964, **60**, 5112.

[190] R. W. Ancrum and A. W. Evans, British 745,051, Feb. 22, 1956, to British Titan Products Co., Ltd.; *Chem. Abs.*, 1956, **50**, 17358.

[191] I. Egami, Japanese 5503, July 26, 1958; *Chem. Abs.*, 1959, **53**, 15829.

[192] E. Wainer, U.S. 2,762,691, Sept. 11, 1956, to Horizons Inc.; *Chem. Abs.*, 1957, **51**, 6103.

[193] L. Aagaard and G. E. Bronson, U.S. 2,608,464, Aug. 26, 1952, to National Lead Co.; *Chem. Abs.*, 1953, **47**, 3531: British 692,901, June 17, 1953, to Titan Co., Inc.; *Chem. Abs.*, 1953, **47**, 12775.

[194] F. V. Schossberger, U.S. 2,857,242, Oct. 21, 1958, to Armour Research Foundation; *Chem. Abs.*, 1959, **53**, 4674: British 838,822, June 22, 1960, Armour Research Foundation of Illinois Institute of Technology; *Chem. Abs.*, 1960, **54**, 23226: German

1,045,996, Dec. 11, 1958, to Armour Research Foundation of Illinois Institute of Technology; *Chem. Abs.*, 1960, **54**, 21679.

[195] O. P. Sprague, U.S. 2,650,873, Sept. 1, 1953, to National Lead Co.; *Chem. Abs.*, 1954, **48**, 3650.

[196] J. Kamlet, U.S. 2,761,760, Sept. 4, 1956, to National Distillers Products Corp.; *Chem. Abs.*, 1957, **51**, 680.

[197] H. Grothe, German 1,050,324, Feb. 12, 1959; *Chem. Abs.*, 1961, **55**, 3022.

[198] R. G. Knickerbocker, E. H. Gorski, H. Kenworthy, and A. G. Starliper, *J. Metals, Trans.*, 1949, **1**, No. 11, 785; *Chem. Abs.*, 1950, **44**, 89.

[199] A. Pechukas, U.S. 2,245,358, June 10, 1941, to Pittsburgh Plate Glass Co.; *Chem. Abs.*, 1941, **35**, 6072. F. J. Cleveland, British 550,750, Jan. 22, 1943, to Pittsburgh Plate Glass Co.; *Chem. Abs.*, 1944, **38**, 1615.

[200] I. E. Muskat and R. H. Taylor, U.S. 2,245,077, June 10, 1941, to Pittsburgh Plate Glass Co.; *Chem. Abs.*, 1941, **35**, 5845.

[201] A. Pechukas, U.S. 2,306,184, Dec. 22, 1942, to Pittsburgh Plate Glass Co.; *Chem. Abs.*, 1943, **37**, 3235.

[202] R. Mas and P. Matthieussent, French 1,115,115, April 19, 1956, to Fabriques de produits chimiques de Thann et de Mulhouse; *Chem. Abs.*, 1959, **53**, 19328.

[203] W. H. Coates, British 806,052, Dec. 17, 1958, to British Titan Products Co., Ltd.; *Chem. Abs.*, 1959, **53**, 12605.

[204] G. T. Mahler, U.S. 2,870,869, Jan. 27, 1959, to New Jersey Zinc Co.; *Chem. Abs.*, 1959, **53**, 8559.

[205] E. W. Nelson, A. Dietz, J. P. Wikswo, and W. E. Trees, U.S. 2,849,083, Aug. 26, 1958, to American Cyanamid Co.; *Chem. Abs.*, 1959, **53**, 2554.

[206] British 853,115, Nov. 2, 1960, to Fabriques de produits chimiques de Thann et de Mulhouse; *Chem. Abs.*, 1961, **55**, 16928.

[207] A. W. Evans and J. D. Groves, U.S. 3,017,254, Jan. 16, 1962, to British Titan Products Co., Ltd.; *Chem. Abs.*, 1962, **56**, 11402: British 724,193, Feb. 16, 1955, to British Titan Products Co., Ltd.; *Chem. Abs.*, 1955, **49**, 12790.

[208] R. J. Mas and A. L. Michaud, U.S. 2,792,077, May 14, 1957, to Fabriques de produits chimiques de Thann et de Mulhouse; *Chem. Abs.*, 1957, **51**, 13332: British 796,967, June 25, 1958, to Fabriques de produits chimiques de Thann et de Mulhouse; *Chem. Abs.*, 1958, **52**, 20946: German 1,085,510, July 21, 1960, to Fabriques de produits chimiques de Thann et de Mulhouse; *Chem. Abs.*, 1961, **55**, 21507.

[209] W. Frey, U.S. 2,675,889, April 20, 1954, to Saurefabrik Schweizerhall; *Chem. Abs.*, 1954, **48**, 7858: Swiss 277,021, Nov. 1, 1951, to Saurefabrik Schweizerhall; *Chem. Abs.*, 1952, **46**, 8818: British 673,426, June 4, 1952, to Saurefabrik Schweizerhall; *Chem. Abs.*, 1952, **46**, 10560.

[210] W. Frey, U.S. 2,675,891, April 20, 1954, to Saurefabrik Schweizerhall; *Chem. Abs.*, 1954, **48**, 11740: British 673,427, June 4, 1952, to Saurefabrik Schweizerhall; *Chem. Abs.*, 1952, **46**, 10561: Swiss 276,715, Oct. 16, 1951, to Saurefabrik Schweizerhall; *Chem. Abs.*, 1952, **46**, 8818.

[211] J. D. Groves, British 872,865, Jan. 16, 1957, to British Titan Products Co., Ltd.; *Chem. Abs.*, 1962, **56**, 169.

[212] I. J. Krchma, U.S. 2,463,396, March 1, 1949, to E. I. du Pont de Nemours and Co.; *Chem. Abs.*, 1949, **43**, 4822.

[213] A. W. Evans and J. D. Groves, German 1,080,989, May 5, 1960, to British Titan Products Co., Ltd.; *Chem. Abs.*, 1961, **55**, 22734.

[214] E. Falkum, Norwegian 91,592, May 5, 1958, to Norske Hydro-Elektrisk Kvaelstofaktieslskab; *Chem. Abs.*, 1959, **53**, 15504.

[215] V. Y. Krumnik, V. I. Borodin, and L. S. Garba, U.S.S.R. 133,469, Nov. 25, 1960; *Chem. Abs.*, 1961, **55**, 13789.

[216] Belgian 634,379, Jan. 2, 1964, to Laporte Titanium Ltd.; *Chem. Abs.*, 1964, **61**, 335.

[217] P. B. Kraus, British 803,432, Oct. 22, 1958, to E. I. du Pont de Nemours and Co.; *Chem. Abs.*, 1959, **53**, 10683.

[218] P. B. Kraus, U.S. 2,718,279, Sept. 20, 1955, to E. I. du Pont de Nemours and Co.; *Chem. Abs.*, 1955, **49**, 16373.

[219] A. C. Mueller, U.S. 2,668,424, Feb. 9, 1954, to E. I. du Pont de Nemours and Co.; *Chem. Abs.*, 1954, **48**, 9028.

[220] B. Simek and L. Jaeger, Czechoslovakian 101,836, Dec. 15, 1961; *Chem. Abs.*, 1963, **58**, 11006.

[221] W. Frey and R. Weber, U.S. 2,675,890, April 20, 1954, to Saurefabrik Schweizerhall; *Chem. Abs.*, 1954, **48**, 7858: British 675,038, to Saurefabrik Schweizerhall; *Chem. Abs.*, 1953, **47**, 9245.

[222] P. I. Miroshnikov, S. T. Guz, V. I. Borodin, V. Y. Kramnik, V. I. Trvetkov, L. S. Garba, V. F. Shipilov, and A. P. Gobov, U.S.S.R. 130,496, Aug. 5, 1960; *Chem. Abs.*, 1961, **55**, 5887.

[223] I. J. Krchma, U.S. 2,463,396, March 1, 1949, to E. I. du Pont de Nemours and Co.; *Chem. Abs.*, 1949, **43**, 4822.

[224] R. J. Mas and P. R. Matthieussent, U.S. 2,815,091, Dec. 3, 1957, to Fabriques de produits chimiques de Thann et de Mulhouse; *Chem. Abs.*, 1958, **52**, 8484: British 789,552, Jan. 22, 1958, to Fabriques de produits chimiques de Thann et de Mulhouse; *Chem. Abs.*, 1958, **52**, 13207.

[225] K. Azuma and S. Goto, *J. Mining Inst. Japan*, 1955, **71**, 687; *Chem. Abs.*, 1956, **50**, 17351.

[226] R. L. Buchanan, U.S. 2,541,495, Feb. 13, 1951, to E. I. du Pont de Nemours and Co.; *Chem. Abs.*, 1951, **45**, 5424.

[227] E. C. Perkins, H. Dolezal, and R. S. Lang, *U.S. Bur. Mines, Rept. Invest.*, 5428, 1958; *Chem. Abs.*, 1959, **53**, 3620.

[228] V. A. Kozhemyakin, A. S. Berengard, and N. A. Filatova, *Tsvetnye Metally*, 1961, No. 9, 70; *Chem. Abs.*, 1962, **56**, 8368.

[229] W. H. Coates and J. Hayden, U.S. 2,943,704, July 5, 1960, to British Titan Products Co., Ltd.; *Chem. Abs.*, 1960, **54**, 25648: German 1,029,812, May 14, 1958, to British Titan Products Co., Ltd.; *Chem. Abs.*, 1960, **54**, 25647.

[230] G. V. Seryakov and A. P. Masterova, U.S.S.R. 116,207, Jan. 19, 1959; *Chem. Abs.*, 1959, **53**, 16904.

[231] J. D. Groves, U.S. 2,940,827, June 14, 1960, to British Titan Products Co., Ltd.; *Chem. Abs.*, 1960, **54**, 20116.

[232] V. A. Niberlin, *U.S. Bur. Mines, Rept. Invest.* No. 5602, 1960; *Chem. Abs.*, 1960, **54**, 15151.

[233] I. J. Krchma, U.S. 2,533,021, Dec. 5, 1950, to E. I. du Pont de Nemours and Co.; *Chem. Abs.*, 1951, **45**, 1738.

[234] I. S. Morozov, S. L. Stefanyuk, and D. Y. Toptygin, U.S.S.R. 133,866, Dec. 10, 1960; *Chem. Abs.*, 1961, **55**, 14846.

[235] A. C. Ellsworth, British 849,172, Sept. 21, 1960, to Columbia-Southern Chemical Corp.; *Chem. Abs.*, 1961, **55**, 23952.

[236] W. Kangro, German 1,011,151, June 27, 1957; *Chem. Abs.*, 1960, **54**, 7510.

[237] K. Kanbayashi, Japanese 2961, April 24, 1959; *Chem. Abs.*, 1960, **54**, 3888.

[238] Swiss 255,404, Jan. 17, 1949, to Saurefabrik Schweizerhall; *Chem. Abs.*, 1949, **43**, 7652: British 653,035, May 9, 1951, to Saurefabrik Schweizerhall; *Chem. Abs.*, 1951, **45**, 8727.

[239] A. S. Berengard and V. A. Kozhemyakin, *Zavodskaya Lab.*, 1960, **26**, 316; *Chem. Abs.*, 1962, **56**, 2296.

[240] S. Mitsui, T. Noda, and K. Nakamura, Japanese 1494, March 4, 1957, to Osaka Titanium Co.; *Chem. Abs.*, 1958, **52**, 10521.

[241] Y. Saheki and K. Funaki, *Nippon Kagaku Zasshi*, 1957, **78**, 754; *Chem. Abs.*, 1958, **52**, 4301.

[242] H. Ishizuka, U.S. 2,889,687, June 9, 1959; *Chem. Abs.*, 1959, **53**, 22791.

[243] Swiss 270,820, Dec. 16, 1950, to Saurefabrik Schweizerhall; *Chem. Abs.*, 1951, **45**, 10519.

[244] W. Frey, U.S. 2,656,011, Oct. 20, 1953, to Saurefabrik Schweizerhall; *Chem. Abs.*, 1954, **48**, 336.

[245] French 1,315,837, Jan. 25, 1963, Dec. 12, 1961, to American Cyanamid Co.; *Chem. Abs.*, 1963, **59**, 1312.

[246] J. F. Wythe, British 866,002, April 26, 1961, to Laporte Titanium Ltd.; *Chem. Abs.*, 1961, **55**, 23952.

[247] C. Ito, K. Kojima, and K. Uotome, Japanese 7414, Oct. 15, 1955, to Osaka Titanium Co.; *Chem. Abs.*, 1959, **53**, 14438.

[248] W. L. Kay and C. E. Ricks, U.S. 2,600,881, June 17, 1952, to E. I. du Pont de Nemours and Co.; *Chem. Abs.*, 1952, **46**, 11605.

[249] S. Mitsui and C. Ito, Japanese 3364, June 6, 1957, to Osaka Titanium Co.; *Chem. Abs.*, 1959, **52**, 20946.

[250] W. H. Coates, British 791,651, March 5, 1958, to British Titan Products Co., Ltd.; *Chem. Abs.*, 1958, **52**, 14999: German 1,058,483, June 4, 1959, to British Titan Products Co., Ltd.; *Chem. Abs.*, 1960, **54**, 21679.


## CHAPTER 20

[1] O. van der Pfordten, *Analen*, 1887, **237**, 201; *J. Chem. Soc.*, 1887, **A**, 337. A. Stahler, *Ber.*, 1905, **38**, 2619; *J. Chem. Soc.*, 1905, **A**, 595.

[2] A. V. Pamfilov and E. G. Shtandel, *J. Gen. Chem. (U.S.S.R.)*, 1937, **7**, 258; *Chem. Abs.*, 1937, **31**, 4609.

[3] N. I. Delarova and T. A. Zavaritskaya, *Tr. V. N. I. A.-M. Inst.; Chem. Abs.*, 1964, **90**, 218.

[4] N. N. Ruban, V. D. Ponomarev, and K. A. Vinogradova, *Tr. I. M. i O. A. N. K. S.S.R.*, 1963, **6**, 1174.

[5] N. N. Ruban, V. D. Ponomarev, and L. G. Romanov, *I. A. N. K. S.S.R., S. M. O. i O.*, 1960, No. 3, 47; *Chem. Abs.*, 1961, **55**, 11048.

[6] J. Moriyama and H. Inagaki, *Nippon Kogyo Kaishi*, 1960, **76**, 101; *Chem. Abs.*, 1961, **55**, 14839.

[7] N. I. Delarova, T. A. Tsekhovolskaya, I. A. Zevakin, and D. I. Tsekhovolskaya, *I. A. N. S.S.S.R., O. T. M i T*, 1960, No 4, 33; *Chem. Abs.*, 1961, **55**, 13782.

[8] T. A. Zavaritskaya and N. I. Delorova, *Tr. V. N.-I. A.-M. Inst.*, 1961, *No.* 47, 96; *Chem. Abs.*, 1962, **57**, 2914.

[9] C. K. Stoddard and E. Pietz, *U.S. Bur. Mines, Rept. Invest.* **4153**, 1947; *Chem. Abs.*, 1947, **42**, 2406.

[10] W. F. Meister, U.S. 2,416,191, Feb. 18, 1947, to National Lead Co.; *Chem. Abs.*, 1947, **41**, 2543. S. S. Cole and W. F. Meister, U.S. 2,396,458, Mar. 12, 1946, to National Lead Co.; *Chem. Abs.*, 1946, **40**, 2948.

[11] W. Noll, German 723,223, June 18, 1942, to I. G. Farbenind., A-G.; *Chem. Abs.*, 1943, **37**, 5204.

[12] British 905,370, Sept. 5, 1962, to National Lead Co.; *Chem. Abs.*, 1963, **58**, 1150.

[13] H. H. Schaumann, U.S. 2,530,735, Nov. 21, 1950, to E. I. du Pont de Nemours and Co.; *Chem. Abs.*, 1951, **45**, 2162.

[14] A. R. Tarsey and W. D. Guthrie, U.S. 2,871,094, Jan. 27, 1959, to Titanium Metals Corp. of America; *Chem. Abs.*, 1959, **53**, 8559.

[15] H. Ishizuka, Japanese 971, Feb. 25, 1954; *Chem. Abs.*, 1954, **48**, 14139.

[16] J. P. Levy, D. H. Pickard, and L. Prickard, U.S. 2,725,350, Nov. 29, 1955; *Chem. Abs.*, 1956, **50**, 4470: British 730,863, June 1, 1955; *Chem. Abs.*, 1956, **50**, 1276.

[17] T. A. Zavaritskaya and G. N. Grigoreva, *Trudy, V. A.-M. Inst.*, 1960, No. 44, 248; *Chem. Abs.*, 1961, **55**, 15850.

[18] C. K. Stoddard and E. Pietz, U.S. Application 706,498, 706,499 and 653,699, 1950, to the United States of America; *Chem. Abs.*, 1961, **45**, 8727.

[19] P. Ehrlich and W. Siebert, *Z. anorg. u. allgem. Chem.*, 1959, **302**, 275; *Chem. Abs.*, 1960, **54**, 8401.

[20] N. K. Druzhinina, *Trans. V. N.-I. A.-M. Inst.*, 1963, 147; *Chem. Abs.*, 1963, **59**, 10974.

[21] V. L. Hansley and S. Schott, U.S. 2,958,574, Nov. 1, 1960, to National Distillers and Chemical Corp.; *Chem. Abs.*, 1961, **55**, 5887.

[22] H. Espenschied, U.S. 2,560,424, July 10, 1951, to National Lead Co.; *Chem. Abs.*, 1951, **45**, 9232.

[23] E. P. Stambaugh, U.S. 2,754,255, July 10, 1956, to National Lead Co.; *Chem. Abs.*, 1957, **51**, 1558: British 732,941, June 29, 1955, to National Lead Co.; *Chem. Abs.*, 1956, **50**, 1276.

[24] W. K. Nelson and H. Espenschied, U.S. 2,555,361, June 5, 1951, to National Lead Co.; *Chem. Abs.*, 1951, **45**, 10523: British 674,315, June 18, 1952, to Titan Co., Inc.; *Chem. Abs.*, 1952, **46**, 11605: German 833,488, March 10, 1952, to Titan Co., Inc.; *Chem. Abs.*, 1955, **49**, 3490.

[25] A. Pechukas, U.S. 2,289,327 and 2,289,328, July 7, 1942, to Pittsburgh Plate Glass Co.; *Chem. Abs.*, 1943, **37**, 237: British 588,657, May 30, 1947, to Pittsburgh Plate Glass Co.; *Chem. Abs.*, 1947, **41**, 6376.

[26] B. de Witt, U.S. 2,370,525, Feb. 27, 1945, to Pittsburgh Plate Glass Co.; *Chem. Abs.*, 1945, **39**, 3132.

[27] W. D. Guthrie, U.S. 2,758,009, Aug. 7, 1956, to Titanium Metals Corporation of America; *Chem. Abs.*, 1956, **50**, 17358.

[28] J. S. Dunn, U.S. 2,819,147, Jan. 7, 1958, to Monsanto Chemical Co.; *Chem. Abs.*, 1958, **52**, 7634.

[29] C. T. Baroch, T. B. Kaczmarek, W. D. Barnes, L. W. Galloway, W. W. Mark, and G. A. Lee, *U.S. Bur. Mines, Rept. Invest.*, 5141, 1955; *Chem Abs.*, 1956, **50**, 131.

[30] C. K. Stoddard and P. J. Maddox, U.S. 2,836,547, May 27, 1958, to Titanium Metals Corp. of America; *Chem. Abs.*, 1958, **52**, 20947.

[31] H. A. S. Bristow, British 866,771, May 3, 1961, to Laporte Titanium Ltd.; *Chem. Abs.*, 1961, **55**, 27810.

[32] H. Espenschied, U.S. 2,560,423, July 10, 1951, to National Lead Co.; *Chem. Abs.*, 1951, **45**, 9231.

[33] H. Espenschied, U.S. 2,598,897, June 3, 1952, to National Lead Co.; *Chem. Abs.*, 1952, **46**, 11604.

[34] W. F. Meister, U.S. 2,344,319, Mar. 4, 1944, to National Lead Co.; *Chem. Abs.*, 1944, **38**, 3428.

[35] H. Espenschied, U.S. 2,598,898, June 3, 1952, to National Lead Co.; *Chem. Abs.*, 1952, **46**, 11605; German 931,104, Aug. 1, 1955, to Titan Co., Inc.; *Chem. Abs.*, 1958, **52**, 20946: British 690,470, April 22, 1953, to Titan Co., Inc.; *Chem. Abs.*, 1953, **47**, 8332.

[36] D. G. Nicholson, U.S. 2,512,807, June 27, 1950; *Chem. Abs.*, 1950, **44**, 9128.

[37] F. Gage, U.S. 2,178,685, Nov. 7, 1939, to Pittsburgh Plate Glass Co.; *Chem. Abs.*, 1940, **34**, 1448.

[38] M. B. Alpert and W. F. Sullivan, U.S. 2,712,523, July 5, 1955, to National Lead Co.; *Chem. Abs.*, 1955, **49**, 14543: German 948,600, Sept. 6, 1956, to Titangesellschaft m. b. H.; *Chem. Abs.*, 1959, **53**, 3949.

[39] D. B. Davis and A. R. Tarsey, U.S. 2,890,100, June 9, 1959, to Titanium Metals Corp. of America; *Chem. Abs.*, 1959, **53**, 19328.

[40] D. G. Nicholson, U.S. 2,457,917, Jan. 4, 1949; *Chem. Abs.*, 1949, **43**, 3158.

[41] H. H. Schaumann, U.S. 2,508,775, May 23, 1950, to E. I. du Pont de Nemours and Co.; *Chem. Abs.*, 1950, **44**, 9163.

[42] E. S. Norwicke, U.S. 2,543,591, Feb. 27, 1951, to Stauffer Chemical Co.; *Chem. Abs.*, 1951, **45**, 5377.

[43] L. G. Jenness and R. L. Annis, U.S. 2,230,538, Feb. 4, 1941, to Vanadium Corp. of America; *Chem. Abs.*, 1941, **35**, 3400.

[44] A. Pechukas, U.S. 2,224,061, Dec. 3, 1940, to Pittsburgh Plate Glass Co.; *Chem. Abs.*, 1941, **35**, 1949.

[45] B. C. Meyers, U.S. 2,412,349, Dec. 10, 1946, to Pittsburgh Plate Glass Co.; *Chem. Abs.*, 1947, **41**, 1072.

[46] W. J. Cauwenberg and A. Dietz, U.S. 2,879,132, March 24, 1959, to American Cyanamid Co.; *Chem. Abs.*, 1959, **53**, 12607.

⁴⁷ W. J. Cauwenberg and A. Dietz, U.S. 2,879,131, March 24, 1959, to American Cyanamid Co.; *Chem. Abs.*, 1959, **53**, 12607.
⁴⁸ E. P. Stambough, U.S. 2,754,256, July 10, 1956, to National Lead Co.; *Chem. Abs.*, 1956, **50**, 16058.
⁴⁹ P. E. Sharr and L. E. Bohl, U.S. 2,920,016, Jan. 5, 1960, to Columbia-Southern Chemical Corp.; *Chem. Abs.*, 1960, **54**, 12517.
⁵⁰ W. Frey and E. Boller, U.S. 2,592,021, April 8, 1952, to Saurefabrik Schweizerhall; *Chem. Abs.*, 1952, **46**, 5276: British 656,098, Aug. 15, 1951, to Saurefabrik Schweizerhall; *Chem. Abs.*, 1952, **46**, 6340: Swiss 262,267, 265,393 and 265,394, Sept. 16, 1949, to Saurefabrik Schweizerhall; *Chem. Abs.*, 1951, **45**, 2639.
⁵¹ Norwegian 88,593, Jan. 21, 1957, to Titan Co. A./S.; *Chem. Abs.*, 1957, **51**, 18509.
⁵² H. A. S. Bristow, British 869,602, May 31, 1961, to Laporte Titanium Ltd.; *Chem. Abs.*, 1961, **55**, 27810.
⁵³ British 743,735, Jan. 25, 1956, to National Lead Co.; *Chem. Abs.*, 1956, **50**, 14191.
⁵⁴ C. K. Stoddard and E. Pietz, U.S. Application 718,727, 1950, to the United States of America; *Chem. Abs.*, 1951, **45**, 8727.
⁵⁵ British 777,539, June 26, 1957, to Montecatini Società generale per l'industria mineraria e chimica; *Chem. Abs.*, 1957, **51**, 15910.
⁵⁶ G. O. R. Milani, German 1,010,514, June 19, 1957, to Montecatini Società generale per l'industria mineraria e chimica; *Chem. Abs.*, 1959, **53**, 14438.
⁵⁷ Swiss 250,071, May 18, 1948; to Saurefabrik Schweizerhall; *Chem. Abs.*, 1949, **43**, 6374.
⁵⁸ K. Nishida, Japanese 1229, Feb. 24, 1955, to Bureau of Industrial Technics; *Chem. Abs.*, 1956, **50**, 17358.
⁵⁹ A. Aradi, *Femipari Kulato Intezet Kozlemenyei*, 1959, **3**, 412; *Chem. Abs.*, 1960, **54**, 13569.
⁶⁰ L. Jaeger and B. Simek, Czechoslovakian 97,317, July 8, 1957; *Chem. Abs.*, 1962, **56**, 8299.
⁶¹ W. L. Kay and C. E. Ricks, British 712,295, July 21, 1954, to E. I. du Pont de Nemours and Co.; *Chem. Abs.*, 1955, **49**, 3489.
⁶² H. Ishizuka, Japanese 4028, June 14, 1955; *Chem. Abs.*, 1957, **51**, 17119.
⁶³ W. S. Clabaugh and R. Gilchrist, U.S. 2,915,364, Dec. 1, 1959, to U.S. Dept. of the Navy; *Chem. Abs.*, 1960, **54**, 6062: *J. Res. Natl. Bur. Standards*, 1955, **55**, 261; *Chem. Abs.*, 1956, **50**, 11867.
⁶⁴ W. S. Clabaugh and R. Gilchrist, U.S. 2,915,364, Dec. 1, 1959, to the United States of America; *Chem. Abs.*, 1960, **54**, 6062.
⁶⁵ H. M. Dees, U.S. 2,927,843, March 8, 1960, to Union Carbide Corp.; *Chem. Abs.*, 1960, **54**, 13579.
⁶⁶ R. Mas, M. Fiquet, and A. Michaud, German 1,085,510, July 21, 1960; *Chem. Abs.*, 1961, **55**, 21507.

## CHAPTER 21

¹ H. Haber and P. Kubelka, U.S. 1,931,381, Oct. 17, 1933, to Krebs Pigment and Color Corp.; *Chem. Abs.*, 1934, **28**, 268.
² I. E. Muskat, U.S. 2,240,343, Apr. 29, 1941, to Pittsburgh Plate Glass Co.; *Chem. Abs.*, 1941, **35**, 5264.
³ A. Pechukas and G. Atkinson, U.S. 2,394,633, Feb. 12, 1946, to Pittsburgh Plate Glass Co.; *Chem. Abs.*, 1946, **40**, 2654.
⁴ W. O. H. Schornstein, British 541,343, Nov. 24, 1941; *Chem. Abs.*, 1942, **36**, 4982.
⁵ J. Heinen, U.S. 2,367,118, Jan. 9, 1945, to Alien Property Custodian; *Chem. Abs.*, 1945, **39**, 2892.

[6] I. E. Muskat, U.S. 2,240,343, Apr. 29, 1941, to Pittsburgh Plate Glass Co.; *Chem. Abs.*, 1941, **35**, 5264: U.S. 2,333,948, Nov. 9, 1943, to Pittsburgh Plate Glass Co.; *Chem. Abs.*, 1944, **38**, 2514: Canadian 404,721, May 12, 1942, to Pittsburgh Plate Co.; *Chem. Abs.*, 1942, **36**, 4982: French 860,768, Jan. 23, 1941, to Pittsburgh Plate Glass Co.; *Chem. Abs.*, 1948, **42**, 6500.

[7] I. E. Muskat and A. Pechukas, U.S. 2,340,610, Feb. 1, 1944, to Pittsburgh Plate Glass Co.; *Chem. Abs.*, 1944, **38**, 4457: British 535,213, Apr. 2, 1941, to Pittsburgh Plate Glass Co.; *Chem. Abs.*, 1942, **36**, 1448: Canadian 404,725, May 12, 1942, to Pittsburgh Plate Glass Co.; *Chem. Abs.*, 1942, **36**, 4982. A. Pechukas, U.S. 2,445,691, July 20, 1948, to Pittsburgh Plate Glass Co.; *Chem. Abs.*, 1948, **42**, 8429.

[8] L. N. Shchegrov, *Titan i ego Splavy*. Akad. Nauk S.S.S.R., Inst. Met., 1961, No. 5, 211; *Chem. Abs.*, 1962, **57**, 13404.

[9] B. N. Melentev, *I. A. N. S.S.S.R., O. T. N., M. i T.*, 1960, No. 4, 69; *Chem. Abs.*, 1961, **55**, 11163.

[10] S. G. Moinov, B. N. Malentev, and V. A. Rezmichenko, *Titan i ego Splavy*, A. N.S.S.S.R., I. M., 1961, No. 5, 205; *Chem. Abs.*, 1962, **57**, 13404.

[11] H. H. Schaumann, U.S. 2,798,819, July 9, 1957, to E. I. du Pont de Nemours and Co.; *Chem. Abs.*, 1957, **51**, 15146.

[12] R. L. Buchanan, U.S. 2,541,495, Feb. 13, 1951, to E. I. du Pont de Nemours and Co.; *Chem. Abs.*, 1951, **45**, 5424.

[13] C. E. Ricks, 2,788,260, April 9, 1957, to E. I. du Pont de Nemours and Co.; *Chem. Abs.*, 1956, **51**, 11673: British 805,570, Dec. 10, 1958, to E. I. du Pont de Nemours and Co.; *Chem. Abs.*, 1959, **53**, 5611.

[14] O. B. Wilcox, U.S. 2,791,490, May 7, 1957, to E. I. du Pont de Nemours and Co.; *Chem. Abs.*, 1957, **51**, 11778.

[15] W. Hughes and B. Harris, U.S. 2,937,928, May 24, 1960, to British Titan Products Co., Ltd.; *Chem. Abs.*, 1960, **54**, 16870.

[16] T. Kitamura, Japanese 7393, June 18, 1960, to Toho Titanium Co., Ltd.; *Chem. Abs.*, 1961, **55**, 20455.

[17] A. Pechukas, U.S. 2,450,156, Sept. 28, 1948, to Pittsburgh Plate Glass Co.; *Chem. Abs.*, 1949, **43**, 1200.

[18] E. Wagner, U.S. 2,990,249, June 27, 1961, to Deutsche Gold und Silber Scheideanstalt vorm. Roessler; *Chem. Abs.*, 1961, **55**, 26392.

[19] A. E. Callow, W. Hughes, and J. D. Groves, British 907,211, Oct. 2, 1962, to British Titan Products Co., Ltd.; *Chem. Abs.*, 1963, **58**, 2179.

[20] W. L. Wilson, U.S. 3,069,281, Dec. 18, 1962, to Pittsburgh Plate Glass Co.; *Chem. Abs.*, 1963, **58**, 6478.

[21] F. W. Lane, U.S. 2,653,078, Sept. 22, 1953, to E. I. du Pont de Nemours and Co.; *Chem. Abs.*, 1954, **48**, 6142.

[22] Swiss 262,553, Oct. 1, 1949, to Saurefabrik Schweizerhall; *Chem. Abs.*, 1951, **45**, 2639.

[23] C. M. Olson and J. N. Tully, U.S. 2,670,275, Feb. 23, 1954, to E. I. du Pont de Nemours and Co.; *Chem. Abs.*, 1954, **48**, 6712: British 764,084, Dec. 19, 1956, to E. I. du Pont de Nemours and Co.; *Chem. Abs.*, 1957, **51**, 8452.

[24] Belgian 624,372, Feb. 14, 1963, to Laporte Titanium Ltd.; *Chem. Abs.*, 1963, **58**, 13583.

[25] Belgian 628,095, Aug. 6, 1963, to British Titan Products Co., Ltd.; *Chem. Abs.*, 1964, **60**, 10256.

[26] H. H. Schaumann, U.S. 2,614,028, Oct. 14, 1952, to E. I. du Pont de Nemours and Co.; *Chem. Abs.*, 1953, **47**, 2948.

[27] Belgian 621,199, Nov. 30, 1962, to Pittsburgh Plate Glass Co.; *Chem. Abs.*, 1963, **58**, 12206.

[28] W. L. Wilson, Belgian 631,070, Aug. 16, 1963, to Pittsburgh Plate Glass Co.; *Chem. Abs.*, 1964, **61**, 333.

[29] H. Zirngible, W. Gutsche, and W. Weidmann, Belgian 636,306, Dec. 16, 1963, to Farbenfabriken Bayer A.-G.; *Chem. Abs.*, 1964, **61**, 14226.

[30] H. Ishizuka, Japanese 3528, May 25, 1955; *Chem. Abs.*, 1957, **51**, 14220.
[31] C. M. Olson and J. N. Tully, German 1,176,630, Aug. 27, 1964, to E. I. du Pont de Nemours and Co.; *Chem. Abs.*, 1964, **61**, 12979.
[32] British 673,782, June 11, 1952, to Saurefabrik Schweizerhall; *Chem. Abs.*, 1952, **46**, 11609.
[33] C. L. Carpenter, Belgian 620,391, Nov. 14, 1962, to Cabot Corp.; *Chem. Abs.*, 1963, **58**, 7704.
[34] A. Mittasch, R. Lucas, and R. Griessbach, U.S. 1,850,286, Mar. 22, 1932, to I. G. Farbenind., A-G.; *Chem. Abs.*, 1932, **26**, 2831.
[35] E. W. Nelson, J. E. Bondurant, and G. C. Marcot, U.S. 2,957,753, Oct. 25, 1960, to American Cyanamid Co.; *Chem. Abs.*, 1961, **55**, 5988.
[36] I. J. Krchma, U.S. 2,512,341, June 20, 1950, to E. I. du Pont de Nemours and Co.; *Chem. Abs.*, 1950, **44**, 8677.
[37] Swiss, 265,192, March 1, 1950, to Saurefabrik Schweizerhall; *Chem. Abs.*, 1950, **44**, 9642.
[38] Swiss 295,398, March 1, 1954, to Saurefabrik Schweizerhall; *Chem. Abs.*, 1955, **49**, 13612.
[39] Swiss 272,991, April 16, 1951, to Saurefabrik Schweizerhall; *Chem. Abs.*, 1952, **46**, 4185.
[40] O. Saladin, German 959,365, March 7, 1957, to Fabriques de produits chimiques de Thann et de Mulhouse; *Chem. Abs.*, 1960, **54**, 842.
[41] Swiss 275,685, Aug. 16, 1951, to Saurefabrik Schweizerhall; *Chem. Abs.*, 1952, **46**, 8340.
[42] British 707,389, April 14, 1954, to Godfrey L. Cabot, Inc.; *Chem. Abs.*, 1954, **48**, 11682.
[43] W. Frey and O. Breig, German 963,462, May 9, 1957, to Fabriques de produits chimiques de Thann et de Mulhouse; *Chem. Abs.*, 1960, **54**, 11418: German 947,788, Sept. 13, 1956, to Saurefabrik Schweizerhall; *Chem. Abs.*, 1958, **52**, 14110.
[44] W. Frey, U.S. 2,779,662, Jan. 29, 1957, to Fabriques de produits chimiques de Thann et de Mulhouse; *Chem. Abs.*, 1957, **51**, 9182.
[45] British 686,568, Jan. 28, 1953, to Saurefabrik Schweizerhall; *Chem. Abs.*, 1953, **47**, 7794.
[46] R. J. Mas and A. L. Michaud, Belgian 619,926, Oct. 31, 1962, to Fabriques de produits chimiques de Thann et de Mulhouse; *Chem. Abs.*, 1963, **58**, 8679.
[47] E. W. Nelson, J. E. Bondurant, and G. C. Marcot, U.S. 3,062,621, Nov. 6, 1962, to American Cyanamid Co.; *Chem. Abs.*, 1963, **58**, 1644.
[48] E. M. Allen, U.S. 3,069,282, Dec. 18, 1962, to Pittsburgh Plate Glass Co.; *Chem. Abs.*, 1963, **58**, 7635: Belgian 620,983, Nov. 30, 1962, to Pittsburgh Plate Glass Co.; *Chem. Abs.*, 1963, **59**, 1309.
[49] E. W. Allen and F. E. Benner, Jr., U.S. 3,105,742, Oct. 1, 1963, to Pittsburgh Plate Glass Co.; *Chem. Abs.*, 1964, **60**, 223.
[50] E. M. Allen and R. R. Palumbo, British 969,619, Sept. 9, 1964, to Pittsburgh Plate Glass Co.; *Chem. Abs.*, 1964, **61**, 14226.
[51] A. Walmsley, U.S. 3,109,708, Nov. 5, 1963, to Laporte Titanium Ltd.; *Chem. Abs.*, 1964, **60**, 2565: British 927,171, May 29, 1963, to Laporte Titanium Ltd.; *Chem. Abs.*, 1963, **59**, 4816.
[52] Swiss 250,369, June 16, 1948, to Saurefabrik Schweizerhall; *Chem. Abs.*, 1949, **43**, 6374: Swiss 250,370, June 16, 1948, to Saurefabrik Schweizerhall; *Chem. Abs.*, 1949, **43**, 6734.
[53] A. W. Evans and W. Hughes, U.S. 2,828,187, March 28, 1958, to British Titan Products Co.; *Chem. Abs.*, 1958, **52**, 10606: British 761,770, Nov. 21, 1950, to British Titan Products Co., Ltd.; *Chem. Abs.*, 1957, **51**, 9182. A. W. Evans, W. Hughes, and J. D. Groves, U.S. 2,964,386, Dec. 13, 1960, to British Titan Products Co., Ltd.; *Chem. Abs.*, 1961, **55**, 10825.
[54] W. Hughes and A. W. Evans, U.S. 3,043,657, July 10, 1962, to British Titan Products Co., Ltd.; *Chem. Abs.*, 1963, **58**, 273.

[55] Belgian 627,965, Aug. 5, 1963, to British Titan Products Co.; *Chem. Abs.,* 1964, **60,** 10256.

[56] J. D. Groves, German 1,170,386, May 21, 1964, to British Titan Products Co., Ltd.; *Chem. Abs.,* 1964, **61,** 3962.

[57] A. W. Evans and K. Arkless, British 948,813, Feb. 5, 1964, to British Titan Products Co., Ltd.; *Chem. Abs.,* 1964, **60,** 10256.

[58] K. Arkless, U.S. 3,097,923, July 16, 1963, to British Titan Products Co., Ltd.; *Chem. Abs.,* 1963, **59,** 12433: British 948,814, May 29, 1963, to British Titan Products Co., Ltd.; *Chem. Abs.,* 1964, **60,** 10256.

[59] H. A. S. Bristow, British 866,363, April 26, 1961, to Laporte Titanium Ltd.; *Chem. Abs.,* 1961, **55,** 23958. R. J. Wigginton and J. Leighton, U.S. 3,073,712, Jan. 15, 1963, to Laporte Titanium Ltd.; *Chem. Abs.,* 1963, **58,** 8678.

[60] J. T. Richmond and H. A. S. Bristow, U.S. 2,760,846, Aug. 28, 1956, to Laporte Titanium Ltd.; *Chem. Abs.,* 1956, **50,** 17362: British 815,891, July 1, 1959, to Laporte Titanium Ltd.; *Chem. Abs.,* 1959, **53,** 22994.

[61] J. T. Richmond, U.S. 3,148,027, Sept. 8, 1964, to Laporte Titanium Ltd.; *Chem. Abs.,* 1964, **61,** 12978.

[62] H. H. Schaumann, U.S. 2,488,439, Nov. 15, 1949, to E. I. du Pont de Nemours and Co.; *Chem. Abs.,* 1950, **44,** 2769.

[62a] H. Pries, Swiss 221,309, Aug. 17, 1942; *Chem. Abs.,* 1949, **43,** 421.

[63] W. Hughes and B. Harris, U.S. 2,937,928, May 24, 1960, to British Titan Products Co., Ltd.; *Chem. Abs.,* 1960, **54,** 16870: British 791,657, March 5, 1958, to British Titan Products Co., Ltd.; *Chem. Abs.,* 1958, **52,** 14192.

[64] H. H. Schaumann, U.S. 2,488,440, Nov. 15, 1949, to E. I. du Pont de Nemours and Co.; *Chem. Abs.,* 1950, **44,** 2769.

[65] A. E. Callow and R. G. Wynne, German 1,121,035, Jan. 4, 1962, to British Titan Products Co., Ltd.; *Chem. Abs.,* 1962, **57,** 16136.

[66] H. H. Schaumann, U.S. 2,502,347, March 28, 1950, to E. I. du Pont de Nemours and Co.; *Chem. Abs.,* 1950, **44,** 9164.

[67] H. H. Schaumann and I. J. Krchma, U.S. 2,691,571, Oct. 12, 1954, to E. I. du Pont de Nemours and Co.; *Chem. Abs.,* 1955, **49,** 2755: British 760,644, Nov. 7, 1956, to E. I. du Pont de Nemours and Co.; *Chem. Abs.,* 1955, **49,** 2755.

[68] W. L. Wilson, U.S. 2,968,529, Jan. 17, 1961, to Columbia-Southern Chemical Corp.; *Chem. Abs.,* 1961, **55,** 9903.

[69] I. J. Krchma and J. E. Booge, U.S. 2,462,978, March 1, 1949, to E. I. du Pont de Nemours and Co.; *Chem. Abs.,* 1949, **43,** 4028.

[70] Swiss 287,877, April 16, 1953, to Saurefabrik Schweizerhall; *Chem. Abs.,* 1954, **48,** 6142.

[71] A. N. Dempster and R. J. Mundy, U.S. 2,760,938, Aug. 28, 1956, to National Lead Co.; *Chem. Abs.,* 1957, **51,** 780.

[72] H. H. Schaumann, U.S. 2,798,819, July 9, 1957, to E. I. du Pont de Nemours and Co.; *Chem. Abs.,* 1957, **51,** 15146: British 839,022, June 29, 1960, to E. I. du Pont de Nemours and Co.; *Chem. Abs.,* 1960, **54,** 25638.

[73] Swiss 287,876, April 16, 1953, to Saurefabrik Schweizerhall; *Chem. Abs.,* 1954, **48,** 6142.

[74] British 686,570, Jan. 28, 1953, to Saurefabrik Schweizerhall; *Chem. Abs.,* 1953, **47,** 7794.

[75] British 689,123, March 18, 1953, to Saurefabrik Schweizerhall; *Chem. Abs.,* 1953, **47,** 9030.

[76] R. J. Wigginton and J. Leighton, U.S. 3,073,712, Jan. 15, 1963, to Laporte Titanium Ltd.; *Chem. Abs.,* 1963, **58,** 8679.

[77] F. Strain, W. L. Wilson, and P. L. Dietz, Jr., U.S. 3,068,113, Dec. 11, 1962, to Pittsburgh Plate Glass Co.; *Chem. Abs.,* 1963, **59,** 2421.

[78] British 922,671, April 3, 1963, to E. I. du Pont de Nemours and Co.; *Chem. Abs.,* 1963, **59,** 4816.

[79] E. W. Nelson and G. C. Marcot, U.S. 2,750,260, June 12, 1956, to American Cyanamid Co.; *Chem. Abs.,* 1957, **51,** 1624.

[80] H. H. Schaumann, U.S. 2,805,921, Sept. 10, 1957, to E. I. du Pont de Nemours and Co.; *Chem. Abs.*, 1958, **52**, 3284: British 822,910, Nov. 4, 1959, to E. I. du Pont de Nemours and Co.; *Chem. Abs.*, 1960, **54**, 6067.

[81] F. W. Lane, U.S. 2,653,078, Sept. 22, 1953, to E. I. du Pont de Nemours and Co.; *Chem. Abs.*, 1954, **48**, 6142: British 764,082, Dec. 19, 1956, to E. I. du Pont de Nemours and Co.; *Chem. Abs.*, 1957, **51**, 8452.

[82] R. D. Nutting, U.S. 2,670,272, Feb. 23, 1954, to E. I. du Pont de Nemours and Co.; *Chem. Abs.*, 1954, **48**, 6712; British 764,083, Dec. 19, 1956, to E. I. du Pont de Nemours and Co.; *Chem. Abs.*, 1957, **51**, 8452.

[83] E. W. Nelson, German 1,119,838, Dec. 21, 1961, to American Cyanamid Co.; *Chem. Abs.*, 1962, **56**, 13812.

[84] C. M. Olson and J. N. Tully, U.S. 2,915,367, Dec. 1, 1959, to E. I. du Pont de Nemours and Co.; *Chem. Abs.*, 1960, **54**, 6067.

[85] P. B. Kraus and H. H. Schaumann, U.S. 2,619,434, Nov. 25, 1952, to E. I. du Pont de Nemours and Co.; *Chem. Abs.*, 1953, **47**, 2948.

[86] Swiss 276,037, Sept. 17, 1951, to Saurefabrik Schweizerhall; *Chem. Abs.*, 1952, **46**, 8339.

[87] C. M. Olson and J. N. Tully, U.S. 2,670,275, Feb. 23, 1954, to E. I. du Pont de Nemours and Co.; *Chem. Abs.*, 1954, **48**, 6712.

[88] J. Diether, German 974,874, June 22, 1961, to Deutsche Gold und Silber Scheideanstalt vorm Koessler; *Chem. Abs.*, 1962, **56**, 13806.

[89] O. Saladin, W. Schornstein, and W. Frey, U.S. 2,657,979, Nov. 3, 1953, to Saurefabrik Schweizerhall; *Chem. Abs.*, 1954, **48**, 3648.

[90] C. E. Ricks, U.S. 2,721,626, Oct. 25, 1955, to E. I. du Pont de Nemours and Co.; *Chem. Abs.*, 1956, **50**, 2132.

[91] I. J. Krchma, U.S. 2,833,627, May 6, 1958, to E. I. du Pont de Nemours and Co.; *Chem. Abs.*, 1958, **52**, 14192.

[92] I. J. Krchma, U.S. 2,508,271, May 16, 1950, to E. I. du Pont de Nemours and Co.; *Chem. Abs.*, 1950, **44**, 6686.

[93] Swiss 272,248, Dec. 15, 1950, to Saurefabrik Schweizerhall; *Chem. Abs.*, 1951, **45**, 10519.

[94] A. Walmsley, British 903,992, Aug. 22, 1962, to Laporte Titanium Ltd.; *Chem. Abs.*, 1962, **57**, 10978.

[95] P. Gregoire, German 1,033,192, July 3, 1958, to Fabriques de produits chimiques de Thann et de Mulhouse; *Chem. Abs.*, 1960, **54**, 25638: British 817,940, Aug. 6, 1959, to Fabriques de produits chimiques de Thann et de Mulhouse; *Chem. Abs.*, 1960, **54**, 10347.

[96] R. J. Mas and A. L. Michaud, French 1,307,280, Oct. 26, 1962, to Fabriques de produits chimiques de Thann et de Mulhouse; *Chem. Abs.*, 1963, **58**, 5290.

[97] J. E. Booge, U.S. 2,508,272, May 16, 1950, to E. I. du Pont de Nemours and Co.; *Chem. Abs.*, 1950, **44**, 6686.

[98] O. Saladin, W. Schornstein, and W. Frey, U.S. 2,657,979, Nov. 3, 1953, to Saurefabrik Schweizerhall; *Chem. Abs.*, 1954, **48**, 3648.

[99] R. J. Mas and A. L. Michaud, Belgian 619,926, Oct. 31, 1962, to Fabriques de produits chimiques de Thann et de Mulhouse; *Chem. Abs.*, 1963, **58**, 8679.

[100] Belgian 621,524, Dec. 14, 1962, to Titangesellschaft m. b. H.; *Chem. Abs.*, 1963, **58**, 13488.

[101] Belgian 621,525, Dec. 14, 1962, to Titangesellschaft m. b. H.; *Chem. Abs.*, 1963, **58**, 13488.

[102] P. B. Kraus, U.S. 2,789,886, April 23, 1957, to E. I. du Pont de Nemours and Co.; *Chem. Abs.*, 1957, **51**, 10091.

[103] T. Kitimura, Japanese 7056, June 14, 1960, to Toho Titanium Co., Ltd.; *Chem. Abs.*, 1961, **55**, 20358.

[104] A. J. Werner, U.S. 2,512,079, June 20, 1950, to E. I. du Pont de Nemours and Co.; *Chem. Abs.*, 1950, **44**, 9164.

[105] K. Arkless and E. Whagman, German 1,121,249, Jan. 4, 1962, to British Titan Products Co., Ltd.; *Chem. Abs.*, 1963, **58**, 12205.

[106] W. Hughes and K. Arkless, German 1,142,340, Jan. 17, 1963, to British Titan Products Co., Ltd.; *Chem. Abs.*, 1963, **58**, 12207.
[107] H. H. Schaumann, U.S. 2,689,781, Sept. 21, 1954, to E. I. du Pont de Nemours and Co.; *Chem. Abs.*, 1955, **49**, 5001.
[108] W. Frey, U.S. 2,739,904, March 27, 1956, to Saurefabrik Schweizerhall; *Chem. Abs.*, 1956, **50**, 7478: British 684,016, Dec. 10, 1952, to Saurefabrik Schweizerhall; *Chem. Abs.*, 1953, **47**, 5697.

## CHAPTER 22

[1] S. S. Svendsen, U.S. 1,911,004, May 23, 1933, to Clay Reductions Co.; *Chem. Abs.*, 1933, **27**, 4037.
[2] S. S. Svendsen, U.S. 1,959,747, May 22, 1934, to Clay Reductions Co.; *Chem. Abs.*, 1934, **28**, 4546: French 643,981, Oct. 24, 1927, to Clay Reductions Co.; *Chem. Abs.*, 1929, **23**, 1727.
[3] F. von Bichowsky, U.S. 2,031,750, Feb. 25, 1936, to Krebs Pigment and Color Corp.; *Chem. Abs.*, 1936, **30**, 2333.
[4] E. Silbermann, French 811,062, Apr. 6, 1937; *Chem. Abs.*, 1937, **31**, 7025.
[5] C. A. Doremus, U.S. 1,501,587, July 15, 1924, to Titanium Pigment Co., Inc.; *Chem. Abs.*, 1924, **18**, 2791: British 206,809, May 24, 1923, to Titanium Pigment Co., Inc.; *Chem. Abs.*, 1924, **18**, 1182.
[6] S. S. Svendsen, U.S. 1,995,334, Mar. 26, 1935, to C. F. Burgess Laboratories, Inc.; *Chem. Abs.*, 1935, **29**, 3122.
[7] S. S. Svendsen, U.S. 2,042,434, May 26, 1936, to Burgess Titanium Co.; *Chem. Abs.*, 1936, **30**, 5006: British 456,058, Nov. 2, 1936, to C. F. Burgess Laboratories, Inc.; *Chem. Abs.*, 1937, **31**, 2368: French 789,815, Nov. 7, 1935, to C. F. Burgess Laboratories, Inc.; *Chem. Abs.*, 1936, **30**, 3175: German 695,633, Aug. 1, 1940, to Sherwin-Williams Co.; *Chem. Abs.*, 1941, **35**, 5730.
[8] S. S. Svendsen, U.S. 2,042,435, May 26, 1936, to Burgess Titanium Co.; *Chem. Abs.*, 1936, **30**, 5006: Canadian 369,914, May 16, 1937, to Burgess Titanium Co.; *Chem. Abs.*, 1938, **32**, 2301.
[9] S. S. Svendsen, U.S. 2,042,436, May 26, 1936, to Burgess Titanium Co.; *Chem. Abs.*, 1936, **30**, 5006: British 456,314, Nov. 2, 1936, to C. F. Burgess Laboratories, Inc.; *Chem. Abs.*, 1937, **31**, 2369.
[10] M. J. Sterba, U.S. 2,121,992, June 28, 1938, to Sherwin-Williams Co.; *Chem. Abs.*, 1938, **32**, 6407: British 495,248, Nov. 8, 1938, to Sherwin-Williams Co.; *Chem. Abs.*, 1939, **33**, 3086: French 833,515, Oct. 24, 1938, to Sherwin-Williams Co.; *Chem. Abs.*, 1939, **33**, 3546.
[11] S. P. Todd and F. C. Verduin, U.S. 2,355,187, Aug. 8, 1944, to Sherwin-Williams Co.; *Chem. Abs.*, 1944, **38**, 6579: Canadian 416,028, Oct. 26, 1943, to Sherwin-Williams Co.; *Chem. Abs.*, 1944, **38**, 1130.
[12] S. P. Todd and E. S. Hays, Jr., Canadian 413,723, July 6, 1943, to Sherwin-Williams Co.; *Chem. Abs.*, 1944, **38**, 1082.
[13] P. E. Mayer, U.S. 2,288,727, July 7, 1942, to Sherwin-Williams Co.; *Chem. Abs.*, 1943, **37**, 237.
[14] F. C. Verduin, U.S. 2,290,922, July 28, 1942, to Sherwin-Williams Co.; *Chem. Abs.*, 1943, **37**, 733.
[15] S. P. Todd, H. F. Saunders, and F. C. Verduin, U.S. 2,326,182, Aug. 10, 1943, to Sherwin-Williams Co.; *Chem. Abs.*, 1944, **38**, 500.
[16] S. S. Svendsen, U.S. 2,165,315, July 11, 1938, to Sherwin-Williams Co.; *Chem. Abs.*, 1939, **33**, 8428: Canadian 373,328, Apr. 26, 1936, to Sherwin-Williams Co.; *Chem. Abs.*, 1938, **32**, 4806.
[17] H. M. Kleiforth, Canadian 373,395, Apr. 26, 1938, to Sherwin-Williams Co.; *Chem. Abs.*, 1938, **32**, 4806.
[18] French 815,184, July 7, 1937, to Burgess Titanium Co.; *Chem. Abs.*, 1938, **32**, 1498.

[19] F. C. Verduin, U.S. 2,232,817, Feb. 25, 1941, to Sherwin-Williams Co.; *Chem. Abs.*, 1941, **35**, 3779.

[20] H. C. Kawecki, U.S. 2,418,073 and 2,418,074, Mar. 25, 1947; *Chem. Abs.*, 1947, **41**, 3590.

[21] French 1,049,140, Dec. 28, 1953, to Fabriques de produits chimiques de Thann et de Mulhouse; *Chem. Abs.*, 1958, **52**, 9628.

[22] R. M. McKinney, U.S. 2,193,563, Mar. 13, 1940, to E. I. du Pont de Nemours and Co.; *Chem. Abs.*, 1940, **34**, 4930.

[23] J. L. Keats, U.S. 2,193,559, Mar. 12, 1940, to E. I. du Pont de Nemours and Co.; *Chem. Abs.*, 1940, **34**, 4930.

[24] P. Kubelka and J. Srbek, U.S. 2,062,133, Nov. 24, 1936, to E. I. du Pont de Nemours and Co.; *Chem. Abs.*, 1937, **31**, 893.

[25] R. Miller, U.S. 2,732,310, Jan. 24, 1956, to Chemical Foundation, Inc.; *Chem. Abs.*, 1956, **50**, 8228.

[26] P. Pipereaut and A. Helbronner, British 207,555, Nov. 26, 1923; *Chem. Abs.*, 1924, **18**, 1554.

[27] M. A. Minot, French 710,732, May 6, 1930; *Chem. Abs.*, 1932, **26**, 1727.

[28] A. M. Thomasen, U.S. 3,042,492, July 3, 1962; *Chem. Abs.*, 1962, **47**, 9461.

[29] K. Viswanathan, *Bull. Central Research Inst. Univ. Travancore*, 1952, II, 106; *Chem. Abs.*, 1953, **47**, 2505.

[30] N. Cugnasco, Italian 447,673, April 23, 1949; *Chem. Abs.*, 1951, **45**, 1740.

[31] A. J. Gaskin and A. E. Ringwood, Australian 222,517, June 29, 1959, to Commonwealth Scientific and Industrial Research Organization; *Chem. Abs.*, 1961, **55**, 15956.

[32] Japanese 155,897, April 7, 1943, to Showa Mineral Refinery, Inc.; *Chem. Abs.*, 1949, **43**, 6796.

[33] D. W. Young, U.S. 2,285,104, June 2, 1942, to General Chemical Co.; *Chem. Abs.*, 1942, **36**, 6822.

[34] D. C. Pease, U.S. 3,012,857, March 23, 1960, to E. I. du Pont de Nemours and Co.; *Chem. Abs.*, 1962, **56**, 11238.

[35] K. L. Berry, U.S. 2,980,510, April 18, 1961, to E. I. du Pont de Nemours and Co.

[36] K. L. Berry, U.S. 3,030,183, April 17, 1962, to E. I. du Pont de Nemours and Co.; *Chem. Abs.*, 1962, **57**, 3082: German 1,082,891, June 9, 1960, to E. I. du Pont de Nemours and Co.; *Chem. Abs.*, 1961, **55**, 23876.

[37] R. G. Russell, W. L. Morgan, and L. F. Scheffter, Belgian 624,547, Feb. 28, 1963, to Owens-Corning Fiberglass Corp.; *Chem. Abs.*, 1963, **59**, 1166.

[38] E. S. Ritter and A. Milch, *J. Appl. Phys.*, 1962, **33**, No. 1, 228; *Chem. Abs.*, 1963, **58**, 1973.

[39] J. H. Peterson, U.S. 2,448,683, Sept. 7, 1948, to E. I. du Pont de Nemours and Co.; *Chem. Abs.*, 1949, **43**, 369.

[40] J. W. Sprague, U.S. 3,092,457, June 4, 1963, to Standard Oil Co.; *Chem. Abs.*, 1963, **59**, 5841.

[41] R. L. Jenkins, U.S. 3,018,186, Jan. 23, 1962, to Monsanto Chemical Co.; *Chem. Abs.*, 1962, **57**, 12108.

[42] W. B. Spencer, W. R. Smith, and A. F. Cosman, U.S. 3,024,089, March 6, 1962, to Cabot Corp.; *Chem. Abs.*, 1962, **56**, 13806.

[43] R. J. Wigginton, U.S. 3,062,673, Nov. 6, 1963, to Laporte Titanium Ltd.; *Chem. Abs.*, 1963, **58**, 2178: British 934,626, Aug. 21, 1963, to Laporte Titanium Ltd.; *Chem. Abs.*, 1963, **59**, 11085.

[44] H. F. Dantro, A. T. Kalinowski, and W. T. Siuta, U.S. 3,091,515, May 28, 1963, to National Lead Co.; *Chem. Abs.*, 1963, **59**, 9617: Belgian 622,065, Dec. 28, 1962, to Titangesellschaft m. b. H.; *Chem. Abs.*, 1963, **59**, 1308.

[45] F. A. A. Gregoire, U.S. 2,905,530, Sept. 22, 1959, to Sociétés des blancs de zinc de la Méditerranée; *Chem. Abs.*, 1960, **54**, 3992: French 1,109,087, Jan. 20, 1956; *Chem. Abs.*, 1959, **53**, 7622: British 791,302, Feb. 26, 1958; *Chem. Abs.*, 1958, **52**, 15091.

[46] Y. Ito, Japanese 9563, Nov. 9, 1956, to Nippon Light Metal Co.; *Chem. Abs.*, 1958, **52**, 17752.

[47] I. E. Ishikawa and N. Sugiyama, *Kogyo Kagaku Zasshi*, 1958, **61**, 395; *Chem. Abs.*, 1961, **55**, 3936.

[48] Japanese 161,700, Feb. 8, 1948, to Osaka Ilmenite Industry Co.; *Chem. Abs.*, 1949, **43**, 3576.

[49] V. J. Cobb, Belgian 625,676, March 29, 1963, to Titangesellschaft m. b. H.; *Chem. Abs.*, 1963, **59**, 1308.

[50] W. F. Washburn and L. Aagaard, U.S. 1,906,730, May 2, 1933, to Titanium Pigment Co., Inc.; *Chem. Abs.*, 1933, **27**, 3626: British 364,112, Sept. 24, 1929, to Titanium Pigment Co., Inc.; *Chem. Abs.*, 1933, **27**, 1733: German 588,230, Nov. 15, 1933, to Titanium Pigment Co., Inc.; *Chem. Abs.*, 1934, **28**, 1555: French 704,044 and 704,585, Sept. 22, 1930, to Titanium Pigment Co., Inc.; *Chem. Abs.*, 1931, **25**, 4723.

[51] F. L. Kingsbury, S. S. Cole, and W. B. Anderson, U.S. 2,369,468, Feb. 13, 1945, to National Lead Co.; *Chem. Abs.*, 1945, **39**, 5093.

[52] R. W. Brickenkamp and C. L. Schmidt, U.S. 2,818,347, Dec. 31, 1957, to National Lead Co.; *Chem. Abs.*, 1958, **52**, 12420.

[53] D. P. Doll and H. H. Volkening, Belgian 615,851, April 13, 1962, to Titangesellschaft m. b. H.; *Chem. Abs.*, 1963, **58**, 4201.

[54] W. Rodgers and C. R. Trampier, U.S. 2,956,859, Oct. 18, 1960, to National Lead Co.; *Chem. Abs.*, 1961, **55**, 5987.

[55] L. E. Barton, U.S. 1,251,170, Dec. 25, 1918; *Chem. Abs.*, 1918, **12**, 535: British 108,805, Jan. 11, 1917, to Titanium Alloy Mfg. Co.; *Chem. Abs.*, 1918, **12**, 230.

[56] R. W. Sullivan, U.S. 2,549,261, April 17, 1951, to E. I. du Pont de Nemours and Co.; *Chem. Abs.*, 1951, **45**, 5946.

[57] J. E. Booge, U.S. 2,046,054, June 30, 1936, to E. I. du Pont de Nemours and Co.; *Chem. Abs.*, 1936, **30**, 5821: French 796,587, April 10, 1936, to Krebs Pigment and Color Corp.; *Chem. Abs.*, 1936, **30**, 6222: British 517,913, Feb. 13, 1940, to E. I. du Pont de Nemours and Co.; *Chem. Abs.*, 1941, **35**, 7219.

[58] R. M. McKinney and C. E. Smith, U.S. 2,062,137, Nov. 24, 1936, to E. I. du Pont de Nemours and Co.; *Chem. Abs.*, 1937, **31**, 893: Reissue 21,427, Apr. 16, 1940, to E. I. du Pont de Nemours and Co.; *Chem. Abs.*, 1940, **34**, 5685.

## CHAPTER 23

[1] N. Heaton, *J. Roy. Soc. Arts*, 1922, **70**, 552; *Chem. Abs.*, 1922, **16**, 3003. J. E. Booge, U.S. 2,224,777, Dec. 10, 1940, to E. I. du Pont de Nemours and Co.; *Chem. Abs.*, 1941, **35**, 2348. C. P. van Hoek, *Farben-Ztg.*, 1929, **34**, 2828; *Chem. Abs.*, 1929, **23**, 5598. R. H. Sawyer, *Am. Soc. Testing Materials, Symp. on Paints*, Mar. 3, 1943; *Chem. Abs.*, 1944, **38**, 267.

[2] D. Wait and I. E. Weber, *J. Oil & Colour Chemists' Assoc.*, 1934, **17**, 257; *Chem. Abs.*, 1934, **28**, 6577.

[3] H. Wagner, *Paint Manuf.*, 1934, **4**, 5-9, 12; *Brit. Chem. Abstracts*, 1934, **B**, 209.

[4] J. R. McGregor, *Paint, Oil, Chem. Rev.*, 1938, **100**, 37; *Chem. Abs.*, 1938, **32**, 4803.

[5] A. Zhukova and A. Sovalova, *Lakokrasochnuyu Ind., Za*, 1934, No. 2, 11; *Chem. Abs.*, 1935, **29**, 3859.

[6] L. V. Lyutin and E. V. Gusyatzkaya, *J. Applied Chem. (U.S.S.R.)*, 1935, **8**, 833; *Chem. Abs.*, 1936, **30**, 4698.

[7] B. A. Rubin, *Byull. Obmena Opyt. Lakokrasoch. Prom.*, 1939, No. **8**, 23; *Chem. Abs.*, 1940, **34**, 8303: *Org. Chem. Ind. (U.S.S.R.)*, 1940, **7**, 223; *Chem. Abs.*, 1941, **35**, 4225.

[8] E. G. Rutter, *J. Oil & Colour Chemists' Assoc.*, 1945, **28**, 187; *Chem. Abs.*, 1946, **40**, 1046.

[9] J. Taylor, *J. Oil & Colour Chemists' Assoc.*, 1955, **38**, 233; *Chem. Abs.*, 1955, **49**, 16456.
[10] T. A. Ermolaeva, *Lakokrasochnye Materialy i ikh Primenenie*, 1961, No. 5, 46.
[11] D. K. Killian, *Paint, Oil and Chem. Rev.*, 1952, **115**, No. 8, 14; *Chem. Abs.*, 1952, **46**, 6401.
[12] A. A. Krasnovskii and T. N. Gurevich, *Doklady Akad. Nauk S.S.S.R.*, 1950, **74**, 569; *Chem. Abs.*, 1952, **46**, 755.
[13] S. Todisco, *Ind. vernice (Milan)*, 1954, **8**, 17; *Chem. Abs.*, 1954, **48**, 7316.
[14] A. G. Vondjidis and W. C. Clark, *Nature*, 1963, **198**, 278.
[15] C. P. van Hoek, *Farben-Ztg.*, 1930, **36**, 267.
[16] C. F. Goodeve and J. O. Kitchener, *Trans. Faraday Soc.*, 1938, **34**, 57 and 902.
[17] C. Renz, *Helv. Chim. Acta.*, 1921, **4**, 961.
[18] A. E. Jacobsen, *Ind. Eng. Chem.*, 1949, **41**, 524; *Chem. Abs.*, 1949, **43**, 4490.
[19] E. Lund and C. F. Weider, *F. A. T. I. P. E. C.*, Second Congress, 1953, 286; *Chem. Abs.*, 1954, **48**, 5518.
[20] A. A. Krasnovskii and V. S. Kiselev, *Org. Chem. Ind. (U.S.S.R.)*, 1940, **7**, 221; *Chem. Abs.*, 1941, **35**, 4225.
[21] R. W. Ancrum and A. G. Oppegaard, U.S. 2,161,755, June 6, 1939, to Titan Co., Inc.; *Chem. Abs.*, 1939, **33**, 7603: British 479,072, Jan. 31, 1938, to British Titan Products Co., Ltd.; *Chem. Abs.*, 1938, **32**, 5237: French 826,216, Feb. 25, 1938, to Titan Co., Inc.; *Chem. Abs.*, 1938, **32**, 6484: British 529,596, Nov. 25, 1940, to British Titan Products Co., Ltd.; *Chem. Abs.*, 1941, **35**, 7739. J. E. Booge, U.S. 2,224,777, Dec. 10, 1940, to E. I. du Pont de Nemours and Co.; *Chem. Abs.*, 1941, **35**, 2348. Advertisement, British Titan Products Co., Ltd.; *J. Oil & Colour Chemists' Assoc.*, 1938, **21**, No. 219.
[22] J. E. Booge, U.S. 2,213,542, Sept. 3, 1940, to E. I. du Pont de Nemours and Co.; *Chem. Abs.*, 1941, **35**, 640.
[23] D. B. Pall, U.S. 2,397,430, Mar. 26, 1946, to American Cyanamid Co.; *Chem. Abs.*, 1946, **40**, 3275.
[24] A. T. McCord and H. F. Saunders, U.S. 2,379,019, June 26, 1941, to Sherwin-Williams Co.; *Chem. Abs.*, 1945, **39**, 4240.
[25] British 580,734, Sept. 18, 1946, to E. I. du Pont de Nemours and Co.; *Chem. Abs.*, 1947, **41**, 1467.
[26] J. E. Booge, U.S. 2,253,551, Aug. 26, 1941, to E. I. du Pont de Nemours and Co.; *Chem. Abs.*, 1941, **35**, 8327. I. E. Weber and A. N. C. Bennett, British 536,208, May 7, 1941, to National Titanium Pigments, Ltd.; *Chem. Abs.*, 1942, **36**, 1508. R. W. Ancrum, British 549,120, Nov. 6, 1942, to British Titan Products Co., Ltd.; *Chem. Abs.*, 1944, **38**, 653.
[27] J. H. Peterson, U.S. 2,369,246, Feb. 13, 1945, to E. I. du Pont de Nemours and Co.; *Chem. Abs.*, 1945, **39**, 5092.
[28] R. W. Ancrum, British 580,809, Sept. 20, 1946, to British Titan Products Co., Ltd.; *Chem. Abs.*, 1947, **41**, 1852.
[29] R. W. Ancrum, British 581,008, Sept. 27, 1946, to British Titan Products Co., Ltd.; *Chem. Abs.*, 1947, **41**, 1853.
[30] J. E. Booge, U.S. 2,273,431, Feb. 17, 1942, to E. I. du Pont de Nemours and Co.; *Chem. Abs.*, 1942, **36**, 3976: British 530,877, Dec. 24, 1940, to E. I. du Pont de Nemours and Co.; *Chem. Abs.*, 1942, **36**, 1200.
[31] G. C. Marcot and G. M. Sheehan, U.S. 2,766,133, Oct. 9, 1956, to American Cyanamid Co.; *Chem. Abs.*, 1957, **51**, 5443.
[32] A. T. Kalinowski, U.S. 2,817,595, Dec. 24, 1957, to National Lead Co.; *Chem. Abs.*, 1958, **52**, 5000.
[33] J. Petit and R. Poisson, French 1,120,600, July 9, 1956, to Fabriques de produits chimiques de Thann et de Mulhouse; *Chem. Abs.*, 1959, **53**, 22994.
[34] H. A. Gardner, U.S. 2,092,838, Sept. 14, 1937; *Chem. Abs.*, 1937, **31**, 8231.
[35] J. Blumenfeld and M. Mayer, U.S. 1,892,693, Jan. 3, 1933, to Krebs Pigment and Color Corp.: U.S. 2,026,862, Jan. 7, 1936, to Krebs Pigment and Color Corp.;

*Chem. Abs.*, 1936, **30**, 1595: British 252,262, Dec. 28, 1924; *Brit. Chem. Abstracts*, 1926, **B**, 680: German 533,236, Aug. 12, 1924, to Verein für Chemische und Metallurgische Produktion; *Chem. Abs.*, 1932, **26**, 609.

³⁶ G. D. Patterson, U.S. 2,138,118, Nov. 29, 1938, to E. I. du Pont de Nemours and Co.; *Chem. Abs.*, 1939, **33**, 1974.

³⁷ H. A. Gardner, *Ind. Eng. Chem.*, 1936, **28**, 1020; *Ind. Eng. Chem.*, 1937, **29**, 640: British 455,717, Oct. 27, 1936; *Chem. Abs.*, 1937, **31**, 2029.

³⁸ *Natl. Paint, Varnish Lacquer Assoc., Sci. Sect., Circ. 546*, 1937, 251; *Chem. Abs.*, 1938, **32**, 6886.

³⁹ W. K. Nelson and A. O. Ploetz, U.S. 2,234,681, Mar. 11, 1941, to National Lead Co.; *Chem. Abs.*, 1941, **35**, 4230: U.S. 2,244,258, June 3, 1941, to National Lead Co.; *Chem. Abs.*, 1941, **35**, 5731.

⁴⁰ J. M. Jarmus and W. W. Plechner, U.S. 2,291,082, July 28, 1942, to National Lead Co.; *Chem. Abs.*, 1943, **37**, 783.

⁴¹ R. D. Nutting, U.S. 2,233,358, Feb. 25, 1941, to E. I. du Pont de Nemours and Co.; *Chem. Abs.*, 1941, **35**, 3840: French 845,591, Aug. 28, 1939, to E. I. du Pont de Nemours and Co.; *Chem. Abs.*, 1941, **35**, 1650.

⁴² French 836,058, Jan. 10, 1939, to E. I. du Pont de Nemours and Co.; *Chem. Abs.*, 1939, **33**, 5208. R. W. Sullivan, Canadian 401,832, Dec. 30, 1941, to E. I. du Pont de Nemours and Co.; *Chem. Abs.*, 1942, **36**, 1794.

⁴³ G. F. New, *J. Oil & Colour Chemists' Assoc.*, 1937, **20**, 352; *Chem. Abs.*, 1938, **32**, 1122.

⁴⁴ J. Blumenfeld, U.S. 1,639,423, Aug. 16, 1927; *Chem. Abs.*, 1927, **21**, 3276. C. Weizmann and J. Blumenfeld, British 256,302, Apr. 22, 1925; *Chem. Abs.*, 1927, **21**, 2992. J. Blumenfeld and M. Mayer, Canadian 265,972, and 265,973, Nov. 23, 1926; *Chem. Abs.*, 1927, **21**, 1020: German 530,211, Apr. 29, 1926, to Verein für Chemische und Metallurgische Produktion; *Chem. Abs.*, 1931, **25**, 5304.

⁴⁵ French 793,526, Jan. 27, 1936, to I. G. Farbenind., A-G.; *Chem. Abs.*, 1936, **30**, 4702.

⁴⁶ C. F. Goodeve, *Trans. Faraday Soc.*, 1937, **33**, 340; *Chem. Abs.*, 1937, **31**, 3307.

⁴⁷ L. O. Kekwick and A. Pass, *Oil Colour Trades J.*, 1937, **92**, 1651; *Chem. Abs.*, 1938, **32**, 1119.

⁴⁸ A. P. Jaeger, U.S. 1,317,164, Sept. 30, 1919; *Chem. Abs.*, 1919, **13**, 3327: Canadian 201,707, July 6, 1920; *Chem. Abs.*, 1920, **14**, 2721.

⁴⁹ L. E. Barton, U.S. 1,313,874, Aug. 26, 1919; *Chem. Abs.*, 1919, **13**, 2768.

⁵⁰ W. J. O'Brien, U.S. 1,864,504, June 21, 1932, to Krebs Pigment and Color Corp.; *Brit. Chem. Abstracts*, 1933, **B**, 478.

⁵¹ French 48,344, Dec. 27, 1937, to I. G. Farbenind., A-G.; *Chem. Abs.*, 1938, **32**, 5647: addition to 787,384, *Chem. Abs.*, 1936, **30**, 1594: British 493,871, Oct. 17, 1938, to I. G. Farbenind., A-G.; *Chem. Abs.*, 1939, **33**, 2743.

⁵² G. D. Patterson, U.S. 2,232,723, Feb. 25, 1941, to E. I. du Pont de Nemours and Co.; *Chem. Abs.*, 1941, **35**, 3839.

⁵³ G. F. New, U.S. 2,242,320, May 20, 1941, to Titan Co., Inc.; *Chem. Abs.*, 1941, **35**, 5730.

⁵⁴ D. H. Dawson, U.S. 2,232,168, Feb. 18, 1941, to E. I. du Pont de Nemours and Co.; *Chem. Abs.*, 1941, **35**, 3468.

⁵⁵ W. K. Nelson, U.S. 2,346,322, Apr. 11, 1944, to National Lead Co.; *Chem. Abs.*, 1944, **38**, 5094: British 557,877, Dec. 9, 1943, to British Titan Products Co., Ltd.; *Chem. Abs.*, 1945, **39**, 5511.

⁵⁶ W. H. Madson, U.S. 2,166,257, July 18, 1939, to E. I. du Pont de Nemours and Co.; *Chem. Abs.*, 1939, **33**, 8428.

⁵⁷ W. T. Hancock, U.S. 2,157,205, May 9, 1939; *Chem. Abs.*, 1939, **33**, 6624.

⁵⁸ British 534,538, Mar. 10, 1941, to E. I. du Pont de Nemours and Co.; *Chem. Abs.*, 1942, **36**, 1508.

⁵⁹ C. J. Kinzie, U.S. 2,099,019, Nov. 16, 1937, to Titanium Alloy Mfg. Co.; *Chem. Abs.*, 1938, **32**, 374.

[60] A. Carpmael, British 448,345, June 8, 1936, to I. G. Farbenind., A-G.; *Chem. Abs.*, 1936, **30**, 7364. K. W. Petersen, German 660,647, May 31, 1938; *Chem. Abs.*, 1938, **32**, 6891.

[61] N. F. Miller, U.S. 2,668,776, Feb. 9, 1954, to New Jersey Zinc Co.; *Chem. Abs.*, 1955, **49**, 11297.

[62] D. G. Nicholson, *Trans. Illinois State Acad. Sci.*, 1936, **29**, 97; *Chem. Abs.*, 1937, **31**, 3307.

[63] P. Farup, U.S. 1,368,392, Feb. 15, 1921; *Chem. Abs.*, 1921, **15**, 1224: Canadian 201,705, July 6, 1920; *Chem. Abs.*, 1920, **14**, 2721.

[64] H. Giese, U.S. 2,260,177, Oct. 21, 1941, to Titan Co., Inc.; *Chem. Abs.*, 1942, **36**, 920.

[65] G. R. Seidel, U.S. 2,284,772, June 2, 1942, to E. I. du Pont de Nemours and Co.; *Chem. Abs.*, 1942, **36**, 6821.

[66] G. R. Seidel, U.S. 2,387,534, Oct. 23, 1945, to E. I. du Pont de Nemours and Co.; *Chem. Abs.*, 1946, **40**, 477.

[67] G. D. Patterson, U.S. 2,187,050, Jan. 16, 1940, to E. I. du Pont de Nemours and Co.; *Chem. Abs.*, 1940, **34**, 3520.

[68] A. Pechukas, U.S. 2,441,225, May 11, 1948, to Pittsburgh Plate Glass Co.; *Chem. Abs.*, 1948, **42**, 5687.

[69] M. L. Hanahan and R. M. McKinney, U.S. 2,212,935, Aug. 27, 1940, to E. I. du Pont de Nemours and Co.; *Chem. Abs.*, 1941, **35**, 640.

[70] French 829,430, June 27, 1938, to Titan Co., Inc.; *Chem. Abs.*, 1939, **33**, 1525.

[71] G. M. Sheehan and E. R. Lawhorne, U.S. 3,086,877, Nov. 1, 1960: French 1,315,236, Jan. 18, 1963, to American Cyanamid Co.; *Chem. Abs.*, 1963, **58**, 14317.

[72] K. W. Petersen, U.S. 2,161,975, June 13, 1939, to I. G. Farbenind., A-G.; *Chem. Abs.*, 1939, **33**, 8042.

[73] A. A. Krasnovskii and V. S. Kiselev, *Chem. Zentr.*, 1943, **1**, 99; *Chem. Abs.*, 1944, **38**, 3147.

[74] M. L. Hanahan and R. M. McKinney, U.S. 2,286,910, June 16, 1942, to E. I. du Pont de Nemours and Co.; *Chem. Abs.*, 1942, **36**, 7334.

[75] J. A. Geddes, U.S. 2,357,101, Aug. 29, 1944, to E. I. du Pont de Nemours and Co.; *Chem. Abs.*, 1945, **39**, 428.

[76] B. W. Allan and W. E. Land, U.S. 2,297,523, Sept. 29, 1942, to American Zirconium Corp.; *Chem. Abs.*, 1943, **37**, 1616: British 544,283, Apr. 17, 1942, to British Titan Products Co., Ltd.; *Chem. Abs.*, 1942, **36**, 6363.

[77] Belgian 366,444, Jan. 31, 1930, to De Keyn Frères; *Chem. Abs.*, 1930, **24**, 4646.

[78] W. R. Whatley, U.S. 2,671,031, March 2, 1954, to American Cyanamid Co.; *Chem. Abs.*, 1954, **48**, 6713.

[79] British 686,570, Jan. 28, 1953, to Saurefabrik Schweizerhall; *Chem. Abs.*, 1953, **47**, 7794.

[80] W. Hughes, British 867,479, May 10, 1961, to British Titan Products Co.; *Chem. Abs.*, 1961, **55**, 22865.

[81] I. J. Krchma and H. H. Schaumann, U.S. 2,559,638, July 10, 1951, to E. I. du Pont de Nemours and Co.; *Chem. Abs.*, 1951, **45**, 9280.

[82] S. C. Marcot, S. P. Todd, and S. A. Lamanna, U.S. 2,824,050, Feb. 17, 1958, to Fabriques de produits chimiques de Thann et de Mulhouse; *Chem. Abs.*, 1958, **52**, 7737.

[83] O. B. Wilcox, U.S. 2,780,558, Feb. 5, 1957, to E. I. du Pont de Nemours and Co.; *Chem. Abs.*, 1957, **51**, 8451. British 803,454, Oct. 22, 1958, to E. I. du Pont de Nemours and Co.; *Chem. Abs.*, 1959, **53**, 7621.

[84] D. W. Robertson, U.S. 2,346,188, Apr. 11, 1944, to National Lead Co.; *Chem. Abs.*, 1944, **38**, 5094: U.S. 2,378,790, June 19, 1945, to National Lead Co.; *Chem. Abs.*, 1945, **39**, 4499.

[85] G. D. Patterson, U.S. 2,296,618, Sept. 22, 1942, to E. I. du Pont de Nemours and Co.; *Chem. Abs.*, 1943, **37**, 1284: British 545,604, June 4, 1942, to British Titan Products Co., Ltd.; *Chem. Abs.*, 1943, **37**, 2200.

[86] C. J. Kinzie, U.S. 2,099,019, Nov. 16, 1937, to Titanium Alloy Mfg. Co.; *Chem. Abs.*, 1938, **32**, 374.

[87] R. W. Ancrum, U.S. 2,576,434, Nov. 27, 1951, to National Lead Co.; *Chem. Abs.*, 1952, **46**, 3298.

[88] British 686,568, Jan. 28, 1953, to Saurefabrik Schweizerhall; *Chem. Abs.*, 1953, **47**, 7794.

[89] O. B. Wilcox, U.S. 2,591,988, April 8, 1952, to E. I. du Pont de Nemours and Co.; *Chem. Abs.*, 1952, **46**, 6405.

[90] C. Renz, *Helv. Chim. Acta*, 1921, **4**, 961; *Chem. Abs.*, 1922, **16**, 1363.

[91] W. O. Williamson, *Nature*, 1937, **140**, 238; *Chem. Abs.*, 1937, **31**, 8129.

[92] W. O. Williamson, *Mineralog. Mag.*, 1940, **25**, 517; *Chem. Abs.*, 1940, **34**, 5348.

[93] W. O. Williamson, *Nature*, 1939, **143**, 279; *Chem. Abs.*, 1939, **33**, 3696.

[94] C. L. Moore, *Rayon Textile Monthly*, 1936, **17**, 792.

[95] G. Zerr, *Farben-Ztg.*, 1929, **34**, 1430; *Chem. Abs.*, 1929, **23**, 3112.

[96] H. Wagner, *Farben-Ztg.*, 1929, **34**, 1243; *Chem. Abs.*, 1929, **23**, 2840.

[97] E. Kiedel, *Farben-Ztg.*, 1929, **34**, 1242; *Chem. Abs.*, 1939, **23**, 2839.

[98] C. F. Goodeve and J. A. Kitchener, *Trans. Faraday Soc.*, 1938, **34**, 570; *Chem. Abs.*, 1938, **32**, 5305.

[99] A. Benrath, *Z. wiss. Phot.*, 1915, **14**, 217; *Chem. Abs.*, 1915, **9**, 1723.

[100] Norwegian 99,110, Dec. 11, 1961, to Titan Co., A./S.; *Chem. Abs.*, 1963, **58**, 3614.

[101] T. A. Ermolaeva, D. L. Abramson, and N. S. Amufrieva, *L. M. i ikh P.*, 1963, No. 1, 36; *Chem. Abs.*, 1963, **58**, 11574.

[102] British 874,511, Aug. 10, 1961, to National Lead Co.; Norwegian 95,212, Dec. 14, 1959, to Titan Co. A./S.; *Chem. Abs.*, 1961, **55**, 6883.

[103] T. N. Gurevich, V. A. Zubchuk, and S. V. Yakubovich, *L. M. i ikh P.*, 1963, No. 1, 55; *Chem. Abs.*, 1963, **58**, 13337.

[104] A. V. Pamfilov, Y. S. Mazurkevich, and N. P. Novalkovskii, *L. M. i ikh P.*, 1963, No. 1, 23; *Chem. Abs.*, 1963, **58**, 12106.

[105] A. G. Vondjidis and W. C. Clark, *Nature*, 1963, **198**, 278; *Chem. Abs.*, 1963, **59**, 14195.

[106] M. L. Meyers and W. R. Whatley, British 909,220, Oct. 31, 1962, to American Cyanamid Co.; *Chem. Abs.*, 1963, **58**, 1644.

[107] J. T. Richmond and G. G. Durant, German 1,023,165, Jan. 23, 1958, to Laporte Titanium Ltd.; *Chem. Abs.*, 1960, **54**, 17911.

[108] L. Csonka, Hungarian 148,370, Jan. 30, 1961; *Chem. Abs.*, 1963, **58**, 9334.

[109] F. R. Tarantine, U.S. 3,067,053, Dec. 4, 1962, to American Cyanamid Co.; *Chem. Abs.*, 1963, **58**, 9334.

[110] D. J. Oliver and S. F. Craddock, U.S. 3,056,689, May 29, 1959, to Laporte Titanium Ltd.; *Chem. Abs.*, 1963, **58**, 2573.

[111] A. J. Werner, U.S. 2,737,460, March 6, 1956, to E. I. du Pont de Nemours and Co.; *Chem. Abs.*, 1956, **50**, 9037.

[112] F. L. Kingsbury, Belgian 617,089, May 15, 1962, to Titangesellschaft m. b. H.; *Chem. Abs.*, 1963, **58**, 680.

[113] G. A. Marcot and J. P. Wikswo, U.S. 2,804,374, Aug. 27, 1957, to American Cyanamid Co.; *Chem. Abs.*, 1958, **52**, 749.

[114] W. H. Holback, W. J. Cauwenberg, and W. R. Whatley, U.S. 2,479,836, Aug. 23, 1949, to American Cyanamid Co.; *Chem. Abs.*, 1940, **44**, 3722.

[115] W. R. Whatley, Belgian 611,552, June 14, 1962, to American Cyanamid Co.; *Chem. Abs.*, 1962, **57**, 12108.

[116] W. Kampfer, *Offic. Dig. Federation Socs. Paint Technol.*, 1960, **32**, 454; *Chem. Abs.*, 1960, **54**, 18981.

[117] M. B. Alpert, A. E. Jacobsen, and P. B. Mitton, *Ind. and Eng. Chem.*, 1963, **2**, No. 4, 264.

[118] W. von Fischer, W. D. Trautman, and J. Friedman, *Am. Paint J.*, 1949, **34**, No. 13, 74; *Chem. Abs.*, 1950, **44**, 1719.

[119] F. B. Steig, Jr., *Offic. Dig., Federation Paint and Varnish Prod. Clubs*, 1957, **29**, 439; *Chem. Abs.*, 1958, **52**, 5851.
[120] A. Goeb, *Farbe u. Lack*, 1951, **57**, 14; *Chem. Abs.*, 1951, **45**, 3165.
[121] E. S. Larsen and H. Berman, *U.S. Geol. Surv., Bull.* No. 848, 1934.

## CHAPTER 24

[1] R. Mach, *Chemik*, 1964, **17**, 96; *Chem. Abs.*, 1964, **61**, 5227.
[2] *European Chemical News*, May 25, 1962, p. 24.
[3] Minerals Yearbook, U.S. Bureau of Mines, yearly volumes.

# Index

Abundance of titanium, 5
Accelerated test for chalking, 538, 539, 540, 541
Acidity, factor of, 300
Active sulfuric acid in ilmenite solution, analysis for, 141, 142
Aerogel
　alumina, to apply alumina coating, 559
　silica, to apply silica coating to pigment, 559, 560, 563
Africa
　ilmenite sand deposits in, 32
　ilmenite-magnetite deposits in, 25
　rutile deposits in, 37
After-treatment of titanium dioxide pigment, 536, 537
Alloying elements in titanium, 184
Alloys of titanium, 184
　chlorination of, 441, 442
Alum, preparation of, containing titanium, 91, 92
Alumina, addition of, to titanium dioxide during thermal splitting, 498, 499
Alumina coating of pigment
　to improve chalking resistance, 552, 553, 554, 555, 556, 557
　to prevent discoloration in films, 552, 553, 554, 555, 556, 557, 558
Amalgams as reducing agents for titanium metal, 162
Ammonium bromotitanate, preparation and properties of, 104
Ammonium pentafluotitanite, preparation of, 102
Ammonium titanofluoride, preparation and properties of, 102
Ammonium titanyl sulfate, preparation and properties of, 90
Analysis
　chemical, methods of, 136
　for free sulfuric acid, 141, 142
　of ilmenite, 44, 45, 46
　of rutile, 46
　of sulfate solution of ilmenite, 301

of titanium compounds, 136
　colorimetric, 141, 143, 144
　continuous process, 142
　gravimetric, 140, 141
　　using cupferon, 136
　volumetric
　　using aluminum, 140
　　using Jones reductor, 137, 138, 139
　　using zinc amalgam, 139
of titanium metal, spectrographic, 142
Anatase
　properties of, 10, 11
　pseudomorphs of, 15
　and rutile, identification of, 142
Anatase and rutile modification, hydrolysis of solutions, 257, 258, 259
Anatase-to-rutile transformation, 258, 259, 335, 336, 337, 338, 339, 340, 341
　prevention of, by stabilizing anions, 338, 339
　studies of, 338, 339, 340, 341
Anhydrite, preparation of, for pigment, 528, 529, 530, 531
Animals, titanium in, 8, 9
Anodizing titanium metal, 180
Arc, high erosion, titanium metal from the dioxide through a, 173
Argentina, ilmenite sand deposits in, 34
Arizonite, properties of, 13, 14
Arkansas, rutile deposits in, 36
Atmosphere, titanium in, 9
Australia
　ilmenite production in, 49, 50
　ilmenite sand deposits in, 31
　rutile deposits in, 38
　rutile resources of, 39
Autoclave
　production of titanium metal in, 150
　removing iron from ilmenite in, 195, 196
Automatic casting process for titanium, 183

675

# INDEX

Ball mill, production of titanium metal in, 158
Barium thiotitanate, preparation of, 115
Barium titanate
  in capacitors, 122
  preparation of, 121, 122, 123, 520, 521
  properties of, 122
Barium titanium sulfide, 114
Bauxite
  concentrate, titania slag from, 209
  red mud from
    pigment from, 229, 230
    selective chlorination of, 445
Bismuth titanate, preparation of, 125
Bleaching hydrolysis product, 320
Boride of titanium, 118
Brazil
  ilmenite and deposits in, 34
  rutile deposits in, 36
Briquets
  binder
    sodium hydroxide, 402
    sulfite liquor, 416
    tar and pitch, 402
  ilmenite, chlorination of, 416, 417, 418
  slag, chlorination of, 425, 426, 427, 428
Bromide process for pigment manufacture, 519
Bromides of titanium, 104
Brookite, properties of, 10, 11

Calcination; *see also* Hydrous titanium dioxide, calcination of
  double, of hydrolysis product, 334
  recycling of exhaust gases, 334
Calcination nuclei, rutile from hydrous oxide, 341, 342, 343, 344, 345
Calcium hydride, reduction of titanium dioxide with, 172
Calcium sulfate-titanium dioxide pigment, 527, 528, 529, 530, 531
Calcium titanate
  preparation of, 123, 124, 125
  properties of, 124
Calcium titanium oxalate, preparation of, 124
Calcium titanium sulfide, 114
Canada
  ilmenite production in, 49
  ilmenite-hematite deposits in Quebec, 21, 22
  ilmenite-magnetite deposits in Alberta, 22
  ilmenite-magnetite deposits in Quebec, 21, 22
  rutile deposits in Quebec, 36

Capacitors from barium titanate, 122
Carbides of titanium, 115
Carbon to titanium bond, compounds characterized by, 133, 134
Centrifuge, production of titanium metal in, 158
Ceylon, ilmenite production in, 49
Chalking
  accelerated test for, 538, 539, 540, 541
  effects of, 533
  experimental studies of, 534, 535, 536, 537, 538, 539
  explanations given for, 534, 535, 536, 537, 538, 539
Chalking resistance
  alumina coating to improve, 552, 553, 554, 555, 556, 557, 558
  calcination
    with added agents to improve, 541, 542, 543, 544, 545
    with antimony oxide to improve, 541, 542, 544
    with zinc oxide to improve, 543, 544, 545, 549
  coating the pigment particles to improve, 545, 546, 547, 548, 549, 550, 551, 552, 558
  organic coating to improve, 547, 548, 550
  of pigment, titanium phthalate added to improve, 546, 547
  silica coating to improve, 558, 559, 560
Chloride process, flow sheet, 482
Chloride solution of ilmenite
  ion exchange extraction of titanium compounds from, 254
  solvent extraction of iron compounds from, 254
Chlorination
  with agents other than chlorine, 445
  of briquets, 401, 402, 403, 404, 405
  chlorine and natural gas as agent in, 417
  with chloroform, 445
  in a fluidized bed, 407
  of ilmenite briquets, 416, 417, 418
  of ilmenite in a fluidized bed, 407, 411, 413, 414, 416, 417, 418, 419, 420, 434
  of ilmenite, manganese oxide catalyst, 423
  of leucoxene, 407
  of ores and concentrates, 400
  of perovskite in a fused salt bath, 439
  of red mud, 437, 438
  of rutile, 402
    with carbon tetrachloride, 446

# INDEX

with chlorine and carbon monoxide, 442, 443, 444
with chlorine and producer gas, 445
in a fluidized bed, 407, 408, 409, 410, 411, 412
with nitrosyl chloride, 448
with phosgene, 442, 443
selective, of ilmenite, 413, 415, 416, 420, 421, 422, 423
of slag, 403, 405
  with chlorine and carbon monoxide, 443
  with chlorine, carbon monoxide, hydrogen chloride, and carbon dioxide, 445, 446
  in a fluidized bed, 409, 412, 418, 428, 429, 430, 431, 432, 433, 434
  in a fused salt bath, 438, 439
  methods of improving, 435
  with phosgene, 443
  studies of, 436, 437
of slag briquets, 425, 426, 427, 428
of sphene, 425
of titaniferous ores with nitrosyl chloride, 448
of titanium alloys, 441, 442
of titanium carbide, 406, 440
of titanium dioxide with aluminum chloride, 448
of titanium dioxide concentrate, 400, 401, 403, 404, 406
of titanium metal, 441, 442
Chlorination furnace, 411, 412
Chlorination gases, waste, titanium tetrachloride from, 462
Chlorination products
  cooling of
    in bed of sand, 454
    with chlorination gas, 453
    in heat exchangers, 459
    with liquid titanium tetrachloride, 448, 449, 450, 451, 452, 453, 454, 455, 456, 460
    in stages, 455, 460
  mechanical separation of, after cooling, 456
  prevention of clogging by, 455
  recovery of chlorine from, 561
  removal of ferric chloride from
    by absorbing in alkali chloride, 458
    with refractory filter, 457
    in rock salt column, 457, 459, 460
  removal of unreacted chlorine from, 461
  selective condensation of, 454, 455
  separation of, 448, 449, 450, 451, 452, 453, 454, 455, 456, 457, 458, 459, 460
  in molten salt bath, 459
  separation of ferric chloride with Cottrell precipitator, 457
  separation of gaseous products from, 461
Chlorine
  removal of
    from chlorination product, 461
    from titanium dioxide, 506, 507
    from titanium dioxide by aeration, 507
  and natural gas as chlorinating agent, 417
Chloroacetate process for titanium dioxide, 523
Cladding metals with titanium, 180, 181
Clarification of sulfate solution of ilmenite or slag, 248, 249
  with glue or colloidal metallic sulfides, 248, 249
Clays, titanium in, 6
Coagulation of dispersed titanium dioxide, 350, 352, 353, 354
Coating agents for titanium dioxide pigments, 545, 546, 547, 548, 549, 550, 551, 552, 558
Cobalt titanate, preparation of, 126
Collector for ilmenite flotation, 40, 41, 42, 43, 44, 45
Colloidal compounds of titanium, 78, 79, 80, 81
Color of enamel, relation of particle size to, 566
Colorado, ilmenite-magnetite deposits in, 21
Colored pigments, 531, 532
Composite pigments
  blended, 530, 531
  coalesced, 527, 528, 529, 530
Concentrate, titanium dioxide, from ores or slag by treatment with hydrochloric acid, 393, 394, 395, 396, 397, 398
Concentration
  of ilmenite ores, 40, 41, 42, 43, 44, 45
  of sulfate solution of ilmenite or slag, 255
Condensation of chlorination products, selective, 454, 455
Constant composition of solution during hydrolysis, 288, 289, 290, 291
Consumption
  of ilmenite, 52
  of rutile, 52

Consumption (*Continued*)
  of titanium dioxide pigments, 576, 577
  of titanium metal, 184
Continuous process
  of hydrolysis, 291, 292, 293
  for titanium metal, 149, 150, 151, 157, 160, 165
Cooling
  chlorination products
    in a bed of sand, 454
    with chlorinating gas, 453
    in a heat exchanger, 459
    with liquid titanium tetrachloride, 448, 449, 450, 451, 452, 453, 454, 455, 456, 460
  titanium dioxide suspension from thermal splitting, 502, 503, 504, 505, 506
Cost
  production, of titanium dioxide pigment, 575
  of pigment plants, 575
Crystal growth of titanium dioxide, minimizing of, during reaction, 481, 486, 487
Crystallization of sulfate solution of ilmenite, 252
Cupferon reagent in analysis of titanium, 136

Digestion
  of ilmenite
    with circulating sulfuric acid, 223
    continuous, with sulfuric acid, 240, 241, 242, 243, 244
    with sulfuric acid, 216, 217, 218, 219, 220, 221, 222
      after leaching with hydrochloric acid, 222
      methods of improving, 224, 225, 243
      to give porous product, 237, 238, 239
      under pressure, 239
      to produce titanyl sulfate, 217
    with waste sulfuric acid, 224
  of ores continuously with sulfuric acid, 240, 241, 242, 243, 244
  of perovskite with sulfuric acid, 227
  of red mud with sulfuric acid, 229
  of slag continuously with sulfuric acid, 243
  of slag with sulfuric acid, 220, 222
  of sphene with sulfuric acid, 227
  of titania slag with sulfuric acid, 231, 232, 233, 234, 236
  of titania slag with waste sulfuric acid, 224
  of titaniferous magnetite with waste sulfuric acid, 224
Digestion product of ilmenite or slag, solution of, 224, 225
Discoloration
  alumina coating of pigment particles to prevent, 552, 553, 554, 555, 556, 557, 558
  coating pigment particles to prevent, 546, 548, 550, 551
  effect of, 533
Discovery of titanium, 3
Dispersibility of pigment in paint vehicles, agents to improve, 562, 563, 564
Dispersing agents for calcined titanium dioxide, 345, 346, 347, 348, 349, 350, 351, 353
  efficiency of, 346
Disproportionation of titanium tetraiodide to produce the metal, 181
Dititanosotitanic carboxide from ilmenite by roasting with carbon, 199
Dual sulfate process for titanium dioxide manufacture, 526

Egypt
  ilmenite production in, 49
  rutile deposits in, 40
Electric field, high-frequency, purification of titanium metal with, 178
Electric furnace, titania slag from, 210, 211
Electrode potentials, fused solutions, 165
Electrolysis to produce titanium metal, 57, 58, 59, 162, 163, 164, 165, 166, 167, 168, 169, 170, 171
  of lower chlorides, 165, 166
  of potassium fluotitanate in a fused bath, 168, 169
  of sodium fluotitanate in a fused bath, 168, 169
  of $Na_2TiCl_4$ in a fused bath, 164
  of the tetrachloride in a fused bath, 164
  of titanium alloys in a fused bath, 171
  of titanium carbide anode, 170
  of titanium carbide in a fused bath, 170
  of titanium dichloride in a fused bath, 166, 167
  of titanium dioxide in a fused bath, 169, 170
  of titanium dioxide with molten zinc cathode, 174

of titanium nitride, 171
of titanium tetrachloride in a fused bath, 162, 163, 164, 165
of titanium tetraiodide in a fused bath, 175
of titanium trichloride, 164, 167
of the trichloride in a fused bath, 167
Electrolysis to purify titanium metal, 175, 176, 177
Electromagnetic separation of titanium minerals, 40, 41, 42, 43, 44, 45
Electroplating titanium
from fused bath, 180
from nonaqueous solution, 179, 180
from water solution, 178, 179
Electroplating titanium metal, 175, 176, 177
Elutriation to effect particle size separation, 350
Esters of titanium, preparation of, 133
Europe, ilmenite sand deposits in, 34; *see also individual countries*
Exsolution, minerals formed by, 14

Ferricyanide of titanium, 118
Ferrocyanide of titanium, 118
Fibrous alkali metal titanates, 121
Fibrous titanium dioxide, 524, 525
Filtration
of hydrous titanium dioxide, 317, 318, 319, 320, 321, 322
media for, 317, 321
of sulfate solution of ilmenite or slag, 251, 252
Finland
ilmenite production in, 49
ilmenite-magnetite deposits in, 25
Flame oxidation of titanium tetrachloride, 488, 489, 490, 491
Florida
ilmenite sand deposits in, 28
rutile deposits in, 36
Flotation
of ilmenite ores, 40, 41, 42, 43, 44, 45
of rutile ores, 40, 41, 42, 43, 44, 45
of titanium minerals, 40, 41, 42, 43, 44, 45
Flow sheet
concentration of ilmenite sand, 27
concentration of ilmenite-magnetite, 19
chloride process of pigment manufacture, 482
sulfate process of pigment manufacture, 214, 217
titanium metal production, 147
Fluid energy milling of calcined titanium dioxide, 359, 360

Fluidized bed
calcination of hydrous titanium dioxide in, 334
chlorination
of ilmenite in, 419
of rutile in, 406
of slag in, 429
oxidation of titanium tetrachloride in, 491, 492, 493
production of titanium metal in, 157
Fluoride process
hydrous titanium dioxide from, calcination of, 518
for pigment manufacture, 509
two-stage, 517
Fluorides of titanium, 101
Fusion, alkali, of ilmenite to form titanium compounds soluble in acid, 196, 197

Geikieilite, 13
Geology of titanium, 10
Gloss of enamel films
factors influencing, 565
methods of improving, 564, 565, 566
perchloric acid or a perchlorate, 564
Greenland, rutile deposits in, 37

Hardening, surface, of titanium metal and alloys, 182
Heat exchanger, cooling chlorination products in, 459
Hexachlorosiloxane, removal from titanium tetrachloride, 478
Hiding power of pigment
relation of particle size to, 566
relation to pigment volume concentration to, 565, 566
Hydrides of titanium, 119
Hydrochloric acid in titanium dioxide produced by thermal splitting, neutralization of, 507
Hydrolysis
constant composition of solution during, 288, 289, 290, 291
continuous process for, 291, 292, 293
of dilute sulfate solution of ilmenite, 260
ilmenite solution added to water, 278, 279, 280, 281, 282, 283, 284, 285, 286, 287, 288
pressure, of sulfate solution of ilmenite, 261, 262, 263, 264
procedure to improve filtering rate of hydrate, 286, 287
of solution of basic sulfate of titanium, 259, 261

# 680　INDEX

Hydrolysis (*Continued*)
　of sulfate solution of ilmenite
　　concentration of crystalloid and colloid, 283
　　indirect methods of, 297, 298, 299, 300, 301
　　studies of methods of, 288, 293, 294, 295, 296, 297
　　trend of the reaction, 281
　of sulfate solution to produce rutile, no nuclei, 313
　of titanium sulfate solution
　　anatase and rutile modifications, 257, 258, 259
　　mechanism of, 256
　of titanium tetrachloride solution, 379, 380, 381, 382, 383, 384, 385, 386, 387, 388, 389
Hydrolysis nuclei
　added to sulfate solution of ilmenite, 264, 265, 266, 267, 268, 269, 270, 271, 272, 273, 274, 275, 276
　coagulation of, 304, 309
　formed in place, 278, 279, 280, 281, 282, 283, 284, 285, 286, 287, 288
　rutile
　　hydrous titanium dioxide, 272, 273, 274
　　by peptizing hydrous titanium dioxide in hydrochloric acid or nitric acid, 304, 305, 306, 307, 308, 309, 311, 315
　　from sodium titanate, 276, 277, 308, 310, 312, 313
　　from titanium tetrachloride, 302, 303, 304, 309, 310, 313, 314
　　from zinc titanate, 313
　for titanium tetrachloride solution, 379, 380, 381, 382, 383, 384, 388, 389
Hydrolysis product
　bleaching, 320
　electrolysis during washing, 322
　filtering, 317, 318, 319, 320, 321, 322
　neutralization of residual acid, 322, 323
　properties of, 333
　treatment with ammonium hydroxide during washing, 321
　washing, 317, 318, 319, 320, 321, 322
　　in the presence of powdered zinc, 319
　　in the presence of a reducing agent, 318, 319, 320, 321
Hydrosols of titanium, 82
Hydrous titanium dioxide; *see also* Hydrolysis product
　calcination of, 324, 325, 326, 327, 328, 329, 330, 331, 332, 333, 334

　　with alkaline earth compounds, 326
　　with an antimony compound, 329
　　with a carbonaceous material, 331
　　changes during, 333, 334
　　in a fluidized bed, 334
　　with a fluoride compound, 328
　　from the fluoride process, 518
　　with a niobium or tantalum compound, 330, 331
　　with a phosphate compound, 330
　　with potassium compounds, 325, 326, 327, 332
　　removal of phosphates before, 339
　　with sodium tungstate, 327
　neutralization of residual acid, 322, 323
　properties of, 333

Ilmenite
　analyses of, 44, 45, 46
　chlorination in a fluidized bed, 407, 411, 413, 414, 416, 417, 418, 419, 420, 434
　chlorination of, with a manganese oxide catalyst, 423
　concentrate from, by heating in autoclave with sulfuric acid and coke, 195, 196
　concentration of, 40, 41, 42, 43, 44, 45
　consumption, 52
　deposits, massive, occurrence of, 16
　digestion of; *see* Digestion of ilmenite
　direct solution in sulfuric acid without solid formation, 223
　dititanosotitanic carboxide from, by roasting with carbon, 199
　extraction of iron with sulfuric acid, 191, 192
　extraction of phosphorus, 190
　flotation of, 40, 41, 42, 43, 44, 45
　　collector for, 40, 41, 42, 43, 44, 45
　fusion
　　with sodium carbonate and coke to form a product soluble in sulfuric acid, 198
　　with sodium carbonate or sodium hydroxide, 187, 188, 189, 193, 194
　　with sodium sulfide, 187, 188
　heating
　　with barium sulfate to form products soluble in hydrochloric acid, 198
　　with sodium sulfate and sulfuric acid to form soluble sulfates, 197, 198, 199

# INDEX

imports of, 51
intergrowths of hematite and magnetite with, 14
leaching; see Leaching ilmenite
oxidation of slag with ferric oxide from, 237
price of, 51
production of, 47
properties of, 12, 13
recovery from beach and buried sands, 27, 28
recovery from ilmenite-magnetite ore, 18, 19
reduction roast in a fluidized bed, 194
removal of iron by electrolyzing in alkaline solution, 190
roasting with sodium carbonate and calcium oxide to form products soluble in sulfuric acid, 197
roasting with sodium sulfate followed by leaching with sulfuric acid, 191, 195
selective chlorination of, 413, 415, 416, 420, 421, 422, 423, 424
selective solution of iron with hydrochloric acid, 200
selective solution of iron with sulfuric acid, 200
smelting, silica as flux, 204
solution
  in hydrochloric acid, 373, 374, 375
  in sulfuric acid, 200
    after alkali fusion, 196, 197
    after reduction roast, 199
titania concentrate from, with hydrochloric acid, 393, 394, 395, 396, 397, 398
titania slag from; see Slag, titania
titanium dioxide concentrate, by roasting with sodium sulfate and coke, 200
titanium dioxide residue after reduction roast, 199
titanium dioxide slag from, 189, 190, 191, 193
weathering of, scheme for, 14
weathering of, steps in process, 14
world production of, 50
Ilmenite ores, degree of alteration, 14
Ilmenite production
  in Australia, 49, 50
  in Canada, 49
  in Ceylon, 49
  in Egypt, 49
  in India, 50
  in Malaysia, 49
  in Republic of South Africa, 49
  in Spain, 49
  in United States, 47, 48, 50
Ilmenite residue, solution in sulfuric acid, 199, 200
Ilmenite sand deposits
  in Alaska, 29
  in Argentina, 34
  in Arkansas, 29
  in Australia, 31
  in Brazil, 34
  in Europe, 34
  in Florida, 28
  in India, 31
  in Japan, 32
  in Latin America, 34
  in Mississippi, 29
  in Montana, 29
  in New Jersey, 29
  in Portugal, 34
  in Union of Soviet Socialist Republics, 30
  in United States, 26
Ilmenite sands, occurrences of, 26
Ilmenite solution, constant composition during hydrolysis, 288, 289, 290, 291
Ilmenite-magnetite
  occurrences of, 18
  recovery of ilmenite from, 18, 19
Ilmenite–sulfuric acid mixture
  heating with steam, 218
  phosphoric acid as dispersing agent, 219
India
  ilmenite production in, 48
  ilmenite sand deposits in, 30
Insecticidal value of titanium dioxide, 135
Interstellar space, titanium in, 7
Iodides of titanium, 105
Ion exchange resin, extraction of titanium compounds with, 254
Iron titanate, preparation of, 125, 127
Irradiation of titanium dioxides to produce the metal, 173
Ivory Coast, ilmenite sand deposits in, 33

Japan, ilmenite sand deposits in, 32

Lanthanum titanate, preparation of, 120, 125
Latin America, ilmenite sand deposits in, 34
Leaching ilmenite
  with ferric chloride, 190

Leaching ilmenite (*Continued*)
  with hydrochloric acid, 188, 189
    before digestion in sulfuric acid, 222
    to remove iron, 188, 189, 195
  with nitric acid, 194
  to remove phosphorus, 190
  with sodium hydroxide, 196
Lead titanate, preparation of, 126, 127
Leucoxene
  chlorination of, 407
  properties of, 14, 15
Lithium thiotitanate, preparation of, 115

Machining, electrochemical, of titanium, 183
Madagascar, rutile deposits in, 38
Magnesium, reduction of titanium tetrachloride with, 148
Magnesium and magnesium chloride
  dissolving in acetone, ketone, or alcohol, 155
  dissolving in methyl alcohol, 154
Magnesium titanate, preparation of, 125
Malaysia, ilmenite production in, 49
Melting titanium metal, 182
Metal, titanium; see Titanium metal
Metatitanic acid, 78
Meteorites, titanium in, 7
Milling
  of calcined titanium dioxide
    coatings added during, 359
    dry, 355, 356, 357, 358, 359, 360
    wet, 351, 352, 354
  chemical, of titanium, 183
  reduction of oil absorption by, 357, 358
Minerals
  titanium, 5, 10, 11, 12, 13, 14
    electromagnetic separation of, 40, 41, 42, 43, 44, 45
    flotation of, 40, 41, 42, 43, 44, 45
  titanium in, 5
Minnesota, ilmenite-magnetite deposits in, 20
Montana, ilmenite-magnetite deposits in, 20
Mozambique, ilmenite sand deposits in, 33

Naming of titanium, 4
New Jersey, ilmenite sand deposits in, 29
New York, ilmenite-magnetite deposits in, 18, 20
Nickel titanate, preparation of, 126
Nitrate process for pigment manufacture, 520, 521

Nitrate process for pigment manufacture, equations for chemical reactions, 521
Nitrogen compounds of titanium, 107
North Carolina, massive ilmenite deposits in, 17
Norway
  ilmenite-magnetite deposits in, 23, 24
  rutile deposits in, 37
Nuclei calcination
  hydrous oxide, to yield rutile, 341, 342, 343, 344, 345
  rutile, for hydrous oxide
    from sodium titanate, 342, 343, 344, 345
    from titanium phosphate, 344
    from titanium tetrachloride, 345
Nuclei for vapor phase oxidation of titanium tetrachloride, 493, 494, 495, 496, 497
Nuclei, hydrolysis; see Hydrolysis nuclei
Nuclei, rutile; see Rutile nuclei
Nuclei, yield seed, for hydrolysis, 276, 277, 278

Octahedrite, properties of, 10, 11
Oil absorption
  determination of, 355
  effect of calcination agents, 356, 357
  of pigments, effect of, 355
  reduction by milling, 357, 358
  regulation of, 355, 356, 357
Organic compounds of titanium, 128
Organometallic compounds of titanium, preparation of, 132
Orthotitanic acid, 78
Oxidation of titanium III in slags, 234, 235, 236
Oxides of titanium, 67

Particle size of pigment, relation of
  to color of enamel, 567
  to hiding power, 566
Perovskite
  chlorination in a fused salt bath, 439
  digestion with sulfuric acid, 227
  pigment from, 227
  properties of, 13
Philippines, ilmenite sand deposits in, 32
Phosphate compounds of titanium, 112
Photosensitivity of pigment
  calcination with added agents to prevent, 561
  cause of, 562
  coating to prevent, 560, 561, 562
    alumina, silica, and ceric oxide coatings, 561

# INDEX

study of, 560, 561
Photosensitivity of rutile, 537
Physiological effects of titanium compounds, 134
Pigment
 bromide process, 519
 composite, blended, 530, 531
 composite, coalesced, 527, 528, 529, 530
 titanium, development of the industry, 568, 569, 570, 571, 574
 titanium dioxide; see Titanium dioxide pigment
Pigment process
 flame oxidation of titanium tetrachloride, 488
 thermal splitting of titanium tetrachloride, 480
 volume concentration, relation to hiding power, 565, 566
Pigments
 colored, 531, 532
 dispersibility in paint vehicle, agents to improve, 562, 563, 564
Pioneer investigations in titanium, 4
Plants, titanium in, 7, 8
Plants, titanium pigment; see Titanium pigment plants
Portugal, ilmenite sand deposits in, 34
Porous digestion product from ilmenite, 237, 238, 239
Potassium chlorotitanate, titanium tetrachloride by thermal decomposition of, 447
Potassium fluotitanate
 preparation of, 101, 102, 103
 properties of, 103
Potassium titanate, fibrous, 121
Potassium titanium chloride, preparation of, 96
Potassium titanium oxalate, preparation of, 91, 133
Preheating titanium tetrachloride and oxygen, 480, 481, 483, 485, 486
 metal heat exchangers, 484
Pressure digestion of ilmenite or slag with sulfuric acid, 239, 244
Pseudobrookite, properties of, 13
Purification
 of sulfate solution of ilmenite or slag, 250, 251
 of titanium metal, 177
 of titanium tetrachloride, 92, 463, 464
  with absorptive carbon, 475, 476, 478
  agents used, 92, 463

  with alkali metal and alkaline earth hydroxides, 471
  with aluminum, 467, 468
  with arsenic sulfide, 469
  with copper, 464, 465, 466, 467, 468, 478
  with esters, 476
  with fatty acids, 477
  with ferrous sulfate, 471
  with hydrides, 474
  with hydrocarbons, 476, 477
  with hydrogen sulfide, 467, 469, 470
  with inorganic compounds, 470, 471, 472
  with iodine, 467, 471
  with iron, 467, 468
  with iron carbonyl, 474
  with lead, 469
  with lead alloy, 469
  with magnesium, 468, 470, 471
  with metallic elements, 462, 463, 464, 465, 466, 467, 468, 469
  with metallic soap, 476
  with organic compounds, 474, 475, 476, 477
  with potassium iodide, 471
  with silica gel, 478
  with sodium, 468
  with sulfides, 469, 470
  with sulfur, 467, 468
  with titanium metal, 469
  with titanium trichloride, 472, 473
  with titanyl chloride, 472, 473
  with zinc, 467, 468
Pyrophanite, 13


Reaction products, thermal splitting of titanium tetrachloride, cooling of, 502, 503, 504, 505, 506
Recovery of waste sulfuric acid, 361, 362, 363, 364, 365, 366, 367, 368, 369, 370
Red mud from bauxite
 chlorination of, 437, 438
  selective, 445
 digestion with sulfuric acid, 229, 230
 titanium dioxide concentrate from, 195
 titanium pigment from, 229, 230
Reduction to produce titanium metal
 of the dioxide, 55, 56, 172, 173, 174
  with calcium carbide, 173
  with calcium hydride, 172
  with calcium-sodium sludge, 155
  with magnesium, 173
  with titanium carbide, 173

# INDEX

Reduction to produce titanium metal (*Continued*)
  of lower salts with sodium in liquid ammonia, 160
  with sodium or magnesium, 160, 161
  of Na$_2$TiCl$_4$ with sodium, 158
  of subchlorides with sodium, 155
  of the sulfides or oxides, 181
  of the sulfide with sodium or potassium, 174
  of the tetrachloride, 54, 145, 147, 148, 155
    in an autoclave, 150
    in a ball mill, 158
    in a bath of molten magnesium chloride, 150, 151
    in a fused bath of lower chlorides, 160
    with hydrogen, 171, 172
    without an inert gas atmosphere, 149
    with lithium, 181
    with magnesium, 56, 145, 146, 147, 148, 149, 150, 151, 152, 153, 154
    with sodium, 54, 155, 156
      in a fluidized bed, 157
      in liquid ammonia, 155
      in a two-stage process, 155, 156, 158
      with sodium amalgam in liquid ammonia, 162
      with a solution of magnesium in zinc, 153
  of the tetrafluoride with silicon, 181, 182
Reduction of sulfate solution of ilmenite or slag, 245, 246, 247
  by electrolysis, 246
  with iron, 246
Republic of South Africa
  ilmenite production in, 49
  ilmenite sand deposits in, 33
  ilmenite-magnetite deposits in, 25
Residue from ilmenite
  after extraction with sulfuric acid, 191, 192
    solution in sulfuric acid, 200
  after fusion with sodium carbonate or sodium hydroxide, 187, 188, 189, 193, 194
    with sodium sulfide, 187, 188
  after leaching with nitric acid, 194
Rhode Island, ilmenite-magnetite deposits in, 20
Rocks, titanium in, 5
Romania, ilmenite-magnetite deposits in, 25

Rutile
  analyses of, 46
  calcination with magnesium oxide to produce titanium compounds soluble in sulfuric acid, 197
  chlorination of; *see* Chlorination of rutile
  flotation of, 40, 41, 42, 43, 44, 45
  hydrolysis
    of titanium nitrate solution, 315, 316
    of titanium sulfate solution, 215, 312, 313, 314
    of titanium tetrachloride solution, 379, 391
  imports of, 51
  occurrences of, 34
  ores, composition of, 50
  price of, 51
  production, 50
  properties of, 10, 11
  solution after fusion, 225, 226
  solution in sulfuric acid after fusion with sodium carbonate, 218
  world resources of, 35
Rutile from anatase, 335, 336, 337, 338, 339, 340, 341
  calcination
    with lithium compounds, 335
    with rutile crystallites, 337
    with rutile nuclei, 335
    with TiOCl$_2$, 336
    with zinc oxide, 335
  double calcination, 335, 336
  studies of the transformation, 338, 339, 340, 341
Rutile consumption in the United States, 52
Rutile deposits
  in Africa, 37
  in Arkansas, 36
  in Australia, 38
  in Brazil, 36
  in Egypt, 40
  in Florida, 36
  in Greenland, 37
  in Madagascar, 38
  in Norway, 37
  in Quebec, 36
  in Senegal, 38
  in Sierra Leone, 37
  in Union of Soviet Socialist Republics, 37
  in Virginia, 35
Rutile from hydrous oxide, calcination nuclei, 341, 342, 343, 344, 345
Rutile from ilmenite, 14

Rutile nuclei
  by electrolytic hydrolysis of chloride solution, 315
  silica-titania gel as, 274
  stabilization
    with glue, 315
    with organic acids, 275, 314
    with sulfuric acid, 314
  hydrolysis, sulfate solution, 302, 303, 304, 305, 306, 307, 308, 309, 310, 311, 312, 313, 314, 315
  hydrolysis, from zinc titanate, 313
Rutile from sulfate solution of ilmenite, 301, 302, 303, 304, 305, 306, 307, 308, 309, 310, 311, 312, 313, 315

Samarium titanate, preparation of, 120
Scale, methods of avoiding titanium dioxide, 499, 500, 501, 502
Sediments, titanium in deep-sea, 7
Seed, hydrolysis; see Nuclei
Seed, yield, for hydrolysis, 276, 277, 278
Senegal
  ilmenite sand deposits in, 32
  rutile deposits in, 38
Selective chlorination
  of ilmenite, 423, 424
  of red mud from bauxite, 445
Separation of chlorination products, 448, 449, 450, 451, 452
  in molten salt bath, 459
Sierre Leone
  ilmenite-magnetite deposits in, 25
  rutile deposits in, 37
Silica coating to improve chalking resistance, 558, 559, 560
Silicides of titanium, 118
Silicoilmenite, properties of, 15
Slag, chlorination of; see Chlorination of slag
  digestion of, with sulfuric acid, 220, 222
    continuous, 243
    under pressure, 239
  titanium dioxide from, 189, 190, 191, 193
Slag, titania
  from bauxite concentrate, 209
  calcium oxide and magnesium oxide as flux, 204, 205
  cooled slowly to improve solubility, 202
  digestion with sulfuric acid, 231, 232, 233, 234, 236
    optimum conditions for, 233
  digestion with waste sulfuric acid, 224
  electric furnace charge pattern, 212
  oxidation of titanium III in, 209, 234, 235, 236
  oxidized with ferric oxide from ilmenite, 237
  pigment from, 232, 233
  purification of, 207
  by smelting ilmenite
    with barium carbonate flux, 207, 208
    in an electric furnace, 210, 211
      magnesia flux, 210, 211
      no flux, 211
    with limestone and sodium oxide, 202, 203
    in a low-shaft blast furnace, 203
    with silica flux, 204, 205, 206
    with sodium carbonate and coke, 204, 205
    with sodium oxide or potassium oxide, 207, 208
    two-step process, 205
  by smelting ores in a cupola, 203
  by smelting with sodium carbonate, 205, 206, 207, 209
  by smelting titanium ore with sodium carbonate, 201
  by smelting titanium ores with calcium oxide and magnesium oxide, 202
  solubility of the oxide, 208
Slag briquets, chlorination of, 425, 426, 427, 428
Sodium, reduction of titanium tetrachloride with, 155
Sodium amalgam as a reducing agent for titanium metal, 162
Sodium fluotitanate, preparation and properties of, 103
Sodium oxyfluopertitanate, preparation of, 103
Sodium titanate
  pigment from, 526, 527
  preparation of, 120, 121
Sodium titanyl sulfate, preparation and properties of, 90
Soils, titanium in, 5, 6
Solubility of impurities in titanium tetrachloride, 463, 464
Solution
  of the digestion product of ilmenite or slag, 244, 245
  direct, of ilmenite in sulfuric acid, 223
  of rutile after fusion, 225, 226
  of sphene in hydrochloric acid, 226, 373, 385, 386
  of sulfate cake, 244, 245

Solvent extraction
   of iron compounds, 253, 254
   of titanium compounds, 253
Spain, ilmenite production in, 49
Sphene
   chlorination of, 425
   digestion with sulfuric acid, 227
   heated with potassium sulfate to form products soluble in sulfuric acid, 198
   leaching with hydrochloric acid followed by digestion with sulfuric acid, 226, 227, 228
   pigment from, 226, 227, 228
   properties of, 15
   solution in hydrochloric acid, 226, 373, 385, 386
Stars, titanium in, 7
Strontium titanium sulfide, 114
Sulfate cake, solution of, 244, 245
Sulfate solution of ilmenite
   analysis of, 301
   basic, hydrolysis of, 259, 261
   continuous hydrolysis, 291, 292, 293
   crystalloid and colloid during hydrolysis, 283
   dilute, hydrolysis of, 260
   hydrolysis, nuclei added, 264, 265, 266, 267, 268, 269, 270, 271, 272, 273, 274, 275, 276
   hydrolysis, nuclei formed in place, 278, 279, 280, 281, 282, 283, 284, 285, 286, 287, 288
   hydrolysis to produce rutile without nuclei, 313
   indirect methods of hydrolysis, 297, 298, 299, 300, 301
   precipitation with alkaline agents, 259
   pressure hydrolysis, 261, 262, 263, 264
   reduction with iron, 246
   removal of ferrous sulfate by crystallization, 252
   rutile from, 301, 302, 303, 304, 305, 306, 307, 308, 309, 310, 311, 312, 313, 314, 315
   studies of hydrolysis of, 293, 294, 295, 296, 297
   trend of hydrolysis reaction, 281
Sulfate solution of ilmenite or slag
   clarification of, 248, 249
   clarification with glue or metallic sulfides, 248, 249
   concentration of, 255
   extraction of iron and titanium compounds from, 253
   filtration of, 251
   purification of, 250, 251
   reduction of, 245, 246, 247
   by electrolysis, 246
Sulfide process for pigment manufacture, 522, 523
Sulfides of titanium, 113
Sulfuric acid
   disposal of, in ocean, 370
   recovery of, 361, 362, 363, 364, 365, 366, 367, 368, 369, 370
   addition of hydrogen chloride, 367, 368
   concentration, 361, 362, 364, 365, 367, 368
   contact process, 363, 364, 369
   cyclic process, 369, 370
   multistage process, 369
   purification by cooling, 366
Sun's atmosphere, titanium in, 7
Sweden, ilmenite-magnetite deposits in, 25

Tanks for digestion of ilmenite with sulfuric acid, 220
Thermite process, titanium metal from the dioxide by, 173, 174
Titanates
   preparation of, 119, 120, 121
   types of, 119
Titania gel, preparation and properties of, 79, 83, 84, 85, 86
Titania slag; see Slag, titania
Titanic acid, 78, 79
Titanic oxalate, preparation of, 128, 130
Titanic phosphate, preparation of, 112, 113
Titaniferous magnetite, digestion of, with waste sulfuric acid, 224
Titanite, properties of, 15
Titanium
   abundance in earth's crust, 5
   alloying elements in, 184
   in animals, 8, 9
   in atmosphere, 9
   in clays, 6
   in deep-sea sediments, 7
   discovery of, 3
   in eggs, 9
   geology of, 10
   in interstellar space, 7
   in meteorites, 7
   in milk, 9
   in minerals, 5
   naming of, 4
   ore deposits of, 16
   in plants, 7, 8
   in rocks, 5

# INDEX

in soil, 5, 6
in stars, 7
in sun, 7
in water, 7
Titanium alcoholates, preparation of, 131
Titanium alloys, 184
  chlorination of, 441, 442
Titanium aluminides, preparation of, 115
Titanium carbide
  chlorination of, 406, 440
  higher, preparation of, 117
  preparation of, 115, 116, 117, 118
  properties of, 116, 117
Titanium to carbon bond, compounds characterized by, 133, 134
Titanium chlorides, 92, 373, 400
Titanium chloroacetate, titanium dioxide from, 523
Titanium chlorosulfonate chlorides, preparation and properties of, 95
Titanium containing slimes, reaction with sulfuric acid, 219
Titanium cyanides, preparation of, 109
Titanium cyanonitride
  preparation of, 109
  properties of, 109, 110
Titanium diboride, preparation of, 118
Titanium dibromide, preparation of, 105
Titanium dichloride
  electrolysis of, in fused bath to produce titanium, 166, 167
  preparation and properties of, 100, 101
Titanium diiodide, 107
  preparation of, 106
Titanium dioxide
  addition of alumina to, in vapor phase, 498, 499
  calcined
    coatings added during milling, 359
    dispersing agents for, 345, 346, 347, 348, 349, 350, 351, 353, 354, 355
    dispersion in water, 345, 346, 347, 350, 351
      with ammonium hydroxide, 346
      with sodium hydroxide, sodium carbonate, sodium silicate, or sodium phosphate, 346, 348
    fluid energy milling, 359, 360
    particle size separation
      by centrifuging, 351
      by decantation, 350
      by elutriation, 350
    wet milling, 351, 352, 354
  chlorination of, 406
  from the chloroacetate, 523
  coarse particle size, 488, 526
  fibrous, 524, 525
  flame oxidation of the tetrachloride, 488, 489, 490, 491
  fluidized bed oxidation of the tetrachloride, 491, 492, 493
  insecticidal value, 135
  nontoxic nature, 134
  nuclei for formation from the tetrachloride, 493, 494, 495, 496, 497
  preparation and properties of, 67, 68, 69, 70, 71, 72
  reduction with calcium-sodium sludge, 155
  separation from reaction products
    in a Cottrell precipitator, 490, 504, 506
    in a cyclone, 490, 503, 506
    in a filter, 503
  small particle size, 525, 526
  spherical particles, 490, 497
  from the sulfate, dual process, 526
  thermal splitting of the tetrachloride, 480
Titanium dioxide concentrate
  chlorination with chloroform, 445
  mechanically from reduced ilmenite, 398, 399
Titanium dioxide gels, preparation and properties of, 83, 84, 85, 86
Titanium dioxide hydrate, particle size and properties of, 221
Titanium dioxide ointment, 134
Titanium dioxide pigment
  bromide process, 519
  consumption of, 576, 577
  cooling suspension of, from thermal splitting, with inert gas, 504
  dispersion in water, 345, 346, 347
    with ammonium hydroxide, 346
    with sodium hydroxide, sodium carbonate, sodium silicate, or sodium phosphate, 346
  fluoride process, 509
  improving by double precipitation, 201
  neutralization of hydrochloric acid in, with alkaline earth hydroxide, 507
  nitrate process, 520, 521
  production of, 576
  production cost of, 575
  removal of chlorine
    by aeration, 507
    with boric acid, 507
    with steam, 507
  separation from chlorine, 502, 503, 504, 505, 506
    in Cottrell precipitator, 503, 504, 505, 506

Titanium dioxide pigment (*Continued*)
  separation from chlorine (*Continued*)
    in a cyclone, 502, 503, 505, 506
    on filters, 503
    in a fluidized bed, 505
  small particle size, 525, 526
  from sodium titanate, 526, 527
  sulfide process, 522, 523
  from titanium fluosilicate, 519
  uses of, 576, 577
Titanium dioxide residue
  leaching ilmenite
    with hydrochloric acid, 393, 394, 395, 396, 397, 398
    with hydrochloric acid in an autoclave, 393
    with titanium tetrachloride, 394
  leaching slag with hydrochloric acid, 394, 396, 397
  treating ilmenite with hydrogen chloride, 393, 396, 397, 398
  treating slag with hydrogen chloride, 397
Titanium dioxide scale, methods of avoiding on walls of reactor and conduit, 499, 500, 501, 502
Titanium dioxide suspension from thermal splitting, cooling of, 502, 503, 504, 505, 506
  with chlorine gas, 505, 507
  in conduits, 503
  with titanium tetrachloride spray, 504
Titanium disulfate, preparation and properties of, 87
Titanium disulfide, preparation and properties of, 113, 114
Titanium fluosilicate, pigment from, 519
Titanium gluconate, preparation of, 130
Titanium hydride
  metal from, by thermal decomposition, 172
  preparation and properties of, 119
Titanium hydroxide, precipitation with alkaline agents, 259
Titanium hydroxide sol, preparation and properties of, 82
Titanium industry, early workers in, 568, 569, 570
Titanium metal
  addition of the tetrachloride to magnesium in stages, 153
  advantages of the sodium reduction process, 155
  alkali metal amalgams as reducing agents, 162
  alkaline earth metal amalgams as reducing agent, 162

aluminothermic process, 55
anodizing to produce coating, 180
arc melting of, 182, 183
automatic casting process, 183
automatic regulation of manufacturing process, 154
casting techniques, 183
chemical milling, 183
chlorination of, 441, 442
cladding metals with, 180, 181
codeposition with copper in fused electrolyte, 181
commercial process for, 145, 148
commercial production of, 145
consolidated shapes, 183
consumption of, 184
continuous process, 160, 165
continuous process, sodium reduction, 157
continuous reduction of the tetrachloride, 149, 150, 151, 157
cost, 184, 186
decomposition of titanium tetraiodide, 174, 175
dehydrogenation of the hydride, 181
from the dioxide by the thermite process, 173, 174
disproportionation of the tetraiodide, 181
electrochemical machining, 183
electrolysis to produce; *see* Electrolysis to produce titanium metal
electroplating
  from a fused bath, 180
  from nonaqueous solutions, 179, 180
  from water solutions, 178, 179
improving, by heat treatment, 161
irradiation of the dioxide, 173
  with dilute acid containing a complexing agent, 154
leaching
  magnesium and magnesium chloride from, 154, 155
  of reaction products with liquid ammonia, 159
liner of titanium metal in the reactor, 154, 158
melting techniques, 182, 183
pioneer investigators of, 53
price of, 184
production of, 53, 184
production in centrifuge, 158
properties of, 59, 60, 61, 62, 63, 64, 65, 66, 185
purification
  by chemical means, 177
  by electrolysis, 175, 176, 177

by high-frequency electric field, 178
with the tetrachloride, 177
reaction of magnesium vapor with titanium tetrachloride, 152, 153
reaction of titanium carbide and iodine, 175
reduction to produce; see Reduction to produce titanium metal
removing magnesium and magnesium chloride from, by dissolving in methyl alcohol, 154
sodium amalgam as reducing agent, 162
sodium or magnesium reducing agent produced in the reactor by electrolysis, 160, 161
surface hardening of, 182
thermal decomposition of the hydride, 172
thermal dissociation of the tetraiodide, 55, 56, 174, 175
thermite process for, 173, 174
titanium carbide anode in fused bath, 170
titanium dioxide-carbon anode, 173
two-stage process, 151, 161
uses of, 185, 186
zinc cathode in electrolytic production of, 165
Titanium monoxide, preparation and properties of, 74, 75, 76
Titanium monosulfate, preparation of, 92
Titanium monosulfide, preparation and properties of, 114
Titanium nitrate, preparation of, 107, 111, 520, 521
Titanium nitrate solution, hydrolysis to produce rutile, 315, 316
Titanium nitride
preparation of, 107, 108, 109
properties of, 107, 108, 109, 110
Titanium oxides, 67
Titanium oxybromide, preparation of, 96, 104
Titanium oxychloride, preparation of, 96
Titanium oxytrisulfate, preparation and properties of, 87
Titanium peroxide, preparation and properties of, 76, 77, 78
Titanium phosphides, preparation of, 113
Titanium phthalate, preparation of, 129, 546, 547
Titanium pigment
flow sheet of chloride process, 482
flow sheet of sulfate process, 214, 216
Titanium pigment industry, development of, 568, 569, 570, 571, 574

Titanium pigment plants, 572
costs of production, 575
history of, 569, 570, 571, 573
of the world, 572, 573
Titanium pigment process, general information, 213, 215
Titanium potassium oxalate, preparation of, 129
Titanium pyrophosphate, preparation and properties of, 113
Titanium selenide, structure of, 115
Titanium silicide, preparation of, 119
Titanium soaps, preparation of, 129
Titanium sulfates, 87, 197, 201, 216, 225, 231
Titanium sulfides
preparation of, 114, 115, 522, 523
roasting to produce the dioxide, 522
Titanium tetrabromide
preparation of, 104, 519
properties of, 104
vapor phase oxidation of, 519
Titanium tetrachloride
action of hydrogen chloride on the nitride, 376
alkali metal chlorides to sulfate solution, 373, 376, 387
anhydrous, solution in water, 377, 378
chlorination; see Chlorination
electrolysis of the dioxide in a fused chloride bath, 446, 447
flame oxidation of, 488, 489, 490, 491
flame oxidation of, gases for auxiliary flame, 488, 489, 490, 491
from titanium dioxide and aluminum chloride, 448
heating
with an auxiliary flame, 487
by direct contact with burning carbon monoxide, 488
with electric resistor elements, 483
of titanium phosphate in molten calcium chloride, 447
hydrolysis of
in an autoclave, 391
cyclic process, 387, 390
hydrolysis nuclei for, 379, 380, 381, 382, 383, 384, 388, 389
solution of, by boiling, 379, 380, 381, 382, 383, 384, 385, 386, 387, 388, 389
with steam, 391, 392
iron chloride separation from by a cyclone, 452
iron chloride separation from by an electrofilter, 452, 453

Titanium tetrachloride (*Continued*)
 mixing with oxygen, method of, 485, 486
 nuclei for vapor phase oxidation, 493, 494, 495, 496, 497
 oxidation of
  apparatus for, 489
  in a fluidized bed, 491, 492, 493
  to form pigment, 480
  prevention of formation of coarse particles, 481, 486, 487
  refractory equipment for, 484
 preparation of, 92
 properties of, 92, 93, 94, 95
 purification of; *see* Purification of titanium tetrachloride
 recovery from ferric chloride sludge, 459
 reduction with magnesium, 145, 148, 149, 150, 151, 152, 153, 154, 155
 reduction with sodium, 155, 156, 157, 158, 159, 160
 removal of hexachlorosiloxane from, 478
 removal of organic matter from, 478
 separation of iron chlorides from, 449, 450, 451, 452, 453
 separation of solid chlorides from, 449, 450, 451, 452, 453, 454, 455, 456, 457, 458, 459, 460
 separation of solid products after cooling, 456
 skin irritation by, 135
 solubility of ferric chloride in, 456
 solubility of impurities in, 463, 464
 solution of ferrotitanium in hydrochloric acid, 373
 solution of ilmenite in hydrochloric acid, 373, 374, 375
 solution of ilmenite or slag in hydrochloric acid after alkali fusion, 375
 solution of rutile in hydrochloric acid after alkali fusion, 374
 solution of sphene in hydrochloric acid, 373
 special methods of preparation, 446
 thermal decomposition of potassium chlorotitanate, 447
 vapor phase oxidation of, cooling products from, 502, 503, 504, 505, 506
 from waste chlorination gases, 462
 water-enriched air for oxidation of, 485
Titanium tetrachloride and oxygen, preheated, filters for, 484
Titanium tetrachloride solution, purification of, 515, 516, 517

Titanium tetrachloride vapor, heating with quartz sand, 487
Titanium tetrachloride–oxygen reaction, studies of, 483
Titanium tetraethylate, preparation of, 130
Titanium tetrafluoride
 preparation of, 102
  from ores, 510, 511, 512, 513, 514, 515
 properties of, 102
 solution of the reaction product, 515
Titanium tetrafluoride solution
 precipitation of the dioxide with ammonia, 511, 512, 513, 514, 515, 516, 517, 518
 purification of, 515, 516, 517, 518
Titanium tetraiodide
 metal by decomposition of, 174, 175
 preparation and properties of, 105, 106
Titanium tetraphenylate, preparation of, 130
Titanium tribromide, preparation of, 105
Titanium trichloride
 preparation of, 96, 97, 98, 99
 properties of, 97, 99, 100
Titanium trifluoride, preparation and properties of, 103
Titanium triiodide, preparation and properties of, 106
Titanium trioxide, preparation and properties of, 73, 74
Titanium trisulfide, preparation and properties of, 114
Titanium tungsten carbide, preparation of, 117
Titanous phosphate, preparation and properties of, 113
Titanyl phosphate, preparation of, 112
Titanyl sulfate
 crystallization from solution, 300
 preparation of, 87, 88, 89, 90, 91
 properties of, 88, 89, 90, 91
 reaction of ilmenite with sulfuric acid, 217
Triethanolamine, improving dispersibility of pigments in paint vehicles with, 563, 564
Two-stage process for titanium metal production, 155, 156, 157, 161


Union of Soviet Socialist Republics
 ilmenite sand deposits in, 30
 ilmenite-magnetite deposits in, 22, 23
 rutile deposits in, 37
 titanium ores, resources of, 24

United States; *see also individual states*
  ilmenite consumption in, 52
  ilmenite deposits in, 16, 17, 18, 20, 21, 26, 28, 29, 30
  ilmenite production in, 47, 48, 50
  resources of titanium ores in, 17, 18, 20, 29, 35
  rutile consumption in, 52
  rutile deposits in, 35, 36
  rutile production in, 50
  titanium metal consumption in, 184
  titanium metal producers in, 184
  titanium metal production in, 184
  titanium pigment consumption in, 576, 577
  titanium pigment producers in, 572
  titanium pigment production in, 576
Uses of titanium dioxide pigments, 576, 577
Uses of titanium metal, 185, 186

Vapor phase oxidation of titanium tetrachloride, 480
Virginia
  massive ilmenite deposits in, 16, 17
  rutile deposits in, 35

Washing hydrolysis product after treatment with ammonium hydroxide, 321
Washing hydrous titanium dioxide, 317, 318, 319, 320, 321, 322
  electrolysis during, 322
  in the presence of powdered zinc, 319
  in the presence of reducing agents, 318, 319, 320, 321
Water, titanium in, 7
Weathering of ilmenite
  chemical changes during, 15
  steps in the process, 14
Wyoming, ilmenite-magnetite deposits in, 20

Yttrium titanate, preparation of, 120

Zinc titanate
  nuclei from, 313
  preparation and properties of, 125
Zircon, separation from ilmenite sand, 40, 43, 45
Zirconium titanate, preparation of, 124